Lecture Notes in Computer Science 15612

Founding Editors

Gerhard Goos
Juris Hartmanis

AF148199

Editorial Board Members

Elisa Bertino, *Purdue University, West Lafayette, IN, USA*
Wen Gao, *Peking University, Beijing, China*
Bernhard Steffen , *TU Dortmund University, Dortmund, Germany*
Moti Yung , *Columbia University, New York, NY, USA*

The series Lecture Notes in Computer Science (LNCS), including its subseries Lecture Notes in Artificial Intelligence (LNAI) and Lecture Notes in Bioinformatics (LNBI), has established itself as a medium for the publication of new developments in computer science and information technology research, teaching, and education.

LNCS enjoys close cooperation with the computer science R & D community, the series counts many renowned academics among its volume editors and paper authors, and collaborates with prestigious societies. Its mission is to serve this international community by providing an invaluable service, mainly focused on the publication of conference and workshop proceedings and postproceedings. LNCS commenced publication in 1973.

Pablo García-Sánchez · Emma Hart ·
Sarah L. Thomson
Editors

Applications of Evolutionary Computation

28th European Conference, EvoApplications 2025
Held as Part of EvoStar 2025, Trieste, Italy, April 23–25, 2025
Proceedings, Part I

 Springer

Editors
Pablo García-Sánchez (ID)
University of Granada
Granada, Spain

Emma Hart (ID)
Edinburgh Napier University
Edinburgh, UK

Sarah L. Thomson (ID)
Edinburgh Napier University
Edinburgh, UK

ISSN 0302-9743   ISSN 1611-3349 (electronic)
Lecture Notes in Computer Science
ISBN 978-3-031-90061-7   ISBN 978-3-031-90062-4 (eBook)
https://doi.org/10.1007/978-3-031-90062-4

© The Editor(s) (if applicable) and The Author(s), under exclusive license
to Springer Nature Switzerland AG 2025

This work is subject to copyright. All rights are solely and exclusively licensed by the Publisher, whether the whole or part of the material is concerned, specifically the rights of translation, reprinting, reuse of illustrations, recitation, broadcasting, reproduction on microfilms or in any other physical way, and transmission or information storage and retrieval, electronic adaptation, computer software, or by similar or dissimilar methodology now known or hereafter developed.
The use of general descriptive names, registered names, trademarks, service marks, etc. in this publication does not imply, even in the absence of a specific statement, that such names are exempt from the relevant protective laws and regulations and therefore free for general use.
The publisher, the authors and the editors are safe to assume that the advice and information in this book are believed to be true and accurate at the date of publication. Neither the publisher nor the authors or the editors give a warranty, expressed or implied, with respect to the material contained herein or for any errors or omissions that may have been made. The publisher remains neutral with regard to jurisdictional claims in published maps and institutional affiliations.

This Springer imprint is published by the registered company Springer Nature Switzerland AG
The registered company address is: Gewerbestrasse 11, 6330 Cham, Switzerland

If disposing of this product, please recycle the paper.

Preface

This volume is one of two which contain the proceedings of the 28th International Conference on the Applications of Evolutionary Computation (2025) – also known as EvoApplications, and previously known as EvoWorkshops. EvoApplications is held as part of EvoStar, the leading European conference in Bio-inspired AI. The other co-occurring conferences relate to combinatorial optimisation (EvoCOP); genetic programming (EuroGP); and evolutionary art and music (EvoMUSART). EvoStar 2025 was held in Trieste, Italy, from 23rd to 25th April.

EvoApplications 2025 received 96 submissions in total. In addition to the general conference, this also included nine special sessions, chaired by world-leading experts in the respective fields. These were: 30 Years of Particle Swarm Optimisation (by Leonardo Vanneschi, Marco S. Nobile, and Vasco Coelho); Analysis of Evolutionary Computation Methods: Theory, Empirics, and Real-World Applications (by Thomas Bartz-Beielstein, Carola Doerr, and Christine Zarges); Bio-inspired Algorithms for Green Computing and Sustainable Complex Systems (Josefa Díaz Álvarez and Juan Luis Jimenez Laredo); Computational Intelligence for Sustainability (Valentino Santucci, Fabio Caraffini, Jamal Toutouh, and Lucia Ballerini); EvoLLMs (Integrating Evolutionary Computing with Large Language Models (LLMs)) (Niki van Stein, Thomas Bäck, Anna V. Kononova); Evolutionary Computation in Edge, Fog, and Cloud Computing (Diego Oliva, Seyed Jalaleddin Mousavirad, and Mahshid Helali Moghadam); Evolutionary Computation in Image Analysis, Signal Processing, and Pattern Recognition (Harith Al-Sahaf, Pablo Mesejo, Ying Bi, and Qurrat Ul Ain); Machine Learning and AI in Digital Healthcare and Personalized Medicine (Stephen Smith and Marta Vallejo); and Soft Computing Applied to Games (Alberto P. Tonda and Antonio M. Mora). From the 96 submissions, after rigorous double-blind review 45 were selected for full oral presentation, while 18 were accepted for short oral presentation. All papers, whether associated with a long or short presentation, appear as full-length papers in this volume. There was also the Evolutionary Machine Learning (EML, organised by João Correia and Mengjie Zhang) special joint track, which was shared between EvoApplications and the co-occurring genetic programming conference: EuroGP. This year, there were ten submissions to EML. From these, five were accepted as long talks and one as a short talk. From those accepted, all of them are in the EvoApplications volumes except one EuroGP paper: "Micro-step Time-series Regression: Insights from System Identification Using Symbolic Regression", which can be found in the EuroGP proceedings.

We would like to extend our earnest appreciation to the following people, whose invaluable efforts allowed EvoApplications to happen.

- Firstly, thank you to the authors for choosing EvoApplications for your works.
- We are very grateful to the reviewers for ensuring the quality of the conference and dedicating their own time to this.
- We would also like to thank Nuno Lourenço (University of Coimbra, Portugal) for his dedicated work as Submission System Coordinator.

- We thank the EvoStar Graphic Identity Team: Sérgio Rebelo and Jéssica Parente (University of Coimbra, Portugal), for their dedication and excellence in graphic design.
- We are grateful to Francisco Chicano and João Correia for their impressive work managing and maintaining the EvoStar website and handling the publicity, respectively.
- We credit the invited keynote speakers: Tea Tušar (Jožef Stefan Institute, Slovenia) and Daniela Besozzi (University of Milano-Bicocca, Italy) for their inspiring and interesting talks.
- We would like to express our gratitude to the Steering Committee of EvoApplications for helping to organize the conference.
- Our immense thanks to the local organisers: Luca Manzoni, Eric Medvet, Giorgia Nadizar, and Gloria Pietropolli (all from the University of Trieste, Italy).
- We are grateful to the support provided by SPECIES, the Society for the Promotion of Evolutionary Computation in Europe and its Surroundings, for the coordination and financial administration.
- Thank you to the coordinators of student affairs: Jéssica Parente and Alina Geiger.
- Lastly, we express our immense gratitude to Anna I. Esparcia-Alcázar (EvoStar coordinator) whose exceptional work and leadership allowed our conference to be a success.

April 2025

Pablo García-Sánchez
Emma Hart
Sarah L. Thomson

Organisation

EvoApplications Conference Chairs

Pablo García-Sánchez   University of Granada, Spain
Emma Hart   Edinburgh Napier University, UK

EvoApplications Publication Chair

Sarah L. Thomson   Edinburgh Napier University, UK

30 Years of Particle Swarm Optimisation Chairs

Leonardo Vanneschi   Universidade NOVA de Lisboa, Portugal
Marco S. Nobile   Ca' Foscari University of Venice, Italy
Vasco Coelho   University of Milano-Bicocca, Italy

Analysis of Evolutionary Computation Methods: Theory, Empirics, and Real-World Applications Chairs

Thomas Bartz-Beielstein   TH Köln, Germany
Carola Doerr   CNRS and Sorbonne Université, France
Christine Zarges   Aberystwyth University, UK

Bio-inspired Algorithms for Green Computing and Sustainable Complex Systems Chairs

Josefa Díaz Álvarez   University of Extremadura, Spain
Juan Luis Jimenez Laredo   University of Granada, Spain

Computational Intelligence for Sustainability Chairs

Valentino Santucci   Università per Stranieri di Perugia, Italy
Fabio Caraffini   Swansea University, UK
Jamal Toutouh   University of Málaga, Spain
Lucia Ballerini   University for Foreigners of Perugia, Italy

EvoLLMs: Intergrating Evolutionary Computing with Large Language Models Chairs

Niki van Stein   Leiden University, Netherlands
Thomas Bäck   Leiden University, Netherlands
Anna V. Kononova   Leiden University, Netherlands

Evolutionary Computation in Edge, Fog, and Cloud Computing Chairs

Diego Oliva   Universidad de Guadalajara, Mexico
Seyed Jalaleddin Mousavirad Hakim Sabzevari University, Iran
Mahshid Helali Moghadam   RISE Research Institutes of Sweden, Sweden

Evolutionary Computation in Image Analysis, Signal Processing, and Pattern Recognition Chairs

Pablo Mesejo   Universidad de Granada, Spain
Harith Al-Sahaf   Victoria University of Wellington, New Zealand
Ying Bi   Zhengzhou University, China
Qurrat Ul Ain   Victoria University of Wellington, New Zealand

Machine Learning and AI in Digital Healthcare and Personalized Medicine Chairs

Stephen Smith   University of York, UK
Marta Vallejo   Heriot-Watt University, UK

Soft Computing Applied to Games Chairs

Alberto P. Tonda — INRAE, France
Antonio M. Mora — University of Granada, Spain

Evolutionary Machine Learning Chairs

João Correia — University of Coimbra, Portugal
Mengjie Zhang — Victoria University of Wellington, New Zealand

EvoApplications Steering Committee

Stefano Cagnoni — University of Parma, Italy
Pedro A. Castillo — University of Granada, Spain
Anna I Esparcia-Alcázar — Universitat Politècnica de València, Spain
Mario Giacobinni — University of Turin, Italy
Paul Kaufmann — University of Mainz, Germany
Antonio Mora — University of Granada, Spain
Günther Raidl — Vienna University of Technology, Austria
Franz Rothlauf — Johannes Gutenberg University Mainz, Germany
Kevin Sim — Edinburgh Napier University, UK
Giovanni Squillero — Politecnico di Torino, Italy
J. L. Jimenez-Laredo — University of Granada, Spain
Honorary member: Cecilia di Chio — King's College London, UK

Program Committee

Harith Al-Sahaf — Victoria University of Wellington, New Zealand
Jacopo Aleotti — University of Parma, Italy
Anca Andreica — Babeş-Bolyai University, Romania
Claus Aranha — University of Tsukuba, Japan
Kehinde Babaagba — Edinburgh Napier University, UK
Jaume Bacardit — Newcastle University, UK
Thomas Bäck — Leiden University, Netherlands
Marco Baioletti — Università degli Studi di Perugia, Italy
Lucia Ballerini — University of Edinburgh, UK
Wolfgang Banzhaf — Michigan State University, USA
Thomas Bartz-Beielstein — TH Köln, Germany
Ying Bi — Zhengzhou University, China

János Botzheim	Eötvös Loránd University, Hungary
Jörg Bremer	University of Oldenburg, Germany
Karina Brotto Rebuli	Università degli Studi di Torino, Italy
Will Browne	Queensland University of Technology, Australia
Doina Bucur	University of Twente, Netherlands
Maxim Buzdalov	Aberystwyth University, UK
Stefano Cagnoni	University of Parma, Italy
Fabio Caraffini	Swansea University, UK
Pedro Castillo	University of Granada, Spain
Paolo Cazzaniga	University of Bergamo, Italy
Ying-Ping Chen	National Yang Ming Chiao Tung University, Taiwan
Francisco Chicano	University of Málaga, Spain
Anders Christensen	University of Southern Denmark, Denmark
Anthony Clark	Pomona College, USA
Vasco Coelho	University of Milano-Bicocca, Italy
José Manuel Colmenar	Universidad Rey Juan Carlos, Spain
Feijoo Colomine D.	Universidad Nacional Experimental del Táchira, Venezuela
Stefano Coniglio	University of Bergamo, Italy
João Correia	University of Coimbra, Portugal
Luís Correia	Universidade de Lisboa, Portugal
Ernesto Costa	University of Coimbra, Portugal
Carlos Cotta	Universidad de Málaga, Spain
Fabio D'Andreagiovanni	University of Modena and Reggio Emilia, Italy, Technical University of Berlin, Germany
Gregoire Danoy	University of Luxembourg, Luxembourg
Bilel Derbel	University of Lille, France
Travis Desell	Rochester Institute of Technology, USA
Laura Dipietro	Massachusetts Institute of Technology, USA
Federico Divina	Pablo de Olavide University, Spain
Carola Doerr	Sorbonne University, France
Bernabe Dorronsoro	University of Cádiz, Spain
Josefa Díaz Álvarez	University of Extremadura, Spain
Tome Eftimov	Jožef Stefan Institute, Slovenia
Abdelrahman Elsaid	University of North Carolina Wilmington, USA
Andres Faina	IT University of Copenhagen, Denmark
Francisco Fernandez De Vega	University of Extremadura, Spain
Antonio Fernández Ares	University of Granada, Spain
Antonio J. Fernández Leiva	Universidad de Málaga, Spain
Francesco Fontanella	Università di Cassino e del Lazio Meridionale, Italy

Pablo García-Sánchez	Universidad de Granada, Spain
Mario Giacobini	University of Turin, Italy
Kyrre Glette	University of Oslo, Norway
Michael Guckert	Mittelhessen University of Applied Sciences, Germany
Maria Habib	University of Granada, Spain
Heiko Hamann	University of Konstanz, Germany
Emma Hart	Edinburgh Napier University, UK
Daniel Hernandez	Tecnológico Nacional Le México/Instituto Tecnológico de Tijuana, Mexico
Ignacio Hidalgo	Universidad Complutense de Madrid, Spain
Rolf Hoffmann	TU Darmstadt, Germany
Giovanni Iacca	University of Trento, Italy
Juanlu Jiménez	University of Granada, Spain
Juan Luis Jiménez Laredo	Université Le Havre Normandie, France
Karlo Knezevic	University of Zagreb, Croatia
Anna V. Kononova	Leiden University, Netherlands
Ana Kostovska	Jožef Stefan Institute, Slovenia
Gurhan Kucuk	Yeditepe University, Türkiye
Waclaw Kus	Silesian University of Technology, Poland
Yuri Lavinas	University of Toulouse, France
Kenji Leibnitz	National Institute of Information and Communications Technology, Japan
Zichao Li	University of Waterloo, Canada
Federico Liberatore	Cardiff University, UK
Fernando Lobo	University of Algarve, Portugal
Michael Lones	Heriot-Watt University, UK
Nuno Lourenço	University of Coimbra, Portugal
Gabriel Luque	University of Málaga, Spain
Evelyne Lutton	INRAE, France
Penousal Machado	University of Coimbra, Portugal
Katherine Malan	University of South Africa, South Africa
Luca Mariot	University of Twente, Netherlands
Eric Medvet	University of Trieste, Italy
David Megías	Universitat Oberta de Catalunya, Spain
Paolo Mengoni	Hong Kong Baptist University, China
Pablo Mesejo	University of Granada, Spain
Krzysztof Michalak	Wrocław University of Economics and Business, Poland
Mahshid Helali Moghadam	RISE Research Institutes of Sweden, Sweden
Salem Mohammed	Mustapha Stambouli University, Algeria
Antonio Mora	University of Granada, Spain

Seyed Jalaleddin Mousavirad	Mid Sweden University, Sweden
Mario Andrés Muñoz	University of Melbourne, Australia
Geoff Nitschke	University of Cape Town, South Africa
Marco S. Nobile	Ca' Foscari University of Venice, Italy
Gustavo Olague	CICESE, Mexico
Diego Oliva	Universidad de Guadalajara, Mexico
Marcos Ortega Hortas	University of A Coruña, Spain
Daniele Maria Papetti	Università degli Studi di Milano-Bicocca, Italy
Anna Paszynska	Jagiellonian University, Poland
Diego Daniel Pedroza-Perez	Universidad de Málaga, Spain
Arkadiusz Poteralski	Silesian University of Technology, Poland
Petr Pošík	Czech Technical University in Prague, Czechia
Raneem Qaddoura	Al Hussein Technical University, Jordan
Elena Raponi	Technical University of Munich, Germany
Quentin Renau	Edinburgh Napier University, UK
José Carlos Ribeiro	Polytechnic Institute of Leiria, Portugal
Luigi Rovito	University of Trieste, Italy
Jose Santos	University of A Coruña, Spain
Valentino Santucci	University for Foreigners of Perugia, Italy
Enrico Schumann	University of Basel, Switzerland
Lennart Schäpermeier	TU Dresden, Germany
Sevil Sen	University of York, UK
Roman Senkerik	TBU in Zlin, Czechia
Chien-Chung Shen	University of Delaware, USA
Sara Silva	Universidade de Lisboa, Portugal
Kevin Sim	Edinburgh Napier University, UK
Anabela Simões	Coimbra Institute of Engineering, Portugal
Stephen Smith	University of York, UK
Maciej Smołka	AGH University of Science and Technology, Poland
Andrea Tangherloni	Bocconi University, Italy
Ernesto Tarantino	ICAR-CNR, Italy
Andrea Tettamanzi	Université Côte d'Azur, France
Sarah L. Thomson	Edinburgh Napier University, UK
Renato Tinós	University of São Paulo, Brazil
Alberto Tonda	INRAE, France
Jamal Toutouh	Massachusetts Institute of Technology, USA
Alexander Turner	University of Nottingham, UK
Qurrat Ul Ain	Victoria University of Wellington, New Zealand
Neil Urquhart	Edinburgh Napier University, UK
Marta Vallejo	Heriot-Watt University, UK
Niki van Stein	Leiden University, Netherlands

Leonardo Vanneschi	Universidade NOVA de Lisboa, Portugal
Diederick Vermetten	Leiden University, Netherlands
Marco Villani	University of Modena and Reggio Emilia, Italy
Rafael Villanueva	Universitat Politècnica de València, Spain
Vanessa Volz	CWI, Netherlands
Jaroslaw Was	AGH University of Science and Technology, Poland
Simon Wells	Edinburgh Napier University, UK
Anil Yaman	Vrije Universiteit Amsterdam, Netherlands
Furong Ye	Leiden University, Netherlands
Ales Zamuda	University of Maribor, Slovenia
Christine Zarges	Aberystwyth University, UK
Mengjie Zhang	Victoria University of Wellington, New Zealand

Contents – Part I

xvi   Contents – Part I

Contents – Part II

Analysis of Evolutionary Computation Methods: Theory, Empirics, and Real-World Applications

Bio-inspired Algorithms for Green Computing and Sustainable Complex Systems

Computational Intelligence for Sustainability

EvoLLMs (Integrating Evolutionary Computing with Large Language Models (LLMs))

Evolutionary Computation in Edge, Fog, and Cloud Computing

EvoApplications

Optimizing Dietary Plans Using Evolutionary Algorithms

Iqra Azfar$^{(\boxtimes)}$ ⓘ, Rabia Shahab ⓘ, Javeria Azfar ⓘ, and Syeda Saleha Raza ⓘ

Habib University, Karachi 75290, Sindh, Pakistan
{ia07614,ra07528,ja07622}@st.habib.edu.pk

Abstract. This paper presents a novel application of Computational Intelligence to optimize dietary plans tailored to athletes' unique nutritional needs across diverse sports disciplines. Leveraging Evolutionary Algorithms (EAs), the system dynamically generates personalized meal plans based on multi-factorial inputs, including age, gender, sport type, and individual food preferences. By integrating multi-objective optimization, the approach ensures precise macronutrient balance for enhanced performance, recovery, and long-term health. The system draws from a structured food database, offering real-time adaptability to dietary restrictions and preferences, and delivering nutrient-dense meal plans optimized for metabolic efficiency. Unlike traditional methods, this solution provides a practical, cost-effective alternative to professional diet planning. The research underscores the transformative potential of EAs in sports nutrition, combining tailored recommendations with scientific accuracy to set a new standard for athlete-focused dietary optimization.

Keywords: Evolutionary Algorithm · Computational Intelligence · Diet Plan Optimization

1 Introduction

Athletes' nutritional requirements are highly specific and complex due to the intensive physical demands and diverse metabolic needs characteristic of different sports disciplines. Extensive research highlights the pivotal role of targeted nutrition in enhancing athletic performance, supporting recovery, and fostering long-term health [1]. However, designing adaptable diet plans that precisely address each athlete's unique needs remains a challenging task. Conventional diet planning methods frequently lack the specificity required to achieve an optimal balance of calories, macro-nutrients (carbohydrates, proteins, dietary fats), and essential micro-nutrients, often overlooking the unique physiological demands imposed by high-intensity activity and adaptive training responses [2].

Despite progress in nutritional science, current methodologies often rely on generalized guidelines, which fail to deliver the precision necessary for athlete-centered dietary planning. These approaches typically do not account for variables such as age, gender, sport-specific requirements, and individual food preferences, which are critical for sustaining performance and recovery. This study

© The Author(s), under exclusive license to Springer Nature Switzerland AG 2025
P. García-Sánchez et al. (Eds.): EvoApplications 2025, LNCS 15612, pp. 3–19, 2025.
https://doi.org/10.1007/978-3-031-90062-4_1

addresses these limitations by employing a Genetic Algorithm (GA) within a computational intelligence framework to optimize dietary planning for athletes [3]. By utilizing two robust datasets-one from Kaggle, containing detailed nutritional profiles of 200 food items, and another outlining sport-specific dietary requirements segmented by age and gender. This system generates comprehensive, customized meal plans as shown in the tables below as well. In addition to offering precise nutrient recommendations, the system provides visualized nutritional breakdowns to support informed dietary decisions. Ultimately, this research aims to offer a scientifically grounded, adaptive tool for athletes, enhancing their capacity to achieve optimal nutrition and peak performance.

Table 1. Nutritional Requirements for Boxers by Age and Gender

Athlete	Gender	Age Group	Protein (g)	Carbohydrates (g)	Fats (g)
Boxers	Female	20–30	60	300	70
Boxers	Female	30–40	65	320	75
Boxers	Female	40–50	70	350	80
Boxers	Female	50+	75	370	85

Table 2. Nutritional Composition of Selected Ingredients

No	Ingredients	Unit (g)	Protein (g)	Fat (g)
0	Milled rice	100	8.4	1.7
1	Young corn	100	5.1	0.7
2	Duck	200	32	57.2
3	Black glutinous rice	100	8	2.3
4	Yardlong bean	100	2.3	0.1
5	Yellow sweet potato	100	0.5	0.4

2   Motivation and Related Work

Research on automated diet planning using computational algorithms has evolved over decades, yet significant gaps remain, especially in addressing athletes' specialized nutritional needs. While systems like MenuGene offered customizable weekly diet plans based on user preferences such as food composition and aesthetics, they lacked the precision to meet athletes' sport-specific nutritional demands [4]. Other approaches, such as Jia et al.'s method, balanced taste preferences and health standards using genetic algorithms but did not focus on the distinct needs of athletes [5]. Similarly, Salloum and Tekli's transportation problem adaptation improved meal planning flexibility, and Ferrario and Gedrich explored machine learning for personalized recommendations. However, neither catered specifically to athletes [6, 7].

Stefanidis et al.'s PROTEIN AI Advisor framework provided recommendations for different user groups, including those with health conditions [8]. Syahputra demonstrated how genetic algorithms could optimize weekly diets for diabetic patients by considering caloric and nutritional needs [9]. Despite these advancements, such methods often focused on single-objective optimization, leaving room for improvement in handling multiple, often conflicting, nutritional goals [10].

Recent advancements, such as the Compromise Differential Evolution (DE) algorithm, address multi-objective optimization in dietary planning. DE effectively identifies Pareto-optimal solutions for complex problems, offering simplicity, speed, and robustness compared to traditional Genetic Algorithms (GA) [11]. For instance, DE's dominance selection operator enables efficient balancing of conflicting objectives, such as optimizing nutritional goals for athletes. Evolutionary algorithms have also been applied to generate entire meal plans based on client preferences while maintaining nutritional standards [10].

However, existing methods often overlook the unique demands of athletes, such as activity-specific metabolic requirements and individual preferences. This research aims to bridge these gaps by employing advanced Computational Intelligence techniques to create tailored dietary plans for athletes. The proposed system focuses on offering athletes optimized ingredients they can directly consume, rather than complete meals.

Evolutionary Algorithms (EAs), particularly Multi-Objective Genetic Algorithms (MOGAs), are ideal for this task due to their ability to handle complex constraints and preferences without extensive reformulation [11,12]. MOGAs enable assigning weights to nutritional objectives such as protein, carbohydrates, and fats, that can be leveraged to develop balanced and personalized diet plans that align with athletes' performance goals [13]. Compared to simpler GA models, MOGAs demonstrate superior stability and accuracy in optimizing intricate problems like nutrition planning [10,12].

To validate this approach, two datasets were analyzed. The EA-based algorithm consistently provided balanced nutritional solutions tailored to athletes' requirements. While promising, additional validation and comparative analysis with traditional methods are required to confirm scalability and robustness [13].

In summary, this research introduces a novel, multi-objective optimization approach for dietary planning that efficiently balances diverse nutritional needs. Leveraging the DE algorithm, the study demonstrates the adaptability of EAs to real-world problems like nutrition planning, paving the way for more personalized and accurate dietary recommendations [2]. Unlike traditional methods, this approach offers a swift, cost-effective alternative to professional nutritionists, making it a practical option for athletes [14]. Future work will explore the effectiveness of alternative EA implementations, such as the Bounded Knapsack approach, to further enhance diet plan optimization.

3   Problem Description

Thus, we also explore the essential nutritional targets and user preferences that our solution must address to create effective, personalized diet plans for athletes. We delve into specific dietary needs, such as macronutrient ratios, energy intake, and micronutrient balance, tailored to athletic performance. Additionally, we consider user preferences, including dietary restrictions, meal timings, and taste preferences, which are critical to user adherence and plan effectiveness. By addressing these factors, we identify the core constraints and challenges that guide the design of our optimization approach, ensuring it aligns with both nutritional requirements and athlete satisfaction.

- **Nutritional Goals**
  - **Targets:** These parameters represent the specific nutritional targets for protein, carbohydrates, and fats that the optimized dietary plans aim to achieve (see Table 1 for detailed requirements based on athlete categories).
  - **Usage:** The genetic algorithm evaluates the fitness of each dietary plan based on its ability to meet these target nutritional goals. Deviations from the targets contribute to the fitness score, guiding the search towards plans that better align with the athlete's nutritional requirements.
- **Preference and All Foods**
  - **Preference:** These variables store information about the preferred ingredients selected by the user and the complete dataset of available food items, respectively.
  - **Usage:** User preferences guide the selection of ingredients considered for dietary plans, ensuring that the generated plans include preferred foods. The nutritional composition of the ingredients used for these plans is provided in Table 2, derived from the All Foods dataset, which contains the necessary nutritional information for all available food items.

4   Comparative Algorithm Implementation for Dietary Optimization

Genetic algorithms are a class of optimization algorithms designed to find the most optimal solution to a computational problem by either maximizing or minimizing a specific objective function. They are rooted in the broader field of evolutionary computation, which simulates biological processes such as reproduction and natural selection to identify the "fittest" solutions. These algorithms blend elements of randomness with controlled operations to efficiently navigate the search space. Genetic algorithms rely on three core genetic operators: selection, crossover, and mutation, applied with specified probabilities to evolve better solutions over successive generations.

4.1 MOGA Inspired Implementation

A new metaheuristic based on Multi-Objective Genetic Algorithm (MOGA) principles was developed, focusing on optimizing multiple nutritional parameters like fats, carbohydrates, and proteins simultaneously. By assigning specific weights to these parameters, the algorithm aimed to generate dietary plans that balanced at least two objectives effectively. Using Pareto optimization, each solution was ranked based on trade-offs, ensuring that improving one objective did not worsen another. The solutions were then normalized with assigned weights to give balanced attention to each nutritional goal, enabling flexible adherence to various dietary constraints and preferences. This approach aimed to ensure a diverse set of well-rounded solutions, catering to individual dietary needs and offering significant improvements in dietary personalization, while also maintaining nutritional balance tailored to specific athletic requirements.

4.2 Limitations of the Pareto-Based MOGA Approach

The Multi-Objective Genetic Algorithm (MOGA), inspired by Pareto optimization, presents several challenges that limit its effectiveness in certain scenarios, particularly when it comes to dietary optimization.

First, the penalty system's simplicity and directness in other approaches often proves advantageous, as it straightforwardly penalizes deviations from nutritional targets. The fitness function in these methods employs a linear penalty mechanism where any excess in nutrient thresholds severely reduces the fitness score. This clear and predictable behavior allows for an intuitive understanding of how solutions are evaluated. In contrast, the Pareto-based MOGA introduces complexity and computational overhead, as it requires multi-step processes, including Pareto dominance calculations, ranking of solutions, and the application of weighted scores. This complexity can complicate the fitness evaluation, making it less transparent and harder to manage. Furthermore, the weights are assigned to variables on a subjective manner, which introduces potential biases, as inappropriate or inaccurate weight assignments may lead to suboptimal outcomes.

While the Pareto approach seeks to identify trade-offs between multiple objectives, it does not guarantee practical dietary outcomes. For instance, the generated plans may meet all the nutritional objectives defined in the scope of this research. Still, they would not account for factors such as allergies, supplementary pills an athlete is taking, or ingredient availability. This research opens the grounds to continue this study and incorporate the factors above to ensure that the dietary plan is both realistic and sustainable for individuals. Moreover, the lack of interpretability in MOGA's multi-layered optimization process can hinder its adoption by diet planners or nutritionists, who may find it difficult to justify the rationale behind selected plans. This limitation becomes particularly significant in scenarios where it is important to consider creating diets that remain appealing and varied over time. Additionally, reliability and robustness are concerns with the Pareto-based method. Although the direct penalty system's strength lies in its simplicity and consistent feedback, the reliance on Pareto scores and composite metrics can sometimes lead to unstable or inconsistent selections, especially if the algorithm is not finely tuned or the population lacks diversity. This instability can compromise the quality and reliability of the generated dietary plans, making MOGA less suitable for cases requiring precise and consistent nutrient optimization. Lastly, this study is currently in its early phases of testing, focusing on a small subset of sports and lacking stratification by factors such as specific dietary preferences of individuals. This limitation restrict the generalizability of this study's findings to broader populations and contexts. Future work will aim to refine the results by addressing individual specific factors such as their food preferences, allergies, alongside exploring the inclusion of real-world considerations such as dietary supplements and long-term meal variability. This will aid in enhancing the practical applicability of the dietary plan results achieved. Further results that were obtained are discussed in the section of result and analysis.

5   Problem Formulation Using Modified Bounded Knapsack

Our approach draws a close parallel between the Knapsack Problem and dietary optimization. In the Knapsack Problem, items are chosen to maximize value without exceeding a weight limit. Similarly, our Evolutionary Algorithm (EA) selects food items based on nutritional value and user preferences, aiming to fulfill dietary requirements while adhering to specific constraints, such as caloric limits and nutrient balance. The Algorithm 1 is shown on the next page.

The process begins by generating an initial population of random chromosomes, each representing a potential dietary plan with various food items. Each chromosome's fitness is assessed through a specialized function that ensures it meets key nutritional goals, such as protein, carbohydrate, and fat intake. Higher-fitness individuals are selected as parents, favoring those that best align with the user's dietary requirements.

To create a new population, genetic material from the parent chromosomes is recombined through crossover, producing diverse offspring. Random mutations are introduced to maintain genetic diversity, helping the algorithm avoid local optima and explore new dietary combinations.

The EA continues iterating, checking a termination criterion such as a set number of generations or a convergence threshold for optimal solutions. Once this criterion is met, the algorithm stops and returns the best dietary plan identified. This approach carefully balances exploration of new possibilities with refinement of promising solutions, ensuring effective, personalized dietary plan optimization.

Algorithm 1. Evolutionary Algorithm for Dietary Plan Optimization

---

Require: Population size N, mutation rate μ, number of generations G, crossover parameter η_c

Ensure: Optimized dietary plan meeting target nutritional goals

    Initialize Parameters:

    Load athlete data and food dataset

    Set target nutritional goals: $target\_protein$, $target\_fat$, $target\_carbs$

    Generate initial population of chromosomes with random food item selections

    for each generation $g = 1$ to G **do**

        Calculate Maximum Units:

        for each food item i **do**

            Compute max_units_i using:

$$max\_units_i = \min\left(\frac{target\_protein}{protein_i}, \frac{target\_fat}{fat_i}, \frac{target\_carbs}{carbs_i}\right)$$

        end for

        Selection:

        Perform `Fitness_Proportional_Selection` to select k chromosomes:

$$P_{select}(chromosome_i) = \frac{fitness(chromosome_i)}{\sum_{j=1}^{N} fitness(chromosome_j)}$$

        Crossover:

        for each pair of selected parents $(parent_1, parent_2)$ **do**

            Generate offspring using `SBX()` function

            Ensure offspring values respect max_units

        end for

        Mutation:

        for each offspring chromosome **do**

            Mutate with probability μ using:

$$chromosome[i] = random(0, max\_units_i) \text{ if } r < \mu, r \sim U(0,1)$$

        end for

        Evaluate Fitness:

        Calculate fitness scores for all chromosomes using `Calculate_Fitness_Score()`.

        Survivor Selection:

        Retain top N chromosomes with highest fitness

        Convergence Check:

        if convergence criteria met **then**

            Break

        end if

    end for

    Output: Return the best chromosome as the optimized dietary plan

5.1   Chromosome Structure

- **Length:** The length of the chromosome corresponds to the total number of available food items or ingredients that can be included in the diet plan.
- **Genes:** Each gene in the chromosome represents the quantity or units of a specific food item to include in the diet. The value of each gene indicates the number of units of the corresponding food item to include.
- **Encoding:** The chromosome is typically encoded as a list or array of integers, where each integer represents the quantity of a specific food item. The index of each integer corresponds to the position of the food item in the list of available food items or ingredients.Example: For example, if the chromosome has a length of 10 and the value of the genes at indices 0, 3, and 7 are 2, 1, and 0 respectively, it means:
  - Include 2 units of the food item at index 0.
  - Include 1 unit of the food item at index 3.
  - Exclude the food item at index 7 from the diet plan
- **Visualization:** Chromosome: [2, 0, 0, 1, 0, 0, 0, 0, 1, 0]
- **Interpretation:** The above chromosome indicates a diet plan that includes:
  - 2 units of the food item at index 0.
  - 1 unit of the food item at index 3.
  - 1 unit of the food item at index 8.

The remaining food items are not included in the diet plan (quantities are set to 0).

5.2   Fitness Function

The fitness function for the Bounded Knapsack optimization evaluates each dietary plan based on its total content of protein, fat, and carbohydrates. These levels reflect the nutritional contributions of the ingredients and form the basis for assessing the quality of the plan. The fitness score starts at 100 and measures how closely a plan meets target values, with a 10% flexibility threshold allowing for minor deviations. Exceeding this threshold sets the score to zero, preventing over-supply of nutrients and maintaining balance.

To ensure that the fitness function aligns with different dietary goals, specific weights are assigned to each macronutrient, reflecting their relative importance. For instance, if the goal is muscle building, protein is given a higher weight, such as 0.4, compared to fats and carbohydrates, which may each be weighted at 0.3. This weighted system ensures that deviations from the target values for protein are penalized more heavily, emphasizing the importance of protein intake in that specific diet plan.

The overall fitness score is then calculated by deducting points for deviations from the target values. To prevent any single nutrient from disproportionately affecting the total score, a maximum deduction of 25 points is set per nutrient. This cap creates a balanced penalty structure, ensuring that even if one macronutrient slightly deviates from its target, it does not excessively dominate or skew the fitness assessment. This approach provides a fair and consistent evaluation across all macronutrients, maintaining a well-rounded diet plan.

The formula for calculating the fitness score is

$$\text{Fitness} = 100 - \left(\frac{|\text{Protein\_Deviation}|}{\text{Target\_Protein}} \times \text{Weight\_Protein} \times 25 \right)$$

$$- \left(\frac{|\text{Fat\_Deviation}|}{\text{Target\_Fat}} \times \text{Weight\_Fat} \times 25 \right)$$

$$- \left(\frac{|\text{Carbs\_Deviation}|}{\text{Target\_Carbs}} \times \text{Weight\_Carbs} \times 25 \right)$$

Deviations are calculated as the difference between actual and target values for protein, fat, and carbohydrates, with larger deviations leading to greater point deductions from the initial score of 100. This approach ensures a positive scoring framework, penalizing nutrient imbalances while preventing negative scores. The weighting system is flexible and adaptable to different dietary goals, such as muscle gain or weight loss, allowing users to prioritize macronutrients as needed. For instance, assigning a higher weight to protein ensures the algorithm focuses more on optimizing protein intake to align with specific dietary objectives.

5.3   Crossover

For this problem, we tried to achieve the appropriate number of steps between exploration and exploitation to prevent the algorithm from getting stuck at a local optimum and at the same time to cover all the potential solutions. We tested four different crossover methods: SBX, Two-Point, Uniform, and a newly introduced New Crossover. Out of all the methods, SBX method simulates binary crossover and the New Crossover method exhibited the finest results.

New Crossover. This technique selects a random crossover point in the parent chromosomes, exchanging genes from one parent with the other after this point. This approach enhances exploration by producing offspring with varied characteristics, covering a broader solution space. However, it can sometimes lead to lower-quality offspring, potentially losing good solutions. By using randomly selected crossover points and controlled gene addition, the method balances exploration and exploitation, allowing the algorithm to search diverse possibilities while retaining solution quality within the search space.

SBX Crossover (Simulated Binary Crossover). The SBX method simulates one-point crossover from binary genetic algorithms, generating two offspring from two parent solutions. It introduces randomness by creating a random number for a gene pair: a value below 0.5 generates offspring close to the parents (exploitation), while a value above 0.5 generates offspring further from the parents (exploration). The parameter 'eta c' adjusts the offspring spread, with higher values favoring exploitation and lower values promoting exploration. This balance helps maintain offspring diversity and adaptability, optimizing both exploration and exploitation in the search space.

SBX Crossover: The SBX (Simulated Binary Crossover) operator generates offspring by simulating the behavior of single-point crossover in binary genetic algorithms. For two parent solutions p_1 and p_2, the offspring solutions c_1 and c_2 are computed as follows:

$$c_1 = 0.5\left[(1+\beta)p_1 + (1-\beta)p_2\right]$$
$$c_2 = 0.5\left[(1-\beta)p_1 + (1+\beta)p_2\right]$$

where:

$$\beta = \begin{cases} (2u)^{\frac{1}{\eta_c+1}}, & \text{if } u \leq 0.5 \\ (1/2(1-u))^{\frac{1}{\eta_c+1}}, & \text{if } u > 0.5 \end{cases}$$

Here: - p_1, p_2: Parent solutions. - c_1, c_2: Offspring solutions. - β: Distribution index controlling the spread of offspring. - $u \sim U(0,1)$: Random number drawn from a uniform distribution. - η_c: Crossover control parameter determining exploration vs. exploitation.

The first brought changes in the chromosomes of offspring necessary for exploration, and the second kept good values from the parents suitable for exploitation. But the Two-Point and Uniform methods were not very effective in exploration since they were probabilistic and at times favorable two-alleles passed on to the offspring. Selecting the crossover method affected the algorithm substantially, particularly between the SBX and the New Crossover offer a perfect combination between exploitation and exploration and hence the superior coverage of the solution space.

5.4   Mutation

During mutation the indices in a chromosome are selected at random a predetermined rate of mutation is provided in the algorithm. If it is to be mutated, a random integer from the allowed range of that index is derived. This helps to bring new individuals into the population and, thus, to diversify the population, which is helpful to explore the solution space. The rate of mutation defines how often mutation happens and that defines the trade-off between exploration and exploitation. Mutation can enhance or diminish the fitness of solutions: changing some indices increases the efficiency of the search for the best solutions.

5.5   Parent and Survivor Selection

The population is then created with potential solutions or chromosomes regarding the user preference and restriction on diets. For parent selection, the following methods are utilized: The Truncation Selection method work by ranking the population according to the fitness of each chromosome, meaning how meeting the dietary needs it is. It then selects the top-performing individuals as parents for the next generation. Fitness Proportional Selection selects individuals from the population with a probability proportional to their fitness. Chromosomes with higher fitness values have a greater chance of being selected as parents. These selection mechanisms ensure that the genetic algorithm explores promising areas of the solution space by favoring solutions that perform well, ultimately guiding the population towards optimal or near-optimal solutions over successive generations.

6   Experiments and Results

Table 3 is referenced for the analysis of the results obtained for each method discussed below. The achieved values were compared with the target values set for each athlete.

Table 3. Nutritional Targets for Athletes.

Athlete	Gender	Age Group	Protein (g)	Carbohydrates (g)	Fats (g)
Boxers	Male	30–40	75	380	90
Runners	Female	40–50	65	320	75
Swimmers	Male	20–30	55	280	65
Cyclists	Female	20–30	45	240	55
...	...	...	...	...	...
Boxers	Male	20–30	70	370	85

For each athlete, a single set of food items was used across all categories. This consistent food set ensured that both the Bounded Knapsack and Pareto methods were tested under identical conditions, allowing for a fair comparison of their effectiveness. The Bounded Knapsack method as discussed ahead, consistently provided solutions that were closer to the target nutritional values for each athlete group, regardless of the food set or category, further showcasing its superior ability to manage conflicting nutritional requirements.

6.1   Pareto Method

The Pareto-based evolutionary algorithm applied for dietary optimization used multi-objective optimization to find solutions that balanced several conflicting nutritional goals, such as protein, carbohydrates, and fats. Theoretically, this method should provide a range of possible solutions, each representing a trade-off between different objectives. However, an in-depth analysis of the Pareto results reveals several shortcomings, particularly in achieving the required nutritional goals for athletes, as shown in Table 4 below.

Table 4. Pareto optimization results for athletes.

Athlete	Gender	Age Group	Protein (g)	Carbohydrates (g)	Fats (g)
Boxers	Male	30–40	37.50	354.80	8.00
Runners	Female	40–50	57.50	210.20	62.20
Swimmers	Male	20–30	25.90	255.50	6.20
Cyclists	Female	20–30	12.30	123.20	2.90
...	...	...	...	...	...
Boxers	Male	20–30	32.00	149.30	57.20

In contrast, the Pareto method exhibited significant imbalances, particularly in key macronutrients such as protein and fats. For example, for male boxers, the Pareto method provided only 37.5 g of protein, a 50% shortfall from the required 75 g. This underperformance extended to fat intake, which reached only 8 g–91% below the target of 90 g. While carbohydrate intake was relatively close to the target, the overall nutrient distribution was heavily skewed, resulting in suboptimal diet plans that fail to meet the basic nutritional requirements of athletes.

This trend was also evident in other athlete categories. For female runners, the Pareto method provided a reasonable fat intake, missing the target by 17.7%, but carbohydrates and protein were under-supplied by 34.3% and 11.5% respectively. For male swimmers, the carbohydrate intake was closer to the target with only an 8.8% deviation, but protein and fat were severely under-supplied, missing the required intake by 52.9% and 90.5%. The most extreme case was for female cyclists, where the protein was 72.7% below the required level, carbohydrates were under-supplied by 48.7%, and fats by 94.7% as seen in Fig. 1 below. Such imbalances make the Pareto approach less viable for optimizing real-world dietary plans, especially for athletes who rely on precise nutrient intake for performance and recovery.

One of the core issues with the Pareto approach is that it does not effectively prioritize critical nutrients such as protein and fats, which are essential for athletes. The method attempts to balance all objectives equally, but this often leads to suboptimal solutions that fail to meet the nutritional needs required for optimal athletic performance. The trade-offs between different objectives result in large gaps in essential nutrients, making the Pareto method less suited for this type of optimization problem. Its theoretical focus on balancing multiple objectives overlooks the critical macronutrients needed for sustained athletic function, highlighting its limitations in practical applications.

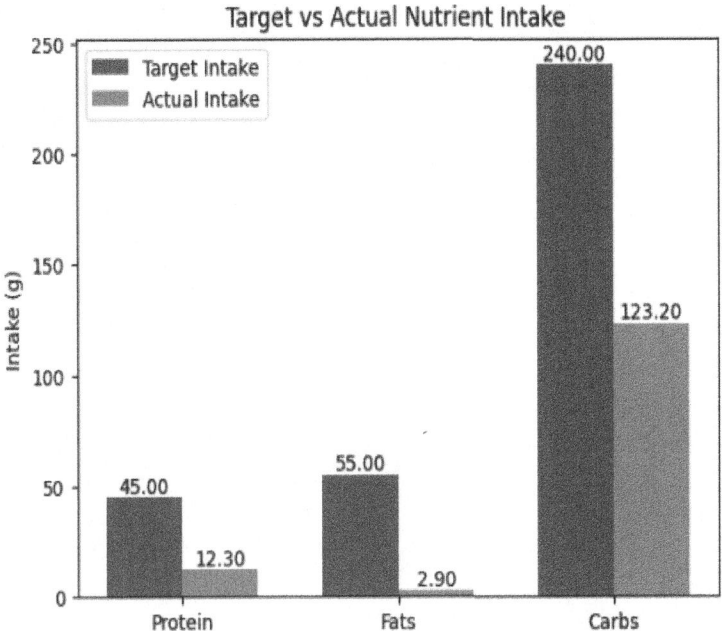

Fig. 1. Nutrient Distribution Achieved for Female Cyclists Using Pareto.

6.2 Modified Bounded Knap Sack

The Bounded Knapsack method provides a highly effective solution for optimizing dietary plans, as it addresses the complex nutritional requirements of athletes in a more structured and targeted way. Unlike the Pareto method, which often sacrifices one nutritional objective for another, the Bounded Knapsack method ensures that each macronutrient; protein, carbohydrate, and fat, is optimized under strict constraints. This structured approach allows for a more precise balance between nutrients, which is essential for athletic performance.

Table 5. Bounded Knapsack optimization results for athletes.

Athlete	Gender	Age Group	Protein (g)	Carbohydrates (g)	Fats (g)
Boxers	Male	30–40	72.60	382.40	69.00
Runners	Female	40–50	64.80	303.40	65.20
Swimmers	Male	20–30	55.30	277.10	62.90
Cyclists	Female	40–50	54.20	229.70	61.70
...	...	...	...	...	...
Boxers	Male	20–30	66.70	323.10	89.20

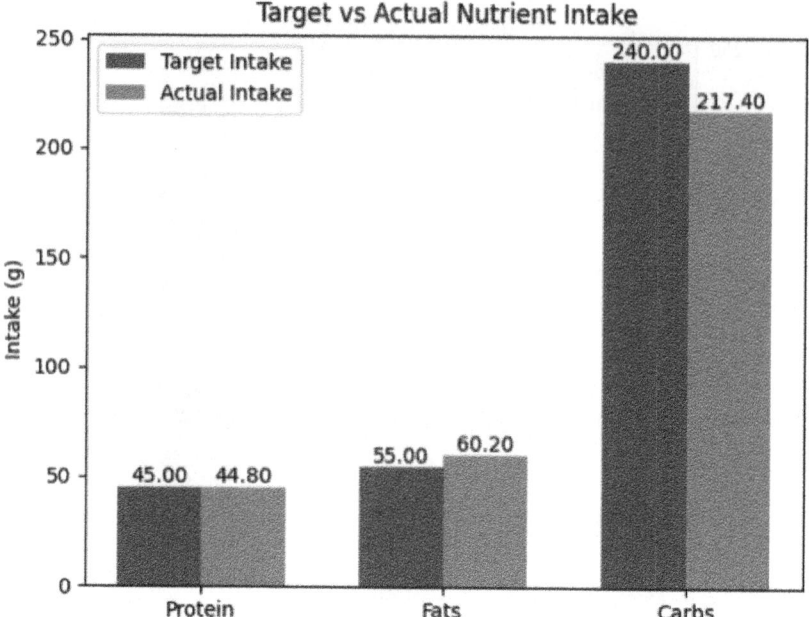

Fig. 2. Nutrient Distribution Achieved for Female Cyclists Using Bounded Knapsack.

In contrast, the Bounded Knapsack method delivered significantly better results as observed in Table 5 and Fig. 2, achieving a well-balanced macronutrient distribution. For example, in the case of male boxers, it provided 72.6 g of protein, just 3.2% below the required 75 g, while carbohydrates slightly exceeded the target at 382.4 g, with only a 0.6% deviation from the goal of 380 g. Fat intake, though still under the target, reached 69 g–23.3% below the required 90 g but significantly better than the Pareto results. Overall, this balanced distribution reflects the method's ability to optimize nutrients without sacrificing any critical macronutrients.

This improvement is seen across other athlete categories. For female runners, protein intake reached 64.8 g, just 0.3% below the target of 65 g, with carbohydrates and fats also close to the targets at 5.2% and 13% deviations, respectively. In male swimmers, the Bounded Knapsack method achieved near-perfect protein levels at 55.3 g (target: 55 g), with carbohydrate and fat intakes slightly off by 1% and 3.2%, respectively. For female cyclists, although carbohydrates fell short by 9.46% as shown in the, the overall nutrient balance was far closer to the required values compared to the Pareto approach, especially for protein and fats, which reached 44.80 g and 60.20 g.

The Bounded Knapsack method demonstrates a key strength in its ability to maintain a structured and balanced nutrient optimization process, ensuring near-target values for all essential macronutrients. Unlike the Pareto method, which often sacrifices one nutrient to balance another, the Bounded Knapsack approach provides a more practical and nutritionally sound solution, particularly suited for athletes who require precision in their dietary intake to support performance and recovery. Its capacity to manage conflicting nutritional requirements without extreme trade-offs makes it highly effective for dietary optimization, delivering well-rounded meal plans that closely align with macronutrient targets.

6.3 Comparison

The comparison between the Pareto and Bounded Knapsack methods reveals significant differences in their ability to balance macronutrients effectively for athletes. The Pareto method consistently struggled to meet essential nutritional targets, particularly for proteins and fats, while often focusing more heavily on carbohydrates. For instance, in the case of male boxers, the Pareto method provided only 50% of the required protein and was 91% below the fat target, leading to a highly imbalanced diet. Similar patterns were seen across other athlete categories, where critical shortfalls in proteins and fats made the dietary plans unsuitable for real-world athletic performance, despite carbohydrates being closer to the target in some cases.

In contrast, the Bounded Knapsack method produced much more balanced results across all categories, coming very close to the required protein and carbohydrate targets, while offering significantly better fat optimization compared to the Pareto method. For example, male boxers received 72.6g of protein, just 3.2% below the target, and the fat intake, though still not perfect, was much closer to the required level at 69g compared to the Pareto method's mere 8g. This trend of achieving near-optimal protein and carbohydrate levels, with reasonable fat intake, was consistent across all athlete types. The Bounded Knapsack approach thus ensured that no macronutrient was severely compromised, providing a far more suitable and realistic nutritional plan for athletes.

In summary, while the Pareto method struggled to achieve balanced diets, particularly in proteins and fats, the Bounded Knapsack method delivered more well-rounded solutions, optimizing all macronutrients much closer to their required values. This makes the Bounded Knapsack approach more practical and reliable for dietary planning, especially in athletic contexts where precision and balance are critical for performance and recovery.

7 Conclusion

Our study highlights the novelty of using the Bounded Knapsack method for dietary optimization, offering a structured and targeted approach that ensures no single nutritional factor dominates the plan. By effectively balancing critical macronutrients, our approach achieves a higher degree of accuracy and reliability compared to traditional methods. This makes it particularly well-suited for real-world applications, such as sports nutrition, where precision is crucial. The structured penalty system of the Bounded Knapsack method consistently ensures that dietary solutions align with exact nutritional targets, leading to more predictable and effective outcomes.

In contrast, the Pareto method often results in trade-offs that lead to significant deviations in essential nutrients like protein and fats, limiting its suitability for performance-oriented dietary planning. Beyond athletes, our method's adaptability makes it beneficial for the general population, providing personalized and nutritionally balanced diet plans that cater to various health and wellness goals. By adjusting nutrient weights, the Bounded Knapsack framework offers a versatile solution for anyone looking to optimize their diet.

As part of our future work, we aim to expand this approach by incorporating more granular dietary preferences and constraints. This includes not only dividing dietary plans across diverse food choices and itineraries but also introducing a feature that prevents the repetition of the same dish within a given timeframe. Furthermore, we plan to enhance personalization by focusing on specific nutrients and ingredients to better align with user preferences and health requirements, including the avoidance of allergens. Such advancements would allow our approach to cater to a broader range of users, including those with specific dietary restrictions or unique nutritional goals. These enhancements will ensure smoother transitions between diet plans, foster higher user satisfaction, and pave the way for a more focused and inclusive dietary optimization framework.

References

1. Spriet, L.L.: Performance nutrition for athletes. Sports Med. **49**(S1), 1–2 (2019)
2. Kaufman, M., Nguyen, C., Shetty, M., Oppezzo, M., Barrack, M., Fredericson, M.: Popular dietary trends impact on athletic performance: a critical analysis review. Nutrients **15**(16), 3511 (2023)
3. R. S., Cs 451 ci project diet optimization plan (2025). Accessed 22 Jan 2025
4. Pinter, B., Vassányi, I., Gaál, B., Mák, E., Kozmann, G.: Personalized Nutrition Counseling Expert System, pp. 957–960. Springer, Heidelberg (2011)
5. Jia, N., Chen, J., Wang, R., Li, M.: Towards automatically generating meal plan based on genetic algorithm. Soft. Comput. **28**(9–10), 6893–6908 (2024)
6. Salloum, G., Tekli, J.: Automated and personalized meal plan generation and relevance scoring using a multi-factor adaptation of the transportation problem. Soft. Comput. **26**(5), 2561–2585 (2021)
7. Ferrario, P.G., Gedrich, K.: Machine learning and personalized nutrition: a promising liaison? Eur. J. Clin. Nutr. **78**(1), 74–76 (2023)
8. Stefanidis, K., et al.: PROTEIN AI advisor: a knowledge-based recommendation framework using expert-validated meals for healthy diets. Nutrients **14**(20), 4435 (2022)
9. Syahputra, M.F., Felicia, V., Rahmat,R.F., Budiarto, R.: Scheduling diet for diabetes mellitus patients using genetic algorithm. J. Phys. Conf. Ser. **801**, 012 033 (2017)
10. Corne, D., Lones, M.A.: Evolutionary algorithms, pp. 1–22 (2018)
11. Chen, X.-C., et al.: Multiobjective evolutionary algorithm with double-level archives for nutritional dietary decision problem. In: 2019 9th International Conference on Information Science and Technology (ICIST) (2019)

12. Pei, Z., Liu, Z.: Nutritional diet decision using multi-objective difference evolutionary algorithm. In: 2009 International Conference on Computational Intelligence and Natural Computing (2009)
13. Springmann, M., Clark, M., Willett, W.: Feedlot diet for Americans that results from a misspecified optimization algorithm. In: Proceedings of the National Academy of Sciences, vol. 115, no. 8 (2018)
14. Nilu, T.Y., Ahmed, S., Ahmed, H.: Analysis of diet choice towards a proper nutrition plan by linear programming. Sci. J. Appl. Math. Stat. **8**(5), 59 (2020)

Building Cross-Sectional Trading Strategies via Geometric Semantic Genetic Programming

Kritpol Bunjerdtaweeporn$^{(\boxtimes)}$ and Alberto Moraglio$^{(\boxtimes)}$

Department of Computer Science, University of Exeter, Exeter, UK
{kb801,a.moraglio}@exeter.ac.uk

Abstract. Cross-sectional trading strategies involves constructing portfolios by comparing expected performance of assets within a group, typically using predicted returns. In this study, we frame the estimation of cross-sectional expected returns as a symbolic regression problem, and investigate the predictive capabilities of geometric semantic genetic programming in developing cross-sectional trading strategies in the U.S. stock market. We employ standard genetic programming and other common methods used for studying cross-sectional returns as baselines for comparison. Our findings indicate that geometric semantic genetic programming provides better forecast accuracy, portfolio performance, and ranking accuracy than standard genetic programming. Furthermore, we show the limitations of errors-based metrics as performance measurement in cross-sectional trading strategies.

Keywords: Geometric Semantic Genetic Programming · Portfolio Construction · Stock Returns Prediction · Stock Selection

1 Introduction

Cross-sectional trading (CST) strategies are a widely used investment approach in asset management. The goal of CST strategies is to build a portfolio by buying assets expected to outperform and selling those expected to underperform within an asset group at a specific point in time. Assets are compared using various criteria, ranging from a single characteristic[1] to output of models that combine multiple characteristics. A common criterion for evaluating assets is their predicted returns. Forming an effective portfolio in CST strategies requires understanding why certain assets are expected to give higher or lower returns than others, rather than focusing on how each asset returns will change over time. The models are found by estimating the cross-sectional expected asset returns, which can be formulated as a symbolic regression problem. The portfolio formed by CST strategies profits from relative differences in asset returns with minimal influenced by

---

[1] In the context of computer science, a characteristic of an asset is analogous to a predictor or feature, terms which will be used interchangeably.

© The Author(s), under exclusive license to Springer Nature Switzerland AG 2025
P. García-Sánchez et al. (Eds.): EvoApplications 2025, LNCS 15612, pp. 20–37, 2025.
https://doi.org/10.1007/978-3-031-90062-4_2

overall market movement [15]. Linear models estimated via ordinary least squares (OLS) have long been the preferred choice in both research and practice for cross-sectional returns forecasting due to its simplicity and effectiveness. In recent years, numerous machine learning (ML) applications [13,16,17] have been proposed to improve predictions over OLS. The improved predictive performance of these ML models is largely attributed to their ability to capture non-linear signals [6]. Geometric semantic genetic programming (GSGP) is a variant of genetic programming (GP) that uses geometric semantic operators (GSOs) to replace the standard genetic operators. GSOs apply precise syntax modification to individuals, resulting in predictable and well-known geometric properties in their semantics. Moreover, the fitness landscape seen by GSOs is unimodal error surface for any supervised ML problem. Given the previous success of ML models in improving prediction accuracy, GSGP presents a promising, yet unexplored, approach for building CST strategies. In this paper, we examine predictive power of GSGP for constructing CST strategies in the U.S. stock market, exploring its potential to enhance forecasting accuracy and portfolio performance. The main contributions of this study are threefold. First, we introduce a novel application of GSGP. Second, we present a comparative analysis of GSGP and the variant in which the optimal mutation step is calculated for geometric semantic mutation operator ($GSGP_o$) [14,29] against standard GP and other common methods used for studying the cross-sectional stock returns, including OLS, LASSO, gradient boosted regression trees (GBRT), random forest (RF), and neural networks (NNs) [7,11,16]. Our results show that both GSGP and $GSGP_o$ offer additional value gains over standard GP, providing better forecasting accuracy, portfolio performance and ranking performance. Comparing to the remaining methods, $GSGP_o$ portfolio tends to show competitive performance across overall metrics, ranking near the top-performing models, although it may not always be the absolute best. Third, we empirically show the limitations of error-based metrics when used for evaluating CST strategies or as an optimisation objective. The rest of this paper is organised as follows: Sect. 2 introduces CST strategies. Section 3 outlines methodology. Section 4 describes experimental setup. Section 5 presents results and discussion. Section 6 concludes and suggests future work.

2   Cross-Sectional Trading Strategies

Cross-sectional trading (CST) strategies involve selecting assets by comparing their relative performance at a specific point in time. In this paper, we consider a set of stocks as the primary asset class for our CST strategy. Essentially, CST strategies focus on buying stocks that are expected to outperform in a long portfolio and selling those that are expected to underperform in a short portfolio. Forming a portfolio in CST strategies typically involves four main steps [37]: *score calculation, score ranking, stock selection,* and *portfolio construction.*

Score Calculation. The score for each individual stock is computed based on its corresponding characteristics. This process produces a vector of scores for all stocks as the final output. For example, an individual stock score could be determined by a single characteristic such as its earnings-to-price ratio [3] or through more sophisticated models that combine several characteristics [16,24,37].

Score Ranking. Score ranking involves sorting the score vector, where stocks with the highest scores are considered the best. The sorting procedure produces a predicted rank list.

Stock Selection. The selection step involves retaining some groups of the stocks based on their rank to form a long-short portfolio. Stocks expected to outperform are included in a long portfolio, those expected to underperform in a short portfolio, while stocks ranked in the middle are excluded.

Portfolio Construction. Finally, weights are assigned to the selected stocks such that the total in both long and short portfolio sums to zero. Some common weight allocation schemes include equal-weighting, where weights are distributed equally to each stock, and characteristic-weighting, which allocates weights based on stock-specific characteristics, such as market capitalisation [11,16] or in proportion to the inverse of their historical volatility [37].

CST strategies contrast with time-series trading (TST) strategies, where trading decisions for each asset are based on expected individual performance. The key difference lies in the threshold for buying or selling assets: TST strategies use a zero-return threshold, while CST strategies employ average return as the benchmark. For example, consider a scenario with 100 stocks, each predicted to yield a positive return. CST strategies would buy a group of stocks with the highest expected return and sell those with the lowest, while TST strategies would buy all 100 stocks.

3   Methodology

This section introduces a general framework for stock returns prediction. We then describe our forecasting models and provide a brief explanation of their implementation in each subsection.

3.1   General Framework

The score for each stock is calculated based on its predicted future returns corresponding to its characteristics. Suppose $\mathbf{S} = \{s_1, \ldots, s_N\}$ represents the set of N stocks, where s_i denotes an individual stock. For simplicity, we assume that the number of stocks remains constant over time $t = 1, \ldots, T$. We consider the relationship between stock returns and their characteristics as a general additive prediction model, as in Gu et al. [16]:

$$r_{i,t+1} = \mathbb{E}_t(r_{i,t+1}) + \epsilon_{i,t+1} \qquad (1)$$

where $r_{i,t+1}$ is the return on stock s_i at time $t + 1$. The cross-sectional expected return $\mathbb{E}_t(r_{i,t+1})$ is assumed to be represented by some *true* underlying function g^*, which takes stock characteristics as an input:

$$\mathbb{E}_t(r_{i,t+1}) = g^*(p_{i,t}) \tag{2}$$

where $p_{i,t} = (p_{i,t}^{(1)}, ..., p_{i,t}^{(K)}) \in \mathbb{R}^K$ denotes a vector of K stock characteristics of stock s_i at time t. The objective is to approximate a function g^* by estimating a function g that maps $p_{i,t}$ into a single real value, $g : \mathbb{R}^K \to \mathbb{R}$, a task that can be viewed as a symbolic regression problem. The form of an estimated function g is left unspecified, allowing it to be either linear or non-linear, as well as parametric $g(p_{i,t}; \theta)$ or non-parametric $g(p_{i,t})$. Although the algorithms used to approximate g^* vary, the output $g(p_{i,t})$ generally aims to predict the *true* stock returns by minimising the mean squared forecast errors (MSFE), defined as:

$$MSFE(g) = \frac{1}{N(T-1)} \sum_{t=1}^{T-1} \sum_{i=1}^{N} \left(r_{i,t+1} - g(p_{i,t}) \right)^2 \tag{3}$$

We use $g(p_{i,t})$ instead of $\hat{r}_{i,t+1}$ to emphasise that scores do not necessarily need to be predicted returns but rather the output of a function. By assuming g^* to be time-invariant, the MSFE function utilises pooled data across the entire panel (see Fig. 1) to estimate the model once, ignoring the time dimension rather than estimating it across each time as in the Fama-MacBeth framework [17]. This approach reduces the computational time required by ML algorithms.

3.2   Forecast Models

Our selection of algorithms to be compared with GSGP are those commonly used in the cross-sectional studies for estimating expected stock returns including nature-inspired algorithms, penalised linear models, and tree-based approaches [7,11,16,24].

Genetic Programming (GP). GP is an evolutionary computation method based on the principle of Darwinian evolution. The population contains the candidate solutions of function g, represented in the form of tree structure. While the advantage of using GP lies in its flexibility to choose fitness function(s) specifically tailored to the nature of the problem [4,5,21,24], we use MSFE as our fitness function to ensure direct comparability with other algorithms and to serve as a baseline for GSGP. Our function set consists of both linear and non-linear operators: $\{+, -, *, aq, sin, tanh\}$. The combination of linear function and function such as *sin* or *tanh* in our function set enables GP to approximate any arbitrary function similar to the universal approximation capability of neural network [33,40]. Analytic quotient (aq) [32] is used to remove discontinuities that can often arise from using unprotected and protected division. Replacing protected division with aq in the function set improves the generalisation ability of the evolved model in symbolic regression, an issue that cannot be resolved through model selection using a validation dataset [33]. Our terminal set consists of stock characteristics and an ephemeral constant that return a value in range $[-1, 1]$. While Liu et al. [24] consider the population size ($npop$) and the maximum number of generations ($ngen$) to be the two most important hyperparameters for a similar problem, they found that $ngen$ plays a more critical role. In our preliminary runs, the training fitness did not vary significantly with different values of $npop$ and when $ngen$ exceeded 40. Therefore, we consider $ngen$ and maximum tree depth ($maxdepth$) to be the two most important hyperparameters, keeping $npop$ fixed. When recombination operators result in an individual that exceed the tree depth limit, we randomly return one of its parents. To further improve the model robustness, we adopt an ensemble approach in our model training [22, 24, 41]. Specifically, we select the five best unique individuals from the final population and form a prediction by taking the arithmetic mean of each individual forecast.

Geometric Semantic Genetic Programming (GSGP). GSGP uses geometric semantic operators (GSOs) to replace the standard syntax-based genetic operators. Given N inputs corresponding to the characteristics of each stock, the semantic of a function g is defined as $s(g) = \big(g(p_1), \ldots, g(p_N)\big)$. This vector can be represented as a point in N-dimensional space, called semantic space. Note that the target vector $\vec{r} = (r_1, \ldots, r_N)$ is also a point in the semantic space. The offspring's semantic produced by GSOs is predictable with well-known geometric properties. Moreover, the fitness landscape seen by GSOs is unimodal for any supervised ML problem. The distance between the target semantic and individual semantic can be used as the fitness function in GSGP, which is MSFE for our problem. Geometric semantic crossover (GSCX) and geometric semantic mutation (GSM) are originally defined in Moraglio et al. [31] as:

Definition 1. *Geometric Semantic Crossover (GSCX) Given two parent functions $g_1, g_2 : \mathbb{R}^K \to \mathbb{R}$, $GSCX(g_1, g_2) = g_r \cdot g_1 + (1 - g_r) \cdot g_2$, where g_r is a random function whose codomain(g_r) $\in [0, 1]$*

Definition 2. *Geometric Semantic Mutation (GSM) Given a parent function $g : \mathbb{R}^K \to \mathbb{R}$, $GSM(g) = g + ms \cdot (g_{r_1} - g_{r_2})$, where ms is a mutation step and g_{r_1}, g_{r_2} are random functions*

GSCX has a nice property that the fitness of an offspring is guaranteed not to be worse than the worst of its parents. However, the offspring semantic can reach the target only if the target semantic lie within the convex hull of its parents [35]. Moreover, applying GSCX increases the node size exponentially, making the operator inapplicable in some cases [39]. While applying simplification helps reduce node size, the evolved function often remains large and computationally intensive [27]. In contrast, GSM increases node size linearly and using only GSM can yield comparable or better results than using both GSOs in the evolutionary process [31, 39]. Although the semantic of random functions generated by the original GSM operator are unbounded, later studies showed that limiting the codomain of random functions to a pre-defined interval using a sigmoid function improves generalisation [14, 39]. However, Nicolau and McDermott [34] found that offspring semantic from unbounded GSM have long-tailed distributions in each dimension due to the lack of a bounded radius, with variance differing according to the distribution of training cases. Thus, applying a sigmoid function blindly may result in poor semantic and unnecessary complexity. For these reasons, we adopt a semantic stochastic hill climber (GSGP with population of size 1) that relies solely on GSM in our evolutionary process. We use a variant of GSM defined in [1] as $GSM(g) = g + ms \cdot \mathcal{N}(g_r)$, where $\mathcal{N}(g_r) = 2 * \frac{g_r - min(s(g_r))}{max(s(g_r)) - min(s(g_r))} - 1$. This variant stabilises the output distribution of random function without requiring a sigmoid function and reduces node size by generating a single random function. A random function is generated using either *grow* or *full* method, randomly chosen with equal probability, with the same primitive set used in GP. To ensure the offspring is at least as good as its parent, if the normalised random function $\mathcal{N}(g_r)$ worsen the offspring, then $-\mathcal{N}(g_r)$ is considered. The new solution is accepted only if it improves upon its parent. Additionally, we consider GSGP with an optimal mutation step $(GSGP_o)$ [14, 29], which uses the GSM operator that selects the mutation step to minimise MSFE of the mutated function. We select the best individual (i.e., the one with the lowest training error) from 10 independent evolutionary processes to account for potential poor choice of initial individual [29] and variations in convergence speed. Our GSGP implementation also incorporates higher-order functions and memoization techniques for improved efficiency [30]. A pseudocode for our GSGP evolutionary process is provided in Algorithm 1.

Algorithm 1. Pseudocode for GSGP Evolutionary Process (GSGP)

```
1:  procedure GSGP(F, P, r⃗)                                    ▷ F: Fitness function
2:      ngen ← 0                                          ▷ P: Set of stock characteristics
3:      Initialise random function g₀              ▷ r⃗: A vector of stock returns (target)
4:      while termination criteria not satisfied do
5:          ngen ← ngen + 1
6:          Generate random function gᵣ       ▷ N(gᵣ) = 2 * (gᵣ−min(s(gᵣ)))/(max(s(gᵣ))−min(s(gᵣ))) − 1
7:          Normalise gᵣ to obtain N(gᵣ)
8:          Calculate ms (pre-determined or based on N(gᵣ))
9:          g_temp ← g_{ngen−1} + ms · N(gᵣ)
10:         if F(g_temp; P, r⃗) is better than F(g_{ngen−1}; P, r⃗) then
11:             g_ngen ← g_temp
12:         else
13:             g_temp ← g_{ngen−1} − ms · N(gᵣ)
14:             if F(g_temp; P, r⃗) is better than F(g_{ngen−1}; P, r⃗) then
15:                 g_ngen ← g_temp
16:             else
17:                 g_ngen ← g_{ngen−1}
18:     return g_ngen
```

Ordinary Least Squares (OLS). OLS is the least complex method in our analysis. It serves as a baseline to assess the additional value gained from using more sophisticated algorithms. Despite its simplicity, OLS is widely used in studies of cross-sectional stock returns [2, 3, 9, 20, 23, 25].

LASSO. LASSO adds an l_1-penalty term to the optimisation problem, encouraging parsimonious models by shrinking some coefficients to zero. This helps mitigate overfitting caused by multicollinearity among predictors, as OLS may potentially captures noise instead signal. We use LASSO since it is analogous to the support vector machine algorithm [19].

Gradient Boosted Regression Trees (GBRT) and Random Forest (RF). Both GBRT and RF are built on regression tree (RT) using the CART algorithm as it is one of the most widely used algorithm among the existing options [26]. Ensemble methods like GBRT and RF aim to tackle overfitting by combining predictions from multiple simple trees into one consensus forecast, rather than building a single complex tree. GBRT is constructed via stagewise additive expansions. The process starts by fitting a simple RT. Next, a second RT is added by fitting prediction errors of the previous RT. The forecast of the second tree is shrunken by some small positive value, called learning rate, to help prevent the model from overfitting the residuals. The process is repeated for a predetermined number of iterations. The final output is an additive model of simple RTs. RF is built based on the idea of bootstrap aggregation or bagging. Bagging involves randomly selecting samples from the training data with replacement. These bootstrapped samples are then used to construct simple RTs. This process is repeated multiple times. RF extend bagged trees by randomly selecting a subset of features at each split, which helps reducing the correlation among trees across different bootstrap samples.

Neural Networks (NNs). We consider the feedforward neural networks architecture, where each node is connected to all nodes in the previous layer and the connections follow a one-way direction, from the input to output layer. Each node in the input layer represents a predictor and the output layer contains a single node producing the score output. We use Rectified Linear Units (ReLU) as our activation function for its computational efficiency and universal approximation capability. We avoid overfitting by adopting multiple regularisation strategy. First, we use stochastic gradient descent to train our NNs. Second, we apply learning rate shrinkage, where the learning rate is divided by 5 each time MSFE failed to decrease training loss or failed to improve the validation samples, retained from the training samples by some pre-determined threshold. Third, we train NNs with multiple random seeds to form an ensemble to improve generalisation ability [18]. Specifically, we train NNs with five independent random seeds. The predictions are then averaged as the final output. Lastly, we use l_2-penalty term to the weight parameters, shrinking the weights toward zero. We consider NNs with up to three hidden layers, denoted as NN_1, NN_2, and NN_3, selected according to the pyramid rule [28], as using more hidden layers could potentially lead to performance deterioration [16].

The implementation of our models utilises well-established frameworks. We employ Scikit-learn [36] for OLS, LASSO, GBRT, RF, and NNs. For GP, GSGP, and $GSGP_o$, the implementations are based on DEAP framework [12]. The hyperparameters are selected from a comprehensive set of parameter specifications (see Table A1), following commonly used choices in the literature [7,11,16,24].

4 Experimental Setup

We begin with a description of the dataset and preprocessing steps. Next, we explain the sample splitting and hyperparameter tuning process. Lastly, we detail the performance evaluation methods used to assess each algorithm.

4.1 Data

The data are sourced from *FactSet financial data and analytics*, obtained with a permission of *Jupiter Asset Management Systematic Equities* team. We consider the latest available values of the data at the end of each month, represented as a panel, shown in Fig. 1. Our stock universe consists of all U.S. firms listed in MSCI North America. The sample spans from January 1990 to December

Month	Predictor 1	...	Predictor K	Return
	$p_{1,1}^{(1)}$	...	$p_{1,1}^{(K)}$	$r_{1,2}$
1	$\vdots$	$\vdots$	$\vdots$	$\vdots$
	$p_{N,1}^{(1)}$	...	$p_{N,1}^{(K)}$	$r_{N,2}$
$\vdots$	$\vdots$	$\vdots$	$\vdots$	$\vdots$
	$p_{1,T-1}^{(1)}$	...	$p_{1,T-1}^{(K)}$	$r_{1,T}$
T-1	$\vdots$	$\vdots$	$\vdots$	$\vdots$
	$p_{N,T-1}^{(1)}$	...	$p_{N,T-1}^{(K)}$	$r_{N,T}$

Fig. 1. An example of dataset presented in a panel format

2022, encompassing approximately 3000 stocks in total, with an average of 650 stocks per month. We consider 12 stock characteristics, as described in Fig. 2. Since each stock characteristic differs in their range of values and unit, it is not straightforward to intuitively compare them. Therefore, we employ a typical transformation to each stock characteristic by converting them into ranks and map into an interval $[-1, 1]$ across each time [7,11,16]. Specifically, we rank each characteristic in ascending order then divided by the number of stocks at each month. The transformed values in the range $[0, 1]$ are multiplied by 2 then subtracted by -1. We use stock returns instead of their excess returns of treasury-bill rate on our analysis to represent the actual performance without any adjustments. The returns are winsorized each month at 1% and 99% during model training to handle outliers, while retaining their original value during portfolio formation. We only include stock observations with complete data on returns and all stock characteristics, i.e., observations with no missing values.

#	Predictor	Definition
1	Short-term reversal [25]	Returns of the prior month
2	Momentum [20]	Returns from 12 months prior to 2 months prior
3	Long-term reversal [9]	Returns from 36 months prior to 13 months prior
4	Size [2]	Market value of an equity at the end of prior month
5	Earnings-to-price [3]	Net income in the prior fiscal year divided by market capitalisation at the end of the prior month
6	Asset turnover	Net sales and revenues in the prior year divided by market capitalisation at the end of prior month
7	Beta	Market beta estimated from daily returns against MSCI North America from the prior 12 months
8	Book-to-price	Book value of equity divided by market value of equity
9	Debt-to-assets	Short term plus long term debts to total assets
10	Volatility	Daily returns standard deviation from the prior 252 days
11	Dividend yield	Dividends per share over the prior 12 months divided by price at the end of the prior month
12	Return on assets	Income before extraordinary items divided by average total assets in the prior year

Fig. 2. Description of all stock characteristics

4.2   Sample Splitting and Hyperparameter Tuning via Validation

The data are divided into three disjoint parts: *training*, *validation*, and *out-of-sample*, while keeping the temporal order of the dataset, following common practices [7, 11, 16]. The training data are used to estimate the model for each combination in the hyperparameter set. The validation data are used for hyperparameter tuning. The out-of-sample data, which have neither been used for training nor hyperparameter tuning, are used to evaluate the method's predictive performance. Specifically, forecasts are generated based on the model trained with training samples. MSFE is then calculated using the validation samples for each combination of hyperparameters. This hyperparameter tuning step, which preserves the temporal ordering of the data, has been shown to be preferable to the standard cross-validation scheme for data with a time dimension [8, 38]. The model with the hyperparameter combination that best minimises MSFE is selected to forecast during out-of-sample periods. The data from the first 12 months are initially used as a training sample and the subsequent 12 months as validation sample, making the out-of-sample period span from January 1992 to December 2022, for a total of 372 months. The models are retrained on an annual basis. The chosen hyperparameter combination of the trained model is fixed, making out-of-sample predictions over the next 12 months. The training sample increases by 12 months while retaining the entire history. The validation sample remains a fixed size and is rolled forward by 12 months. Figure 3 illustrates the sample splitting and model retraining process.

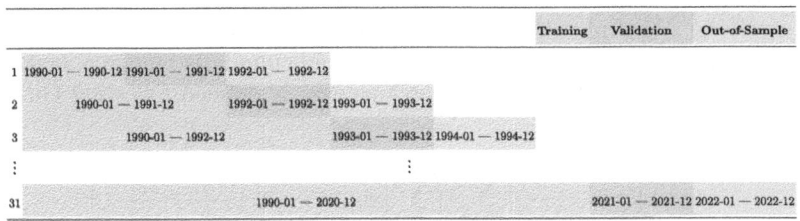

				Training	Validation	Out-of-Sample
1	1990-01 — 1990-12 1991-01 — 1991-12 1992-01 — 1992-12					
2	1990-01 — 1991-12	1992-01 — 1992-12 1993-01 — 1993-12				
3	1990-01 — 1992-12	1993-01 — 1993-12 1994-01 — 1994-12				
⋮		⋮				
31	1990-01 — 2020-12				2021-01 — 2021-12 2022-01 — 2022-12	

Fig. 3. Sample Splitting and Model Retraining

4.3   Performance Evaluation

The performance of each method is evaluated using various type of metrics, including prediction accuracy, portfolio performance, and ranking accuracy. All calculations are made during out-of-sample period.

We assess the statistical significance of the differences in forecast errors among models using a modified Diebold-Mariano test [10], as defined in Gu et al. [16]. The Diebold-Mariano test is adapted by comparing average prediction errors over time rather than comparing the errors of each individual forecast. Specifically, to test for the forecast performance of method (1) against method (2), the test statistic DM_{12} is defined as:

$$DM_{12} = \frac{\bar{d}_{12}}{\hat{\sigma}_{\bar{d}_{12}}} \tag{4}$$

where,

$$d_{12,t} = \frac{1}{N} \sum_{i=1}^{N} \left(\left(r_{i,t} - g_1(p_{i,t}) \right)^2 - \left(r_{i,t} - g_2(p_{i,t}) \right)^2 \right) \tag{5}$$

$g_1(p_{i,t})$ and $g_2(p_{i,t})$ denote the output score for stock s_i at time t using each method. $\bar{d}_{12}$ and $\hat{\sigma}_{\bar{d}_{12}}$ denote the mean and Newey-West standard error of $d_{12,t}$ over the out-of-sample period. We compare the predictive performance against a naive prediction using R^2_{oos}, defined in Gu et al. [16] as:

$$R^2_{oos} = 1 - \frac{\sum_{(i,t)\in\mathcal{T}_{oos}} (r_{i,t} - g(p_{i,t}))^2}{\sum_{(i,t)\in\mathcal{T}_{oos}} r_{i,t}^2} \tag{6}$$

$\mathcal{T}_{oos}$ indicates that the predictions are only evaluated during the out-of-sample period. R^2_{oos} pools prediction errors across stocks and over time into one panel-level assessment of each model.

We form a long-short portfolio at each point in time during the out-of-sample period, following the procedures described in Sect. 2. Specifically, we divide stocks into ten groups according to their predicted return (i.e., their score). Next, we buy stocks in the top decile to include them in a long portfolio and sell those in the bottom decile in a short portfolio. The selected stocks are weighted equally, ensuring that the total weight sums to 1 in a long portfolio and -1 in a short portfolio. We report the long-short portfolio monthly return mean and its standard deviation, annualised Sharpe ratio, final cumulative returns, maximum drawdown, and turnover. The portfolio turnover is defined following Gu et al. [16] as:

$$Turnover = \frac{1}{|\mathcal{T}_{oos}|} \sum_{t \in \mathcal{T}_{oos}} \left(\sum_{i \in \mathbf{S}} \left| w_{i,t} - \frac{w_{i,t-1}(1 + r_{i,t})}{1 + \sum_j w_{j,t-1} r_{j,t}} \right| \right) \tag{7}$$

We also assess the correctness of the predicted rank list by computing the average of Spearman's rank correlation coefficient across each time, denoted as ρ_{oos}.

5   Results and Discussion

We report the run closest to the median Sharpe ratio out of 30 simulation runs, as it aligns towards the investment objective and likely presents how each method will perform in practice. All calculations are made based on out-of-sample period.

5.1   Results

Table 1 reports pairwise comparison of the test statistics of modified Diebold-Mariano test. Negative value indicates that the row method outperform the column method by having lower average forecast errors and vice versa. Bold font indicates that the difference is significant at 5% level. Asterisk sign indicates that the difference is significant after conservative Bonferroni adjustment.

Table 1. This table reports pairwise modified Diebold-Mariano test statistics comparing the average forecast errors during the out-of-sample period. Negative value indicates that the row method outperforms the column method, i.e., it has lower average forecast errors and vice versa. Bold font indicates that the difference is significant at the 5% level. Asterisk sign indicates that the difference is significant after Bonferroni adjustment.

	OLS	LASSO	GBRT	RF	NN_1	NN_2	NN_3	GP	GSGP	$GSGP_o$
OLS	–	**5.81***	0.78	**3.79***	**2.38**	0.60	1.06	**−10.15***	**−4.11***	**−5.92***
LASSO		–	−0.98	1.04	**−5.28***	**−3.07**	**−3.06**	**−13.23***	**−7.42***	**−9.07***
GBRT			–	2.06	−1.78	−0.51	−0.32	**−7.94***	**−2.51**	**−3.71***
RF				–	**−4.98***	**−3.24***	**−3.28***	**−11.81***	**−5.84***	**−7.37***
NN_1					–	**2.91**	**2.92**	**−8.26***	−1.59	**−3.47***
NN_2						–	0.45	**−10.17***	**−3.86***	**−5.52***
NN_3							–	**−11.41***	**−4.29***	**−6.07***
GP								–	**8.28***	**6.91***
GSGP									–	**−2.32**
$GSGP_o$										–

The forecast errors of GSGP and $GSGP_o$ are significantly less than those of standard GP, with GSGP considered best. However, after applying the conservative Bonferroni adjustment, there is no significant difference between GSGP and $GSGP_o$. Nonetheless, their forecast errors remain relatively high compared to the remaining methods.

Figure 4 shows the cumulative returns of the decile long-short portfolio as selected by the models during out-of-sample period. The portfolio performance and R^2_{oos} for each method are summarised in Table 2. The numbers are presented without transaction cost, highlighting the raw predictive ability. The results show that both GSGP and $GSGP_o$ achieve higher R^2_{oos} than GP, with GSGP slightly higher than $GSGP_o$. Nonetheless, their R^2_{oos} are still relatively lower than the rest of the methods. $GSGP_o$ achieves the best portfolio performance among the GP-based approaches. Although GP portfolio has lower volatility, its risk-adjusted return, Sharpe ratio, is significantly lower at 0.06, which is more than four times lower. The final cumulative return of GP portfolio is the only one that is less than 1, indicating losses at the end of the period. $GSGP_o$ ranked behind OLS, NN_1, and NN_2 in terms of return mean, Sharpe ratio, and final cumulative return but had a lower maximum drawdown compared to those methods, except for NN_1.

Figure 5 illustrates stock characteristics importance across each method. We calculate the importance of each stock characteristic following the approach in [7,16]. Specifically, the importance of a stock characteristic $p^{(k)}_{i,t}$ is measured by the reduction in R^2_{oos} when it is set to 0 for all predictions at each time period, while other characteristics remain unchanged. The value for each characteristic are normalised to sum to 1 for relative interpretation. The colour gradient within each column indicates the importance of the characteristics for a particular method, with darker colours representing greater importance. The figure

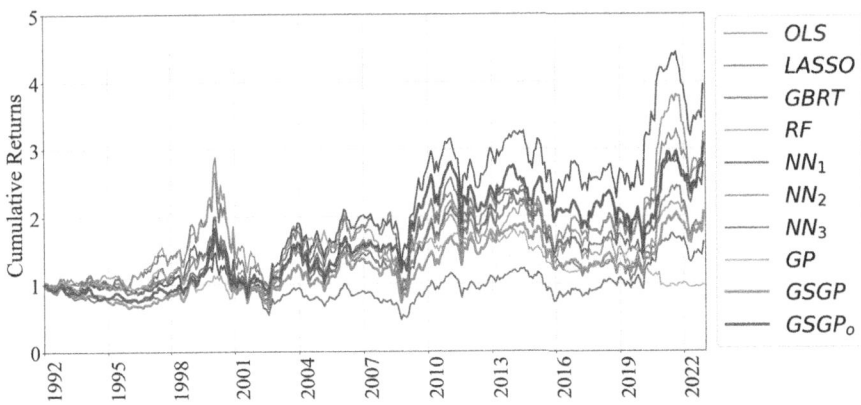

Fig. 4. This figure shows the cumulative returns of decile long-short portfolio, as selected by the models, during out-of-sample period of a run closest to the median Sharpe ratio out of 30 simulation runs for each method

Table 2. This table compares the performance of decile long-short portfolio during out-of-sample periods across methods. The numbers are presented for equal-weighting scheme with monthly rebalance excluding transaction cost, highlighting the raw predictive ability of the models.

	OLS	LASSO	GBRT	RF	NN_1	NN_2	NN_3	GP	GSGP	$GSGP_o$
Return Mean (%)	0.55	0.43	0.33	0.41	0.56	0.50	0.46	0.06	0.39	0.49
Return Std (%)	6.86	7.11	6.39	6.88	6.20	6.29	6.15	3.47	6.37	6.26
Sharpe Ratio	0.28	0.21	0.18	0.21	0.31	0.28	0.26	0.06	0.21	0.27
Final CumRet	3.24	1.97	1.62	1.98	3.95	3.14	2.79	0.97	2.06	3.08
MaxDD (%)	59.83	66.10	72.13	68.12	49.75	62.15	63.42	47.68	64.17	56.37
Turnover (%)	103.12	82.84	96.55	89.44	103.81	105.69	109.76	82.24	101.09	98.25
R^2_{oos} (%)	0.71	0.75	0.73	0.77	0.68	0.71	0.72	0.44	0.65	0.61
ρ_{oos} (%)	0.59	0.41	0.30	0.01	0.69	0.30	0.27	0.02	0.19	0.57

suggests that linear models like OLS and LASSO consider *Beta* to be the most important characteristic, while GBRT, NN_1, and $GSGP_o$ prioritise *Volatility*; other methods prioritise different characteristics.

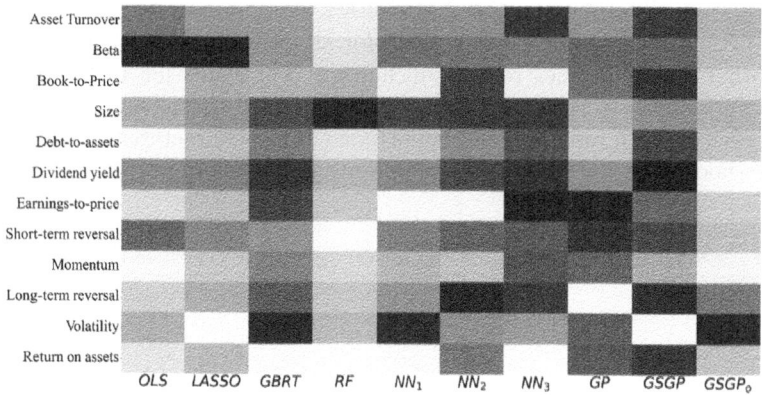

Fig. 5. This figure shows stock characteristics importance across methods. The importance of a characteristic is calculated as the reduction in R^2_{oos} when its values are set to 0, while other characteristics remain fixed. The values are normalised to sum to 1 for relative interpretation. The colour gradient within each column indicates the importance of the characteristics of a particular method, with darker colours representing greater importance.

5.2   Discussion

Our results show that both GSGP and $GSGP_o$ offer advantages over standard GP in terms of forecasting accuracy, with improvements in both forecast errors (see Table 1) and R_{oos}^2 (see Table 2), contrary to previous findings suggesting GSGP limited benefits in financial applications [29]. These improvements likely result from periodic model retraining and hyperparameter updates with new data applied here, a strategy not fully explored in prior studies. Additionally, GSGP and $GSGP_o$ achieve better portfolio performance and higher ρ_{oos} than standard GP, with $GSGP_o$ generally outperforming GSGP. The differences between their portfolios performance are likely due to the distinct signals each method capture (see Fig. 5). When compared to the remaining methods, the forecast errors of GSGP and $GSGP_o$ remains relatively high (see Table 1) and their values of R_{oos}^2 are somewhat lower (see Table 2). Nonetheless, $GSGP_o$ portfolio tends to show competitive performance across non-error-based metrics, ranking near the top-performing models in each aspect, though it might not always come out on top.

Interestingly, OLS portfolio performs well and is difficult to beat, ranking third in Sharpe ratio, with final cumulative return and ρ_{oos} just behind NN_1. This could be partly explained by the smaller predictor set used in this study, as previous studies has shown that the additional values provided by ML algorithms often comes from larger predictor sets, e.g., 920 predictors in Gu et al. [16] and over 100 predictors in Cakici et al. [7]. However, only 12 predictors, which likely contain fewer non-linear signals, are considered here. Nonetheless, this explanation does not account for all observed outcomes, particularly the inconsistencies in portfolio performance and forecasting errors across many method pairs e.g., $GSGP_o$ against the remaining methods, except GP and LASSO against NNs.

A more plausible explanation for these discrepancies lies in the limitation of MSFE as an optimisation objective. Although perfectly minimised MSFE would ideally capture the correct relative order among stock returns, even low but non-zero MSFE values can still lead to large ranking errors. Because CST strategies prioritise the relative performance of assets over absolute accuracy, MSFE fails to account for important aspect like the order of asset returns. As a result, the portfolio performances likely depends more on the signals each method capture rather than their forecast errors. Ranking metrics such as ρ_{oos} better capture the effectiveness of CST strategies compared to traditional error-based metrics like the modified Diebold-Mariano test and R_{oos}^2. These findings suggest that the current definition of semantics in semantics GP framework overlooks cases where the relative order of outputs is more critical than their precise values.

6   Conclusion and Future Work

We introduce an application of GSGP in developing CST strategies in the U.S. stock market. The predictive power of GSGP and its variant in which the optimal mutation step is used in GSM operator ($GSGP_o$) are compared against standard GP and other common methods used for studying cross-sectional stock returns including OLS, LASSO, GBRT, RF, and NNs. The results show that both GSGP and $GSGP_o$ offer additional value gains over standard GP, providing better forecasting errors, R_{oos}^2, ρ_{oos}, and portfolio performance. Part of the success is attributed to periodic model retraining and hyperparameter updates with new data, previously overlooked. GSGP and $GSGP_o$ forecast errors remain high compared to the remaining methods. However, the portfolio performance of $GSGP_o$ often ranked among the top across metrics, though it may not always be the winner. The findings further suggest that MSFE, commonly used in GSGP and other regression models, may be suboptimal as an optimisation objective for building CST strategies as it overlooks the relative order of asset returns. In future work, we aim to explore the incorporation of semantics that consider order structure among outputs within GSGP framework, potentially leading to more accurate portfolio formation and improved performance in CST strategies.

Acknowledgments. We thank Dr. Linquan Chen, Matus Mrazik, and Jupiter Systematic Equities team for their helpful comments.

Disclosure of Interests. The first author is a PhD student partially supported by Jupiter Asset Management. The views and findings expressed in this paper are solely those of the authors and do not necessarily reflect the opinions or positions of Jupiter Asset Management or its subsidiaries.

A   Hyperparameter Tuning

Table A1. This table describes hyperparameter combinations for tuning with validation samples, with all possible combination are formed from a Cartesian product. For example, for an algorithm with hyperparameters $A = \{a_1, a_2\}$ and $B = \{b_1, b_2\}$, the pairs (a_1, b_1), (a_1, b_2), (a_2, b_1), (a_2, b_2) are considered.

	Hyperparameter	Specification	Definition
OLS	none		
LASSO	alpha	$\{10^{-5}, 10^{-4}, 10^{-3}\}$	Constant that multiplies the l_1 penalty term
	tol	$\{10^{-5}, 10^{-4}, 10^{-3}\}$	If the updates are smaller than *tol*, checks the dual gap for optimality and continues until it is smaller than *tol*
GBRT	n_estimators	$\{50, 100\}$	Number of boosting stages to perform
	learning_rate	$\{10^{-2}, 10^{-1}\}$	The contribution of each tree
	maxdepth	$\{1, 2, 3, 4, 5\}$	Maximum depth of each individual tree
RF	n_estimators	$\{50, 100, 200\}$	Number of trees in the forest
	maxdepth	$\{1, 2, 3, 4, 5\}$	Maximum depth of each individual tree
	max_samples	0.75	Proportion of samples to draw to train each base estimator
	max_features	sqrt	Number of features to consider when looking for the best split
NN_1-NN_3[a]	solver	sgd	Solver for weight optimisation
	learning_rate_init	$\{10^{-2}, 10^{-1}\}$	Initial learning rate
	batch_size	500	Batch size
	validation_fraction	0.2	Proportion of training data to set aside as validation set for early stopping
	tol	$\{10^{-6}, 10^{-5}, 10^{-4}\}$	When the loss or validation score is not improving by at least *tol*, the current learning rate is divided by 5
	alpha	$\{10^{-5}, 10^{-4}, 10^{-3}\}$	Strength of l_2 regularisation term
	max_iter	400	Number of epochs
	ensemble	5	Number of independent random seeds used for model training
GP	npop	200	Number of population
	ngen	$\{10, 20, 40\}$	Number of maximum generation
	maxdepth	$\{10, 20, 30\}$	Maximum depth of an individual after applying genetic operators
GSGP	npop	1	Number of population
	ngen	$\{20, 50, 100\}$	Number of maximum generation
	max_gen_depth	$\{3, 5, 7\}$	Maximum depth for each tree generation
	ms	$\{10^{-3}, 10^{-2}, 10^{-1}\}$	Mutation step
	nrun	10	Number of independent runs
$GSGP_o$	npop	1	Number of population
	ngen	$\{20, 50, 100\}$	Number of maximum generation
	max_gen_depth	$\{3, 5, 7\}$	Maximum depth for each tree generation
	ms	optimal	Mutation step
	nrun	10	Number of independent runs

[a] The hidden layers for neural network with 1, 2 and 3 hidden layers are (32), (32, 16), and (32, 16, 8) respectively

References

1. Bakurov, I., et al.: Geometric semantic genetic programming with normalized and standardized random programs. Genet. Program Evolvable Mach. **25**(1), 6 (2024)
2. Banz, R.W.: The relationship between return and market value of common stocks. J. Financ. Econ. **9**(1), 3–18 (1981)
3. Basu, S.: Investment performance of common stocks in relation to their price-earnings ratios: a test of the efficient market hypothesis. J. Financ. **32**(3), 663–682 (1977)
4. Becker, Y.L., Fei, P., Lester, A.M.: Stock selection: an innovative application of genetic programming methodology. In: Genetic Programming Theory and Practice IV, pp. 315–334 (2007)
5. Becker, Y.L., Fox, H., Fei, P.: An empirical study of multi-objective algorithms for stock ranking. In: Genetic Programming Theory and Practice V. Springer (2008)
6. Bonne, G., Wang, J., Zhang, H.: Machine learning factors: capturing nonlinearities in linear factor models. MSCI Research (2021)
7. Cakici, N., Fieberg, C., Metko, D., Zaremba, A.: Machine learning goes global: cross-sectional return predictability in international stock markets. J. Econ. Dyn. Control **155**, 104725 (2023)
8. Cerqueira, V., Torgo, L., Smailović, J., Mozetič, I.: A comparative study of performance estimation methods for time series forecasting. In: 2017 IEEE International Conference on Data Science and Advanced Analytics (DSAA), pp. 529–538. IEEE (2017)
9. De Bondt, W.F., Thaler, R.: Does the stock market overreact? J. Financ. **40**(3), 793–805 (1985)
10. Diebold, F.X., Mariano, R.S.: Comparing predictive accuracy. J. Bus. Econ. Stat. **13**(3), 253–263 (1995). https://ideas.repec.org/a/bes/jnlbes/v13y1995i3p253-63.html
11. Drobetz, W., Otto, T.: Empirical asset pricing via machine learning: evidence from the European stock market. J. Asset Manag. **22**, 507–538 (2021)
12. Fortin, F.A., De Rainville, F.M., Gardner, M., Parizeau, M., Gagné, C.: DEAP: evolutionary algorithms made easy. J. Mach. Learn. Res. **13**(1), 2171–2175 (2012)
13. Giglio, S., Kelly, B., Xiu, D.: Factor models, machine learning, and asset pricing. Annu. Rev. Financ. Econ. **14**(1), 337–368 (2022)
14. Gonçalves, I., Silva, S., Fonseca, C.M.: On the generalization ability of geometric semantic genetic programming. In: Genetic Programming: 18th European Conference, EuroGP 2015, Copenhagen, Denmark, April 8-10, 2015, Proceedings 18, pp. 41–52. Springer (2015)
15. Goyal, A., Jegadeesh, N.: Cross-sectional and time-series tests of return predictability: what is the difference? Rev. Fin. Stud. **31**(5), 1784–1824 (2018)
16. Gu, S., Kelly, B., Xiu, D.: Empirical asset pricing via machine learning. Rev. Fin. Stud. **33**(5), 2223–2273 (2020)
17. Han, Y., He, A., Rapach, D.E., Zhou, G.: Cross-sectional expected returns: new Fama-MacBeth regressions in the era of machine learning. Rev. Fin. **28**(6), 1807–1831 (2024)
18. Hansen, L.K., Salamon, P.: Neural network ensembles. IEEE Trans. Pattern Anal. Mach. Intell. **12**(10), 993–1001 (1990)
19. Jaggi, M.: An equivalence between the LASSO and support vector machines. In: Regularization, Optimization, Kernels, and Support Vector Machines, pp. 1–26 (2013)

20. Jegadeesh, N., Titman, S.: Returns to buying winners and selling losers: implications for stock market efficiency. J. Financ. **48**(1), 65–91 (1993)
21. Kim, M., Becker, Y.L., Fei, P., O'Reilly, U.M.: Constrained genetic programming to minimize overfitting in stock selection. In: Genetic Programming Theory and Practice VI, Genetic and Evolutionary Computation, pp. 179–195 (2008)
22. Kotanchek, M., Smits, G., Vladislavleva, E.: Trustable symbolic regression models: using ensembles, interval arithmetic and pareto fronts to develop robust and trust-aware models. In: Genetic Programming Theory and Practice V, pp. 201–220 (2008)
23. Lewellen, J.: The cross-section of expected stock returns. Crit. Fin. Rev. **4**(1), 1–44 (2015)
24. Liu, Y., Zhou, G., Zhu, Y.: Maximizing the Sharpe ratio: a genetic programming approach. SSRN Electr. J. **3726609** (2020). https://doi.org/10.2139/ssrn
25. Lo, A.W., MacKinlay, A.C.: When are contrarian profits due to stock market overreaction? Rev. Fin. Stud. **3**(2), 175–205 (1990)
26. Loh, W.Y.: Classification and regression trees. Wiley Interdisc. Rev. Data Min. Knowl. Discov. **1**(1), 14–23 (2011)
27. Martins, J.F.B., Oliveira, L.O.V., Miranda, L.F., Casadei, F., Pappa, G.L.: Solving the exponential growth of symbolic regression trees in geometric semantic genetic programming. In: Proceedings of the Genetic and Evolutionary Computation Conference, pp. 1151–1158 (2018)
28. Masters, T.: Practical Neural Network Recipes in C++. Morgan Kaufmann (1993)
29. McDermott, J., Agapitos, A., Brabazon, A., O'Neill, M.: Geometric semantic genetic programming for financial data. In: Esparcia-Alcázar, A.I., Mora, A.M. (eds.) EvoApplications 2014. LNCS, vol. 8602, pp. 215–226. Springer, Heidelberg (2014). https://doi.org/10.1007/978-3-662-45523-4_18
30. Moraglio, A.: An efficient implementation of GSGP using higher-order functions and memoization. In: Semantic Methods in Genetic Programming, Workshop at Parallel Problem Solving from Nature (2014)
31. Moraglio, A., Krawiec, K., Johnson, C.G.: Geometric semantic genetic programming. In: Coello, C., Cutello, V., Deb, K., Forrest, S., Nicosia, G., Pavone, M. (eds.) PPSN 2012. LNCS, vol. 7491, pp. 21–31. Springer, Heidelberg (2012). https://doi.org/10.1007/978-3-642-32937-1_3
32. Ni, J., Drieberg, R.H., Rockett, P.I.: The use of an analytic quotient operator in genetic programming. IEEE Trans. Evol. Comput. **17**(1), 146–152 (2012)
33. Nicolau, M., Agapitos, A.: Choosing function sets with better generalisation performance for symbolic regression models. Genet. Program Evolvable Mach. **22**(1), 73–100 (2021)
34. Nicolau, M., McDermott, J.: Genetic programming symbolic regression: what is the prior on the prediction? In: Genetic Programming Theory and Practice XVII, pp. 201–225 (2020)
35. Pawlak, T.P., Krawiec, K.: Competent geometric semantic genetic programming for symbolic regression and Boolean function synthesis. Evol. Comput. **26**(2), 177–212 (2018)
36. Pedregosa, F., et al.: Scikit-learn: machine learning in Python. J. Mach. Learn. Res. **12**, 2825–2830 (2011)
37. Poh, D., Lim, B., Zohren, S., Roberts, S.: Building cross-sectional systematic strategies by learning to rank. J. Fin. Data Sci. **3**(2), 70–86 (2021)
38. Schnaubelt, M.: A comparison of machine learning model validation schemes for non-stationary time series data, Technical report, FAU Discussion Papers in Economics (2019)

39. Vanneschi, L., Silva, S., Castelli, M., Manzoni, L.: Geometric semantic genetic programming for real life applications. In: Genetic Programming Theory and Practice XI, pp. 191–209 (2014)
40. Yao, X.: Universal approximation by genetic programming. In: Foundations of Genetic Programming, pp. 66–67 (1999)
41. Zhang, Y., Bhattacharyya, S.: Genetic programming in classifying large-scale data: an ensemble method. Inf. Sci. **163**(1–3), 85–101 (2004)

Adjacent Distance Matrix-Based Competitive Swarm Optimizer

Yang Cao[3] , Rui Zhong[1(✉)] , Jun Yu[2] , and Masaharu Munetomo[1]

[1] Information Initiative Center, Hokkaido University, Sapporo, Japan
{zhongrui,munetomo}@iic.hokudai.ac.jp
[2] Institute of Science and Technology, Niigata University, Niigata, Japan
yujun@ie.niigata-u.ac.jp
[3] Graduate School of Information Science and Technology, Hokkaido University, Sapporo, Japan
yang.cao.y4@elms.hokudai.ac.jp

Abstract. This paper improves the global exploration capability of the Competitive Swarm Optimizer (CSO) by introducing an adjacent distance matrix selection strategy and proposes an Adjacent distance Matrix-based CSO (AMCSO). The traditional CSO updates offspring individuals by randomly selecting two particle individuals for competition, which ignores the differences and similarities of particle individuals. To fully utilize the knowledge of and fitness landscape and particle swarm to guide the selection mechanism in competition, our proposed AMCSO adopts an adjacent distance matrix based on the Euclidean distance between every pairwise particle individual, which allows the selection of particle individuals with closer distance for competition and further ensures population diversity. Comprehensive numerical experiments were conducted on 10-D and 20-D CEC2022 benchmark functions to investigate the performance of AMCSO against the original CSO and four well-known metaheuristic algorithms (MAs), and experimental results and statistical analysis confirm the effectiveness of the integration of the adjacent distance matrix and demonstrate a statistically significant improvement of AMCSO.

Keywords: Competitive Swarm Optimizer · Adjacent Distance Matrix · Metaheuristic Algorithms (MAs) · Numerical Optimization

1 Introduction

With the advancement of computers and information technology, systems and industrial products are becoming increasingly complex, making the optimization of these systems and artifacts more challenging. The demand for solving optimization problems within limited time frames is also growing [1]. Metaheuristic Algorithms (MAs), inspired by biological, physical, or population-based phenomena [2], include popular methods such as Genetic Algorithm (GA) [3], Particle Swarm Optimization (PSO) [4], Differential Evolution (DE) [5], and CSO [6].

© The Author(s), under exclusive license to Springer Nature Switzerland AG 2025
P. García-Sánchez et al. (Eds.): EvoApplications 2025, LNCS 15612, pp. 38–51, 2025.
https://doi.org/10.1007/978-3-031-90062-4_3

These algorithms are designed to find optimal approximate solutions, significantly reducing computational time and resource requirements. Additionally, MAs have the advantage of not needing gradient information for the problem being solved [7].

The CSO proposed by Cheng and Jin in 2015 [6], has gained significant attention due to its simplicity and efficiency. Since its introduction, CSO has been widely applied across various fields, with numerous studies further improving and utilizing the algorithm. Although the CSO has potential in large-scale single-objective optimization problems, it overly relies on the "loser-to-winner learning" paradigm and neglects the winner determination mechanism, making it difficult to escape local optima. To address this issue, Gao et al. [8] proposed a flexible ranking-based CSO for large-scale multi-objective optimization problems. This approach introduces a novel winner-determination strategy and competitive mechanism to improve search efficiency. Simulation results show that the algorithm significantly enhances the exploration and exploitation capabilities of the traditional CSO and performs well in benchmark tests and application cases. Li et al. [9] proposed an improved CSO to address the issues of low solution precision and premature convergence in the traditional CSO algorithm, which only relies on the winners to guide population evolution. First, a super particle obtained through a cumulative learning strategy provides a promising direction for population evolution. Then, a weight-based dynamic omnidirectional strategy is employed to enhance the population's exploration ability. Experimental results on CEC2017 benchmark problems demonstrate that the proposed algorithm achieves a better balance between exploration and exploitation, making it highly competitive. Meanwhile, CSO has also been applied to address real-world problems. Xing et al. [10] proposed a regional integrated energy system model that combines electric and thermal energy. By considering the operational constraints of electric-thermal coupling, the study employs the CSO to optimize the allocation of distributed power sources and loads within the system. The results demonstrate that the CSO algorithm significantly enhances the economic performance of the integrated energy system, achieving optimal operational modes, improving energy utilization efficiency, and ensuring economic and stable system operation.

Although many improved CSO variants have been proposed, there is still a gap in enhancing its low-dimensional global exploration capability. Furthermore, considering the Proximal Optimality Principle (POP) [11] that the closer points may share structural similarities, and this redundancy will consume an additional computational budget to evaluate the fitness, thus, we integrate an Adjacent distance matrix-based competition into CSO and propose Adjacent distance Matrix-based CSO (AMCSO), which increases the probability of selecting closer points as competitors and maintain the population diversity during optimization. To validate our proposed AMCSO, we conducted experiments on 10- and 20-D (D = dimensional) CEC2022 benchmark functions, comparing our improved algorithm with the original CSO and four other well-known MAs. The comprehensive experimental results demonstrate that the proposed AMCSO achieves superior performance on several benchmark functions against competitor algorithms.

The remainder of this paper is structured as follows: Sect. 2 introduces the CSO. Section 3 details the proposed AMCSO. The numerical experiments are introduced in Sect. 4. Section 5 analyzes the experimental results. Finally, Sect. 6 concludes our work and discusses future directions.

2   Competitive Swarm Optimizer (CSO)

CSO is an MA proposed in 2015 [6]. It is an improved variant of PSO and incorporates a competitive mechanism to enhance optimization performance. Unlike PSO, CSO doesn't rely on global best information. CSO emphasizes competition within particle individuals to ensure the survival of better individuals and update the worse individuals, which allows CSO to converge faster and perform better in complex, high-D problems.

The core idea of CSO is that particles compete in pairs based on their fitness values, and the particle individual with worse fitness moves towards the better one. Specifically, the velocity and position update of CSO are as follows:

- **Velocity Update:** The position and velocity of the winner and loser in the k-th round of competition during generation t as $X_{w,k}(t)$, $X_{l,k}(t)$, and $V_{w,k}(t)$, $V_{l,k}(t)$, respectively, where $k = 1, 2, \ldots, m/2$. Consequently, after the k-th competition, the loser's velocity is updated according to the following learning strategy:

$$
\begin{aligned}
V_{l,k}(t+1) = {} & R_1(k,t)V_{l,k}(t) \\
& + R_2(k,t)(X_{w,k}(t) - X_{l,k}(t)) \\
& + \varphi R_3(k,t)(\bar{X}_k(t) - X_{l,k}(t))
\end{aligned}
\tag{1}
$$

- **Position Update:** As a result, the position of the loser can be updated with the new velocity:

$$
X_{l,k}(t+1) = X_{l,k}(t) + V_{l,k}(t+1)
\tag{2}
$$

where $R_1(k,t)$, $R_2(k,t)$, and $R_3(k,t) \in [0,1]^n$ are three randomly generated vectors following the k-th competition and learning process in generation t, $\bar{X}_k(t)$ represents the mean position of the relevant particles, and φ is a parameter that controls the influence of $\bar{X}(t)$.

Through these updates, CSO gradually leads the population of particles to converge to an optimal or near-optimal solution, making it effective for solving complex optimization problems. CSO has been successfully applied in various domains, including global optimization and high-dimensional optimization tasks.

3   Our Proposal: AMCSO

The original CSO algorithm randomly selects two individuals for win/loss comparison, which can eliminate promising solutions that are far apart and carry diverse information, which may reduce the diversity of the search space. To address this issue, we introduce an Adjacent distance matrix-based competition selection, where the matrix constructed based on the Euclidean distances between different n individuals is demonstrated in Fig. 1.

	x_0	x_1	x_2	x_3	......	x_n
x_0	d_{00}	d_{01}	d_{02}	d_{03}	......	d_{0n}
x_1	d_{10}	d_{11}	d_{12}	d_{13}	......	d_{1n}
x_2	d_{20}	d_{21}	d_{22}	d_{23}	......	d_{2n}
x_3	d_{30}	d_{31}	d_{32}	d_{33}	......	d_{3n}
......	......	......	......	......		......
x_n	d_{n0}	d_{n1}	d_{n2}	d_{n3}	......	d_{n3}

Euclidean distances: $d_{ij} = \sqrt{(x_{i1} - x_{j1})^2 + (x_{i2} - x_{j2})^2 + \ldots + (x_{in} - x_{jn})^2}$

Fig. 1. Adjacent Distance Matrix.

According to the Proximal Optimality Principle (POP), individuals who are closer in distance tend to share similar local information. By selecting individuals that are close in distance for comparison, we reduce the risk of converging to a local optimum, enhance the diversity of the population, and increase the likelihood of finding a globally optimal solution, The differences in the competition mechanism between CSO and AMCSO are demonstrated in the Fig. 2.

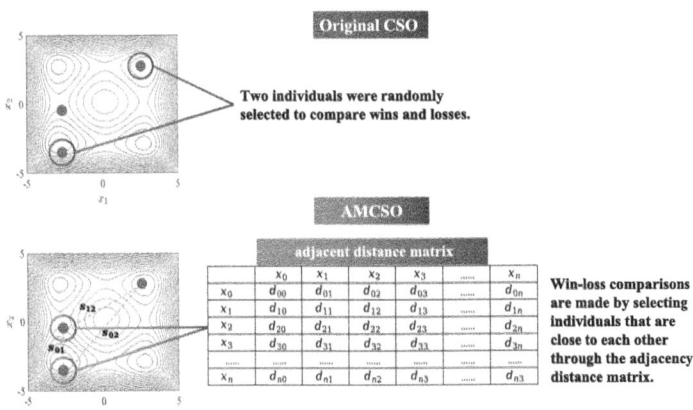

Fig. 2. The differences between CSO and AMCSO.

We first construct the Adjacent distance matrix by calculating Euclidean distances to achieve this. Then, we randomly select one individual, identify the k closest individuals based on the matrix, and randomly select one of these k individuals for the win/loss comparison. The winning individual remains unchanged, while the losing is updated using the Eq. (1) and Eq. (2). Briefly, Algorithm 1 presents the pseudocode of AMCSO.

Algorithm 1: AMCSO

---

1 Population size:N, Dimension:D, Max. iteration:T;
2 $t = 0$, $X^{(t)} \leftarrow$ **initial**(N, D);
3 **while** $t < T$ **do**
4   Calculate the fitness of all particles in $X(t)$;
5      $U = X(t)$, $X(t+1) = \emptyset$;
6   **while** $U \neq \emptyset$ **do**
7   Constructing the Adjacent distance matrix by calculating the Euclidean distance between different individuals;
8   Randomly choose one particle $X_1(t)$ from U;
9   Using the Adjacent distance matrix, an individual $X_2(t)$ is randomly selected from the k individuals closest to $X_1(t)$;
10   **if** $f(X_1(t)) \leq f(X_2(t))$ **then**
11              $X_w(t) = X_1(t), X_l(t) = X_2(t)$;
12   **else**
13              $X_w(t) = X_2(t), X_l(t) = X_1(t)$;
14   **end**
15   Add $X_w(t)$ into $X(t+1)$;
16   Update $X_l(t)$ using Eqs. (1) and Eqs. (2);
17   Add the updated $X_l(t+1)$ to $X(t+1)$;
18   Remove $X_1(t), X_2(t)$ from U;
19   **end**
20      $t = t + 1$;
21 **end**

---

4 Numerical Experiments

This section presents an overview of the numerical experiments and details the experimental setup and results to enable a fair comparison in evaluating the performance of the proposed AMCSO.

4.1 Experimental Settings

This experiment adopts 12 functions from CEC2022 to evaluate the performance of the proposed AMCSO. Five well-known MAs are employed as competitors: Evolution Strategy (ES) [12], GA, DE, Coral Reefs Optimization (CRO) [13], and the original CSO, which are implemented using the MEALPY library[1]. The experiments are conducted in both 10 and 20 dimensions. The maximum

---

[1] https://github.com/thieu1995/mealpy.

number of fitness evaluations (FEs) is set to $1000 \times D$, where D represents the dimension size, for the CEC2022. Additionally, each algorithm is repeated 20 times to reduce the impact of randomness. The parameter settings for CSO are summarized in Table 1 and adopt the optimal configurations of competitors as reported in relevant publications [14].

Table 1. Parameters selection.

Parameters	Values
Population size (N)	200
Dimension size (D)	10, 20
parameter φ (Phi)	0.1
The number of selected nearest individuals (K)	50

4.2 Experimental Results

Tables 2 and 3 present the experimental results on 10-D and 20-D CEC2022, displaying the mean and standard deviation for each algorithm across the 12 test functions from CEC2022. A smaller mean indicates that the algorithm is able to find solutions closer to the optimal value over multiple runs, reflecting better performance. Similarly, a smaller standard deviation suggests that the algorithm produces more stable solutions, implying better convergence and more consistent solution quality. Conversely, a larger standard deviation indicates that the solution distribution is more dispersed, suggesting that the algorithm may be unstable or exhibit poor convergence, potentially requiring further tuning or improvement.

In the table, "+" indicates that the mean value of AMCSO is better than the comparison algorithm, with a p-value less than 0.05. "−" indicates that the mean value of AMCSO is worse, also with a p-value less than 0.05. "≈" indicates that the p-value between AMCSO and the comparison algorithm is greater than 0.05, meaning no significant difference. In optimization problems, a p-value less than 0.05 signifies a statistically significant difference in the performance of the two algorithms, indicating that the observed differences in solution quality or convergence are not due to chance.

Figures 3 and 6 show the convergence plots of each algorithm on 12 CEC2022 test functions in 10 and 20 dimensions. These plots illustrate the algorithms' convergence behavior as the number of iterations increases. Figures 4 and 7 show the boxplots of the algorithms for 10 and 20 dimensions across the 12 test functions of CEC2022. In optimization problems, boxplots help evaluate the quality and stability of an algorithm's solutions:

- **Median**: A smaller median indicates better solution quality.
- **Box and Whiskers**: Shorter boxes and whiskers refl ct a more concentrated distribution, suggesting greater algorithm stability.

- **Outliers**: Outliers indicate the algorithm may occasionally get stuck in local optima or be influenced by unfavorable factors.

By analyzing boxplots, one can gain a deeper understanding of the performance of different optimization algorithms. Figures 5 and 8 display the ranking radar charts for each algorithm in 10 and 20 dimensions on CEC2022.

Table 2. Experimental results and statistical analysis in 10-D CEC2022.

Func.		ES	GA	DE	CRO	CSO	AMCSO
f_1	mean	1.655e+03 +	3.992e+02 +	6.486e+03 +	6.273e+02 +	3.012e+02 +	**3.010e+02**
	std	1.352e+03	1.088e+02	1.884e+03	3.985e+02	4.503e−01	1.841e+00
f_2	mean	4.738e+02 +	4.258e+02 +	4.531e+02 +	4.278e+02 +	4.075e+02 +	**4.036e+02**
	std	4.866e+01	2.702e+01	1.294e+01	2.674e+01	5.323e+00	3.990e+00
f_3	mean	6.000e+02 +	6.000e+02 +	6.000e+02 +	6.000e+02 +	6.000e+02 +	**6.000e+02**
	std	1.948e−02	5.535e−05	9.238e−03	2.441e−05	3.938e−07	2.668e−09
f_4	mean	8.563e+02 +	8.491e+02 +	9.041e+02 +	8.613e+02 +	8.386e+02 +	**8.175e+02**
	std	1.903e+01	1.728e+01	1.565e+01	2.367e+01	9.001e+00	1.007e+01
f_5	mean	9.018e+02 +	9.004e+02 +	9.028e+02 +	9.005e+02 +	**9.000e+02** ≈	9.000e+02
	std	1.193e+00	2.717e−01	5.506e−01	3.959e−01	4.102e−05	1.883e−04
f_6	mean	5.120e+04 +	2.031e+05 +	8.052e+04 +	2.806e+04 +	1.961e+04 +	**1.501e+04**
	std	1.420e+04	2.814e+05	2.576e+04	1.600e+04	4.844e+03	3.444e+03
f_7	mean	2.085e+03 ≈	2.030e+03 +	2.063e+03 +	**2.026e+03** +	2.032e+03 +	2.027e+03
	std	5.031e+01	3.697e+00	9.336e+00	5.150e+00	2.057e+00	1.819e+00
f_8	mean	2.686e+03 +	2.672e+03 +	**2.253e+03** ≈	2.734e+03 +	2.259e+03 +	2.256e+03
	std	4.990e+02	4.592e+02	7.402e+00	5.061e+02	3.426e+01	1.965e+01
f_9	mean	2.669e+03 +	2.589e+03 +	2.641e+03 +	2.577e+03 +	**2.300e+03** ≈	2.304e+03
	std	1.707e+02	1.667e+02	4.960e+01	1.807e+02	2.997e−01	1.658e+01
f_{10}	mean	2.652e+03 +	2.639e+03 −	2.647e+03 +	2.645e+03 +	**2.630e+03** −	2.639e+03
	std	5.284e+01	3.779e+01	6.274e+00	4.062e+01	2.390e+01	5.009e+00
f_{11}	mean	2.681e+03 +	2.605e+03 +	2.629e+03 +	2.610e+03 +	2.600e+03 +	**2.600e+03**
	std	2.067e+02	9.146e+00	4.535e+00	9.307e+00	2.389e−02	3.474e−02
f_{12}	mean	2.933e+03 +	2.869e+03 ≈	2.868e+03 −	2.873e+03 +	**2.867e+03** ≈	2.869e+03
	std	3.987e+01	2.980e+00	6.786e−01	5.091e+00	5.744e−01	1.950e+00
+/≈/−:		11/1/0	10/1/1	10/1/1	12/0/0	8/3/1	−
Avg. ranks:		5.33	3.42	4.75	3.83	1.92	**1.75**

5 Discussion

First, we analyzed the results of the algorithm on the 10-dimensional CEC2022 test functions:

- **Experimental Results Table Analysis** (Table 2): The "+" symbol appears 51 times, the "−" symbol appears 3 times, and the "≈" symbol appears 6 times. This indicates that AMCSO achieves solutions closer to the optimal in most cases, with significant differences compared to other algorithms, showing that the results are not due to chance.

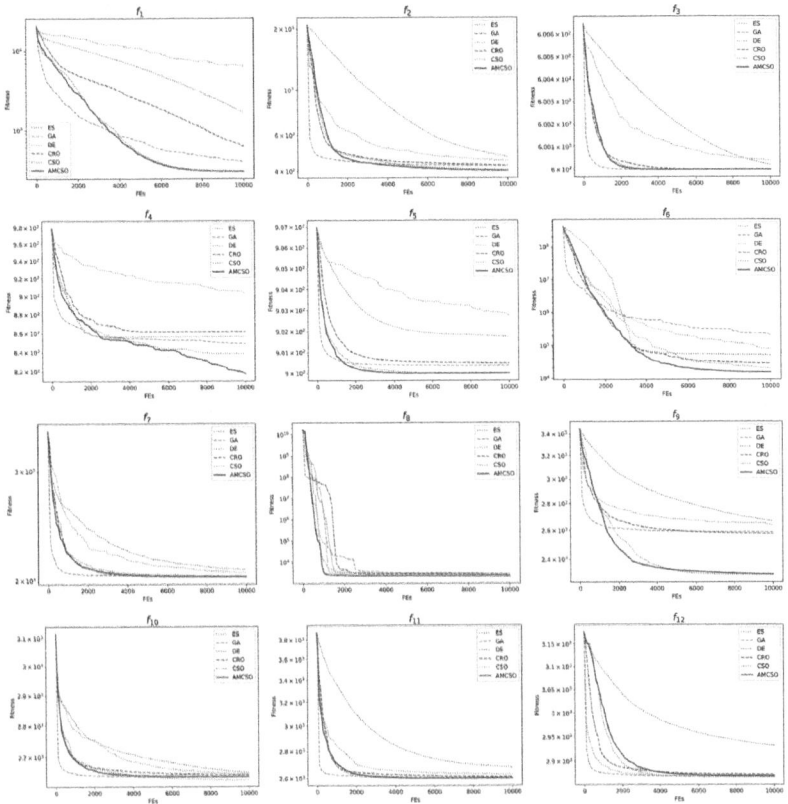

Fig. 3. Convergence plot in 10-D CEC2022.

- **Convergence Analysis** (Fig. 3): Except for f_{12}, AMCSO demonstrates faster convergence on most functions. The solutions stabilize and approach the optimal in the later stages, with particularly strong performance on f_4 and f_6.
- **Boxplot Analysis** (Fig. 4): The boxplots show that AMCSO has a more concentrated solution distribution and smaller standard deviations across most functions, indicating higher stability.
- **Radar Chart Analysis** (Fig. 5): The radar chart shows that AMCSO ranks first in f_1, f_2, f_3, f_4, f_6, and f_{11}, while maintaining strong rankings in the other functions as well.

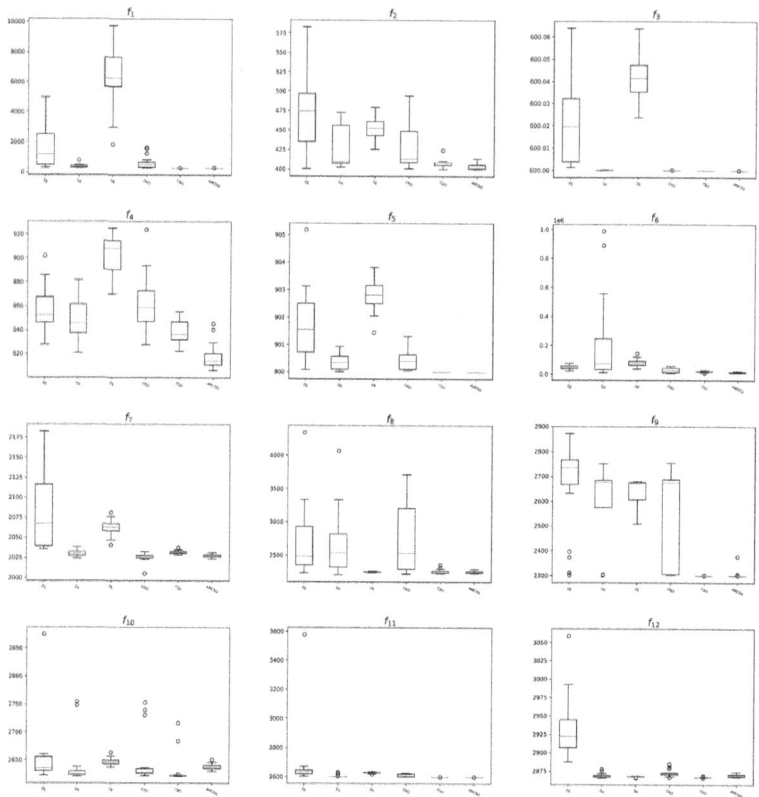

Fig. 4. Boxplot in 10-D CEC2022.

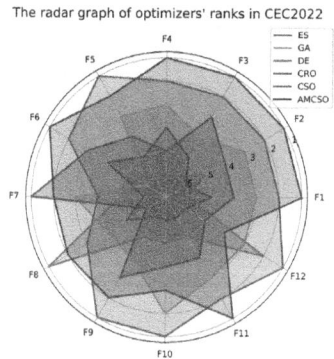

Fig. 5. Radar graph of optimizers' ranks in 10-D CEC2022.

– **Average Ranking Analysis:** AMCSO's average rank is 1.75, outperforming CSO's 1.92, making it the top performer overall, with other algorithms ranking significantly lower.

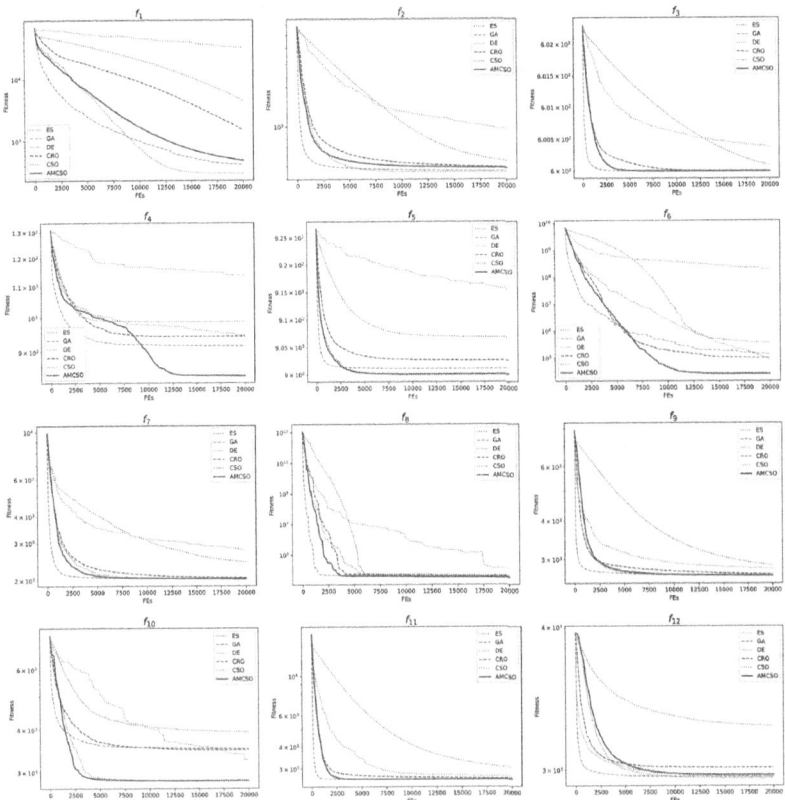

Fig. 6. Convergence plot in 20-D CEC2022.

Next, we analyzed the results of the algorithm on the 20-D CEC2022 test functions:

- **Experimental Results Table Analysis** (Table 3): The "+" symbol appeared 42 times, the "−" symbol appeared 12 times, and the "≈" symbol appeared 6 times. Although the number of "+" symbols decreased compared to the 10-D results, overall, AMCSO still achieved solutions closer to the optimal in most cases, excluding randomness.
- **Convergence Analysis** (Fig. 6): While the algorithm showed weaker convergence on f_1, f_2, and f_{12}, it exhibited faster convergence on most other functions and was able to stably approach the optimal solution, with particularly strong performance on f_4 and f_6.
- **Boxplot Analysis** (Fig. 7): In 20 dimensions, AMCSO's solution distribution was concentrated with smaller standard deviations across almost all functions, demonstrating excellent stability.

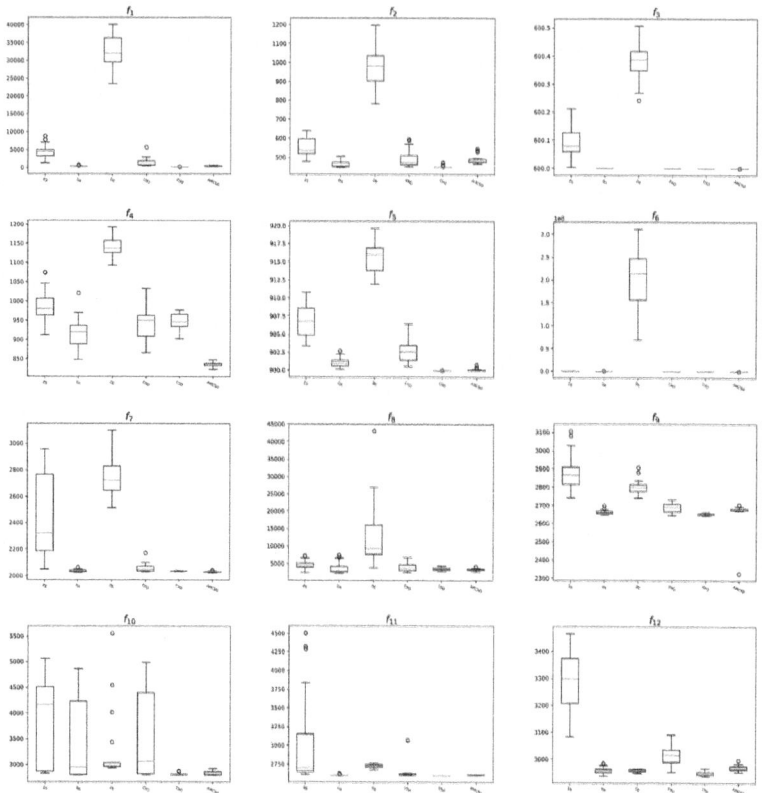

Fig. 7. Boxplot in 20-D CEC2022.

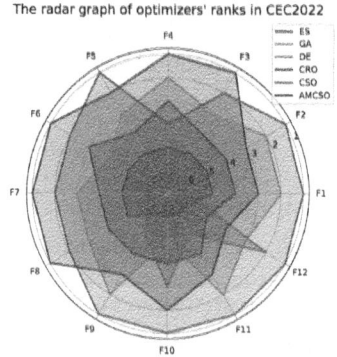

Fig. 8. Radar graph of optimizers' ranks in 20-D CEC2022.

- **Radar Chart Analysis** (Fig. 8): The radar chart shows that AMCSO ranked first on f_3, f_4, f_6, f_7, and f_8, while maintaining relatively high rankings on other functions.

Table 3. Experimental results and statistical analysis in 20-D CEC2022.

Func.		ES	GA	DE	CRO	CSO	AMCSO
f_1	mean	4.426e+03 +	4.194e+02 −	3.251e+04 +	1.526e+03 +	**3.003e+02** −	4.837e+02
	std	1.913e+03	1.284e+02	4.354e+03	1.252e+03	1.589e−01	1.054e+02
f_2	mean	5.510e+02 +	4.628e+02 −	9.727e+02 +	4.932e+02 +	**4.523e+02**	4.883e+02
	std	5.017e+01	1.515e+01	1.045e+02	4.527e+01	7.169e+00	2.246e+01
f_3	mean	6.001e+02 +	6.000e+02 +	6.004e+02 +	6.000e+02 +	6.000e+02 +	**6.000e+02**
	std	5.864e−02	2.997e−05	6.363e−02	5.799e−05	7.294e−09	3.433e−10
f_4	mean	9.878e+02 +	9.176e+02 +	1.140e+03 +	9.442e+02 +	9.469e+02 +	**8.355e+02**
	std	3.984e+01	3.853e+01	2.544e+01	4.557e+01	2.261e+01	6.741e+00
f_5	mean	9.069e+02 +	9.011e+02 +	9.156e+02 +	9.027e+02 +	**9.000e+02** +	9.001e+02
	std	2.178e+00	6.263e−01	2.060e+00	1.543e+00	7.488e−06	2.136e−01
f_6	mean	3.791e+05 +	1.411e+05 +	2.025e+08 +	9.836e+04 +	8.681e+04 +	**2.526e+04**
	std	1.376e+05	7.448e+04	6.820e+07	2.929e+04	3.491e+04	3.764e+03
f_7	mean	2.439e+03 +	2.037e+03 ≈	2.757e+03 +	2.058e+03 +	2.036e+03 +	**2.033e+03**
	std	2.901e+02	9.204e+00	1.630e+02	3.349e+01	2.152e+00	4.132e+00
f_8	mean	4.637e+03 +	3.701e+03 ≈	1.316e+04 +	3.979e+03 +	3.549e+03 ≈	**3.416e+03**
	std	1.188e+03	1.547e+03	9.398e+03	1.176e+03	3.926e+02	2.981e+02
f_9	mean	2.878e+03 +	2.662e+03 −	2.803e+03 +	2.690e+03 ≈	**2.651e+03** −	2.665e+03
	std	9.852e+01	1.160e+01	3.848e+01	2.534e+01	4.678e+00	7.800e+01
f_{10}	mean	3.943e+03 +	3.528e+03 +	3.268e+03 +	3.492e+03 +	**2.818e+03** −	2.840e+03
	std	7.889e+02	7.888e+02	6.623e+02	7.764e+02	3.255e+01	4.165e+01
f_{11}	mean	3.049e+03 +	2.604e+03 ≈	2.730e+03 +	2.637e+03 +	**2.600e+03** −	2.604e+03
	std	6.379e+02	6.864e+00	2.273e+01	1.000e+02	5.001e−03	3.616e+00
f_{12}	mean	3.279e+03 +	2.959e+03 +	2.958e+03 −	3.016e+03 +	**2.948e+03** −	2.970e+03
	std	1.164e+02	1.268e+01	4.308e+00	3.494e+01	7.469e+00	1.005e+01
+/≈/−:		12/0/0	5/3/4	11/0/1	10/2/0	4/1/7	−
Avg. ranks:		5.33	2.83	5.25	3.92	**1.58**	2.08

- **Average Ranking Analysis:** AMCSO's average rank was 2.08, second only to CSO's 1.58. This indicates that while AMCSO's performance slightly decreased as the dimension increased, it still outperformed other algorithms. Particularly on f_4 and f_6, AMCSO consistently demonstrated a significant advantage.

The AMCSO algorithm demonstrates exceptional performance and stability in the CEC2022 benchmark, significantly outperforming the original CSO algorithm and other algorithms in 10-D test functions. AMCSO ranked first on average in the 10-D tests, fully showcasing its ability to enhance the original CSO algorithm by escaping local optima more effectively and achieving the global optimum. AMCSO underperforms compared to the original CSO in 20 dimensions, likely due to the increased complexity of the search space at higher dimensions, which makes finding the global optimum more challenging. Furthermore, AMCSO's parameters and mechanisms may require fine-tuning to better handle high-dimensional problems, while its adjacency distance selection strategy could constrain global exploration in such scenarios. Additionally, AMCSO excels across various categories of challenging functions, including shifted and rotated functions (e.g., F1 Zakharov Function, F2 Rosenbrock Function), complex landscapes with high modality (e.g., F3 Expanded Schaffer's F7 Function, F4 Non-Continuous Rastrigin Function), multimodal problems, and hybrid and composition functions (e.g., F6 Hybrid Function 1, F7 Hybrid Function 2, F8 Hybrid Function 3). These functions often involve complex search spaces, asymmetry, and diverse landscape properties.

6 Conclusion

The proposed AMCSO algorithm improves the original CSO algorithm by enhancing the win-loss comparison mechanism and introducing an Adjacent distance matrix based on the Euclidean distance between exploring individuals during the selection process. By increasing the diversity of individuals, AMCSO effectively enhances the algorithm's ability to escape local optima and find the global optimum. Experimental results show that AMCSO achieves excellent performance on the CEC2022 test functions, fully validating its global optimization capability and effectiveness.

Although AMCSO outperforms other algorithms, particularly showing significant advantages when dealing with complex functions, its performance decreases as the dimensionality increases. This suggests that high-D problems impose greater demands on the algorithm's adaptability. In future research, we plan to further enhance AMCSO's adaptive capability to address the challenges posed by high-dimensional optimization problems, improving its stability and performance across different dimensions.

Acknowledgement. This work was supported by JST SPRING, Grant Number JPMJSP2119 and JSPS KAKENHI Grant Number 21A402 and 24K15098.

References

1. Liu, L., Fei, T., Zhu, Z., Wu, K., Zhang, Y.: A survey of evolutionary algorithms. In: 2023 4th International Conference on Big Data, Artificial Intelligence and Internet of Things Engineering (ICBAIE), pp. 22–27 (2023)
2. Felipe Coello Castillo, C., Coello, C.A.: A survey of applications of multi-objective evolutionary algorithms in biotechnology. In: 2024 IEEE Congress on Evolutionary Computation (CEC), pp. 1–8 (2024)
3. Zhong, R., Yu, J.: A novel evolutionary status guided hyper-heuristic algorithm for continuous optimization. Clust. Comput. **27**, 12209–12238 (2024)
4. Kumar, K.V., Merugu, B., Devalapalli, S.S., Mekala, M., Lakavath, J.S.: Sizing of DGS for minimization of power losses in distribution system using particle swarm optimization. In: 2024 IEEE International Conference on Electronics, Computing and Communication Technologies (CONECCT), pp. 1–6 (2024)

5. Zhong, R., Yu, J.: Dea^{2}h^2: differential evolution architecture based adaptive hyper-heuristic algorithm for continuous optimization. Cluster Comput. 1–28 (2024)
6. Cheng, R., Jin, Y.: A competitive swarm optimizer for large scale optimization. IEEE Trans. Cybern. **45**(2), 191–204 (2015)
7. Cao, Y., Zhong, R., Yu, J., Munetomo, M.: Optimization of electricity consumption forecasting models via hyper-heuristic algorithm. In: 2024 6th International Conference on Data-driven Optimization of Complex Systems (DOCS), pp. 114–120 (2024)
8. Gao, X., Song, S., Zhang, H., Wang, Z.: A flexible ranking-based competitive swarm optimizer for large-scale continuous multi-objective optimization. IEEE Trans. Evol. Comput. 1–1 (2024)
9. Li, W., Gao, Y., Wang, L.: An improved competitive swarm optimizer with super-particle-leading. Neural Process. Lett. **55**, 1–33 (2023)
10. Xing, Y., et al.: Research on optimal control of regional integrated energy system based on competitive swarm optimizer. In: 2021 IEEE 5th Conference on Energy Internet and Energy System Integration (EI2), pp. 1525–1530 (2021)
11. Yaguchi, K., Tamura, K., Yasuda, K., Ishigame, A.: Basic study of proximate optimality principle based combinatorial optimization method. In: 2011 IEEE International Conference on Systems, Man, and Cybernetics, pp. 1753–1758 (2011)
12. Rimcharoen, S., Leelathakul, N.: Adaptive evolution strategy for symbolic regression. In: 2023 IEEE 6th International Conference on Computer and Communication Engineering Technology (CCET), pp. 38–42 (2023)
13. Asghari, A., Sohrabi, M.K.: Multiobjective edge server placement in mobile-edge computing using a combination of multiagent deep q-network and coral reefs optimization. IEEE Internet Things J. **9**(18), 17503–17512 (2022)
14. Van Thieu, N., Mirjalili, S.: Mealpy: an open-source library for latest meta-heuristic algorithms in python. J. Syst. Archit. (2023)

The More the Merrier: On Evolving Five-Valued Spectra Boolean Functions

Claude Carlet[1,2], Marko Đurasević[3], Domagoj Jakobovic[3(⊠)], Luca Mariot[4], and Stjepan Picek[5]

[1] Department of Mathematics, Université Paris 8, 2 Rue de la Liberté, 93526 Saint-Denis Cedex, France
[2] University of Bergen, Bergen, Norway
[3] Faculty of Electrical Engineering and Computing, University of Zagreb, Unska 3, Zagreb, Croatia
{marko.durasevic,domagoj.jakobovic}@fer.hr
[4] Semantics, Cybersecurity and Services Group, University of Twente, Drienerlolaan 5, 7522 Enschede, NB, The Netherlands
l.mariot@utwente.nl
[5] Digital Security Group, Radboud University, PO Box 9010, Nijmegen, The Netherlands
stjepan.picek@ru.nl

Abstract. Evolving Boolean functions with specific properties is an interesting optimization problem since, depending on the combination of properties and Boolean function size, the problem can range from very simple to (almost) impossible to solve. Some problems are more interesting as there may be only a few options for generating the required Boolean functions. This paper investigates one such problem: evolving five-valued spectra Boolean functions, the functions whose Walsh-Hadamard coefficients can only take five distinct values. We experimented with three solution encodings, two fitness functions, and 12 Boolean function sizes and showed that the tree encoding is superior to other choices, as we can obtain five-valued Boolean functions with high nonlinearity.

Keywords: Boolean Functions · Plateaudness · Evolutionary Algorithms · Five-valued Functions

1 Introduction

Evolving Boolean functions with specific cryptographic properties is an interesting optimization problem. From one side, already a vast body of work showed that evolutionary algorithms (EAs) can generate Boolean functions with excellent properties, see, e.g., [1–3] that even rival algebraic constructions.[1] On the

---

[1] In some cases, unfortunately, rare ones, heuristics have been even more successful than at-the-moment state-of-the-art algebraic techniques. For instance, Kavut et al. used heuristics to generate Boolean functions in 9 variables having nonlinearity 241, which was an open problem for several decades [4].

© The Author(s), under exclusive license to Springer Nature Switzerland AG 2025
P. García-Sánchez et al. (Eds.): EvoApplications 2025, LNCS 15612, pp. 52–67, 2025.
https://doi.org/10.1007/978-3-031-90062-4_4

other hand, most EA works, while successful, consider problems where we know a number of algebraic constructions that reach excellent (or even optimal) results. As such, the practical relevance of such new, metaheuristics-based results is questionable. Naturally, this does not mean that evolutionary algorithms should not be used to evolve Boolean functions with specific properties, but that one should carefully select a problem where there are not many known approaches to solve it or the known approaches provide peculiar functions not suited for applications. One such problem is the design of five-valued spectra Boolean functions (whose Walsh-Hadamard spectra contain only five values) [5].

As discussed in Sect. 2, five-valued spectra functions are very interesting from a cryptographic perspective. They can be balanced and have excellent nonlinearity values. It was shown that a concatenation of two properly chosen $(n-1)$-variable bent functions from the Maiorana-McFarland class can give functions with five valued Walsh-Hadamard spectra [5]. These resulting functions can possess good cryptographic properties, be again used in constructing bent functions, or have the same extended Walsh-Hadamard spectrum as the only known Almost Perfect Nonlinear (APN) permutation in dimension 6 [6]. Unfortunately, there is not much interesting work on the design of five-valued spectra functions. To our knowledge, there are few algebraic constructions [5–9] and no computational or heuristic approaches. What is more, as we discuss later in this paper, the spectral inversion approach [10] used to evolve plateaued functions (i.e., three-valued spectra Boolean functions) cannot be straightforwardly adapted for five-valued spectra functions, making the problem more complex.

This work explores how evolutionary algorithms can evolve five-valued spectra functions. We conduct experiments with three solution encodings—including tree encoding with genetic programming (GP) as it generally shows excellent performance when evolving Boolean functions [11]—two fitness functions, and 12 Boolean function sizes. Our results show that for the smallest sizes (dimensions 5 and 6), all approaches work well and obtain five-valued functions. However, already for dimension 7, only the tree encoding can consistently find five-valued functions. From dimension 10, we see more difference between the two fitness functions, where the first one managed to reach significantly higher nonlinearity values. Our main contributions are:

1. We are the first to consider the design of five-valued spectra Boolean functions with evolutionary algorithms. This problem is interesting from both the evolutionary and mathematical perspectives.
2. Our experimental results show that the tree encoding is superior to the truth table one. Moreover, we can obtain five-valued functions with high nonlinearity for all tested Boolean function sizes. Additionally, for several (odd) sizes, we reach five-valued spectra functions that match the best-known nonlinearity.
3. We explain why the spectral inversion principle commonly used when evolving plateaued functions cannot be straightforwardly adapted for five-valued spectra functions.

2   Preliminaries

In this section, we recall the background notions related to Boolean functions and their cryptographic properties used in the remainder of the paper. For a systematic treatment of the subject, we refer the reader to Carlet's book [12].

2.1   Notation

In what follows, we endow the binary set $\{0, 1\}$ with the structure of a finite field, and we denote it by $\mathbb{F}_2$. In particular, sum and multiplication in $\mathbb{F}_2$ amount respectively to the XOR (denoted as $\oplus$) and logical AND (denoted by concatenation) of two bits. For all $n \in \mathbb{N}$, the set $\mathbb{F}_2^n$ denotes the n-dimensional vector space over $\mathbb{F}_2$, with vector sum being defined as the bitwise XOR of two n-tuples of bits $x, y \in \mathbb{F}_2$, while multiplication by a scalar $a \in \mathbb{F}_2$ of a vector $x \in \mathbb{F}_2$ corresponds to the logical AND of a with each coordinate of x. The vector space $\mathbb{F}_2^n$ is equipped with the inner product defined as $x \cdot y = \bigoplus_{i=1}^n x_i y_i$ for all $x, y \in \mathbb{F}_2^n$. The Hamming weight $w_H(x)$ of a vector x, where $x \in \mathbb{F}_2^n$, is the number of non-zero positions in x. The Hamming distance $d_H(x, y)$ between two vectors $x, y \in \mathbb{F}_2^n$ is the number of positions where x and y differ, and it is equivalent to the Hamming weight of their bitwise XOR.

2.2   Boolean Functions

A Boolean function of $n \in \mathbb{N}$ variables is a mapping of the form $f : \mathbb{F}_2^n \to \mathbb{F}_2$, which can be uniquely represented by its truth table. Formally, the truth table of f is the 2^n-bit vector that lists all output values $f(x)$, for $x \in \mathbb{F}_2^n$, where some total order has been chosen on $\mathbb{F}_2^n$ (most commonly, the lexicographic order). While the truth table representation is "human-friendly", there is not much that can be directly deduced from it, except for the Hamming weight of the function. Other cryptographic properties are more easily characterized through the Walsh-Hadamard transform $W_f : \mathbb{F}_2^n \to \mathbb{Z}$, which measures the correlation between f and the linear function $a \cdot x$:

$$W_f(a) = \sum_{x \in \mathbb{F}_2^n} (-1)^{f(x) \oplus a \cdot x}. \tag{1}$$

The vector of all Walsh-Hadamard coefficients $W_f(a)$ for all $a \in \mathbb{F}_2^n$ is also called the Walsh-Hadamard spectrum of f.

A Boolean function f is balanced if it takes the value one exactly the same number 2^{n-1} of times as value zero when the input ranges over $\mathbb{F}_2^n$. Considering the Walsh-Hadamard transform representation, a function is balanced if and only if $W_f(\underline{0}) = 0$. Balancedness is a fundamental criterion for Boolean functions used in cryptographic applications since unbalanced functions have a statistical bias that can be exploited in distinguishing attacks.

The minimum Hamming distance between the truth table of f and the truth tables of all affine functions (i.e., functions of the form $a \cdot x \oplus b$, where $b \in \mathbb{F}_2$) is called the nonlinearity of f. This value can be expressed in terms of the Walsh-Hadamard coefficients as follows:

$$nl_f = 2^{n-1} - \frac{1}{2} \max_{a \in \mathbb{F}_2^n} |W_f(a)|. \tag{2}$$

The Parseval relation states that $\sum_{a \in \mathbb{F}_2^n} W_f(a)^2 = 2^{2n}$ for every Boolean function of n variables. The above relation implies that the arithmetic mean of the squared Walsh-Hadamard spectrum $W_f(a)^2$ equals 2^n. The maximum of $W_f(a)^2$ being equal to or larger than its arithmetic mean, we can deduce that $\max_{a \in \mathbb{F}_2^n} |W_f(a)|$ must be larger than or equal to $2^{\frac{n}{2}}$. Therefore, the nonlinearity of every n-variable Boolean function is bounded above by the so-called covering radius bound:

$$nl_f \leq 2^{n-1} - 2^{n/2-1}. \tag{3}$$

The nonlinearity $2^{n-1} - 2^{\frac{n-1}{2}}$ is called the quadratic bound since for n odd, it is a tight upper bound on the nonlinearity of Boolean functions with algebraic degree at most two. The quadratic bound is the best value of the nonlinearity that can be reached for $n \leq 7$ while for $n \geq 9$, better nonlinearity exists, see [12]. A Boolean function can be considered highly nonlinear if its nonlinearity is close to the covering radius bound.

Another way to uniquely represent a Boolean function f on $\mathbb{F}_2^n$ is as a multivariate polynomial in the quotient ring $\mathbb{F}_2[x_1,...,x_n]/(x_1^2 \oplus x_1,...,x_n^2 \oplus x_n)$, which is called the Algebraic Normal Form (ANF), and is defined as:

$$f(x) = \bigoplus_{a \in \mathbb{F}_2^n} h(a) \cdot x^a, \tag{4}$$

where $h(a)$ is defined by the binary Möbius inversion principle:

$$h(a) = \bigoplus_{x \preceq a} f(x), \text{ for any } a \in \mathbb{F}_2^n. \tag{5}$$

Here, $x \preceq a$ means that a covers x (alternatively, x precedes a), that is $x_i \leq a_i$ for all $i \in \{0,...,n-1\}$. The Möbius transform is an involution, meaning that if we apply it to the ANF of f (i.e. we swap $h(a)$ with $f(x)$ in the above equation), we obtain the original truth table of f.

2.3   Bent Boolean Functions

The functions whose nonlinearity equals the maximal value $2^{n-1} - 2^{n/2-1}$ are called bent. In particular, any bent function $f : \mathbb{F}_2^n \to \mathbb{F}_2$ satisfies $W_f(a) = \pm 2^{\frac{n}{2}}$ for all $a \in \mathbb{F}_2^n$, due to the Parseval relation. Hence, bent functions are not balanced (since $W_f(\underline{0} = \pm 2^{\frac{n}{2}} \neq 0$), and they exist only for n even (since the Walsh-Hadamard transform can only take integer values).

Bent functions are interesting mathematical objects with diverse usages. For instance, they are used in coding theory with Kerdock codes and to build bent function sequences for telecommunications. Bent Boolean functions are rare, and we know the exact number of bent Boolean functions for $n \leq 8$ only [13].

When designing cryptographic functions, a common problem is how to fulfill the requirements of multiple conflicting criteria. The best-known example is designing a balanced function with high nonlinearity, high algebraic degree, and an unconstrained number of input bits. Bent functions achieve the highest possible nonlinearity, but they are not balanced, have an algebraic degree of at most $n/2$, and exist only when the number of variables is even. On the other hand, some other related classes of Boolean functions allow to reach high nonlinearity while being balanced and existing for every number of input variables. We discuss such examples next.

2.4   Plateaued Boolean Functions

A Boolean function $f : \mathbb{F}_2^n \rightarrow \mathbb{F}_2$ is said to be plateaued if its Walsh-Hadamard spectrum takes at most three values: 0 and $\pm\alpha$, where necessarily $\alpha = 2^\lambda$ with $\lceil \frac{n}{2} \rceil \leq \lambda \leq n$. The value α is also called the amplitude of the function [12].

When $\alpha = \frac{n}{2}$, a plateaued function corresponds to a bent function. A Boolean function $f : \mathbb{F}_2^n \rightarrow \mathbb{F}_2$ is called semi-bent if its Walsh-Hadamard transform satisfies $W_f(a) \in \{0, \pm 2^{\frac{n+2}{2}}\}$ for all $a \in \mathbb{F}_2^n$ when n is even, or $W_f(a) \in \{0, \pm 2^{\frac{n+1}{2}}\}$ for all $a \in \mathbb{F}_2^n$ when n is odd. In the latter case, a semi-bent function is also called near-bent. The nonlinearity of a semi-bent function f equals $2^{n-1} - 2^{\frac{n-1}{2}}$ when n is odd, and $2^{n-1} - 2^{\frac{n}{2}}$ when n is even. More in general, the nonlinearity value of a plateaued function is determined by the size of the Boolean function n and its amplitude 2^λ: $nl_f = 2^{n-1} - 2^{\lambda-1}$.

2.5   Five-Valued Spectra Boolean Functions

Five-valued spectra Boolean functions of n variables have a Walsh-Hadamard spectrum whose values range in $\{0, \pm 2^{\lambda_1}, \pm 2^{\lambda_2}\}$ where $\lceil \frac{n}{2} \rceil \leq \lambda_1, \lambda_2 < n$. The nonlinearity of a five-valued spectrum function $f : \mathbb{F}_2^n \rightarrow \mathbb{F}_2$ depends on its largest amplitude. Without loss of generality, let us assume that $\lambda_1 > \lambda_2$. Then, the nonlinearity of f equals: $nl_f = 2^{n-1} - 2^{\lambda_1-1}$.

Such functions can achieve very good cryptographic properties. For instance, for n odd, having $\lambda_1 = \frac{n-1}{2}$ and $\lambda_2 = \frac{n+1}{2}$ ensures the same nonlinearity as those of semi-bent functions. Similarly, for n even $\lambda_1 = \frac{n}{2}$ and $\lambda_2 = \frac{n+2}{2}$ yields the nonlinearity of a semi-bent function. Bent functions in n variables can be viewed through (four) restrictions to the cosets of some $(n-2)$-dimensional linear subspace. These restrictions are either bent, semi-bent, or five-valued spectra functions. Thus, five-valued spectra functions can be used to build bent functions, a technique known as the 4-bent decomposition [6].

3   Related Work

In this section, we give an overview of some of the works related to designing Boolean functions with good cryptographic properties via evolutionary algorithms and metaheuristics. Our treatment of the subject is inevitably limited; for a complete survey, we refer the reader to [11].

Historically, most research focused on adopting the truth table representation, which naturally lends itself to genetic algorithms (GA), where the genomes correspond to fixed-length bitstrings. For example, Millan et al. [14] proposed a GA where the individuals' genotypes are 2^n-bit strings that encode the truth tables of balanced Boolean functions. The authors developed specific crossover and mutation operators to preserve balancedness and evolved Boolean functions up to $n = 12$ variables under a fitness function that optimized a combination of high nonlinearity, correlation immunity, and strict avalanche criterion. Next, Clark et al. [15] investigated a two-stage optimization approach where simulated annealing is first used to reach a promising area of the search space, and then hill climbing is applied to exploit that area by finding highly nonlinear functions with low autocorrelation. Later works further explored the truth table-based encoding by focusing on different classes of Boolean functions (such as bent functions [16]), multi-objective optimization approaches [17], balancedness-preserving operators [18] or combining EA with local search [19].

A second research strand considered the representation of candidate Boolean functions as syntactic trees evolved through Genetic Programming (GP). Picek et al. [20] compared the performances of GP and GA to optimize different cryptographic properties of Boolean functions, investigating the influence of the underlying genetic operators in the process. The main finding was that GP obtained the best performance. This fact has been corroborated in later research [21, 22], which generally shows that evolving syntactic trees with GP yields Boolean functions with better cryptographic properties than using GA with a truth table-based encoding. Other works in this research line explored the use of GP variants, such as Cartesian GP to evolve bent [23] or balanced, highly nonlinear functions [24], or linear GP to evolve correlation-immune functions [25].

The tree encoding approach discussed above is used as an indirect representation of the function's truth table. A different angle is to represent candidate solutions as Walsh-Hadamard spectra, the rationale being that several cryptographic properties of interest are characterized in terms of the Walsh-Hadamard transform, as discussed in Sect. 2. Under this perspective, an EA or metaheuristic optimization algorithm is used to tweak a constrained spectrum, then compute the inverse Walsh-Hadamard transform and check how far is the resulting pseudo-Boolean function $f : \mathbb{F}_2^n \to \mathbb{R}$ from being an actual Boolean function. When an optimal solution is found, the corresponding Boolean function "by design" satisfies a good combination of properties, given the constraints imposed on the spectrum (e.g., low maximum absolute value for high nonlinearity). This so-called *spectral inversion approach* was pioneered by Clark et al. [10], who employed simulated annealing to permute three-valued spectra with the objective of finding plateaued Boolean functions. Following this idea, Mariot and Leporati [26] developed a GA to evolve the spectra of plateaued functions. More recently, Rovito et al. [27] used GP to evolve Walsh-Hadamard spectra, with the objective of finding highly nonlinear Boolean functions under the spectral inversion method.

A final research avenue worth mentioning is the application of EA and meta-heuristics to optimize *algebraic constructions* of Boolean functions. Instead of directly optimizing single Boolean functions, the idea here is to design a deterministic procedure yielding a class of functions with specific properties. These procedures can be classified into primary and secondary constructions [12]. A primary construction leverages other combinatorial structures (such as permutations or partial spreads) to construct from scratch new Boolean functions with specific properties, such as high nonlinearity. On the other hand, secondary constructions start from existing functions and extend them into new functions (usually on a larger number of variables) that satisfy analogous properties. So far, all works in this area address the design of secondary constructions via GP. Picek and Jakobovic [1] employed GP to evolve secondary constructions of bent functions. More recently, Carlet et al. [2] considered using GP to evolve constructions for highly nonlinear balanced functions. Interestingly, the authors remarked that GP tends to discover the well-known indirect sum construction, concealing it under different syntactic versions of the evolved trees. Finally, Mariot et al. [28] used Evolutionary Strategies (ES) to investigate a secondary construction of semi-bent functions based on cellular automata (CA).

4   Methodology

4.1   Solution Encodings and Operators

Truth Table Representation (TT). The most common option for representing a Boolean function is the truth table representation (denoted by TT) [11]. For a Boolean function with n inputs, the truth table is coded as a bit string with a length of 2^n. The bit string represents the Boolean function upon which the algorithm operates directly. In each evaluation, the truth table is transformed into the Walsh-Hadamard spectrum, after which the nonlinearity and other desired properties are calculated.

For the mutation operators, we use the simple bit mutation and the shuffle mutation. The simple bit mutation inverts a randomly selected bit, and the shuffle mutation shuffles the bits within a randomly selected substring. We used one-point and uniform crossover operators. The one-point crossover combines a new solution from the first part of one parent and the second part of the other parent with a randomly selected breakpoint. The uniform crossover randomly selects one bit from both parents at each position in the child bitstring that is copied. Each time the evolutionary algorithm invokes a crossover or mutation operation, one of the previously described operators is randomly selected.

Algebraic Normal Form Representation (ANF). From the standpoint of the evolutionary algorithm, both TT and ANF representation use the same genotype encoding, which is a bit string of length 2^n. Therefore, the same set of genetic operators is used for TT as well as for ANF representation. As discussed in

Sect. 2, the Möbius transform is an involution, which means that we can apply it to an ANF genotype to obtain the truth table representation of the function as phenotype, over which we can compute the desired cryptographic properties.

Symbolic Encoding (GP). The third approach in our experiments uses tree-based GP to represent a Boolean function in its symbolic form. In this case, we represent a candidate solution by a tree whose leaf nodes correspond to the input variables $x_1, \cdots, x_n \in \mathbb{F}_2$. The internal nodes are Boolean operators that combine the inputs received from their children and forward their output to the respective parent nodes. The output of the root node is the output value of the Boolean function. The corresponding truth table of the function $f : \mathbb{F}_2^n \to \mathbb{F}_2$ is determined by evaluating the tree over all possible 2^n assignments of the input variables at the leaf nodes. Each GP individual is thus evaluated according to its truth table.

In our experiments, we use the following function set: OR, XOR, AND, AND2, XNOR, IF, and function NOT that takes a single argument. The function AND2 behaves the same as the AND function but with the second input inverted. The function IF takes three arguments and returns the second one if the first one is evaluated as true and the third one otherwise. This function set is common when dealing with the evolution of Boolean functions with cryptographic properties [29].

The genetic operators used in our experiments with tree-based GP are simple tree crossover, uniform crossover, size fair, one-point, and context preserving crossover [30] (selected at random), and subtree mutation. The option to use multiple genetic operators was based on the initial experiments.

Since the search size grows rapidly with the number of inputs, we expect the bit string encoding, using both the truth table and ANF representation, to perform much worse than the GP, which is in accordance with most previous works. However, we include all representations for completeness and a more reliable estimate of the problem's difficulty.

Walsh-Hadamard Spectra and Spectral Inversion. We conclude this section by discussing why adapting the spectral inversion approach set forth by Clark et al. in [10] is more complex for five-valued spectra functions than for plateaued functions (i.e., 3-valued spectra functions). Consequently, we did not consider Walsh-Hadamard spectra as an additional encoding in our experiments.

As mentioned in Sect. 3, the main idea underlying spectral inversion is to encode a candidate solution as a Walsh-Hadamard spectrum satisfying specific constraints, such as having only three spectral values $-2^\lambda, 0, 2^\lambda$ to enforce a plateaued function. The problem is that, however, not every Walsh-Hadamard spectrum corresponds to an actual Boolean function. More specifically, if one starts from a random spectrum and applies the inverse Walsh-Hadamard transform, the result will most likely be a pseudo-Boolean function $f : \mathbb{F}_2^n \to \mathbb{Z}$. Therefore, the optimization problem becomes to permute the values in a spectrum until an actual Boolean function $f : \mathbb{F}_2^n \to \mathbb{F}_2$ is found via the inverse Walsh-Hadamard transform. Further, one needs to determine beforehand the multiplicities of each spectral value for the profile of the desired Boolean function, which is relatively easy to do in the case of a plateaued function. In fact, plateaued functions are characterized by a single amplitude λ with $\lceil n/2 \rceil \leq \lambda \leq n$. Moreover, one can exploit a) the Parseval relation to enforce that the sum of the squared Walsh-Hadamard spectrum must be equal to 2^{2n} for every Boolean function, and b) reduce the search space by half by setting $f(\underline{0}) = 0$. As described in [26], these two observations give rise to a linear system of two equations and two unknowns, where the latter represent respectively the multiplicities of -2^λ and $+2^\lambda$:[2]

$$\begin{cases} \#2^\lambda + \# - 2^\lambda & = \frac{2^{2n}}{2^{2\lambda}} \\ \#2^\lambda - \# - 2^\lambda & = 2^n \end{cases} \quad . \tag{6}$$

However, a 5-valued spectrum is characterized by *two* amplitudes λ_1, λ_2 with $\lceil n/2 \rceil \leq \lambda_1, \lambda_2 \leq n$. Therefore, in this situation, there are four multiplicities to be determined, namely those of $-2^{\lambda_1}, -2^{\lambda_2}, +2^{\lambda_1}$ and $+2^{\lambda_2}$. On the other hand, the constraints are always determined by the Parseval relation and setting the value of the function on the null vector equal to zero. Consequently, one ends up with an undetermined system with four unknowns and two equations, which has multiple solutions. In practice, adapting the spectral inversion technique for five-valued spectra functions would add an additional parameter tuning step since not all solutions of the system might yield an area of the search space containing actual Boolean functions. For this reason, we decided not to adopt this encoding in our experiments.

4.2   Fitness Functions

We designed two fitness functions to evolve five-valued spectra Boolean functions in different numbers of variables. Both fitness functions enforce balancedness in the following way: if the Boolean function is not balanced, it receives a negative penalty equal to the number of bits in the truth table that must be changed to reach a balanced function. Only if the function is balanced will it receive an additional score based on the Walsh-Hadamard spectrum values.

The first fitness function counts the number of distinct values in the Walsh-Hadamard spectrum; if the number of distinct values is not equal to 5, it receives the score of $\frac{1}{1+|\#values-5|}$, where $\#values$ denotes the number of the Walsh-Hadamard spectrum values. Next, only if the number of the Walsh-Hadamard spectrum values equals 5 does the function receive a score based on its nonlinearity. Note that the nonlinearity value is always larger than the score obtained when not having 5 values in the Walsh-Hadamard spectrum. However, rather than taking just the nonlinearity value nl_f, we consider the whole Walsh-Hadamard spectrum to provide approximate gradient information. In this case, the number of occurrences of the maximal absolute value in the spectrum is calculated

---

[2] The multiplicity of 0 can be determined separately by considering further cryptographic criteria such as correlation immunity, which force to zero some coefficients in the spectrum.

and denoted with #max_values. As higher nonlinearity corresponds to a lower maximal absolute value in the Walsh-Hadamard spectrum, we try to direct the search to as few occurrences of the maximal value as possible to facilitate reaching the next nonlinearity value. Finally, only if the function is balanced and has 5 values in the Walsh-Hadamard spectrum, it receives the score defined with

$$fitness_1(f) = nl_f + \frac{2^n - \#max\_values}{2^n}. \tag{7}$$

Note that the second term never reaches the value of 1 since, in that case, we effectively reach the next nonlinearity level.

The second fitness function defines a (generic) penalty factor by counting the number of coefficients whose values deviate from the five allowed ones. More precisely, given $f : \mathbb{F}_2^n \rightarrow \mathbb{F}_2$, we define the penalty of f as follows:

$$pen(f) = \begin{cases} |\{a \in \mathbb{F}_2^n : W_f(a) \notin \{0, \pm 2^{\frac{n-1}{2}}, \pm 2^{\frac{n+1}{2}}\}\}| & , \text{ if n is odd,} \\ |\{a \in \mathbb{F}_2^n : W_f(a) \notin \{0, \pm 2^{\frac{n}{2}}, \pm 2^{\frac{n+2}{2}}\}\}| & , \text{ if n is even.} \end{cases}$$

Then, a possible fitness function that optimizes both the nonlinearity of f and its "five-valuedness" is the following:

$$fitness_2(f) = \frac{nl_f}{1 + pen(f)}.$$

This fitness function optimizes the number of the Walsh-Hadamard spectrum values and nonlinearity concurrently, so a high fitness value does not indicate that the resulting function has only the desired values in the spectrum. However, this might allow the search algorithm to reach the desired property by traversing a wider portion of the search space. Note that it is also possible to obtain functions having only a subset of distinct values in the spectrum, but this is dealt with in post-processing, and solutions that are not five-valued spectra functions are penalized. The fitness values defined in this manner assume values differing in orders of magnitude; however, since we do not use fitness proportional selection (but a ranking one), only the relationship between the two values is relevant.

4.3 Algorithms and Parameters

We employ the same evolutionary algorithm for both bitstring and symbolic encoding: a steady-state selection with a 3-tournament elimination operator (denoted SST). In each iteration of the algorithm, three individuals are chosen at random from the population for the tournament, and the worst one in terms of fitness value is eliminated. The two remaining individuals in the tournament are used with the crossover operator to generate a child individual, which then undergoes mutation with individual mutation probability $p_{mut} = 0.5$. Finally, the mutated child replaces the eliminated individual in the population.

The experiments are executed in 30 repetitions for every problem size, fitness function, and representation. The termination condition has been set to 10^6 evaluations; this proved to be more than enough for all the variants to converge.

5 Experimental Results

The experiments were performed for three representations (TT, ANF, symbolic/GP), using two fitness functions and for Boolean functions from 5 to 16 variables (since these sizes represent the common choice in related works [11]). The general remarks can be summarized as follows: for 5 and 6 variables, all representations and fitness functions result in a five-valued spectra Boolean function with the maximal nonlinearity value found in every algorithm run.

Already for 7 variables, there are significant differences: both bitstring representations (TT and ANF) fail to find a five-valued function in almost every run. In fact, out of 60 runs for TT and ANF with $fitness_1$, only a single run managed to find a Boolean function with 5 values in the spectrum (Fig. 1a). The situation is somewhat better for $fitness_2$, where multiple optimal solutions (with a nonlinearity value of 56) for that size were found using TT and ANF representations (Fig. 1b). However, the GP's performance is clearly superior since it obtained the perfect result in almost every run.

For sizes larger than $n = 7$, the genetic algorithm with TT or ANF representation could not find even a single five-valued function in any of the experiments. In these cases, the nonlinearity value is irrelevant since the primary objective was not fulfilled in the first place. On the other hand, GP easily finds five-valued functions for all problem sizes; it also manages to optimize the nonlinearity value further while keeping this constraint satisfied. For this reason, we exclude the bitstring (TT and ANF) results from further analysis and show only the results obtained with GP.

Table 1. Results of fitness values for various problem sizes, GP representation.

Size	$fitness_1$			$fitness_2$		
	avg	stdev	max	avg	stdev	max
5	12.63	0.00	12.63	12.62	0.02	12.63
6	24.94	0.01	24.94	24.94	0.00	24.94
7	56.00	1.52	56.63	55.10	8.33	56.63
8	112.96	0.01	112.97	112.96	0.01	112.97
9	235.68	6.21	240.63	67.04	79.01	240.63
10	480.98	0.01	480.98	148.30	169.41	480.98
11	985.13	11.97	992.63	110.17	3.20	115.40
12	1984.97	0.02	1984.99	219.51	8.05	230.60
13	3996.07	50.47	4032.63	419.61	5.51	422.60
14	8064.99	0.02	8065.00	838.17	13.33	870.60
15	16183.77	56.77	16256.60	1645.43	11.51	1664.20
16	31958.31	2990.04	32513.00	3292.20	23.32	3328.20

Table 1 outlines the fitness values obtained by GP for both fitness function definitions. Since both fitness functions do not necessarily reveal the nonlinearity

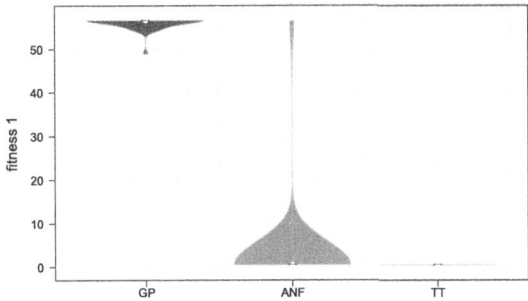

(a) Results for $fitness_1$.

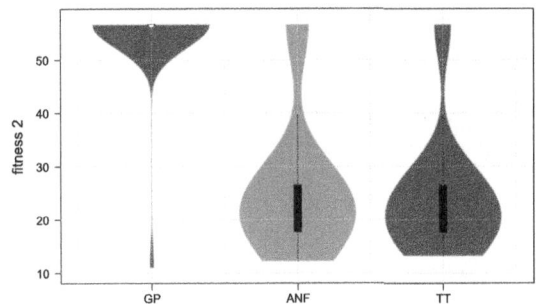

(b) Results for $fitness_2$.

Fig. 1. Violin plot representation of the results for problem size 7.

value in all cases (due to their definition), we present the best obtained nonlinearity values in all experiments, depending on problem size and fitness function, in Table 2. Moreover, we provide the best-known nonlinearities (taken from [12] and [2]) as a reference for the quality of the obtained solutions.[3] Although the two used fitness functions are not directly comparable, already from Table 1, we can see that $fitness_2$ provides inferior results when the number of variables is larger than $n = 10$. This is evident from Table 2, where the obtained nonlinearity values differ greatly.

Let us assess the results with $fitness_1$ in more detail. Observe that for odd sizes, we generally obtain better results. In fact, we reach functions that match the best-known nonlinearity for sizes 5, 7, 9, and 11. Furthermore, for size 13, the nonlinearity we obtained is very close to the best-known nonlinearity. For even values, we could consider that up to size 10, we reach nonlinearity similar (but slightly worse) to the best-known one. Interestingly, we can see that for odd sizes,

---

[3] Note that the best-known nonlinearity is given for a general case of balanced functions and not necessarily five-valued ones.

we generally (the exception is size 16, where the standard deviation is two orders of magnitude higher than for any other case) have a higher standard deviation, which indicates that while the obtained nonlinearity values are closer (or match the best-known ones), it is not necessarily trivial to reach those values. On the other hand, for even sizes, the standard deviation is extremely small, indicating that the algorithm gets stuck in local optima very easily or that the obtained nonlinearity values are the best possible ones for even-sized five-valued spectra Boolean functions.

Table 2. Nonlinearity values obtained for the two fitness functions for different problem sizes.

Size	$fitness_1$ max NL	$fitness_2$ max NL	best-known nonlinearity
5	**12**	**12**	12
6	24	24	26
7	**56**	**56**	56
8	112	112	116
9	**240**	240	240
10	480	480	492
11	**992**	114	992
12	1984	230	2010
13	4032	422	4036
14	8064	870	8120
15	16256	1664	16272
16	32512	3328	32638

Figure 2 outlines the convergence of the GP algorithm for two selected problem sizes: 10 and 16. The problem of size 10 is selected since it represents the largest size for which both fitness functions reach the same maximum value, and size 16 was selected since it represents the largest problem size considered. For size 10, we see that the convergence between the two fitness functions is quite similar, although the algorithm converges slightly faster when $fitness_1$ is used. However, this difference can be considered almost negligible since it is restricted to only a few generations. On the other hand, we see a large difference between the algorithm's performance for the two fitness functions for size 16. In this case, with $fitness_2$, the algorithm stagnates throughout the optimization process and does not obtain any better solution. For $fitness_1$, we see that the algorithm starts with a worse nonlinearity value but soon jumps to a much better one. However, from that point on, the algorithm mostly does not improve any further for the rest of the run, confirming that the selected number of generations is more than enough.

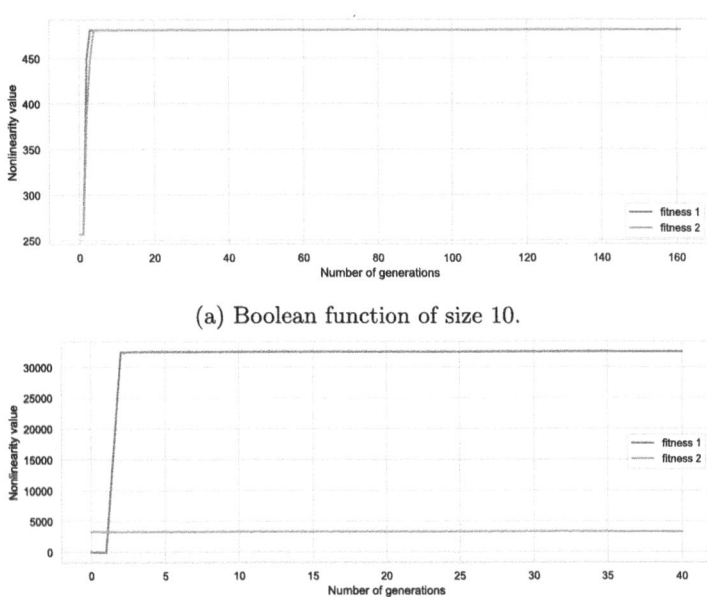

(a) Boolean function of size 10.

(b) Boolean function of size 16.

Fig. 2. Convergence of the algorithm.

6 Conclusions and Future Work

This paper explores the evolution of five-valued spectra Boolean functions. This problem has significant practical relevance since there are only a few known ways how to construct such functions. We experiment with three solution encodings, two fitness functions, and 12 Boolean functions sizes. We show it is possible to construct five-valued functions for each dimension when using GP. Moreover, we evolve five-valued functions that are highly nonlinear and for odd sizes up to dimension 11, matching the best-known nonlinearities. Next, we observe that a simpler fitness function reaches better nonlinearity values, indicating a more elaborate penalty function may be needed.

In future work, it would be relevant to design better fitness functions and experiment with Cartesian Genetic Programming. Next, one could compare the difficulty of evolving bent Boolean functions (since they have only two values in their spectra), 3-valued functions, and five-valued functions. Finally, considering that the results obtained in this work rival state-of-the-art obtained with algebraic constructions, it could be interesting to try to evolve constructions of five-valued spectra functions.

References

1. Picek, S., Jakobovic, D.: Evolving algebraic constructions for designing bent Boolean functions. In: Proceedings of the Genetic and Evolutionary Computation Conference 2016. GECCO '16, pp. 781–788. Association for Computing Machinery (2016)
2. Carlet, C., Djurasevic, M., Jakobovic, D., Mariot, L., Picek, S.: Evolving constructions for balanced, highly nonlinear Boolean functions. In Fieldsend, J.E., Wagner, M. (eds.) GECCO '22: Genetic and Evolutionary Computation Conference, Boston, Massachusetts, USA, July 9 - 13, 2022, pp. 1147–1155. ACM (2022)
3. Mariot, L., Picek, S., Jakobovic, D., Djurasevic, M., Leporati, A.: Evolutionary construction of perfectly balanced Boolean functions. In: IEEE Congress on Evolutionary Computation, CEC 2022, Padua, Italy, July 18–23, 2022, pp. 1–8. IEEE (2022)
4. Kavut, S., Maitra, S., Yucel, M.D.: Search for Boolean functions with excellent profiles in the rotation symmetric class. IEEE Trans. Inf. Theory **53**(5), 1743–1751 (2007)
5. Maitra, S., Sarkar, P.: Cryptographically significant Boolean functions with five valued Walsh spectra. Theoret. Comput. Sci. **276**(1–2), 133–146 (2002)
6. Hodžić, S., Pasalic, E., Zhang, W.: Generic constructions of five-valued spectra Boolean functions. IEEE Trans. Inf. Theory **65**(11), 7554–7565 (2019)
7. Cao, X., Hu, L.: Two Boolean functions with five-valued Walsh spectra and high nonlinearity. Int. J. Found. Comput. Sci. **26**(05), 537–556 (2015)
8. Mesnager, S., Zhang, F.: On constructions of bent, semi-bent and five valued spectrum functions from old bent functions. Adv. Math. Commun. **11**(2), 339–345 (2017)
9. Xu, G., Cao, X., Xu, S.: Several classes of Boolean functions with few Walsh transform values. Appl. Algebra Eng. Commun. Comput. **28**(2), 155–176 (2016)
10. Clark, J.A., Jacob, J.L., Maitra, S., Stanica, P.: Almost Boolean functions: the design of Boolean functions by spectral inversion. Comput. Intell. **20**(3), 450–462 (2004)
11. Djurasevic, M., Jakobovic, D., Mariot, L., Picek, S.: A survey of metaheuristic algorithms for the design of cryptographic Boolean functions. Cryptogr. Commun. **15**(6), 1171–1197 (2023)
12. Carlet, C.: Boolean Functions for Cryptography and Coding Theory. Cambridge University Press, Cambridge (2021)
13. Langevin, P., Leander, G.: Counting all bent functions in dimension eight 99270589265934370305785861242880. Des. Codes Cryptogr. **59**(1–3), 193–205 (2011)
14. Millan, W., Clark, A., Dawson, E.: Heuristic design of cryptographically strong balanced boolean functions. In: Advances in Cryptology - EUROCRYPT '98, pp. 489–499 (1998)
15. Clark, J.A., Jacob, J.L.: Two-stage optimisation in the design of Boolean functions. In: Dawson, E.P., Clark, A., Boyd, C. (eds.) ACISP 2000. LNCS, vol. 1841, pp. 242–254. Springer, Heidelberg (2000). https://doi.org/10.1007/10718964_20
16. Fuller, J., Dawson, E., Millan, W.: Evolutionary generation of bent functions for cryptography. In: The 2003 Congress on Evolutionary Computation, 2003. CEC '03, vol. 3, pp. 1655–1661 (2003)
17. Aguirre, H., Okazaki, H., Fuwa, Y.: An Evolutionary multiobjective approach to design highly non-linear Boolean functions. In: Proceedings of the Genetic and Evolutionary Computation Conference GECCO'07, pp. 749–756 (2007)

18. Manzoni, L., Mariot, L., Tuba, E.: Balanced crossover operators in genetic algorithms. Swarm Evol. Comput. **54**, 100646 (2020)
19. Behera, P.K., Gangopadhyay, S.: An improved hybrid genetic algorithm to construct balanced Boolean function with optimal cryptographic properties. Evol. Intell. **15**(1), 639–653 (2022)
20. Picek, S., Jakobovic, D., Golub, M.: Evolving cryptographically sound Boolean functions. In: GECCO (Companion), pp. 191–192 (2013)
21. Picek, S., Jakobovic, D., Miller, J.F., Batina, L., Cupic, M.: Cryptographic Boolean functions: One output, many design criteria. Appl. Soft Comput. **40**, 635–653 (2016)
22. Picek, S., Carlet, C., Guilley, S., Miller, J.F., Jakobovic, D.: Evolutionary algorithms for Boolean functions in diverse domains of cryptography. Evol. Comput. **24**(4), 667–694 (2016)
23. Hrbacek, R., Dvorak, V.: Bent function synthesis by means of cartesian genetic programming. In: Bartz-Beielstein, T., Branke, J., Filipič, B., Smith, J. (eds.) PPSN 2014. LNCS, vol. 8672, pp. 414–423. Springer, Cham (2014). https://doi.org/10.1007/978-3-319-10762-2_41
24. Picek, S., Jakobovic, D., Miller, J.F., Marchiori, E., Batina, L.: Evolutionary methods for the construction of cryptographic Boolean functions. In: Machado, P., et al. (eds.) Genetic Programming, pp. 192–204. Springer, Cham (2015)
25. Husa, J.: Designing correlation immune Boolean functions with minimal hamming weight using various genetic programming methods. In: López-Ibáñez, M., Auger, A., Stützle, T., eds.: Proceedings of the Genetic and Evolutionary Computation Conference Companion, GECCO 2019, Prague, Czech Republic, July 13-17, 2019, pp. 342–343. ACM (2019)
26. Mariot, L., Leporati, A.: A genetic algorithm for evolving plateaued cryptographic Boolean functions. In: Dediu, A.-H., Magdalena, L., Martín-Vide, C. (eds.) TPNC 2015. LNCS, vol. 9477, pp. 33–45. Springer, Cham (2015). https://doi.org/10.1007/978-3-319-26841-5_3
27. Rovito, L., Lorenzo, A.D., Manzoni, L.: Discovering non-linear Boolean functions by evolving Walsh transforms with genetic programming. Algorithms **16**(11), 499 (2023)
28. Mariot, L., Saletta, M., Leporati, A., Manzoni, L.: Heuristic search of (semi-)bent functions based on cellular automata. Nat. Comput. **21**(3), 377–391 (2022)
29. Carlet, C., Durasevic, M., Gasperov, B., Jakobovic, D., Mariot, L., Picek, S.: A new angle: on evolving rotation symmetric Boolean functions. In Smith, S.L., Correia, J., Cintrano, C. (eds.) Applications of Evolutionary Computation - 27th European Conference, EvoApplications 2024, Held as Part of EvoStar 2024, Aberystwyth, UK, April 3-5, 2024, Proceedings, Part I. Lecture Notes in Computer Science, vol. 14634, pp. 287–302. Springer (2024)
30. Poli, R., Langdon, W.B., McPhee, N.F.: A field guide to genetic programming (2008). Published via http://lulu.com and freely http://www.gp-field-guide.org.uk

Search Trajectory Networks Applied to a Real-World Parallel Batch Scheduling Problem

Francesca Da Ros[1]([✉])(iD), Luca Di Gaspero[1](iD), Marie-Louise Lackner[2](iD),
Nysret Musliu[2](iD), and Michael Soprano[1](iD)

[1] Università degli Studi di Udine, Udine, Italy
{francesca.daros,luca.digaspero,michael.soprano}@uniud.it
[2] Christian Doppler Laboratory for Artificial Intelligence and Optimization for
Planning and Scheduling, TU Wien, Vienna, Austria
{marie-louise.lackner,nysret.musliu}@tuwien.ac.at

Abstract. We investigate solution methods for the Oven Scheduling Problem (OSP), a parallel batch scheduling optimization problem in semiconductor manufacturing, using Search Trajectory Networks (STNs). STNs are a recently introduced tool to analyze and compare the behavior of metaheuristic algorithms concerning their exploration ability w.r.t. single problem instances. We consider two state-of-the-art algorithms for the OSP, a Simulated Annealing (SA) and a Large Neighborhood Search (LNS), and instances from the literature. The STNs enable us to draw the following conclusions: (i) The two algorithms' trajectories overlap especially at the beginning of the trajectories, as revealed by a search space partitioning based on Hierarchical Agglomerative Clustering; (ii) The fitness landscape of many instances is multi-modal, with several high-quality solutions scattered in the search space; (iii) SA trajectories are longer, but the number of locations visited by SA and LNS is similar.

Keywords: Oven scheduling problem · Algorithmic behavior · Empirical analysis · Search space partitioning · Large neighborhood search · Simulated annealing · Algorithm visualization · Explainability

1 Introduction

Comparing different solution approaches for an optimization problem is essential to find the most suitable algorithm for a given class of instances sharing similar characteristics. To this end, it is common practice for optimization practitioners to collect quantitative data on the algorithm's behavior (e.g., best objective function value, number of iterations, time, etc.) and to display them in tables, box plots, and line plots. While these analyses are useful, they fail to represent the complete behavior of the algorithms and focus only on the final step of the optimization process. Recently, Search Trajectory Networks (STNs) have

© The Author(s), under exclusive license to Springer Nature Switzerland AG 2025
P. García-Sánchez et al. (Eds.): EvoApplications 2025, LNCS 15612, pp. 68–85, 2025.
https://doi.org/10.1007/978-3-031-90062-4_5

emerged as a powerful tool for enhancing the interpretability and explainability of algorithmic outcomes [33]. STNs offer a graphical representation of algorithmic behavior by representing the trajectories taken by algorithms in the search space as a directed graph. These networks are particularly useful for visualizing and analyzing the paths that search algorithms traverse when applied to the same problem instance.

The Oven Scheduling Problem (OSP) is a real-world NP-hard parallel batch scheduling problem arising in the electronic component manufacturing industry [27]. The OSP aims to group jobs in batches and to devise a schedule that minimizes the cumulative processing time of the batches, the tardiness, and the setup costs. Feasible schedules must respect a set of constraints, including job compatibility, availability times, due and release dates, etc. Many methods have been proposed in the literature to solve the OSP, including Integer Linear Programming (ILP) [27], Constraint Programming (CP) [27], Local Search (LS) [13,14,28], Large Neighborhood Search (LNS) [12], and a construction heuristic [27]. Classical analyses of stochastic algorithms, such as *gap analysis* and *critical difference plots*, demonstrated that LNS and Simulated Annealing (SA) are the most successful algorithms [14]. However, these findings are primarily based on evaluating the final outcomes, with a limited exploration of how these algorithms behave within the search space.

This paper aims to analyze the behavior of two state-of-the-art algorithms, SA and LNS, in solving a real-world parallel batch scheduling problem using STNs. Through this analysis, we contribute to the study of algorithms in parallel batch scheduling by providing deeper insights into the solution landscape of problem instances and its relationship with algorithm performance. Additionally, we highlight the value of STNs in algorithm analysis, with the goal of encouraging their adoption as a standard component in comparative algorithm studies.

The remainder of this paper is structured as follows. Section 2 reports the definition of the problem at hand. Section 3 provides the necessary background on STNs. Section 4 summarizes related work. Section 5 describes our methodology and Sect. 6 reports the results. Finally, Sect. 7 concludes and outlines future research directions.

2   Oven Scheduling Problem

In this section, we provide an informal description of the combinatorial optimization at hand, the OSP. For the formal ILP model, we forward the interested reader to the original work by Lackner et al. [27].

2.1   Instance Parameters

The key elements of the problem include a set of machines (or ovens), a set of jobs to be scheduled, and a set of attributes (i.e., temperature at which the jobs should be processed).

Each oven has a maximum capacity and a schedule of available and unavailable intervals (e.g., for downtime, maintenance, etc.). The initial state of each oven is described by a specified attribute (i.e., temperature setting). Each job is defined by an attribute (i.e., processing temperature), a release date (i.e., the earliest time the job can be processed), a due date, and a size. Furthermore, its processing time is constrained within a minimum and maximum range and each job may only be assigned to a subset of eligible machines. Attribute dependent setup times and costs are incurred for adjusting a machine's settings before processing the first batch and between consecutive batches on the same machine, as these adjustments ensure the required attribute setting for the incoming jobs.

2.2   Objective Function

The goal is to minimize the cumulative batch processing time (p), the cumulative setup costs (sc), and the number of tardy jobs (t). The objective function obj is a linear combination of these components and is defined as follows:

$$\text{obj} = (w_p \cdot \tilde{p} + w_t \cdot \tilde{t} + w_{sc} \cdot \tilde{sc})/(w_p + w_t + w_{sc}) \quad \in [0,1] \subset \mathbb{R} \qquad (1)$$

where $\tilde{p}$, $\tilde{t}$, and $\tilde{sc}$ are the normalized versions of p, t, and sc, respectively, and w_p, w_{sc}, and w_t are the associated weights. We consider the objectives in a lexicographic manner: minimizing processing time takes precedence over minimizing tardiness, which, in turn, is prioritized over minimizing setup costs. This is achieved by setting $w_p = w_t \cdot (n+1)$, $w_t = n \cdot \max_{SC} + 1$, and $w_{sc} = 1$, where n is the number of jobs and $\max_{SC}$ is the maximal setup cost.[1]

2.3   Constraints

The aim of the OSP is to assign each job to a machine and devise an optimal schedule respecting the following constraints: (i) Each job is assigned to exactly one batch on one of its eligible machines. (ii) Each batch must be processed within the scheduling horizon. (iii) All jobs in one batch have the same start and end time. (iv) All jobs assigned to one batch must share the same attribute. (v) The batch size must not exceed the capacity of the machine it is assigned to. (vi) A batch may not start before the release time of any job in that batch. (vii) The processing time of each batch must lie between the minimum and maximum processing times of all assigned jobs. Preemption is not allowed. (viii) Batches on the same machine may not overlap, and setup times must be accounted for between consecutive batches. (ix) Both the batch processing time and the preceding setup time must lie entirely within one of the machine's availability intervals.

---

[1] See Lackner et al. [27, Use Case 2].

3   Search Trajectory Networks

STNs are visualizations of directed graphs used to analyze and compare the search behavior of stochastic algorithms such as metaheuristics. STNs are based on the following notions, as introduced by Ochoa et al. [33]:

- *Representative solution*: A solution to the problem at hand at a given point in time. It represents the state of the search algorithm (e.g., the new best solution found by SA during the search).
- *Location*: A non-empty subset of representative solutions coming from a partition of the search space (note that if no search space partitioning is applied, then the set of locations and the set of representative solutions coincide, see also Sect. 5.5). Each representative solution belongs to exactly one location.
- *Search trajectory*: A sequence of locations created by substituting each representative solution in a sequence with its corresponding location, based on the order in which the solutions are encountered during the search.
- *Node*: A location in the search trajectory. The set of nodes is denoted as N.
- *Edges*: Directed edges connecting two consecutive locations in the search trajectory. Their weight is proportional to the number of transitions between the two locations. The set of edges is denoted as E.
- *Search Trajectory Network (STN)*: The directed weighted graph STN = $G(N, E, w)$.

STNs are available both as a standalone[2] and a web-based tool [6]. STNs have been recently extended to the multi-objective case [29] and integrated with Large Language Models (LLMs) [8]. Recent literature works applied STNs to complex problems, like parameter tuning [35], neuroevolution [36], social networks [5], and genetic programming [20]. However, to the best of our knowledge, STNs have neither been applied to scheduling problems (see also Sect. 4.2).

4   Related Work

4.1   Parallel Batch Scheduling Problems

Batch scheduling problems with and without parallelization of machines play an important role in modern manufacturing, as they aim to optimize resource usage by grouping and processing together compatible jobs. They have been applied to several fields, including semiconductor [9,30], composite [42], steel [46], and pharmaceutical [24] industry. Batch scheduling problems consider different objectives and constraints, including makespan [10,16], tardiness [23,42], and flowtime [3]. The OSP differs from these formulations because of the different mix of objectives and constraints, i.e., it includes more real-world details than other parallel batch scheduling problems.

Solution methods for batch scheduling problems span from exact methods, like Constraint Programming [42] and Mixed-Integer Programming (MIP) [23], to heuristic and metaheuristics methods, such as SA [11,15,31], Tabu Search (TS) [4,24], and Genetic Algorithms [17,45].

---

[2] See https://github.com/gabro8a/STNs-v2.

4.2 Analysis of Solution Methods for Combinatorial Optimization

In the last decade, there has been a growing interest in analyzing algorithms and in interpreting the optimization process in combinatorial problems [40,41]. Tools included advanced statistical methodologies, such as meta-analysis [43], and visualization frameworks, such as Instance Space Analysis (ISA), Local Optima Networks (LONs), and STNs. Smith-Miles and Muñoz [39] introduced the ISA methodology, which links algorithm performance to instance features, allowing for the identification of regions where algorithms perform well or poorly. This instance space is depicted in a $2D$ projection. Ochoa et al. [34] and Verel et al. [44] proposed to model the fitness landscape through networks called LONs. Each node is a local optimum and the edges are the possible transitions between them. STNs differ from LONs as they focus on the search behavior, whereas the latter focus on the structure of the fitness landscape.

Considering the application of STNs to combinatorial optimization problems, four studies could be discerned.[3] Akbay and Blum [1] demonstrated the utility of STNs by applying them to two instances each of two combinatorial optimization problems, the Minimum Positive Influence Dominating Set Problem (MPIDS) and the Electric Vehicle Routing Problem with Time Windows and Simultaneous Pickup and Delivery (EVRP-TW-SPD). The authors showcased the potential of the methodology; however, the procedure was not fully integrated within the broader context of algorithm analysis. Meanwhile, Alza et al. [2] examined significant implications of rotating the landscape within combinatorial domains, employing STNs as part of their approach. The work by Narvaez-Teran et al. [32] closely resembles our study, as the authors analyzed STNs across multiple instances for two state-of-the-art algorithms; however, their focus is restricted to a problem with limited real-world applicability, the Cyclic Bandwidth Sum Problem (CBSP). The only work that incorporated STNs directly into the evaluation of algorithm performance is the one on Influence Maximization (IM), a graph-based benchmark problem, by Chacon-Sartori and Blum [5].

5 Methodology

In this section, we outline the methodology followed in this work, as summarized in Fig. 1. We apply two state-of-the-art algorithms for the OSP (Sect. 5.1) on a subset of problem instances (Sect. 5.2). Representative solutions are collected (Sect. 5.3, Sect. 5.4) and the locations are identified through search space partitioning (Sect. 5.5). Finally, we represent the STNs as graphs and extract descriptive metrics for analysis (Sect. 5.6).

---

[3] We consider the 51 literature works that cite Ochoa et al. [33] as per Scopus (28-10-2024).

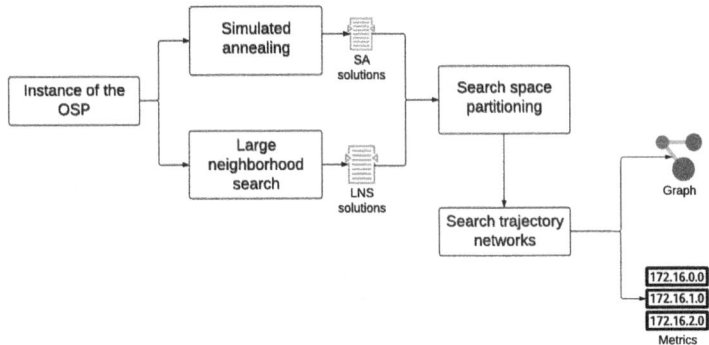

Fig. 1. Methodology followed in this work.

5.1 Algorithms

We consider the two state-of-the-art algorithms for the OSP (SA and LNS).

Simulated Annealing. We consider the multi-neighborhood SA proposed by Da Ros et al. [14]. The algorithm uses the Metropolis acceptance criterion, a geometric cooling scheme, and a cutoff mechanism to reduce computational costs during the early stages of the search [18,22]. The initial solution is computed using the construction heuristic by Lackner et al. [27]. The search space is explored using several neighborhoods: swapping two batches, creating new unary or larger batches, and adding jobs to an existing batch. We use the code available in public repositories of the original work.[4] The code, written in C++, is compiled using Clang++18.

Large Neighborhood Search. We consider the LNS proposed by Da Ros et al. [12]. LNS iteratively destroys and repairs parts of an existing solution [38]. The initial solution is computed using the construction heuristic by Lackner et al. [27]. The procedure uses three destroy methods: destroying random batches, destroying random batches with the same attribute, and destroying random batches on the same machine. The repair operator is based on a CP model utilizing interval variables and two levels for the optimality tolerance. The size of the destruction, i.e., the number of destroyed batches, is increased after several idle iterations and reset to the initial size whenever an improvement in the objective function is found. We use the code available in public repositories of the original work.[5] The control structure and the destroy operators were implemented in Python (v.3.10.12) and the repair operator is run with CP Optimizer (v.22.1.0) [25].

Experimental Setting. The experiments are executed on a machine featuring 2x Intel Xeon Platinum 8368 2.4 GHz 38C, 8×64 GB RDIMM. Each algorithm is executed on a single thread 5 times, each time with a different seed (i.e., from 42 to 46). Both methods are stopped with a 1-h time limit or after 10000 idle

---

[4] See https://github.com/iolab-uniud/osp-ls.

[5] See https://github.com/iolab-uniud/osp-lns.

iterations. The parameters of the algorithms were set as in the original papers, given that the machines' characteristics are comparable.

5.2 Instances

We consider the OSP benchmark dataset [13, 27] available on Zenodo [26]. Out of the 120 benchmark instances, we randomly select 10. Their characteristics are reported in Table 1, together with the algorithm that finds the solution with the best objective (label "winner"). The roughly estimated search space size is given by n^{n+k}, where n is the number of jobs and k the number of machines.

Table 1. Characteristics of the selected instances. STNs for instances marked with a "◇" are presented in Fig. 3 on page 14.

	code	jobs	machines	attributes	horizon	search space size	winner
	inst_45	50	2	2	15403	$\approx 10^{89}$	LNS
	inst_50	50	2	5	3021	$\approx 10^{89}$	LNS
◇	inst_61	100	2	2	1267	$\approx 10^{204}$	LNS
	inst_62	100	2	2	10304	$\approx 10^{204}$	SA
	inst_70	100	2	5	32861	$\approx 10^{204}$	LNS
	inst_72	100	5	2	3748	$\approx 10^{210}$	SA
◇	inst_77	100	5	5	1135	$\approx 10^{210}$	SA & LNS
	inst_79	100	5	5	29816	$\approx 10^{210}$	SA
	inst_82	250	2	2	8224	$\approx 10^{605}$	LNS
◇	inst_91	250	5	2	7605	$\approx 10^{612}$	SA

5.3 Sampling Process

The first step in implementing STNs is to track the representative solutions and their transitions. In our approach, we log any changes in the best solution encountered. We avoid using frequency-based logging (e.g., recording after a given number of steps) since the recording process itself does not impact the algorithm's runtime – our as SA method is iteration-based and LNS logging procedures require less than 0.0001 s to register a new solution. Additionally, frequency-based logging risks missing important information, as only a subset of the representative solutions may be captured using this approach.

5.4 Solution Encoding

A solution to the OSP is represented by two arrays of size equal to the number of jobs. The first array stores, for each job j, the machine m on which it is processed. The second array stores, for each job j, the position b in the schedule of machine

m at which it is processed. This representation is sufficient because processing times, batch start times, and other elements of the schedule can be calculated deterministically based on the job's processing order, assigned machine, and the machine's availability and setup times, following an earliest-possible scheduling procedure. This is how both SA and LNS deal with solutions when evaluating their objective value.

STNs require all solutions to be in string format, with all the strings having the same length. For this reason, we concatenate the above-mentioned arrays in a single data structure and convert the numbers to bitstrings.

Figure 2 illustrates an example of the solution encoding for a random instance with 10 jobs, 2 machines, and 2 attributes. At the top, we display a human-readable solution representation in the form of a Gantt chart for both machines. The horizontal axis represents time, while the vertical axis denotes machine capacity. The machine's maximum capacity is indicated by a dashed red line, and unavailable time intervals are shown in black. Each job is represented by a rectangle, where the color corresponds to the attribute and the height corresponds to the job's size. Jobs processed in the same batch are stacked, so their combined height reflects the batch's total size. Jobs within a batch begin and end processing simultaneously, with the processing time set by the longest minimum processing time among the jobs. The middle section of Fig. 2 encodes the solution using two arrays: one specifying the machine assigned to each job and another indicating the processing order. Finally, these two arrays are concatenated into a single array, with colors distinguishing the machine assignment segment from the processing order segment. The bottom portion of Fig. 2 illustrates this concatenated array in its bitstring form.

Preliminary experiments considering additional elements in the solution representation (e.g., the processing start time, processing duration, etc.) provided results comparable to this minimal solution representation and are therefore not included here.

5.5   Search Space Partitioning

STNs built solely from representative solutions (where each location corresponds to a unique solution) often become cluttered and hard to interpret. This issue is especially pronounced in real-world problems where algorithmic paths explore a large number of solutions. To address this, search space partitioning can be employed. This involves mapping similar solutions to a smaller set of locations, based on some measure of proximity or similarity between solutions. In this work, we consider two partitioning strategies, one based on Shannon Entropy (SE) [33] and the other on Hierarchical Agglomerative Clustering (HAC) [7].

Shannon Entropy. The core idea of SE-based search space partitioning is to merge solutions by removing variables from the search space. The procedure is as follows (see [33, Section 5.4] for theoretical details): First, the SE [37] of each decision variable is calculated, based on the values they assume in the

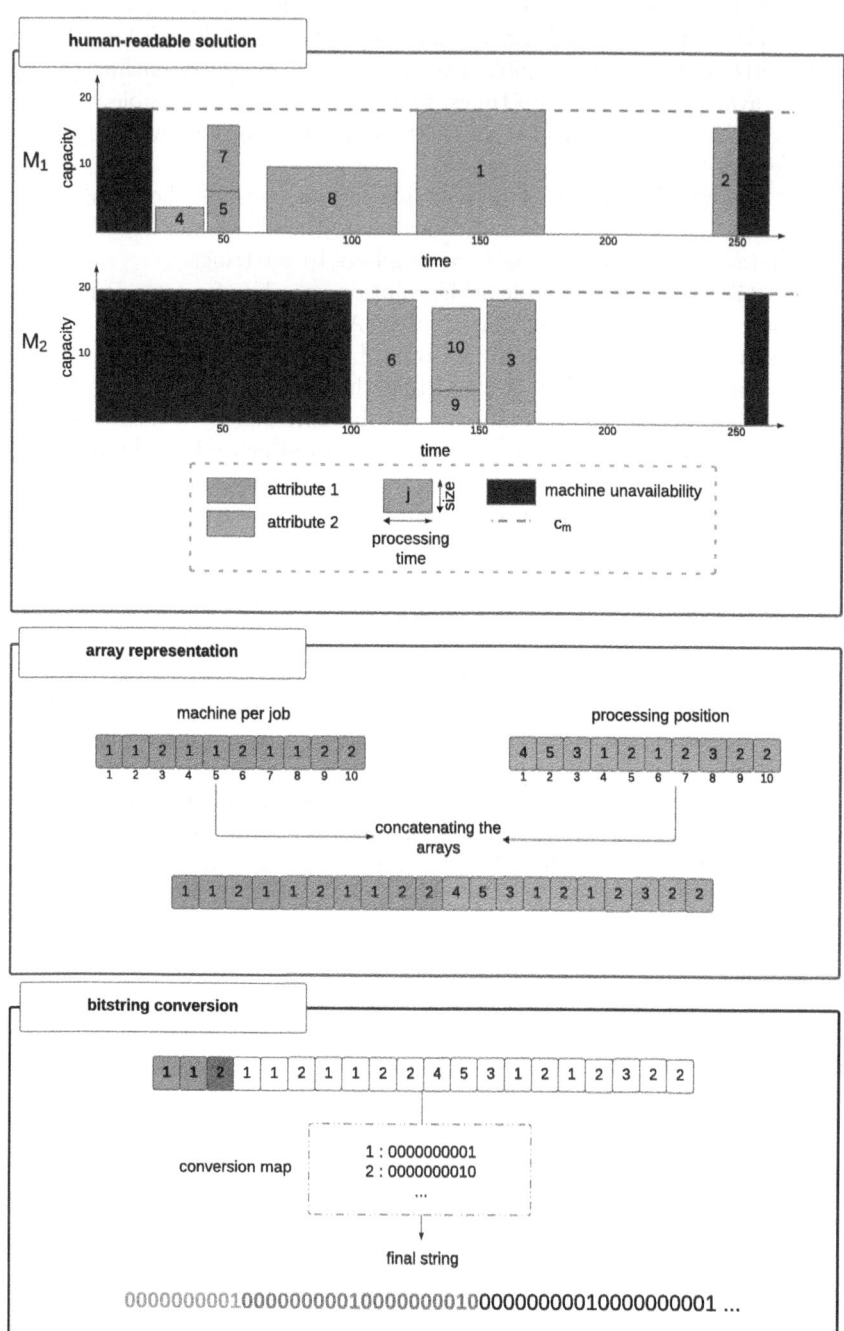

Fig. 2. Solution representation.

list of representative solutions so that variables that present the same values in multiple solutions have lower SE values; this means that they are not so informative. Second, decision variables with lower SE values are removed. The number of variables removed is based on a percentage threshold set by the user.

Hierarchical Agglomerative Clustering. HAC-based search partitioning is a bottom-up approach. The implementation uses single-linkage and is extended w.r.t. traditional HAC with a mechanism controlling the size (i.e., the percentage of all solutions a cluster can maximally contain) and volume (i.e., the percentage of the covered search space volume the solutions of a cluster span) of the clusters. Both these control values are set by the user (as a percentage w.r.t. the number of representative solutions, see [7, Section 3] for theoretical details). The procedure is as follows: Initially, each data point (i.e., representative solution) is considered as a cluster of size 1. Based on a distance measure that can be specified by the user, the two closest clusters are determined and, if possible in terms of resulting cluster volume and size, are merged into a new cluster. This step is repeated iteratively.

Experimental Setting. We report the experiments conducted with and without search partitioning. When dealing with partitioning, we cannot choose the same values for reduction, volume size, or cluster size for all instances as they present different properties (e.g., number of solutions, similarity between solutions, etc.) [33]. In the case of Shannon Entropy, we consider a reduction between 50% and 90% (with 10% steps). In the case of HAC, we consider the Euclidian distance[6] and cluster size and volume between 1% and 5% (with 1% steps). For each instance, the choice was made through a full factorial analysis of the possible combinations, choosing the setting with the most expressive visualization.

5.6   Search Trajectory Networks Analysis

Graphical Representation. We interchangeably use networks generated by the Kamada-Kawai [21] and Fruchterman-Reingold [19] algorithms. Both algorithms produce force-directed layouts with a roughly even distribution of nodes, minimized edge crossings, nearly uniform edge lengths, and preserved symmetries.

Metrics. In this work, we evaluate the following set of STN metrics: (i) *nodes*: Number of nodes. (ii) *edges*: Number of edges. (iii) *best*: Number of nodes with the best-found evaluation. (iv) *ends*: Number of nodes at the end of the trajectories other than the nodes with the best evaluation. (v) *shared*: Number of nodes visited by more than one algorithm. (vi) *strength*: Incoming weighted degree of best nodes normalized by the total number of runs. We do not consider the

---

[6] The only other available option in the stnweb tool is the Manhattan distance, which is not suitable in this a case.

number of nodes at the beginning of the trajectories (*initial*) since all algorithms start from the same initial solution.

Experimental Setting. To plot STN graphs and extract the metrics, we used the web tool available at the public GitHub repository `stnweb`.[7] The tool was run on a MacBook Air M3 with 16GB of memory and macOS Sonoma 14.3.

6 Experimental Results

In this section, we report the results of the experimental analysis, considering the metrics and the visual representation as described in the previous section. Table 2 reports the STN metrics distinguishing the instances (label "code") and the search space partitioning strategy (label "partitioning", the percentages between brackets correspond to the chosen settings of the related parameters).

In all instances, STNs with no space partitioning (indicated with "–") do not identify areas explored by both algorithms, as the number of shared nodes is always 1 (since all runs start at the same initial solution). When applying SE, the number of nodes decreases slightly; however, the aggregation into locations happens only within representative solutions of the same algorithm (i.e., only representative solutions explored by SA or only representative solutions explored by LNS, therefore *shared*=1). We also conducted experiments increasing the partitioning size (up to 99%), but the results were not different w.r.t. those of lower partitioning percentages. Conversely, when applying HAC as a search partitioning strategy, it is possible to reduce significantly the number of locations and to identify locations visited by both algorithms: for all instances, the number of shared nodes is greater than 1. The difference in behavior between the two partitioning strategies may be attributed to the overestimation inherent in the solution representation using bitstrings. Specifically, HAC shows greater robustness to the zero-padding of strings, while SE is more sensitive to this characteristic.

Considering the number of best nodes, four instances (`inst_50`, `inst_61`, `inst_72`, and `inst_77`) have more than one best solution. Only in one case (`inst_61`), all best solutions belong to different locations of the search space (i.e., before and after the HAC partitioning the number of best solutions does not decrease). In the other three cases, the best solutions do not belong to a single location either, since the number of best nodes is always greater than 1 after partitioning. This is an indicator of a multi-modal landscape with high-quality solutions scattered across it. The strength values of best solutions are higher in the case of search space partitioning with HAC, meaning that the best nodes have many connections with other nodes. In this case, we can assume that the location where the best nodes lies is linked to several other locations and is therefore relatively easy to reach.

---

[7] See https://github.com/camilochs/stnweb.

Table 2. Search trajectory network metrics. The STNs for instances marked with a "◇" are presented in Fig. 3 on page 14.

	code	partitioning	nodes	edges	shared	best	end	strength
		–	361	360	1	1	9	0.1
	inst_45	SE (90%)	358	357	1	1	9	0.1
		HAC (1%,1%)	131	184	12	1	9	0.1
		–	276	275	1	5	5	0.5
	inst_50	SE (90%)	275	274	1	5	5	0.5
		HAC (1%,1%)	139	177	14	4	4	0.7
		–	836	835	1	3	7	0.3
◇	inst_61	SE (90%)	830	829	1	3	7	0.3
		HAC (1%,1%)	113	246	10	3	5	0.8
		–	1169	1168	1	1	9	0.1
	inst_62	SE (90%)	1163	1162	1	1	9	0.1
		HAC (1%,1%)	118	242	15	1	8	0.4
		–	1789	1788	1	1	9	0.1
	inst_70	SE (90%)	1765	1764	1	1	9	0.1
		HAC (2%,2%)	59	173	4	1	6	0.4
		–	828	827	1	5	5	0.5
	inst_72	SE (90%)	824	823	1	5	5	0.5
		HAC (1%,1%)	119	180	13	3	5	1.1
		–	837	836	1	5	5	0.5
◇	inst_77	SE (90%)	833	832	1	5	5	0.5
		HAC (1%,2%)	121	215	10	3	4	1.4
		–	950	949	1	1	9	0.1
	inst_79	SE (90%)	940	939	1	1	9	0.1
		HAC (1%,1%)	169	220	9	1	8	0.6
		–	553	552	1	1	9	0.1
	inst_82	SE (90%)	553	552	1	1	9	0.1
		HAC (1%,1%)	117	247	7	1	9	0.1
		–	1391	1390	1	1	9	0.1
◇	inst_91	SE (90%)	1385	1384	1	1	9	0.1
		HAC (2%,2%)	57	104	3	1	9	0.4

Due to space limitations, we present only the network representations for a subset of the selected instances (Fig. 3). The networks are constructed with distinct colors, shapes, and sizes to convey specific information: (i) Each algorithm is represented by a unique color; the trajectories of SA appear in gray, while those of LNS appear in a lighter shade od gray. (ii) Nodes visited by both algorithms are indicated by black circles. (iii) Nodes representing the best solu-

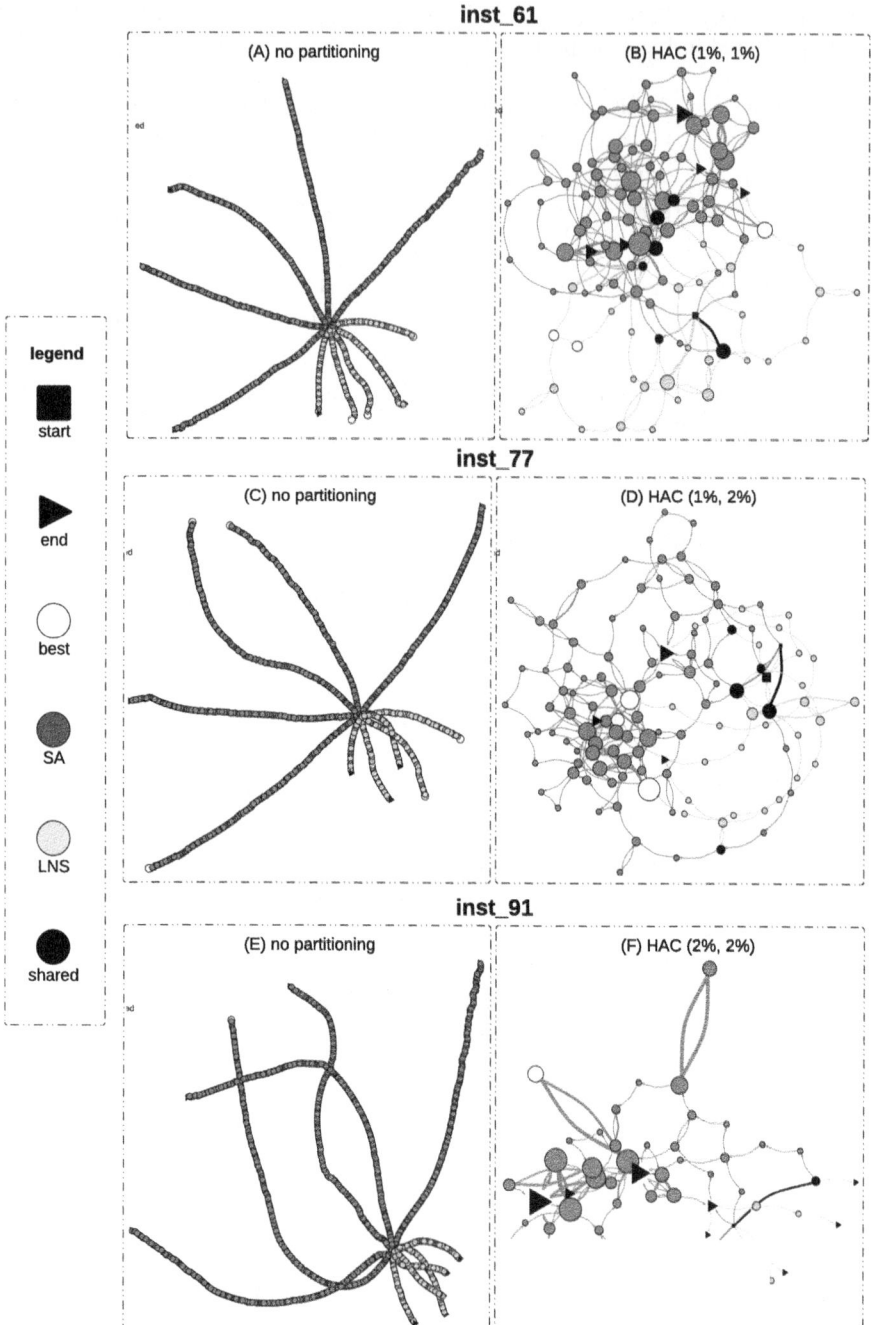

Fig. 3. Search trajectory networks.

tions are indicated by white circles. (iv) Starting nodes are represented by black squares. (v) End nodes are represented by black triangles. (vi) The size of a node is proportional to the number of visits paid to that node.

The selected instances present different characteristics: one where one algorithm outperforms the other, but both algorithms visit the locations of the best solutions according to HAC space partitioning (inst_61); one where both algorithms reach optimal solutions (inst_77); one where one algorithm is superior and only it visits the best solution locations (inst_91). We consider the networks without search space partitioning (Fig. 3, images A, C, and E) and those partitioned using HAC (Fig. 3, images B, D, and F). Results partitioned with SE are omitted, as it does not identify locations shared between the two algorithms.

In all cases, SA trajectories appear longer than those of LNS: this is justified by the algorithm procedures themselves, since SA randomly explores the search space with small and quick changes, whereas LNS considers sub-problems iteratively, performing fewer but larger changes on the current solution.

HAC partitioning highlights that SA tends to visit more representative solutions within the same locations (i.e., the nodes are larger); on the contrary, LNS tends to visit representative solutions in different locations of the space. The majority of the shared locations are within the initial part of the search, this is justified by the fact that both algorithms start from the same solution (i.e., both algorithms construct the initial solution using the algorithm proposed by Lackner et al. [27]). Finally, the end nodes also appear to be visited multiple times (i.e., they are larger in size than other nodes), suggesting that the algorithms may become trapped in sub-optimal regions, where good but not optimal solutions are located.

7   Conclusions

In this work, we applied STNs to analyze the behavior of two state-of-the-art algorithms, SA and LNS, for solving a real-world parallel batch scheduling problem, the OSP. We considered a subset of 10 instances taken from the literature and represented STNs with and without search space partitioning. Considering the search partitioning strategies, the HAC can uncover the behavior of the algorithms better, as it is able to detect shared locations between the algorithms.

The search space of the instances is extensive (up to 1391 unique visited nodes) and displays signs of multi-modality with high-quality solutions appearing in different space locations. While the algorithms explore different areas within the space, most shared locations occurring early in the search; SA frequently revisits the same locations multiple times, whereas LNS visits fewer representative solutions, typically situated in distinct locations. This observation indicates that it may be beneficial to combine both algorithms in a portfolio or hybrid method.

Despite providing useful information on the underlying mechanisms of metaheuristics, STNs also have some limitations. First of all, constructing meaningful STNs that enable new insights is a highly human-intensive task that is only somewhat prone to automatization. For instance, the evaluation of search partitioning is conducted mainly through visual analysis of the produced network. This approach is time-consuming and limits the application of STNs to many instances, requiring the user to focus the analysis on a smaller subset of them.

While HAC appears as the best search partitioning strategy, it may show scalability issues towards a higher number of solutions due to the extensive computations required (e.g., the distance matrix is computed and stored for all combinations of solutions).

Future research directions include additional studies on the differences between algorithms considering ISA and LONs. Furthermore, we plan to understand how HAC can perform with a higher number of solutions (e.g., calculating the distance matrix with heuristics like Approximate Nearest Neighbor (ANN)).

Acknowledgements. We thank C. Blum and C. Chacón Sartori for answering our questions about Search Trajectory Networks and the related web tool. We thank K. Roitero for his precious suggestions on the topic of clustering. We thank the SPECIES Society for funding a visiting scholarship for Francesca Da Ros at TU Wien (Austria). The financial support from the Austrian Federal Ministry of Labour and Economy, the National Foundation for Research, Technology and Development and the Christian Doppler Research Association is gratefully acknowledged.

Data Availability Statement. Relevant material is available at the following link: https://github.com/francesdaros/stn-osp.

References

1. Akbay, M.A., Blum, C.: Two examples for the usefulness of STNWeb for analyzing optimization algorithm behavior. In: Sevaux, M., Olteanu, A.L., Pardo, E.G., Sifaleras, A., Makboul, S. (eds.) Metaheuristics, pp. 341–346. Springer, Cham (2024). https://doi.org/10.1007/978-3-031-62922-8_25

2. Alza, J., Bartlett, M., Ceberio, J., McCall, J.: Analysing the fitness landscape rotation for combinatorial optimisation. In: Rudolph, G., Kononova, A.V., Aguirre, H., Kerschke, P., Ochoa, G., Tušar, T. (eds.) Parallel Problem Solving from Nature – PPSN XVII, pp. 533–547. Springer, Cham (2022). https://doi.org/10.1007/978-3-031-14714-2_37

3. Azizoglu, M., Webster, S.: Scheduling a batch processing machine with incompatible job families. Comput. Ind. Eng. **39**(3), 325–335 (2001). https://doi.org/10.1016/S0360-8352(01)00009-2

4. Çelik, C., Saricicek, I.: Tabu Search for Parallel Machine Scheduling with Job Splitting. In: 2009 Sixth International Conference on Information Technology: New Generations, pp. 183–188 (2009). https://doi.org/10.1109/ITNG.2009.271

5. Chacon-Sartori, C., Blum, C.: Boosting a genetic algorithm with graph neural networks for multi-hop influence maximization in social networks. In: 2022 17th Conference on Computer Science and Intelligence Systems (FedCSIS), pp. 363–371 (2022). https://doi.org/10.15439/2022F78

6. Chacon-Sartori, C., Blum, C., Ochoa, G.: Search trajectory networks meet the web: a web application for the visual comparison of optimization algorithms. In:

Proceedings of the 2023 12th International Conference on Software and Computer Applications. ICSCA '23, New York, NY, USA, pp. 89–96. Association for Computing Machinery (2023). https://doi.org/10.1145/3587828.3587843

7. Chacon-Sartori, C., Blum, C., Ochoa, G.: An extension of STNWeb functionality: on the use of hierarchical agglomerative clustering as an advanced search space partitioning strategy. In: Proceedings of the Genetic and Evolutionary Computation Conference. GECCO '24, New York, NY, USA, pp. 151–159, Association for Computing Machinery (2024). https://doi.org/10.1145/3638529.3654084

8. Chacón Sartori, C., Blum, C., Ochoa, G.: Large Language Models for the Automated Analysis of Optimization Algorithms. In: Proceedings of the Genetic and Evolutionary Computation Conference, pp. 160–168, GECCO '24, Association for Computing Machinery, New York, NY, USA (2024). https://doi.org/10.1145/3638529.3654086

9. Chang, P.Y., Damodaran, P., Melouk, S.: Minimizing makespan on parallel batch processing machines. Int. J. Prod. Res. 42(19), 4211–4220 (2004). https://doi.org/10.1080/00207540410001711863

10. Cheng, B., Wang, Q., Yang, S., Hu, X.: An improved ant colony optimization for scheduling identical parallel batching machines with arbitrary job sizes. Appl. Soft Comput. 13(2), 765–772 (2013). https://doi.org/10.1016/j.asoc.2012.10.021

11. Chou, F.D.: Minimising the total weighted tardiness for non-identical parallel batch processing machines with job release times and non-identical job sizes. Eur. J. Ind. Eng. 7(5), 529–557 (2013). https://doi.org/10.1504/EJIE.2013.057380

12. Da Ros, F., Di Gaspero, L., Lackner, M.L., Musliu, N.: Reducing energy consumption in electronic component manufacturing through large neighborhood search. In: Proceedings of the Genetic and Evolutionary Computation Conference Companion. GECCO '24 Companion, New York, NY, USA, pp. 1706–1714. Association for Computing Machinery (2024). https://doi.org/10.1145/3638530.3664132

13. Da Ros, F., Di Gaspero, L., Lackner, M.L., Musliu, N., Winter, F.: Local search algorithms for the oven scheduling problem. In: Proceedings of the Genetic and Evolutionary Computation Conference Companion, GECCO '24 Companion, New York, NY, USA, pp. 191–194. Association for Computing Machinery (2024). https://doi.org/10.1145/3638530.3654158

14. Da Ros, F., Di Gaspero, L., Lackner, M.L., Musliu, N., Winter, F.: Multi-neighborhood simulated annealing for the oven scheduling problem (2024). https://doi.org/10.2139/ssrn.4998899

15. Damodaran, P., Srihari, K., Lam, S.S.: Scheduling a capacitated batch-processing machine to minimize makespan. Robot. Comput.-Integr. Manuf. 23(2), 208–216 (2007). https://doi.org/10.1016/j.rcim.2006.02.012

16. Damodaran, P., Vélez-Gallego, M.C.: A simulated annealing algorithm to minimize makespan of parallel batch processing machines with unequal job ready times. Expert Syst. Appl. 39(1), 1451–1458 (2012). https://doi.org/10.1016/j.eswa.2011.08.029

17. Fowler, J.W., Phojanamongkolkij, N., Cochran, J.K., Montgomery, D.C.: Optimal batching in a wafer fabrication facility using a multiproduct G/G/c model with batch processing. Int. J. Prod. Res. 40(2), 275–292 (2002). https://doi.org/10.1080/00207540110081489

18. Franzin, A., Stützle, T.: Revisiting simulated annealing: a component-based analysis. Comput. Oper. Res. 104, 191–206 (2019). https://doi.org/10.1016/j.cor.2018.12.015

19. Fruchterman, T.M.J., Reingold, E.M.: Graph drawing by force-directed placement. Software Pract. Exper. **21**(11), 1129–1164 (1991). https://doi.org/10.1002/spe.4380211102

20. Hu, T., Ochoa, G., Banzhaf, W.: Phenotype search trajectory networks for linear genetic programming. In: Pappa, G., Giacobini, M., Vasicek, Z. (eds.) Genetic Programming, pp. 52–67. Springer, Cham (2023). https://doi.org/10.1007/978-3-031-29573-7_4

21. Kamada, T., Kawai, S.: An algorithm for drawing general undirected graphs. Inf. Process. Lett. **31**(1), 7–15 (1989). https://doi.org/10.1016/0020-0190(89)90102-6

22. Kirkpatrick, S., Gelatt, D., Vecchi, M.P.: Optimization by simulated annealing. Science **220**(4598), 671–680 (1983). https://doi.org/10.1126/science.220.4598.671

23. Kosch, S., Beck, J.C.: A New MIP model for parallel-batch scheduling with non-identical job sizes. In: Simonis, H. (ed.) Integration of AI and OR Techniques in Constraint Programming, pp. 55–70, Springer, Cham (2014). https://doi.org/10.1007/978-3-319-07046-9_5

24. Krim, H., Zufferey, N., Potvin, J.-Y., Benmansour, R., Duvivier, D.: Tabu search for a parallel-machine scheduling problem with periodic maintenance, job rejection and weighted sum of completion times. J. Sched. **25**(1), 89–105 (2021). https://doi.org/10.1007/s10951-021-00711-9

25. Laborie, P., Rogerie, J., Shaw, P., Vilím, P.: IBM ILOG CP optimizer for scheduling. Constraints **23**(2), 210–250 (2018). https://doi.org/10.1007/s10601-018-9281-x

26. Lackner, M.L., Mrkvicka, C., Musliu, N., Walkiewicz, D., Winter, F.: Benchmark instances and models for the Oven Scheduling Problem [Dataset] (2022). https://doi.org/10.5281/zenodo.7456937

27. Lackner, M.L., Mrkvicka, C., Musliu, N., Walkiewicz, D., Winter, F.: Exact methods for the oven scheduling problem. Constraints **28**(2), 320–361 (2023). https://doi.org/10.1007/s10601-023-09347-2

28. Lackner, M.L., Musliu, N., Winter, F.: Solving an industrial oven scheduling problem with a simulated annealing approach. In: Proceedings of the 13th International Conference on the Practice and Theory of Automated Timetabling, pp. 115–120 (2022)

29. Lavinas, Y., Aranha, C., Ochoa, G.: search trajectories networks of multiobjective evolutionary algorithms. In: Jiménez Laredo, J.L., Hidalgo, J.I., Babaagba, K.O. (eds.) Applications of Evolutionary Computation, pp. 223–238. Springer, Cham (2022). https://doi.org/10.1007/978-3-031-02462-7_15

30. Mazumdar, C.S., Mathirajan, M., Gopinath, R., Sivakumar, A.I.: Tabu search methods for scheduling a burn-in oven with non-identical job sizes and secondary resource constraints. Int. J. Oper. Res. **3**, 119–139 (2008). https://doi.org/10.1504/IJOR.2008.016157

31. Melouk, S., Damodaran, P., Chang, P.Y.: Minimizing makespan for single machine batch processing with non-identical job sizes using simulated annealing. Int. J. Prod. Econ. **87**(2), 141–147 (2004). https://doi.org/10.1016/S0925-5273(03)00092-6

32. Narvaez-Teran, V., Ochoa, G., Rodriguez-Tello, E.: Search trajectory networks applied to the cyclic bandwidth sum problem. IEEE Access **9**, 151266–151277 (2021). https://doi.org/10.1109/ACCESS.2021.3126015

33. Ochoa, G., Malan, K.M., Blum, C.: Search trajectory networks: a tool for analysing and visualising the behaviour of metaheuristics. Appl. Soft Comput. **109**, 107492 (2021). https://doi.org/10.1016/j.asoc.2021.107492

34. Ochoa, G., Tomassini, M., Vérel, S., Darabos, C.: A study of NK landscapes' basins and local optima networks. In: Proceedings of the 10th Annual Conference on Genetic and Evolutionary Computation. GECCO '08, New York, NY, USA, pp. 555–562. Association for Computing Machinery (2008). https://doi.org/10.1145/1389095.1389204

35. Riveros, M., Rojas-Morales, N., Montero, E., Ochoa, G.: Understanding search trajectories in parameter tuning. In: Proceedings of the Genetic and Evolutionary Computation Conference. GECCO '24, New York, NY, USA, pp. 778–786. Association for Computing Machinery (2024). https://doi.org/10.1145/3638529.3654146

36. Sarti, S., Adair, J., Ochoa, G.: Neuroevolution trajectory networks of the behaviour space. In: Jiménez Laredo, J.L., Hidalgo, J.I., Babaagba, K.O. (eds.) Applications of Evolutionary Computation, pp. 685–703. Springer, Cham (2022). https://doi.org/10.1007/978-3-031-02462-7_43

37. Shannon, C.E.: A mathematical theory of communication. Bell Syst. Tech. J.l **27**(3), 379–423 (1948). https://doi.org/10.1002/j.1538-7305.1948.tb01338.x

38. Shaw, P.: Using constraint programming and local search methods to solve vehicle routing problems. In: Proceedings of the 4th International Conference on Principles and Practice of Constraint Programming, pp. 417–431, CP '98. Springer, Heidelberg (1998). https://doi.org/10.1007/3-540-49481-2_30

39. Smith-Miles, K., Muñoz, M.A.: Instance space analysis for algorithm testing: methodology and software tools. ACM Comput. Surv. **55**(12) (2023). https://doi.org/10.1145/3572895

40. Sörensen, K.: Metaheuristics–the metaphor exposed. Int. Trans. Oper. Res. **22**(1), 3–18 (2015). https://doi.org/10.1111/itor.12001

41. Swan, J., et al.: Metaheuristics "In the Large". Eur. J. Oper. Res. **297**(2), 393–406 (2022). https://doi.org/10.1016/j.ejor.2021.05.042

42. Tang, T.Y., Beck, J.C.: CP and hybrid models for two-stage batching and scheduling. In: Hebrard, E., Musliu, N. (eds.) Integration of Constraint Programming, Artificial Intelligence, and Operations Research, pp. 431–446. Springer, Cham (2020). https://doi.org/10.1007/978-3-030-58942-4_28

43. Turkeš, R., Sörensen, K., Hvattum, L.M.: Meta-analysis of metaheuristics: quantifying the effect of adaptiveness in adaptive large neighborhood search. Eur. J. Oper. Res. **292**(2), 423–442 (2021). https://doi.org/10.1016/j.ejor.2020.10.045

44. Verel, S., Ochoa, G., Tomassini, M.: Local optima networks of NK landscapes with neutrality. IEEE Trans. Evol. Comput. **15**(6), 783–797 (2011). https://doi.org/10.1109/TEVC.2010.2046175

45. Wang, H.M., Chou, F.D.: Solving the parallel batch-processing machines with different release times, job sizes, and capacity limits by metaheuristics. Expert Syst. Appl. **37**(2), 1510–1521 (2010). https://doi.org/10.1016/j.eswa.2009.06.070

46. Zhao, Z., Liu, S., Zhou, M., Guo, X., Qi, L.: Decomposition method for new single-machine scheduling problems from steel production systems. IEEE Trans. Autom. Sci. Eng. **17**(3), 1376–1387 (2020). https://doi.org/10.1109/TASE.2019.2953669

Optimizing the Logistics Operations of Distribution Network Operators from a Multinational Electric Utility Company

Diego Dantas Almeida[1], Mariana Azevedo[1], Victor Vieira[1],
Nelson Ion de Oliveira[1], Anna Giselle Câmara Dantas Ribeiro Rodrigues[1],
Leonardo C. T. Bezerra[1,2(✉)], Lucas Nunes[3], Thaís Alves de Mendonça[3],
and Rodrigo Manfredini[3]

[1] Federal University of Rio Grande do Norte (UFRN), Natal, RN, Brazil
{diego.dantas.064,mariana.brito.110,victor.vieira.700}@ufrn.edu.br,
{nelson.oliveira,anna.ribeiro,leonardo.bezerra}@ufrn.br
[2] University of Stirling, Stirling, Scotland
leonardo.bezerra@stir.ac.uk
[3] Neoenergia, Rio de Janeiro, Brazil
{lucas.nunes,thais.alves,rodrigo.manfredini}@neoenergia.com

Abstract. Distribution network operators (DNOs) share a significant responsibility regarding the assurance of electrical energy supply quality and continuity. In detail, DNOs are legally required to: (i) address electrical emergency occurrences quickly, especially to restore electricity supply, and; (ii) ensure the efficiency of service concerning commercial occurrences. In this work, we propose a solution to optimize the logistics operations of the Spanish multinational electric utility company Iberdrola. Our work scope is Neoenergia, the Brazilian subsidiary controlling five different DNOs. In our work, we follow the CRISP-DM data science framework to address the allocation of operations bases. The solution was developed and successfully deployed in collaboration with the analytics team of Neoenergia. In detail, we model the problem as a knapsack and tackle it with an iterated greedy metaheuristic. Results show a decrease in the distances between bases and occurrences when compared to the current approach adopted by Neoenergia. Our approach also reduces travel times, contributes to the improvement of supply continuity indices, and better meets company business requirements. Importantly, we provide a simulation tool to recommend future base allocation, which comprises valuable input to planning.

Keywords: electrical energy · heuristic optimization · data science

Supplementary Information The online version contains supplementary material available at https://doi.org/10.1007/978-3-031-90062-4_6.

© The Author(s), under exclusive license to Springer Nature Switzerland AG 2025
P. García-Sánchez et al. (Eds.): EvoApplications 2025, LNCS 15612, pp. 86–102, 2025.
https://doi.org/10.1007/978-3-031-90062-4_6

1   Introduction

Electrical energy is one of the most critical resources for the social and economical development of a country, with clean and affordable energy being one of the United Nations sustainable development goals. In this context, countries and companies have heavily invested in both renewable energy sources and in optimizing costs associated with the generation, transmission, and distribution of energy. Iberdrola, the Spanish multinational electric utility company, has research and innovation initiatives targeting these issues. The Brazilian subsidiary of Iberdrola in Brazil, Neoenergia, controls five distribution network operators (DNOs) across the country, and also operates in generation, transmission, and commercialization of electrical energy. Altogether, Neoenergia DNOs total over 16 million clients and over 37 million consumers.

Electrical energy distribution in Brazil is regulated by the National Agency for Electrical Energy (ANEEL), which establishes policies for product, service, and commercial quality, including supply continuity indicators that are used to evaluate the performance of distribution companies. Importantly, ANEEL regulates compensation for consumers under certain circumstances, e.g. when indicators fall below pre-determined thresholds or when deadlines are not met. The Neoenergia Cosern DNO, for instance, paid over $1.5 million in consumer compensation in 2022, despite having ranked second nation-wide in terms of energy supply continuity. From 2018 to 2023, Neoenergia DNOs have paid a combined yearly average of $22 million in consumer compensation.

To improve its operational efficiency and mitigate the compensation loss, Neoenergia has been actively collaborating with Brazilian universities for research and innovation. Among the most pressing logistics operations tasks that Neoenergia needs to address is the allocation of operational bases. From a general perspective, managing operational bases is a complex optimization problem that involves (i) determining how many and where the bases will be deployed; (ii) how many teams will be available in each base and their roster, and; (iii) planning daily occurrence routes for each team to address. The analytics team of Neoenergia addresses these problems separately, and in this work we follow the same approach and focus on (i). A further business requirement concerns distribution territorial units (DTUs). In detail, each of Neoenergia DNOs splits the area under its coverage into DTUs, and occurrences within a given DTU must be addressed by an operational base from that DTU. In this context, the underlying optimization problem is to allocate a pre-determined number of operational bases to locations within DTUs such that each DTU has at least one operational base and the total time taken to address occurrences is minimized. Two additional business requirements are that (i) locations selected to host an operational base present a good overall infrastructure to support the base daily activities, and; (ii) occurrences that affect more clients (or clients with higher business priority) be addressed more quickly, as these potentially affect continuity indicators more.

In this work, we follow the CRISP-DM data science framework [15] to address the allocation of operational bases at Neoenergia DNOs. Initially, we identified that the company's most recent model is an unsupervised learning (UL) approach proposed for the allocation of operational bases at the Neoenergia Brasília DNO. Given that all other Neoenergia DNOs operate at a state-level rather than at a city-level, this could heavily affect model assumptions and performance. Next, we propose a general knapsack modeling [12] suitable for all of Neonergia DNOs. By modeling operational bases allocation as a knapsack, we are able to meet all the business requirement set by Neoenergia, including DTU-adherence, location infrastructure, and client priority. Finally, we propose an iterated greedy (IG) algorithm [11] to optimize the knapsack problem. Importantly, because the IG algorithm can use different parameter configurations, we are able to generate a number of alternative allocations that are aggregated as recommendations of locations that would be preferable for hosting an operational base. We empirically demonstrate the benefits of our approach using Neoenergia Cosern DNO data, identifying base allocations that would reduce the median time to address occurrences by 18.53% w.r.t. the UL approach.

The remainder of this work is organized according to the CRISP-DM framework, as follows. Section 2 further deepens business understanding and briefly discusses related work funded by ANEEL in Brazil. Next, Sect. 3 details the data made available by Neoenergia, and provides an exploratory analysis that is instrumental for the modeling proposed in Sect. 4. The evaluation of the proposed model is given in Sect. 5 and the development and deployment of the recommendations dashboard is detailed in Sect. 6. Finally, we conclude and discuss future work in Sect. 7.

2   Business Context

Electrical energy distribution is often a monopolized market, demanding indirect incentives for companies to invest in their service. ANEEL adopts a *price-cap* model, setting an initial level for fees that is increased periodically according to a pre-selected official consumer price index. A percentage of that increase is deducted by ANEEL based on whether expected efficiency and productivity gains were reached. As such, distribution companies can only increase their profits by optimizing their operating costs [5,6]. In addition, ANEEL regulates consumer compensation based on supply continuity indicators, so preventing and solving electrical emergency interruptions becomes critical. In this section, we initially define the main indicators of interest to our work, and then discuss the administrative organization of the Neoenergia Cosern DNO, which is illustrative of the remaining Neoenergia DNOs. Finally, we detail business requirements and overview ANEEL-funded related work in Brazil.

2.1   Supply Quality Indicators

To monitor supply continuity, ANEEL establishes two collective and two individual quality indicators. Our focus is on the former, namely the (i) *equivalent interruption duration per consumer unit* (DEC) and; (ii) *equivalent interruption frequency per consumer unit* (FEC). Each indicator has pre-defined thresholds that companies must not violate over a given period of time for any set of consumer units. If a violation occurs, consumer units are automatically compensated via their utility bill. However, interruptions that last up to three minutes are not included for violation assessment. For this reason, the total time taken to solve an interruption is monitored by ANEEL and companies alike through measures of the individual stages of the electrical emergency occurrences:

Average Preparation Time (APT): average time required for teams to be dispatched to solve electrical emergency occurrences.

Average Travel Time (ATT): average time required between teams dispatch and arrival at electrical emergency occurrence locations.

Average Execution Time (AET): average time required to solve electrical emergency occurrences once teams are on-site.

Average Emergency Resolution Time (AERT): average time required for teams to solve electrical emergency occurrences (includes preparation, travel, and execution times).

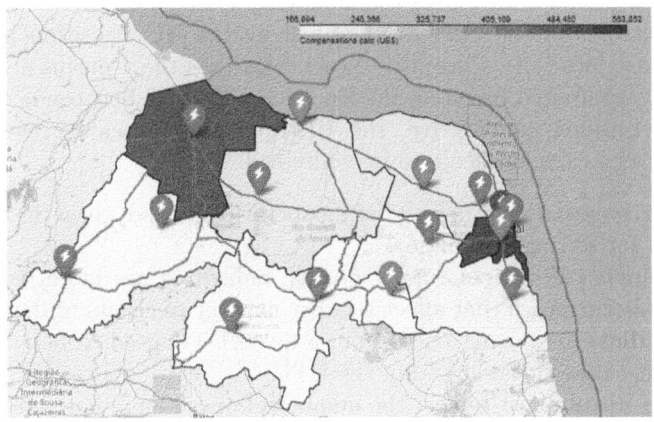

Fig. 1. Distribution territorial units (DTUs) spatial discretization adopted by Neoenergia Cosern. Darker DTU colors mean more consumer compensation. The color of an operational base (marker) depicts set up date: up to 2019 (green); between 2020 and 2022 (orange), or; from 2023 on (red). (Color figure online)

2.2   Administrative Organization

As discussed above, ANEEL defines supply continuity indicators based on sets of consumer units. In turn, Neoenergia DNOs further group those sets into distribution territorial units (DTUs). For Neoenergia Cosern, the 56 sets of consumer units are grouped into eight DTUs, illustrated in Fig. 1 with DTU geographical limits and markers indicating the locations of operations bases. Details for each

DTU are given as supplementary material, for brevity. As a business requirement, occurrences within a given DTU need to be addressed by an operational base from the same DTU. In addition, occurrences can differ as to being (i) *emergency*, meaning they are not planned for and must be addressed as quickly as possible in the case of an interruption, and; (ii) *technical-commercial*, which are planned interventions that are also regulated by ANEEL. Both emergency and technical-commercial occurrences can lead to consumer compensation, as follows. Regarding the former, compensation is due when supply interruptions last longer than three minutes, as previously discussed. Concerning the latter, each type of intervention has a prescribed deadline, and failing to meet the deadline incurs on a corresponding consumer compensation fee.

2.3   Business Requirements

The solution in use at Neoenergia for the allocation of operational bases was originally developed for the Neoenergia Brasília DNO. As previously discussed, the solution models base allocation as an unsupervised learning (UL) problem. In detail, the target geographic area is considered to be a bi-dimensional Euclidean space where the x and y coordinates correspond to latitude and longitude. In this approach, the analyst inputs the number of DTUs and the occurrence data, and the centroids of the clusters identified by the UL algorithm become the suggested base locations. Currently, Neoenergia uses the k-means algorithm, so centroids are computed based on the Euclidean distance between occurrences.

From a business perspective, the UL approach presents drawbacks that Neoenergia set as requirements to be addressed.

1. **Existing bases:** do not move existing operational bases given the economical overhead for setting up a base;
2. **Occurrence relevance:** supply quality indicators are more heavily influenced by occurrences that affect a larger number of clients or demand;
3. **DTU adherence:** occurrences from a given DTU need to be addressed by operational bases located within the DTU;
4. **Flexible number of bases:** the number of operational bases should be set independently of the number of existing DTUs;
5. **Optimized DTU allocation:** when more bases than DTUs are given, the allocation of bases to DTUs should optimize supply quality indicators;
6. **Road infrastructure:** allocation should consider real-world routes and distances, and federal routes should be preferred due to their infrastructure;
7. **Location infrastructure:** bases should be allocated to locations that correspond to cities, which need to present a minimally viable infrastructure.

2.4   Related Work

ANEEL funds projects related to electrical energy in Brazil. These are traditionally industry-academia collaborations, and the data about the projects conducted between 2008 and 2023 is publicly available [1]. Nearly half of the projects funded by ANEEL in the period concern electrical energy distribution, and the two most recurring themes are the (i) supervision, control, and protection of electrical energy systems, and; (ii) alternative sources for electrical energy generation. Most of the projects concern applied research and over recent years have often delivered information technology solutions. However, seldom works involve optimization and/or artificial intelligence.

We then conducted a systematic review of the publications associated with the projects funded by ANEEL, to identify works related to ours that also fit an applied research industry-academia collaboration. Among the projects that employ computational intelligence techniques, the works based on machine, deep, or statistical learning usually focus on modeling continuity indicators [7,8], forecasting energy demand [4], or fraud detection [10]. By contrast, works that employ metaheuristic algorithms tackle a range of different optimization problems, such as the allocation of remote control switches [9,13], maintenance planning [14], or consumer set gerrymandering [2]. Regarding metaheuristic techniques, the techniques employed range from variable neighborhood search [9] to artificial immune systems [14], but evolutionary algorithms are the most recurring algorithm [2,13]. To our knowledge, ANEEL has not yet funded an industry-academia collaboration addressing the allocation of operational bases.

3 Data Understanding

Logistics operations at Neoenergia uses multiple information systems (IS). In addition, the different companies that comprise the Neoenergia economical group were acquired over the years, and hence use legacy IS that are being gradually replaced. For the context of emergency occurrences, Neoenergia developed a system dubbed GSE, built on ArcGIS, which has already been deployed by most of the Neoenergia DNOs. In turn, technical-commercial occurrences are

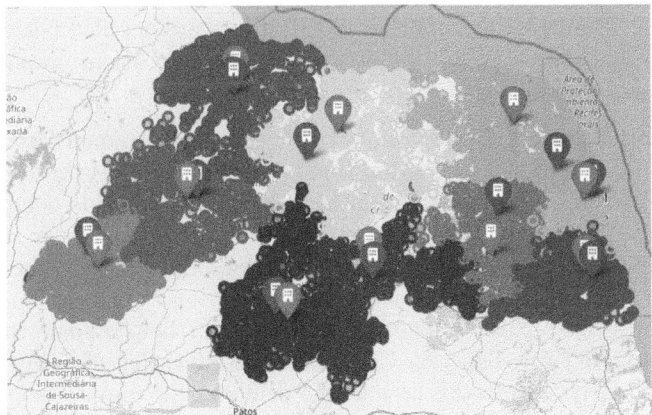

Fig. 2. Spatial distribution of the electrical emergency occurrences, colored by the addressing 2019 operational bases (red markers). Blue markers: UL approach allocation. (Color figure online)

managed by a commercial software that has also been deployed by most DNOs. Given the similarity in IS across DNOs, in this work we use data provided by the Neoenergia Cosern DNO as representative of the remaining DNOs. Importantly, Neoenergia Cosern operates at a state-level, similarly to all other Neoenergia DNOs – the only exception is Neoenergia Brasília, as discussed.

In this section, we discuss insights we obtain from an exploratory analysis of the data using the currently employed unsupervised learning (UL) approach devised by Neoenergia Brasília. Details on data cleaning and enrichment stages are given as supplementary material. Figure 2 illustrates the geographical distribution of electrical emergency occurrences, where color indicates the operational base that addressed a given occurrence. Notice that in very few situations there is a mix in colors, confirming that occurrences are mostly addressed by teams from the nearest operational base. In addition, we observe that some operational bases appear considerably shifted from the centroids of the occurrences they have addressed. This insight is what potentially motivated the analytics team at Neoenergia to propose an UL approach to operational base allocation.

An important remark about the data provided by Neoenergia concerns the effect of the COVID-19 pandemic. In detail, the digital transformation rushed by the pandemic meant that the efforts of the IS sectors were concentrated on transitioning the company from in-place to remote working environments. As a result, some IS such as GSE became inconsistent with developments occurred during the 2020–2021 period, most notably the addition of novel operational bases. In this context, we remark that the bases given in Fig. 1 in orange and red were not registered in the occurrences datasets provided. Hence, we differentiate the three time periods depicted when assessing the data below.

Prior to modeling, we conduct an exploratory analysis of the current operational bases allocation of the Neoenergia Cosern DNO. To do so, we employ the Neoenergia Brasília UL solution, which also evidences the limitations that the business requirements provided by Neoenergia seek to address. For instance, Euclidian space representation violates the (6) road infrastructure requirement by definition. The violations of other requirements are detailed below, comprising four different scenarios. For brevity, most of the allocations are illustrated in the supplementary material.

(1) Existing bases. Figure 2 compares the 2019 operational base allocation (red markers) with the UL allocation (blue markers). Most of the 10 then existing operational bases are relatively near the cluster centroids computed from the 2019–2022 data, with the exception of three. Importantly, two of the three alternative centroids match the operational bases that Neoenergia Cosern set up between 2019–2022, and are justified by the absence of operational bases in the north coast up to 2019. Altogether, these insights help validate the planning currently adopted by Neoenergia Cosern for the allocation of their operational bases. Importantly, we use this scenario to evidence the importance of business requirement (1) *existing bases*, as this allocation incurs in base relocation.

(2) Occurrence relevance. When weighted UL is employed, many significant changes to the 2019 allocation are observed. Two of the most important concern:

(i) an operational base at the north of the state would be replaced by two novel bases, and (ii) two operational bases at the south of the state would be replaced by a single novel base. These insights evidence that the existing base allocation could be improved if occurrence importance were factored, but the clustering approach provides this at the cost of business requirement (1) *existing bases*.

(3) DTU adherence. We compare the 2022 allocation with a DTU-adherent UL allocation. To do so, we split the original dataset into DTU-exclusive datasets, and run the UL approach taking each DTU-exclusive dataset and their 2022 existing number of operational bases. The centroids identified match most of the existing operational bases, once again validating the allocation planning of Neoenergia Cosern. However, one of the centroids is located in a region where no cities are eligible to host an operational bases, violating business requirement (7) *location infrastructure*. Furthermore, requiring a predefined number of bases per DTU violates business requirement (5) *optimized DTU allocation*.

(4) Flexible number of bases. When the number of clusters exceeds the number of existing operational bases, the centroids differ considerably from the 2022 allocation. Among the most relevant modifications are (i) replacing three operational bases with a single, novel base; (ii) replacing an existing base with two novel bases, which the model recommends in three separate regions. Altogether, this allocation further reinforces the difficulty to meet business requirement (i) *existing bases* when using an UL approach.

4   Data Preparation and Modeling

The operational bases allocation problem addressed in this work can be naturally modeled as an optimization problem. Depending on scale, the problem can be addressed through exact methods or metaheuristics. In this work, we opt for the second option as the different Neoenergia DNOs present different numbers of operational bases and locations that would be eligible to host them. More importantly, the stochastic nature of metaheuristics enable the analysis of alternative scenarios that can be aggregated as recommendations for stakeholders. In this section, we initially discuss how we use data preparation to meet part of the business requirements from Neoenergia. Later, we detail our proposed problem formulation, solution representation, and algorithmic approach.

4.1   Data Preparation

To complement the datasets provided by Neoenergia, we additionally engineer a real-world distance matrix to meet four business requirements, namely (2) *occurrence relevance*; (3) *DTU adherence*; (6) *road infrastructure*, and; (7) *location infrastructure*. Below, we detail the three-stage enrichment procedure.

Location Infrastructure. We identify cities with at least 10 000 residents in 2023 according to official projections. This threshold was selected given the demographics of the cities currently hosting Neoenergia Cosern bases. In total, 74 cities are eligible to host an operational bases according to this criterion.

Road Infrastructure. We preprocess real-world distances between occurrences and eligible cities using the osmnx [3] real-world routing Python library. For feasibility, we spatially-aggregated occurrences prior to computing real-world distances, as follows. First, we associated each occurrence o_i to its nearest city c_j. For each city c_j, we then computed the centroid of its associated occurrences o_i using their latitude and longitude information. In practice, centroids roughly matched city centers. Finally, we computed real-world distances between the 74 eligible cities e_k and all 167 cities c_j that Neoenergia Cosern is responsible for, resulting on a 167×74 real-world distance matrix. Effectively, this matrix enables us to identify for each city c_j the nearest eligible city c_k in a given allocation. We further enriched this matrix with three columns computed from the occurrences o_i associated with city c_j, namely (i) n_j, the number of occurrences; (ii) r_j, the number of clients affected by the occurrences (a proxy for **occurrence relevance**), and; (iii) m_j, the median distance between city c_j and the occurrences.

DTU Adherence. We set distance matrix entries $d_{jk} = \infty$ for every combination of city j and eligible city k that do not belong to the same DTU, effectively discarding them from the analysis.

4.2 Modeling

We model the operational bases allocation problem as a knapsack problem, where each eligible city is considered a unit-weight item and the knapsack capacity equals the total number of bases to be allocated. For a given eligible city e_k, item profit is computed as a function of the cities that a base located in e_k would address. As a result, item profit is dynamic, since the addition or removal of a base from the knapsack affect city-base association. Next, we detail the objective function, our iterated greedy algorithmic approach, and the heuristic function used by the greedy constructive procedure.

Objective Function. We compute the objective function using an auxiliary distance matrix, as follows. Given each city c_j, the objective function first computes city cost $t_j = n_j \times (m_j + d_{jk})$, where k is the index of the eligible city $e_k \in S$ nearest to city c_j. Effectively, t_j is a proxy for the total distance a team departing from eligible city e_k would travel to address all occurrences from city c_j. In addition, we remark that n_j may be replaced by r_j to meet (2) *occurrence relevance*. The next stage of the objective function is to compute the cost z_k of each eligible city $e_k \in S$, which equals $z_k = \sum t_j$, for all cities c_j associated with e_k. Finally, the objective function computes the total cost $f(S) = \sum z_k$, for all of eligible cities $e_k \in S$.

Algorithm 1. Iterated greedy with tabu list pseudocode

Require: B: existing base set; E: eligible city set; n: number of bases to be allocated; d: ruin degree;
t: tabu tenure; $\mathcal{R}$: relaxed acceptance criterion

1: $T \leftarrow \emptyset$	▷ tabu list initialization
2: $S \leftarrow generateSolution(n, B, E, T)$	▷ incumbent solution creation
3: $S^* \leftarrow S$	▷ best-so-far solution
4: $E' \leftarrow E \setminus S$	▷ working set of eligible cities
5: **repeat**	▷ ruin-and-recreate loop
6: **if** *current iteration* $> t$ **then**	
7: $E' \leftarrow E \cup T.pop(d)$	▷ oldest tabu city becomes eligible
8: $S' \leftarrow S$	▷ candidate solution
9: $T.append(S'.popRandom(d))$	▷ solution ruin and tabu list update
10: $S' \leftarrow generateSolution(d, B, E', T, S')$	▷ solution recreation
11: $E' \leftarrow E' \setminus S'$	▷ update working set of eligible cities
12: **if** $f(S') < f(S')$ **then**	
13: $S \leftarrow S'$	▷ update incumbent solution
14: **if** $f(S') < f(S^*)$ **then**	
15: $S^* \leftarrow S$	▷ update best-so-far solution
16: **else if** $\mathcal{R}(S', S)$ **then**	
17: $S \leftarrow S'$	▷ update incumbent solution
18: **until** *stopping criterion met*	

Ensure: S_b

Algorithmic Approach. We adopt an iterated greedy coupled with a tabu list approach including a relaxed acceptance criterion, depicted in Algorithm 1. The algorithm requires the (i) existing base set B; (ii) eligible city set E; (iii) number of bases to be allocated n; (iv) ruin degree d; (v) tabu tenure t, and; (vi) relaxed acceptance criterion r, to accept a tentative solution S' even if it is not better than the incumbent solution S. Notice that argument (iii) ensures business requirement (4) *flexible number of bases*, which combined with data preprocessing DTU-adherence also ensures (5) *optimized DTU allocation*.

The algorithm first initializes the tabu list T (Line 1) and generates an incumbent solution by adding the existing base set and filling the best remaining spots with the best eligible cities from E (Line 2), which is then recorded as the best-so-far solution (Line 3). Next, a working eligible city set E' is created by excluding cities in S (Line 4). The ruin-and-recreate loop is given in Lines 5–18. Initially, Lines 6–7 update the working set of eligible cities and tabu list once the number of iterations is greater than the tabu tenure t. Next, a candidate solution S' is created from the incumbent solution S (Line 8), and the ruin procedure removes d random eligible cities e_k from S', which become tabu (Line 9). Recreation is given in Line 10, when procedure *generateSolution* adds to S' the best d available eligible cities $e_k \in E' \setminus T$. The cities added to the solution are then removed from the working eligible city set E' (Line 11). Finally, the candidate solution S' may replace incumbent (Lines 12–13) and best-so-far solutions (Lines 14–15). S' may also replace S based on the relaxed acceptance criterion $\mathcal{R}$ (Lines 16–17).

The greedy procedure *generateSolution* either generates a solution from scratch or completes a partial solution, depending on the number of arguments provided. The procedure requires four arguments, namely (i) remaining solution capacity, (ii) existing base set; (iii) a set of eligible cities (excluding existing bases), and; (iv) a tabu list. When no additional argument is provided, the algorithm creates a solution from scratch, initially populating the solution with the existing bases to respect business requirement (1) *existing bases*. Next, eligible cities are greedily selected to fill the remaining solution capacity. Conversely, if a partial solution is provided as fourth argument, procedure *generateSolution* does not use the existing base set and only greedily fills the remaining solution capacity. In both cases, cities present in the tabu list cannot be selected.

Heuristic Function. The heuristic value of a given eligible city e_k is computed in two steps, as follows. First, we compute for each city c_j its value $v_j = n_j \cdot m_j$, where n_j and m_j are respectively the precomputed number of occurrences associated with city c_j and the median distance between those occurrences and city c_j. For a weighted analysis, the number of clients r_j replaces n_j. Finally, the heuristic value h_k of a given eligible city e_k equals $h_k = \sum v_j$ for the 10 cities c_j that are nearest to e_k.

5   Evaluation

The proof-of-concept evaluation conducted in this section concerns two major perspectives. First, we evaluate the heuristic optimization (HO) approach for travel distance, as a proxy to benefits to supply quality indicators. Since the dataset provided by Neoenergia does not include all the existing operational bases for the period considered, we take the unsupervised learning (UL) approach from the Neoenergia Brasília DNO as baseline. Later, we discuss proposed allocations in light of the business requirements set by Neoenergia.

5.1   Travel Distance

We assess illustrative 15-base allocations adopting the parameter configuration given in the supplementary material for brevity. In detail, the allocations considers the 2019 existing operational bases given in Fig. 1, and five new bases must be allocated. Results are given in Fig. 3, which also gives baseline results from: (i) the 2019 existing bases, and; (ii) the different UL scenarios discussed in Sect. 3. However, UL scenarios with $n = 10$ use instead $n = 14$ to increase comparability with HO allocations. The real-world distances for the different allocations are given as boxplots (clipped at 150km).

Overall, no statistical significant difference is observed between approaches, given the overlap in boxplots. However, compared to the 15-base UL allocation, the HO unweighted allocation presents lesser first and second quartile, equivalent average, greater third quartile, and a longer tail. Two insights indicate that this is a consequence of having to respect DTUs, namely (i) a similar difference in distributions between the two 14-base unweighted UL allocations, and; (ii) the similarity between the boxplot of the HO allocation and the boxplot of the 14-base DTU-adherent UL allocation. Regarding (ii), Table 1 shows that the median and mean distances for the HO approach are, respectively, 15.4% and 2.24% lower than for the UL approach, likely a result of having an additional base.

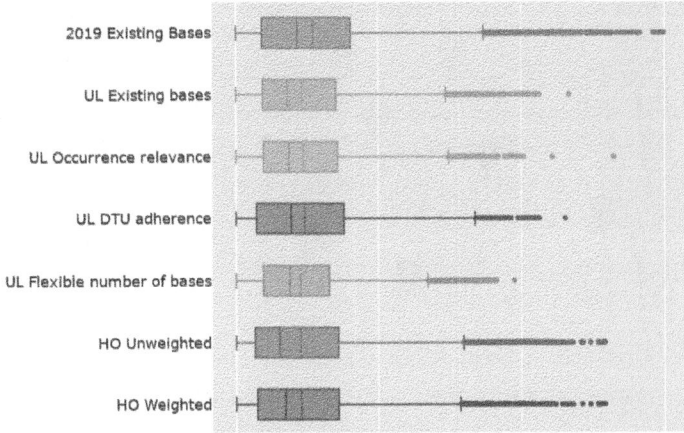

Fig. 3. Boxplot of real-world travel distances (in km) using different operational base allocations. Dashed lines depict averages; only data points lesser than 150km are given.

Table 1. Real-world distances (km) for different allocations.

Allocation	Bases	Mean	Median
2019 Existing bases	10	27.05	21.59
UL Existing bases	14	23.17	**18.18**
UL Occurrence relevance	14	23.62	18.67
UL DTU adherence	14	24.09	19.36
UL Flexible number of bases	15	**22.68**	18.88
HO Unweighted	15	**22.65**	**15.38**
HO Weighted	15	22.82	17.36

Nonetheless, when both 15-base allocations are compared, the median distance for the HO allocation is 18.53% lower than for the UL allocation. Finally, we remark the the addition of five bases by the HO allocation respectively improves the mean and median distances of the 2019 existing bases by 16.27% and 28.76%.

When we compare the weighted and unweighted HO allocations, we notice from Fig. 3 that the former presents quartiles that are slightly greater than the latter. Indeed, Table 1 shows that means are very much identical between the HO allocations, whereas the median distance for the weighted allocation is 12.87% worse. These results follow a similar pattern observed between UL 14-base weighted and unweighted allocations, with quartiles for the former between slightly greater than for the latter.

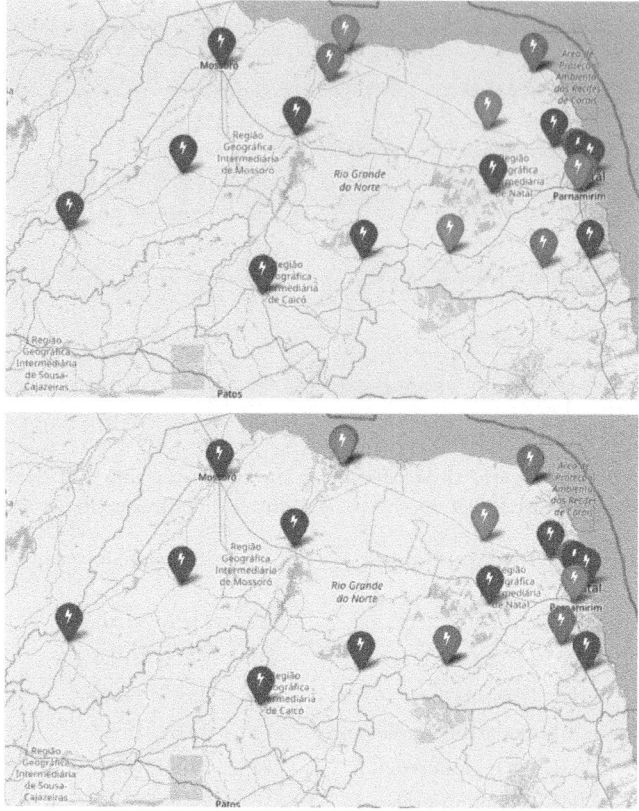

Fig. 4. Unweighted (top) and weighted (bottom) allocations proposed by the HO approach. Markers are colored as a function status. Green: existing at 2019; orange: deployed between 2020–2022 (Color figure online); red: deployed in 2023, and; blue: recommended.

5.2 Discussion

We conclude the evaluation discussing the illustrative operational bases allocations previously proposed. Figure 4 depicts the allocations for the unweighted (top) and weighted (bottom) analyses.

Unweighted Analysis. The proposed allocation partially validates the planning of Neoenergia Cosern, as three out of the five recommended bases either match or strongly approximate bases that were deployed after 2019. By contrast, the remaining two bases greatly differ from the ones that were set up by the company. Interestingly, both suggestions favor the coast over the countryside, either becoming closer to the northeastern or the southeastern coast. Importantly, the former suggestion matches the deployment planned by Neoenergia.

Weighted Analysis. When occurrence relevance is factored in, the planning of Neoenergia Cosern is further validated. In detail, three of the suggested bases exactly match the bases that have been set up since 2019. Furthermore, the northeastern coast base under consideration by Neoenergia is once again recommended by the HO approach. The most significant difference between the weighted and unweighted assessment comprises the southeastern base suggestion, which becomes closer to the capital city (located at the middle of the eastern coast) to account for occurrence relevance.

6   Deployment

The analytics team at Neoenergia develops dashboards and elaborates reports for the different DNOs with recommendations about alternative scenarios they may consider when planning their operational bases allocations. To meet this need, we implemented and deployed simulation and recommendation dashboards not disclosed for confidentiality. In this section, we give a high-level description of the dashboards, discuss the insights from recommendation, and detail the benefits to Neoenergia DNOs since deployment.

6.1   Dashboards

The simulation dashboard depicts plots of an HO approach run taking as input: (i) an occurrence dataset; (ii) an existing bases dataset; (iii) algorithmic parameters, and; (iv) whether to consider occurrence relevance. In turn, the recommendation dashboard ranks eligible cities based on the frequency with which they appear in different allocations. These allocations are the results obtained from running the 108 different parameter configurations of the HO approach given as supplementary material, for brevity.

Five of the eight most recommended eligible cities in the dashboard are cities where Neoenergia Cosern had set up operational bases after 2019. An additional recommended base was also under consideration for deployment, namely the northeastern coast base given in Fig. 4. These results serves as further validation that the planning department of Neoenergia Cosern has conducted sound assessments prior to allocating operational bases. More importantly, they help validate the HO approach modeling that we propose. In complement, the remaining most recommended eligible cities are located in the central and mid-southern regions, which the company had not yet considered, providing the analytics teams with novel insights for their planning.

6.2   Monitoring

Since deployment at the end of 2022, Neoenergia has experienced a number of benefits resulting from the dashboards we developed. Below, we discuss the benefits observed for Neoenergia DNOs, starting with Neoenergia Cosern.

Benefits to Neoenergia Cosern. In addition to validating the planning of the existing bases adopted by the company, the insights produced in this work provided (i) further evidence for Neoenergia Cosern to set up operational bases that were under analysis, and; (ii) indications of the regions where the addition of novel operation bases would bring most benefits. In 2023, the company set up a novel operational base alongside the southern coast recommended in Fig. 4 (right). The supply quality indicator thresholds for those regions were respected in 2023, despite adverse climatic conditions that traditionally trigger an increased number of occurrences.

Benefits to Neoenergia DNOs. The dashboards developed in this work are being instrumental for holistic assessments of all five DNOs in the Neoenergia economical group. operational bases locations are revised periodically based on travel distances and team productivity. In this context, the proposed dashboards are being coupled with existing solutions at Neoenergia, thus extending the benefits to DNOs other than Neoenergia Cosern. In 2023, for instance, the analytics team used the suite of solutions to recommend a novel operational base for the Neoenergia Brasília DNO. Along 2024, the Neoenergia Coelba DNO expanded their operational bases, which address around 6 million clients and an area of circa $570,000km^2$. Importantly, the same demographic profile adopted for identifying the 74 eligible cities for Neoenergia Cosern results in 341 eligible cities for Neoenergia Coelba, reinforcing the need for scalable solutions. In this context, data science projects such as the one conducted in this work improve the analytics capacity of Neoenergia and lead to more accurate and optimized planning of the required investment for base allocation.

7   Conclusion

Neoenergia, the Brazilian subsidiary of the multinational Iberdrola electrical utility company, is an economical group active in the distribution, generation, transmission, and commercialization energy markets. In total, Neoenergia companies operate in 18 different Brazilian states, as well as in the federal district. Altogether, the five distribution network operators (DNO) of Neoenergia serve around 16.5 million consumer units, equivalent to a population of 40 million people approximately. Given its scale, Neoenergia has paid an yearly average $22 million dollars in consumer compensation in the 2018–2022 period, even with DNOs like Neoenergia Cosern ranking among the best in the country.

In this work, we have addressed operational bases allocation, one of the most pressing logistic operations problems that the analytics and planning teams of Neoenergia and their DNOs attempt to optimize. In detail, for this work we have followed the approach adopted by Neoenergia of addressing allocation independent of roster scheduling and route planning. Nonetheless, the heuristic optimization (HO) modeling we proposed demonstrated its benefits in comparison to unsupervised learning model currently adopted by the company. Besides significant reductions of nearly 20% in median travel times for the Neoenergia Cosern DNO, the proposed modeling also meets all the business requirements set by Neoenergia. Importantly, the solution is general and scalable, being applicable to the context of all five Neoenergia DNOs.

The contributions of this work extend beyond modeling. Concretely, we have developed and deployed simulation and recommendation dashboards to assess planning operational bases allocation. Simulation allows analysts to evaluate custom runs of the HO solution, whereas recommendation aggregates results from several runs using alternative parameter configurations. The ranking of eligible cities provided by the recommendation algorithm is instrumental for Neoenergia DNOs, helping to validate planning regarding (i) existing and (ii) under analysis operational bases. More importantly, recommended eligible cities often provide novel insights that planning and analytics teams had not yet antecipated, as observed in the context of Neoenergia Cosern.

The practical benefits to Neoenergia DNOs already observed since dashboard deployment are encouraging. In detail, two of Neoenergia DNOs have already deployed novel operational bases with the aid of the proposed dashboards, as well as a significant expansion in 2024 for the largest DNO. Importantly, the supply quality indicators for Neoenergia Cosern were positively affected by the implementation of the recommendations provided in this work, even in an year of adverse climatic conditions. Future work will address the different aspects of operational bases management, with the goal of developing a solution that integrates allocation, scheduling, and routing. Critically, the work envisioned will be eligible for research funding from the ANEEL Brazilian regulatory agency, which will enable the development of yet more robust solutions. In addition, this work is also applicable to the context of DNOs from other subsidiaries of Iberdrola, which often promote international knowledge exchange workshops. We then expect to expand the current academia-industry collaboration to other countries, e.g. Scottish Power DNOs in the United Kingdom.

References

1. Agência Nacional de Energia Elétrica: Projetos de P&D em energia elétrica (2023). https://dadosabertos.aneel.gov.br/dataset/projetos-de-p-d-em-energia-eletrica.
2. Araújo, R.J.P.d.: Otimização de desempenho de indicadores de continuidade do serviço em concessionárias de distribuição utilizando algoritmos evolutivos. Ph.D. thesis, Universidade de São Paulo (2011)
3. Boeing, G.: OSMnx: new methods for acquiring, constructing, analyzing, and visualizing complex street networks. Comput. Environ. Urban Syst. **65**, 126–139 (2017). https://doi.org/10.1016/j.compenvurbsys.2017.05.004
4. De Medeiros, R.Á.O., et al.: Previsão de demanda no médio prazo utilizando redes neurais artificiais em sistemas de distribuição de energia elétrica. Master's thesis, Universidade Federal da Paraíba (2016)
5. Filho, J.B.S.A., Bacic, M.J.: Modelos de custos na regulação da indústria de distribuição de energia elétrica. In: XV Congresso Brasileiro de Custos. Curitiba, PR (Nov 2008)
6. Littlechild, S.C.: Privatisation, competition and regulation. The 29th Wincott Lecture delivered on 14th October 1999. Forward by Sir Geoffrey Owen., United Kingdom: The Wincott Foundation (1999)
7. Magalhães, E.F.A.: Modelagem e simulação de indicadores de continuidade: Ferramenta auxiliar para a manutenção em redes de distribuição de energia elétrica. Master's thesis, Universidade Federal da Bahia (2017)

102 D. Dantas Almeida et al.

8. Mattos, J.B.: Aplicação de redes neurais artificiais na estimação de indicadores de continuidade do fornecimento de energia elétrica. Master's thesis, Departamento de Engenharia Elétrica, Centro Tecnológico, UFES, Vitória, ES (2018)
9. Nunes, F.S.: Alocação ótima de chaves em sistemas de distribuição de energia elétrica utilizando a metaheurística VNS. Undergraduate dissertation. Universidade Federal do Ouro Preto (2018)
10. Pulz, J.: Abordagem alternativa para cálculo regulatório de perdas não-técnicas do sistema de distribuição e técnicas de machine learning para detecção de fraude na baixa tensão. Ph.D. thesis, Universidade de São Paulo (2022)
11. Ruiz, R., Stützle, T.: A simple and effective iterated greedy algorithm for the permutation flowshop scheduling problem. European journal of operational research (2007)
12. Salkin, H.M., De Kluyver, C.A.: The knapsack problem: a survey. Naval Res. Logist. Quart. **22**(1), 127–144 (1975)
13. Tenfen, D.: Alocação ótima de chaves telecomandadas em redes de distribuição com multi-objetivo via algoritmos genéticos de Pareto. Master's thesis, Universidade Federal de Santa Catarina (2011)
14. Trentini, C.: Planejamento da manutenção de sistemas de distribuição com foco em custos e confiabilidade. Master's thesis, Universidade Federal de Juiz de Fora (2019)
15. Wirth, R., Hipp, J.: CRISP-DM: Towards a standard process model for data mining. In: Proceedings of the 4th International Conference on the Practical Applications of Knowledge Discovery and Data Mining. vol. 1, pp. 29–39. Manchester (2000)

Analysis of Illicit Drug Mixtures at Festivals Using Portable Near-Infrared Spectroscopy with Genetic Programming

Steven Dockter[1]([✉]) [iD], Deepak Karunakaran[2] [iD], Qi Chen[1] [iD],
and Yongshi Deng[2] [iD]

[1] School of Engineering and Computer Science, Victoria University of Wellington,
Wellington 6400, New Zealand
{Steven.Dockter,Qi.Chen}@ecs.vuw.ac.nz
[2] Institute of Environmental Science and Research, Auckland, New Zealand
{Deepak.Karunakaran,Yongshi.Deng}@esr.cri.nz
https://www.wgtn.ac.nz/ecs , https://www.esr.cri.nz

Abstract. Illicit drugs are often mixtures containing cutting agents and other adulterants. The adulterants can cause harm on their own or when consumed in conjunction with the illicit drug. Drug checking services mitigate the risks associated with illicit drugs by providing substance composition and other harm prevention information. A widely embraced delivery model for drug checking involves offering the service at events and festivals. To help reduce drug-related harm at festivals, it is beneficial to have substance identification methods that are timely, safe, portable, and easy to use. Near-infrared spectroscopy (NIRS) is a well-known method for identifying and quantifying a range of substances, including the analysis of illicit drug mixtures, particularly when using high-end bench-top instruments. Providing mixture analysis by a portable NIRS solution is advantageous, however, there are still challenges due to device limitations. Consequently, a variety of methods have been utilised on portable NIRS data of various mixtures. Nevertheless, evolutionary computation methods, known to be robust for solving complex combinatorial and optimisation problems, are not as frequently reported in NIRS. This paper proposes a genetic programming-based approach to develop models for analysing illicit drug mixtures using portable NIRS. The experiment results indicate that the proposed approach can effectively provide a model for the analysis of mixtures that provides an accurate identification of components within a drug sample that is comparable to traditional linear-based models.

Keywords: Near-infrared spectroscopy · Genetic programming · Mixture analysis · Drug harm reduction

1 Introduction

Illicit drugs are detrimental to physical and mental health [25]. Existing outside quality regulations, illicit drugs often contain adulterants or cutting-agents to

© The Author(s), under exclusive license to Springer Nature Switzerland AG 2025
P. García-Sánchez et al. (Eds.): EvoApplications 2025, LNCS 15612, pp. 103–118, 2025.
https://doi.org/10.1007/978-3-031-90062-4_7

increase profit [14]. Drug related harm reduction services have reported that only about 68% of illicit drug samples tested contain the expected substance and a high proportion contain fillers [12]. Adulterants can be more harmful than the illicit drug itself or can become a hazard when consumed in conjunction with the illicit drug. The reliable identification of illicit drugs and harmful adulterants is a key component in the prevention of drug-related harm.

With a vested interest in public safety, front-line police officers in New Zealand are equipped with a portable NIRS solution that can identify common illicit drugs. This solution, known as Lumi™ Drug Scan, can quickly, safely, and accurately detect three highly prevalent street drugs: 3,4-Methylenedioxymethamphetamine (MDMA), cocaine, and methamphetamine (meth) [16]. The rapid in situ identification of a suspected illicit substance enables law enforcement to make more informed decisions at the point of inter-action [26].

Portable NIRS has proved to be a capable technology for the analysis of mix-tures in food and agricultural quality assurance [2, 17, 28]. Although the detection of illicit drugs using portable NIRS has been previously explored [14], approaches specifically focused at detecting harmful adulterants in such mixtures are scarce. Detecting harmful adulterants in mixtures is of particular importance at music festivals where a disproportionately high number of people use drugs and inci-dents of drug-related deaths are a major concern [20]. High-end bench-top solu-tions are effective at analysing mixtures by identifying and quantifying mixture components; however, these devices are expensive, bulky, and require specialised expertise to operate, adding constraints to their large-scale deployment in point-of-care drug-checking services, particularly at festivals where high-throughput screening is essential. Portable spectrometers are lightweight, low cost, and easy to use but suffer from technical limitations like limited wavelength range and poor selectivity. More sophisticated approaches [9] are required for portable spec-trometers to analyse mixtures effectively. However, these approaches need to be highly efficient as rapid analysis is a key requirement in field application scenarios such as festival drug testing.

Genetic programming (GP) [4] is a machine learning method that does not presuppose any relationships within the data. As a subset of evolutionary com-putation, GP employs techniques inspired by natural evolution to solve complex problems [1]. A typical outcome of GP is a program that is often depicted as a tree structure, commonly referred to as a program tree. The GP process begins with a random assortment of potential solutions (programs) which are refined over time through minor adjustments and the preservation of beneficial traits. This iterative evolution sharpens the programs' ability to address problems that were not explicitly defined for the algorithm. Moreover, once evolved, the pro-gram trees are very efficient to compute during inference. Another benefit of program trees are that they are easily visualised. In addition, by not relying on any assumptions about the data, such as having a linear relationship require-ment, GP can determine on its own the best way to model the data be it lin-ear or non-linear. Therefore, GP based approaches to develop a model without assumptions could be efficient and explainable for the analysis of mixtures with illicit drugs in conjunction with portable NIRS.

The primary goal of this work is to develop a GP-based approach that effectively models the relationships between individual components and mixtures of illicit drugs, enabling the prediction of specific components in unknown mixtures. This novel approach offers the added benefit of generating interpretable models and results. Furthermore, we integrate the proposed GP method into a workflow designed for portable, real-time analysis, facilitating the detection of harmful adulterants in illicit drug samples.

2 Background

Near-infrared (NIR) is a sub-region of infrared electromagnetic radiation with wavelengths from 780 to 2500 nanometre (nm) [10]. Since every structurally different molecule produces a unique infrared spectrum [18], every substance has its own distinct spectrum in NIR. When substances are combined together they form a mixture. If the components of a mixture do not interact chemically, then the absorbance of NIR radiation of the individual component substances contribute proportionately to the overall absorbance of the mixture [24]. Therefore, it follows that the mixture NIR absorbance spectrum contains information of the components that constitute the makeup of the mixture.

Lumi™ Drug Scan is a portable NIRS solution [23] that has been deployed at scale by the NZ Police, enabling effective detection of street drugs. The solution consists of (i) a handheld spectrometer (Lumi™ Nano NIR device; see Fig. 1), (ii) the Lumi™ Analysis mobile app, and (iii) a secure cloud service. Lumi™ Drug Scan employs state-of-the-art machine learning algorithms to develop an effective Drug Detection Model (DDM) to detect the primary drug component of a street drug sample.

One of the challenges with portable NIRS is poor spectral reproducibility. This results in high variability in the spectra generated from identical drug mixture samples. Advanced preprocessing methods and complex feature engineering approaches would support development of robust models in the face of high variability in the data. The continued success of the Lumi™ solution in field use for more than two years confirms the robustness of its drug detection models. For detecting harmful adulterants in drug mixtures, a similar modelling approach is necessary.

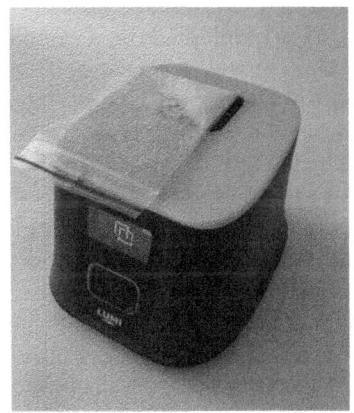

Fig. 1. Lumi™ Nano NIR device

3   The Proposed Method

Motivated by the merits of GP, particularly its ability to generate human readable programs and the low inference cost of computing the program tree which is suitable for an application with strict real-time requirements, we consider using GP for mixture analysis of illicit drugs. One of the effective approaches used in infrared spectroscopy to detect harmful adulterants is to subtract the spectra of the majority component(s) [8] which have already been identified. However, in many cases, due to the issue of spectral residuals [5] left after subtraction of multiple reference spectra from that of a multi-component mixture spectra, the analysis is not accurate. Thus, adapting the spectral subtraction [22] method with GP models for portable NIRS could improve the analysis performance. Therefore, we consider the approaches of spectral subtraction and spectral addition in combination with GP to generate tree based models which could detect harmful components in a drug mixture. As such, the analysis is carried out by two complementary methods. The first is a subtraction model that is well suited for determining an adulterant component that may be present and the second is an addition model that provides correlating evidence when the predicted solution contains the correct mixture components.

3.1   The Overall Workflow for Mixture Analysis

The overall workflow for mixture analysis is shown in Fig. 2. We assume that a primary DDM, similar to the one developed for Lumi™ Drug Scan, serves as the first stage of inference on a sample. The DDM is a calibrated classification model [27] that provides interpretable class probabilities. Based on the class outcome from the DDM and the associated probabilities, we determine whether mixture analysis is necessary. If the DDM detects no drug or if the class probability for a detected drug exceeds a certain threshold, further analysis is not required. Otherwise a mixture analysis is initiated, using the class outcome to extract the reference spectrum of the identified drug from a limited library of drug spectra.

A GP Mixture Analysis by Spectral Subtraction (MASS) model is then employed by using the drug reference spectrum and the sample spectrum as inputs. The GP-based MASS model outputs a set of candidate spectra which are matched with a library of reference spectra of harmful adulterants such as eutylone, fentanyl, etc. The best adulterant match is extracted from the library of adulterant reference spectra. Another GP evolved model, the Mixture Analysis by Spectral Addition (MASA) model, is then used to validate the result. The matched adulterant identified by the MASS model, along with the drug reference spectrum, is input to the MASA model to generate a set of candidate spectra. These spectra are scored against the original sample spectrum, providing a result that indicates the presence of the adulterant in the sample mixture.

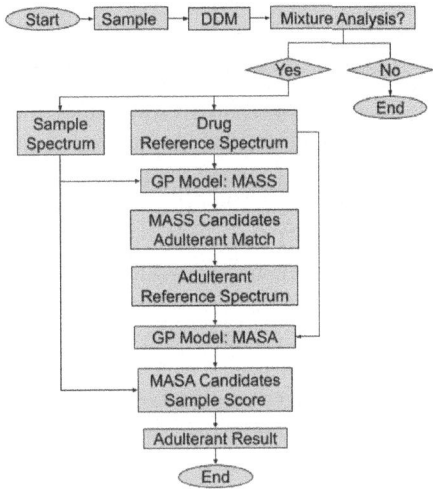

Fig. 2. Overall flowchart for mixture analysis.

3.2   Evolving MASS and MASA Models with GP

GP is an evolutionary computation technique that uses an initial population of individuals (randomly generated solutions), represented as program trees, to evolve more adept candidate solutions to solve optimisation and regression problems. Genetic processes like selection, mating and mutation are employed to evolve an optimised solution by keeping the best traits and introducing small changes. Steps in applying GP involve determining the terminal set, function set, fitness function, control parameters, and termination criteria [11]. A run of a GP algorithm puts the evolutionary processes into action to find the best-fit candidate solutions measured against a target outcome. The evolutionary process is repeated with each iteration referred to as a generation. The evolution's duration is set by stopping criteria such as a lack of significant change in fitness or reaching a preset number of generations, or by whichever criterion is satisfied first.

A particular application of GP is in the realm of symbolic regression (GPSR) [6,19] where a solution is found in the form of the best evolved program that produces a desired target outcome [1]. Strengths of GPSR include the ability to explore a vast solution space without any prior knowledge or assumptions, ease of implementation, and the natural selection of the most important inputs [13].

Evolving Programs for Mixture Analysis. Our proposed tree based model for mixture analysis is composed of a set of evolved programs. GPSR evolves the programs in a way that best relates components and mixtures of those components. Specifically, the relationship is made between reference spectra of the components and the spectrum of the mixture.

Preprocessing spectral data is common for many analysis scenarios by applying data transformations to correct for baseline shifts, noise, or for light scattering effects for example [21]. In addition to no preprocessing (none) of the spectral data, GPSR with mean centred (MC), standard normal variate (SNV), and smoothed gradient (SG) transformations are considered in this work. For SG, a Savitzky-Golay filter with a window size of 15 and poly order of 2 was applied to the first derivative transformed data.

Programs are evolved by GPSR for a given mixture concentration. The best evolved program from a GPSR run becomes a candidate for that mixture concentration. GPSR runs are repeated to obtain a set of candidate programs from a set of training mixture concentrations. Collectively, the set of best candidate programs, one from each training mixture concentration, form a model for making inference on new data. Learning from the training mixture series, two models are produced: a spectral addition model and a spectral subtraction model.

GP-Based Mixture Analysis by Spectral Addition. The MASA model is assembled with GPSR by relating pure component spectra to mixtures consisting of the same components at known concentrations. A single program is a combination of operators and constants that when applied to the input spectra produces a good reconstruction of the target mixture spectrum. In this manner a set of n programs, $[p_1, p_2, ..., p_n]$ are created from a set of n training mixtures $[m_1, m_2, ..., m_n]$ corresponding to n component concentration combinations $[c_1, c_2, ..., c_n]$. The set of training mixtures used for evolving the GPSR programs is designed to be a representative set of varied component concentration combinations.

GP-Based Mixture Analysis by Spectral Subtraction. The MASS model is constructed with GPSR according to the assumed first component spectral subtraction relationship, where, excluding noise, a component is the difference when all other components are subtracted from a mixture. For the case of a binary mixture, MASS GPSR takes the first component reference spectrum and the mixture spectrum as inputs and finds a program to make a good reconstruction of the second component reference spectrum. As with the MASA model, the MASS model is formed from a set of programs that are produced from the training set of mixtures.

4   Experiment Setup

4.1   Parameter Settings for GPSR

GPSR was employed to evolve programs that best describe the relationship between a drug mixture spectrum and component reference spectra. The GPSR implementation makes use of the distributed evolutionary algorithms in python (DEAP) framework [7]. The parameter settings for the evolutionary process

Table 1. Parameter settings for GPSR.

Parameter	Value	Parameter	Value
Terminal set	spectra, constants	Function set	$+, -, *, \%$
Initialisation	half and half	Population size	1000
Selection method	tournament	Tournament size	7
Mating probability	0.9	Mutation probability	0.1
Fitness function	MSE	Maximum tree depth	7
Generations	25	Number of runs	30

were set according to the values summarised in Table 1. Most of them follow the common settings in GPSR.

For terminals, a set of constants ranging from 0 to 1 in increments of 0.1 was used. The input spectra are also included in the terminal set. The function set consists of the basic arithmetic operators: addition, subtraction, multiplication, and protected division (a variation of division that avoids division by zero errors). Mean squared error (MSE) is used as the fitness function to measure the performance of the candidate solution while selection is performed by the tournament method with a size of seven. The initial pool of individuals was generated using the half-and-half method with a population size of 1000.

4.2 Quantifying Spectral Goodness of Fit

In this paper, goodness of fit refers to how well a predicted NIR absorbance spectrum represents a target spectrum. This will be assessed using the statistical measure R^2. Known as the coefficient of determination, R^2 represents the proportion of the variation in the target spectrum that can be explained by the model's predicted spectrum. If the predicted absorbance value at the i-th wavelength is $\hat{y}_i$, and the target absorbance value at the i-th wavelength is y_i, with the mean of the target spectrum denoted as $\bar{y}$, the proportion of the total variation in the target spectrum explained by the model is given by the complement of the ratio of the residual sum of squares (RSS) to the total sum of squares (TSS). Thus, the coefficient of determination R^2 can be written as Eq. 1.

$$R^2 = 1 - \frac{\sum_i (y_i - \hat{y}_i)^2}{\sum_i (y_i - \bar{y})^2} = 1 - \frac{RSS}{TSS} \tag{1}$$

The best possible R^2 score is 1 for when the predicted values exactly match the target values. A model that always predicts $\bar{y}$ will receive an R^2 score of 0. It is possible to obtain negative R^2 values when the predicted values are worse than predicting $\bar{y}$.

4.3 Benchmarks

Benchmark Dataset Near-Infrared Absorbance Spectra. The substances meth, methylsulfonylmethane (MSM), and caffeine were used to form two binary mixture series, each with meth as the first component: [meth, MSM] and [meth, caffeine]. Each series consists of mixtures of meth combined with either MSM or caffeine at varying weight-based concentrations. For example, the [meth, MSM] series ranges from 5% meth and 95% MSM (denoted as 5:95) to 95% meth and 5% MSM (95:5) in 5% increments: 5:95, 10:90, 15:85, ... , 85:15, 90:10, and 95:5. Additionally, 100% pure samples of each component are included in each series as references.

The [meth, MSM] and [meth, caffeine] datasets consist of near-infrared absorbance spectra obtained from the Lumi™ Nano NIR device, each based on measurements of the respective mixture series. The device provides 228 absorbance values over a wavelength range of 900 to 1700 nm. However, due to noise and anomalies at the extreme ends of this range, a restricted wavelength range was used for model training and testing in this work. The training set of mixtures consist of a selected subset from the [meth, MSM] dataset, labelled as meth:MSM % w/w, as defined in Eq. 2.

$$s \subset [\text{meth, MSM}], s = \{10:90, 25:75, 40:60, 55:45, 70:30, 85:15\} \qquad (2)$$

The unselected mixtures from the [meth, MSM] dataset, along with all mixtures from the [meth, caffeine] dataset, are reserved for testing.

Benchmark Model - Partial Least Squares Regression. Partial least squares regression (PLSR) is one of the most popular algorithms for model creation on NIRS data [3]. Its popularity is due to its robustness in quantitative analysis [15] by providing a linear modelling method that builds the relationship between independent and dependent variables [28]. In this study, PLSR serves as the benchmark model against which the results of the GP-based models are compared. The number of latent variables of the PLSR model was matched to the number of components. This approach leverages all available component spectra, enabling the model to capture the linear relationship of the components to the mixture spectrum.

5   Results

5.1   The Training Performance

Programs were evolved by GPSR on the training set of mixtures. A MASA model was developed and then GPSR was repeated to create the MASS model. GPSR training performance was compared to PLSR models trained on the same set of training mixtures. Table 2 presents a comparison of the best training R^2 scores for the GPSR and PLSR MASA and MASS models. The results are tabulated by scoring the calculated spectrum to the actual spectrum (mixture and component two for MASA and MASS respectively) at each training mixture concentration across preprocessing methods. As evidenced by the low variation across preprocessing methods, the GPSR evolved programs show little sensitivity to the data

Table 2. GPSR and PLSR MASA and MASS model training R^2 values.

Model Type	Pre-proc	MASA Model Training						MASS Model Training					
		10:90	25:75	40:60	55:45	70:30	85:15	10:90	25:75	40:60	55:45	70:30	85:15
GPSR	none	.9868	.9757	.9438	.9298	.8546	.8483	.9868	.9767	.9540	.9012	.8263	.7817
PLSR	none	−0.84	−13.5	−13.2	−37.9	−7.21	−16.9	−0.74	−15.2	−14.0	−43.0	−5.05	−14.0
GPSR	MC	.9878	.9789	.9588	.9242	.8700	.8682	.9880	.9723	.9481	.8980	.8411	.7928
PLSR	MC	.9870	.9768	.9507	.9184	.8600	.8568	.9868	.9722	.9391	.8793	.8056	.7776
GPSR	SNV	.9898	.9855	.9613	.9317	.8859	.8837	.9895	.9799	.9559	.9112	.8541	.8324
PLSR	SNV	.9870	.9768	.9507	.9184	.8600	.8568	.9868	.9722	.9391	.8793	.8056	.7776
GPSR	SG	.9888	.9750	.9611	.9285	.8792	.8317	.9888	.9723	.9408	.8684	.7084	.5399
PLSR	SG	.9884	.9741	.9550	.9105	.8508	.7945	.9883	.9687	.9360	.8377	.6638	.4760

transformations. In general, the models exhibit good performance in predicting a target spectrum. Overall, the ability of the models to achieve a good representation of the training mixture decreases as the concentration of component one (meth) increases and component two (MSM) decreases. The SG transformed data shows a significantly lessened ability to create a good representation of the training mixture at high component one/low component two concentrations, this is especially true with the MASS models.

Preprocessing of the data is found to be a requirement with the PLSR models when comparing the calculated to actual spectrum. This is highlighted by the large negative scores with the unprocessed data. The large negative values arise from the independent measurements of the component spectra and the mixture spectrum in which the differences in baseline levels (baseline offset) can be significant. The result is a large difference between the predicted absorbance value and the target value.

As with the GPSR models, the same trend is found on the PLSR models with lower representation ability as component one increases and component two concentrations decrease. The GPSR models provide a more accurate representation of the actual spectra compared to the PLSR models across all preprocessing methods, as indicated by the higher scores achieved by the GPSR models. The GPSR model predictions also have less variation than their PLSR model counterparts across training mixtures.

An example of a GPSR MASS evolved program from the training mixture concentration meth:MSM 85:15 % w/w can be viewed in Fig. 3. The input spectra are represented by $X0$ and $X1$. The result of evaluating the program tree is a spectrum derived from the operations and constants applied to the input spectra. Figure 3 reveals the full variety of operators from the function set along with several select constants that essentially transform the input spectra according to the learnt relationship to produce a good fit to the regression target spectrum.

5.2 The Testing Performance

Due to their consistency and relatively better performance on the training data, the best GPSR models generated from the SNV transformed data were used to

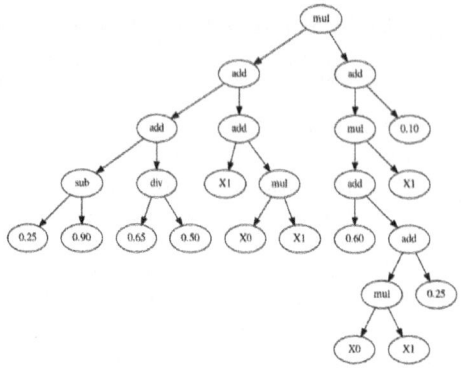

Fig. 3. An example GPSR MASS program tree from training mixture meth:MSM 85:15 % w/w.

make inference on the test data. The same SNV transformation was also applied to the test data, with no additional preprocessing performed.

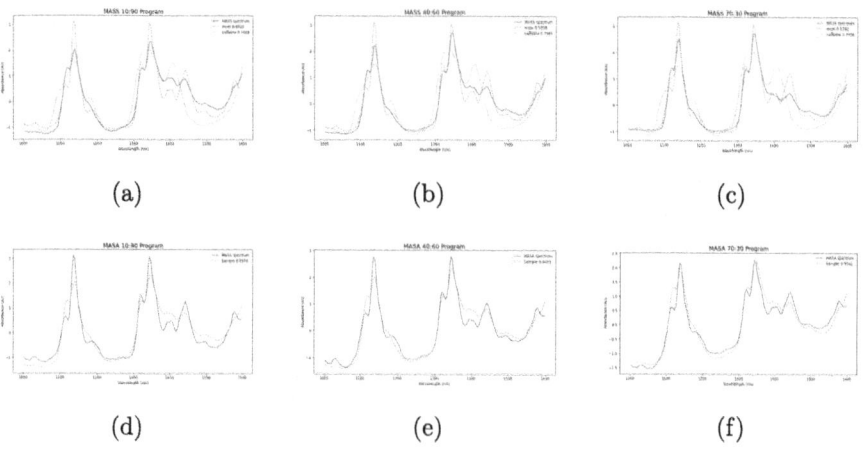

Fig. 4. Example GPSR MASS and MASA predicted spectra from a meth:MSM 50:50 % w/w test sample. GPSR MASS programs (a) 10:90, (b) 40:60, (c) 70:30. GPSR MASA programs (d) 10:90, (e) 40:60, and (f) 70:30.

Predict Second Component Spectra by MASS. The GPSR MASS model is used to predict second component spectra on test mixture samples when given the first component reference and the sample spectrum. From those two inputs a set of second component predicted spectra are obtained. Figure 4(a-c) depict GPSR MASS predicted spectra for a meth:MSM 50:50 % w/w test sample. A comparison is made to a set of two substances, MSM and caffeine. The model predictions (solid lines) match more closely to the MSM reference (dashed lines) than the caffeine reference (dotted lines). The best score among the predicted spectra comes from the 70:30 program ($R^2 = 0.9392$) for a correct match to the substance MSM.

Predict Mixture Spectra by MASA. The GPSR MASA model is used to predict mixture spectra that are scored against the test sample. Given the first and the second component references, the GPSR MASA model predicts a set of mixture spectra. Using the same meth:MSM 50:50 % w/w test sample as before, the drug reference remains the same and the second component reference is appointed from the MASS model best match result. For the GPSR MASA predicted spectra see Fig. 4(d-f). The GPSR MASA model has predicted very good reconstructions of the test sample. The best score comes from the 70:30 program ($R^2 = 0.9542$). The MASA result gives corroborating evidence that the test sample is indeed a mixture of meth and MSM.

Discriminate the Positive and Negative Classes. The GPSR models are used to provide a classification prediction for adulterant detection in two scenarios. Either an adulterant is present in the mixture or an adulterant is absent from the test mixture. The datasets, which have a common first component and different second components, are suited for testing the differentiating ability of the GPSR models. This can be accomplished by first testing for the substance MSM (a single adulterant) on the two binary mixture series [meth, MSM] and [meth, caffeine]. The second test is then arranged for detecting the presence of caffeine given the same sets of mixtures. When looking for an adulterant of interest the MASA and MASS models should give a low prediction value on samples that do not contain the adulterant. Conversely, a high prediction value is expected for mixture samples in which the adulterant under test is present. First it is assumed that the DDM has identified each sample as containing the drug meth. Then the test on samples that do and do not contain an adulterant is carried out by the GPSR mixture analysis models.

The highest prediction scores from the GPSR models are given for each test sample in Fig. 5. The ability of the models to distinguish between mixtures containing MSM and those without is correlated to the distance between their respective scores. At low meth/high second component mixtures the distance is greatest. When testing for MSM, the [meth, MSM] mixtures are detected near a score of 1.0 while the [meth, caffeine] mixtures are below 0.75 as shown in Fig. 5(a). As the concentrations change to high meth/low second component mixtures, the ability of the model to distinguish lessens. At the 95:5 mixtures (mostly meth) the model makes virtually no distinction between the two samples with respect to their MSM content. At the pure meth sample (no MSM) the scores fall lower as expected for a sample without the adulterant. The GPSR MASA and MASS models perform comparably, however, throughout most of the range of mixtures a slightly larger area of discernment is noted with the

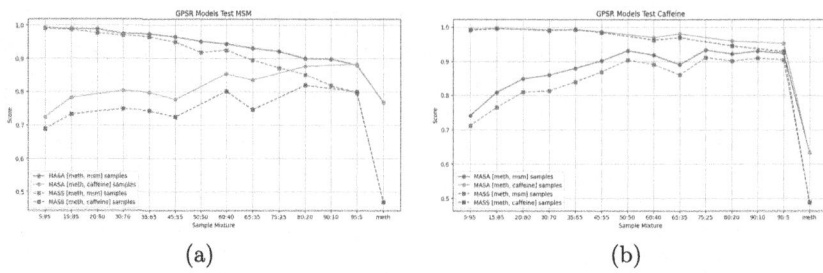

Fig. 5. Prediction scores from GPSR models. Test for MSM (a) and test for caffeine (b) on [meth, MSM] and [meth, caffeine] mixtures.

MASS model. In addition, a favourably lower score is given to the all meth/no MSM sample by the MASS model compared to the MASA model.

The highest prediction values from testing for caffeine by the GPSR MASA and MASS models on the [meth, MSM] and [meth, caffeine] samples is given in Fig. 5(b). Like with the previous test for MSM, high second component/low meth concentration samples give the largest margin of discernment. As the mixtures change to high meth/low second component concentrations, the distance between positive and negative classes decreases but a noticeable gap remains between samples with and without the adulterant of interest (caffeine). As with the test for MSM, at the pure meth sample (no caffeine), and in line with expectations, the scores tally considerably lower.

A few noteworthy commonalities from the two tests is the wider discrimination distance of the GPSR MASS model for most of the mixture range and the better discrimination of classes at the most extreme high component one concentrations by the GPSR MASA model. Therefore, the two models working in tandem is a good strategy for implementing a mixture analysis workflow.

5.3   Comparison to Linear Models on the Test Data

The performance of the GPSR models was compared to the PLSR models on the test data. The PLSR MASA results, test for MSM, have been added to the plot with the GPSR MASA results as shown in Fig. 6(a). Likewise, the GPSR and PLSR MASS models, test for MSM, results are presented in Fig. 6(b). The GPSR and PLSR models have comparable outcomes when discriminating the positive and negative classes. The GP mixture analysis approach is validated by demonstrating similar performance to the benchmark PLSR models.

5.4   GP Mixture Analysis Sample Workflow

A mixture analysis is initiated when the DDM results meet specific conditions: a drug is detected, and its class probability falls below a certain threshold (see Fig. 2). The reference spectrum of the drug identified by the DDM, along with the test sample spectrum, are then input into the GPSR MASS model to identify

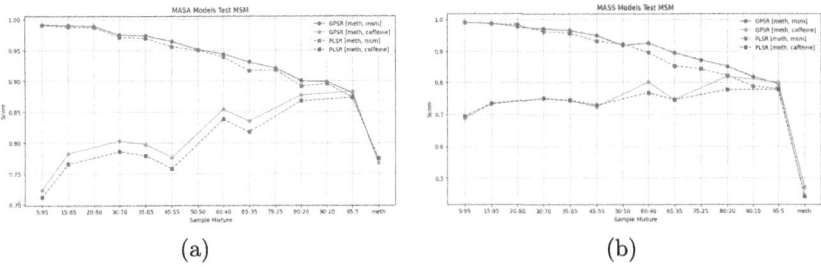

(a)   (b)

Fig. 6. Prediction scores from GPSR and PLSR models when testing for MSM. MASA models (a) and MASS models (b) prediction scores from [meth, MSM] and [meth, caffeine] test sample mixtures.

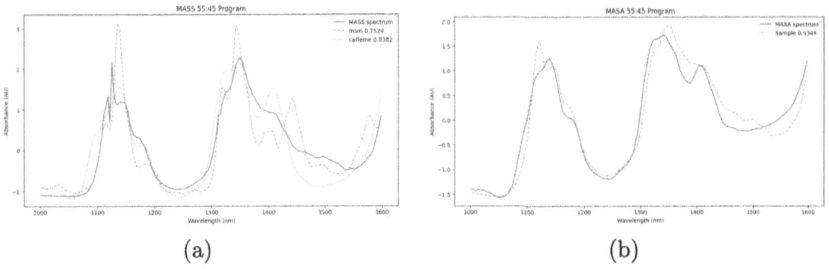

(a)   (b)

Fig. 7. The GPSR MASS model predicted spectrum (a) with adulterant spectra and matching scores. The best GPSR MASA spectrum (b) with comparison to the actual sample mixture spectrum.

a best-match adulterant. Next, the reference spectra of both the identified drug and the best-matched adulterant from the GPSR MASS model are input into the GPSR MASA model. The MASA model predicts spectra, which are scored against the test sample to either confirm the presence of the adulterant in the mixture or conclude otherwise.

For a demonstration of the workflow, an example test sample is analysed. The DDM has already identified meth as the main drug, with a class probability that warrants further mixture analysis. The meth reference and the sample spectra are given to the GPSR MASS model to match to a potential adulterant. The result of the GPSR MASS model best score against a library of two adulterants is shown in Fig. 7(a). The conclusion of the GPSR MASS model is a best matched adulterant of the substance caffeine with a score of 0.8382. The score on the other adulterant analysed, MSM, is 0.7524.

To substantiate the GPSR MASS prediction result, the matched adulterant and the drug reference are input to the GPSR MASA model to examine how well the predicted components can recreate the test sample. The resulting plot of the best GPSR MASA produced spectrum is depicted in Fig. 7(b). The GPSR MASA prediction reveals a convincing representation of the test sample with a score of 0.9344. The true identity of the example test sample is a meth:caffeine 80:20 % w/w mixture of meth and caffeine. Mixture analysis by GP has correctly identified the adulterant caffeine present in the example test sample mixture.

6 Conclusions

This work proposes using genetic programming for symbolic regression (GPSR) to provide an analysis of components in a mixture such as adulterants in illicit drugs. The GPSR models were found to effectively model the relationship between a mixture and its components. The models can also be easily visualised and understood as program trees. The results indicate that GPSR models are comparable to partial least squares regression (PLSR) models when employed to detect components in a mixture. A proposed workflow for illicit drug mixture analysis at music festivals using portable spectroscopy was introduced and its effectiveness was demonstrated by example for the successful identification of an adulterant in an illicit drug sample.

Both the PLSR and GPSR models exhibit decreasing performance at higher concentrations of component one/lower concentrations of component two. Recognising a theoretical detection limit, future work to address this issue may involve feature selection techniques or additional training instances at the concentrations of concern, or a combination thereof. Mixtures of two components were presented in this work as the basis for an extension into the analysis of mixtures with more than two components. The same line of reasoning applies to the expansion of the library of adulterants. Another area for exploration is in an additional machine learning approach to the information as a whole given by the GPSR MASS and MASA models for improving prediction confidence.

References

1. Afzal, W., Torkar, R.: On the application of genetic programming for software engineering predictive modeling: a systematic review. Expert Syst. Appl. **38**(9), 11984–11997 (2011). https://doi.org/10.1016/j.eswa.2011.03.041
2. Aykas, D.P., Menevseoglu, A.: A rapid method to detect green pea and peanut adulteration in pistachio by using portable FT-MIR and FT-NIR spectroscopy combined with chemometrics. Food Control **121**, 107670 (2021). https://doi.org/10.1016/j.foodcont.2020.107670
3. Balabin, R.M., Safieva, R.Z., Lomakina, E.I.: Comparison of linear and nonlinear calibration models based on near infrared (NIR) spectroscopy data for gasoline properties prediction. Chemom. Intell. Lab. Syst. **88**(2), 183–188 (2007). https://doi.org/10.1016/j.chemolab.2007.04.006
4. Banzhaf, W.: Artificial intelligence: genetic programming. In: Smelser, N.J., Baltes, P.B. (eds.) International Encyclopedia of the Social & Behavioral Sciences, pp. 789–792. Pergamon, Oxford (Jan 2001).https://doi.org/10.1016/B0-08-043076-7/00557-X

5. Bradley, M., Izzia, F., Nunn, S.: Analysis of Mixtures by FT-IR: Spatial and Spectral Separation of Complex Samples. Spectroscopy -Springfield then Eugene then Duluth- **0**, 20–+ (Aug 2008). https://www.spectroscopyonline.com/view/analysis-mixtures-ft-ir-spatial-and-spectral-separation-complex-samples, publisher: MJH Life Sciences
6. Chen, Q., Xue, B., Browne, W., Zhang, M.: Evolutionary regression and modelling. In: Banzhaf, W., Machado, P., Zhang, M. (eds.) Handbook of Evolutionary Machine Learning, pp. 121–149. Springer Nature Singapore, Singapore (2024). https://doi.org/10.1007/978-981-99-3814-8_5
7. Fortin, F.A., De Rainville, F.M., Gardner, M.A.G., Parizeau, M., Gagné, C.: DEAP: evolutionary algorithms made easy. J. Mach. Learn. Res. **13**(1), 2171–2175 (2012), publisher: JMLR.org
8. Gonclaves, R., et al.: Suitability of infrared spectroscopy for drug checking in harm reduction centres. Int. J. Drug Policy **88**, 103037 (2021). https://doi.org/10.1016/j.drugpo.2020.103037
9. Gozdzialski, L., Wallace, B., Stege, U., Hore, D.: Linear programming for spectral mixture analysis (2021)
10. Hong, F.W., Chia, K.S.: A review on recent near infrared spectroscopic measurement setups and their challenges. Measurement **171**, 108732 (2021), publisher: Elsevier
11. Khan, M.W., Alam, M.: A survey of application: genomics and genetic programming, a new frontier. Genomics **100**(2), 65–71 (2012), publisher: Elsevier
12. KnowYourStuffNZ: Know Your Stuff NZ Webpage (2024). https://knowyourstuff.nz/results-reports/. Accessed 24 Oct 2024
13. Kotanchek, M., Smits, G., Kordon, A.: Industrial strength genetic programming. In: Riolo, R., Worzel, B. (eds.) Genetic Programming Theory and Practice, pp. 239–255. Springer US, Boston, MA (2003). https://doi.org/10.1007/978-1-4419-8983-3_15
14. Kranenburg, R.F., Ramaker, H.J., Sap, S., van Asten, A.C.: A calibration friendly approach to identify drugs of abuse mixtures with a portable near-infrared analyzer. Drug testing and analysis **14**(6), 1089–1101 (2022), publisher: Wiley Online Library
15. Liu, C.M., Han, Y., Min, S.G., Jia, W., Meng, X., Liu, P.P.: Rapid qualitative and quantitative analysis of methamphetamine, ketamine, heroin, and cocaine by near-infrared spectroscopy. Forens. Sci. Int. **290**, 162–168 (2018), publisher: Elsevier
16. New Zealand Police: New Zealand Police Webpage (Jul 2022). https://www.police.govt.nz/news/release/police-rolls-out-lumi-drug-scan-tool-nationwide-following-successful-pilot. Accessed 24 Oct 2024
17. Oliveira, M., Cruz-Tirado, J., Roque, J., Teófilo, R., Barbin, D.: Portable near-infrared spectroscopy for rapid authentication of adulterated paprika powder. J. Food Compos. Anal. **87**, 103403 (2020), publisher: Elsevier
18. Pavia, D., Lampman, G., Kriz, G., Vyvyan, J.: Introduction to Spectroscopy. Cengage Learning (2008)
19. Poli, R., Langdon, W., Mcphee, N.: A field guide to genetic programming. Lulu Press **10**(2), 229–230 (2008)
20. Santamarina, R., Caldicott, D., Fitzgerald, J., Schumann, J.L.: Drug-related deaths at Australian music festivals. Int. J. Drug Policy **123**, 104274 (2024). https://doi.org/10.1016/j.drugpo.2023.104274
21. Shi, H., Lei, Y., Prates, L.L., Yu, P.: Evaluation of near-infrared (NIR) and Fourier transform mid-infrared (ATR-FT/MIR) spectroscopy techniques combined

with chemometrics for the determination of crude protein and intestinal protein digestibility of wheat. Food Chem. **272**, 507–513 (2019), publisher: Elsevier

22. Smith, B.: Spectral Subtraction. Publisher: MJH Life Sciences **36**, 14–19 (May 2021). https://www.spectroscopyonline.com/view/spectral-subtraction. Accessed 24 Oct 2024

23. Spectral Engines, The Nynomic Group [formerly m.u.t AG]: TactiScan Brochure (2024). https://f.hubspotusercontent10.net/hubfs/4905262/Assets/Brochures/TactiScan_Brochure_en_v03_web.pdf. Accessed 24 Oct 2024

24. Szabadai, Z., Vlaia, V., Țăranu, I., Țăranu, B.O., Vlaia, L., Popa, I.: Multivariate data processing in spectrophotometric analysis of complex chemical systems. Edited by Jamal Uddin p. 291 (2012)

25. The New Zealand Drug Foundation: New Zealand Drug Foundation Webpage (2024). https://drugfoundation.org.nz/articles/report-drug-use-in-aotearoa-202223

26. The New Zealand Herald: The New Zealand Herald Webpage (Jul 2022). https://www.nzherald.co.nz/nz/lumi-the-new-police-drug-testing-tool-hitting-the-streets/V5J2HP3NXDG22GQEPGKWXYPD3U/. Accessed 24 Oct 2024

27. Wenger, J., Kjellström, H., Triebel, R.: Non-parametric calibration for classification. In: International Conference on Artificial Intelligence and Statistics, pp. 178–190. PMLR (2020)

28. Zhou, L., Tan, L., Zhang, C., Zhao, N., He, Y., Qiu, Z.: A portable NIR-system for mixture powdery food analysis using deep learning. Lwt **153**, 112456 (2022), publisher: Elsevier

Hybrid Optimization of Horizontal Alignments in European Terrains: A Comparative Study

Ane Espeseth[1]([✉]) [iD], Martin Juříček[2] [iD], Harald Michael Ludwig[3] [iD],
and Tea Tušar[4,5] [iD]

[1] Center for Computing in Science Education, University of Oslo, Oslo, Norway
anekes@uio.no
[2] Faculty of Mechanical Engineering, Brno University of Technology, Brno, Czechia
200543@vutbr.cz
[3] Johannes Kepler University, Linz, Austria
ludwig@csh.ac.at
[4] Department of Intelligent Systems, Jožef Stefan Institute, Ljubljana, Slovenia
tea.tusar@ijs.si
[5] Jožef Stefan Postgraduate School, Ljubljana, Slovenia

Abstract. Path planning across terrain is a fundamental challenge in civil engineering, with applications ranging from transportation infrastructure to urban development. Recent advances in computational methods have enabled automated route optimization, particularly in horizontal alignment problems that balance construction costs with terrain constraints. However, standardized comparisons of optimization approaches across diverse geographical contexts remain limited, hindering the development of reliable automated planning systems. Here we show through a systematic comparative study across three European landscapes that A* significantly outperforms RRT* in initial path generation, with better computational efficiency and terrain adaptation, while PSO demonstrates superior optimization capabilities compared to CMA-ES and DE in refining these paths against roadway construction criteria. Through extensive parameter validation, we find these performance advantages remain consistent across different geographical contexts and topographical challenges, with the hybrid A*-PSO approach achieving significantly better results than applying optimization algorithms to straight-line paths alone. These findings provide a comprehensive comparison of key algorithms in infrastructure planning optimization, demonstrating the relative strengths of different approaches in horizontal alignment tasks. This comparative analysis offers practical guidance for algorithm selection while highlighting opportunities for further development through the incorporation of real-world engineering constraints.

Keywords: Horizontal alignment optimization · Path planning · Comparative study

A. Espeseth, M. Juíček and H. M. Ludwig—Contributed equally to this work.

© The Author(s), under exclusive license to Springer Nature Switzerland AG 2025
P. García-Sánchez et al. (Eds.): EvoApplications 2025, LNCS 15612, pp. 119–136, 2025.
https://doi.org/10.1007/978-3-031-90062-4_8

1   Introduction

Horizontal Alignment Optimization (HAO) is a core optimization problem in civil engineering and transportation planning, commonly applied in the design of linear infrastructure such as roadways and railways. It involves finding an optimal path across a horizontal plane, represented by a 2D or 3D map. The goal is to minimize construction costs, factoring in land acquisition, terrain features, environmental constraints, safety, and path curvature, while ensuring compliance with geometric standards such as maximum curvature, path length, and safety requirements [41].

Aligning paths across complex terrains with multiple constraints demands high computational effort, particularly when dealing with non-differentiable cost functions. Early approaches used traditional pathfinding algorithms to generate feasible paths in controlled environments with predictable obstacles. Global optimization techniques were also introduced early on, at first in the form of numerical methods [47] and dynamic programming [46], but soon also through Genetic Algorithms (GAs) in order to address non-linear, large-scale problems in infrastructure planning [18,22]. Over time, a variety of strategies has been developed to improve efficiency, accuracy, and complexity management in HAO. Problem-specific challenges have also been identified and addressed with tailored solutions.

However, this diversity has introduced a challenge: with tailored solutions and evaluations dominating the literature, cross-comparison becomes difficult. As a result, generalization of methods across HAO contexts is limited.

To address the need for standardized comparison, we conduct a structured evaluation using simplified HAO scenarios based on publicly available maps. We employ widely recognized algorithms—A-Star (A*) [16], Rapidly-exploring Random Tree Star (RRT*) [23], Particle Swarm Optimization (PSO) [24], Covariance Matrix Adaptation Evolution Strategy (CMA-ES) [15], and Differential Evolution (DE) [43]—without domain-specific customizations to ensure generalizability and facilitate meaningful comparisons across optimization techniques in static horizontal alignment. We compare the algorithms across three structured scenarios: graph-based algorithms, optimization algorithms within a straight corridor, and a hybrid approach that initializes the optimization algorithms with the path created by A* (the best-performing graph-based algorithm). This study thereby provides an initial reference point for evaluating the effectiveness of standard optimization and graph-based algorithms in infrastructure planning.

2   Related Work

The Horizontal Alignment Optimization Task. There are several variations of HAO tasks, ranging from graph-based path optimization to 2D-corridor selection, parametric tuning of curves, and complex 3D problems combining horizontal and vertical alignment. The objective is often to optimize a single alignment with multiple constraints [22], though recent work emphasizes generating diverse, feasible alternatives to aid infrastructure planners [34].

Graph-Based Optimization Algorithms for Corridor Selection. Graph-based least-cost optimization algorithms have long played an essential role in infrastructure planning. They are easily adopted for minimal distance and obstacle avoidance, features of both preliminary alignment and adaptive planning tasks [1, 7, 38]. Among modern methods, A* and RRT* stand out due to their effectiveness in static and dynamic problems, respectively. A* is effective in predictable environments, while RRT* is adapted for complex, dynamic scenarios such as mobile robot trajectory planning [17, 29]. Perhaps due to the similarities between mobile planning and horizontal planning tasks, RRT* and its variations are also widely applied in infrastructure planning [33, 44, 49].

Optimization-Only Techniques. Algorithms such as PSO, CMA-ES, and DE are robust for single-objective optimization tasks in infrastructure planning, including adjusting paths for energy efficiency and crowd safety [53], timetable synchronization [12, 27], and construction constraints in challenging terrains [21]. Methods from other disciplines have also been tested. Among the most successful ones is mathematical optimization, particularly Mixed Integer Linear Programming (MILP) and its variants, and the Distance Transform (DT) algorithm, first introduced by [39] in 2006. In later years, methods like Deep Reinforcement Learning [14] have also been used with some success.

In engineering publications, optimization methods are often customized to address the unique needs of specific alignment problems. GAs in particular are frequently modified to handle project-specific requirements, creating intricate genetic representations and operations specific to the problem (see, e.g., [19, 26]). These customizations, while effective in specific scenarios, reduce the generalizability of findings and limit direct comparisons between different methods.

Hybrid Strategies: Pathfinding + Optimization. In recent years, more general-purpose optimization techniques are finding a place as *refinement tools* for initial paths in static alignment tasks. Bi-level optimization strategies use two different optimization algorithms in sequence, e.g., to find an optimal horizontal corridor to use for vertical optimization [30, 50]. While graph-based least-cost algorithms are not ideal for complex objective functions, they can efficiently find promising paths which then drastically reduce the search space of more sophisticated optimization algorithms. For instance, [49] uses RRT* to initialize a DE optimization stage. In [40], a four-step hybrid approach is used which mixes Dijkstra with PSO to optimize for a complex railway alignment problem. Another paper, [32], initializes with DT and optimizes with PSO. In [45], a variant of RRT* is used alongside Ant Colony Optimization. Hybrid methods like these are gradually gaining popularity, but a 2023 review of HAO [41] still identified their exploration as a gap in the current literature.

Solution Encoding. In infrastructure optimization, physical constraints like path curvature are critical. While dynamic path planning often incorporates curves directly into the solution representation (see, e.g., [6,52]), HAO typically uses lightweight representations. Solutions are stored as point lists, which are transformed into smooth paths at evaluation [41] (for a comprehensive review of these methods, see [36]).

Comparison Studies. Despite much modern progress in optimization and pathfinding, comprehensive studies that evaluate and compare these methods in standardized static alignment contexts remain limited: in [2], a comparison was made between bi-objective optimization techniques for vertical alignment, including three scalarization methods and two GAs, and [48] compares the use of GAs, PSO and Nonlinear Optimization with Mesh Adaptive Direct Search (NOMAD) for initializing the Sequential Quadratic Programming (SQP) optimizer on a 3D problem. Research is shifting towards hybrid and multi-objective approaches that better use the strong points of each optimization technique, but consistent comparisons are still needed to make these advancements applicable across alignment problems.

3   Methods

3.1   Path Planning Algorithms

A*. The A* algorithm [16] is a classic and widely used algorithm for pathfinding and graph traversal that combines the properties of Dijkstra's and Uniform-Cost-Search [37]. As a heuristic search algorithm, it dynamically expands toward the goal under the guidance of a heuristic function, continually seeking the most efficient path between the start and end point. A* and its variations are particularly applicable in fields such as mobile robot navigation [5,9,10] and computer games [3,25,51].

The algorithm uses an evaluation function, $f(n)$, defined as:

$$f(n) = g(n) + h(n),$$

where $f(n)$ represents the total cost estimate for node n, $g(n)$ is the actual cost of the path from the start to n, and $h(n)$ is the heuristic estimate of the cost from n to the target. The heuristic must be admissible: $h(n) \leq h^*(n)$, where $h^*(n)$ is the actual optimal path cost from node n to the goal. Common heuristics include Manhattan distance, Euclidean distance, and diagonal distance. For each current node v_n, the cost $g(v'_n)$ for neighboring nodes v'_n is calculated using the formula:

$$g(v'_n) = g(v_n) + c(v'_n),$$

where $c(v'_n)$ is the distance between the nodes. The algorithm systematically explores nodes by selecting the most promising candidate from the open list, then moves it to the closed list as part of the path. It expands neighboring nodes and updates the candidate list until the target node is added to the closed list, at which point a complete path has been found.

Adjusted slightly for HAO, the implementation accounts for elevation differences using an extended cost function, allows for diagonal movement, and operates with an extended neighborhood.

RRT*. RRT*, a cousin of A*, also incrementally builds a tree in a defined environment [23]. Its key feature, which adapts it to dynamic or uncertain environments, is its ability to continuously reevaluate and improve paths. If a new branch offers a shorter path to the goal, the tree is "rewired" to best incorporate it, increasing the likelihood of finding an optimal trajectory. RRT* can be easily extended with various heuristics and constraints, and has a prominent presence in the fields of robotics, such as motion planning for mobile robots [35], drones [11] and manipulators [28]. It guarantees asymptotic optimality, improving path quality as computation time increases.

In each iteration, the algorithm randomly samples a point x_{rand} from the configuration space and finds the nearest existing node x_{nearest} using a distance metric $d(x_1, x_2)$. It then attempts to connect a new point x_{new} at a step size δ in the direction of x_{rand}. The best parent for x_{new} is identified within a neighborhood radius, ensuring the path improves progressively. The path cost $c(x)$ is calculated as:

$$c(x) = c(x_{\mathrm{parent}}) + d(x_{\mathrm{parent}}, x).$$

After adding x_{new}, the algorithm attempts to rewire existing nearby nodes to shorten the overall path, ensuring that the tree converges toward the optimal path as $n \to \infty$. The condition for rewiring can be expressed as:

$$c(x_{\mathrm{new}}) + d(x_{\mathrm{new}}, x_{\mathrm{near}}) < c(x_{\mathrm{near}}).$$

This enables RRT* to improve paths iteratively, making it highly effective for complex path-planning tasks in high-dimensional spaces.

3.2   Optimization Algorithms

PSO. In Particle Swarm Optimization, a population of particles iteratively searches for the optimal solution by adjusting their positions within a multidimensional search space based on individual experiences and the collective knowledge of the swarm [13, 20, 24]. The key mechanisms of the PSO algorithm are described in two equations. The position $\mathbf{x}_i^{(t)}$, of particle i at iteration t is updated as:

$$\mathbf{x}_i^{(t+1)} = \mathbf{x}_i^{(t)} + \mathbf{v}_i^{(t+1)},$$

where $\mathbf{v}_i^{(t+1)}$ is the particle's velocity at the next iteration. The velocity is updated based on four factors: the particle's current position $\mathbf{x}_i^{(t)}$, its previous velocity $\mathbf{v}_i^{(t)}$, its previous best position $\mathbf{p}_i^{(t)}$ and the global best position found by the swarm $\mathbf{g}^{(t)}$:

$$\mathbf{v}_i^{(t+1)} = w \cdot \mathbf{v}_i^{(t)} + c_1 \cdot r_1 \cdot (\mathbf{p}_i^{(t)} - \mathbf{x}_i^{(t)}) + c_2 \cdot r_2 \cdot (\mathbf{g}^{(t)} - \mathbf{x}_i^{(t)}),$$

here w is the inertia weight, c_1 and c_2 are acceleration coefficients, and r_1 and r_2 are random numbers drawn from a uniform distribution.

CMA-ES. The Covariance Matrix Adaptation Evolution Strategy is an evolutionary optimization algorithm that generates new solutions by sampling from an evolving probability distribution [15]. A key feature of CMA-ES is its ability to learn correlations between variables, enabling efficient search even for complex and non-separable functions. CMA-ES adapts two components during optimization: the covariance matrix and the global step size. The covariance matrix is updated based on evolutionary paths and the differences between the best individuals of consecutive generations. The global step size is adjusted by monitoring the length of consecutive steps: it increases when steps are larger than expected and decreases if the steps are smaller. The algorithm begins with initialization, followed by population generation and fitness evaluation. After sorting and selecting the best individuals, the covariance matrix and step size are updated, and the process is repeated until a stopping criterion is met.

DE. Differential Evolution is an iterative algorithm that creates new candidate solutions by blending information from randomly selected individuals within the current population [43]. Candidates are then compared to existing solutions, and if they show better performance, they replace them. Its key operations are *mutation*, *crossover* and *selection*.

Mutation (differential variant): At each iteration t, new solutions $\mathbf{v}$ are created by combining randomly selected individuals $\mathbf{x}$ from the population:

$$\mathbf{v}_i^{(t+1)} = \mathbf{x}_{r_1}^{(t)} + F \cdot (\mathbf{x}_{r_2}^{(t)} - \mathbf{x}_{r_3}^{(t)}),$$

where $i = 1, 2, \ldots, N$, and r_1, r_2, and r_3 are distinct random indices. Here, F is the mutation factor that scales the differential solution.

Crossover (exponential variant): Trial solutions $\mathbf{u}_i^{(t+1)} = (u_{i,1}^{(t+1)}, \ldots, u_{i,D}^{(t+1)})$ are created by mixing elements from $\mathbf{v}$ and $\mathbf{x}$. For each element j in the solution:

$$u_{i,j}^{(t+1)} = \begin{cases} v_{i,j}^{(t+1)} & \text{if rand} \leq CR \text{ or } j = j_{\text{rand}}, \\ x_{i,j}^{(t)} & \text{otherwise}, \end{cases}$$

where CR is the crossover constant, rand is a random value between 0 and 1, and j_{rand} is a random index, ensuring at least one element from V is included.

Selection: The better solution between the trial solution and the current solution is retained, ensuring elitism where $f(\cdot)$ is the fitness function:

$$\mathbf{x}_i^{(t+1)} = \begin{cases} \mathbf{u}_i^{(t+1)} & \text{if } f(\mathbf{u}_i^{(t+1)}) < f(\mathbf{x}_i^{(t)}), \\ \mathbf{x}_i^{(t)} & \text{otherwise}, \end{cases}$$

where $f(\cdot)$ is the fitness function.

3.3   Experiment Design

The Python code used in this work is available on Github[1].

Geographical Information Model. To base this study on real-world data, we used digital elevation maps from the OpenDEM project, produced using Copernicus data and information funded by the European Union – EU-DEM layers [31]. Our experiments focused on regions in Slovenia, Austria, and Italy, selected for their diverse topographical features[2]. In Slovenia, we examined the area between Deskle and Ljubljana; in Austria, the section between Lienz and Villach; and in Italy, the path between Terni and L'Aquila. The elevation data for these regions was downloaded in GeoTIFF format from OpenDEM, then normalized and converted into 2D array format for analysis.

Path Planning Comparison. In the first stage, we compared the A* and RRT* algorithms for finding paths across all maps. The goal was to identify a foundational planning algorithm for subsequent path optimization. The comparison of these algorithms is then made in the context of path length (Euclidean distance), time computation and elevation differences along the path. The RRT* algorithm typically works with constraints or collision objects that it must avoid, and in our case, we introduced an altitude constraint set to 800 m to improve computational efficiency by limiting the search space. For our *baseline* comparison we connected the start and goal points with a line segment, which is then optimized in the next phase based on the objective function.

Path Optimization. In the second stage, paths produced by the selected path planning algorithm and the baseline approach are further refined because they initially fail to meet all horizontal alignment constraints—such as a clothoidal shape and minimum turning radius. To address these limitations, optimization algorithms like PSO, CMA-ES, and DE are applied to this constrained optimization problem. The steps of this stage are illustrated in Fig. 1.

Starting with an initial path P_{init}, we first simplify it through downsampling with the Douglas-Peucker method[3], yielding a more manageable path P with n points. The downsampling precision is controlled by the parameter ε. Since both P_{init} and P follow the map grid and are rugged, we further smooth P using splines of clothoid curves, with their smoothness controlled by the parameter τ. The clothoid curves are defined by control points, which can be placed on perpendicular cutting planes along the path P. The length of cutting planes is determined by the Cutting Plane Factor (CPF) parameter.

---

[1] https://github.com/Steigner/HorizAligns-Hybrid-Optimization.

[2] The specific maps were **Slovenia**: N255E460, N255E465; **Austria**: N260E450, N260E455, N260E460; **Italy**: N210E455, N210E460, N215E460, N215E455, N215E450, N210E450.

[3] The Douglas-Peucker method is a widely used algorithm for simplifying curves, polygons or paths by eliminating points that minimally affect the overall shape [8].

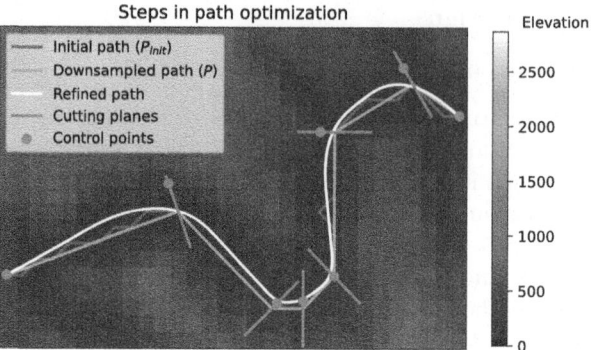

Fig. 1. The path optimization process consists of two steps: 1) The initial path P_{init} is simplified to generate a downsampled path P, which establishes cutting planes— line segments designated for the placement of clothoid control points. 2) Optimization algorithms search for the optimal positioning of these control points along the cutting planes, yielding an optimized path trajectory that satisfies the given constraint.

Optimizing the path trajectory means finding the positions of clothoid control points that minimize the objective function:

$$\sum_{i=1}^{n-1} \|P_{i+1} - P_i\| \cdot (1 + w_g + w_t),$$

where $\|P_{i+1} - P_i\|$ represents the Euclidean distance between consecutive points in the path $P = \{P_1, P_2, \ldots, P_n\}$. The weights w_g and w_t penalize steep terrain gradients, the mitigation of which would demand additional construction costs, and expensive digging of tunnels, respectively:

$$w_g = \begin{cases} 2, & \text{if } G > 8\,\% \text{ or } G < -8\,\% \\ 0, & \text{otherwise} \end{cases}, \quad w_t = \begin{cases} 5, & \text{if } T > 800\,\text{m} \\ 0, & \text{otherwise} \end{cases}.$$

The constants used in these formulations were determined by a domain expert. Additionally, to adhere to safety constraints, the solution requires that the turning radius $R > 100\,\text{m}$ along the entire path. We verify this using the differential curvature method [42].

4   Results

Graph-Based Algorithms. Our investigation revealed a striking hierarchy of efficiency among path-planning approaches, with implications for real-world infrastructure optimization. The comparison between A* and RRT* algorithms proved particularly illuminating: A* demonstrated remarkable computational efficiency, executing approximately 78 times faster than RRT* while simultaneously discovering paths that more naturally conformed to terrain features

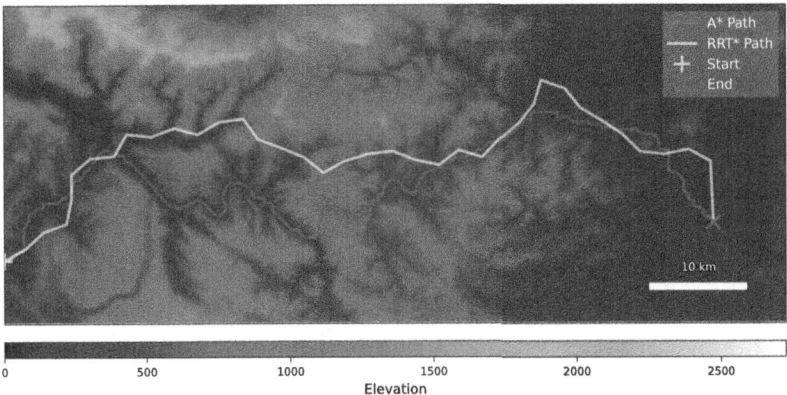

Fig. 2. Comparison of paths generated by A* and RRT* algorithms on the Slovenian map from Deskle to Ljubljana city center, overlaid on elevation data (meters). The A* path (red) demonstrates more direct routing compared to the RRT* path (white). (Color figure online)

(Fig. 2). This dramatic performance difference stems from A*'s ability to leverage terrain-aware heuristics better than RRT*, enabling it to prioritize paths through valleys and avoid unnecessary elevation changes. Furthermore, A* exhibits favorable computational scaling with terrain resolution. Our experiments show that, on average, a 10-fold increase in resolution results in only a 6.2-fold increase in computation time, from 0.8 s to 5 s, demonstrating efficient performance.

Hyperparameter Search. This initial finding guided our hybrid optimization strategy. Rather than pursuing parallel development of both algorithms, we leveraged A*'s superior performance as a foundation for a deeper exploration of path refinement. Using Weights & Biases' Bayesian optimization framework [4], we conducted an extensive hyperparameter search across 128 preliminary runs on the Slovenian map, systematically exploring the parameter space for three distinct optimization algorithms: PSO, CMA-ES, and DE.

To validate whether these parameters, optimized for the Slovenian terrain, would generalize to different geographical contexts, we performed a comprehensive grid search comprising 246 runs across three distinct geographical regions (Slovenia, Austria, and Italy), three algorithms, three random seeds, and multiple parameter combinations (Table 1). Notably, the parameter values that performed well in Slovenia proved robust across all three terrains, suggesting that our optimization approach captures fundamental aspects of the path-planning problem rather than terrain-specific features. The results of this exploration were unequivocal: PSO emerged as the consistently superior approach across all tested scenarios, maintaining its performance advantage even in substantially different geographical contexts.

Table 1. Parameter ranges explored during Bayesian optimization using A*-initialized paths.

Parameter	Sweep Range	Most Promising Values	Selected Values
Cutting Plane Factor	[1, 5]	1, 3, 5	1
ε	[0.0, 1.0]	0.6, 0.8, 1.0	1.0
τ	[0.1, 1.0]	0.2, 0.4, 0.75	0.4
Population Size	[20, 100]	60	60
Generation Size	[5, 50]	50	50

Table 2. Fitness mean for the best CPF and across CPFs for the three stand-alone optimization algorithms on each map. Apart from *, which used CPF 3, all top runs used CPF 1.

Map	Best CFP ($n = 3$)			All CFPs ($n = 9$)		
	PSO	CMA-ES	DE	PSO	CMA-ES	DE
Slovenia	487	490	**482**	501	593	604
Italy	**883**	895	896	984	1012	1121
Austria	1365*	**1190***	1443	1428	1419	1581
Overall	912	**858**	940	971	1008	1102

Optimization on a Straight Corridor. We next created a baseline for the three algorithms PSO, CMA-ES, and DE by optimizing over the straight path from *start* to *goal*. To enable a fair comparison with the hybrid strategy, we used the hyperparameters identified in the previous section, but allowed a re-tuning of the CPF, as the optimal path needed to deviate quite far from the straight line. A grid search over CPF values 1, 3 and 5 nevertheless favoured CPF 1, i.e., the paths tightly following the original line, as can be seen in Table 2. The overall best-performing algorithm was PSO, but the top algorithm differed for each map, showing how the HAO task can have strong sensitivity to the choice of algorithm when it operates on its own. Figure 3 demonstrates this descrepancy between the best paths for each CPF on the Austrian map.

Hybrid Optimization Results. Next, we compared the performance of PSO, CMA-ES and DE when optimizing the A* path. The optimization convergence plots (Fig. 4) reveal several key insights. First, PSO consistently outperformed both CMA-ES and DE across all geographical contexts, achieving better solutions with fewer generations. This superiority manifested not just in final fitness values but also in the reliability of convergence, as evidenced by the tighter interquartile ranges in the PSO trials. Second, the relative difficulty of optimization varied significantly across regions, with the Slovenian terrain proving most amenable to optimization while the Italian landscape presented the greatest challenge. This variation appears to be driven by topographical constraints:

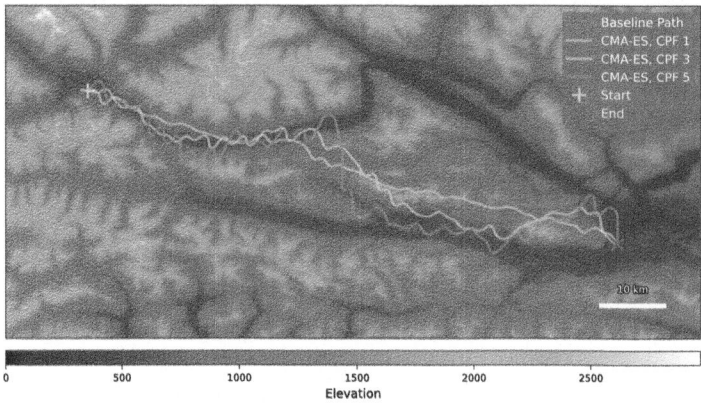

Fig. 3. Comparison of CMA-ES paths optimized with three different CPFs on the Austrian map. As CPF increases, curves become more extreme, leading to lower fitness. For the Austrian map, CPF 3 is just large enough to benefit from the nearby valleys, and outperforms CPF 1.

the Italian test case includes a path where approximately half the optimal route lies in a very narrow corridor of low altitude. In this region, even slight deviations from the optimal path result in substantial fitness penalties, leading to a more challenging optimization landscape. Notably, in these topographically constrained sections, the optimized path closely follows the initial A* solution, suggesting that the initial path-planning phase had already identified near-optimal routing through these challenging areas. Third, all algorithms showed rapid initial improvement followed by diminishing returns, but PSO maintained progress longer than its competitors. When compared to the baseline results, the hybrid optimization consistently outperforms the baseline optimization, as demonstrated in Table 3.

Implications. The practical implications of these findings are visualized in Fig. 5, where we can observe how the optimized paths deviate from their A*-initialized predecessors. The optimized routes demonstrate smoother transitions and better adaptation to terrain features while maintaining feasible construction constraints. This improvement is particularly evident in areas where the original A* path made sharp turns or traversed challenging elevation changes. Since the optimization process adjusts the cutting planes, additional turns can be introduced along the optimized path.

Our results suggest that a two-phase approach—initial path planning with A* followed by PSO-based refinement—represents a robust strategy for real-world infrastructure planning. This combination effectively balances computational efficiency with solution quality, providing a practical framework for addressing complex routing challenges in varied geographical contexts.

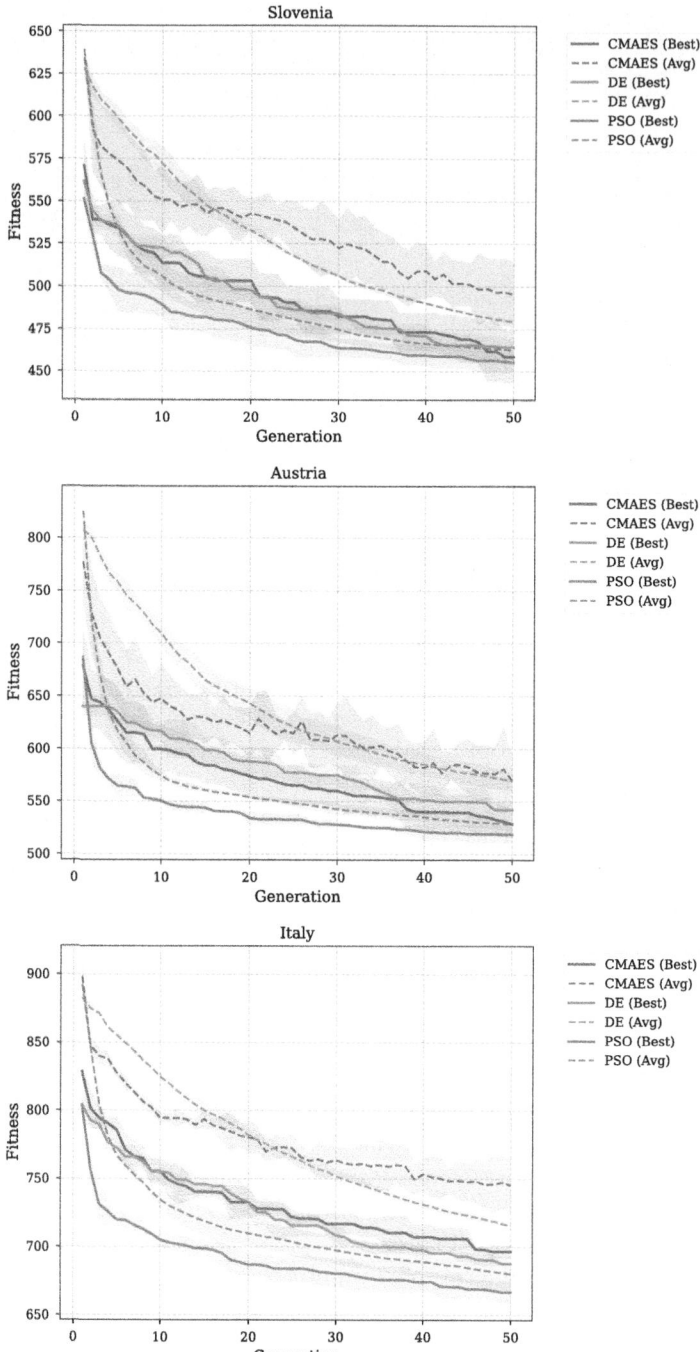

Fig. 4. Fitness progressions for optimization algorithms across different geographical maps. Solid lines represent best fitness values per generation, while dashed lines show average fitness. Shaded areas indicate the inter-quartile range across three independent runs.

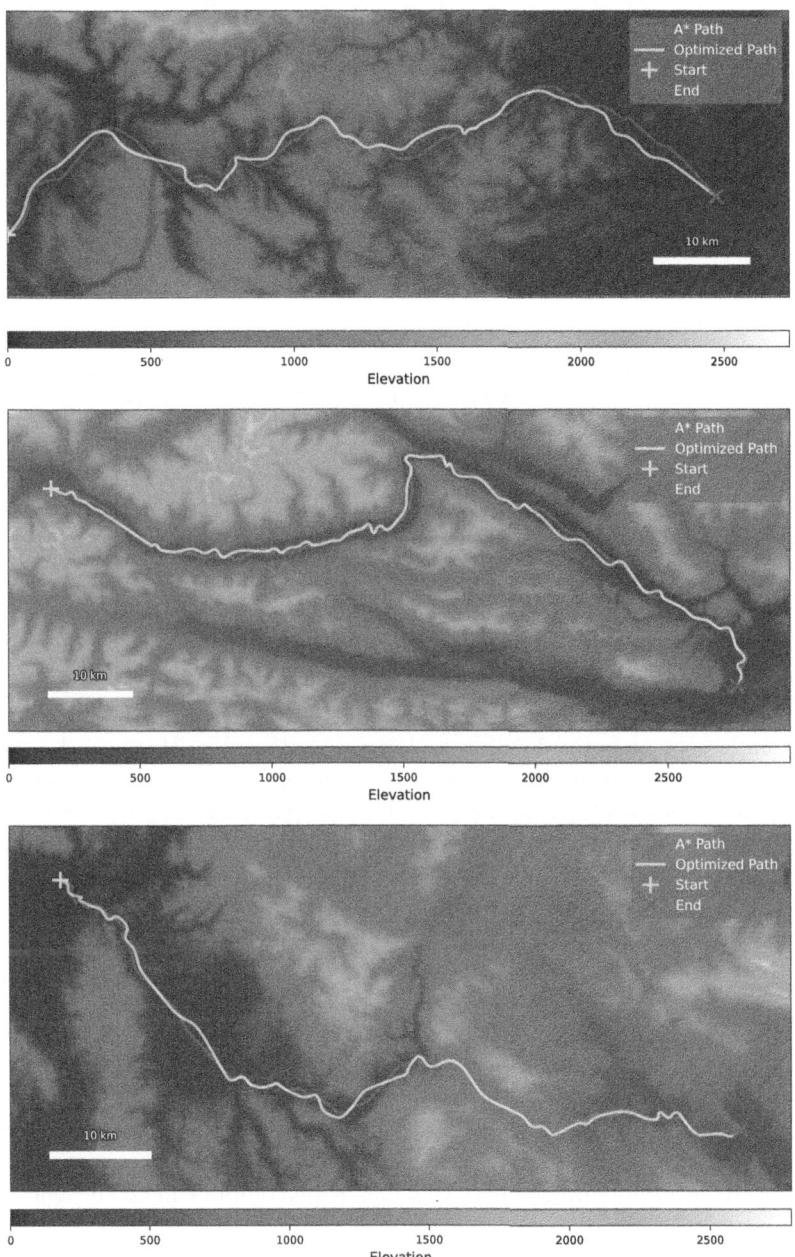

Fig. 5. Comparison of initial A* paths (red) and final PSO optimized paths (white) across three different geographical regions, showing how optimization improves path smoothness and terrain adaptation while maintaining construction constraints. (Color figure online)

Table 3. Comparison of the final mean fitness values from optimization on a A* path versus on the straight corridor. T-tests for the overall scores of the two approaches showed them to be highly significant ($p \lll 0.01$ for distinct maps, $p < 0.05$ overall).

Map	Algorithm	Hybrid	Baseline
Slovenia	PSO	**455** ± 5.9	487 ± 13.0
	CMA-ES	**459** ± 15.3	491 ± 4.8
	DE	**465** ± 0.5	482 ± 7.5
	Overall	**460** ± 9.1	487 ± 8.8
Austria	PSO	**519** ± 8.5	1365 ± 49.7
	CMA-ES	**529** ± 13.6	1190 ± 105
	DE	**542** ± 6.6	1443 ± 6.8
	Overall	**530** ± 13.2	1351 ± 150
Italy	PSO	**667** ± 8.2	884 ± 5.7
	CMA-ES	**697** ± 4.7	895 ± 11.3
	DE	**688** ± 0.8	896 ± 10.5
	Overall	**684** ± 14.0	892 ± 10.2
Overall		**558** ± 96	910 ± 369

5   Conclusions

Our comparison of path-planning and optimization algorithms for horizontal alignment has yielded significant insights for infrastructure planning. The clear superiority of A* over RRT* for initial route planning suggests that graph-based approaches may be more suitable for static environments. Furthermore, our systematic evaluation across diverse European terrains revealed that PSO consistently outperforms both CMA-ES and DE, with parameter settings that proved robust across different geographical contexts.

However, important limitations must be acknowledged. Our study focused on relatively small geographical regions and simplified cost functions compared to real infrastructure projects. The computational requirements of our approach may increase significantly for larger-scale projects or when incorporating more complex terrain features and constructions constraints.

Despite these limitations, our two-phase approach—combining A* for initial path generation with PSO for refinement—offers a promising direction for practical infrastructure planning, effectively balancing computational efficiency with solution quality. Future work could explore the method's scalability to larger geographical areas, incorporate multi-objective optimization for competing priorities such as environmental impact and social factors, and explore more diverse landscapes.

Acknowledgments. Special thanks to the Species Society and the organizers of the Species Summer School 2024 for connecting us thus making this research possible.

Tea Tušar acknowledges financial support from the Slovenian Research and Innovation Agency (research core funding No. P2-0209, and project No. N2-0254 "Constrained Multiobjective Optimization Based on Problem Landscape Analysis"). Martin Juríček was supported by the project GACR No. 24-12474S "Benchmarking derivative-free global optimization methods".

Disclosure of Interests. The authors have no competing interests to declare that are relevant to the content of this article.

References

1. Aguiar, M.O., et al.: Optimizing forest road planning in a sustainable forest management area in the Brazilian Amazon. J. Environ. Manage. **288**, 112332 (2021). https://doi.org/10.1016/j.jenvman.2021.112332
2. Akhmet, A., Hare, W., Lucet, Y.: Bi-objective optimization for road vertical alignment design. Comput. Oper. Res. **143**, 105764 (2022). https://doi.org/10.1016/j.cor.2022.105764
3. Álvaro, V., Ortega, J., Serrano, J.E., Acuña, L.C., Martinez-Santos, J.C.: Path planning for non-playable characters in arcade video games using the wavefront algorithm. In: 2020 IEEE Games, Multimedia, Animation and Multiple Realities Conference (GMAX), pp. 1–5 (2020). https://doi.org/10.1109/GMAX49668.2020.9256835
4. Biewald, L.: Experiment tracking with weights and biases (2020). https://www.wandb.com/, software available from wandb.com
5. Chatzisavvas, A., Dossis, M., Dasygenis, M.: Optimizing mobile robot navigation based on A-star algorithm for obstacle avoidance in smart agriculture. Electronics **13**(11) (2024). https://doi.org/10.3390/electronics13112057
6. Dolgov, D., Thrun, S., Montemerlo, M., Diebel, J.: Path planning for autonomous vehicles in unknown semi-structured environments. Int. J. Robot. Res. **29**(5), 485–501 (2010). https://doi.org/10.1177/0278364909359210
7. Douglas, D.H.: Least-cost path in GIS using an accumulated cost surface and slope-lines. Cartographica: the International Journal for Geographic Information and Geovisualization **31**(3), 37–51 (1994). https://doi.org/10.3138/D327-0323-2JUT-016M
8. Douglas, D.H., Peucker, T.K.: Algorithms for reducing the number of points needed to represent a digitized line or its simplified version. Cartographica **10**(2), 112–122 (1973). https://doi.org/10.3138/FM57-6770-U75U-7727
9. Ducho, F., et al.: Path planning with modified a star algorithm for a mobile robot. Proc. Eng. **96**, 59–69 (2014). https://doi.org/10.1016/j.proeng.2014.12.098
10. Erke, S., Bin, D., Yiming, N., Qi, Z., Liang, X., Dawei, Z.: An improved A-star based path planning algorithm for autonomous land vehicles. Int. J. Adv. Rob. Syst. **17**(5), 1729881420962263 (2020). https://doi.org/10.1177/1729881420962263
11. Flores-Caballero, G., Rodríguez-Molina, A., Aldape-Pérez, M., Villarreal-Cervantes, M.G.: Optimized path-planning in continuous spaces for unmanned aerial vehicles using meta-heuristics. IEEE Access **8**, 176774–176788 (2020). https://doi.org/10.1109/ACCESS.2020.3026666
12. Fournier, D., Martinez, T., Fages, F., Mulard, D.: Metro Energy Optimization through Rescheduling: Mathematical Models and Heuristic Algorithm Compared to MILP and CMA-ES. Ph.D. thesis, Inria Saclay Ile de France (2016)

13. Gad, A.G.: Particle swarm optimization algorithm and its applications: a systematic review. Arch. Comput. Methods Eng. **29**(5), 2531–2561 (2022). https://doi.org/10.1007/s11831-021-09694-4
14. Gao, T., et al.: A deep reinforcement learning approach to mountain railway alignment optimization. Comput.-Aided Civil Infrastruct. Eng. **37**, 73–92 (2021). https://doi.org/10.1111/mice.12694
15. Hansen, N., Ostermeier, A.: Completely derandomized self-adaptation in evolution strategies. Evol. Comput. **9**(2), 159–195 (2001). https://doi.org/10.1162/106365601750190398
16. Hart, P., Nilsson, N., Raphael, B.: A formal basis for the heuristic determination of minimum cost paths. IEEE Trans. Syst. Sci. Cybern. **4**, 100–107 (1968). https://doi.org/10.1109/TSSC.1968.300136
17. Jiang, F., Ma, L., Broyd, T., Li, J., Jia, J., Luo, H.: Systematic framework for sustainable urban road alignment planning. Transp. Res. Part D: Transp. Environ. **120**, 103796 (2023). https://doi.org/10.1016/j.trd.2023.103796
18. Jong, J.C., Jha, M.K., Schonfeld, P.: Preliminary highway design with genetic algorithms and geographic information systems. Comput.-Aided Civil Infrastruct. Eng. **15**(4), 261–271 (2000). https://doi.org/10.1111/0885-9507.00190
19. Jong, J.C., Schonfeld, P.: An evolutionary model for simultaneously optimizing three-dimensional highway alignments. Transport. Res. Part B: Methodol. **37**(2), 107–128 (2003). https://doi.org/10.1016/S0191-2615(01)00047-9
20. Juíček, M., Parák, R., Kdela, J.: Evolutionary computation techniques for path planning problems in industrial robotics: a state-of-the-art review. Computation **11**, 245 (2023). https://doi.org/10.3390/computation11120245
21. Kaleybar, H.J., Davoodi, M., Brenna, M., Zaninelli, D.: Applications of genetic algorithm and its variants in rail vehicle systems: A bibliometric analysis and comprehensive review. IEEE Access, pp. 68972–68993 (2023). https://doi.org/10.1109/ACCESS.2023.3292790
22. Kang, M.W., Schonfeld, P.: Artificial intelligence in highway location and alignment optimization: applications of genetic algorithms in searching, evaluating, and optimizing highway location and alignments. World Sci. (2020). https://doi.org/10.1142/11059
23. Karaman, S., Frazzoli, E.: Sampling-based algorithms for optimal motion planning. Int. J. Robot. Res. **30**(7), 846–894 (2011). https://doi.org/10.1177/0278364911406761
24. Kennedy, J., Eberhart, R.: Particle swarm optimization. In: Proceedings of ICNN'95 – International Conference on Neural Networks. vol. 4, pp. 1942–1948 vol.4 (1995)
25. Khantanapoka, K., Chinnasarn, K.: Pathfinding of 2D & 3D game real-time strategy with depth direction A* algorithm for multi-layer. In: 2009 Eighth International Symposium on Natural Language Processing, pp. 184–188 (2009). https://doi.org/10.1109/SNLP.2009.5340922
26. Kim, E., Jha, M.K., Son, B.: Improving the computational efficiency of highway alignment optimization models through a stepwise genetic algorithms approach. Transport. Res. Part B: Methodol. **39**(4), 339–360 (2005). https://doi.org/10.1016/j.trb.2004.06.001
27. Kwan, C.M., Chang, C.S.: Timetable synchronization of mass rapid transit system using multiobjective evolutionary approach. IEEE Transactions on Systems, Man, and Cybernetics, Part C (Applications and Reviews) **38**(5), 636–648 (2008). https://doi.org/10.1109/TSMCC.2008.923872

28. Liu, Z., et al.: An optimal motion planning method of 7-DOF robotic arm for upper limb movement assistance. In: 2019 IEEE/ASME International Conference on Advanced Intelligent Mechatronics (AIM), pp. 277–282 (2019). https://doi.org/10.1109/AIM.2019.8868594

29. Loganathan, A., Ahmad, N.S.: A systematic review on recent advances in autonomous mobile robot navigation. Eng. Sci. Technol. Int. J. **40**, 101343 (2023). https://doi.org/10.1016/j.jestch.2023.101343

30. Mondal, S., Lucet, Y., Hare, W.: Optimizing horizontal alignment of roads in a specified corridor. Comput. Oper. Res. **64**, 130–138 (2015). https://doi.org/10.1016/j.cor.2015.05.018

31. Open digital elevation model (OpenDEM) the portal for sharing the 3rd dimension. https://www.opendem.info/index.html

32. Pu, H., et al.: Mountain railway alignment optimization using stepwise and hybrid particle swarm optimization incorporating genetic operators. Appl. Soft Comput. (2019). https://doi.org/10.1016/j.asoc.2019.01.051

33. Pu, H., Wan, X., Song, T., Schonfeld, P., Peng, L.: A 3D-RRT-star algorithm for optimizing constrained mountain railway alignments. Eng. Appl. Artif. Intell. **130**, 107770 (2024). https://doi.org/10.1016/j.engappai.2023.107770

34. Pushak, Y., Hare, W., Lucet, Y.: Multiple-path selection for new highway alignments using discrete algorithms. Eur. J. Oper. Res. **248**(2), 415–427 (2016). https://doi.org/10.1016/j.ejor.2015.07.039

35. Rapalski, A., Dudzik, S.: Energy consumption analysis of the selected navigation algorithms for wheeled mobile robots. Energies **16**(3) (2023). https://doi.org/10.3390/en16031532

36. Ravankar, A., Ravankar, A.A., Kobayashi, Y., Hoshino, Y., Peng, C.C.: Path smoothing techniques in robot navigation: state-of-the-art, current and future challenges. Sensors **18**(9), 3170 (2018). https://doi.org/10.3390/s18093170

37. Russell, S., Norvig, P.: Artificial Intelligence: A Modern Approach. Prentice Hall Press, USA, 3rd edn. (2009). https://doi.org/10.5555/1671238

38. Saha, A.K., Arora, M.K., Gupta, R.P., Virdi, M., Csaplovics, E.: GIS-based route planning in landslide-prone areas. Int. J. Geogr. Inf. Sci. **19**(10), 1149–1175 (2005). https://doi.org/10.1080/13658810500105887

39. Smith, M.: Determination of gradient and curvature constrained optimal paths. Comput. Aided Civ. Infrastruct. Eng. **21**, 24–38 (2006). https://doi.org/10.1111/j.1467-8667.2005.00414.x

40. Song, T., et al.: Mountain railway alignment optimization integrating layouts of large-scale auxiliary construction projects. Comput.-Aided Civil Infrastruct. Eng. **38**(4), 433–453 (2023). https://doi.org/10.1111/mice.12839

41. Song, T., Schonfeld, P., Pu, H.: A review of alignment optimization research for roads, railways and rail transit lines. IEEE Trans. Intell. Transp. Syst. **24**, 4738–4757 (2023). https://doi.org/10.1109/TITS.2023.3235685

42. Spivak, M.: A Comprehensive Introduction to Differential Geometry. No. 1 in A Comprehensive Introduction to Differential Geometry, Publish or Perish, Incorporated (1999)

43. Storn, R., Price, K.: Differential evolution - a simple and efficient heuristic for global optimization over continuous spaces. J. Global Optim. **11**(4), 341–359 (1997). https://doi.org/10.1023/A:1008202821328

44. Sushma, M., Maji, A.: A modified motion planning algorithm for horizontal highway alignment development. Comput.-Aided Civil Infrastruct. Eng. **35**(8), 818–831 (2020). https://doi.org/10.1111/mice.12534

45. Sushma, M., Roy, S., Maji, A.: Optimizing points of intersection for highway and railway alignment—Using path planner method and ant algorithm-based approach. In: Recent Advances in Transportation Systems Engineering and Management: Select Proceedings of CTSEM 2021, pp. 127–143. Springer (2022). https://doi.org/10.1007/978-981-19-2273-2_9
46. Trietsch, D.: A family of methods for preliminary highway alignment. Transp. Sci. **21**(1), 17–25 (1987). https://doi.org/10.1287/trsc.21.1.17
47. Turner, A.K.: Route selection. Photogramm. Eng. Remote. Sens. **44**(12), 1561–1576 (1978)
48. Vázquez-Méndez, M.E., Casal, G., Santamarina, D., Castro, A.: A 3D model for optimizing infrastructure costs in road design. Comput.-Aided Civil Infrastruct. Eng. **33**(5), 423–439 (2018). https://doi.org/10.1111/mice.12350
49. Yang, D., He, Q., Yi, S.: Underground metro interstation horizontal-alignment optimization with an augmented rapidly exploring random-tree connect algorithm. J. Transport. Eng., Part A: Syst. **146**(11), 04020129 (2020). https://doi.org/10.1061/JTEPBS.0000454
50. Yang, D., He, Q., Yi, S.: Bilevel optimization of intercity railway alignment. Transp. Res. Rec. **2675**(11), 985–1002 (2021). https://doi.org/10.1177/03611981211023756
51. Yao, J., Lin, C., Xie, X., Wang, A.J., Hung, C.C.: Path planning for virtual human motion using improved A* star algorithm. In: 2010 Seventh International Conference on Information Technology: New Generations, pp. 1154–1158 (2010). https://doi.org/10.1109/ITNG.2010.53
52. Zhang, T., et al.: 3D constrained hybrid A*: Improved vehicle path planning algorithm for cost-effective road alignment design. Autom. Constr. **166**, 105645 (2024). https://doi.org/10.1016/j.autcon.2024.105645
53. Zhong, J., Li, D., Cai, W., Chen, W.N., Shi, Y.: Automatic crowd navigation path planning in public scenes through multiobjective differential evolution. IEEE Trans. Comput. Social Syst. **11**(1), 905–918 (2022). https://doi.org/10.1109/TCSS.2022.3217417

Facial Geometric Feature Extraction for Dimensional Emotion Analysis Using Genetic Programming

Wenlong Fu, Qi Chen$^{(\boxtimes)}$, Bing Xue, and Mengjie Zhang

The Center for Data Center and AI and School of Engineering and Computer
Science, Victoria University of Wellington, Wellington 6140, New Zealand
Qi.Chen@ecs.vuw.ac.nz

Abstract. Geometric features derived from single static images have the
potential to be highly effective for facial emotion analysis, as shape, structure, and spatial relationships are key factors. However, these aspects
are rarely explored in existing research. In this paper, we propose a
novel approach that utilizes Genetic Programming (GP) to automatically extract geometric features for more effective emotional representation. The proposed GP system uses various evaluation strategies, evolving either a single feature per run or multiple features within a single
run. These GP-evolved features capture critical angular and distance-
based relationships between facial landmarks, which are then integrated
with an existing deep learning model to enhance performance. The
results show that the proposed method achieves improved performance
in dimensional emotion analysis, providing a more comprehensive understanding of emotional expressions in static images. In addition, our approach is effective in improving the accuracy of emotion predictions, establishing a foundation for more precise facial emotion analysis using geometric information.

Keywords: Genetic Programming · Dimensional Emotion · Feature
Extraction

1 Introduction

Dimensional emotion analysis has garnered increasing attention and plays an
important role in affective computing, where emotions are represented as points
in a continuous multi-dimensional space, typically characterised by dimensions such as valence and arousal [21,29]. Unlike categorical approaches that
group emotions into distinct categories, e.g., happiness, disgust, and fear [25],
dimensional emotion models offer a more refined representation of emotional
states [21,29]. It makes them particularly well-suited for applications requiring the capture of subtle emotional changes, such as human-computer interaction and mental health monitoring [1].

Most existing studies on representation learning have focused on extracting features for categorical emotion recognition [25]. However, there has been

© The Author(s), under exclusive license to Springer Nature Switzerland AG 2025
P. García-Sánchez et al. (Eds.): EvoApplications 2025, LNCS 15612, pp. 137–153, 2025.
https://doi.org/10.1007/978-3-031-90062-4_9

limited research specifically addressing the extraction of geometric facial features for dimensional emotion analysis, particularly from static images. While geometric facial features have proven effective for facial emotion classification, their significance for dimensional emotion analysis from a single image is still unclear. This challenge is further compounded by the significant imbalance in dimensional emotion datasets, leaving geometric feature extraction for dimensional emotions from static images a relatively under-explored area.

Genetic programming (GP) has demonstrated significant potential in learning facial features for emotion classification across various domains, owing to its capacity to evolve complex, human-readable models [6]. GP has the ability to automatically discover features that are optimal for a given task, making it a promising method for feature extraction in dimensional emotion analysis. Despite this promise, very few studies have investigated its potential in automatically extracting geometric features from facial images for dimensional emotion analysis.

The AffectNet dataset, one of the largest datasets for facial expression analysis, provides annotations for both categorical labels e.g., happy, sad and angry and dimensional labels including valence and arousal [21]. AffectNet has been widely used for developing and evaluating emotion recognition models due to its diversity and rich annotation. However, the inherent class imbalance poses significant challenges [29], particularly for methods like GP, which typically require balanced data distributions for effective learning.

Extracting geometric features from single static facial images involves analysing the shape and position of key facial landmarks like the eyes, eyebrows, nose, and mouth. By measuring their distances, angles, and relationships, we can capture unique facial patterns linked to different emotions. This method offers valuable insights for emotion analysis, but it has been less explored compared with other facial recognition techniques. This paper aims to propose a GP approach to extracting geometric features from single static facial images for dimensional emotion analysis on the imbalanced AffectNet dataset. The proposed GP approach automatically evolves multiple facial features which are well-suited for valence and arousal predictions. To identify effective features, GP is employed with various fitness functions, allowing the extraction of features tailored to different objectives. Each fitness function guides the GP to select features that are most relevant for specific emotional dimensions, such as valence and arousal. These extracted features are then used as inputs for a neural network forecasting valence and arousal.

To start with, GP evolves a single feature with a single fitness function, generating a set of features from different fitness functions across multiple runs, referred to as *SFGP* (single features from multiple runs). This method operates as a sequential approach, where each run focuses on a unique fitness function to evolve a single feature at a time. Over multiple runs, this approach accumulates a diverse set of features, each optimized for a specific objective. *Then,* multiple objective functions are applied concurrently within a single run to evaluate and select the most optimal GP individuals, resulting in the *BFsGP* (the best features from different objective functions in a single run). This approach gathers the best features identified by each objective function, integrating them into a single solution. *Moreover*, a non-dominated sorting strategy on multiple objective functions is used to select Pareto-optimal solutions, which represent the extracted features in the *MOGP* (non-dominated sorting Multi-Objective GP) method. The specific objectives of this paper are as follows:

- whether the proposed three GP methods can evolve/learn features to enable neural networks to outperform the existing models in predicting valence and arousal;
- whether batch feature generation methods, *BFsGP* or *MOGP*, that produce a set of features in a single run lead to better performance compared to the sequential feature generation approach *SFGP* which produces multiple features over multiple iterations; and
- whether *MOGP*, a multiobjective method that generates features from the non-dominated individuals on the Pareto front outperfoms *BFsGP*, which uses various fitness functions to construct features, for valence and arousal prediction.

2   Background

2.1   Facial Feature Extraction

Facial emotion analysis is a crucial area of research in affective computing and computer vision, with applications ranging from human-computer interaction to psychological studies [1,7,19]. The process involves capturing and interpreting facial expressions to identify underlying emotions. Facial expressions are rich sources of information, offering critical insights into a person's emotional state.

Emotion analysis primarily relies on two main types of features: appearance-based features and geometric features, which involve manually extracting texture information or geometric data from detected facial landmarks [7,19]. Appearance-based features focus on the texture and shading information across the face. Facial emotion features have been extracted by traditional methods, such as the Scale-Invariant Features Transformation (SIFT), Histogram of Oriented Gradients (HoG), and Local Binary Patterns (LBP) [6,8,14]. On the other hand, geometric features involve identifying key facial landmarks such as the corners of the eyes, mouth, and eyebrows [8,18,23]. These landmarks serve as reference points for calculating the distances and angles between specific points, forming a set of features that describe facial expressions in a high-dimensional space. In addition, action units, which are moved based on key facial landmarks, have been used for extracting facial features in image sequences [7,9,15].

Deep learning techniques have been successfully applied to facial emotion analysis as well [24,25,29]. Neural networks can automatically extract facial features from static images and image sequences. Convolutional Neural Networks (CNNs) are widely used in facial emotion analysis due to their outstanding performance in image-based tasks. CNNs automatically learn spatial hierarchies of features from images, making them suitable for analyzing facial expressions.

Notable CNN-based architectures, such as VGGNet [26] and ResNet [17], have been successfully adapted for facial emotion recognition [29]. Since facial expressions are dynamic, effective emotion recognition significantly hinges on capturing temporal dependencies. Recurrent Neural Networks (RNNs) and Long Short-Term Memory (LSTM) networks are well-suited for this task [25]. They process sequences of frames to capture the temporal evolution of facial expressions. Recent approaches have combined CNNs for spatial feature extraction with RNNs or LSTMs to capture temporal patterns in videos, improving performance in emotion recognition from video data [32]. Multimodal approaches for facial emotion recognition combine multiple data sources, such as facial expressions, voice, body gestures, and physiological signals, to improve the accuracy of emotion detection [13]. The transformer architecture, originally proposed for natural language processing, has recently shown promise in vision tasks, including facial emotion analysis. Vision Transformers (ViTs) [12] capture long-range dependencies between different face regions, capturing more global information than CNNs.

Facial emotion recognition approaches typically employ categorical models that classify emotions into discrete categories such as happiness and sadness. However, dimensional emotion models provide a more nuanced representation by placing emotions on continuous dimensions. The two most widely used dimensions are *valence*, ranging from positive to negative, and *arousal*, ranging from calm to excited [9,13]. This model allows emotions to be represented as points in a two-dimensional space. For example, joy is characterized by high valence and high arousal, whereas sadness has low valence and low arousal. These models offer a more flexible and scalable approach to emotion representation, especially when dealing with complex or mixed emotional states. For static image recognition, the pre-trained models from EfficientNetv2 [27] and MaxVit-Tiny [28] were used for analysing facial emotion [29] to obtain the start-of-the-art predictions on valence and arousal.

2.2 Genetic Programming for Facial Feature Extraction

For emotion classification, extensive research has been conducted on various aspects, such as feature selection [14], feature extraction [6] and evolving classifiers [2]. GP has been employed to select features from a predefined texture attributes and geometric characteristics to construct binary classifiers [14]. Additionally, GP has been used to extract texture information [3] and geometric characteristics [16]. Also, features extracted by Principal Component Analysis (PCA) were selected to evolve GP classifiers [2]. GP composite operators were proposed to construct Gabor filter-based features for used in Bayesian classifiers [3] as well. Furthermore, a set of texture features (such as HoG and LBP) are fused by GP for expression classification [30]. Via using two convolution operators to extract image features, GP evolved trees including base learners with their hyperparameters, such as K-nearest neighbor (KNN) with the number of neighbors [33]. Texture information were automatically extracted based on selected region [4–6]. All above studies leverage GP for facial expression classification

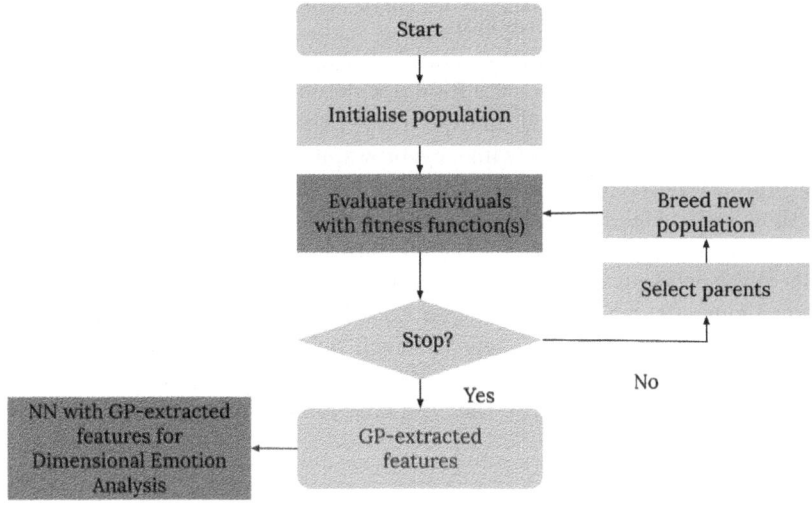

Fig. 1. The overall framework

tasks. However, there is a lack of its use in continuous analysis on arousal and valences.

In the realm of dimensional emotion analysis, action units on key points were proposed for evolving GP regression formulae to express the degrees of happiness, anger, and fear [20]. This is likely the first investigation into dimensional emotion analysis using GP, albeit with a limited dataset and features extracted from sequence images rather than static images.

While predefined geometric features from key points on static images have been selected to construct GP classifiers, there is still a need to further explore potential geometric features for dimensional emotion analysis, especially on large and unbalanced datasets.

3   The Proposed Approach

This section describes the proposed GP methods for extracting geometric features for dimensional emotion analysis. The overall framework of the system is shown in Fig. 1. The framework generally follows the standard evolutionary process of GP, starting with the initialisation of the GP population. This population then undergoes an iterative process of evaluating GP individuals, selecting parents, and breeding a new population, until some predefined stopping criterion is met. In this work, GP is employed to extract features with a pretrained neural network (NN) model used for dimensional emotion analysis. During evaluation, GP individuals, which represent the extracted features, are evaluated using a single fitness function or multiple different fitness functions. The GP extracted features are integrated with the NN model to analyse valance and arousal.

3.1   GP for Feature Extraction

The Terminal Set: To extract geometric features from static images, this work leverages geometric characteristics from a widely used set of 68 facial landmarks in computer vision and facial analysis [25]. These landmarks outline essential facial features, including the jawline, eyebrows, nose, eyes, and mouth, capturing the overall geometric structure of the face. By enabling precise feature extraction, they are crucial for applications such as facial expression analysis, emotion detection, and face alignment. For the GP terminals, this work utilises distances, specifically Euclidean and absolute distances and angles calculated among sets of two to four of these key points. The Euclidean distances ed and the absolute distance ad are in Eq. (1) through (4). Each point p_i is represented by its horizontal location $p_{i,x}$ and vertical location $p_{i,y}$. Angles θ are defined in Eqs. (7) and (8). It is important to note that the coordinates are normalized within the range of 0 to 1.

$$ed(p_1, p_2) = \sqrt{(p_{1,x} - p_{2,x})^2 + (p_{1,y} - p_{2,y})^2} \tag{1}$$

$$ed(p_1, p_2, p_3, p_4) = ed\left(\frac{p_1 + p_2}{2}, \frac{p_3 + p_4}{2}\right) \tag{2}$$

$$ad_x(p_1, p_2) = |p_{1,x} - p_{2,x}| \tag{3}$$

$$ad_y(p_1, p_2) = |p_{1,y} - p_{2,y}| \tag{4}$$

$$cc(p_1, p_2, p_3) = (p_{2,x} - p_{1,x})(p_{3,x} - p_{2,x}) + (p_{2,y} - p_{1,y})(p_{3,y} - p_{2,y}) \tag{5}$$

$$cc(p_1, p_2, p_3, p_4) = (p_{2,x} - p_{1,x})(p_{4,x} - p_{3,x}) + (p_{2,y} - p_{1,y})(p_{4,y} - p_{3,y}) \tag{6}$$

$$\theta(p_1, p_2, p_3) = \arccos \frac{cc(p_1, p_2, p_3)}{ed(p_1, p_2)ed(p_2, p_3)} \tag{7}$$

$$\theta(p_1, p_2, p_3, p_4) = \arccos \frac{cc(p_1, p_2, p_3, p_4)}{ed(p_1, p_2)ed(p_3, p_4)} \tag{8}$$

In summary, there are $\binom{68}{2} = 2,278$ unique pairs (p_1, p_2), $\binom{68}{3} = 50,116$ unique triplets (p_1, p_2, p_3), and $\binom{68}{4} = 814,385$ unique quadruplets (p_1, p_2, p_3, p_4), derived from the 68 landmarks. Each unique pair contributes three types of features: the Euclidean distance in Eq. (1), the horizontal absolute distance in Eq. (3), and the vertical absolute distance in Eq. (4). Each unique triplet provides one type of feature, i.e., angle in Eq. (7), while each unique quadruplet adds two types of features, Euclidean distance between midpoints in Eq. (2), angle in Eq. (8). Altogether, this results in a total of $2278 \times 3 + 50,116 \times 1 + 814,385 \times 2 = 1,685,720$ features in the terminal set.

The Function Set: The function set in the proposed GP system comprises standard operators $\{+, -, \times, \div, log, sin, cos, exp,, max, min\}$. The division operator $\div$ is protected, returning 1 when divided by 0 occurs. Similarly, log is protected, returning 0 when the input is less than a tolerance level of 10^{-12}. The exponential operator exp is capped, returning e^{30} for inputs greater than 30. The max and min operators take two arguments to compare values.

The Set of Objective Functions: In our GP for feature learning approaches, individuals are evaluated based on their outputs (the outputs of each GP program on single static images) in relation to valance or arousal. Each GP program represents one individual feature. The GP approaches consider two objectives: the Mean of Squared Error (MSE) and the Concordance Correlation Coefficient (CCC) as referenced in [29,31] and shown in Eqs. 9 and 10, respectively.

$$MSE = \frac{1}{n} \sum_{i=1}^{n} (y_i - \hat{y}_i)^2 \tag{9}$$

where n is the number of observations, y_i is the actual value, and $\hat{y}_i$ is the predicted value.

$$CCC = \frac{2 \, \text{cov}(y, \hat{y})}{\sigma_y^2 + \sigma_{\hat{y}}^2 + (\mu_y - \mu_{\hat{y}})^2} \tag{10}$$

where $\text{cov}(y, \hat{y})$ is the covariance between the actual values y and the predicted values $\hat{y}$, σ_y^2 and $\sigma_{\hat{y}}^2$ are the variances of y and $\hat{y}$, respectively, and μ_y and $\mu_{\hat{y}}$ are the means of y and $\hat{y}$, respectively.

These two objectives capture different but complementary aspects of model performance. Specifically, MSE measures the average squared difference between predicted and true values, focusing on minimising overall prediction error. In contrast, CCC assesses the agreement between predicted and true values, accounting for both precision and accuracy by considering the correlation and bias between the two sets.

In addition, MSE and CCC can sometimes conflict. For instance, a model with a low MSE might have a low CCC if there is a systematic bias or scaling issue between predicted and actual values [10]. Conversely, a model with a high CCC might not necessarily have a lower MSE if it sacrifices some error to achieve better agreement. Combining MSE and CCC into a single objective function, fit_{com}, using a weight parameter α (as defined in Eq. (11)), provides a more comprehensive evaluation of model performance which ensures that the model minimizes prediction errors while also maintaining strong agreement with the true values, thereby addressing potential shortcomings of using either metric alone. The weight parameter α is determined through experimental investigation and is set to 3. This choice of α indicates a greater emphasis on minimising prediction error.

$$fit_{com} = \alpha * MSE - CCC \tag{11}$$

Including both the individual objectives, MSE and CCC, and the combined function fit_{com} allows for a more nuanced search in the solution space. Therefore, the proposed GP algorithms evaluate six objective functions, applying MSE, CCC, and fit_{com} separately to both arousal and valence, to optimize performance across both dimensions effectively.

Algorithm 1. Single Feature GP (*SFGP*)

Input: GP Settings including generation N_{gen}
Output: Multiple GP-Extracted Features f_{gp}
1: Initialise the learnt feature set $f_{gp} = \{\}$
2: **for each** $fit \in$ the fitness function set **do**
3:     Initialise GP population
4:     **for each** $g \in N_{gen}$ **do**
5:         Evaluate and Select parents with the fitness function fit
6:         Create new population using crossover, mutation and reproduction
7:         Replace the current population by the new population
8:     **end for**
9:     Select the best individual into the GP feature set f_{gp}
10: **end for**
11: **return** f_{gp}

Algorithm 2. Best Features GP (*BFsGP*)

Input: GP settings including generation N_{gen}
Output: Multiple GP-Extracted Features f_{gp}
1: Initialise GP population and learnt feature set $f_{gp} = \{\}$
2: **for each** $g \in N_{gen}$ **do**
3:     Evaluate and Select parents with a *random fitness function*
4:     Create new population using crossover, mutation and reproduction
5:     Collect the best individual based on each objective function to form f_{gp}
6:     Add f_{gp} into the new population
7:     Replace the current population by the new population
8: **end for**
9: **return** f_{gp}

The Three GP Methods: Three GP algorithms, *SFGP*, *BFsGP*, and *MOGP* are proposed in this work. Algorithm 1 outlines the sequential feature generation approach, *SFGP*, which constructs features by evolving them with different objective functions. In each iteration, one of the six objective functions is selected to evolve a new GP feature, ensuring diversity and coverage across different performance criteria.

To reduce the computational cost, a batch feature generation method, *BFsGP*, is developed as shown in Algorithm 2. Different from *SFGP*, a single execution of *BFsGP* generates a diverse set of GP features. During the tournament selection process in *BFsGP*, a randomly chosen objective function is used to compare individuals ensuring varied selection pressure. At the end of the evolutionary process, the constructed GP features are evaluated and selected based on each of the specified objective functions. In both *SFGP* and *BFsGP*, each single objective function is used to select the best GP program. As a result, the final number of GP features is the same as the number of objective functions.

Furthermore, a multi-objective GP method, *MOGP*, is proposed in Algorithm 3 motivated by the inherent trade-offs in optimising competing goals of

---

Algorithm 3. Multi-Objective GP (*MOGP*) f_{gp}

Input: GP settings including generation N_{gen}
Output: Multiple GP-Extracted Features f_{gp}
1: Initialise GP population
2: **for each** $g \in N_{gen}$ **do**
3:     Evaluate and Rank the population with the *six fitness functions*
4:     Create new population using crossover, mutation and reproduction
5:     Replace the current population with the new population
6: **end for**
7: Choose the solutions in the front as the GP feature set f_{gp}.
8: **return** f_{gp}

---

minimising MSE and maximising agreement metrics CCC. $MOGP$ ranks the individuals using the Non-dominated Sorting strategy [11]. The six objective functions—namely, MSE, CCC, and fit_{com} for both arousal and valence—act as multiple objectives, allowing $MOGP$ to simultaneously evaluate diverse aspects of feature performance. At the end of the evolution, $MOGP$ outputs a feature set that lies on the Pareto front, representing non-dominated solutions that cannot be improved in one objective without sacrificing performance in another. Additionally, unlike the fixed number of features produced by $BFsGP$, $MOGP$ provides a diverse feature set with varying numbers, offering greater flexibility and adaptability for different modeling tasks.

The extracted geometric features may be less effective for dimensional emotion analysis when derived from single images alone. Therefore, to address this potential limitation, and inspired by the success of CNN models [29], the outputs of arousal and valence, o_{aro} and o_{val}, from the trained neural network models are combined with the GP-evolved geometric features to train the final model. Once GP extracts geometric features, the selected features are combined with o_{aro} and o_{val} to create a new training dataset. This dataset is then used to train a neural network model comprising two linear layers, with the loss function defined by Eq. (11). The outputs of this neural network model provide predictions for both arousal and valance.

4   Design of the Experiments

4.1   Datasets

The AffectNet dataset is one of the largest available datasets for facial expression analysis, including around 0.4 million images with manual annotations for both categorical and dimensional emotions [21]. AffectNet labels emotions using both discrete categories, e.g., fear and surprise, and continuous dimensions, e.g., valence and arousal. This can be shown in a 2D Cartesian coordinate system where the x-axis and y-axis represent the valence and arousal, respectively, as some example images and the related arousal and valence shown in Fig. 2. The annotations are within the circumplex of a range of $[-1, 1]$.

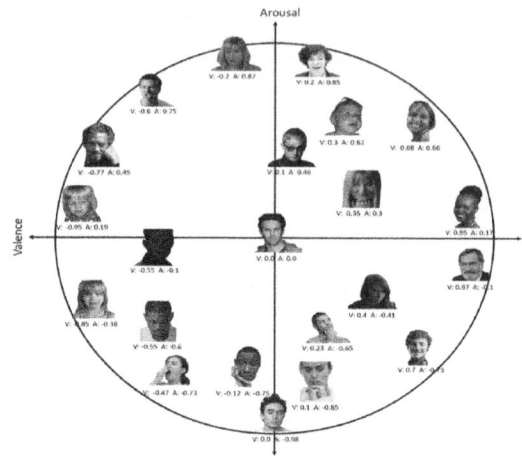

Fig. 2. Examples of images in AffectNet with labels.

The expanded AffectNet-8 dataset builds on AffectNet-7 by introducing a new category. This paper focuses on AffectNet-8 as it contains the whole set of AffectNet-7 data while presenting a more complex challenge. AffectNet-8 includes around 0.4 million pre-defined training images. Specifically, it provides dimensional emotion annotations alongside geometric landmarks, which facilitate the extraction of geometric features, such as distances, angles, and areas between key facial points. These features are important for predicting emotional dimensions but they pose challenges due to the imbalance in the dataset [29], where certain valence and arousal values are underrepresented. The skewed distribution makes it difficult for machine learning models to generalize effectively [21,29]. The independent training images are utilized as the training dataset, while the remaining images are reserved exclusively for testing. To address the imbalance issue, the sampling technique proposed in [29] is applied during training. More details on the AffectNet dataset can be found in [21,29].

4.2   Parameter Settings

Each GP method is executed for 30 independent runs with a population size of 500 over 200 generations. The crossover probability was set to 0.65, the mutation probability was set to 0.3, and the elitism (reproduction) to 0.05. For the pretrained NN model, EfficientNetv2s [27] was selected due to its comparable performance to MaxViT [28] but with fewer parameters, making it more parameter-efficient for the scope of this study.

5   Results

This section presents the results of the GP system, along with visual examples, followed by discussions and analysis of the findings.

Table 1. The Overall Test Performance on AffectNet-8

	MSE	MAE	RMSE	CCC
MaxViT	0.1021	0.2351	0.3196	0.7840
EfficientNetv2s	0.1028	0.2387	0.3206	0.7816
$SFGP$	**0.0972 ± 0.0008**	0.2285 ± 0.0015	**0.3118 ± 0.0012**	0.7898 ± 0.0021
$BFsGP$	0.0973 ± 0.0005	**0.2283 ± 0.0010**	0.3120 ± 0.0008	**0.7899 ± 0.0015**
$MOGP$	0.0977 ± 0.0005	**0.2283 ± 0.0013**	0.3125 ± 0.0008	0.7881 ± 0.0017

Table 2. Results on the Statistic Significant Test on AffectNet-8 with the four metrics

	MaxViT	EfficientNetv2s	$SFGP$	$BFsGP$	$MOGP$
MaxViT		↑ ↑ ↑ ↑	↓ ↓ ↓ ↓	↓ ↓ ↓ ↓	↓ ↓ ↓ ↓
EfficientNetv2s	↓ ↓ ↓ ↓		↓ ↓ ↓ ↓	↓ ↓ ↓ ↓	↓ ↓ ↓ ↓
$SFGP$	↑ ↑ ↑ ↑	↑ ↑ ↑ ↑		= = = =	= ↑ = ↑ ↑
$BFsGP$	↑ ↑ ↑ ↑	↑ ↑ ↑ ↑	= = = =		↑ = ↑ ↑
$MOGP$	↑ ↑ ↑ ↑	↑ ↑ ↑ ↑	↓ = ↓ ↓	↓ = ↓ ↓	

Each cell presents the comparison results in terms of the following order: MSE, MAE, $RMSE$, and CCC. "↓" indicates the results in the corresponding row position are significantly worse than the results from the relevant column, while "↑" means significantly better. "=" denotes no significant difference.

5.1 Overall Test Performance

Table 1 presents the overall test performances (combining arousal and valance) on the AffectNet 8 dataset. Performance metrics include MSE, the mean absolute error (MAE), the root mean square error (RMSE) and CCC, which are the same as [29]. The results for MaxVit and EffcientNet2s are sourced directly from [29]. The best test results for each metric are highlighted in bold. Table 2 provides the compared results using the Tukey multiple comparison tests, based on analysis of variance (ANOVA) [22]. It is noted that the results from Table 1 exhibit characteristics of normal distributions.

There are three interested key observations from the comparisons of the overall test performance. First, the three GP methods have significantly better results than MaxVit and EfficeintNetv2s as shown on the comparisons in Table 2. This suggests that the geometric features extracted using GP significantly enhance the prediction performance metrics, including all the four metrics of MSE, MAE, RMSE and CCC, for dimensional emotion metrics including arousal and valence. Second, $SFGP$ and $BFsGP$ have the best results across the comparisons, particularly in terms of MSE, RMSE and CCC. Notably, there are no significant differences between the results from $SFGP$ and $BFsGP$. This highlights the effectiveness of the proposed $BFsGP$ in evolving diverse features from multiple objective functions in a single run, as opposed to relying on separate runs with a single-objective fitness function. Third, the results from $MOGP$ are not

148   W. Fu et al.

Table 3. Test Performances of Valence and Arousal. The results of SVR and CNN (AlexNet) are from [21], and $MaxVit_{comb}$ from [29].

	Arousal		Valance	
	CCC	RMSE	CCC	RMSE
$BFsGP\ (CCC)$	**0.645 ± 0.001**	0.309 ± 0.003	**0.724 ± 0.001**	0.343 ± 0.004
$BFsGP\ (fit_{com})$	0.638 ± 0.002	**0.300 ± 0.001**	0.716 ± 0.002	**0.328 ± 0.001**
$MOGP\ (CCC)$	0.639 ± 0.001	0.302 ± 0.001	0.723 ± 0.001	0.342 ± 0.005
$MOGP\ (fit_{com})$	0.643 ± 0.001	0.310 ± 0.004	0.717 ± 0.002	0.330 ± 0.001
SVR	0.199	0.400	0.340	0.494
CNN (AlexNet)	0.450	0.402	0.541	0.394
$MaxVit_{comb}$	0.642	0.305	0.716	0.331

obliviously different from the results from $SFGP$ and $BFsGP$ on MAE but significantly worse on MSE, RMSE and CCC. It is noted that all results from the three GP methods are stable and the standard deviations of MSE, MAE, RMSE and CCC are very small. The number of the features evolved by $MOGP$ is 105.03 ± 25.68. While $SFGP$ and $BFsGP$ only have the predefined six features extracted by GP. Future work is required to refine the extraction of GP features from Pareto front solutions more effectively.

5.2 Detailed Results on Arousal and Valance

Table 3 presents the test performances on arousal and valance for $BFsGP$ (using CCC and fit_{com} to select the best model, following [29]), $MOGP$, Support Vector Regression (SVR) [21], a CNN model (AlexNet) [21], and $MaxVit_{comb}$ which refers to the best results on arousal and valance from [29]. Note that $MaxVit_{comb}$ combined CCC and MSE as the loss function, which is different from the loss function in $MaxVit$ [29].

As shown in Table 3, $BFsGP$ has the best test performance on arousal and valance, in terms of CCC. However, the RMSE values of $BFsGP$ are higher (worse) than the results from $MaxVit_{comb}$. $MOGP$ has worse performance than $MaxVit_{comb}$ in terms of the average of RMSE on arousal, although $MOGP$ is slightly better than $MaxVit_{comb}$ when CCC is used. Besides, $MOGP$ has slightly better results than $MaxVit_{comb}$ on valance when fit_{com} is used. Given that AffectNet-8 dataset is highly unbalanced, effective model selection criteria for CCC and RMSE, especially in arousal prediction, remain an open challenge.

5.3 Analysis on the Example of Learnt Features

Due to the complexity of the full tree of a GP-evolved geometric feature, only a sub-tree example is provided in Fig 3. In the sub-tree, c_1 denotes a constant coefficient, while θ_i (where $i = 1, 2, 3, 6, 7, 8, 9$) correspond to the angles formed by three specific landmark points. Additionally, $\theta_{1,0}$ represents an angle formed

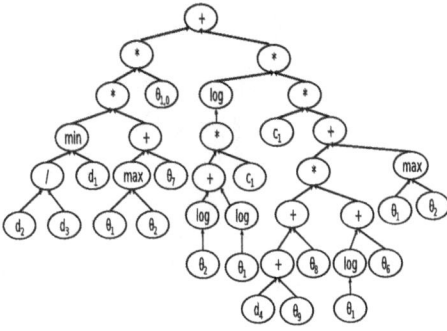

Fig. 3. Example of a sub-tree from a GP evolved feature.

using four landmark points. The terms d_i (where $i = 2, 3, 4$) refer to distances between pairs of landmark points. As observed, the sub-tree highlights several angle attributes, suggesting that geometric angle features may encode information pertinent to dimensional emotion analysis.

To visually demonstrate these GP-evolved geometric features, Fig. 4 (a)-(d) illustrate examples of distances and angles constructed from selected facial landmark points. Figure 4 (a) and (c) show how variations in mouth movement, such as the contrast between open and closed mouth positions, correspond to emotional dimensions across different facial expressions. For instance, the distance between two points on the mouth in Fig. 4 (a), representing the expression of anger, is longer than the corresponding distance in Fig. 4 (c), which shows the expression of disgust. Figure 4 (b) and (d) mainly focus on the angles constructed from mouth, eyes, and eyebrows. Similar to the use of action units in facial expression analysis, the GP-evolved sub-tree extracts key geometric features from regions associated with expressive cues, specifically, the mouth, eyes, and eyebrows. These features capture essential movements and shapes that align closely with human-defined action units, facilitating a structured analysis of facial expressions through precise, interpretable metrics. Unlike action units, however, these key points are extracted by GP from static images, which may also reflect variations in dimensional emotion.

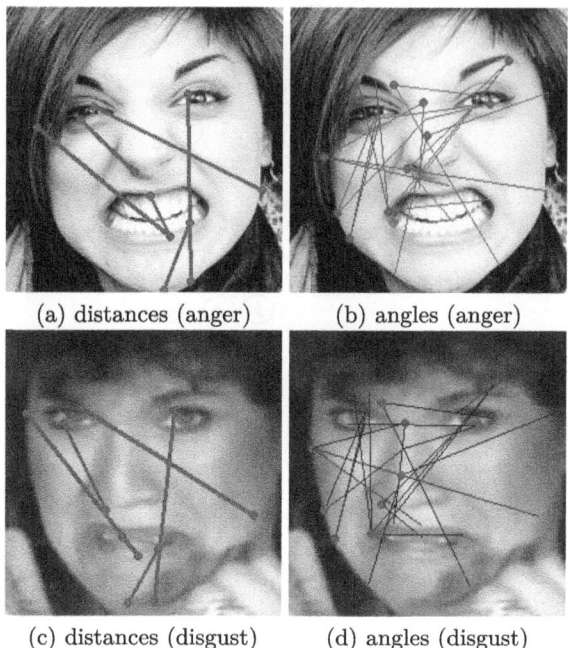

(a) distances (anger)   (b) angles (anger)

(c) distances (disgust)   (d) angles (disgust)

Fig. 4. Examples of geometry distances and angles extracted from selected points.

6   Conclusions

This paper introduced a novel approach to facial dimensional emotion analysis by leveraging GP to automatically extract geometric features from static images, which has been rarely explored. The GP system evolved a robust set of features via utilising angular and distance-based relationships between facial landmarks. Those features were integrated with a deep learning model to significantly enhance the accuracy of emotion prediction. The proposed GP methods *BFsGP* and *MOGP*, efficiently extracted multiple geometric features in a single run, achieving comparable performance to *SFGP* which requires multiple runs to attain similar results. There are no significant differences between *BFsGP* and *SFGP* in valence and arousal prediction. Furthermore, *BFsGP* outperforms than *MOGP* in the extracted geometric features from the AffectNet dataset for dimensional analysis, in terms of MSE, RMSE and CCC.

Our experimental results demonstrate that the inclusion of GP-evolved geometric features leads to improved performance in dimensional emotion analysis, validating the potential of GP for feature extraction in this domain. This work not only offers a new direction for integrating geometric and deep learning techniques but also lays the groundwork for future research into more precise and interpretable emotional representation from static facial images.

For future work, it is worth expanding the GP system to handle dynamic emotion recognition from image sequences which could provide more comprehensive emotional insights by incorporating temporal information. Additionally, exploring the integration of other types of geometric features, such as curvature or texture-based descriptors, could further enhance the representation of emotional expressions.

References

1. Barrett, L.F., Adolphs, R., Marsella, S., Martinez, A.M., Pollak, S.D.: Emotional expressions reconsidered: challenges to inferring emotion from human facial movements. Psychol Sci Public Interest **20**(3), 165–166 (2019)
2. Behzad Bozorgtabar, G.A.R.R.: A genetic programming-PCA hybrid face recognition algorithm. J. Signal Inform. Process. **2**, 170–174 (2011)
3. Bhanu, B., Yu, J., Tan, X., Lin, Y.: Feature synthesis using genetic programming for face expression recognition. In: Deb, K. (ed.) GECCO 2004. LNCS, vol. 3103, pp. 896–907. Springer, Heidelberg (2004). https://doi.org/10.1007/978-3-540-24855-2_103
4. Bi, Y., Xue, B., Zhang, M.: Automatically extracting features for face classification using multi-objective genetic programming. In: Proceedings of the 2020 Genetic and Evolutionary Computation Conference Companion, pp. 117—-118 (2020)
5. Bi, Y., Xue, B., Zhang, M.: Genetic programming-based feature learning for facial expression classification. In: 2020 IEEE Congress on Evolutionary Computation (CEC), pp. 1–8 (2020)
6. Bi, Y., Xue, B., Zhang, M.: Multi-objective genetic programming for feature learning in face recognition. Appl. Soft Comput. **103**, 107152 (2021)
7. Canal, F.Z., et al.: A survey on facial emotion recognition techniques: a state-of-the-art literature review. Inf. Sci. **582**, 593–617 (2022)
8. Chen, J., Chen, D., Gong, Y., Yu, M., Zhang, K., Wang, L.: Facial expression recognition using geometric and appearance features. In: Proceedings of the 4th International Conference on Internet Multimedia Computing and Service, pp. 29—-33 (2012)
9. Chen, Y., Li, J., Shan, S., Wang, M., Hong, R.: From static to dynamic: adapting landmark-aware image models for facial expression recognition in videos. IEEE Transactions on Affective Computing, pp. 1–15 (2024)
10. Chicco, D., Warrens, M.J., Jurman, G.: The coefficient of determination R-squared is more informative than SMAPE, MAE, MAPE, MSE and RMSE in regression analysis evaluation. Peerj Comput. Sci. **7**, e623 (2021)
11. Deb, K., Pratap, A., Agarwal, S., Meyarivan, T.: A fast and elitist multiobjective genetic algorithm: NSGA-II. IEEE Trans. Evol. Comput. **6**(2), 182–197 (2002)
12. Dosovitskiy, A., et al.: An image is worth 16x16 words: transformers for image recognition at scale. In: Proceedings of the 9th International Conference on Learning Representations (2021)
13. Geetha, V., Mala, T., Priyanka, D., Uma, E.: Multimodal emotion recognition with deep learning: advancements, challenges, and future directions. Inform. Fusion **105**, 102218 (2024)
14. Ghazouani, H.: A genetic programming-based feature selection and fusion for facial expression recognition. Appl. Soft Comput. **103**, 107173 (2021)

15. Ghimire, D., Lee, J.: Geometric feature-based facial expression recognition in image sequences using multi-class adaboost and support vector machines. Sensors **13**(6), 7714–7734 (2013)
16. Ibrahem, H., Nasef, M., Emam, M.: Genetic programming based face recognition. Int. J. Comput. Appl. **69**(27), 1–6– (May 2013)
17. He, K., Zhang, X., Ren, S., Sun, J.: Deep residual learning for image recognition. In: Proceedings of the IEEE Conference on Computer Vision and Pattern Recognition, pp. 770–778 (2016)
18. Kalyta, O., Krak, I., Barmak, O., Wójcik, W., Radiuk, P.: Method of facial geometric feature representation for information security systems. In: Proceedings of the 3rd International Workshop on Intelligent Information Technologies & Systems of Information Security, Khmelnytskyi, Ukraine, March 23-25, 2022. vol. 3156, pp. 319-328 (2022)
19. Li, S., Deng, W.: Deep facial expression recognition: a survey. IEEE Trans. Affect. Comput. **13**(3), 1195–1215 (2022)
20. Loizides, A., Slater, M., Langdon, W.B.: Measuring Facial Emotional Expressions Using Genetic Programming, pp. 545–553. Springer London (2002)
21. Mollahosseini, A., Hasani, B., Mahoor, M.H.: Affectnet: a database for facial expression, valence, and arousal computing in the wild. IEEE Trans. Affect. Comput. **10**, 18–31 (2017)
22. Montgomery, D.C.: Design and Analysis of Experiments. John Wiley & Sons, Inc (2019)
23. Murugappan, M., Mutawa, A.: Facial geometric feature extraction based emotional expression classification using machine learning algorithms. PloS One **16**(2) (2021)
24. Saeed, S., Shah, A.A., Ehsan, M.K., Amirzada, M.R., Mahmood, A., Mezgebo, T.: Automated facial expression recognition framework using deep learning. J. Healthcare Eng. **2022**(1), 5707930 (2022)
25. Sajjad, M., et al.: A comprehensive survey on deep facial expression recognition: challenges, applications, and future guidelines. Alexandria Eng. J. **68**, 817–840 (2023)
26. Simonyan, K., Zisserman, A.: Very deep convolutional networks for large-scale image recognition. arXiv preprint arXiv:1409.1556 (2014)
27. Tan, M., Le, Q.V.: Efficientnetv2: Smaller models and faster training. In: Meila, M., Zhang, T. (eds.) Proceedings of the 38th International Conference on Machine Learning, ICML, vol. 139, pp. 10096–10106 (2021)
28. Tu, Z., et al.: Maxvit: Multi-axis vision transformer. In: Avidan, S., Brostow, G., Cissé, M., Farinella, G.M., Hassner, T. (eds.) Computer Vision – ECCV, pp. 459–479. Springer Nature Switzerland (2022)
29. Wagner, N., Mätzler, F., Vossberg, S.R., Schneider, H., Pavlitska, S., Zöllner, J.M.: Cage: circumplex affect guided expression inference. In: 2024 IEEE/CVF Conference on Computer Vision and Pattern Recognition Workshops (CVPRW), pp. 4683–4692 (2024)
30. Wu, M., Li, M., He, C., Chen, H., Wang, Y., Li, Z.: Facial expression recognition based on genetic programming learning CCA fusion. In: 2022 5th International Conference on Pattern Recognition and Artificial Intelligence (PRAI), pp. 526–532 (2022)
31. Xu, J., Chen, Q., Xue, B., Zhang, M.: A new concordance correlation coefficient based fitness function for genetic programming for symbolic regression. In: 2024 IEEE Congress on Evolutionary Computation (CEC), pp. 01–08 (2024). https://doi.org/10.1109/CEC60901.2024.10611932

32. Zhang, H., Huang, B., Tian, G.: Facial expression recognition based on deep convolution long short-term memory networks of double-channel weighted mixture. Pattern Recogn. Lett. **131**, 128–134 (2020)
33. Zhang, T., Ma, L., Liu, Q., Li, N., Liu, Y.: Genetic programming for ensemble learning in face recognition. In: Advances in Swarm Intelligence. pp. 209–218. Springer International Publishing (2022)

Methodology for Designing Injection Molds: Data Mining and Multi-objective Optimization

António Gaspar-Cunha$^{(\boxtimes)}$ (iD), João Melo, Tomás Marques, and António Pontes (iD)

IPC-Institute for Polymers and Composites, University of Minho, Guimarães, Portugal
{agc,pontes}@dep.uminho.pt, {pg50466,pg50789}@alunos.uminho.pt

Abstract. Injection molding is a complex process where effective mold design is essential to avoid defects like incomplete parts, flash, sink marks, weld lines, air bubbles, warping, and shrinkage. Proper mold design helps minimize these defects, but it is challenging due to the number of variables across stages such as plasticization, filling, packing, and cooling. This study focuses on optimizing the cooling phase of injection molding using numerical simulation tools (Moldex3D) to evaluate the impact of design variables on process performance. Due to the complexity and multi-objective nature of the problem, Artificial Intelligence (AI) techniques, including data mining, Artificial Neural Networks (ANN), and Multi-Objective Evolutionary Algorithms (MOEAs), were used to explore solutions effectively. The optimization aimed to improve the thermal efficiency of conformal cooling systems, critical for ensuring part quality and minimizing cycle time. A case study with a cylindrical part and multiple cooling systems initially defined 34 objectives, including temperature gradients, cycle time, and defects. These objectives were reduced to four using Principal Component Analysis (PCA) to facilitate optimization. The results showed significant improvements in temperature uniformity and defect reduction, confirming the effectiveness of integrating AI-based optimization techniques with numerical modeling for advanced cooling system design in injection molds.

Keywords: Data Mining · Artificial Neural Networks · Multi-Objective Evolutionary Algorithms · Injection Molding

1 Introduction

The injection molding process is widely used for manufacturing plastic components due to its versatility and mass production capability. However, product quality is heavily influenced by the uniformity of cooling, which is crucial for reducing defects like warping, residual stresses, and shrinkage. These defects arise from non-uniform heat dissipation during cooling, leading to dimensional inconsistencies (Rosato et al. 2000; Kazmer 2007; Menges 2008). To ensure high-quality parts, homogeneous and efficient cooling solutions are essential. Conventional optimization of processing parameters, such as temperature and pressure, has been common. However, Conformal Cooling Channels (CCC) have recently revolutionized thermal control. CCC adapts to the part's geometry, promoting uniform cooling and eliminating hotspots typical in conventional systems (Kanbur et al. 2020; Feng et al. 2021a; Silva et al. 2022).

© The Author(s), under exclusive license to Springer Nature Switzerland AG 2025
P. García-Sánchez et al. (Eds.): EvoApplications 2025, LNCS 15612, pp. 154–170, 2025.
https://doi.org/10.1007/978-3-031-90062-4_10

CCC offers several benefits, including more uniform temperature distribution, enhanced cooling efficiency, reduced cycle times, and improved part quality. Studies show that CCC significantly reduces cycle time and defects compared to conventional channels (Kanbur et al. 2020; Feng et al. 2021a; Silva et al. 2022). Their feasibility has been greatly aided by advancements in Additive Manufacturing (AM), such as Direct Metal Laser Sintering (DMLS), which enables the creation of complex, efficient cooling geometries (Feng et al. 2021.b, Silva et al. 2022).

This paper proposes a methodology to deal with the design of CCC using numerical modeling to evaluate the solutions, data mining techniques to analyze the data generated, and multiobjective optimization algorithms to search for improved solutions. The results of the methodology will be assessed using a practical example.

The proposed methodology consists of several key steps: i) definition of design variables, identifying parameters like cooling channel geometry; ii) modeling, developing a simulation of the injection molding process; iii) applying PCA to reduce dimensionality and focus on impactful objectives; iv) using an ANN as a surrogate model to replace costly simulations during MOEA iterations, accelerating convergence; and v) executing the multi-objective optimization with MOEA to find the optimal mold design, focusing on minimizing defects, improving cooling efficiency, and reducing cycle time.

Numerical modeling simulations will be conducted using Moldex3D, a simulation tool for injection molding processes. Moldex3D enables detailed analysis of the mold filling, cooling, packing, and warpage behavior. The software allows the simulation of complex phenomena like shrinkage and the impact of cooling channel design on part quality (Chen 2018). Moldex3D provides temperature profiles during cooling, which are used to optimize cooling channel configurations. By analyzing these profiles, areas of uneven cooling can be identified and adjusted for more uniform temperature distribution, ultimately improving part quality by minimizing defects such as shrinkage, warping, and residual stresses (Chen & Chen 2020; Huang et al. 2021).

AI-driven optimization combined with surrogate models like artificial neural networks addresses existing limitations. AI optimizes design and reduces the computational burden of iterative simulations. Neural networks can predict the thermal performance of mold designs, reducing the need for exhaustive numerical simulations and accelerating the design process (Sun et al. 2023). PCA reduces the problem dimensionality by transforming correlated variables into fewer uncorrelated components that retain most of the variance. This simplifies the optimization process, enhances computational efficiency, and improves model robustness by filtering out noise and irrelevant information (Jolliffe & Cadima 2065; Abdi & Williams 2010).

Multi-objective optimization algorithms consider multiple performance metrics, such as temperature profiles, shrinkage, warpage, Von Mises stresses, cycle time, and cooling efficiency (Deb 2001; Coello et al. 2007). These metrics ensure the mold design balances structural integrity, dimensional stability, and efficient production. Temperature profiles affect shrinkage and warpage, impacting part accuracy (Kazmer 2007). Von Mises stresses provide insight into mold strength during the injection cycle, ensuring operational reliability (Suresh et al. 2019). Cycle time and cooling efficiency evaluate productivity and energy consumption, affecting cost-effectiveness (Thiriez & Gutowski 2006). These algorithms help navigate trade-offs to achieve a well-balanced mold design.

This work's main contribution is the development of a methodology for designing CCC for injection molding based on PCA analysis to select the most relevant objectives to be optimized, ANN as a surrogate model, and MOEA to optimize.

The remainder of this text is organized as follows: Sect. 2 presents a literature review on designing CCC, Sect. 3 describes the methodology proposed, Sect. 4 presents and discusses the results obtained, and Sect. 6 presents the conclusion accomplished.

2 Literature Review

Injection molding is indispensable in many manufacturing processes, allowing the efficient production of parts with complex geometries at reduced costs (Rosato et al., 2007). A key advantage of injection molding is its superior dimensional accuracy and repeatability, which are crucial in sectors like automotive, where part consistency and quality are imperative for safety and functionality. The injection molding process involves five primary stages (Fig. 1): i) Mold Closing – the mold halves close to form the cavity into which the polymer will be injected; ii) Filling – molten polymer is introduced into the mold cavity at high velocity and pressure; iii) Packing – after filling, pressure is maintained to compensate for material shrinkage as the polymer cools; iv) Cooling – the molten polymer solidifies within the mold cavity as it cools; v) Mold Opening and Ejection – once the polymer has sufficiently solidified, the mold halves are separated and the solidified part is ejected from the mold using ejector pins or other mechanisms. Simultaneously, the solid polymer is plasticized for the next injection cycle during cooling (Rosato et al. Rosato et al. 2000).

Each stage's duration depends on the material properties and specific part design, consequently, the optimization of these stages is fundamental to enhancing production efficiency, see Fig. 1 (Rosato et al. 2000; Herausgegeben et al. 2007). Cooling is often the most time-consuming stage of the cycle, representing between 50% and 80% of the overall cycle time (Silva et al. 2022). Thus, efficient heat transfer is necessary to minimize cooling time while maintaining part quality and dimensional accuracy.

Gaspar-Cunha et al. (2022a, 2022b) provide an overview of optimization methodologies applied to polymer processing, focusing on extrusion and molding technologies. These reviews analyze optimization strategies, categorizing them according to application areas and challenges, from heuristic approaches to advanced computational techniques like MOEA and AI-based models. Surrogate modeling techniques such as Response Surface Methodology (RSM) and ANN reduce computational burden by constructing predictive models from limited simulations, facilitating efficient solution exploration. Hybrid approaches combining EAs with gradient-based methods have also shown faster convergence and superior solutions (Gaspar-Cunha et al. 2022a).

Injection molding, in particular, offers significant potential for optimization, aiming to minimize material usage, enhance product quality, and improve efficiency. Optimization of injection molding includes the design of the cavity, gate, runner system, and cooling channels, as well as defining operating conditions. Many studies focus on single-objective and multi-objective optimization techniques. For example, Alam and Kamal (2003, 2004, 2005) optimized runner dimensions and operating conditions to minimize system volume, cycle time, and shrinkage. Gaspar-Cunha et al. (2005) used a MOEA

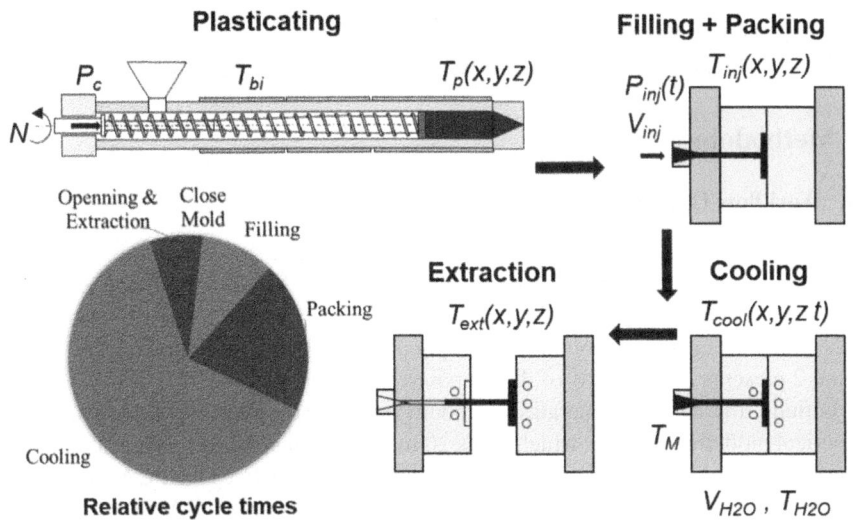

Fig. 1. Injection molding cycle: stages and relative cycle times.

linked to a numerical modeling code to determine conditions inducing specific morphological characteristics. Fernandes et al. (2010, 2012) applied an MOEA to optimize cooling channel designs and operating conditions to minimize temperature differentials, cavity pressure, shrinkage, warpage, and cycle time. Xu et al. (2012) employed a multi-objective particle swarm optimization algorithm minimizing part weight, shrinkage, and flash using an ANN trained with experimental data from the Taguchi method. These studies demonstrate the effectiveness of AI-driven optimization techniques for complex molding challenges.

Conformal cooling channels have emerged as a significant innovation in mold design, tailored to follow the intricate geometry of the molded part, thereby enabling faster and more uniform cooling, which can result in a reduction of the total cycle time by approximately 47% (Kitayama et al. 2018.a). The advent of additive manufacturing technologies, such as DMLS, has facilitated the fabrication of these complex cooling channels. CCC also improves cooling uniformity, thereby mitigating internal stresses and distortions, which leads to parts with greater dimensional stability. Nevertheless, injection molding remains challenged by defects such as flow marks, bubbles, sink marks, warpage, and weld lines, which predominantly result from suboptimal process parameters or inadequate cooling distribution (Farooque et al., 2020; Kitayama et al. 2018).

Computational simulation tools are increasingly pivotal in optimizing the injection molding process. Software platforms such as Moldflow™ and Moldex3D™ enable comprehensive mold filling simulations and the prediction of defects before physical production, thus allowing for adjustments that ensure part quality and reduce development time (Osswald & Hernández-Ortiz 2006). These tools allow the evaluation of various cooling configurations enabling the optimization of mold design without the need for costly physical prototypes. Designing CCC exemplifies the synergy between mold engineering and additive manufacturing technologies, providing a pathway that combines enhanced productivity with improved part quality (Nguyen et al. 2023).

3 Methodology

3.1 Workflow Overview

Given the high dimensionality of the objective space and the computational expense inherent to the numerical modeling process, PCA and ANN are employed to enhance efficiency. Specifically, PCA reduces the dimensionality of the objective space, while ANN evaluates candidate solutions throughout successive generations of the MOEA. Figure 2 presents an overview of the proposed methodology.

Initially, the MOEA generates a random population directly evaluated using the numerical modeling routine, which takes 20 min for each solution. This evaluation has two key purposes: first, to identify the most significant objectives via PCA, and second, to train an ANN based on these selected objectives. The trained ANN subsequently facilitates the evaluation process in the remaining generations, significantly reducing computational costs.

In the present study, the complexity is further increased by the possibility of injecting the part at two distinct locations. Preliminary analyses indicated that a single ANN model could not adequately capture the behaviors associated with the different gate locations. Consequently, the initial dataset includes samples from both gate types, and after the application of PCA, two separate ANNs are trained, one for each gate type. During the MOEA optimization, both gate types are considered by introducing an additional integer decision variable, which can assume 1 or 2, representing the respective gate types.

This methodology enables the MOEA to explore the design space effectively while maintaining computational efficiency, ensuring that the dedicated ANN model can appropriately manage the two gate types. Similar approaches integrating PCA and ANN into optimization frameworks have been investigated in various studies, yielding significant improvements in computational efficiency and solution quality (Jin et al. 2001; Jin 2011).

3.2 Principal Component Analysis (PCA)

Principal Component Analysis (PCA) is a statistical technique for reducing the dimensionality of complex datasets, making interpretation easier by revealing latent structures. PCA transforms variables to maximize variance while retaining critical features, reducing the number of objectives, and enhancing optimization efficiency.

PCA is useful for high-dimensional datasets, such as injection molding processes. Reducing variables simplifies models, optimizes computational resources, and accelerates analysis. PCA transforms original variables into orthogonal principal components, ranked by the variance they capture, prioritizing the most informative components and reducing complexity (Hotelling 1933; Jolliffe 2002; Abdi and Williams 2010).

PCA involves centering data by subtracting the mean, computing the covariance matrix, and decomposing it into eigenvalues and eigenvectors. Eigenvalues represent the variance explained by each principal component, while eigenvectors define component directions.

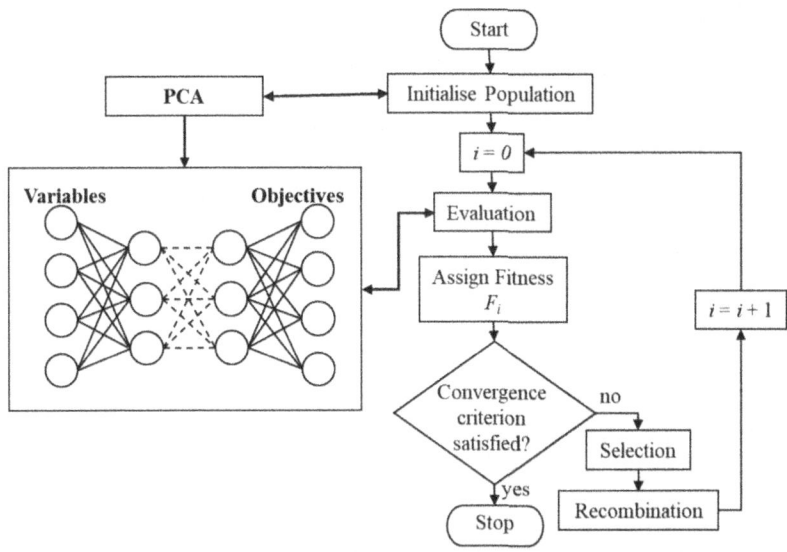

Fig. 2. MOEA/PCA/ANN global methodology proposed.

The transformation of original variables into principal components (PC_k) is represented by:

$$PC_k = \phi_{1k}X_1 + \phi_{2k}X_2 + \cdots + \phi_{pk}X_p \qquad (1)$$

Here, ϕ_{ik} represents the coefficient (loading) of variable x_i in the k-th principal component, and p denotes the number of components.

Principal components are selected based on eigenvalues, typically retaining those that explain at least 80% of the total variance. This reduces data complexity while retaining most information. The transformed dataset yields a reduced-dimensional representation that preserves most variance, aiding analysis (Hotelling 1933).

To capture the nonlinearities present in the data, PCA was compared with Nonlinear Principal Component Analysis (NL-PCA), which was designed to handle complex data structures with nonlinear relationships. The Multi-Dimensional Scaling (MDS) variant was adopted. This technique is used for dimensionality reduction, primarily focused on preserving the relative distances or similarities between data points when they are mapped to a lower-dimensional space (Borg & Groenen 2005).

In injection molding, where objectives include minimizing temperature gradients, defects, and cycle times, PCA optimizes processes by establishing a hierarchy of objectives. Focusing on components with significant variance makes multi-objective optimization more efficient. In the example studied, PCA reduced thirty-four principal components to four or three, explaining over 80% of the total variance. This reduction allows clearer analysis, improving mold design and the injection process.

3.3 Surrogate Modeling with ANN

Using MOEA to optimize complex processes like polymer injection molding requires evaluating many candidate solutions, each involving computational simulations to assess multiple objectives. Injection molding depends on various parameters, such as temperature, pressure, and material properties, influencing product quality, cycle time, and energy consumption. Given the high computational cost, evaluating all solutions is impractical, making surrogate models an effective alternative (Simpson et al. 2001; Forrester & Keane 2009).

Surrogate models based on Artificial Neural Networks (ANN) provide an efficient substitute for high-fidelity simulations in MOEA. ANNs can learn complex, non-linear relationships between the injection molding process's input and output parameters. Using ANN to approximate numerical model outputs reduces computational costs, allowing MOEA to explore more solutions and improve the chances of finding optimal results.

ANNs consist of layers of interconnected neurons (see Fig. 2), processing inputs through activation functions. They are trained using datasets to minimize prediction errors, enabling them to approximate numerical simulations effectively (Goodfellow et al. 2016). Implementing ANN-based surrogates involves generating a dataset of simulation results through techniques. This dataset is used to train the ANN, which can then predict the performance of new designs. During optimization, the trained ANN replaces full-scale simulations, significantly speeding up the process (Jin 2011).

In injection molding optimization, surrogate models can address objectives like minimizing cycle time, reducing material usage, and improving product quality. ANNs capture the intricate relationships within the process, allowing thousands of evaluations per second and enabling MOEA to converge toward an optimal Pareto front more efficiently (Gao et al. 2023; Konuskan et al. 2024). Thus, this methodology reduces optimization time enabling simultaneously more extensive exploration of the design space, improving process performance and product quality.

3.4 Multi-objective Optimization Procedure

Multi-objective evolutionary algorithms are typically categorized based on underlying principles and mechanisms. The primary categories of MOEA include dominance-based, decomposition-based, indicator-based, and hybrid-based approaches. These algorithms have been developed to address specific challenges in multi-objective optimization, with each method having unique strengths suitable for different problems.

Dominance-based MOEAs use Pareto dominance relations to determine the fitness of individuals in the population. NSGA-II (Non-dominated Sorting Genetic Algorithm II) is a well-known example of a dominance-based MOEA, where individuals are ranked based on Pareto dominance and crowdedness to promote diversity (Deb et al. 2001, 2002).

Decomposition-based MOEAs, such as MOEA/D (Multi-objective Evolutionary Algorithm based on Decomposition), transform the multi-objective problem into several scalar optimization subproblems, which are then solved simultaneously (Zhang & Li 2007).

Indicator-based MOEAs employ performance indicators, such as hypervolume, to drive the search process by considering both convergence and diversity. SMS-EMOA (S-Metric Selection Evolutionary Multi-objective Optimization Algorithm) is a prominent example that uses the hypervolume metric to select solutions (Beume et al. 2007). Hypervolume-based indicators provide information on the convergence to the Pareto front and the spread of solutions, which makes indicator-based algorithms particularly appealing for complex problems.

Hybrid-based MOEAs combine elements from multiple MOEA types to take advantage of their respective strengths. An example is NSGA-III (Non-dominated Sorting Genetic Algorithm III), which integrates dominance-based approaches with reference vector-based techniques to guide the search process and address many-objective optimization problems (Deb et al. 2014).

In this work, NSGA-III was chosen because of its capability to maintain a good balance between convergence and diversity and its effectiveness in handling many-objective optimization problems. The NSGA-III parameters are: number of generations is 50, the population size is 100, number of offspring is 100, the duplicates are eliminated, float random sampling, tournament selection, SBX crossover, and PM mutation are adopted.

4 Case Study

4.1 Experimental Setup

The proposed optimization methodology was applied to determine the optimal placement and geometry of the CCC for a specific plastic cup. This straightforward example enables assessing the results grounded in existing knowledge and expertise. Nevertheless, the ultimate aim is to establish a robust methodology addressing more intricate real-world challenges. For that purpose, two different situations are analyzed, one considering two circular rings (Fig. 3 left) and the second three circular rings (Fig. 3 right). As can be seen, the design variables identified in each case allowed the rings to adapt to the geometry of the cup using the concept of CCC. The aim is to define the diameter and the distances between the rings and between the rings and the cup. Simultaneously, there is the possibility of injecting the part either in the center of the cup bottom (gate = 1) or laterally (gate = 2). Table 1 shows the range of variation of the design variables used.

The objectives of this study are categorized into six domains: temperature differences, warpage, shrinkage, cycle time, density, and cooling efficiency (Table 2).

The first focus is on temperature differences within the part during cooling, which significantly affect the internal microstructure and, ultimately, the part's performance. The objective is to minimize temperature differentials across the part, specifically to achieve uniform temperature distribution throughout the wall, thereby mitigating distortion and enhancing the uniformity of the internal microstructure. However, achieving this is inherently challenging, as multiple solutions may yield similar average temperature differences but exhibit distinct distribution patterns.

Table 1. Decision variables and range of variation (dimensions in mm).

Designation	Limits (2 rings)	Limits (3 rings)
d'1	[8, 20]	[8, 20]
d'2	[8, 20]	[8, 20]
d1	[0, 25]	[0, 10]
d2	[8, 20]	[8, 20]
D1	[6, 15]	[6, 12]
D2	[6, 15]	[6, 12]
d'3	--	[8, 20]
d3	--	[8, 20]
D3	--	[6, 12]
Gate location	[1 or 2]	[1 or 2]

To overcome this difficulty, for each temperature set of temperature differences identified two values are calculated, the average difference between all points and the average difference for the 20% higher differences. Temperature differences are divided into three groups: (1) variation between the surface and the mid-thickness, (2) variation between the two surfaces, and (3) variation between the mid-thickness and the opposite surface. Each group's average value and standard deviation are computed for all observed temperature variations.

Regarding warpage and shrinkage, sixteen objectives were defined and categorized into two primary groups: warpage and shrinkage. For each group, the maximum and minimum values along the X, Y, and Z axes are computed independently, and the overall value across all three axes.

During the modeling phase, an initial cycle time value must be defined in Moldex3D, and updated throughout the simulation process by making the final temperature after cooling coincide with the ejection temperature, designated by tcdiff. This value is directly proportional to the cycle time. In the present case, the values defined are: filling is 1.13 s, packing is 7.09 s and cooling time is 16.5 s, giving a total of 24.72 s. After calculations, the cycle time is tcdiff plus 24.72 s.

Density variations across different part regions can lead to differential shrinkage, uneven hardening, or even internal defects such as residual stresses and distortions. Thus, the average and standard deviation of density at all mesh points of the part are calculated and used as objectives. This results in 34 objectives that must be reduced using PCA to an acceptable number to be used in the MOEA to optimize the process, as shown in Table 2. All these objectives are to be minimized.

Table 2. Objectives.

Name	Unities	Description
aDTsm	°C	Average difference in temperature between the inner and outer surfaces and the centre
sDTsm	°C	Standard deviation of aDTsm
aDTsmH	°C	Average difference of 20% higher variation temperatures between the inner and outer surfaces and the centre
sDTsmH	°C	Standard deviation of aDTsmH
aDTms	°C	Average difference in temperature between the centre and outer surface
sDTms	°C	Standard deviation of aDTms
aDTmsH	°C	Average difference of 20% higher variation temperatures between the centre and outer surface
sDTmsH	°C	Standard deviation of aDTmsH
aDTss	°C	Average difference in temperature between the outer and inner surfaces
sDTss	°C	Standard deviation of aDTss
aDTssH	°C	Average difference of the 20% higher variation temperatures between the outer and inner surfaces
sDTssH	°C	Standard deviation of aDTssH
maxWaT	mm	Maximum total warpage
maxWaX	mm	Maximum warpage in X direction
maxWaY	mm	Maximum warpage in Y direction
maxWaZ	mm	Maximum warpage in Z direction
minWaT	mm	Minimum total warpage
minWaX	mm	Minimum warpage in X direction
minWaY	mm	Minimum warpage in Y direction
minWaZ	mm	Minimum warpage in Z direction
maxSrT	mm	Maximum total shrinkage
maxSrX	mm	Maximum shrinkage in X direction
maxSrY	mm	Maximum shrinkage in Y direction
maxSrZ	mm	Maximum shrinkage in Z direction
minSrT	mm	Minimum total shrinkage
minSrX	mm	Minimum shrinkage in X direction
minSrY	mm	Minimum shrinkage in Y direction
minSrZ	mm	Minimum shrinkage in Z direction
tcdiff	s	Difference between cycle time imposed and cycle time predicted by Moldex3D Studio

(continued)

Table 2. (*continued*)

Name	Unities	Description
sdDens	g/cm^3	Standard deviation of density
coolef	%	Difference between heat flux transfer from core cooling system and cavity cooling systems
maxVMS	MPa	Max value of Von Mises Stress
aVMS	MPa	Average value of Von Mises Stress
sdVMS	MPa	Standard deviation of Von Mises Stress

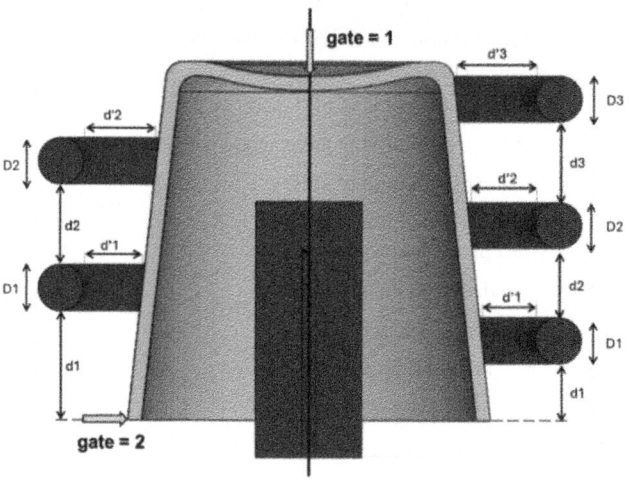

Fig. 3. Decision variables for a mold considering two separated cooling systems, i.e., with two (left) and three (right) circular rings as CCC.

5 Results and Discussion

5.1 Dimensional Reduction Results

Dimensionality reduction using PCA and NL-PCA was conducted in a systematic four-step process. Initially, the 34 objectives were reduced to 20. Subsequently, these 20 objectives were reduced to 10, then further to 5, and ultimately to 4. At each stage, PCA and NL-PCA were applied to the reduced set of objectives derived from the preceding step, with selection based on Eigenvalue analysis. Given the sensitivity of the final results to the order of the data, a final evaluation was performed by randomizing the row order and calculating the average Eigenvalues. Using this approach, the selected objectives were aDTmsH, maxWaX, tcdiff, and minSrY. The optimization outcomes were assessed using an additional bi-objective optimization run focusing on the two objectives minSrY and tcdiff.

5.2 ANN Model Performance

A different ANN is generated for each set of objectives and gate location. As described above, Fig. 4 shows the error produced after the ANN's automatic training for gate1 and gate2. This is necessary because a previous analysis of the objectives scatter plots showed that the solutions are separated into two distinct regions. Bayesian optimization was used in the automatic training of the networks to define hyperparameters. This method is usually employed to optimize objective functions that are expensive to evaluate as it uses a surrogate model to predict the performance of hyperparameter configurations and an acquisition function to select new configurations for evaluation, balancing exploration of the hyperparameter space with exploitation of known promising regions. The hyperparameters determined during this process after 1000 evaluations are the number of layers and units, hidden layer and output layer activation functions, dropout, optimizer, learning rate, number of epochs, and batch size. As shown in Fig. 4, the error obtained is small, 0.0258 and 0.0412 for gates 1 and 2, respectively. Anyway, special care must be taken during the optimization using ANN as a surrogate since the algorithm can conduct the solutions outside the initial region defined by the solutions used during the train.

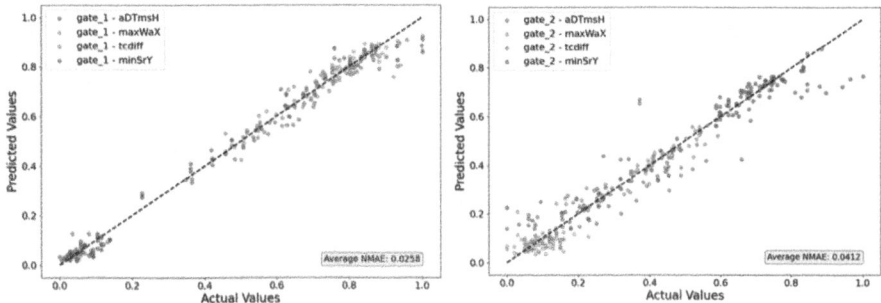

Fig. 4. ANN evaluation assessement.

5.3 Optimization Results

Figures 5, 6 and 7 present the Pareto fronts for the three optimization cases, including the initial populations. These figures offer a visual representation of the optimization trajectory, showing both initial and final generations. Selected solutions are graphically represented for direct comparison, aiding in interpreting improvements and trade-offs achieved.

Figure 5 illustrates optimization results involving two objectives: minimizing shrinkage in the Y direction (minSrY) and minimizing cooling time differential (tcdiff). The figure reveals two clusters of solutions for each gate, highlighting the distribution of viable solutions. Comparing the initial population (generation 1) and final population (generation 100) shows significant improvements, demonstrating effective optimization. For gate 1, transitioning from solution 1 to 2 increases cycle time due to extended cooling but reduces shrinkage, aligning with material behavior during cooling. By moderating

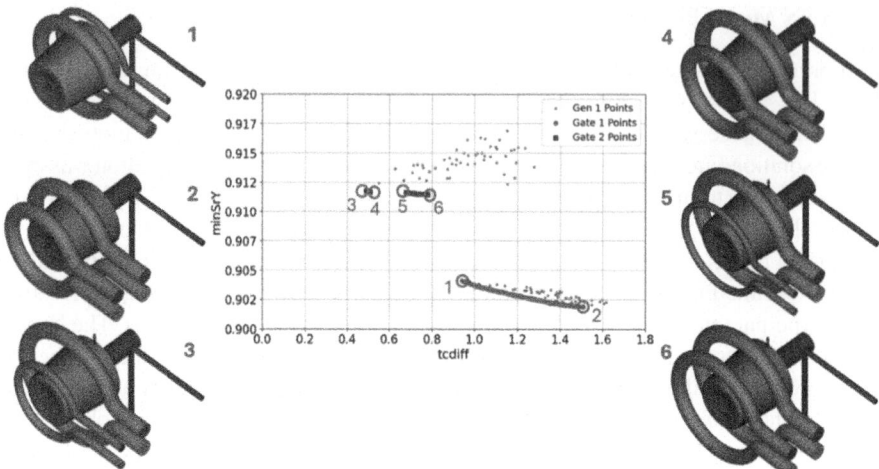

Fig. 5. Results with two objectives for two rings: minSrY and tcdiff.

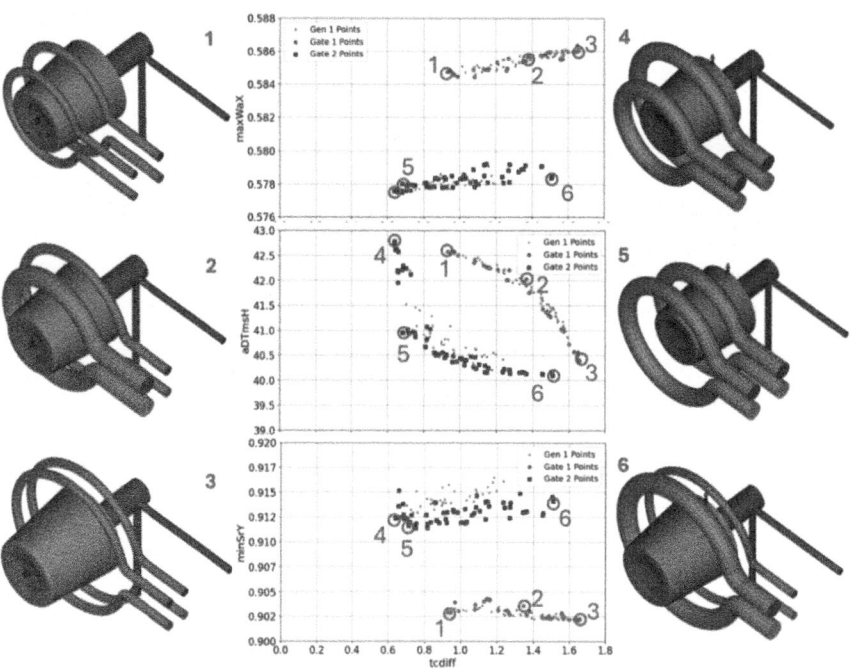

Fig. 6. Results with four objectives for two rings: aDTmsH, maxWaX, tcdiff and minSrY.

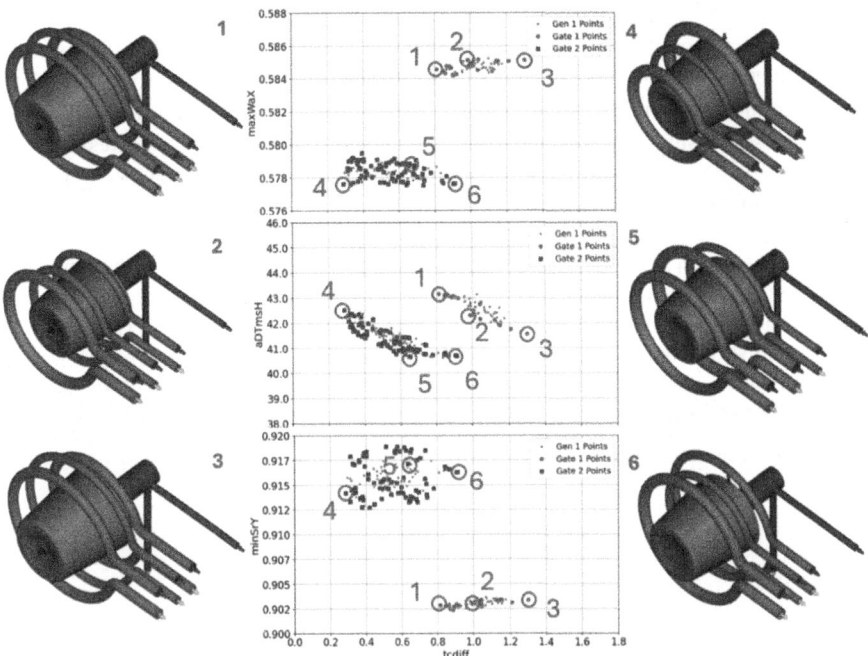

Fig. 7. Results with four objectives for three: aDTmsH, maxWaX, tcdiff and minSrY.

cooling rates, internal stresses are reduced, thereby decreasing shrinkage. This trade-off between cycle time and shrinkage enables balanced solutions that meet quality requirements with feasible production times. The trend is less pronounced for solutions 3 and 4 of gate 2, but solutions 5 and 6 reveal a similar pattern to gate 1. This indicates that the fundamental relationship between cooling rate and shrinkage is consistent across gate configurations, ensuring reliable optimization outcomes.

For the scenario with four objectives and two cooling rings (Fig. 6), changes in minSrY are less pronounced, while variations in tcdiff are more substantial due to trade-offs with additional objectives: minimizing the average difference of 20% higher variation temperatures between the center and outer surface (aDTmsH) and minimizing warpage in X direction (maxWaX). The inclusion of these objectives adds complexity, requiring balance among competing goals. The behavior of maxWaX mirrors minSrY, while aDTmsH is more sensitive to cooling channel geometry. Consequently, cooling channel geometry undergoes significant modifications, as shown in Fig. 6 for solutions 1 to 6, reflecting the need to balance the added objectives. These results are aligned with thermomechanical behavior of the process and show the complexity of the process.

6   Conclusion

This work presents a methodology for designing conformal cooling channels (CCC) in injection molds by integrating numerical modeling, data mining, and optimization algorithms. PCA simplified optimization by focusing on key objectives like temperature uniformity, warpage, and shrinkage. ANNs replaced high-fidelity simulations, significantly reducing computational costs. The NSGA-III algorithm balanced convergence and solution diversity by improving the objectives selected by PCA. In summary, this methodology integrates numerical modeling, machine learning, and optimization to address the complexities of designing efficient cooling systems for injection molding. Results show improved efficiency, reduced cycle times, and lower defect rates, enhancing product quality and advancing mold design.

Acknowledgments. The authors acknowledge the funding by FEDER funds COMPETE 2020 Program and National Funds through FCT projects UID-B/05256/2020, and UID-P/05256/2020, and "Agendas para a Inovação Empresarial" (Project n° 49, acronym "INOV.AM", with reference PRR/49/INOV.AM/EE, operation code 02/C05-i01.01/2022.PC644865234–00000004), supported by the RRP - Recovery and Resilience Plan and by the European Funds NextGeneration EU. http://www.recuperarportugal.gov.pt/.

Disclosure of Interests. The authors have no competing interests to declare related to the content of this article.

References

Abdi, H., Williams, L.J.: Principal component analysis. Wiley Interdisciplinary Reviews: Computational Statistics **2**(4), 433–459 (2010)

Alam, K., Kamal, M.R.: A Genetic Optimization of Shrinkage by Runner Balancing. SPE Annual Technical Conference, vol. 5, pp. 639–641. Society of Plastics Engineers (SPE), Nashville, TN, USA (2003)

Alam, K., Kamal, M.R.: Runner balancing by a direct genetic optimization of shrinkage. Polym. Eng. Sci. **44**, 1949–1959 (2004)

Alam, K., Kamal, M.R.: A robust optimization of injection molding runner balancing. Comp. Chem. Eng. **29**, 1934–1944 (2005)

Beume, N., Naujoks, B., Emmerich, M.: SMS-EMOA: multiobjective selection based on dominated hypervolume. Eur. J. Oper. Res. **181**, 1653–1669 (2007)

Borg, I., Groenen, P.J.F.: Modern Multidimensional Scaling: Theory and Applications, 2nd edn. Springer, Boston, MA (2005)

Chen, C.S., Chen, J.H.: Advanced injection molding simulations using Moldex3D for optimizing process conditions. J. Manuf. Process. **56**, 112–118 (2020)

Chen, W.T.: Injection Molding: Integration of Theory and Modeling Methods. Springer, Boston, MA (2018)

Coello, C.A.C., Lamont, G.B., Van Veldhuizen, D.A.: Evolutionary Algorithms for Solving Multi-Objective Problems. Springer, Boston, MA (2007)

Deb, K.: Multi-objective Optimization Using Evolutionary Algorithms. Wiley, Chichester (2001)

Deb, K., Pratap, A., Agarwal, S., Meyarivan, T.: A fast and elitist multi-objective genetic algorithm: NSGA-II. IEEE Trans. Evol. Comput. **6**(2), 182–197 (2002)

Deb, K., Pratap, A., Agarwal, S., Meyarivan, T.: A fast elitist non-dominated sorting genetic algorithm for multi-objective optimization: NSGA-II. In: Schoenauer, M., et al. (eds.) Parallel Problem Solving from Nature (PPSN) VI, vol. 1917, pp. 849–858. Springer, Paris, France (2000). https://doi.org/10.1007/3-540-45356-3_83

Deb, D., Jain, A., Singh, R.K.: An evolutionary many-objective optimization algorithm using reference-point-based nondominated sorting approach, part i: solving problems with box constraints. IEEE Trans. Evol. Comput. **18**(4), 577–601 (2014)

Farooque, R., Asjad, M., Rizvi, S.J.A.: A current state of the art applied to injection moulding manufacturing process - A review. Materials Today: Proceedings **43**, 441–446 (2020)

Feng, Q., Wang, C., Li, S.: A review of conformal cooling for injection molding. Polymers **13**(6), 945 (2021)

Feng, J., Kanbur, Z., Silva, M.: Optimization of cooling channels in injection molding. Polym. Eng. Sci. **61**(5), 1234–1245 (2021)

Fernandes, C., Pontes, A.J., Viana, J.C., Gaspar-Cunha, A.: Using multiobjective evolutionary algorithms in the optimization of operating conditions of polymer injection molding. Polym. Eng. Sci. **50**, 1667–1678 (2010)

Fernandes, C., Pontes, A.J., Viana, J.C., Gaspar-Cunha, A.: Using multi-objective evolutionary algorithms for optimization of the cooling system in polymer injection molding. Int. Polym. Proc. **27**, 213–223 (2012)

Forrester, A.I., Keane, A.J.: Recent advances in surrogate-based optimization. Prog. Aerosp. Sci. **45**(1–3), 50–79 (2009)

Gao, Z., Dong, G., Tang, Y., Zhao, Y.F.: Machine learning aided design of conformal cooling channels for injection molding. J. Intell. Manuf. **34**(3), 1183–1201 (2023)

Gaspar-Cunha, A., Viana, J.: Using multi-objective evolutionary algorithms to optimize mechanical properties of injection molded part. Int. Polym. Process. **20**, 274–285 (2005)

Gaspar-Cunha, A., Covas, J.A., Sikora, J.: Optimization of polymer processing: a review (part i—extrusion). Materials **15**(1), 384 (2022)

Gaspar-Cunha, A., Covas, J.A., Sikora, J.: Optimization of polymer processing: a review (part ii-molding technologies). Materials **15**(3), 1–20 (2022)

Goodfellow, I., Bengio, Y., Courville, A.: Deep Learning. MIT Press (2016)

Hotelling, H.: Analysis of a complex of statistical variables into principal components. J. Educ. Psychol. **24**(6), 417 (1933)

Huang, C.H., Chen, M.S., Li, Y.R.: Application of Moldex3D in injection molding simulation for enhancing product quality. Polym. Eng. Sci. **61**(8), 2100–2108 (2021)

Jin, Y., Chen, Y., Sendhoff, B.: A framework for modeling and optimization using surrogate models and its application to evolutionary optimization. IEEE Trans. Syst. Man Cybernet. Part C (Appli. Rev.) **31**(1), 88–102 (2001)

Jin, Y.: Surrogate-assisted evolutionary computation: recent advances and future challenges. Swarm Evol. Comput. **1**(2), 61–70 (2011)

Jolliffe, I.T., Cadima, J.: Principal component analysis: a review and recent developments. Philos. Trans. Royal Soc. A: Math. Phys. Eng. Sci. **374**(2065) (2016)

Jolliffe, I.T.: Principal Component Analysis, 2nd edn. Springer, Boston, MA (2002)

Kanbur, B.B., Suping, S., Duan, F.: Design and optimization of conformal cooling channels for injection molding: a review. Int. J. Adv. Manuf. Technol. **106**(7), 3253–3271 (2020)

Kazmer, D.: Injection Mold Design Engineering. Hanser, Munich (2007)

Kitayama, S., Tamada, K., Takano, M., Aiba, S.: Numerical optimization of process parameters in plastic injection molding for minimizing weldlines and clamping force using conformal cooling channel. J. Manuf. Process. **32**, 782–790 (2018)

Kitayama, S., Yamazaki, Y., Takano, M., Aiba, S.: Numerical and experimental investigation of process parameters optimization in plastic injection molding using multi-criteria decision making. Simul. Model. Pract. Theory **85**, 95–105 (2018)

Konuskan, Y., Yılmaz, A.H., Tosun, B., Lazoglu, I.: Machine learning-aided cooling profile prediction in plastic injection molding. Int. J. Adv. Manuf. Technol. **130**, 2957–2968 (2024)

Menges, G., Michaeli, W., Mohren, P.: How to Make Injection Molds. Hanser, Munich (2008)

Nguyen, V.T., et al.: Weld line strength of polyamide fiberglass composite at different processing parameters in injection molding technique. Polymers **15**(20), 4102 (2023)

Osswald, T.A., Hernández-Ortiz, J.P.: Polymer Processing. Hanser, Munich (2006)

Osswald, T., Turng, L.-S., Gramann, P.: Injection Molding Handbook. Hanser, Munich (2007)

Rosato, D.V., Rosato, D.V., Rosato, M.G.: Injection Molding Handbook. Springer, Boston, MA (2000)

Silva, T., Sousa, A.C., Gaspar, F.: Additive manufacturing for conformal cooling channels in injection molds. Proc. Manufact. **55**, 177–183 (2022)

Simpson, T.W., Peplinski, J.D., Koch, P.N., Allen, J.K.: Metamodels for computer-based engineering design: Survey and recommendations. Eng. Comput. **17**(2), 129–150 (2001)

Sun, Y., Chen, J., Zhu, Y.: Surrogate modeling with neural networks for thermal optimization in mold design. J. Manuf. Process. **81**, 335–347 (2023)

Suresh, S., Sampath, K.A., Rajesh, K.: Finite element analysis in mold design. Eng. Anal. Boundary Elem. **104**, 287–294 (2019)

Thiriez, A., Gutowski, T.G.: An environmental analysis of injection molding. Inter. J. Sustainable Manufact. **1**(1/2), 230–245 (2006)

Xu, G., Yang, Z., Long, G.: Multi-objective optimization of MIMO plastic injection molding process conditions based on particle swarm optimization. Int. J. Adv. Manuf. Technol. **58**, 521–531 (2012)

Zhang, Q., Li, H.: MOEA/D: a multiobjective evolutionary algorithm based on decomposition. IEEE Trans. Evol. Comput. **11**(6), 712–731 (2007)

Climbing the Tower of Meta-mutations - The Role of Higher-Order Mutations

Bruno Gašperov[(✉)] and Branko Šter

Faculty of Computer and Information Science, University of Ljubljana,
Ljubljana, Slovenia
{bruno.gasperov,branko.ster}@fri.uni-lj.si

Abstract. Recently, the field of meta-learning, often described as "learning how to learn", has been gaining significant research attention. Its generalization is given by higher-order meta-learning, where multiple levels (orders) of meta-parameters are stacked, leading to systems that learn how to learn how to learn, and so on. Higher-order mutation rates represent an instance of this paradigm used within evolutionary computation. Under such a scheme, the mutation rate of the k-th order mutation rate is determined by the $(k + 1)$-th order mutation rate, and so forth, continuing up to the top order n. In the self-referential variant, the top meta-mutation rate is mutated by itself, thereby removing the need for an additional hyperparameter. While initial experiments employing higher-order mutation rates have yielded promising results, especially in dynamic and adversarial settings, a comprehensive analysis is so far lacking. To address this gap, we provide an empirical study with a focus on interpreting the behavior of higher-order mutation rates, including self-referential ones, under varying selective pressure dynamics (i.e., fitness functions).

Keywords: higher-order meta-learning · genetic algorithm · self-adaptive mutation rates · automated machine learning

1 Introduction

In recent years, the field of meta-learning has garnered increasing research interest in the machine learning community and beyond it [1]. This line of research originally draws inspiration from nature, where meta-learning is prevalent and manifests across multiple scales of learning [2]. While standard machine learning focuses on finding models that achieve high performance on a given task, the aim of meta-learning [2,3] (colloquially described as *"learning how to learn"*) is to identify models that are capable of quickly and sample-efficiently learning novel, previously unseen tasks. This is done by acquiring some meta-knowledge [4] ω which facilitates further learning. Depending on the particular approach, ω can represent the initial model parameters, models themselves, hyperparameters, data points, algorithms, metaheuristics, hyperheuristics, objective functions, mutation rates, step sizes, or any other aspect of the learning procedure.

© The Author(s), under exclusive license to Springer Nature Switzerland AG 2025
P. García-Sánchez et al. (Eds.): EvoApplications 2025, LNCS 15612, pp. 171–186, 2025.
https://doi.org/10.1007/978-3-031-90062-4_11

Meta-learning is considered successful if the meta-learner's *"performance at each task improves with experience and with the number of tasks"* [3]. It can be performed through either gradient-based or gradient-free optimizers; in the case of the latter, evolutionary computation (EC) is commonly leveraged [5,6] due to its simplicity and the capacity of EC metaheuristics to serve as black-box function optimizers. Standard meta-learning approaches stop at the first level above learning; they perform first-order meta-learning without involving higher levels (orders), such as meta-meta-learning (learning how to learn how to learn) and so on. Its generalization, *higher-order* (or higher-level) *meta-learning*, involves the use of multiple stacked levels of meta-parameters, with each level acquiring some meta-knowledge about the level below it, in principle leading to an infinite hierarchy of levels. Put differently, the meta-knowledge encoded in ω depends on some meta-meta-knowledge ω', which is in turn affected by ω'', etc. Very recently, an instance of this paradigm - higher-order mutation rates - has been explored in combination with population-based EC methods. In such a setup, each population member is assigned its own mutation rate, along with the mutation rate of the mutation rate (the meta-mutation rate), and so on, until the final, n-th order meta-mutation rate. Generally, the k-th order (meta)-mutation rate defines the mutation rate of the $(k-1)$-th order (meta)-mutation rate. The top (n-th) mutation rate is either equal to itself (*self-referential*), or some fixed rate. Self-referential mutations are particularly attractive, offering a step toward automated learning, as they bypass the need for the top mutation rate hyperparameter. The n orders (levels) of meta-learning altogether form a "tower" of meta-mutations. Higher-order mutation rates have been successfully applied to very rudimentary [7] as well as more sophisticated settings [8] (non-stationary and competitive). Their use is also suitable in the context of online and continuous learning [9,10], where novel tasks are provided sequentially, without prior knowledge of the task distribution.

Our paper contributes to the body of research on higher-order meta-learning [7,8], self-adaptive mutation rates [11–17], and self-adaptivity [18,19] in general, in the context of EC. To a certain extent, particularly through the analysis of self-referential mutation rates, it also enriches the literature on automated machine learning (auto-ML) [20]. The main contributions are as follows:

1. We perform numerical experiments to shed new light on the behavior of Gaussian higher-order mutations, including those introduced in [7,8]. As part of this, we also consider mutation schemes with an instantaneous propagation of mutations in a top-down fashion, enabling an immediate evaluation of higher-order mutation rates (i.e., meta-parameters).
2. We investigate the use of different higher-order mutation schemes on different synthetic fitness functions, including dynamic ones. Such fitness functions are especially relevant in the context of EC in dynamic (non-stationary) environments. Moreover, we particularly study the extent to which selective pressure permeates higher (meta)-mutational orders under such circumstances.
3. We observe the effect of catastrophic divergence, happening for higher n (order) values on longer time periods even under stabilizing selective pressure.

2   Related Research

2.1   (First-Order) Self-adaptive Mutation Rates

The existing research on self-adaptive mutation rates[1] in the areas of EC and artificial life is relatively rich [11,12,14–17,21,22]. Bäck's work [21] was among the first attempts at self-adaption of mutation rate settings. Instead of having a global, fixed mutation rate, Bäck incorporates mutation rates into the individual's genotypes, hence enabling self-organizing behavior of the population. Clune *et al.* [12] demonstrate that evolution fails at optimizing mutation rates for long-term adaptation when rugged fitness landscapes are involved. Wilke *et al.* [17] show that high mutation rates lead to the survival of solutions located in flatter regions of the fitness surface, due to their increased robustness to mutations. Bedau *et al.* [11] find that mutation rates adapt as the evolutionary demands for novelty vary. However, it remains an open question *whether* and *how* the introduction of higher-order mutation rates would affect all these results.

2.2   Higher Order Mutation Rates

While still rare in the literature, several recent publications study the use of higher-order mutation rates, including self-referential ones [7,8], in various EC and evolutionary reinforcement learning contexts.

Among the pioneers of such research, Nellis [23] discusses the use of arbitrarily high "towers" of (meta)-meta-parameters within an artificial chemistry setting, and the ensuing problem of determining the optimal height of such towers.

Lu *et al.* [7] use a simple setting to empirically demonstrate that population-based evolution implicitly optimizes meta-mutation rates of arbitrarily high order. The authors use an additive mutation scheme, meaning that the mutation rate at the i-th order is updated by adding the $(i+1)$-th order meta-mutation rate, along with some Gaussian noise. However, the employed environment (Numeric Fitness World from [24])) is peculiar in the sense that it sets the fitness of a genome to the value of its first gene; therefore, whatever mutational mechanism can cause this value to diverge the fastest is likely to prevail. On the theoretical side, the authors also prove that, under some circumstances, evolution selects for arbitrarily high orders of meta-learning. While the result holds in principle, its impact on a broader range of practical problems remains somewhat unclear.

In light of this, Coward *et al.* [8] show that self-adaptive and self-referential mutations improve performance in a variety of practical machine learning settings, including non-stationary, adversarial, and multi-agent tasks. More concretely, such mutations are shown to lead to improved hyperparameter robustness in non-stationary population-based training tasks and unsupervised environment design for curriculum generation, as well as to complex adaptations in competitive multi-agent game-theoretic settings. The importance of self-referential mutation

---

[1] For a more comprehensive overview of self-adaptation in evolutionary algorithms see [19].

rates for achieving hyperparameter-free systems is strongly accentuated. Finally, the theoretical results from [7] are generalized to real-vector input objective functions and mutation operators.

3   Methodology

3.1   Higher-Order Mutation Scheme

Gaussian Scheme. In a Gaussian case, the following higher-order (meta-)mutation scheme is used [8]:

$$\sigma'_n \sim \mathcal{N}\left(\sigma_n; \sigma^2_{\text{meta}}\right),$$
$$\sigma'_j \sim \mathcal{N}\left(\sigma_j; \sigma^2_{j+1}\right), \quad 1 \le j < n, \tag{1}$$
$$\boldsymbol{\theta}' \sim \mathcal{N}\left(\boldsymbol{\theta}; \text{diag}(\sigma^2_1)\right),$$

where $\boldsymbol{\theta}$ denotes a solution (here assumed to be a vector but possibly a scalar), $\mathcal{N}(\mu, \sigma^2)$ a Gaussian (normal) mutation operator parametrized by the mean μ and variance σ^2 (or in the multi-variate case, $\mathcal{N}(\boldsymbol{\mu}, \boldsymbol{\Sigma})$, with the mean vector $\boldsymbol{\mu}$ and the covariance matrix $\boldsymbol{\Sigma}$), $\text{diag}(x)$ a diagonal matrix where all the diagonal elements are equal to x, σ_j the j-th (meta)-mutation rate, and n the meta-learning order. The fixed top meta-mutation rate is given by σ_{meta}. Prime symbols ($'$) are used to denote parameters associated with the next generation (i.e., the offspring). All solution elements share the same (meta)-mutation rates. Note that it takes j generations for the changes on the j-th level (σ_j) to propagate to the solution level ($\boldsymbol{\theta}$) and hence affect the fitness value. In the pathological case $n = 0$, the mutation rate is constant. Equation 1 represents the n-th order **self-adaptive** case; the **self-referential** case is obtained by setting the meta-mutation rate of the top (n-th) meta-mutation rate equal to itself, recursively, i.e.: $\sigma'_n \sim \mathcal{N}\left(\sigma_n; \sigma^2_n\right)$. Besides Eq. 1, we also consider the scheme:

$$\sigma'_n \sim \mathcal{N}\left(\sigma_n; \sigma^2_{\text{meta}}\right),$$
$$\sigma'_j \sim \mathcal{N}\left(\sigma_j; (\sigma'_{j+1})^2\right), \quad 1 \le j < n, \tag{2}$$
$$\boldsymbol{\theta}' \sim \mathcal{N}\left(\boldsymbol{\theta}; \text{diag}(\sigma'^2_1)\right).$$

Unlike in Eq. 1, the new higher-order mutation rates are now used instantly to sample new lower-level mutation rates. To achieve this, it is necessary to start from the highest-order rate and proceed to the lowest, mutating the solution only at the very end, using the new mutation rate σ'_1 instead of σ_1. Using this scheme, the assumptions made in [7,8] about the delayed effects of meta-mutations are no longer valid. The scheme is known to lead to a much quicker evaluation of the effect of (meta)-mutation rates [25,26], as changes on the top level immediately affect the solution (and hence the fitness) through an instantaneous chain of effects, with propagation taking place within a single generation. This is particularly relevant given that evolutionary exploration is myopic [27] and prefers short-term over longer-term fitness. Self-referentiality can again be introduced as before. In what follows, the mutation scheme contained in Eq. 1 (resp., Eq. 2) is referred to as mutation with **delayed** (resp., **instantaneous**) **propagation**.

3.2   Genotype Representation and Algorithmic Settings

Each individual x is assigned its own set of meta-parameters (i.e., mutation rates), represented by a genotype vector of length $n + 1$:

$$x = \{\theta, \sigma_1, \ldots, \sigma_n\}. \tag{3}$$

Consequently, meta-parameter values vary both across evolutionary generations and within the population. With such representation, population-based techniques can easily be used to perform higher-order meta-learning, eliminating the need for a global mutation rate, similarly to the setup from [21]. A simple generational genetic algorithm (GA), without recombination/crossover, is employed. Unless noted otherwise, elitism is used in the GA.

4   Experimental Results

4.1   Random Walk

We first consider the case without any selective pressure, where each individual produces a single offspring. Thus, each lineage (a sequence of successors of a solution from the zeroth generation) represents a random walk in the space of meta-parameters defined by a mutation scheme. Our preliminary experimentation indicates that, in the absence of selective pressure, the schemes from Eq. 1 and Eq. 2 lead to very similar results, so we consider only the latter. We discard solutions θ and focus only on (meta)-mutation rates.

Self-referential Higher-Order Mutation Rates. The experimental results for $n = 4$ and the use of self-referentiality with Eq. 2 are shown in Fig. 1, for a population size of $N_{\text{pop}} = 10^5$ and for $N_{\text{gen}} = 2000$. First note that the top meta-mutation rates ($\sigma_n = \sigma_4$) are affected only by their own previous values, as it holds that $\sigma_n^{(i+1)} \sim \mathcal{N}(\sigma_n^{(i)}; (\sigma_n^{(i)})^2)$, where i denotes evolutionary generation. It follows that $\sigma_n^{(i+1)} = \sigma_n^{(0)} \prod_{k=1}^{i+1}(1 + z^{(k)})$, where $z^{(k)}$ are independent realizations of the standard normal distribution, and $\sigma_n^{(0)}$ the initialization value. The process $(\sigma_n^{(i+1)})$ seems to converge to 0 as $i \to \infty$. This happens rapidly; among N_{pop} lineages encountered in our simulation, 95.4% had $|\sigma_4| < 10^{-3}$ already after one hundred generations. After two hundred generations, this percentage was 99.6%. The mutation rates right beneath (σ_3) tend to fluctuate shortly in the beginning, until the corresponding σ_4 values converge. When this happens, σ_3 meta-parameters of different individuals get stuck at different values, remaining constant afterward. Due to such behaviors of the top two mutation rates (σ_n and σ_{n-1}), suboptimal performance when using self-referentiality in combination with low n (order) values is anticipated. Furthermore, since σ_3 values serve as mutation rates of σ_2 rates, the latter move in a random walk fashion, starting from the initialization value and then meandering, driven first by a variable and then a roughly constant mutation rate (following the convergence of

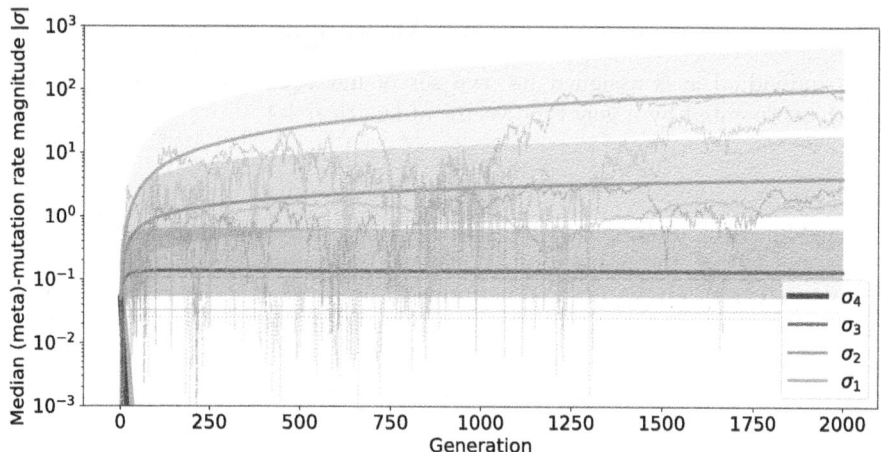

Fig. 1. Random walk trajectories generated by a self-referential Gaussian higher-order mutation scheme with $n = 4$. Median mutation rate magnitudes along with the 25% and 75% percentiles (shaded regions) are shown. A random uniform initialization $\mathcal{U}(0, 0.1)$ is used. Moreover, mutation rate magnitudes for three randomly chosen lineages are illustrated as lighter fluctuating lines of the corresponding colors and styles.

the associated σ_4 value). Assuming that a particular σ_3 locks in on some value v during generation j, it can be easily shown that $\mathbb{V}(\sigma_2^{(j+k)}) \approx kv^2$ for $k > 0$, due to normality. Hence, its variance increases linearly with evolutionary time. Furthermore, σ_1 meta-parameters exhibit even more complex behavior, as they have stochastic variances equal to σ_2 meta-parameters, whose own variances increase. The erratic behavior of σ_1 is seen in Fig. 1, indicating very large median mutation magnitudes and high variances within individual lineages. Setting the order n to a larger value ($n > 4$) would result in the lowest mutation rates becoming even more fickle, with the mutation rate of their mutation rate dynamically changing, and so on. Therefore, we observe cascading variation, with mutations propagating from higher to lower levels in a top-down manner, leading to the compounding of variation at the bottom of the hierarchy. We conclude that the increase in the order n improves variability in the lowest-level rates, with potential implications for the exploration-exploitation dilemma.

Higher-Order Mutation Rates with the Fixed Top Meta-mutation Rate. First note that, since there is now no self-referentiality involved, $\sigma_n^{(i)}$ is a sum of i normally distributed variables and is as such itself normally distributed with $\mathbb{E}(\sigma_n^{(i)}) = \sigma_n^{(0)}$ (the initialization value) and $\mathbb{V}(\sigma_n^{(i+1)}) = (i + 1)\sigma_{meta}^2$. Hence, its variance increases linearly with evolutionary time, as seen in the self-referential case with σ_2 (or more generally σ_{n-2}). Due to the hierarchical nature of the meta-mutational scheme, there is again a significant propagation of variation from higher to lower levels.

4.2 Performance on Synthetic Functions

In this section, we introduce selective pressure through fitness functions. The next generation is populated by the offspring of the top-k individuals (randomly sampled for reproduction) from the current generation, as well as the top-k individuals themselves (if elitism is used). We investigate how optimizing different fitness functions, including dynamic ones, is reflected in higher-order mutation rates under different mutation schemes, and to what extent one can interpret the evolution of these rates. We particularly study the extent to which selection, exerted directly on the solution ($\boldsymbol{\theta}$) level through some fitness function (f), reaches higher-order mutation rates (σ_1, ..., σ_n). This question also holds significance in the related fields of artificial life and evolutionary and theoretical biology. While it has been shown [7,8] that in principle selection can work on arbitrarily high levels of meta-learning, it is unclear what constitutes a beneficial higher-order meta-parameter under dynamic schemes such as those presented in Eq. 1 and 2. We use the following fitness functions to maximize:

$$f_1(\boldsymbol{\theta}) = -\sum_{j=1}^{D} \theta_j^2, \tag{4}$$

$$f_2^i(\boldsymbol{\theta}) = -\sum_{j=1}^{D} (\theta_j - \delta_j^i)^2, \tag{5}$$

$$f_3^i(\boldsymbol{\theta}) = \begin{cases} -|\boldsymbol{\theta}|, & i \bmod 2 = 0 \text{ and } i \in [200k, 200k + 100] \text{ for } k \in \mathbb{Z}_{\geq 0}, \\ -|\boldsymbol{\theta} - \mathbf{1}_D|, & i \bmod 2 = 1 \text{ and } i \in [200k, 200k + 100] \text{ for } k \in \mathbb{Z}_{\geq 0}, \\ -|\boldsymbol{\theta}|, & i \bmod 2 = 0 \text{ and } i \in [200k + 100, 200k + 200] \text{ for } k \in \mathbb{Z}_{\geq 0}, \\ -|\boldsymbol{\theta} - \mathbf{10}_D|, & i \bmod 2 = 1 \text{ and } i \in [200k + 100, 200k + 200] \text{ for } k \in \mathbb{Z}_{\geq 0}, \end{cases} \tag{6}$$

where D denotes the dimensionality of the solution, and $\boldsymbol{\delta}^i = (\delta_1^i, \dots, \delta_D^i)$ is a global optimum. It moves every τ generations to a new location sampled from a random uniform distribution, i.e., $\boldsymbol{\delta}^{k\tau} \sim \mathcal{U}([-u, u]^D)$ for $k \in \mathbb{N}, u \in \mathbb{R}^+$. Also, $\mathbf{1}_D$ (resp., $\mathbf{10}_D$) is d-dimensional vector with all elements equal to 1 (resp., 10).

The (static) function f_1 (sphere function) promotes a mutation rate that is close to zero (once the optimum is reached). The optimization of f_2 (dynamic sphere function) requires a mutation rate that occasionally jumps (at breakpoints $k\tau$). Lastly, f_3 promotes a mutation rate that oscillates between a smaller and a larger value, depending on the regime. In all cases, the search space is given by $[-u, u]^D$, and a uniform random initialization from this range is used for initializing θ_i values. A random uniform initialization $\mathcal{U}(0, 0.1)$ is employed for the meta-parameters. We refer to a run of a simulation as successful if it manages to find a solution for which the absolute value of the fitness function is below a threshold called the *value to reach* (VTR). Moreover, the *function evaluations to success* (FES) metric is defined as the number of function evaluations needed to discover a successful solution. Lastly, the *expected running time* (ERT) is given as:

$$\text{ERT} = \frac{\mu\,\overline{\text{FES}} - (1 - \mu)\,\text{FE}_{\max}}{\mu}, \tag{7}$$

where $\overline{\text{FES}}$ is the mean FES, $\text{FE}_{\max}$ the maximum number of function evaluations, and μ the success rate, given as the ratio of successful runs to total runs, following definitions from [28]. When using the dynamic fitness function f_2, the evolutionary period is split into segments that match the frequency of changes (τ). The metrics are then calculated separately per segment, and the values are finally averaged out.

We start with the f_1 function. Due to the use of uniform initialization over a large search space, we automatically test the capacity of higher-order mutations to address poor initialization. Since the goal boils down to converging to zero as quickly as possible, the problem is the opposite of that in [7], in which divergence of solutions is promoted. The number of (independent) experiments performed is set to $N_{exp} = 100$, the population size to $N_{pop} = 100$, and the number of generations to $N_{gen} = 200$, yielding $\text{FE}_{\max} = 20000$. Additionally, $k = 10$, $u = 100$, and VTR $= 0.01$. The experimental results are depicted in Fig. 2.

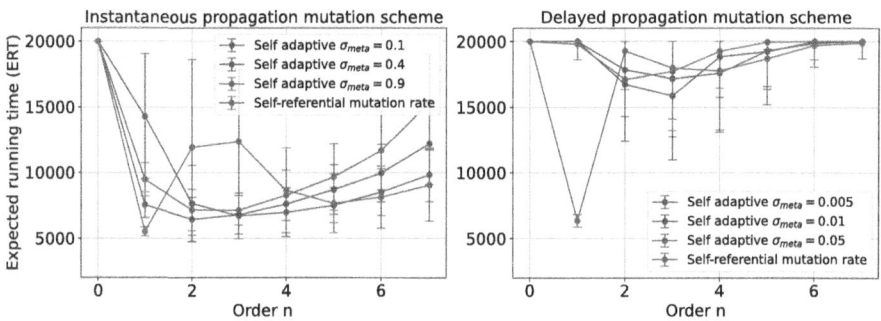

Fig. 2. Expected running times (ERT) on f_1 with standard deviations for mutation schemes with different orders n, types (self-referential vs self-adaptive, delayed vs instantaneous), and fixed top mutation rates σ_{meta}.

Overall, instantaneous propagation results in substantially smaller ERT values. The reason behind this likely lies in its immediate evaluation of the meta-parameters. Curiously, the best results are attained for the self-referential mutation rate with $n = 1$; this holds across propagation types, with mean ERTs ranging between 5000 and 7500. Under this scheme, the same mutation rate is used to mutate itself and the solution[2]. As n is increased to $n = 2$, a strong decline in performance for self-referentiality is seen. This stems from the fact that the top two mutation rates tend to either zero (σ_n) or constant values (σ_{n-1}),

---

[2] This dominance of $n = 1$ is not expected to hold generally, as finding optimal solutions and mutation rates represent two different types of problems [23].

as elaborated in 4.1. While these rates can grow when affected by the selective pressure from the bottom level, the mechanism is not efficient enough to propel rapid changes in θ. This is ameliorated by further increases in n (up to $n = 5$ for the instantaneous and $n = 4$ for the delayed scheme), which generally improve performance with self-referentiality, although still below the level of $n = 1$. Further increases in n decrease performance, especially under delayed propagation, where practically all runs are unsuccessful already for $n = 6$. With many orders, noise from various meta-learning levels accumulates, making it increasingly difficult for selection to distinguish between better and worse higher-order meta-parameters when using feasible population sizes. Under self-adaptive schemes, performance seems to follow U-shaped trajectories, with optima most commonly located at $n = 2$ or $n = 3$, depending on the σ_{meta} value. While the performance of such schemes is relatively competitive with the self-referential variant with $n = 1$ under instantaneous propagation, this is not the case when using delayed propagation mutations. We next analyze the behavior of higher-order mutation rates in the case of self-referentiality, based on the same experimental results, as shown in Fig. 3. With $n = 1$, both the solution and the mutation rate assume values close to zero quickly, for both types of propagation. While advantageous for convergence, this is potentially problematic in the context of tracking optima in dynamic (non-stationary) environments, where exploration should be done continuously. Clearly, the selective pressure reaches both the solution and mutation (σ_1) levels, although there is more noise and a lower magnitude when delayed propagation is used. In the $n = 5$ case, higher-order mutation rates react more swiftly under instantaneous propagation, while under delayed propagation, median (meta)-mutation rates remain approximately constant even after the solutions are close to the VTR, with generally much less responsiveness.

We also study what happens for even larger n values, under the same experimental settings as before, except that now $N_{gen} = 1000$, $n = 15$, and elitism is not used. Interestingly, the use of large n values over longer evolutionary periods can lead to divergent behavior, even in the presence of selective pressure pointing towards convergence. What happens is that individuals with deleterious values of higher-order meta-parameters, but beneficial or good-enough lower-order meta-parameter values, sometimes populate the entire new generation with their offspring. This can result in the gene pool being overwhelmed by genes that lead to long-term fitness reduction (in this case divergence of solutions), or, in the biological realm, to entire species going extinct. Since evolution is myopic, there is no direct mechanism to prevent this. An example of such catastrophic divergence is given in Fig. 4, where uncontrollable growth in higher-order mutation rates afflicts lower-order rates in a cascading manner. Such pernicious events bear some resemblance to the evolutionary concepts of mutational meltdown [29] and evolutionary suicide [30]. Clearly, if elitism is used at least some solutions will not diverge. Our experimentation also shows that this effect requires smaller n values under the delayed scheme, in which the evaluation of higher-order rates is deferred, and hence it is easier for deleterious values to propagate.

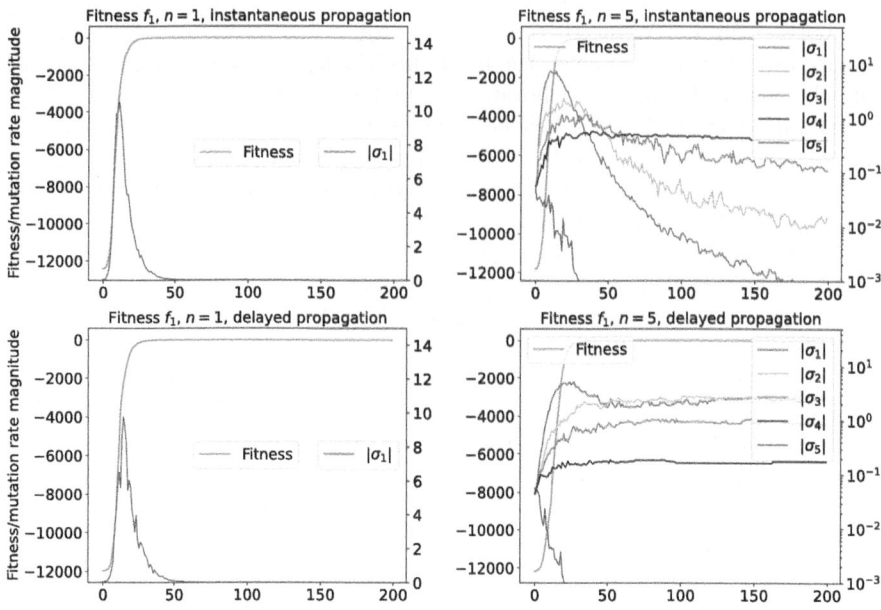

Fig. 3. The behavior of (meta)-mutation rates on f_1 for different mutation schemes, with self-referentiality. The median (meta-)mutation rate magnitudes and solutions for the best-of-generation individual are shown. Note the dual y-axes: one for the fitness values and another one for the (meta-)mutation rate magnitudes. In the sub-graphs on the right-hand side, log scales are used for the secondary y-axes.

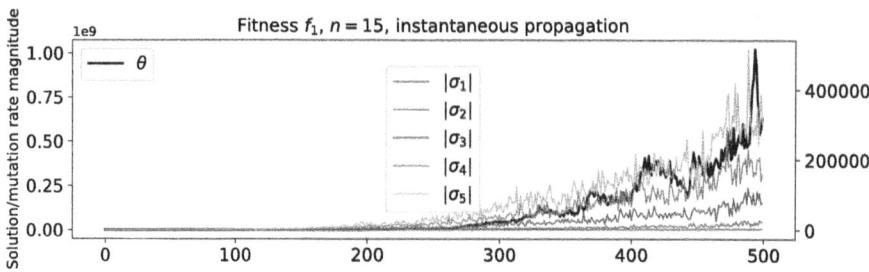

Fig. 4. Divergence of solutions and mutation rates (only the lowest five (meta)-mutation rates are shown) with f_1, $n = 15$, instantaneous propagation, with self-referentiality. As before, median magnitudes are shown.

We now turn to the second fitness function (f_2), which generalizes f_1 to circumstances where the optimum shifts. The same settings (including initialization) are used as with f_1, except that now $N_{gen} = 1600$. Additionally, we set $\tau = 400$, resulting in three shifts during the evolutionary period.

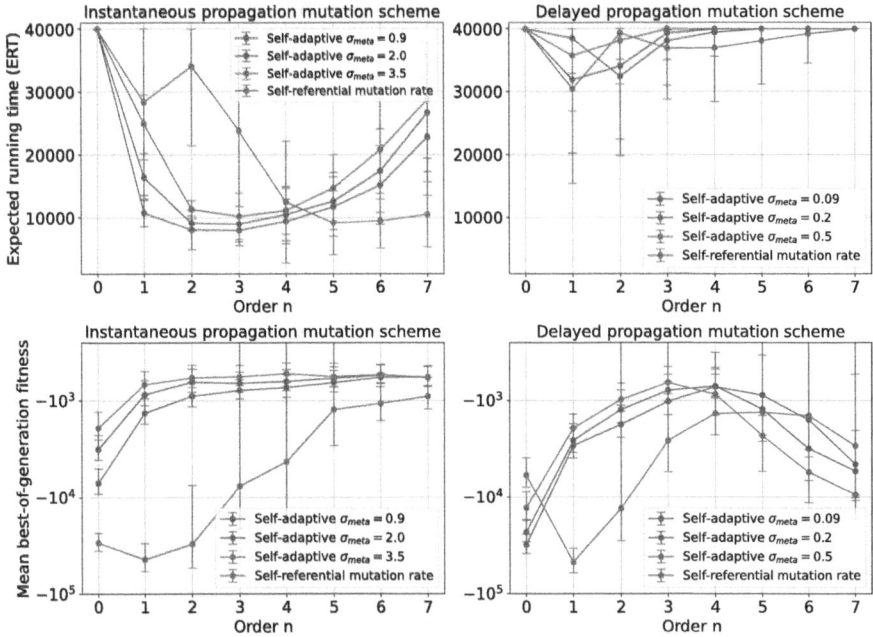

Fig. 5. Expected running time and mean best-of-generation fitness on f_2 averaged out over all generations with standard deviations for different mutation schemes. Again, lower σ_{meta} values are used under a delayed propagation mutation scheme.

The results are shown in Fig. 5. As before, note the substantial differences in performance, with the delayed propagation scheme failing to attain low ERT (or large fitness) values. Unlike the case with f_1, the use of higher n values with self-referential schemes, up to some order, generally leads to better performance under instantaneous propagation. While the lowest ERT values are attained already for $n = 1$ (delayed propagation), higher n values lead to significantly higher mean best-of-generation fitnesses and, generally, to solutions of higher quality, meaning that the movements of optima are tracked better, although VTRs are not reached. Self-adaptive mutation rates significantly outperform self-referential ones with respect to both metrics. Again, optimal n values depend on σ_{meta} and tend to be lower for ERT's than for mean fitness, especially in the instantaneous case. The behavior of higher-order mutation rates (self-referential case) is shown in Fig. 6 for $n = 1$ and $n = 5$. Note that, for $n = 1$, with either type of propagation, the initial optima are successfully found, and the mutation rate quickly plummets after the initial spike. For this reason, the process remains stuck, and shifts in the environment are unable to be tracked, as the mutation rate remains extremely low. This is not the case with the use of higher-order mutation rates where tracking is much more successful, as seen in the figure. For example, with the combination ($n = 5$, instantaneous), the spikes in the mutation rates clearly coincide with breakpoints, as jumps are required. Higher-order mutation rates ($\sigma_2, \sigma_3, \ldots$) also exhibit cyclic patterns, similar to those of σ_1, but with lower maximum magnitudes and ranges. This is in line with the previous observations regarding the amplification of mutations at the lowest levels. With delayed propagation and $n = 5$, the patterns are again less pronounced due to a lack of instantaneous meta-parameter evaluation, and the effect of environmental changes on higher meta-mutational levels is much weaker.

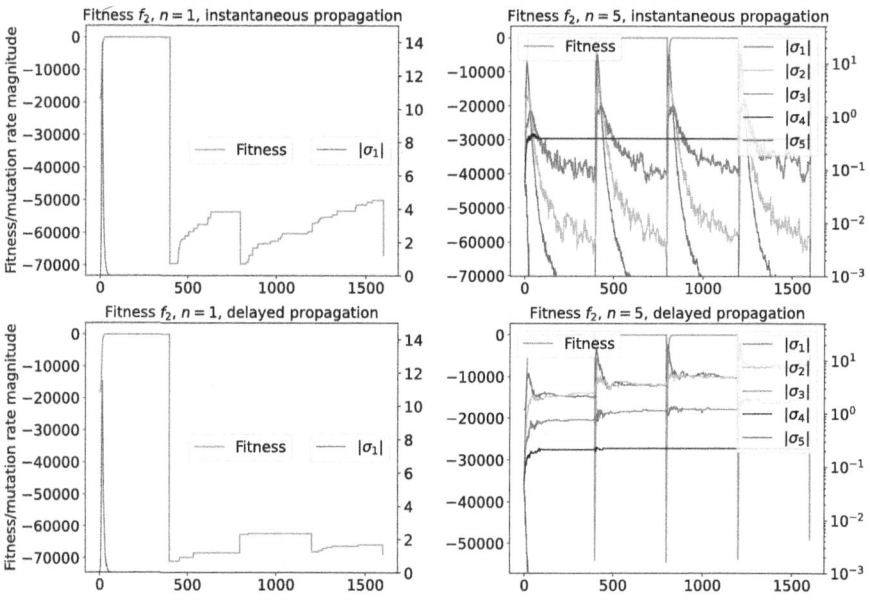

Fig. 6. The results for different mutation schemes on f_2, all with self-referentiality. The median (meta-)mutation rate magnitudes and solutions for the best-of-generation individuals are shown. The same remarks as in Fig. 3 apply.

We finally turn to f_3, using the same experimental settings as before, except $N_{gen} = 500$. The results are shown in Fig. 7. We again have performance peaking at $n = 2$ or $n = 3$ for all the considered σ_{meta} values, and the underperformance of self-referential rates. This highlights the benefits of higher-order mutation rates in scenarios where sudden, discontinuous jumps are required, as is the case with f_3. Roughly U-shaped performance of self-referential rates is noticeable on the considered range of orders, with minima attained for $n = 2$, as before. The behavior of mutation rates, including higher-order ones, is shown in Fig. 8

In all cases, optima are tracked relatively successfully, except for the ($n = 5$, delayed) combination, with solutions oscillating from a small value to 1 (or 10, depending on the regime). We focus on instantaneous propagation. The magnitude of the lowest mutation rate (σ_1) shows a serrated pattern, undergoing jumps after each regime change, thereby successfully self-adapting to the rate of change of the underlying dynamic fitness function. The activity of σ_2 meta-parameters,

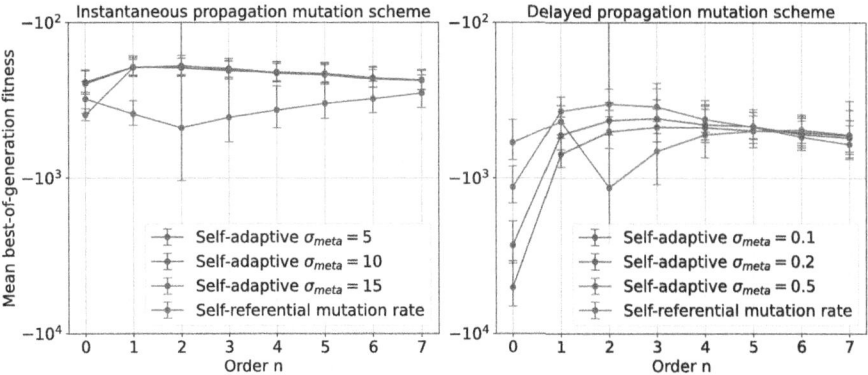

Fig. 7. Mean best-of-generation fitness on f_3 averaged out over all generations with standard deviations for different mutation schemes. Log y-axes are used.

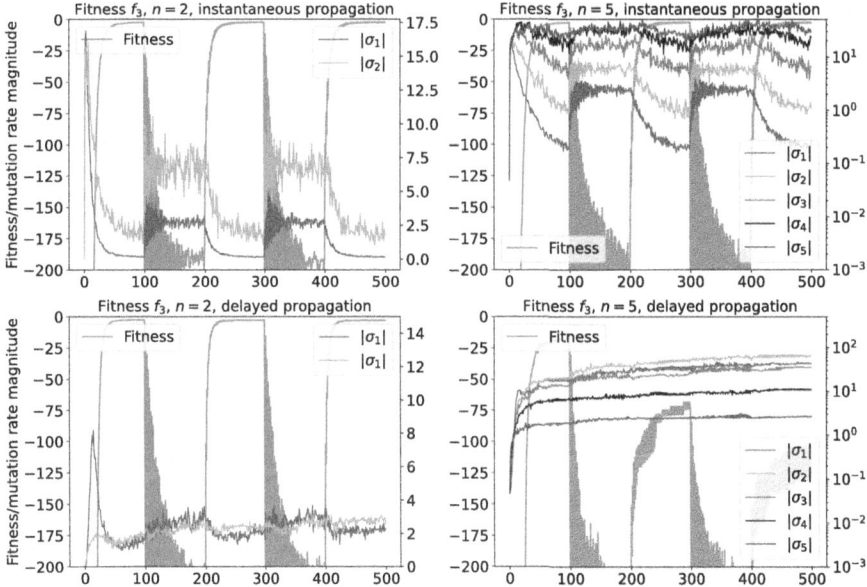

Fig. 8. The behavior of (meta)-mutation rates for different mutation schemes on f_3, with self-referentiality, for different mutation schemes. The median (meta-)mutation rate magnitudes and solutions for the best-of-generation individuals are shown. $\sigma_{\mathrm{meta}} = 15$ (delayed), $\sigma_{\mathrm{meta}} = 0.5$ (instantaneous)

which determine the mutation rates of σ_1, is similar to that of σ_1, while exhibiting more noise and less pronounced spikes, and a lower magnitude. The reason for this is clear: the selective pressure on this level is more indirect and hence weaker. For the same reason, higher-order mutation rates ($n = 5$) exhibit even less regularity, playing no particular role and even deteriorating performance.

In summary, the behavior of higher-order mutation rates significantly varies between the delayed and instantaneous mutation propagation schemes, with our results pointing to significant limitations of the former. In both cases, higher-order rates exhibit a certain degree of interpretability and can contribute to improving performance. However, especially for the delayed propagation scheme, most information is contained already within the first two orders, with higher orders being largely unaffected by the selective pressure, and failing to serve a substantial purpose. The optima are mostly found for $n \leq 3$. This is well anticipated - the pressure to learn suitable meta-mutation rates might also decrease as we ascend the hierarchy, as the influence gets diluted due to multiple noisy samplings, thereby obscuring potentially beneficial higher-order mutations. Adaptive benefits also diminish as most opportunities for improvement are already covered within lower levels. As we ascend the hierarchy, the selective pressure gets more and more indirect, and hence weaker.

5   Conclusion

In this paper, we study the use of self-referential higher-order mutations under different selective pressure dynamics. Our main takeaway is that the instantaneous propagation scheme shows good performance under different fitness functions, even in the self-referential case, which requires no hyperparameter tuning. However, it is still in most cases outperformed by self-adaptive mutation rates with fixed σ_{meta} hyperparameters, with the results exhibiting substantial sensitivity to their values. We also find that, for higher n values, divergence in mutation rates happens, leading to catastrophic performance. This happens even in the presence of selective pressure because of the propagation of mutations. Therefore, even if self-referentiality is used, the order n still remains a hyperparameter to tune, and a further step towards auto-ML would be to sidestep its use. With this in mind, in the following work, we plan to allow the order n to co-evolve with the solutions and meta-parameters (meta-mutation rates), thereby enabling the order of meta-learning to change according to the environmental conditions. The height of the tower of meta-mutations would therefore dynamically change instead of being predetermined. This would introduce another element of self-adaptation to the corresponding GA. The same holds true for the top-k hyperparameter which quantifies the current level of selective pressure, which could also be encoded into the individual's gene pool. Lastly, other mutation schemes could be studied, including those employing non-normal distributions (especially heavy-tailed ones like Cauchy or Student's t).

Acknowledgments. This publication is supported by the European Union's Horizon Europe research and innovation programme under the Marie Skłodowska-Curie Postdoctoral Fellowship Programme, SMASH co-funded under the grant agreement No. 101081355.

References

1. Peng, H.: A comprehensive overview and survey of recent advances in meta-learning. arXiv preprint arXiv: 2004.11149 (2020)
2. Wang, J.X.: Meta-learning in natural and artificial intelligence. Current Opinion Behav. Sci. **38**, 90–95 (2021)
3. Thrun, S., Pratt, L.: Learning to learn: Introduction and overview. In: Learning to Learn, pages 3–17. Springer (1998). https://doi.org/10.1007/978-1-4615-5529-2_1
4. Hospedales, T., Antoniou, A., Micaelli, P., Storkey, A.: Meta-learning in neural networks: a survey. IEEE Trans. Pattern Anal. Mach. Intell. **44**(9), 5149–5169 (2021)
5. Houthooft, R., et al.: Evolved policy gradients. Adv. Neural Inform. Process. Syst. **31** (2018)
6. Abraham, A.: Meta learning evolutionary artificial neural networks. Neurocomputing **56**, 1–38 (2004)
7. Lu, C., Towers, S., Foerster, J.: Arbitrary order meta-learning with simple population-based evolution. In: ALIFE 2023: Ghost in the Machine: Proceedings of the 2023 Artificial Life Conference. MIT Press (2023)
8. Coward, S., Lu, C., Letcher, A., Jiang, M., Parker-Holder, J., Foerster, J.N.: Higher order and self-referential evolution for population-based methods. In: Automated Reinforcement Learning: Exploring Meta-Learning, AutoML, and LLMs (2024)
9. Cesa-Bianchi, N., Lugosi, G.: Prediction, learning, and games. Cambridge university press (2006)
10. Finn, C., Rajeswaran, A., Kakade, S., Levine, S.: Online meta-learning. In: International Conference on Machine Learning, pp. 1920–1930. PMLR (2019)
11. Bedau, M.A., Packard, N.H.: Evolution of evolvability via adaptation of mutation rates. Biosystems **69**(2–3), 143–162 (2003)
12. Clune, J., Misevic, D., Ofria, C., Lenski, R.E., Elena, S.F., Sanjuán, R.: Natural selection fails to optimize mutation rates for long-term adaptation on rugged fitness landscapes. PLoS Comput. Biol. **4**(9), e1000187 (2008)
13. Kaneko, K., Ikegami, T.: Homeochaos: dynamics stability of a symbiotic network with population dynamics and evolving mutation rates. Physica D **56**(4), 406–429 (1992)
14. Riechmann, T.: Learning how to learn-improved mutation within ga learning. IFAC Proc. Vol. **31**(16), 123–127 (1998)
15. Dang, D.-C., Lehre, P.K.: Self-adaptation of mutation rates in non-elitist populations. In: Handl, J., Hart, E., Lewis, P.R., López-Ibáñez, M., Ochoa, G., Paechter, B. (eds.) PPSN 2016. LNCS, vol. 9921, pp. 803–813. Springer, Cham (2016). https://doi.org/10.1007/978-3-319-45823-6_75
16. Beyer, H.-G., Schwefel, H.-P.: Evolution strategies-a comprehensive introduction. Nat. Comput. **1**, 3–52 (2002)
17. Wilke, C.O., Wang, J.L., Ofria, C., Lenski, R.E., Adami, C.: Evolution of digital organisms at high mutation rates leads to survival of the flattest. Nature **412**(6844), 331–333 (2001)
18. Rechenberg, I.: Evolutionsstrategie: Optimierung technischer systeme nach prinzipien der biologischen evolution. frommann-holzbog, stuttgart, 1973. Step-Size Adaptation Based on Non-Local Use of Selection Information. In Parallel Problem Solving from Nature (PPSN3) (1994)

19. Meyer-Nieberg, S., Beyer, H.-G.: Self-adaptation in evolutionary algorithms. In: Parameter Setting in Evolutionary Algorithms, pp. 47–75. Springer (2007). https://doi.org/10.1145/3583131.35904
20. Weng, Z.: From conventional machine learning to automl. J. Physics: Conf. Ser. 1207, 012015 (2019)
21. Bäck, T., et al.: Self-adaptation in genetic algorithms. In: Proceedings of the First European Conference on Artificial Life, pp. 263–271. MIT press Cambridge (1992)
22. Kumar, A., Liu, B., Miikkulainen, R., Stone, P.: Effective mutation rate adaptation through group elite selection. In: Proceedings of the Genetic and Evolutionary Computation Conference, pp. 721–729 (2022)
23. Nellis, A.: Towards meta-evolution via embodiment in artificial chemistries. PhD thesis, University of York (2012)
24. Frans, K., Witkowski O.: Population-based evolution optimizes a meta-learning objective. arXiv preprint, arXiv: 2103.06435 (2021)
25. Simon, D.: Evolutionary optimization algorithms. John Wiley & Sons (2013)
26. Katona, A.: Evolution of evolvability for neuroevolution. PhD thesis, University of York (2023)
27. Watson, R.A., Szathmáry, E.: How can evolution learn? Trends Ecol. Evolution 31(2), 147–157 (2016)
28. Tang, K., Li, X., Suganthan, P.N., Yang, Z., Weise, T.: Benchmark functions for the cec'2010 special session and competition on large-scale global optimization. Nat. Inspired Comput. Appli. Laboratory, USTC, China 24, 1–18 (2007)
29. Lynch, M., Bürger, R., Butcher, D., Gabriel, W.: The mutational meltdown in asexual populations. J. Heredity 84(5), 339–344 (1993)
30. Parvinen, K.: Evolutionary suicide. Acta. Biotheor. 53(3), 241–264 (2005)

Evolutionary Reinforcement Learning for Interpretable Decision-Making in Supply Chain Management

Stefano Genetti[1], Alberto Longobardi[2], and Giovanni Iacca[1]([⊠])

[1] University of Trento Via Sommarive 9, 38123 Povo (Trento), Italy
{stefano.genetti,giovanni.iacca}@unitn.it
[2] Adige BLM Group Via per Barco 11, 38056 Levico Terme (Trento), Italy
alberto.longobardi@blmgroup.it

Abstract. In the context of Industry 4.0, Supply Chain Management (SCM) faces challenges in adopting advanced optimization techniques due to the "black-box" nature of most AI-based solutions, which causes reluctance among company stakeholders. To overcome this issue, in this work, we employ an Interpretable Artificial Intelligence (IAI) approach that combines evolutionary computation with Reinforcement Learning (RL) to generate interpretable decision-making policies in the form of decision trees. This IAI solution is embedded within a simulation-based optimization framework specifically designed to handle the inherent uncertainties and stochastic behaviors of modern supply chains. To our knowledge, this marks the first attempt to combine IAI with simulation-based optimization for decision-making in SCM. The methodology is tested on two supply chain optimization problems, one fictional and one from the real world, and its performance is compared against widely used optimization and RL algorithms. The results reveal that the interpretable approach delivers competitive, and sometimes better, performance, challenging the prevailing notion that there must be a trade-off between interpretability and optimization efficiency. Additionally, the developed framework demonstrates strong potential for industrial applications, offering seamless integration with various Python-based algorithms.

Keywords: Supply Chain Management · Interpretable Reinforcement Learning · Grammatical Evolution · Decision Trees

1 Introduction

Disruptive technological innovations have consistently driven industrial revolutions, transforming production paradigms to enhance productivity [1]. In line with this trend, recent technological enhancements such as the Internet of Things, big data analytics, cloud computing, robotics, artificial intelligence, and augmented reality are motivating and enabling a *fourth industrial revolution*,

© The Author(s), under exclusive license to Springer Nature Switzerland AG 2025
P. García-Sánchez et al. (Eds.): EvoApplications 2025, LNCS 15612, pp. 187–203, 2025.
https://doi.org/10.1007/978-3-031-90062-4_12

commonly referred to as *Industry 4.0* [2]. Since its introduction, the subject has received increasing attention from both academics and companies which are increasingly compelled to adopt new technologies to maintain their competitiveness [3]. Notably, competition is increasingly supply chain-based rather than company-specific [4], with supply chains defined as interconnected networks that produce and deliver value [5]. In this domain, *Supply Chain Management* (SCM) has emerged as a key element of Industry 4.0 [6]. SCM involves the strategic coordination of business functions across companies to enhance both individual and collective performance [5]. Traditional decision-making approaches in SCM, often based on intuition and limited data, struggle in today's dynamic environment. On the other hand, modern decision-making, supported by *business intelligence*, leverages data and analytics for better performance and competitive advantage. While AI techniques have proven effective for this task, they often lack transparency, making their decision processes hard to interpret [7,8]. This misaligns with the interests of company stakeholders, who are hesitant to trust and deploy black-box AI-based solutions [9]. Instead, they tend to prefer solutions that reveal causal relationships and are perceived as more trustworthy and maintainable [10]. To address this issue, *interpretable* and *explainable* AI (IAI/XAI) AI solutions may provide a better alternative.

In this work, we adapt the IAI methodology from [11], in the following referred to as ELDT (Evolutionary Learning Decision Trees), for supply chain optimization. We integrate this approach within a simulation-based optimization framework which allows us to model complex, uncertain supply chain scenarios. We compare the performance of ELDT against alternative approaches based on heuristics, metaheuristics, and Reinforcement Learning (RL) across two case studies, including a real-world application in laser-cutting machine production. The results show that ELDT consistently provides solutions comparable to, and in some cases surpass, those of its competitors.

The rest of the paper is structured as follows. The next section reviews the background principles and the main related works. Section 3 describes the proposed approach. Section 4 defines the two supply chain optimization problems used to assess the approach. Section 5 details the experimental setup. Section 6 presents the experimental results. Finally, Sect. 7 concludes this work.

2   Background and Related Work

Recent advancements, including affordable computational power, have led to the adoption of AI in industrial processes, including SCM. Solving complex optimization problems in real-world organizations using exact methods is often infeasible, leading to the emergence of heuristics, metaheuristics, and deep learning approaches as a viable alternative [12]. While these models can effectively support decision-making, they are often "black-box" [13], meaning their internal workings are not visible or comprehensible, which makes it difficult to trace how decisions are made. This lack of transparency hinders their broader adoption [7], as managers and stakeholders remain accountable for outcomes produced by these models [8]. For instance, a deep neural network might perform well during training, but it can struggle when deployed, necessitating extensive testing for trustworthiness [7]. Still, if an error occurs, it would be highly impractical to identify the root cause within the model, potentially resulting in substantial financial losses and safety hazards [14–16]. Moreover, the lack of transparency becomes a critical issue when the algorithms must be audited or when regulatory authorities, such as the European Union, mandate that outcomes must be understandable. Ultimately, understanding causality is essential for decision-makers, enabling them to grasp cause-effect relationships and make informed choices.

The literature identifies two primary solutions to address this limitation: *Explainable AI* (XAI) and *Interpretable AI* (IAI) [10]. XAI refers to techniques and methods that make the behavior of AI systems more understandable, providing insights into how a model reaches its decisions and thereby enhancing trust and transparency. IAI focuses on creating inherently understandable models, such as decision trees (DTs), which allow for direct inspection and assessment of security and safety. Furthermore, interpretable models enable users to uncover new correlations, identify relevant features, and leverage domain expertise to optimize model performance. Despite their relevance for industrial decision-making, there is a noticeable lack of studies in the literature that propose XAI or IAI techniques for industrial optimization [17], particularly in SCM [18]. One contributing factor is the widely accepted notion that there is a trade-off between interpretability and performance [14,19], though this trade-off remains unproven [20]. Additionally, the relatively novel adoption of AI in supply chains has led researchers to prioritize exploring its potential performance improvements rather than focusing on explaining its outputs and decisions [18].

One of the few directions that have been explored concerning the use of IAI for SCM regards demand forecasting: for instance, in [21] the authors employ a *neuro-symbolic* AI approach which combines neural networks with symbolic reasoning, enhancing interpretability by expressing neural decisions through logical rules [18]. In [22], the authors focus instead on supply chain risk management and utilize Ant Colony Optimization (ACO) to tackle the opacity of neural networks. In this case, the algorithm extracts intuitive rules (e.g., if-then-else statements) effectively representing the knowledge embedded in the black-box model. One notable disadvantage of these implementations is that they introduce explainability by approximating or simulating the behavior of a given black-box model being used, rather than representing the model *exactly*, which is instead the primary feature of IAI models. In [15], the authors introduce a framework using DTs, which are inherently interpretable, to classify supply chain suppliers based on their resilience capacity. Similarly, [14] emphasizes the synergy between AI models and supply chain experts in risk prediction. They compare a non-interpretable support vector machine (SVM) with an interpretable DT-based model for predicting delivery delays. The results show that while DTs achieve acceptable results, their displayed performance is inferior with respect to SVMs. The authors conclude that if precision is critical (e.g., where late deliveries incur significant costs) the weaker performance of the interpretable model may be unacceptable. Conversely, if understanding the factors behind delays is prioritized, a slight reduction in predictive accuracy might be acceptable.

3   Proposed Approach

This work builds on the approach proposed in [11], leveraging a two-level optimization scheme that combines evolutionary computation with RL to train interpretable policies in the form of DTs. In this method, the input to a DT, namely a one-dimensional feature vector, traverses the tree from the root through decision nodes until it reaches a leaf, which outputs the model's decision. Two key challenges arise in designing a DT for decision-making: ① determining the conditions for splits at non-terminal nodes, and ② assigning the appropriate action at each leaf. In the aforementioned two-level optimization scheme, the outer loop uses Grammatical Evolution [23] to evolve the DT structure, searching for optimal state space partitioning. The inner loop, instead, applies Q-learning to map each aggregated state (i.e., all states that lead to a particular leaf) to an appropriate action. Figure 1 shows a block diagram of the algorithm.

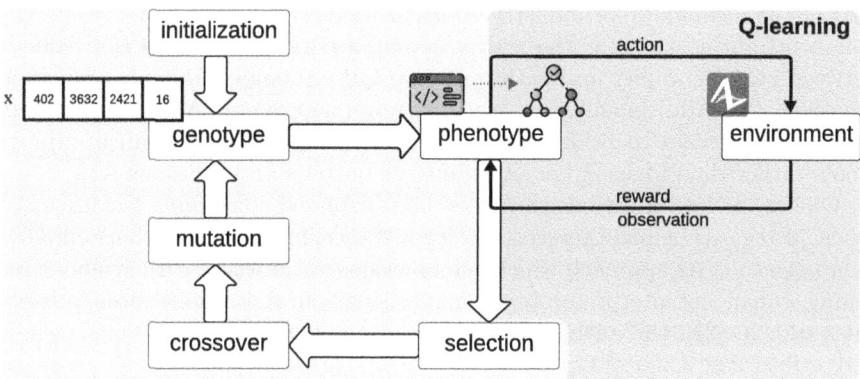

Fig. 1. Conceptual scheme of the proposed algorithm's internal workings. Blue blocks indicate components related to the evolutionary loop, while red blocks represent elements of the Q-learning loop. Adapted from [11].

Individual Encoding. Each individual in the population is represented by a fixed-length list of integers, where each integer ranges from 0 to g_{max}. The parameter g_{max} is set much higher than the total number of productions in the Backus-Naur Form (BNF) grammar that governs the evolutionary process, ensuring uniform selection of all productions. Each genotype has a fixed length, and the initial population is generated randomly.

Mutation Operator. To promote population diversity, uniform mutation is applied to each individual, such that each gene has a probability m_p of being replaced with a random value between 0 and g_{max}.

Crossover Operator. To generate new offspring by combining the genotypes of individuals across generations, one-point crossover is used. In this method, a random point is chosen along the parent genotypes, splitting them into two parts. The offspring are then created by exchanging, after this point, the segments between the two parents. This crossover is applied to selected parent pairs in each generation, with uniform probability.

Replacement of the Individuals. Each generation, a new population replaces the current one using steady-state selection, ensuring that strong individuals are preserved throughout the evolutionary process.

Fitness Evaluation. The list of integers contained in each individual's genotype is translated into its corresponding phenotype, i.e., a DT-based policy, by applying the production rules of a problem-dependent grammar, following the procedure outlined in [11]. If the genotype lacks sufficient genes during this translation, the individual is penalized with a low fitness score. Once the DT is created, its leaves are initialized with a set of actions, which are assigned a random value between -1 and 1. This DT defines the policy π, which is used by an RL agent to interact with an environment implemented using OpenAI Gym[1].

In each of the e episodes, π takes a state as input and selects an action at a leaf. To balance exploration and exploitation, action selection follows an ϵ-greedy strategy: with probability $1 - \epsilon$, the action with the highest value is chosen, otherwise a random action is selected. Feedback from the environment is used to update the corresponding leaf using Bellman's equation for Q-learning:

$$Q(s,a) \leftarrow (1 - \alpha)Q(s,a) + \alpha(r + \gamma \max_{a'}\{Q(s',a')\})$$

where $Q(s,a)$ is the value of action a at state s, α is the learning rate, r is the reward, s and a refer to the current state and action, and s' and a' refer to the next state and action, respectively. The fitness assigned to the individual is the mean of the returns obtained across the e episodes:

$$f(\pi) = \frac{1}{e}\sum_{i=1}^{e} R_i(\pi) = \frac{1}{e}\sum_{i=1}^{e}\sum_{k=1}^{T} r(s_k^i, \pi(s_k^i))$$

Here, $f(\pi)$ is the fitness of the agent encoding the policy π, $R_i(\pi)$ is the return obtained in the i-th episode (i.e., the sum of the rewards in the i-th episode), T is the number of steps per episode, r is the reward function, s_k^i is the k-th state in the i-th episode, and $\pi(s_k^i)$ is the action taken by the policy in the state s_k^i.

3.1   Simulation Based Optimization

This work aims to apply optimization/RL methods for SCM. Due to the inherent complexities of supply chains such as uncertainties, stochastic behaviors, and high dimensionality, creating an analytical solution for diverse industrial

---

[1] https://github.com/openai/gym.git.

applications is impractical. To address this, we propose a flexible simulation-optimization framework consisting of two key components: a simulation module, and an optimization module. The main design goals are applicability across various industrial scenarios, seamless integration with optimization/RL algorithms, and reasonable runtime performance.

For the simulation module, we adopt AnyLogic[2], a Java-based simulation software widely used to develop high-fidelity system simulations across various domains, including supply chains, business processes, transportation, and healthcare. Its established role in recent research further supports our choice [24–26]. The optimization module leverages algorithms written in Python, chosen for its broad adoption and availability of state-of-the-art implementations.

Communication between the AnyLogic's Java simulation environment and the Python-based optimizer is enabled by the open-source package ALPypeOpt[3]. Specifically, communication is achieved by implementing two methods from the `ALPypeOptClientController` interface: `setupAndRun`, which executes simulations using the decision variables provided by the optimizer, and `getModelOutput`, which allows the retrieval of the simulation results, such as revenue or production costs. Through an `AnyLogicModel` object, the optimizer interacts with AnyLogic, specifying inputs, triggering simulations, and retrieving outputs. In the optimization loop, the optimizer explores the solution space by proposing decision variable assignments, sends these to AnyLogic, and uses the simulation output to evaluate the objective function and guide the search.

4   Use Cases

We evaluated the proposed simulation-based optimization framework outlined in Sect. 3 in two supply chain problems, namely a make-or-buy decision problem and a hybrid flow shop scheduling problem, which are detailed below.

Make-or-Buy Decision Problem. The make-or-buy decision, also known as the *outsourcing decision*, is a common and critical task in SCM and industrial optimization [27]. In recent years, there has been growing recognition of the benefits associated with shifting from traditional in-house production to an outsourcing approach, where certain operations previously performed internally are delegated to external suppliers [28]. The ability to accurately determine which activities to retain in-house and which to outsource can significantly impact a company's profitability and competitive standing [29]. Outsourcing certain production stages instead of handling them internally can lead to reduced production times, savings in physical space, and simplification of business processes. However, the decision of what to make and what to buy is complex, as evidenced by the extensive literature proposing various frameworks to guide this process [30]. While outsourcing can offer flexibility and help companies respond to rapid market changes, it also carries risks, such as unexpected cost increases

---

[2] https://www.anylogic.com/.

[3] https://github.com/MarcEscandell/ALPypeOpt.git.

and heightened dependency on a broad range of suppliers [31]. As pointed out in [32], many companies lack a solid foundation for making these decisions, often relying on overhead costs as the primary factor guiding their strategy. A more holistic view that considers additional contributing aspects is necessary for making sustainable, long-term business decisions. In view of this, the use of interpretable AI can provide a significant contribution.

In this study, we examine an instance of the make-or-buy decision problem which can be further adapted to address more specific scenarios. Our focus is on the supply chain of a fictional company aiming to fulfill customer orders efficiently. Each order requires the assembly of a specified number of components A, B, and C, and has a given deadline. In the supply chain, four plants are involved: Plant A produces component A, Plant B produces component B, Plant C produces component C, and Plant D assembles the components to complete orders. The company can fulfill orders using its own plants or outsource the production. Outsourcing ensures timely completion but comes with a significant cost of 30 (generic cost units) per order. The primary objective of this single-objective optimization problem is to maximize the total revenue R, which is defined as:

$$R = 100 \times n_{ordersOnTime} + 50 \times n_{ordersOutOfTime} - 30 \times n_{ordersOutsourced}.$$

On-time deliveries earn the highest reward (100) for customer satisfaction. Late deliveries still contribute (50), acknowledging the value of completion despite delays. Outsourcing is penalized (-30) due to added costs and reduced control. This set of coefficients prioritizes efficient order completion, encouraging timely, internal fulfillment while minimizing delays and outsourcing. The AnyLogic simulation models uncertainties and stochastic behaviors, reflecting real-world variability in production and logistics that cannot be easily addressed through a purely analytical approach. These uncertainties are modeled by allowing production times at Plants A, B, and C to be not fixed but follow probability distributions. A truck cyclically visits these companies to transport components to Plant D (assembly unit). If no components are ready for pickup, the truck skips the stop. Loading, unloading, and travel times are also stochastic, with values sampled from uniform distributions within given ranges.

Given a number of orders n_{orders}, the decision variables are a set of binary values $\mathbf{x} = [x_1, x_2, \ldots, x_{n_{orders}}]$, where $x_i = 1$ indicates outsourcing an order, and $x_i = 0$ indicates fulfilling it using the internal supply chain. The objective is to find the optimal set of binary decisions $\mathbf{x}^*$ that maximizes the total revenue $R = f(\mathbf{x})$ obtained from the AnyLogic simulation given the outsourcing decisions. Even for small problem instances, the number of possible realizations of $\mathbf{x}$ is considerably large. For example, with an input of 100 orders, there are 2^{100} possible arrangements. This renders brute-force methods computationally infeasible, further justifying the need for AI-based solutions to explore this vast solution space and identify regions that yield good revenues.

Hybrid Flow Shop Scheduling Problem. Scheduling problems are ubiquitous across several application domains like manufacturing, services, and project management, each with its unique constraints and goals. In its simplest form, scheduling deals with assigning jobs to machines in a specific order to optimize a desired objective function. In more depth, the literature refers to *flow shop scheduling* as the optimization problem in which multiple jobs need to be processed on a series of machines, where each job must pass through all machines in a predetermined order. *Hybrid flow shop* (HFS) scheduling introduces an additional layer of complexity by allowing multiple machines at each stage of the production process. Unlike in a traditional flow shop, where each stage is served by a single machine, in HFS each stage can have multiple parallel machines. This reflects real-world scenarios where a production stage might have several identical or different machines working in parallel in order to introduce flexibility, increase capacities, and avoid bottlenecks in some operations that are more time-consuming [33]. As a result, HFS is a widely used production system that is applicable across various industrial sectors. For example, in automotive manufacturing, different parts of a vehicle might be assembled in parallel before being brought together in the final assembly line.

In its classical formulation, an HFS scheduling problem [34] involves a set of n input jobs $(j_1, j_2, \ldots, j_n)$, that must be processed sequentially through a workshop consisting of m stages. Each stage i may contain $m_i \geq 1$ machines operating in parallel and for at least one stage $m_i > 1$. The production flow is the same for all jobs: they proceed in order through stage 1, stage 2, and so on until stage m. All jobs and machines are available for processing from time zero, ensuring that no delays are caused by initial availability constraints. Within each stage, machines are identical in function but may have different processing times for different jobs. Each machine can process only one job at a time, and preemption is not allowed, meaning that once a job starts processing on a machine, it must continue until completion without interruptions. After finishing at one stage, a job is immediately available for processing at the next stage. Although jobs may wait between stages, the buffer space is assumed to be unlimited, so jobs can remain in the buffer until a machine becomes available. Given these constraints, the problem is to find a schedule that optimizes a given objective function. The most commonly used objective functions [35] for the HFS problem, as identified in the literature, include the makespan, the total flow time, the maximum tardiness, the total completion time, the total tardiness. Makespan refers to the total time required to complete all jobs, i.e., the completion time of the last job. Total flow time is the sum of the time each job spends in the system, from its release to its completion. Total completion time represents the sum of the completion times of all jobs. Lastly, total tardiness measures the total sum of delays for all jobs beyond their respective due dates. According to the results presented in [36], the HFS problem is NP-hard. This implies that it is infeasible to develop an algorithm that solves the problem with polynomial time complexity, even in the simplified scenario where there are two stages, one with two machines and the other with a single machine.

In this study, we consider a real-world application of the HFS problem. Our focus is on a supply chain scenario involving an Italian manufacturing company, a global leader in the production of laser-cutting machines for tubes. The objective of the supply chain is to produce and deliver laser-cutting machines in response to customer demand. The organization manufactures laser-cutting machines, categorized into two main product families with similar production cycles within each group. The "LT7 family" includes 4 machines, identified as LT7, LT7p, LT7 INS., and LT7p INS., while the "LT8 family" comprises 8 machines, identified as LT8, LT8p, LT8 ULA, LT8p ULA, LT8 12, LT8p 12, LT8 12 ULA, and LT8p 12 ULA. Each machine type has unique production stages with varying durations and specific human resource requirements, classified as M, E, and R. The company has multiple human resources for each stage, so it is possible to work on more than one laser-cutting machine at the same time. The manufacturing facility has 20 assembly areas, allowing for simultaneous work on up to 20 machines. Completed machines are transported by truck to the logistics centers, with transportation time being a significant factor.

Within this scenario, it is crucial to assign proper priorities to customer laser-cutting machine orders to efficiently allocate human resources to machines on a daily basis, thereby optimizing the overall supply chain. From this perspective, we can view this situation as an instance of the HFS problem applied to a real-world supply chain scenario. An instance of the HFS problem in this context involves a sequence (potentially with more than 20 entries) of laser-cutting machine orders (the jobs to be processed). Each order is characterized by the following attributes: ① order ID; ② machine type (mt); ③ due date (dd); ④ basement arrival date (db); ⑤ date of electrical panel arrival (de). The basement arrival date marks when the first M phase can be processed. In other words, it is not possible to start the sequence of tasks with people belonging to stage M before this date. Similarly, the electrical panel arrival date indicates when the first E phase can begin, preventing E tasks from starting before this date. Additionally, the problem instance specifies the total number of available human resources for each type (M, E, R). The objective of the optimizer is to determine the optimal job schedule. We adopt an indirect encoding scheme that represents solutions as job permutations, omitting some decision variables required for constructing an HFS schedule. Paired with surrogate heuristics, this approach reduces the search space and yields better results than direct encoding [33,37]. Job permutations are decoded into schedules using a *list scheduling* technique [33]. Among the common metrics used in the literature to evaluate the output of an HFS solver, we focus on minimizing the makespan, which intuitively reflects the efficiency of other metrics as well.

5   Experimental Setup

Computational Setup. For our experiments, we used the free Personal Edition of AnyLogic, which does not support exporting a simulation model as a standalone application. As a result, the simulations had to be executed directly through the AnyLogic GUI. The experiments were conducted on a Windows 11 laptop with a 14-core Intel i7-12700H CPU @ 2.30 GHz and 32GB of RAM.

The total execution time for all experiments was ca. 400 CPU core hours. Our AnyLogic simulation model and Python code are publicly available on GitHub.[4].

Algorithm Settings. The optimization algorithms we implemented are categorized into two groups, namely:

- *Schedule-as-a-whole approaches.* These methods operate on the entire list of orders as a whole. We implemented a Random Search (RS) baseline and included optuna[5] (OPTUNA), a popular optimization framework, configured with its default sampler (Tree-structured Parzen Estimator [38]) and pruner (Median Pruner). Additionally, we incorporated two metaheuristics, namely a Genetic Algorithm (GA) and Ant Colony Optimization (ACO), both implemented using the inspyred Python library[6]. For the HFS problem, we also considered a greedy heuristic, GREEDY, which prioritizes jobs by the earliest due dates.
- *Policy-generating approaches.* These methods focus on training a policy that processes one order at a time and makes decisions based on its features. In this category, we included a deep RL method (RL) based on Proximal Policy Optimization (PPO) [39] using the rllib[7] implementation. We also incorporated an IAI solution that evolves DTs using Genetic Programming (GP), implemented using the deap Python library[8]. Finally, our proposed approach (ELDT), as described in Sect. 3, belongs to this policy-generating category.

For the implementation details, readers are referred to our GitHub codebase.

To ensure a fair comparison, each algorithm was allocated 5000 AnyLogic simulation executions. Hyperparameters affecting the computational budget were adjusted accordingly, while other parameters were set to their library defaults. For ELDT, we used the settings from [11]. Due to space constraints, detailed hyperparameter descriptions are reported in the public GitHub repository.

Datasets. Because of the specificity of our scenarios, suitable benchmarks for evaluating the implemented algorithms are not readily available. As a result, for both problems, we generated datasets through stochastic procedures (which are also made available in the GitHub repository). For the HFS, we also included a real-world dataset with data provided by the company.

As regards the make-or-buy decision problem, we handcrafted a dataset consisting of 100 order entries. Each order requires a specified quantity of three components: type A, type B, and type C. The quantity of each component is sampled uniformly at random from the range $[0, 20]$. Additionally, each order's deadline is generated stochastically by adding a random number of days, uniformly sampled between 800 and 1500, to the starting simulation date of June

---

[4] https://github.com/DIOL-UniTN/eldt-scm.
[5] https://github.com/optuna/optuna.git.
[6] https://github.com/aarongarrett/inspyred.git.
[7] https://docs.ray.io/en/latest/rllib/index.html.
[8] https://github.com/DEAP/deap.git.

19, 2024. Due to the stochastic nature of the generation process, to ensure a robust and reliable analysis, we generated three independent instances of this dataset, referred to as list1, list2, list3.

For the HFS scheduling problem, we generated five test instances using the procedures described below. In all experiments, the number of human resources of types M, E, and R available to work on laser-cutting machine orders is fixed at $M = E = R = 5$, a value manually calibrated to avoid trivial solutions: too few or too many human resources would result in similar performance across all algorithms. ① Dataset d1 includes a list of orders for all 12 laser-cutting machines in both families. The arrival dates for both the basement component and the electrical panel are independently and uniformly distributed between day 1 and day 20. This relatively narrow range is calibrated by hand to ensure that the close arrival times of the components make scheduling decisions more impactful, thereby highlighting differences in algorithm performance. The delivery date for each order is set to the basement arrival date plus a random number between 20 and 50, which is consistent with the lead time of the company. This dataset contains 100 laser-cutting machine orders. ② Dataset d2 is generated using the same procedure as dataset d1, but it is limited to machines from the "LT7 family". ③ Dataset d3, instead, only includes laser-cutting machines from the "LT8 family". ④ Dataset d4, similarly to d1, includes orders for all 12 laser-cutting machines in both families. However, differently from d1, d4 simulates an annual production plan divided into five groups, G1 (days 1–73), G2 (days 74–146), G3 (days 147–219), G4 (days 220–292), and G5 (days 293–365). Each machine is randomly assigned to one of these groups. For each order in the dataset, the basement arrival date is randomly chosen within the corresponding group's day range, as is the electrical panel arrival date. The due date is set to the basement arrival date plus a random number between 20 and 50. The dataset consists of 100 entries. ⑤ Dataset d5 is populated with 345 orders directly provided by the company from 2021 to 2024, consisting of LT7 and LT8 machines.

6   Results

To ensure reliable results, we aggregated data from 10 independent runs, accounting for the algorithms' stochasticity. For the GREEDY algorithm, a single run is sufficient due to its deterministic nature.

Regarding the make-or-buy decision problem, Fig. 2 shows the fitness trend achieved by the optimization algorithms on dataset list1 (Fig. 2a), along with the best DTs found by GP (Fig. 2c) and ELDT (Fig. 2b). Table 1 provides a quantitative comparison of the results across the three datasets.

For the HFS scheduling problem, Fig. 3 shows the fitness trends of the algorithms on datasets d1, d4, and d5. Figure 4 presents the best DTs obtained by ELDT for the same datasets. Table 2 provides the detailed results.

Note that, for illustration purposes, the DTs shown in the figures have undergone a post-processing step. This step is necessary when the trees contain subtrees that are never executed. While it would be possible to integrate a mechanism to prevent the generation of such non-executable branches within the algorithms, doing so would increase the computational complexity. Furthermore, retaining this redundancy in the population can promote greater diversity. The results demonstrate the effectiveness and flexibility of our simulation-based optimization framework. In particular, comparing the two categories of methods (schedule-as-a-whole and policy-generating), it emerges that neither consistently outperforms the other. However, one potential advantage of policy-generating approaches is that the policy obtained can be continuously refined based on feedback from the simulator as new orders arrive dynamically. In contrast, schedule-as-a-whole methods need to restart the optimization process each time the order list is updated.

Table 1. Results for the make-or-buy decision problem. Revenue R (mean $\pm$ std. dev. across 10 runs) of the different algorithms is reported. The best for each dataset is highlighted in bold. Asterisk indicates a statistically significant difference w.r.t. ELDT results (Wilcoxon test, $\alpha = 0.05$); ns denotes no significance.

Dataset	Schedule-as-a-whole approaches				Policy-generating approaches		
	RANDOM	OPTUNA	GA	ACO	RL	GP	ELDT
list1	$7313.00 \pm 76.60^*$	$\mathbf{7868.00 \pm 50.51}^*$	$7866.00 \pm 25.91^*$	$7733.00 \pm 37.73^*$	7691.00 ± 26.44^{ns}	7650.00 ± 149.00^{ns}	7666.20 ± 67.66
list2	$7397.00 \pm 35.61^*$	$7802.00 \pm 102.93^*$	$\mathbf{7807.00 \pm 45.72}^*$	7704.00 ± 34.38^{ns}	7698.00 ± 32.25^{ns}	7734.00 ± 62.40^{ns}	7712.60 ± 61.03
list3	$7286.00 \pm 58.73^*$	$7848.00 \pm 65.29^*$	$\mathbf{7858.00 \pm 28.60}^*$	$7746.00 \pm 43.51^*$	7719.00 ± 35.73^{ns}	7752.00 ± 102.83^{ns}	7662.80 ± 79.71

Table 2. Results for the HFS scheduling problem. Average $\pm$ std. dev. of best makespan over 10 runs across five datasets. Note that the GREEDY algorithm is deterministic. Bold highlights the best result per dataset. Asterisk indicates a statistically significant difference w.r.t. ELDT (Wilcoxon test, $\alpha = 0.05$).

Dataset	Schedule-as-a-whole approaches					Policy-generating approaches		
	RANDOM	GREEDY	OPTUNA	GA	ACO	RL	GP	ELDT
d1	$452.60 \pm 0.52^*$	462.00^*	$\mathbf{450.90 \pm 0.32}^*$	$451.50 \pm 0.53^*$	$452.60 \pm 0.52^*$	$451.70 \pm 0.48^*$	$451.70 \pm 0.48^*$	454.78 ± 0.74
d2	$436.90 \pm 0.32^*$	445.00^*	$\mathbf{436.00 \pm 0.00}^*$	$436.30 \pm 0.48^*$	$437.20 \pm 0.79^*$	$436.80 \pm 0.42^*$	$436.20 \pm 0.79^*$	439.08 ± 0.44
d3	$458.20 \pm 0.42^*$	461.00^*	$\mathbf{457.10 \pm 0.32}^*$	$457.60 \pm 0.52^*$	$458.00 \pm 0.67^*$	$457.40 \pm 0.52^*$	$457.70 \pm 0.48^*$	460.88 ± 0.53
d4	$518.60 \pm 4.48^*$	482.00^*	$477.30 \pm 1.77^*$	$476.70 \pm 1.16^*$	$481.20 \pm 2.62^*$	$478.50 \pm 1.35^*$	$\mathbf{475.40 \pm 0.70}^*$	489.48 ± 5.27
d5	$2055.40 \pm 10.34^*$	$\mathbf{1510.00}^*$	$1918.12 \pm 10.83^*$	$1851.10 \pm 23.86^*$	$1818.20 \pm 19.01^*$	$1872.90 \pm 44.09^*$	$1564.80 \pm 24.10^*$	1737.30 ± 36.65

It is worth noticing that, for the make-or-buy decision problem, the large standard deviation values in Table 1 reflect the high variability and uncertainty inherent in the simulated supply chain for the make-or-buy decision problem. On the other hand, for the case of HFS, see Table 2, it should be noted that for dataset d2, OPTUNA likely achieved the optimal solution across all runs, as indicated by its zero standard deviation.

Notably, on both problems ELDT achieved results comparable to (or in some cases better than) competitors within the given budget while producing simple, interpretable policies. In most cases, pairwise comparisons (Wilcoxon rank-sum test, $\alpha = 0.05$) show that differences between ELDT and the competitor solutions are statistically significant. In general, interpretable DTs approaches (ELDT

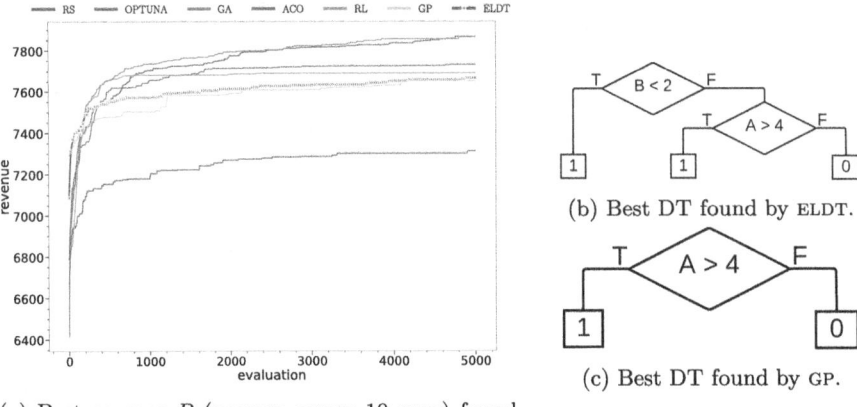

(a) Best revenue R (average across 10 runs) found by each algorithm across evaluations.

(b) Best DT found by ELDT.

(c) Best DT found by GP.

Fig. 2. Results for the make-or-buy decision problem on list1.

and GP) perform on par with, and sometimes outperform, black-box optimizers, particularly excelling on the real-world dataset (note that GREEDY is not black-box). On dataset d5, GREEDY outperforms other algorithms, likely because of its longer time frame (approximately three years) compared to the other datasets, which span only a single year. Over such an extended time frame, prioritizing orders strictly by their deadlines appears to be the most effective approach, especially under practical constraints like limited human resources. The analysis of the fitness curve in Fig. 2a however suggests that, with more evaluations, other optimization algorithms may achieve performance levels comparable to GREEDY. Further investigation is needed to validate these observations.

Interestingly, the policies found by ELDT sometimes resemble those achieved by GP, suggesting convergence towards similar policies. For instance, looking at Fig. 2b and Fig. 2c it can be seen that both methods recommending outsourcing for $A > 4$ and advising no outsourcing below this threshold. Future work should investigate the possible convergences between GP and ELDT DTs.

The DTs from ELDT for the HFS problem also provide useful insights. For dataset d1, the tree (Fig. 4a) prioritizes orders for machines of type "LT8p", which have the shortest manufacturing time. For d5 (Fig. 4c), the root node emphasizes delivery dates, hence mirroring the GREEDY baseline which prioritizes orders by deadline. A future enhancement could involve comparing the policies found by ELDT with the scheduling strategies adopted by company practitioners.

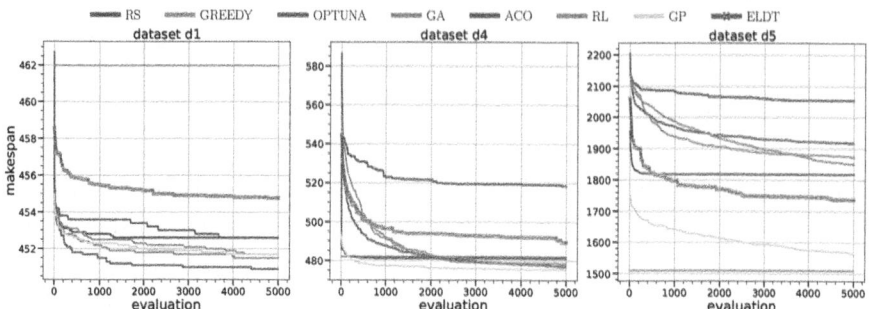

Fig. 3. Results for the HFS scheduling problem (datasets d1, d4, and d5). The y-axis displays the best makespan (average across 10 runs) found by each algorithm across evaluations. The result of the GREEDY policy is reported for reference.

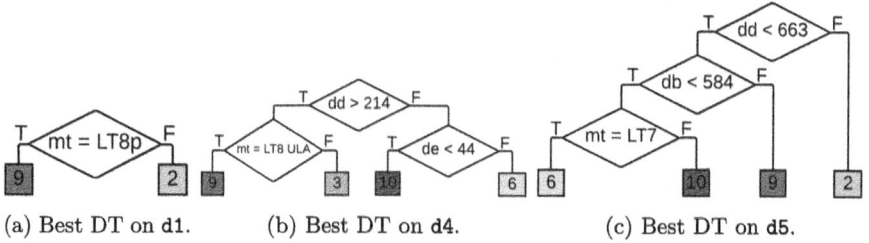

(a) Best DT on d1.   (b) Best DT on d4.   (c) Best DT on d5.

Fig. 4. Results for the HFS scheduling problem. Best DTs found by ELDT on datasets d1, d4, and d5 across 10 runs, with leaf nodes colored on a gradient from light (low priority for input order) to dark red (high priority for input order), indicating how the simulation model prioritizes scheduling.

7   Conclusions

In this work, we applied optimization/RL algorithms to enhance decision-making in supply chain management, a key element of Industry 4.0. We addressed two supply chain optimization problems using a versatile simulation-based optimization framework that successfully integrated various Python-based algorithms. By adopting AnyLogic to model the supply chain, we simplified development by capturing real-world complexities like uncertainties and stochasticity. This approach enables non-experts in AI to contribute to model design and visualize optimization results, reducing the risks and costs of implementation.

We classified the tested methods into two groups: schedule-as-a-whole and policy-generating approaches. Among the latter, we considered, in particular, an instance of Genetic Programming along with the method proposed in [11], which combines Grammatical Evolution with RL to generate interpretable DTs. Results show no clear winner between the two groups of methods, though policy-generating approaches have the potential advantage of continuously refining the decision-making process as new orders arrive dynamically, unlike schedule-as-a-whole approaches, which must reprocess the entire list for each new order.

To the best of our knowledge, this is the first attempt to apply a simulation-based optimization to generate interpretable models (based on DTs) for supply chain optimization. Remarkably, the interpretable DTs performed on par with, and sometimes better than, black-box algorithms, challenging the notion of a trade-off between performance and interpretability. These findings highlight the need for further research into interpretable solutions to meet industry demands. Unlike black-box approaches, which limit users to evaluate model outputs, interpretable models can enable interaction with their internal workings. For instance, integrating ELDT with a user interface could allow domain experts to adjust DT values, fostering human-in-the-loop optimization. Additionally, incorporating Large Language Models into the proposed framework could further enhance interpretability providing natural language explanations of decision-making [40].

References

1. Yin, Y., Stecke, K.E., Li, D.: The evolution of production systems from Industry 2.0 through Industry 4.0. Int. J. Prod. Res. **56**(1-2), 848–861 (2018)
2. Lasi, H., Fettke, P., Kemper, H.-G., Feld, T., Hoffmann, M.: Industry 4.0. Bus. Inf. Syst. Eng. **6**, 239–242 (2014)
3. Göran, A., Lihui, W., Magnus, H., Philip, M.: Cloud manufacturing - a critical review of recent development and future trends. Int. J. Comput. Integr. Manuf. **30**(4–5), 347–380 (2017)
4. Lambert Douglas, M., Enz Matias, G.: Issues in supply chain management: progress and potential. Ind. Mark. Manage. **62**, 1–16 (2017)
5. Mentzer, J.T., et al.: Defining supply chain management. J. Bus. Logist. **22**(2), 1–25 (2001)
6. Brettel, M., Friederichsen, N., Keller, M., Rosenberg, M.: How virtualization, decentralization and network building change the manufacturing landscape: an Industry 4.0 perspective. Int. J. Inf. Commun. Eng. **8**(1), 37–44 (2014)
7. Carter, A., Imtiaz, S., Naterer, G.F.: Review of interpretable machine learning for process industries. Process Saf. Environ. Prot. **170**, 647–659 (2023)
8. Monteiro, W.R., Reynoso-Meza, G.: A multi-objective optimization design to generate surrogate machine learning models in explainable artificial intelligence applications. EURO J. Decis. Processes **11**, 100040 (2023)
9. Došilović, F.K., Brčić, M., Hlupić, N.: Explainable artificial intelligence: a survey. In: International Convention on Information and Communication Technology, Electronics and Microelectronics, pp. 0210–0215. IEEE (2018)
10. Arrieta, A.B., et al.: Explainable artificial intelligence (XAI): concepts, taxonomies, opportunities and challenges toward responsible AI. Inf. Fus. **58**, 82–115 (2020)
11. Custode, L.L., Iacca, G.: Evolutionary learning of interpretable decision trees. IEEE Access **11**, 6169–6184 (2023)
12. Paternina-Arboleda, C.D., Montoya-Torres, J.R., Fabregas-Ariza, A.: Simulation-optimization using a reinforcement learning approach. In: Winter Simulation Conference, pp. 1376–1383. IEEE (2008)
13. Hong, S.R., Hullman, J., Bertini, E.: Human factors in model interpretability: industry practices, challenges, and needs. Proc. ACM Hum.-Comput. Interact. **4**(CSCW1), 1–26 (2020)

14. George, B., Samir, D., Grigoris, A.: Predicting supply chain risks using machine learning: the trade-off between performance and interpretability. Futur. Gener. Comput. Syst. **101**, 993–1004 (2019)
15. Heydarbakian, S., Spehri, M.: Interpretable machine learning to improve supply chain resilience, an Industry 4.0 recipe. IFAC-PapersOnLine **55**(10), 2834–2839 (2022)
16. Samia, Z., Clovis, F., Dieudonné, T., Samuel, F.-W., Bernard, K.-F.: The viability of supply chains with interpretable learning systems: the case of COVID-19 vaccine deliveries. Glob. J. Flex. Syst. Manag. **24**(4), 633–657 (2023)
17. Ahmed, I., Jeon, G., Piccialli, F.: From artificial intelligence to explainable artificial intelligence in Industry 4.0: a survey on what, how, and where. IEEE Trans. Ind. Inf. **18**(8), 5031–5042 (2022)
18. Kosasih, E.E., Papadakis, E., Baryannis, G., Brintrup, A.: A review of explainable artificial intelligence in supply chain management using neurosymbolic approaches. Int. J. Prod. Res. **62**(4), 1510–1540 (2024)
19. Massimo, B., Davide, M., Mattia, N., Francesco, Z.: Machine Learning for industrial applications: a comprehensive literature review. Expert Syst. Appl. **175**, 114820 (2021)
20. Cynthia, R.: Stop explaining black box machine learning models for high stakes decisions and use interpretable models instead. Nat. Mach. Intell. **1**, 206–215 (2019)
21. Yang-Byung, P., Sung-Joon, Y., Jun-Su, Y.: Development of a knowledge-based intelligent decision support system for operational risk management of global supply chains. Euro. J. Ind. Eng. **12**(1), 93–115 (2018)
22. Sin-Jin, L., Ming-Fu, H.: Incorporated risk metrics and hybrid AI techniques for risk management. Neural Comput. Appl. **28**, 3477–3489 (2017)
23. Ryan, C., Collins, J.J., Neill, M.O.: Grammatical evolution: evolving programs for an arbitrary language. In: Genetic Programming: First European Workshop, pp. 83–96. Springer (1998)
24. Ivanov, D.: Operations and supply chain simulation with AnyLogic 7.2: decision-oriented introductory notes for master students. Berlin School of Economics and Law (2016)
25. Mikhail, A., Dmitry, P., Dmitry, K., Hadi, D., Anna, S.: System modeling in solving mineral complex logistic problems with the AnyLogic software environment. Transp. Res. Procedia **68**, 483–491 (2023)
26. Wartelle, A., Dauzère-Pérès, S., Yugma, C., Christ, Q., Roussel, R.: A study on the impact of lot priorities on cycle times in semiconductor manufacturing. In: Winter Simulation Conference, pp. 2266–2275. IEEE (2023)
27. McIvor Ronan, T., Humphreys Paul, K.: A case-based reasoning approach to the make or buy decision. Integr. Manuf. Syst. **11**(5), 295–310 (2000)
28. Cánez Laura, E., Platts Ken, W., Probert David, R.: Developing a framework for make-or-buy decisions. Int. J. Oper. Prod. Manag. **20**(11), 1313–1330 (2000)
29. Paul Yoon, K., Naadimuthu, G.: A make-or-buy decision analysis involving imprecise data. Int. J. Oper. Prod. Manag. **14**(2), 62–69 (1994)
30. Serrano, R.M., Ramírez, M.R.G., Gascó, J.L.G.: Should we make or buy? An update and review. Eur. Res. Manag. Bus. Econ. **24**(3), 137–148 (2018)
31. McIvor Ronan, T., Humphreys Paul, K., McAleer, W.E.: A strategic model for the formulation of an effective make or buy decision. Manag. Decis. **35**(2), 169–178 (1997)
32. Blaxill Mark, F., Hout Thomas, M.: The fallacy of the overhead quick fix. Harv. Bus. Rev. **69**(4), 93–101 (1991)

33. Yu, C., Quirico, S., Andrea, M.: A genetic algorithm for the hybrid flow shop scheduling with unrelated machines and machine eligibility. Comput. Oper. Res. **100**, 211–229 (2018)
34. Rubén, R., Vázquez-Rodríguez, J.A.: The hybrid flow shop scheduling problem. Eur. J. Oper. Res. **205**(1), 1–18 (2010)
35. Li, X., Chehade, H., Yalaoui, F., Amodeo, L.: A new method coupling simulation and a hybrid metaheuristic to solve a multiobjective hybrid flowshop scheduling problem. In: Conference of the European Society for Fuzzy Logic and Technology, pp. 1082–1089. Atlantis Press (2011)
36. Gupta Jatinder, N.D.: Two-stage, hybrid flowshop scheduling problem. J. Oper. Res. Soc. **39**(4), 359–364 (1988)
37. Thijs, U., Rubén, R., Serifoglu, F.S.: Genetic algorithms with different representation schemes for complex hybrid flexible flow line problems. Int. J. Metaheuristics **1**(1), 30–54 (2010)
38. Watanabe, S.: Tree-structured Parzen estimator: understanding its algorithm components and their roles for better empirical performance. arXiv preprint arXiv:2304.11127 (2023)
39. Schulman, J., Wolski, F., Dhariwal, P., Radford, A., Klimov, O.: Proximal policy optimization algorithms. arXiv preprint arXiv:1707.06347 (2017)
40. Serafim, P.B., Crescenzi, P., Gezici, G., Cappuccio, E., Rinzivillo, S., Giannotti, F.: Exploring large language models capabilities to explain decision trees. In: HHAI Hybrid Human AI Systems for the Social Good, pp. 64–72. IOS Press (2024)

Grammatical Feature Construction for Enhanced Interpretability in Breast Cancer Classification

Yumnah Hasan[1]([⊠]), Allan de Lima[1], Darian Reyes Fernández de Bulnes[2], Douglas Mota Dias[3], and Conor Ryan[1,4]

[1] Lero, University of Limerick, Limerick, Ireland
{yumnah.hasan,allan.delima,conor.ryan}@ul.ie
[2] University of Galway, Galway, Ireland
darian.reyes@universityofgalway.ie
[3] Atlantic Technological University, Galway, Ireland
douglas.motadias@atu.ie
[4] Limerick Digital Cancer Research Centre, University of Limerick, Limerick, Ireland

Abstract. Recent advances in Artificial Intelligence have yielded significant progress in developing medical and clinical diagnosis techniques. Machine learning algorithms are among the most promising methods for detection and classification problems. Despite their inherent robustness, the primary challenge in employing these approaches lies in their opaque behaviour, a critical factor in medical diagnosis. Establishing trust between clinicians and patients requires an explainable model. This paper presents a two-stage approach to improving explainability: the first stage, Grammatical Feature Construction (GFC), uses Grammatical Evolution (GE) to perform feature construction. These features are interpretable as they are generated from the original features by applying simple arithmetic operations to the original data. These features are independent of the model/algorithm that will be used for classification, so any classification algorithm could be used in the second stage; we focus here on Linear Discriminant Analysis (LDA) to create GFC/LDA, which provides greater explainability than using LDA alone while maintaining comparable performance. To evaluate the effectiveness of GFC/LDA, we conducted a comprehensive comparative analysis against methods including GE as a classifier and LDA using all original features in two Breast Cancer datasets, the Digital Database for Screening Mammography and the Wisconsin Breast Cancer dataset. The results demonstrate that the GFC/LDA approach yields comparable with the other methods but produces more interpretable models.

Keywords: Breast Cancer · Feature Construction · Grammatical Evolution · Interpretability · Machine Learning

1 Introduction

In recent years, artificial intelligence (AI) has significantly propelled advances in medical imaging through cutting-edge techniques such as machine learning (ML)

© The Author(s), under exclusive license to Springer Nature Switzerland AG 2025
P. García-Sánchez et al. (Eds.): EvoApplications 2025, LNCS 15612, pp. 204–220, 2025.
https://doi.org/10.1007/978-3-031-90062-4_13

and convolutional neural networks (CNN). These algorithms have been instrumental in image recognition, disease detection, and medical diagnosis, ushering in a transformative period in the field [32]. However, the rapid progress in ML and CNN models has led to a major challenge in interpreting the intricate decision-making processes behind their predictions [45]. Despite their effectiveness, the lack of interpretability of these models has raised concerns about their reliability and acceptance within the medical community.

Amidst these challenges, Grammatical Evolution (GE) [37] has emerged as a promising alternative. GE does not necessitate extensive data for effective training, offering a more feasible approach in the context of limited medical datasets [18] , and its symbolic expression representation enhances interpretability, addressing opaque decision-making processes in advanced ML models, and it has been successfully used to improve explainability in medical diagnosis [7,24]. This paper aims to bridge the gap between accuracy and interpretability in medical imaging by using GE to create more transparent solutions.

This study explores GE to construct a concise set of features from a larger set. Our approach, Grammatical Feature Construction (GFC), aims to enhance the interpretability of an ML model, namely, Linear Discriminant Analysis (LDA), without compromising its classification accuracy compared to when it employs the entire original feature set. GFC individuals perform two functions: first, they construct features, and second, they train an ML algorithm on those features. This paper employs LDA as the ML algorithm, and we refer to the system as GFC/LDA. The proposed GFC/LDA approach follows a two-stage process for classification: the first stage is model-agnostic, while the user chooses the model in the second stage.

We evaluated the performance of GFC/LDA on two Breast Cancer (BC) datasets. Additionally, we examined nine different data augmentation methods to address the class imbalance issue that both datasets suffer from. The evaluation involved the features constructed by GFC to train an ML model to make predictions on validation data. Since we need to train an ML model and make predictions by each GE individual in every generation, we chose the LDA classifier because of its low computational cost. The key contributions of this research are the ability of GFC/LDA to develop a small set of features that facilitates a more understandable interpretation of ML solutions while maintaining a good Area Under the Curve (AUC) and, secondly, its applicability to small and highly imbalanced datasets, demonstrating comparable performance to ML classifiers.

This paper is organised as follows: Sect. 2 reviews existing literature, Sect. 3 explores applied methods, and Sect. 4 details the experimentation process. Results and analysis are presented in Sect. 5. Section 6 concludes and outlines future research.

2   Literature Review

ML methods have found widespread application in real-world challenges, with notable performance observed in tasks such as classification [21] and regression

[27] being observed. In classification tasks, ML techniques aim to learn classifiers that effectively categorize unseen instances (test data) based on observed instances (training data), each represented by a set of attributes and a label [2]. The quantity, quality, and representation of the data significantly impact the performance of these classifiers. Recognizing the challenges of many features, such as increased computational complexity and potential overfitting, a crucial need arises for developing an efficient data construction pipeline to feed the model. Feature Construction (FC) enhances the training process by creating high-level features that combine original features, resulting in new ones with more meaningful representations. This can lead to better model performance [28]. FC is a more intricate process than simple feature extraction that involves feature selection and linear and non-linear transformations of the primitive set [13].

In scenarios where the learning algorithm fails to achieve a desired level of accuracy using an existing set of features, FC algorithms can improve performance by reducing the dimensionality of the problem without compromising the expressiveness of the models [25]. Evolutionary Computation methods have shown promising results when used to generate diverse features. Further discussion of these methods follows.

In [41], three Genetic Programming (GP) algorithms and random search were used for a concise feature construction method, benefiting five ML algorithms. Findings across 21 classification and regression datasets showed that generating just two compact features could rival the performance of the entire original set, with the GP Gene-pool Optimal Mixing Evolutionary Algorithm (GOMEA) standing out as the most effective algorithm. An embedded GP-based approach [1] was designed to construct multiple high-level features from low-level features for binary and multi-class spectroscopy data problems. The results showed enhanced classification accuracy and increased biomarker detection rates in most cases. Another work [23] introduced an efficient feature construction method in image segmentation with Filter GP (FGP) and Embedded GP (EGP) methods. Tests on diverse datasets revealed that EGP outperformed FGP's enhanced segmentation performance.

The use of GE to generate newly constructed features has emerged as a useful approach, producing comparable and often superior results to original features. For example, some results [13,22] demonstrated that newly generated features yield more interpretable models, even though fewer features than were present in the original set were used. GE was used to create features that were subsequently assessed by a Radial Basis Function network (RBF) [40] before being used to train an Artificial Neural Network (ANN) training. Comparative experiments with other ANN models showed substantial error reduction for COVID-19 cases and mortality predictions, particularly with increased artificial features.

While FC enhances ML algorithms, considering the interpretability of ML models is crucial for transparency, user trust, and informed decision-making. Different techniques such as Feature Importance (FI) [6], Partial Dependence Plot (PDP) [12], Local Interpretable Model-agnostic Explanations (LIME) [34], and Shapley Additive explanations (SHAP) [26] developed to provide insights into model behaviour, feature reliance, and decision influences.

However, each of these approaches has limitations. For example, SHAP values may pose computational challenges, and interpretation may be complex [5] due to the exponential growth in the number of feature combinations as the feature count increases, which makes the computation of Shapley values more difficult [14], while LIME exhibits sensitivity to sample selection and hyperparameter choices [3]. Feature Importance methods can struggle with correlation and scale issues [38]. PDP assumes independence and may oversimplify complex interactions [29]. Understanding these limitations is essential for the accurate interpretation of models.

This study introduces a novel feature construction approach called GFC/LDA. In this approach, GE constructs diverse features, and LDA classifies breast instances as normal or abnormal, offering more straightforward and interpretable solutions. Data augmentation is employed to address imbalanced class distribution. Notably, the method efficiently handles small and imbalanced datasets without extensive training data.

3   Methodology

GFC/LDA was evaluated on the Digital Database for Screening Mammography (DDSM) [20], a dataset comprising normal and abnormal mammographic images, and the Wisconsin Breast Cancer (WBC) dataset [43], which primarily consists of breast mass features obtained through Fine Needle Aspiration (FNA).

The $DDSM$ dataset includes mammographic images with two views: Cranioaudal (CC) and Medial Lateral Oblique (MLO). Our evaluation focuses on one cancer volume (02) and three volumes of normal images (01–03), intentionally selected low positive and high negative samples for a more realistic scenario.

In the preprocessing phase, images undergo median filtering for noise removal. Subsequently, Otsu thresholding eliminates artefacts such as labels (CC or MLO). Segmentation, based on the approach described in [11], divides the full image ($\mathbf{I}$) into top (I_T), mid (I_M), and bottom (I_B) segments.

Four distinct setups are designed using the $DDSM$ dataset, incorporating features from both segmented and non-segmented images. For feature extraction, the Gray-Level Co-occurrence Matrix (GLCM) is employed in four different orientations, which are $0°$, $45°$, $90°$, and $135°$, utilizing two diagonal and two neighbouring pixel angles. $0°$ is used for horizontally adjacent pixels, and $90°$ is used for vertically adjacent pixels, representing the co-occurrence of grey-level pairs. On the other hand, angles $45°$ and $135°$ are used for diagonal directions considering pairs of diagonally adjacent pixels, capturing co-occurrence from top-left to bottom-right and top-right to bottom-left, respectively. This configuration results in the calculation of 13 Haralick features [16] per orientation. A total of 52 features are obtained per segment/image. The $DDSM$ setups include S_{CC} (segmented images from CC view), S_{MLO} (MLO view images), S_{CC+MLO} (both CC and MLO segmented images), and F_{CC+MLO} (features from non-segmented

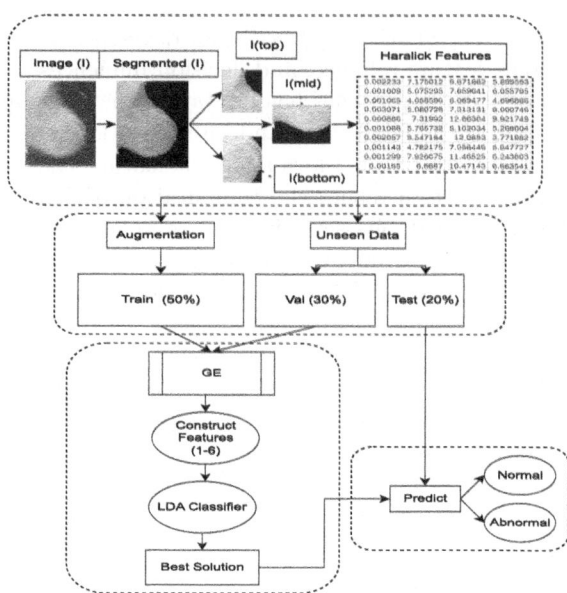

Fig. 1. Proposed workflow of GFC/LDA approach. Here, I is used for the Image segmented into three top, mid, and bottom segments.

images contain both views). On the other hand, the *WBC* dataset has 30 features, which include measurements like radius, texture, perimeter, area, smoothness, compactness, concavity, concave points, symmetry, and fractal dimension. They are divided into normal (benign) and abnormal (malignant) categories.

A severely imbalanced dataset, in which one class considerably outnumbers the others, presents challenges to ML models, resulting in biased predictions. To address this issue in our data setups, nine different augmentation techniques, outlined in Sect. 4, are applied to the training sets. No data augmentation is performed on validation and test data.

Following data augmentation, we used GE incorporating LDA to train models. Each individual used a GE-style mapping to create new features using the original set. GE defines its search space through a grammatical definition. We leverage this to specify how new features can be constructed and how many should be created. GE employs a BNF grammar comprising terminals and non-terminals; the terminals are the elements that will ultimately compose the phenotypic structure, and the non-terminals are the semantic guides to link terminals. Listing 1.1 shows our grammar: the production rule for the non-terminal <start> has six different choices, each having a different number of expressions to be mapped, which allows the total number of constructed features to be from one to six. To maintain the balance between complexity and performance six is selected as the maximum value. These features are constructed through arithmetic operations, and terminals include the original features. In this way, the output solution (best GE individual) uses up to six constructed features composed of a selection of the original features. The number of features available in the non-terminal <x> is 52 (13 Haralick features for each of four orientations) for the *DDSM* scenarios and 30 for the *WBC* scenarios.

$$\langle\text{start}\rangle ::= \texttt{CF[0] = } \langle e \rangle$$
$$| \ \texttt{CF[1] = } \langle e \rangle; \ \texttt{CF[0] = } \langle e \rangle$$
$$| \ \texttt{CF[2] = } \langle e \rangle; \ \texttt{C[F1] = } \langle e \rangle; \ \texttt{CF[0] = } \langle e \rangle$$
$$| \ \texttt{CF[3] = } \langle e \rangle; \ \texttt{CF[2] = } \langle e \rangle; \ \texttt{CF[1] = } \langle e \rangle;$$
$$\texttt{CF[0] = } \langle e \rangle$$
$$| \ \texttt{CF[4] = } \langle e \rangle; \ \texttt{CF[3] = } \langle e \rangle; \ \texttt{CF[2] = } \langle e \rangle$$
$$\texttt{CF[1] = } \langle e \rangle; \ \texttt{CF[0] = } \langle e \rangle$$
$$| \ \texttt{CF[5] = } \langle e \rangle; \ \texttt{CF[4] = } \langle e \rangle; \ \texttt{CF[3] = } \langle e \rangle;$$
$$\texttt{CF[2] = } \langle e \rangle; \ \texttt{CF[1] = } \langle e \rangle; \ \texttt{CF[0] = } \langle e \rangle$$
$$\langle e \rangle ::= \langle 0 \rangle (\langle e \rangle, \langle e \rangle) \ | \ \langle x \rangle$$
$$\langle 0 \rangle ::= \texttt{add} \ | \ \texttt{sub} \ | \ \texttt{mul} \ | \ \texttt{pdiv}$$
$$\langle x \rangle ::= \texttt{x[0]} \ | \ \texttt{x[1]} \ | \ \texttt{x[2]} \ | \ ... \ | \ \texttt{x[51]}$$

Listing 1.1. Proposed GFC Grammar: Individuals create up to six features, each formed from any number of original features.

Afterwards, GE invokes the LDA classifier[1] to train a model with the training set using the constructed features and generate predictions for both training and validation sets. The model's AUC is a performance measure for this problem. The fitness score for each model (a GE individual) is the average between the AUC on the training set (using cross-validation with five folds) and the AUC on the validation set. We designed the fitness function like that to assess the performance of unseen data and guide the evolution in a more general way to avoid overfitting. In addition, we calculated the average of these two scores to achieve more robust results. We noted that AUC on the training set, even with cross-validation, is usually much higher than on the validation set because the training set contains augmented data. In contrast, the validation set consists of the original unbalanced data, similar to the test set.

Next, the features constructed by GFC are fed into an ML classifier. In this work, LDA serves as the classifier. The trained model is then employed to make predictions on an unseen test set, showcasing the practical application of GFC in FC and classification tasks. The workflow of GFC/LDA is illustrated in Fig. 1.

A key focus of this study is to emphasize the simplicity and interpretability of solutions derived from GFC/LDA. As illustrated in Fig. 2, an example outcome

---

[1] Although, as noted earlier, any appropriate classification algorithm could be used.

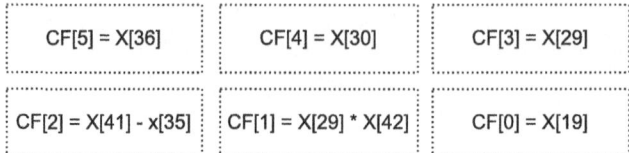

| CF[5] = X[36] | CF[4] = X[30] | CF[3] = X[29] |
| CF[2] = X[41] - x[35] | CF[1] = X[29] * X[42] | CF[0] = X[19] |

Fig. 2. Depiction of one actual solution from GFC/LDA selected among the 30 runs of S_{CC} setup using ADA augmentation approach. The cross-validation training AUC is 0.932, the validation AUC is 0.936, and the test AUC is 0.922. The expression of each constructed feature is presented where X[n] is used for the original features, and CF[n] shows the constructed features.

from our S_{CC} setup reveals that the generated solution comprises merely eight original features, contrasting the initial set of 52 features. The six constructed features are formed through arithmetic operations applied to a core set of 52 features, presenting a streamlined representation that diverges significantly from the original dataset's complexity. This highlights the potential of GFC to distil essential information and derive simplified yet effective solutions.

4   Experimental Details

This study compares our proposed approach, GFC/LDA, with two alternative methods, LDA and GE, as classifiers that used original features. The LDA classifier, GE classifier, and GFC/LDA robustness are assessed on two datasets, *DDSM* and *WBC*.

The *DDSM* and *WBC* datasets are divided into train, validation, and test at a 50:30:20 ratio, respectively. In setups S_{CC}, S_{MLO}, S_{CC+MLO} the positive and negative samples are 6% and 94%, whereas there are 12% abnormal and 88% normal cases present in the F_{CC+MLO} scenario. Similarly, the *WBC* dataset has a class ratio of 37:63 for positive to negative instances. The details of the datasets used in this study are shown in Table 1. Following the approaches detailed in Sect. 3, the training set is augmented, resulting in new and different samples.

The augmentation techniques used in this work include Synthetic Minority Oversampling Technique (SMOTE) [8], ADASYN (ADA) [19], Borderline SMOTE (BSMOTE) [15], SMOTE Nominal and Continous (S-NC) [8], Support Vector Machine SMOTE (SVM-S) [30], SMOTE Edited Nearest Neighbour (S-ENN) [42], SMOTE-Tomek (S-Tomek) [4], Mixup [44] and STEM [17].

AUC is helpful in imbalanced binary classification problems, as typical accuracy measurements may not provide insights into a model's performance. AUC considers sensitivity (true positive rate) and specificity (true negative rate) across different thresholds determined by predicted probabilities. These thresholds typically range from 0 to 1, and the number of thresholds used depends on the dataset. Since it's difficult to specify the exact number of thresholds, it's sufficient to refer to them as a range from 0 to 1. This makes AUC robust, particularly

Table 1. Details of the number of samples used for training, validation, and testing sets. Here, **Pos Neg** indicates positive and negative samples, respectively.

Setups	Train		Validation		Test	
	Pos	Neg	Pos	Neg	Pos	Neg
S_{CC}	59	921	26	395	12	165
S_{MLO}	62	928	27	398	12	167
S_{CC+MLO}	122	1849	53	793	23	331
F_{CC+MLO}	82	575	36	247	16	104
WBC	118	199	51	86	22	37

Table 2. Experimental parameters used in GFC and GE.

Parameter type	Parameter value	
	GFC	GE
Number of runs	30	30
Population size	100	200
Number of generations	50	100
Initialization method	Sensible [36]	Sensible [36]
Mutation probability	0.01	0.01
Crossover probability	0.80	0.80
Elitism size	1	1
Max. tree depth	17	35

when unequal class distributions could lead to bias. Therefore, we used AUC to measure the performance of the proposed method.

All LDA experiments were executed using the scikit-learn library [31]. The GRAPE framework [9] was used for GE experiments both for GFC/LDA and the GE Classifier. To conduct rigorous statistical analysis, the R library pROC [35] was employed to assess the performance of the proposed and implemented approaches. Detailed access to our code is provided in the supplementary material[2] for further reference.

The experiments in the proposed GFC/LDA approach, which involve new FC as described in Sect. 3, utilized the grammar specified in Listing 1.1. LDA is trained on these constructed features for classification. Early experimentation with a smaller maximum tree depth from 3 to 15 led to poor results, prompting a decision to set the maximum tree depth at 17 to capture more complex feature interactions.GE's larger population and generation sizes allowed for more evaluations. However, GFC, despite requiring LDA executions each time and undergoing fewer evaluations to conserve resources, delivered comparable results. The parameters used to perform these experiments are presented in Table 2.

---

[2] https://github.com/yumnahhasan/GFC.

Table 3. Comparative Analysis of the AUC between GFC/LDA, GE Classifier, and LDA using different augmentation approaches for both datasets.

Setups	Methods	ADA	BSMOTE	S-ENN	SMOTE	S-NC	S-Tomek	SVM-S	Mixup	STEM
S_{CC}	GFC/LDA	**0.929**	**0.929**	0.919	**0.929**	**0.929**	0.922	**0.929**	**0.929**	0.916
	GE	0.907	0.905	0.893	0.907	0.905	0.905	0.906	0.907	0.900
	LDA	0.922	0.922	0.912	0.922	0.922	0.926	0.922	0.922	0.912
S_{MLO}	GFC/LDA	0.907	0.902	0.889	0.907	0.907	0.908	0.901	0.910	0.897
	GE	0.897	0.874	0.875	0.876	0.894	0.896	0.876	0.894	0.892
	LDA	0.930	0.917	0.929	0.929	0.924	0.931	0.911	**0.932**	0.926
S_{CC+MLO}	GFC/LDA	0.922	0.930	0.931	0.923	0.922	0.923	0.929	0.928	**0.934**
	GE	0.911	0.912	0.920	0.911	0.916	0.917	0.916	0.908	0.921
	LDA	0.927	0.931	0.931	0.928	0.933	0.929	0.929	0.927	0.931
F_{CC+MLO}	GFC/LDA	0.935	0.939	0.940	0.937	0.940	0.939	0.938	0.937	0.944
	GE	0.930	0.931	0.907	0.928	0.932	0.928	0.929	0.932	0.937
	LDA	0.933	0.933	0.936	0.934	0.936	0.936	0.928	**0.958**	0.946
WBC	GFC/LDA	0.977	0.980	0.990	0.985	0.981	0.979	0.990	0.988	**0.992**
	GE	0.985	0.987	0.991	0.983	0.989	0.987	0.989	0.988	0.991
	LDA	0.951	0.951	0.968	0.943	0.953	0.943	0.953	0.972	0.955

The experiment used GE as a classifier with all original features to classify samples. The grammar, described in the supplementary material, evolves solutions using basic arithmetic operations, including addition, subtraction, multiplication, division, and real constants for diverse mathematical expressions.

Moreover, we assessed the performance of a single ML model, an LDA linear classifier, using the original features. The experiments are carried out in all scenarios. Before training the LDA, we standardized the data. The experiment ran for a single iteration, and we recorded an essential performance metric, AUC, on the same test data as the other experiments.

5   Results and Analysis

GFC/LDA is compared with the other two methods, LDA and GE, as classifiers. The evaluation is performed on four setups from the *DDSM* dataset, constructed based on the view, segment, and unsegmented images. Additionally, one setup is used from the *WBC* dataset. The nine augmentation techniques listed in Sect. 4 are applied to maintain the class ratio in the training data.

GFC/LDA demonstrates superior performance relative to GE while showing comparable performance to LDA. In Table 3, the test AUCs of all the methods implemented in this work are compared using nine augmentation methods. The average AUCs are reported for GFC/LDA and GE, while a single-run AUC is shown for LDA. In S_{CC}, GFC/LDA outperforms the other two methods, achieving a 0.929 AUC using ADA, BSMOTE, S-ENN, SMOTE, S-NC, SVSM-S, and Mixup. In the second setup, S_{MLO}, LDA obtained the highest AUC of all

Table 4. Average number of features constructed for each setup.

Setups	ADA	BSMOTE	S-ENN	SMOTE	S-NC	S-Tomek	SVM-S	Mixup	STEM
S_{CC}	5.80	5.80	5.70	5.80	5.80	5.83	5.80	5.80	5.73
S_{MLO}	5.60	5.50	5.70	5.63	5.83	5.63	5.63	5.70	6.00
S_{CC+MLO}	5.76	5.73	5.86	5.86	5.73	5.86	5.73	5.76	5.83
F_{CC+MLO}	5.50	5.43	5.66	5.56	5.66	5.53	5.26	5.70	5.30
WBC	5.83	5.70	5.43	5.70	5.76	5.46	5.60	5.70	5.50

approaches, scoring 0.932 utilizing Mixup. In the S_{CC+MLO} scenario, applying STEM, GFC/LDA secured the maximum AUC of 0.934. In the final setup of the *DDSM* dataset, F_{CC+MLO}, LDA obtained the highest AUC of 0.958 using Mixup. GFC/LDA with STEM achieves an AUC of 0.992 on the *WBC* problem, which is the best of all methods.

Table 4 shows the average number of features constructed for each scenario over 30 runs. Even though the number of features allowed in the search space ranged from 1 to 6, most of the runs presented six features in the best solution. We suspect this is because LDA can handle the complexity of a higher number of features, but future work will investigate higher possible numbers.

The double dendrogram cluster heatmaps are presented in Fig. 3. The dendrograms follow a tree-like structure, where nodes represent clusters, features serve as the leaves, and the height of the branches indicates the distance between clusters, with shorter branches representing closer relationships. Hierarchical clustering is performed using the *average* linkage method, which calculates the average distance between clusters. Correlation analysis is conducted between the original and constructed features selected from the single individual closest to the average test AUC obtained after applying STEM augmentation. The cluster heatmap arranges the features such that those with similar correlation patterns are placed closer together.

The heatmaps employ a colour gradient ranging from blue to red to illustrate the strength and direction of correlations, with red and blue indicating strong positive and negative correlations, respectively. The heatmaps suggest that the constructed features generally show a lower correlation with the original features. Higher correlations are observed in cases where a constructed feature is derived primarily from a single original feature. In contrast, the correlation appears weaker when a constructed feature incorporates multiple original features using operators. It presents that the constructed features capture different or more complex information not simply derived from any original feature.

An additional analysis of the constructed features for each setup reveals that they are not highly correlated. This indicates that each constructed feature captures unique information from the original features, contributing independently to the model's overall performance. The four individuals closest to the mean test

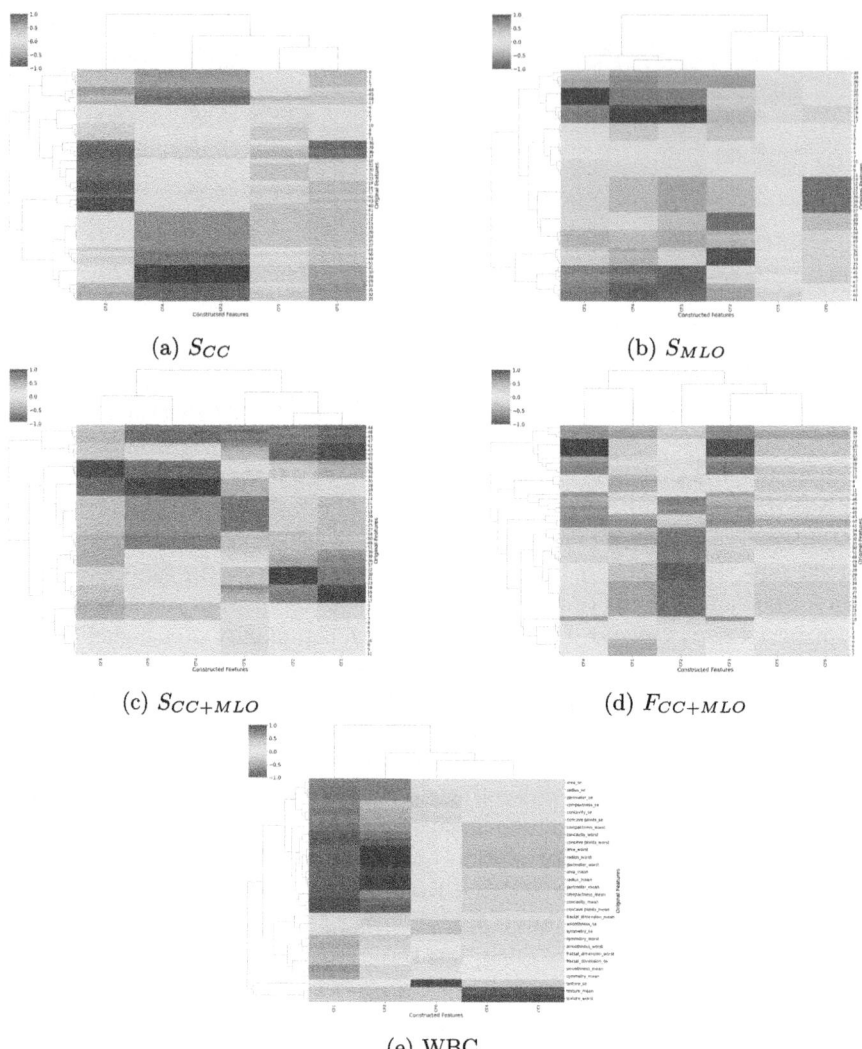

(a) S_{CC}

(b) S_{MLO}

(c) S_{CC+MLO}

(d) F_{CC+MLO}

(e) WBC

Fig. 3. The cluster heatmaps for each setup illustrate the correlations between the constructed features (CF) and the original features. The correlation values range from 1 (highest) to -1 (lowest), indicating the strength of the relationship.

AUC are selected from all 30 runs. Results for each setup are provided in the supplementary material referred to in Sect. 4.

5.1   Significance Test

To compare the AUCs of the GE, LDA and GFC/LDA methods, the DeLong test [10] was performed across all five setups with the significance level set to

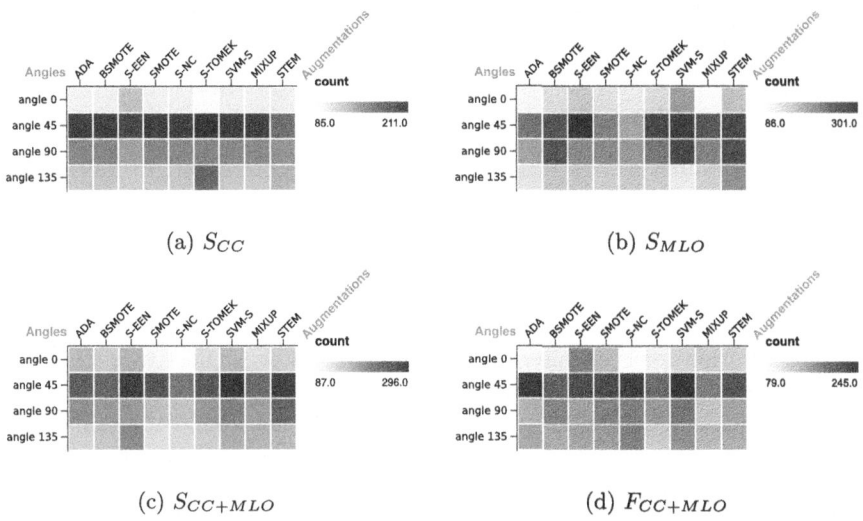

Fig. 4. The counts of features acquired from angles $0°$, $45°$, $90°$, and $135°$ are presented using a heatmap for all the setups of the *DDSM* dataset. Nine augmentation approaches are used here for each setup. The most frequently appearing features observed in each data setup category are those obtained from angle $45°$.

$\alpha=0.05$, as presented in Table 5. The results demonstrate that while the proposed GFC/LDA approach significantly improves over GE as a standalone classifier in one experimental setup S_{MLO}, no statistically significant difference is observed compared to LDA across all setups. This suggests that GFC/LDA retains the classification performance of LDA while enhancing model interpretability through its structured feature construction process, which GE enables. GFC/LDA allows for more insightful feature engineering that can reveal underlying patterns in the data, unlike GE alone, which may generate more complex and less interpretable results. Therefore, even in scenarios where GFC/LDA does not exhibit statistically significant differences in AUC compared to LDA, it still

Table 5. The table summarizes the statistical significance of differences between three models (LDA, GFC/LDA, and GE) across five experimental setups S_{CC+MLO}, S_{CC}, S_{MLO}, F_{CC+MLO}, and WBC. Where, "S" indicates a significant difference ($\alpha <=0.05$), while "NS" denotes no significant difference ($\alpha >0.05$).

Comparison	S_{CC}	S_{MLO}	S_{CC+MLO}	F_{CC+MLO}	WBC
LDA vs GFC/LDA	NS	NS	NS	NS	NS
LDA vs GE	NS	NS	S	NS	NS
GFC/LDA vs GE	NS	NS	S	NS	NS

offers the advantage of increased model interpretability without compromising predictive performance.

5.2   Interpretability of Models

The interpretability of ML models has come under considerable scrutiny, particularly in the medical sector, where confidence in decision-making processes is crucial [33]. Despite various techniques introduced to enhance model interpretability, each approach has limitations, as discussed in Sect. 2, restricting their applicability on a broader scale [39].

The advantage of transparency becomes particularly notable when considering the limitations of LDA interpretability. While LDA is effective in classification tasks, it may face challenges in providing clear insights into decision-making due to its reliance on complex mathematical transformations. The linear combinations of features used by LDA can result in transformed dimensions that are less intuitive and harder to interpret, especially with an increasing number of features. Additionally, LDA assumes that features follow a normal distribution and have identical covariance matrices, limiting its applicability in scenarios with non-linear or heterogeneous data.

In contrast, interpretability becomes more straightforward when LDA is used with the GFC approach. The symbolic representation provided by GFC allows for a clearer understanding of evolved solutions, offering enhanced interpretability. This combination is a valuable alternative when the complexities of ML models pose transparency challenges, making it easier to discern the features contributing more towards correct predictions.

GFC/LDA provides valuable insights into selecting the most informative features used in constructing new features, as shown in Fig. 2. This solution is generated using the S_{CC} setup with ADA augmentation applied. The reason for choosing this solution is to illustrate how GFC/LDA constructs simple expressions. It represents one of our best and most straightforward solutions, achieving a cross-validation training AUC of 0.932, a validation AUC of 0.936, and a test AUC of 0.922. A total of six features are constructed by using only eight original features from the set of fifty-two.

Furthermore, in the heatmap depicted in Fig. 4, we visualize the frequency of features extracted by a GLCM matrix in four different orientations, including 0°, 45°, 90°, and 135°, which appear in the newly constructed features. The counts depicted in the heatmaps provide a clear and interpretable representation, indicating that, in all setups of the $DDSM$ dataset, features acquired from the 45° angle are predominantly utilised. This suggests that texture patterns captured at this orientation are more informative or relevant for the task, possibly because the structural or directional information in the data (e.g., texture variation) aligns more with the 45° direction, providing more distinguishing characteristics than other angles. This highlights the understanding of feature importance in the context of the novel feature construction method, GFC.

In the $DDSM$ setup, the most commonly appearing features in all scenarios are "Inverse Difference Moment" (IDM) (Feature 19) and "Entropy" (Feature 35).

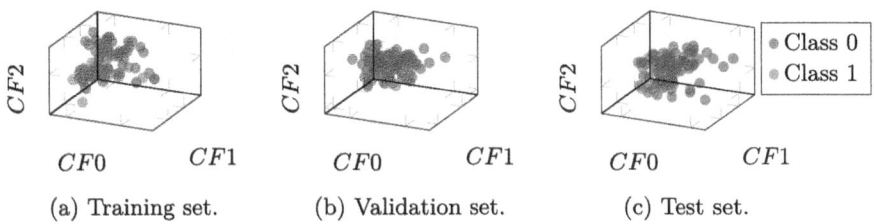

Fig. 5. Example of a sample best individual from a representative GFC/LDA run using the S_{CC} setup augmented by S-ENN. The best solution was constructed with only three features. To improve clarity, only 100 random samples from each set were used to plot the graphs.

IDM and Entropy indicate the difference and randomness in the grey-level pixels. On the other hand, "Radius" (Feature 7) and "Concavity" (Feature 21) emerge as the most prominent features used in the *WBC* dataset. The radius indicates the average distance from the centre point of a cell nucleus, while the irregularity of the concave portion is denoted by concavity. These observations can help health experts understand the importance of these features in the diagnosis.

Figure 5 shows an example where the best solution presented only three features, allowing the representation of the classes in a 3D graph. This example came from a run in the approach with CC segments augmented by S-ENN. For simplicity, we plotted the graphs using only 100 random samples from each set. We can see a clear separation between classes in the training set, and even though we cannot state the same for the validation and test sets, we still can see a concentration of the minority class in a region of the graph. This figure also depicts the high difficulty involved in handling imbalanced datasets.

6   Conclusion and Future Work

This work introduced an interpretable approach, GFC, based on GE, that addresses the challenge of opacity in medical diagnosis associated with ML. By leveraging the newly constructed features generated by GE, an LDA classifier is incorporated to distinguish between cancerous and normal samples. This two-stage approach, GFC/LDA, aimed to provide a more transparent and comparable model. The first stage is not tied to any specific model. In contrast, the user chooses the model in the second stage. The comparative analysis across two breast cancer datasets, DDSM and WBC, involved three approaches, GFC/LDA, GE, and LDA, as standalone classifiers applied directly to the original features. The results reveal that GFC/LDA produces simple and explainable solutions while retaining the AUC similar to that of LDA using the original features.

We envision broadening GFC's applicability by integrating it with other ML classifiers for future work. Furthermore, we aim to investigate the outcomes of linear and non-linear classifiers using this study's most frequently employed features and analyse their influence on predictions.

Acknowledgements. This study was funded by the Taighde Éireann CRT-AI (Grant 18/CRT/6223) and Lero, SFI Centre for Software (Grant 13/RC/2094_2).

References

1. Ahmed, S., Zhang, M., Peng, L., Xue, B.: Multiple feature construction for effective biomarker identification and classification using genetic programming. In: Proceedings of the 2014 Annual Conference on Genetic and Evolutionary Computation, pp. 249–256 (2014)
2. Al-Sahaf, H., et al.: A survey on evolutionary machine learning. J. R. Soc. N. Z. **49**(2), 205–228 (2019)
3. Bansal, N., Agarwal, C., Nguyen, A.: Sam: the sensitivity of attribution methods to hyperparameters. In: Proceedings of the IEEE/CVF Conference on Computer Vision and Pattern Recognition, pp. 8673–8683 (2020)
4. Batista, G.E., Bazzan, A.L., Monard, M.C., et al.: Balancing training data for automated annotation of keywords: a case study. Wob **3**, 10–8 (2003)
5. Białek, J., Bujalski, W., Wojdan, K., Guzek, M., Kurek, T.: Dataset level explanation of heat demand forecasting ANN with SHAP. Energy **261**, 125075 (2022)
6. Breiman, L.: Random forests. Mach. Learn. **45**, 5–32 (2001)
7. Cavaliere, F., Cioppa, A.D., Marcelli, A., Parziale, A., Senatore, R.: Parkinson's disease diagnosis: towards grammar-based explainable artificial intelligence. In: 2020 IEEE Symposium on Computers and Communications (ISCC), pp. 1–6 (Jul 2020), ISSN: 2642-7389
8. Chawla, N.V., Bowyer, K.W., Hall, L.O., Kegelmeyer, W.P.: Smote: synthetic minority over-sampling technique. J. Artifi. Intell. Res. **16**, 321–357 (2002)
9. de Lima, A., Carvalho, S., Dias, D.M., Naredo, E., Sullivan, J.P., Ryan, C.: GRAPE: grammatical algorithms in python for evolution. Signals **3**(3), 642–663 (2022)
10. DeLong, E.R., DeLong, D.M., Clarke-Pearson, D.L.: Comparing the areas under two or more correlated receiver operating characteristic curves: a nonparametric approach. Biometrics, 837–845 (1988)
11. Fitzgerald, J.M., Ryan, C., Medernach, D., Krawiec, K.: An integrated approach to stage 1 breast cancer detection. In: Proceedings of the 2015 Annual Conference on Genetic and Evolutionary Computation, pp. 1199–1206 (2015)
12. Friedman, J.H.: Greedy function approximation: a gradient boosting machine. Annals Statist., 1189–1232 (2001)
13. Gavrilis, D., Tsoulos, I.G., Dermatas, E.: Selecting and constructing features using grammatical evolution. Pattern Recogn. Lett. **29**(9), 1358–1365 (2008)
14. Hamilton, R.I., Papadopoulos, P.N.: Using shap values and machine learning to understand trends in the transient stability limit. IEEE Trans. Power Syst. **39**(1), 1384–1397 (2024)
15. Han, H., Wang, W.-Y., Mao, B.-H.: Borderline-SMOTE: a new over-sampling method in imbalanced data sets learning. In: Huang, D.-S., Zhang, X.-P., Huang, G.-B. (eds.) ICIC 2005. LNCS, vol. 3644, pp. 878–887. Springer, Heidelberg (2005). https://doi.org/10.1007/11538059_91
16. Haralick, R.M., Shanmugam, K., Dinstein, I.H.: Textural features for image classification. IEEE Trans. Syst. Man Cybernet., 610–621 (1973)

17. Hasan, Y., Amerehi, F., Healy, P., Ryan, C.: Stem rebalance: a novel approach for tackling imbalanced datasets using smote, edited nearest neighbour, and mixup. In: 2023 IEEE 19th International Conference on Intelligent Computer Communication and Processing (ICCP), pp. 3–9. IEEE (2023)

18. Hasan, Y., Lima, A.d., Amerehi, F., Bulnes, D.R.F.d., Healy, P., Ryan, C.: Interpretable solutions for breast cancer diagnosis with grammatical evolution and data augmentation. In: International Conference on the Applications of Evolutionary Computation (Part of EvoStar), pp. 224–239. Springer (2024). https://doi.org/10.1007/978-3-031-56852-7_15

19. He, H., Bai, Y., Garcia, E.A., Li, S.: Adasyn: adaptive synthetic sampling approach for imbalanced learning. In: 2008 IEEE international joint conference on neural networks (IEEE world congress on computational intelligence), pp. 1322–1328. IEEE (2008)

20. Heath, M., et al.: Current status of the digital database for screening mammography. In: Digital Mammography: Nijmegen, 1998, pp. 457–460. Springer (1998). https://doi.org/10.1007/978-94-011-5318-8_75

21. Heenaye-Mamode Khan, M., et al.: Multi-class classification of breast cancer abnormalities using deep convolutional neural network (CNN). PLoS ONE **16**(8), e0256500 (2021)

22. Kröll, J.P., Eickhoff, S.B., Hoffstaedter, F., Patil, K.R.: Evolving complex yet interpretable representations: Application to alzheimer's diagnosis and prognosis. In: 2020 IEEE Congress on Evolutionary Computation (CEC), pp. 1–8. IEEE (2020)

23. Liang, Y., Zhang, M., Browne, W.N.: Feature construction using genetic programming for figure-ground image segmentation. In: Leu, G., Singh, H.K., Elsayed, S. (eds.) Intelligent and Evolutionary Systems. PALO, vol. 8, pp. 237–250. Springer, Cham (2017). https://doi.org/10.1007/978-3-319-49049-6_17

24. de Lima, A.D., Lopes, A.J., do Amaral, J.L.M., de Melo, P.L.: Explainable machine learning methods and respiratory oscillometry for the diagnosis of respiratory abnormalities in sarcoidosis. BMC Med. Inform. Decision Making **22**(1), 274 (2022)

25. Liu, H., Motoda, H.: Feature extraction, construction and selection: A data mining perspective, vol. 453. Springer Science & Business Media (1998)

26. Lundberg, S.M., Lee, S.I.: A unified approach to interpreting model predictions. Adv. Neural Inform. Process. Syst. **30** (2017)

27. Mahesh, B.: Machine learning algorithms-a review. Inter. J. Sci. Res. (IJSR) **9**(1), 381–386 (2020)

28. Miquilini, P., Barros, R.C., de Melo, V.V., Basgalupp, M.P.: Enhancing discrimination power with genetic feature construction: A grammatical evolution approach. In: 2016 IEEE Congress on Evolutionary Computation (CEC), pp. 3824–3831. IEEE (2016)

29. Moosbauer, J., Herbinger, J., Casalicchio, G., Lindauer, M., Bischl, B.: Explaining hyperparameter optimization via partial dependence plots. Adv. Neural. Inf. Process. Syst. **34**, 2280–2291 (2021)

30. Nguyen, H.M., Cooper, E.W., Kamei, K.: Borderline oversampling for imbalanced data classification. Inter. J. Knowl. Eng. Soft Data Paradigms **3**(1), 4–21 (2011)

31. Pedregosa, F., et al.: Scikit-learn: machine learning in Python. J. Mach. Learn. Res. **12**, 2825–2830 (2011)

32. Rana, M., Bhushan, M.: Machine learning and deep learning approach for medical image analysis: diagnosis to detection. Multimedia Tools Appli. **82**(17), 26731–26769 (2023)

33. Rasheed, K., Qayyum, A., Ghaly, M., Al-Fuqaha, A., Razi, A., Qadir, J.: Explainable, trustworthy, and ethical machine learning for healthcare: a survey. Comput. Biol. Med., 106043 (2022)
34. Ribeiro, M.T., Singh, S., Guestrin, C.: " why should i trust you?" explaining the predictions of any classifier. In: Proceedings of the 22nd ACM SIGKDD International Conference on Knowledge Discovery and Data Mining, pp. 1135–1144 (2016)
35. Robin, X., et al.: pROC: an open-source package for r and s+ to analyze and compare ROC curves. BMC Bioinform. **12**, 1–8 (2011)
36. Ryan, C., Azad, R.M.A.: Sensible initialisation in grammatical evolution. In: Barry, A.M. (ed.) GECCO 2003: Proceedings of the Bird of a Feather Workshops, Genetic and Evolutionary Computation Conference, pp. 142–145. AAAI, Chigaco (11 Jul 2003)
37. Ryan, C., Collins, J.J., Neill, M.O.: Grammatical evolution: evolving programs for an arbitrary language. In: Banzhaf, W., Poli, R., Schoenauer, M., Fogarty, T.C. (eds.) EuroGP 1998. LNCS, vol. 1391, pp. 83–96. Springer, Heidelberg (1998). https://doi.org/10.1007/BFb0055930
38. Saarela, M., Jauhiainen, S.: Comparison of feature importance measures as explanations for classification models. SN Appli. Sci. **3**(2), 1–12 (2021). https://doi.org/10.1007/s42452-021-04148-9
39. Stiglic, G., Kocbek, P., Fijacko, N., Zitnik, M., Verbert, K., Cilar, L.: Interpretability of machine learning-based prediction models in healthcare. Wiley Interdisciplinary Rev. Data Mining Knowl. Dis. **10**(5), e1379 (2020)
40. Tsoulos, I.G., Tzallas, A.T., Tsalikakis, D.: Prediction of covid-19 cases using constructed features by grammatical evolution. Symmetry **14**(10), 2149 (2022)
41. Virgolin, M., Alderliesten, T., Bosman, P.A.: On explaining machine learning models by evolving crucial and compact features. Swarm Evol. Comput. **53**, 100640 (2020)
42. Wilson, D.L.: Asymptotic properties of nearest neighbor rules using edited data. IEEE Trans. Syst. Man Cybernet., 408–421 (1972)
43. Wolberg, W.H., Street, W.N., Mangasarian, O.L.: Breast cancer wisconsin (diagnostic) data set [uci machine learning repository] (1992)
44. Zhang, H., Cisse, M., Dauphin, Y.N., Lopez-Paz, D.: Mixup: Beyond Empirical Risk Minimization (Apr 2018)
45. Zhang, Y., Tiňo, P., Leonardis, A., Tang, K.: A survey on neural network interpretability. IEEE Trans. Emerging Topics Comput. Intell. **5**(5), 726–742 (2021)

Designing Hardware-Friendly Hash Functions for Network Security Using Cartesian Genetic Programming

Mujtaba Hassan[1]([envelope]) [iD], Jo Vliegen[1] [iD], Stjepan Picek[3] [iD], and Nele Mentens[1,2] [iD]

[1] ES&S, COSIC, ESAT, KU Leuven, Leuven, Belgium
{mujtaba.hassan,jo.vliegen,nele.mentens}@kuleuven.be
[2] LIACS, Leiden University, Leiden, The Netherlands
[3] Digital Security Group, Radboud University, Nijmegen, The Netherlands
stjepan.picek@ru.nl

Abstract. In this study, we propose a novel, hardware-efficient non-cryptographic (NC) hash function, developed using Cartesian Genetic Programming (CGP), to optimize the processing of network flows (e.g., 96-bit vectors in IPv4). With the advent of high-speed terabit Ethernet technologies such as 800G, cybercriminals are increasingly exploiting these networks to launch various attacks, including distributed denial-of-service (DDoS) attacks. To counter these threats, network security applications often employ probabilistic data structures (PDS), such as Bloom filters and Count Min sketches, to monitor network flows. These applications are often deployed on Field Programmable Gate Arrays (FPGAs) to meet real-time processing demands. The performance of PDS depends significantly on the efficiency of NC-hash functions. By utilizing avalanche metrics (avalanche dependence, avalanche weight, and entropy) as the fitness function, CGP-hash ensures robust performance without the need for dataset-specific training. While Genetic Programming (GP)-based NC-hash functions have demonstrated superior computational efficiency on FPGAs in terms of operating frequency, throughput, and latency, they are often less resource-efficient compared to state-of-the-art NC-hash functions. Thus, our hypothesis in this paper is that CGP, with its compact representation, leads to NC hashes with high computational efficiency and fewer resources on an FPGA. Our experimental results confirm that CGP-hash improves computational efficiency by at least 7.3% while improving area efficiency by 4.5× compared to state-of-the-art NC-hash functions. This makes CGP-hash more suitable for FPGA implementations than other bio-inspired and handcrafted hash functions. Moreover, the 48-bit hash output can be extended by evolving additional hash functions and concatenating their outputs, all while maintaining superior resource efficiency and computational speed.

Keywords: Non-cryptographic hash function · Cartesian Genetic Programming · Bloom filters · Network flow detection

© The Author(s), under exclusive license to Springer Nature Switzerland AG 2025
P. García-Sánchez et al. (Eds.): EvoApplications 2025, LNCS 15612, pp. 221–237, 2025.
https://doi.org/10.1007/978-3-031-90062-4_14

1   Introduction

Cybercriminals are taking advantage of high-speed terabit-scale Ethernet networks (bandwidths exceeding 100 Gb/s), leveraging advanced tools such as generative AI and developer copilot technologies to craft increasingly sophisticated code [1]. Consequently, the threat landscape has expanded dramatically. For instance, in 2023, Cloudflare mitigated its largest-ever hyper-volumetric distributed denial-of-service (DDoS) attack, which peaked at 201 million requests per second (rps), nearly eight times greater than the previous record of 26 million rps set in 2022 [2]. A common approach to countering such attacks involves deploying network monitoring and filtering systems (NMFS), which typically rely on machine learning, pattern matching, or large flow detection methods [3]. In this work, our target application is large flow detection. All network packets sharing the same flow ID are considered part of the same flow. The flow ID generally comprises five fields: source and destination IP addresses, source and destination ports, and the transmission protocol, though alternative combinations are also possible. For IPv4 networks, this can be represented with 96 bits [4].

One of the primary challenges in real-time detection and monitoring of network packets on Terabit Ethernet networks is ensuring that the detection rate matches the line rate. Failure to do so risks missing a significant number of packets. To address this, hardware accelerators, such as Field Programmable Gate Arrays (FPGAs), are crucial for enabling parallel packet processing at line rate [5]. Major cloud providers, including Amazon and Microsoft Azure, have integrated FPGAs for hardware acceleration to improve throughput and minimize latency in network applications [6,7]. The increasing adoption of FPGAs for cloud computing and network security is further underscored in [8,9]. A recent deployment of Intel's Agilex FPGA for Network Intrusion Detection achieved an impressive throughput of 400 Gbps [10].

In NMFS, both lookup and counting processes can be implemented using exact data structures, such as dictionaries and associative arrays, or through probabilistic data structures (PDS). PDS, such as Count Min (CM) sketches for frequency counting and Bloom filters for item detection, are commonly used for real-time detection and measurement of large data flows in computer networks. These structures offer faster query and update times and reduced memory consumption compared to exact methods, although with a slight trade-off in accuracy. Performance analysis in the previous studies [11–15] indicates that their efficiency is heavily influenced by the execution speed of the underlying non-cryptographic (NC) hash functions (explained in the next section). Consequently, optimizing NC-hash function execution times can lead to substantial improvements in overall system performance [11–13].

The main objective of this work is to optimize the performance of NC-hash functions on FPGAs, specifically with respect to the computational efficiency (throughput, latency, and operating frequency), while reducing the utilization of FPGA resources such as Look-Up Tables (LUTs) in comparison to state-of-the-art NC-hash functions. This optimization is crucial to meet the growing demands of high-speed terabit Ethernet networks, as the industry anticipates Ethernet speeds reaching 3.2 Tb/s by 2030 [16]. This research builds on the work of [17], which employed Genetic Programming (GP) to develop hardware-efficient NC-hash functions capable of processing 96-bit flow IDs for PDS. Their findings demonstrated that their proposed NC-hash functions surpassed Evolutionary Computing (EC)-based and other state-of-the-art NC-hash functions by at least 36% in operating frequency, 30% in throughput, and achieved a 27% reduction in latency on FPGAs. However, they remained behind in terms of number of occupied resources.

In the present study, we propose utilizing Cartesian Genetic Programming (CGP) to design NC-hash functions. CGP offers a distinct advantage over GP as it allows multiple uses of nodes within the design, resulting in a more compact output representation compared to tree-based GP. We expect that this approach can produce more efficient and FPGA-compatible designs than both GP and other non-EC methodologies.

The contributions of our research can be summarized as follows.

- We introduce a novel 48-bit NC hash function, CGP-hash, designed for 96-bit inputs using CGP. Our CGP-hash demonstrates superior efficiency on FPGAs in terms of throughput, latency, and operating frequency, surpassing current state-of-the-art NC hash functions that rely on the Avalanche Criterion.
- Our proposed CGP-hash achieves a 4.5× improvement in hardware resource efficiency compared to all evaluated state-of-the-art NC hash functions.
- By optimizing key parameters and settings, our approach facilitates the evolution of additional NC-hash functions. This scalability enables the concatenation of hash outputs, effectively extending the overall output size, which is a significant advantage for applications such as Bloom filters in PDS that require larger hash values or multiple independent hash functions.

2   Preliminaries

2.1   Hash Functions

Hash functions convert messages of arbitrary length into a fixed-length output, commonly known as a hash digest or hash code. Mathematically, a hash function can be defined as $h = H(m)$, where h represents the hash code, m is the input message, and H denotes the hash function. Hash functions fall into two primary categories: cryptographic hash functions and NC-hash functions. Regardless of type, all hash functions must exhibit the following essential properties [11,18,19]:

- **Determinism:** This ensures that the same input will always produce the same output.
- **Uniform distribution:** All possible outputs within a given range should be equally likely, preventing clustering and reducing the likelihood of collisions.
- **Avalanche:** A small change in the input should lead to a large change in the output, ensuring high unpredictability and randomness.

- **Speed:** Hash functions should generate outputs quickly and efficiently.

Furthermore, cryptographic hash functions must provide strong security assurances, including pre-image resistance, second pre-image resistance, and collision resistance [20,21]. These additional security features require multiple processing rounds (a series of iterative steps) that increase computational cost, rendering cryptographic hash functions less efficient for applications demanding rapid searches.

2.2   Bloom Filters

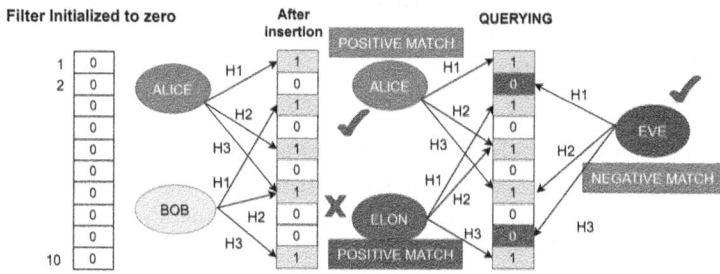

Fig. 1. A 10-bit Standard Bloom filter with three hash functions, where three distinct hash functions ($H1$, $H2$, and $H3$) are used to insert the names "ALICE" and "BOB". After the insertion, three queries are done. The query for "ALICE" results in a correct positive match, the query for "EVE" results in a correct negative match, while the query for "ELON" results in a false positive match due to hash collisions.

A Bloom filter is a PDS designed to efficiently support membership queries using multiple fast, independent, and uniform hash functions [13,19,22,23]. Unlike conventional data structures, it does not store actual set elements, allowing it to represent a large number of elements in a highly space-efficient manner. The process of inserting an element or checking for membership typically operates in constant time, $\mathcal{O}(n)$, where n represents the number of hash functions. Due to the independence of these hash functions, such operations can be parallelized in hardware, further enhancing performance. While the Bloom filter offers significant advantages, particularly in terms of speed and space efficiency, it is inherently probabilistic. This probabilistic nature introduces the possibility of false positives, where an element not present in the set is incorrectly identified as present. This limitation arises from hash collisions, instances where different inputs produce the same output. Consequently, the Bloom filter cannot guarantee perfect detection accuracy. The false positive rate (FPR) is a critical measure of accuracy, influenced by factors such as the number and quality of hash functions, the bit vector size, and the load factor [11,12,19,24]. Despite the risk of false positives, the Bloom filter guarantees no false negatives, ensuring that any element reported as absent is for sure not present. This property is particularly advantageous in applications where negative results predominate, minimizing unnecessary memory accesses. The full working of the Standard Bloom Filter (SBF) is illustrated in Fig. 1.

3   Related Work

The performance of PDS is highly dependent on the efficiency of the underlying NC-hash functions, as discussed in Sect. 1. Additionally, for PDS implementation on FPGAs, it is crucial that these NC-hash functions are hardware-optimized. In a study by [11], a novel approach was introduced for processing 96-bit network flow IDs. Rather than relying on multiple independent NC-hash functions, the authors generated a single hash value and partitioned it into multiple segments, each serving as an independent hash output. To enhance compatibility with hardware, they reduced the number of rounds in the cryptographic permutation XOODOO [25] while maintaining its avalanche properties. This optimized variant, termed XOODOO-NC, was integrated into an FPGA-based Bloom filter, where it outperformed other implementations in both computational efficiency and resource usage. However, the authors did not detail the specific performance metrics of XOODOO-NC itself.

Building on this work, Claesen et al. [15] and Sateesan et al. [13] introduced NC versions of several widely used symmetric-key ciphers, including AES, Pyjamask, GIFT, SKINNY, SPECK, TinyJAMBU, PHOTON, and Sparkle. Of these, GIFT-NC and PHOTON-NC achieved an impressive low hardware latency, with 2.70 ns and 2.6 ns, respectively. XOODOO-NC had a slightly higher latency at 2.75 ns but demonstrated superior efficiency in hardware resource usage. In a separate study [12], the authors reported an execution delay of 1.5 ns for XOODOO-NC. This result reflected the initial two rounds, during which the avalanche properties were not fully satisfied [26]. In [19], another study implementing XOODOO-NC (96-bit input/output) on FPGA recorded a latency of 2 ns. All NC hash functions mentioned above complete their output computation within a single clock cycle on FPGA, thus avoiding the need for complex pipelining techniques.

Cyclic Redundancy Checks (CRCs) are also capable of functioning as NC hash functions and can produce outputs up to 128 bits. However, CRCs are not recommended for PDS applications on FPGAs due to high resource consumption and substantial latency [27–29]. Similarly, lightweight cryptographic hash functions encounter significant limitations. For instance, FPGA-based implementations of Photon and Sponge hash functions exhibit minimum cycle counts of 12 and 45, respectively [30–33]. Each cycle represents a complete period of the system clock governing FPGA operations, and these high cycle counts contribute to increased latency and decreased throughput, making them unsuitable for high-speed applications such as Terabit Ethernet networks.

In recent years, there has been growing interest in applying EC methodologies to develop NC hash functions. A series of studies by Hassan et al. [17,19,27] applied GP with avalanche metrics to design NC hash functions specifically for processing 96-bit network flow IDs and generating high-entropy hash outputs for PDS. Their proposed hash function (GPNCH) [19] outperformed the best FPGA implementations available, achieving at least 19% improvement in operating frequency, throughput, and latency. In later work [27], they refined their approach by exploring various logical operators and parameter settings, identifying XOR and rotation as essential operators, and recommended smaller tree heights to reduce function size and improve software simulation times. This refined NC hash function, termed E-hash, achieved an 8.4% increase in operating frequency and throughput and a 7.74% reduction in latency compared to other hardware-friendly NC hash functions. In [17], the authors introduced a 96-bit NC-hash function (NCGPH96-concat) by concatenating three evolved 32-bit NC hash functions, achieving a highly uniform distribution. When integrated into an SBF, these proposed NC hash functions demonstrated comparable FPR to widely used hash functions like FNV1a, with FPGA evaluations showing at least a 27% improvement in operating frequency, throughput, and latency over 17 other state-of-the-art NC hash functions.

Across their studies, Hassan et al. offered a comprehensive review and critical analysis of various bio-inspired methodologies such as GP, CGP, and Linear Genetic Programming (LGP) approaches for NC hash functions [34–39] and [4,18,40–42]. Three significant limitations emerge in the application of these methods for FPGA-based PDS. First, most NC hash functions use collision rate as a fitness function, performing well only on trained datasets, making them sensitive to input variability [17–19,27,42]. Second, these methods heavily depend on the multiplication operator to introduce entropy, which is a resource-intensive operation generally avoided in FPGA hash function design due to its negative impact on operating frequencies and hardware resource consumption [12,17,36,40,43]. Third, many existing NC hash functions produce only 16-bit or 32-bit outputs, which are inadequate for PDS applications requiring larger hash outputs or multiple independent hash functions.

Motivated by the highly efficient GP-based approach presented in [17] for designing NC hash functions, we propose a method for developing 48-bit NC hash functions using CGP with avalanche metrics (see next section) as the fitness criterion. This approach is anticipated to produce FPGA-compatible designs that are more efficient than those generated by GP and other evolutionary or non-evolutionary methods, owing to CGP's compact structure, which permits multiple uses of nodes. Similar to the method in [17], the 48-bit output can be concatenated to create larger hash outputs required for PDS while ensuring high computational efficiency and optimal resource utilization.

4   Avalanche Metrics

Studies [17–19,27,42] demonstrate that the avalanche property is independent of input data, meaning it does not require measurement in relation to specific datasets, unlike collision resistance and output distribution, which are highly data-dependent. For example, Saez et al. [42] demonstrated that the collision rates of various state-of-the-art hash functions differ across four real and four synthetic datasets, indicating a dependence on input type. A more stringent criterion, known as the Strict Avalanche Criterion (SAC), specifies that altering a single bit in the input should result in each output bit having a 50% probability of changing. Research on NC hash functions has shown that achieving SAC inherently leads to optimal collision rates and uniform output distribution [11, 13,17,18]. To assess the avalanche property, three metrics avalanche dependence (D_{av}), weight (W_{av}), and entropy (E_{av}), as presented in the work of Daemen et al. [25] are frequently applied.

Avalanche Dependence: This property quantifies the number of output bits that may change when a single input bit is altered:

$$D_{av} = m - \sum_{1}^{m} f(pv[t]). \tag{1}$$

It is defined using a probability vector pv, where each element $pv[t]$ represents the probability that the t-th output bit will flip. Here, m represents the total number of output bits. The function $f(pv)$ yields 1 if pv equals 0, and 0 otherwise. This property is satisfied only when D_{av} is approximately equal to m across all outputs resulting from a one-bit change in the input, thereby generalizing full diffusion.

Avalanche Weight: This refers to the expected number of bits that will vary between two output values resulting from a change in a single input bit. Mathematically, it is represented as follows:

$$W_{av} = \sum_{1}^{m} pv[t]. \tag{2}$$

To satisfy the condition, 50% of the output bits should be changed for a single bit change, i.e., $W_{av} \simeq m/2$ for Hamming weight = 1. The avalanche weight generalizes the avalanche criteria.

Avalanche Entropy: This measurement can be employed to assess the level of randomness in the output:

$$E_{av} = \sum_{1}^{m} (-pv[t] * log_2 pv[t] - (1 - pv[t]) * log_2(1 - pv[t])). \tag{3}$$

Full entropy describes the state in which the output produced by the hash function reaches maximum randomness, showing no discernible patterns or meaningful structure within the data.

5   Methodology

5.1   Cartesian Genetic Programming (CGP)

This research employs CGP to model NC hash functions. CGP is a variant of GP in which solutions are represented as a two-dimensional grid of nodes, providing several advantages over the more common tree-based GP approach. These benefits include a smaller required population size, the ability to reuse nodes, the presence of non-coding genes, compact representation (improving interpretability), and a straightforward Evolutionary Algorithm (EA) [44]. Given that the primary objective of this study is to optimize existing NC hash functions by developing more compact designs that satisfy avalanche metrics, we anticipate that these advantages will result in improvements over GP-based NC hash functions.

5.2   Encoding

The proposed NC hash function is specifically designed for processing 96-bit inputs. Thus, the input size is set to 96 bits. This input is divided into two 48-bit segments, referred to as input 0 and input 1, which together are used to produce a 48-bit output. Our approach evolves a single 48-bit hash function as a proof of concept, which can then be replicated to evolve multiple 48-bit hash functions. By concatenating these outputs, we can generate 96-bit or larger hash values suitable as required by PDS. In contrast to the method in [17], which concatenates three 32-bit hash functions to achieve a 96-bit output, our approach requires only two evolved 48-bit hash functions. The study in [17] demonstrated that generating smaller hash outputs and concatenating them to produce a larger output is more advantageous in terms of computational efficiency, including operating frequency, throughput, and latency, on FPGAs. This approach requires far fewer nodes in the design compared to designs that produce all 96 bits at once [19,27].

5.3   Fitness Function

$$fitness function = \sum (D_{av}, W_{av}, E_{av}).   \tag{4}$$

Avalanche metrics are calculated as outlined in Sect. 4. With a 48-bit output, the ideal values for D_{av} and E_{av} are 48 bits, while for W_{av}, it is 24 bits. Following the methodology of [17,19,27], these metrics are combined into a composite fitness score with a theoretical maximum of 120, which serves as the target for the proposed NC hash function. Since these metrics derive from a single probability vector pv, a multi-objective fitness function was not employed. To manage computational demands, a Monte Carlo simulation [17–19,27] is used in a two-phase process. In the initial phase, the population is evaluated on a reduced input subset, with 9600-bit flips (96 × 100). Output differences are incrementally calculated to capture worst-case values, ensuring robustness across various possible inputs. In the second phase, the most promising candidates from phase one are re-evaluated using 24 million bit flips to expand input coverage, as recommended in [11,15,17,25]. Reported results are based on this comprehensive second phase.

5.4 Primitive Set and Parameters

The experimental setup began with an analysis of the recommended primitive set (a combination of functions and terminals) for NC hash functions, as proposed in [17,19,27]. We utilized the same primitive set with one modification: constants were excluded (reason explained in Sect. 6.5). Additionally, we incorporated a set of ten left and ten right circular rotation operators, designed to rotate bits by increments of 1 through 10, linking the most significant bits (MSB) with the least significant bits (LSB) in each direction. Due to differences in output size, three custom swap operators were arbitrarily incorporated, swapping the 24, 12, and 4 MSB with LSB.

In CGP, many parameter settings, such as population size, number of nodes, selection scheme, number of generations, mutation rate, and mutation type, can be explored to achieve optimal results. We experimented with a diverse range of settings, and the best configurations are summarized in Table 1.

Table 1. Primitive Set and general design parameters

Kind	Value
Evolutionary Strategy	(1+12)-ES
No. of Inputs & Outputs	2 & 1
Input/Output Size	96/48 bits
Primitive Set	{XOR, AND, OR, NOT, ROT, SWAP}
Mutation Type	probabilistic
Selection Scheme	Fittest
No. of Generations	20000
Max. Allowed Nodes	3000
Population Size	5
Mutation Rate	0.1
Fitness Function	Avalanche metrics
Stopping Conditions	either max. generations or target fitness is reached

5.5 Stopping Conditions

The evolutionary process terminates under either of the following conditions:

- The optimal fitness value of 120 is attained.
- The number of generations reaches 20k, as explained in Sect. 6.1.

5.6   Software and Hardware Tools

The CGP simulations leverage a cross-platform library developed in C, as documented in [45]. To enhance the computational efficiency of the evolutionary process, the simulations are executed on 24 CPU cores. For the hardware implementation, the Virtex Ultrascale+ xcvu7p-flvb2104-2-i FPGA chip is chosen, with synthesis performed using Vivado 2022.1. This FPGA selection is intentional, as it matches the chip used in compared studies, ensuring consistency in performance evaluation.

6   Results and Discussion

6.1   Fitness Evaluation

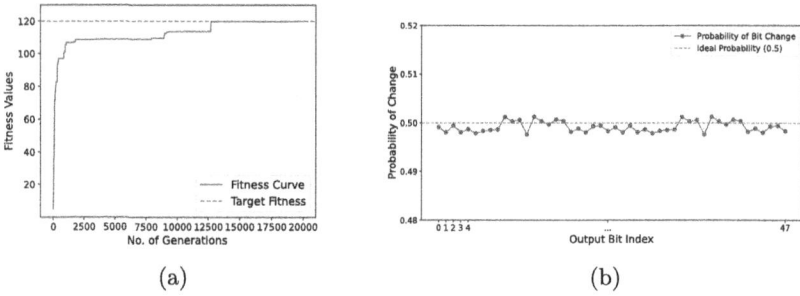

(a)   (b)

Fig. 2. (a) Population fitness growth across generations; (b) Output bit probabilities after 24 million single-bit input flips.

Figure 2a illustrates a rapid increase in population fitness during early generations, reaching around 110 by generation 2500. Afterward, fitness stabilizes, with minor improvements around generations 10000 and 13000, eventually reaching the target fitness of 120. In CGP, active and inactive genes play distinct roles: nodes contributing to the phenotype are termed active, while others remain inactive. Although 3000 nodes are permitted, the final output (Fig. 3) includes only 26 active nodes. Thus, while Fig. 2a may appear stable from generations 2500 to 10000, mutations continue, primarily among inactive nodes, which can become active if linked to the output path, explaining the observed fitness jumps.

Figure 2b shows the probability of each output bit changing upon a single input bit flip (SAC) over 24 million flips of a 96-bit input. The probabilities remain near 0.5, resulting in a D_{av} of 48, W_{av} of 23.96, and E_{av} of 48, achieving a near-perfect fitness score of 119.96. As discussed in Sect. 4, meeting these metrics ensures a uniform output distribution, which is crucial for PDS like Bloom filters.

During the training and testing phases, several challenges were encountered (see Sect. 6.5). After identifying an optimal configuration (detailed in Table 1), the experiment was repeated 210 times (justification given in Sect. 6.3), yielding

44 valid designs that utilized both inputs. Of these, only two designs successfully passed large-scale testing with 24 million bit flips. Interestingly, in both cases, inputs 1 and 2 were used in similar quantities (47 and 56 times, respectively), pointing to the importance of balanced input usage to avoid bias and ensure robust large-scale performance. However, we need more experiments to draw conclusions about this behavior.

For FPGA implementation, the first design was chosen due to its compact structure of 26 active nodes, compared to 28 in the alternative. As demonstrated in [17], designs with fewer nodes typically achieve superior computational efficiency and resource utilization on FPGA.

6.2   Final Design

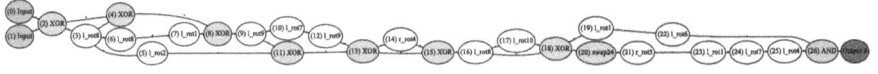

Fig. 3. The optimal hash function designed using CGP with only eight logical operators.

Figure 3 presents the final NC hash function design, which appears compact compared to GP-based designs in [17,19,27], using only 26 nodes. Unlike GP-based designs, this CGP configuration allows node reuse, where a node's output can feed into multiple others, reducing redundancy. This offers an advantage, as GP-based designs require repeated logic to achieve a similar effect. The CGP-based design is also more interpretable than the complex tree structures typical of GP methods.

Table 2. CGP vs. GP: Logic gates comparison of NC hash functions.

Evolutionary NC-Hash	Total Nodes	Total Logic Gates	XOR	OR	AND	NOT	Constants	ROT
NCGPH32-I [17]	300	100	79	21	0	0	0	98
E-hash [27]	432	117	107	01	05	4	6	201
GPNCH-V [19]	550	125	109	02	04	10	12	293
CGP-hash (ours)	352 (tree form)	111	110	0	01	0	0	128

To gain further insights and facilitate comparison with GP-based NC-hash function designs in [17,19,27], the CGP structure was converted into a tree form, with results in Table 2. In this form, it requires far fewer nodes than GP-based hash functions E-hash and GPNCH-V for 96-bit inputs, though their output sizes are 96 bits, compared to the CGP-hash's 48-bit output. Compared to the 32-bit NCGPH32-I, CGP-hash generates a larger output with a similar number of logical elements. Most importantly, to generate a large 96-bit output, only two CGP-hash instances are needed versus three for NCGPH32-I, which would triple FPGA resource use.

The final design contains numerous rotation operators, which incur minimal cost on FPGA and employs only XOR and AND gates from the primitive set.

However, this does not imply that other gates are unnecessary; our experiments confirm that including other gates in the primitive set is essential for guiding evolution toward optimal solutions. In terms of logical and rotation operator usage, both GP and CGP approaches perform comparably.

6.3   Runtime of the Software Simulation

In [27], the authors reported that a complete evolutionary run of a GP-based NC hash function that satisfies avalanche metrics requires between 2 to 4 d. In contrast, a full run of the CGP-based NC hash function averages only 2.5 h (based on 210 repeated experiments). This significant reduction in runtime enables a broader exploration of parameter settings. Based on our experiments, this speedup is attributable to two primary factors. First, our implementation utilizes a C-based library [45], whereas the GP-based studies in [17,19,27] rely on a Python-based library [46], and C generally executes faster than equivalent Python code [47,48]. Second, CGP employs smaller population sizes [44] and leverages Neutral Genetic Drift–the inheritance of non-contributing (inactive) genes across generations–to achieve high fitness. Because mutations frequently occur in these inactive genes, the phenotype often remains unchanged, reducing the need for repeated evaluations and significantly accelerating execution times. However, for large-scale testing (24 million-bit flips), each design requires approximately 20 h to complete, which limits the scope for additional experimentation.

6.4   FPGA Performance Comparison with the State-of-the-Art

Table 3. Comparison with related work.

No.	NC hash function	Block Size	Max. operating frequency (MHz)	Throughput (Tp) (Gbps)	No. of LUTs	Latency (ns)	Tp/Area (Mbps/LUT)
1	Murmur3 [13]	96/64	120.6	2.57	567	24.87	4.53
2	XOODOO-NC (2.5 rounds) [19]	96/96	500	48	480	2.00	100
3	SPECK-NC [15]	96/96	171.6	16.47	432	5.82	38.12
4	Pyjamask-NC [15]	96/96	279.4	26.82	832	3.58	32.23
5	TinyJAMBU-NC [13]	96/96	138.9	13.33	1520	7.2	8.77
6	GPNCH-V [19]	96/96	595.23	59.95	1446	1.68	41.46
7	SipHash [4,13]	96/16	182.8	1.463	1061	21.88	1.38
8	XORHash [4,13]	96/16	627.3	2.868	291	11.13	9.86
9	NSGAHash-VII [4,13]	96/16	184.1	1.473	80	21.72	18.41
10	FNV-1a [13]	96/128	122.9	0.92	567	130.08	1.63
11	GIFT-NC [15]	96/128	369.7	35.49	546	2.70	65
12	AES-NC [15]	96/128	255.9	24.56	2225	3.91	11.04
13	SKINNY-NC [15]	96/128	202.4	19.44	2176	4.93	8.93
14	Sparkle-NC [13]	96/128	133.3	12.79	811	7.2	15.78
15	PHOTON-NC [13]	96/144	384.6	36.92	1338	2.6	27.59
16	CRC-128 [28]	128/32	312.5	40	3511	3.2	11.39
17	NCGPH96-concat [17]	96/96	813	78.04	1683	1.23	46.37
18	E-hash [27]	96/96	645.16	61.94	1542	1.55	40.17
19	CGPhash (ours)	96/48	877.43	84.23	187	1.14	450.43

To benchmark our CGP-based approach against state-of-the-art, we utilized the results from [17], as shown in Table 3, with an additional focus on throughput per LUT (Tp/LUT) to assess FPGA area efficiency. This metric indicates how efficiently the design uses the available LUT resources to achieve a certain level of throughput. A higher value means the hash function can process more data using fewer FPGA resources.

NCGPH96-concat [17] previously demonstrated the highest computational efficiency. However, our CGP-based method improves operating frequency and throughput by 7.92% and reduces latency by 7.3%. While the CGP-hash output is 48 bits, we present this as a proof of concept, demonstrating the viability of the approach. With further experiments, additional hash functions could be concatenated to generate larger outputs, similar to the method proposed in [17]. Given FPGA's capacity for parallel processing, concatenated hash functions should maintain comparable latency, allowing us to conclude that CGP appears to have a slight advantage over GP-based NC hash functions in FPGA performance comparison in terms of operating frequency, throughput, and latency.

However, GP-based studies [17,19,27] were unable to conduct a large number of experiments due to the resource-intensive nature of the methodology. Therefore, it is plausible that GP-based NC hash functions could match the computational efficiency of CGP-based designs on FPGA if larger-scale experiments were conducted.

A comparison of LUT utilization indicates that Tp/LUT, a critical measure of area efficiency, places our method significantly ahead, surpassing XOODOO-NC by 4.5×. For a fair comparison, an equivalent CGP-hash design for 96-bit output, using approximately double the LUTs, would still offer 2.25× higher area efficiency.

GP-inspired hash functions, such as E-hash, GPNCH, and NCGP32-concat, outperform handcrafted, reduced-round functions like XOODOO-NC, GIFT-NC, and PHOTON-NC in operating frequency, throughput, and latency. However, they face challenges with area efficiency. For instance, NCGPH96-concat has an area efficiency of 46.37, roughly half that of XOODOO-NC. In our work, CGP-based node reuse significantly reduces circuit size, providing a distinct advantage in FPGA implementations over GP-based designs.

Lastly, in comparison to widely used hash functions like Murmur3 and Fnv1-a, our CGP-hash approach significantly outperforms them in both computational speed and resource efficiency. While FPGA implementations of SipHash, XORhash, and NSGAHash-VII are on Virtex 7 boards and CRC-128 on a Virtex 6 board, slightly older platforms than those used here, the observed variations in latency and throughput are substantial.

6.5   Limitation and Challenges when using CGP

In our design, we divide the 96-bit input into two 48-bit inputs to produce a 48-bit output. The CGP method allows flexibility in selecting inputs contributing to the output, enabling solutions to evolve using only the inputs deemed beneficial during evolution [45]. However, in our case, the absence of either input results in invalid designs, as both inputs are equally crucial. To address this, we consider three primary solutions.

The first approach involves applying a constraint that discards any solutions not utilizing both inputs, which is a method demonstrated in [39]. In their study, the authors processed a 288-bit IPv6 flow ID by splitting it into five 64-bit blocks. The drawback of this approach is that many potentially strong candidates are excluded due to constraint violations, and this intervention introduces bias in a natural process, so we only considered this as a last resort and did not apply it in our study.

The second approach seeks to increase fitness function complexity, making it more challenging for the algorithm to produce high-quality solutions without using all inputs. We attempted an alternating bit-flip strategy across inputs, but this did not yield the desired effects. Given the 48-bit output size, avalanche metrics can still be satisfied with repeated usage of a single input.

The third approach increases the total number of nodes in the design. Although adding nodes moderately affects software simulation time, a larger number of nodes improves the likelihood that both inputs will be utilized. In experiments averaging 200 runs with node counts set to 1000, 2000, and 3000, we observed an increase in valid designs from 5.6% to a satisfactory 21%. The 95% confidence interval (CI) for the latter value ranged from 15.35% to 26.65%, indicating a statistically significant improvement. Based on these results, we determined 3000 nodes sufficient for large-scale testing. It is worth noting that increasing nodes to around 10000 led to diminished training phase performance, likely due to the fixed 20k generation limit, which may be insufficient for a design of this scale.

Studies indicate that incorporating constants in the design introduces non-linearity and enhances output randomness [18,19,27]. However, due to potential input omissions, as explained above, we excluded constants from the primitive set. Our results demonstrate that evolving hash functions with high entropy is achievable even without constants.

7   Conclusions and Future Work

This research introduces a novel 48-bit, hardware-optimized NC hash function (CGP-hash) developed using a bio-inspired CGP approach. CGP-hash is evolved specifically to process 96-bit IPv4 network flow identifiers, with applications in approximate structures like Bloom filters for real-time monitoring of network traffic for malicious flows. Optimizations are performed to meet the current and future demands of high-speed Terabit Ethernet networks on FPGA. Our methodology surpasses all known FPGA implementations of state-of-the-art NC hash functions, which rely on avalanche metrics. CGP-hash achieves a computational efficiency improvement of at least 7.3% in terms of operating frequency, throughput, and latency while improving area efficiency (Tp/LUT) by 4.5×. This efficiency enables our approach to scale to larger hash outputs through result concatenation, supporting network applications that require extended hash sizes without significant FPGA resource consumption.

When comparing CGP with GP for FPGA implementation of NC hash functions, both methodologies demonstrate almost comparable computational efficiency. However, CGP's unique representation and smaller population size offer advantages in algorithm training time. Additionally, CGP's reusable node structure significantly reduces FPGA resource occupancy, a critical consideration for resource-constrained applications.

Future work will adapt this approach to design hash functions for processing IPv6 network flow IDs, which have a 288-bit size. A distinctive feature of CGP is its capacity to evolve solutions by selectively disregarding certain inputs when they are unnecessary, though this may introduce design biases in certain scenarios. Exploring techniques to manage this input dependency without bias in the evolutionary process presents an intriguing direction for future research.

Acknowledgments. This work is supported by Cybersecurity Research Flanders (VR20192203) and the Higher Education Commission, Pakistan.

References

1. OpenAI: Influence and cyber operations: an update, Accessed 15 Nov 2024
2. Omer Yoachimik, J.P.: DDoS threat report for 2023 Q4, Accessed 01 Oct 2024
3. Le Jeune, L., Sateesan, A., Rabbani, M.M., Goedemé, T., Vliegen, J., Mentens, N.: SoK - network intrusion detection on FPGA. In: Batina, L., Picek, S., Mondal, M. (eds.) SPACE 2021. LNCS, vol. 13162, pp. 242–261. Springer, Cham (2022). https://doi.org/10.1007/978-3-030-95085-9_13
4. Grochol, D., Sekanina, L.: Fast reconfigurable hash functions for network flow hashing in FPGAs. In: 2018 NASA/ESA Conference on Adaptive Hardware and Systems (AHS), pp. 257–263. IEEE (2018)
5. Kekely, M., Kekely, L., Kořenek, J.: General memory efficient packet matching fpga architecture for future high-speed networks. Microprocess. Microsyst. **73**, 102950 (2020)
6. Sharwood, S.: Microsoft adds FPGA-powered network accelerator to Azure, Accessed 02 Oct 2024
7. Amazon: Amazon EC2 F1 Instances:Enable faster FPGA accelerator development and deployment in the cloud, Accessed 02 Oct 2024
8. Babu, P., Parthasarathy, E.: Reconfigurable fpga architectures: a survey and applications. J. Institution Eng. (India): Ser. B **102**, 143–156 (2021)
9. Bobda, C., et al.: The future of FPGA acceleration in datacenters and the cloud. ACM Trans. Reconfigurable Technol. Syst. **15**(3) (2022)
10. Košař, V., Šišmiš, L., Matoušek, J., Kořenek, J.: Accelerating ids using tls pre-filter in FPGA. In: 2023 IEEE Symposium on Computers and Communications (ISCC), pp. pp. 436–442 (2023)
11. Sateesan, A., Vliegen, J., Daemen, J., Mentens, N.: Novel Bloom filter algorithms and architectures for ultra-high-speed network security applications. In: 2020 23rd Euromicro Conference on Digital System Design (DSD), pp. 262–269. IEEE (2020)
12. Sateesan, A., Vliegen, J., Daemen, J., Mentens, N.: Hardware-oriented optimization of Bloom filter algorithms and architectures for ultra-high-speed lookups in network applications. Microprocess. Microsyst. **93**, 104619 (2022)

13. Sateesan, A., Biesmans, J., Claesen, T., Vliegen, J., Mentens, N.: Optimized algorithms and architectures for fast non-cryptographic hash functions in hardware. Microprocess. Microsyst. **98**, 104782 (2023)

14. Sateesan, A., Vliegen, J., Scherrer, S., Hsiao, H.C., Perrig, A., Mentens, N.: Speed records in network flow measurement on fpga. In: 2021 31st International Conference on Field-Programmable Logic and Applications (FPL), pp. 219–224 (2021)

15. Claesen, T., Sateesan, A., Vliegen, J., Mentens, N.: Novel non-cryptographic hash functions for networking and security applications on FPGA. In: 2021 24th Euromicro conference on digital system design (DSD), pp. 347–354. IEEE (2021)

16. Alliance, E.: 2024 Ethernet Roadmap, Accessed 01 Oct 2024

17. Hassan, M., Sateesan, A., Vliegen, J., Picek, S., Mentens, N.: A genetic programming approach for hardware-oriented hash functions for network security applications. Appl. Soft Comput. **165**, 112078 (2024)

18. Estébanez, C., Saez, Y., Recio, G., Isasi, P.: Automatic design of noncryptographic hash functions using genetic programming. Comput. Intell. **30**(4), 798–831 (2014)

19. Hassan, M., Sateesan, A., Vliegen, J., Picek, S., Mentens, N.: Evolving non-cryptographic hash functions using genetic programming for high-speed lookups in network security applications. In: Applications of Evolutionary Computation, Cham, Springer Nature Switzerland, pp. 302–318 (2023)

20. Mittelbach, A., Fischlin, M.: The theory of hash functions and random oracles. Springer Nature, An Approach to Modern Cryptography, Cham (2021)

21. Stallings, W.: Cryptography and network security principles and practice (2014)

22. Ramakrishna, M.: Practical performance of bloom filters and parallel free-text searching. Commun. ACM **32**(10), 1237–1239 (1989)

23. Bloom, B.H.: Space/time trade-offs in hash coding with allowable errors. Commun. ACM **13**(7), 422–426 (1970)

24. Qiao, Y., Li, T., Chen, S.: Fast bloom filters and their generalization. IEEE Trans. Parallel Distrib. Syst. **25**(1), 93–103 (2014)

25. Daemen, J., Hoffert, S., Van Assche, G., Van Keer, R.: The design of Xoodoo and Xoofff. IACR Trans. Symmetric Cryptol., 1–38 (2018)

26. Sateesan, A.: FPGA design for large flow detection in high-speed networks 26 Jan (2024)

27. Hassan, M., Vliegen, J., Picek, S., Mentens, N.: A systematic exploration of evolutionary computation for the design of hardware-oriented non-cryptographic hash functions. In: Proceedings of the Genetic and Evolutionary Computation Conference. GECCO 2024, pp. 1255–1263. Association for Computing Machinery, New York (2024)

28. Bajarangbali, Anand, P.A.: Design of high speed crc algorithm for ethernet on FPGA using reduced lookup table algorithm. In: 2016 IEEE Annual India Conference (INDICON), pp. 1–6 (2016)

29. Clark Shen, Q., Vega, J.C., Chow, P.: Parallel CRC on an FPGA at terabit speeds. In: 2022 International Conference on Field-Programmable Technology (ICFPT), pp. 1–6 (2022)

30. Al-Shatari, M.O.A., Hussin, F.A., Abd Aziz, A., Witjaksono, G., Tran, X.T.: FPGA-based lightweight hardware architecture of the photon hash function for iot edge devices. IEEE Access **8**, 207610–207618 (2020)

31. Lara-Nino, C.A., Morales-Sandoval, M., Diaz-Perez, A.: Small lightweight hash functions in FPGA. In: 2018 IEEE 9th Latin American Symposium on Circuits & Systems (LASCAS), pp. 1–4. IEEE (2018)

32. Nalla Anandakumar, N., Peyrin, T., Poschmann, A.: A very compact FPGA implementation of LED and PHOTON. In: Meier, W., Mukhopadhyay, D. (eds.) INDOCRYPT 2014. LNCS, vol. 8885, pp. 304–321. Springer, Cham (2014). https://doi.org/10.1007/978-3-319-13039-2_18

33. Jungk, B., Lima, L.R., Hiller, M.: A systematic study of lightweight hash functions on FPGAS. In: 2014 International Conference on ReConFigurable Computing and FPGAs (ReConFig 2014), pp. 1–6. IEEE (2014)

34. Dobai, R., Korenek, J., Sekanina, L.: Adaptive development of hash functions in fpga-based network routers. In: 2016 IEEE Symposium Series on Computational Intelligence (SSCI), pp. 1–8 (2016)

35. Dobai, R., Korenek, J.: Evolution of non-cryptographic hash function pairs for FPGA-based network applications. In: 2015 IEEE Symposium Series on Computational Intelligence, pp. 1214–1219. IEEE (2015)

36. Grochol, D., Sekanina, L.: Evolutionary design of fast high-quality hash functions for network applications. In: 2016 Proceedings of the Genetic and Evolutionary Computation Conference, pp. 901–908 (2016)

37. Grochol, D., Sekanina, L.: Multi-objective evolution of hash functions for high speed networks. In: 2017 IEEE Congress on Evolutionary Computation (CEC), pp. 1533–1540. IEEE (2017)

38. Grochol, D., Sekanina, L.: Multi-objective evolution of ultra-fast general-purpose hash functions. In: Castelli, M., Sekanina, L., Zhang, M., Cagnoni, S., García-Sánchez, P. (eds.) EuroGP 2018. LNCS, vol. 10781, pp. 187–202. Springer, Cham (2018). https://doi.org/10.1007/978-3-319-77553-1_12

39. Grochol, D., Sekanina, L.: Evolutionary design of hash functions for ipv6 network flow hashing. In: 2020 IEEE Congress on Evolutionary Computation (CEC), pp. 1–8. IEEE (2020)

40. Kidoň, M., Dobai, R.: Evolutionary design of hash functions for IP address hashing using genetic programming. In: 2017 IEEE Congress on Evolutionary Computation (CEC), pp. 1720–1727. IEEE (2017)

41. Karasek, J., Burget, R., Morský, O.: Towards an automatic design of non-cryptographic hash function. In: 2011 34th International Conference on Telecommunications and Signal Processing (TSP), pp. 19–23. IEEE (2011)

42. Saez, Y., Estebanez, C., Quintana, D., Isasi, P.: Evolutionary hash functions for specific domains. Appl. Soft Comput. **78**, 58–69 (2019)

43. Hua, N., Norige, E., Kumar, S., Lynch, B.: Non-crypto hardware hash functions for high performance networking ASICs. In: 2011 ACM/IEEE Seventh Symposium on Architectures for Networking and Communications Systems, pp. 156–166 (2011)

44. Miller, J.F.: Cartesian genetic programming: its status and future. Genet. Program Evolvable Mach. **21**(1), 129–168 (2020)

45. Turner, A.J., Miller, J.F.: Introducing a cross platform open source cartesian genetic programming library. Genet. Program Evolvable Mach. **16**, 83–91 (2015)

46. Fortin, F.A., De Rainville, F.M., Gardner, M., Parizeau, M., Gagné, C.: DEAP: evolutionary algorithms made easy. J. Mach. Learn. Res. **13**(1), 2171–2175 (2012)

47. Rysak, P.: Comparative analysis of code execution time by c and python based on selected algorithms. J. Comput. Sci. Instit. **26**, 93–99 (2023)

48. Zehra, F., Javed, M., Khan, D., Pasha, M.: Comparative analysis of c++ and python in terms of memory and time (2020)

Understanding Trade-Offs in Classifier Bias with Quality-Diversity Optimization: An Application to Talent Management

Catalina M. Jaramillo[✉][ID], Paul Squires[ID], and Julian Togelius[ID]

New York University, New York, NY, USA
cmj383@nyu.edu

Abstract. Fairness, the impartial treatment towards individuals or groups regardless of their inherent or acquired characteristics [20], is a critical challenge for the successful implementation of Artificial Intelligence (AI) in multiple fields like finances, human capital, and housing. A major struggle for the development of fair AI models lies in the bias implicit in the data available to train such models. Filtering or sampling the dataset before training can help ameliorate model bias but can also reduce model performance and the bias impact can be opaque. In this paper, we propose a method for visualizing the biases inherent in a dataset and understanding the potential trade-offs between fairness and accuracy. Our method builds on quality-diversity optimization, in particular Covariance Matrix Adaptation Multi-dimensional Archive of Phenotypic Elites (MAP-Elites). Our method provides a visual representation of bias in models, allows users to identify models within a minimal threshold of fairness, and determines the trade-off between fairness and accuracy.

Keywords: Fairness · Bias · CMA-ME · Human Capital · Talent Management · Quality-Diversity · Evolution

1 Introduction

Bias exists in various forms throughout society and can be learned by AI based automated decision-making systems, which learn to recognize patterns and consequently replicate biases ingrained in those patterns [12]. While not inherently negative, biases in AI models can exacerbate negative impacts on individuals and society, eroding trust in AI [16,20,24]. Numerous examples highlight how biased AI technologies in areas like hiring, healthcare, and criminal justice have caused harm and raised concerns about maintaining societal biases [19,26]. Research has found how discrimination based on gender, race, age, and other characteristics can be perpetuated by the use of biased AI models [4,15,29].

Any dataset allows for the learning of an almost arbitrarily large number of models with similar accuracy. This situation has been defined as multiplicity. Meyer [21], describes dataset multiplicity as the multiplicity introduced in the model by data inaccuracies (like errors in data collection, data being non-representative of the population, bias, etc.). Black et al. [3] shows how models

© The Author(s), under exclusive license to Springer Nature Switzerland AG 2025
P. García-Sánchez et al. (Eds.): EvoApplications 2025, LNCS 15612, pp. 238–253, 2025.
https://doi.org/10.1007/978-3-031-90062-4_15

with similar performance can have different results from the fairness perspective, and how such multiplicity can be caused by factors like feature selection, training sampling, model class, parameters initialization. Watson et al. [35] add the target output selection as another possible cause.

Algorithm and dataset biases combine in complex ways to create biased learned models. When applying ML to real world problems, bias can manifest in undesirable ways [20]. The debate about the societal effects of biased models is usually oriented towards model bias induced by datasets [14]. Technically, bias can also come from e.g. algorithmic details or architectural choices. For example, algorithms that predict job success and are biased towards a particular gender or ethnicity are clearly problematic. For example, if an algorithm that is used to screen job applicants or provide decision support to a hiring manager is trained on hiring decisions that were made in a biased way (such as favoring men over women), its screening or recommendations will likely incorporate those biases. It becomes extremely important to find ways to not only identify such biases, but also selectively counteract them while minimizing the impact on model performance, as measured by accuracy [19].

A method can be developed to find sets of trained models with good accuracy, but that are different according to their expression of specific biases. Such method could be used to identify models that perform well while minimizing these biases. Such a set of models could also help us understand the underlying trade-offs, if any, between model performance and biases. If the underlying dataset is biased, the best model of data is very likely also biased; we may instead want the best model that is unbiased according to some measure, or that has no more than a certain amount of bias. We could call this the bias-performance trade-off. Visualizing and understanding this trade-off, however, requires that the set of models is identified so that it becomes easy to find the relevant models.

In this paper, we introduce a new method for understanding classifier biases induced by data, and helping to select a model that has the right trade-off between performance and bias. Our method is based on quality diversity (QD) search, a relatively recent family of algorithms that finds a large number of solutions that vary in systematic ways. In other words, we embrace model multiplicity, and provide a way of helping select between models.

2   Background

Our method builds on quality-diversity search and is situated within a set of approaches to address bias in machine-learned models; in this paper, it is applied to a talent management problem. We give a brief background on each topic.

2.1   Bias in Talent Management and the 80% Rule

Our main example here is taken from talent management, specifically promotion, using a dataset of promotion decisions. Here, we provide some background on bias in promotion and the need for visualizing balance and finding sufficiently fair yet performant models.

The issue of bias and fairness toward workers has been of keen interest to employers for many decades. In the US, a greater urgency and attention to bias and fairness was ushered in by the passage of the Civil Rights laws of the 1960s and 1970s and subsequent extensions and revisions. In the years following the passage of these laws there has been a great deal of litigation at all levels of the judicial system, including many US Supreme Court cases (Griggs vs Duke Power, Albemarle vs Moody, Watson vs Fort Worth Bank, Connecticut vs Teal) [1,6, 11,34]. The litigation clarified the law and established "rules" for assessing bias and fairness. One such rule is the 80% or $4/5$ rule. According to the EEOC's Uniform Guidelines on Employee Selection Procedures (1978) the 80% rule is a criterion to identify adverse impact and is defined as, "A selection rate for any race, sex, or ethnic group which is less than four-fifths ($4/5$) (or eighty percent) of the rate for the group with the highest rate will generally be regarded by the Federal enforcement agencies as evidence of adverse impact, while a greater than four-fifths rate will generally not be regarded by Federal enforcement agencies as evidence of adverse impact."

Employers then are challenged with deploying employee selection and promotion procedures and cutoff scores (tests, GPAs, interview ratings, etc.) in a way that accurately identifies talented and deserving individuals while at the same time avoids adverse impact, that is, failing to meet the 80% rule. The problem may be framed as optimizing the balance between fairness and hiring and rewarding top performers. So for example if an employer wishes to set a cut-off of 75% on an employment test in which higher scores predict better job performance and the pass rate for females is less than 80% of the pass rate for male, what should the employer do? The complexity of the problem becomes apparent when considering cutoff scores for multiple screening procedures and their potential adverse impact for multiple groups (gender, ethnicity, age, etc.) [5].

2.2 Methods for Handling Bias in Machine Learning

Research has explored different possible sources of bias and its negative impact in the application of the models. Suresh et al. [28] show how harm can be introduced in different steps of the ML cycle, going from data collection and preparation, model development and evaluation to results deployment.

Multiple approaches have been proposed to mitigate unfairness. Chen et al. [4] evaluate 17 bias mitigation methods, including Optimized Pre-processing, Learning Fair Representation, Disparate Impact Remover, Adversarial Debiasing, Meta Fair Classifier, and Reject Option Classification, among others. Their findings show that bias fairness is improved in about 46% of the studied cases, but there is also a significant negative impact in model accuracy in 66% of the scenarios, and actually in 25% of scenarios fairness is negatively affected by the mitigation method. Zhang et al. [36–38] proposed a multi-objective evolutionary method where learning uses objectives that combine accuracy and fairness metrics; later, they included ensemble learning and a dynamic adaptative selection of used fairness metrics.

Another important factor to consider is the difficulty to understand and measure the trade-off between accuracy and fairness [27] and the complexity of different approaches suggested to manage its impact on models [18]. By finding a set of models with diverse results from the bias perspective, instead of a single model optimization approach as proposed by Zhang et al. [36–38], our method allows for a better understanding of the mentioned trade-off and the selection of the model that better fits the given restrictions.

2.3   Quality-Diversity Search

The quality-diverse (QD) framework is a somewhat recent family of stochastic search/optimization algorithms, where optimization in favor of an objective or fitness function is considered to be only one part of the story [8,25]. QD algorithms instead find diverse sets of solutions, according to some diversity measure or set of diversity measures. In particular, we will focus our investigation on a version of the Map-Elites algorithm, which is a stochastic illumination algorithm that finds a set of solutions to cover a map defined by one or more descriptors [22]. Functionally, Map-Elites can be thought of as an evolutionary algorithm where the population is spread out on a map according to what descriptor values each individual has. The map is divided into cells, where each cell contains the best solution found so far within a particular descriptor interval. This makes it easy to find good solutions that exhibit desired behaviors. For example, if searching for robot gaits, Map-Elites can generate a map of gaits where the descriptors correspond to which legs are used, and a good robot gait that fits a specific leg configuration can easily be found [7].

Map-Elites, like other forms of evolutionary computation, is a stochastic search/optimization algorithm [8]. This means that it is at efficiency disadvantage compared to methods for model optimization tailored to specific learner representations, such as gradient descent (for neural network training) or information gain maximization (for decision tree learning). However, it has the advantage that stochastic search can be applied to optimize anything that has parameters and an objective function. Common learner representations such as neural networks and decision trees can be evolved with good results [33], though this process is often less computationally efficient than using model-specific learning methods.

The particular version of Map-Elites we use here is Covariance Matrix Adaptation MAP-Elites (CMA-ME), which allows Map-Elites to search continuous spaces more efficiently as shown in Algorithm 1 [10]. CMA-ME adapts the covariance matrix adaptation method from Covariance Matrix Adaptation Evolution Strategies to a QD setting [13].

Algorithm 1. Covariance Matrix Adaptation MAP-Elites

1: **Input:** An evaluation function *evaluate* which computes a behavior characterization and fitness, and a desired number of solutions n.
2: **Result:** Generate n solutions storing elites in a map M.
3: Initialize population of emitters E.
4: **for** $i = 1$ to n **do**
5:    Select emitter e from E which has generated the least solutions out of all emitters in E.
6:    $x_i \leftarrow$ generate_solution(e)
7:    $(\beta_i, \text{fitness}) \leftarrow evaluate(x_i)$
8:    return_solution($e, x_i, \beta_i, \text{fitness}$)
9: **end for**

---

3   Methodology

In this paper, we propose the use of QD methods, specifically Map-Elites, to find sets of models that vary systematically in selected bias dimensions. We operationalize biases as the overrepresentation of certain features among the predicted instances of a model, and use degrees of biases as descriptors for CMA-ME.

Additionally, we use a fairness definition based on the predicted outcomes of the classifier, focusing on group fairness, which requires that all groups within a protected attribute have a similar probability of being assigned to the positive predicted class [32].

The first step is to obtain a baseline model: a simple multi-layer neural network is created, using hyper-parameter optimization and 10-fold stratified cross-validation, to identify the best model. The resulting configuration (number of layers and number of nodes) is implemented in the neural network used in the experiments.

Next, the weights of a set of neural networks with the defined architecture are evolved, and their fitness values are calculated as the accuracy of the network forward pass prediction against the real outcome for the whole dataset.

The above-mentioned accuracy is used as the output value to be maximized, while the ratio of positive classification between the two groups present in each protected attribute are used as descriptors of the model (for example the ratio of female positive prediction rate vs male positive prediction rate for gender attribute, or the ratio of young positive rate vs old positive rate for age.) See Algorithm 2.

Algorithm 2. Fitness Function

1: **Input:** `weights, nn_architecture, cases, cases_xa, cases_xb, cases_ya, cases_yb, labels.`
2: **Output:** `accuracy, ratio_x, ratio_y.`
3: **Step 1: Compute Overall Accuracy**
4: Perform a forward pass on `cases` with `weights` and **nn_architecture**.
5: Compute accuracy:

$$\text{accuracy} \leftarrow \frac{\text{correct predictions}}{\text{total cases}}$$

6: **Step 2: Compute Group Ratios**
7: For each group (`cases_xa, cases_xb, cases_ya, cases_yb`), compute mean predictions:

$$\text{mean\_xa}, \text{mean\_xb}, \text{mean\_ya}, \text{mean\_yb}$$

8: Compute fairness ratios:

$$\text{ratio\_x} \leftarrow \frac{\text{mean\_xa}}{\text{mean\_xb} + \epsilon}, \quad \text{ratio\_y} \leftarrow \frac{\text{mean\_ya}}{\text{mean\_yb} + \epsilon}$$

9: **Step 3: Return Fitness Results**
10: **Output:** (`accuracy`, (`ratio_x, ratio_y`))

A two-dimensional map was built, showing the best accuracy reached for each cell. The cells in the map are delimited by the ratios of two protected attributes present in the dataset.

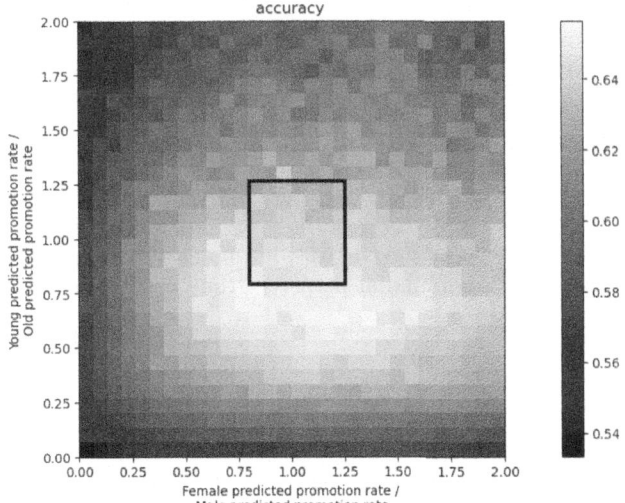

Fig. 1. Fair zone within the map

From the map, we identified the model with best accuracy, and its protected attributes' positive ratios; along with the best accuracy for a model within the "fair zon", and the respective positive ratios. The "fair zone" is defined by the area between 0.8 and 1.25 in the positive ratios based on the 80% rule described before, as shown in Fig. 1. A model that is located closer to (1,1) (where positive prediction is equally likely along both groups within each demographic dimension) has a lower learned bias. The bigger the distance from (1,1), the higher the bias. We used Euclidean distance between (1,1) and the coordinates given by the model descriptors to compare the bias of different models.

Two datasets were used in this research. In the first one we apply the method to predicting promotions in a company using the Promotion Dataset [23], a task with high relevance for ML applications in human resources. In the second case, we used the Adult Dataset [2], a dataset well known for fairness research.

For each case we used two protected attributes in the dataset and selected the best model with an acceptable degree of bias, given by the 80% rule. By comparing the selected model with the most accurate one in the whole map, the method allows to visualize the trade-off between bias and model performance. The map also reflects how the bias in the data impacts the results, by simply displaying the array of most accurate models and color-coding each cell for accuracy.

Table 1. Methods

	Promotion Dataset	Adult Dataset
N	54808	48842
Protected attributes	Gender, Age	Gender, Race
Number features	14	55
Baseline Accuracy	0.6628	0.8416
NN Architecture		
NN Hiden layers	2	2
NN Nodes	[35,15]	[64,32]
NN Activation function	Leaky Relu	Leaky Relu

The used code relies on the pyribs [31] library. Details about the CMA-ME parameters and code can be found at https://github.com/catajara/Classif-Bias-Tradeoffs-QD.

3.1 Data

The Promotion Dataset, is an open source dataset consisting of individuals that were either promoted or not promoted and a set of features including department in the organization, region, education, gender, training, age, performance, and tenure. Categorical variables were encoded using binary dummy variables. We used sampling to generate different scenarios from the data, to evaluate how the presence or absence of bias in the data impacts the results. Random sampling was used to obtain 4 datasets with a defined bias. In all cases the total percentage of promoted individuals is set close to 50%. Given the age distribution in the data, a 34 years old threshold was used to define the young vs. old groups. Rows 1 to 4 in Table 2 show the data used, including the number of cases for each group and Row 1 to 4 in Table 3 show the promotion rate for the group:

1. Promotion Unbiased dataset, with 50% promotion rate for all the categories.
2. Promotion dataset with same proportion for male and female, and for young and old; with higher promotion rate for male cases in both age groups.
3. Promotion dataset with higher sample size for male cases, and similar proportion for young and old groups; with higher promotion rate for male in both age groups.
4. Promotion dataset with same proportion for male and female cases, and similar proportion for young and old; with higher promotion rate for male and young cases (67% compared with 33% for male and old), and higher promotion rate for female and old (67%) vs. female and young cases (33%).

Adult Dataset, predicts annual income exceeding of \$50K/yr using census data. Features include age, gender, race, education, occupation, among others. This dataset has bias for both protected attributes (gender and race in this case). Random sampling was used to obtain a stratified sample with similar distribution to the original dataset and a unbiased sample. Rows 5 and 6 in Tables 2 and 3 show the sample details for Adult dataset:

5. Adult unbiased dataset, with 50% promotion rate for all the categories.
6. Adult stratified sample, with higher number cases for male and for white, and higher positive rate for male and for white.

Table 2. Dataset Information

No.	Dataset	Total (n)	Male (n)	Female (n)	Young/ White (n)	Old Other (n)
1	Unbiased	5728	2864	2864	2864	2864
2	Male Biased	4384	2192	2192	2192	2192
3	Higher Male Sample	6576	4384	2192	3288	3288
4	Cross Biased	4508	2254	2254	2254	2254
5	Adult Unbiased	1680	840	840	840	840
6	Adult Stratified	13565	9224	4341	11666	1899

Table 3. Positive Rate by Subgroup

No.	Dataset	All	Male	Female	Young/White	Old/Other
1	Unbiased	0.50	0.50	0.50	0.50	0.50
2	Male Biased	0.50	0.65	0.35	0.50	0.50
3	Higher Male Sample	0.56	0.67	0.35	0.56	0.56
4	Cross Biased	0.50	0.50	0.50	0.50	0.50
5	Adult Unbiased	0.50	0.50	0.50	0.50	0.50
6	Adult Stratified	0.25	0.31	0.11	0.26	0.16

4   Results

For each experiment, we recorded several metrics. First, we generated a heat map displaying the best accuracy achieved for attributes' bias descriptors combination. The heat map was organized into 30 bins ranging from 0 to 2. Additionally, we compiled a table containing various metrics:

– Accuracy
– Ratio of predicted positive rates for both groups withing each protected attribute
– Deviation -Euclidian distance from (1,1)- for the model with the highest accuracy among all models in the map (best model)
– Deviation for the model with the highest accuracy located within the fair zone, which ranged from 0.8 to 1.25 for both descriptors (best fair).

Following we present the metrics collected for each experiment:

4.1   Unbiased Sample (Figure 2, Table 4)

In this sample (see Row 1 Table 3), we have the same number of promoted individuals in all the groups: male, female, old and young individuals, making the promotion rate 50% for all cases.

The accuracy is 0.0106 lower for the best model in the fair zone vs. the best model in the map, while the distance from (1,1) is reduced from 0.2872 to 0.2255.

4.2   Male Biased Sample (Figure 3, Table 5)

While the number of male and female, and the number of young and old individuals is equally split, the proportion of males in the sample that are promoted is 65% and 35% for females. Regarding age, both groups have a similar promotion rate of 50% (Row 2 Table 3.)

The best model is 0.0401 points more accurate than the best model in the fair zone. We observe a higher distance from (1,1) for the most accurate model (0.7761) and that in that model the promotion rate for women is the 24% of

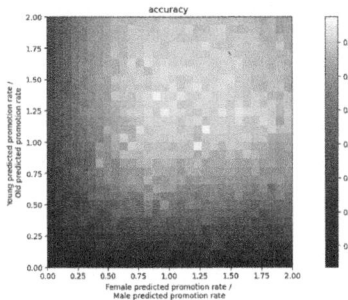

Fig. 2. Unbiased sample map

Table 4. Unbiased sample results

	Best model	Best fair
Accuracy	0.6550	0.6444
Fem/Male	1.2695	0.8365
Young/Old	1.0994	1.1553
Deviation	0.2872	0.2255

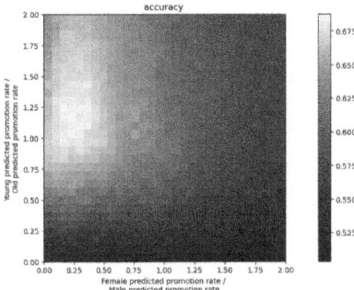

Fig. 3. Male biased sample map

Table 5. Male biased sample results

	Best model	Best fair
Accuracy	0.6877	0.6476
Fem/Male	0.2400	0.8344
Young/Old	1.1573	1.0990
Deviation	0.7761	0.1929

the promotion rate for men. By using the best fair model, distance from (1,1) is reduced to 0.1929 and the ratio between women and men promotion rate is increased to 0.8344.

4.3   Male Biased Sample, Male Oversampling (Figure 4, Table 6)

For this case, we wanted to observe the effect of having a higher promotion rate for males and a higher number of male cases represented in the dataset (Row 2 Tables 2 and 3), a condition found in many real life situations. Both age groups have a similar promotion rate.

The distribution of models is similar to the distribution obtained in the second experiment. In this case, the trade-off in accuracy between the best model and the best fair one is 0.0431, and the ratio for female vs male promotion rate is 0.2575. The distance from (1,1) is 0.7220. This distance is reduced to 0.1863 when we use the best fair model and the female vs male promotion ratio is increased to 0.8202.

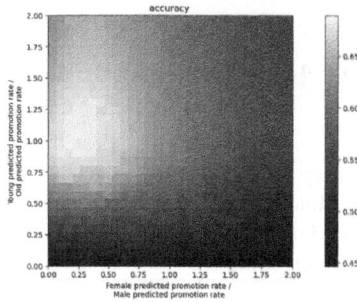

Fig. 4. Male biased and oversampled sample map

Table 6. Male biased and oversampled sample results

	Best model	Best fair
Accuracy	0.6892	0.6461
Fem/Male	0.2575	0.8202
Young/Old	0.8808	1.0491
Deviation	0.7520	0.1863

4.4 Cross Biased Sample, Male Young and Female Old (Figure 5, Table 7)

In this experiment we used a dataset that doesn't have bias considering each dimension independently (male vs. female or young vs old). But, when we combined the dimensions, there is bias: individuals that are male and old have a higher promotion rate, so do individuals that are female and young (Row 4 Table 2)

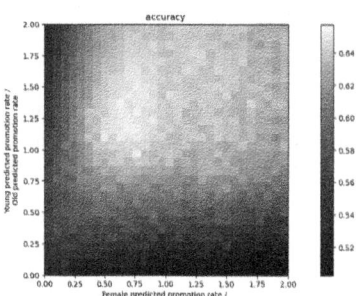

Fig. 5. Cross biased sample map

Table 7. Cross biased sample results

	Best model	Best fair
Accuracy	0.6657	0.6586
Fem/Male	0.8173	0.9313
Young/Old	1.3600	1.0616
Deviation	0.4037	0.0922

The accuracy difference between the best model and the best fair model is 0.0071. And the distance is reduced from 0.4037 to 0.0922.

4.5 Adult Dataset, Unbiased Sample (Figure 6, Table 8)

In this experiment we used a unbiased sample from the Adult dataset (Row 5 Table 3).

The best model in the map is within the "fair zone".

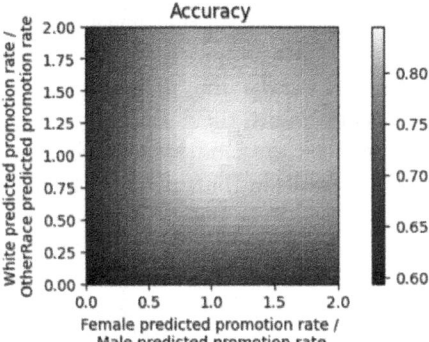

Fig. 6. Adult unbiased sample map

Table 8. Adult unbiased sample results

	Best model	Best fair
Accuracy	0.8452	0.8452
Fem/Male	0.9341	0.9341
White/Other	1.0853	1.0853
Deviation	0.1078	0.1078

4.6 Adult Dataset, Stratified Sample (Figure 7, Table 9)

In this experiment we used a stratified sample from the Adult dataset that corresponds to the distribution of positive outcomes for both groups in the protected attributes (gender and race). See Row 6 Tables 2 and 3.

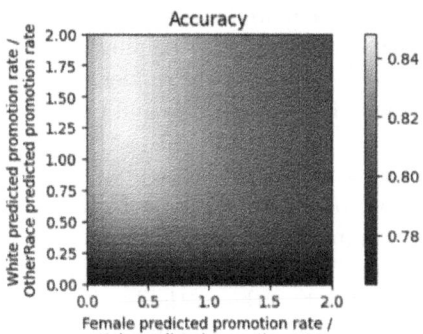

Fig. 7. Adult stratified sample map

Table 9. Adult stratified sample results

	Best model	Best fair
Accuracy	0.8481	0.8331
Fem/Male	0.3443	0.8014
White/Other	1.5121	1.1930
Deviation	0.8320	0.2769

The most accurate models are distributed in a similar way than the bias in the dataset. The accuracy difference between the best model and the best fair model is 0.015. And the distance is reduced from 0.8320 to 0.2769.

We observe that the distribution of the accuracy of the models reflects the bias distribution in the datasets. Notice that for the unbiased datasets (see experiments 4.1, 4.4, and 4.5) the area in the map for models with higher accuracy (lighter color) is wider compared with the area found in the cases where the dataset is biased (experiments 4.2, 4.3, and 4.6.) For the cases where the data is biased, the most accurate models are concentrated in the area of the map that mirrors the bias in the data (that is, the area closer to the cell that has a positive predicted ratio similar to that present in the data for each protected attribute.)

The method allowed to generate a set of models and compare how the accuracy and the bias distribution changed in connection with the data used to train the model. Additionally, in cases where the most accurate model is located out of the fair zone, we were able to identify a best fair solution that increases fairness to the minimum threshold. Also, we could understand the trade-off in accuracy from selecting the best fair model instead of the most accurate one. While the accuracy dropped by an average of 2% points between the most accurate model and the best model within the fair zone, the average distance to (1,1) was reduced by 30%. This, assuring in all cases a model that complies with the $^4/_5$ rule.

5   Discussion

The use of classifier systems and other algorithmic approaches for human resources tasks such as recruitment and promotion is contentious. It could be argued that algorithms should play no part in these decisions, and that they should be solely down to human judgment. At the same time, humans vary widely between them in how they interpret and apply given guidelines, and human judgment can be every bit as biased as algorithmic judgment. One difference is that classifiers (and other algorithms) can be more readily tested because of temporal consistency. We see our work as contributing by helping prospective users to visualize the biases in their classifiers, and better understand the bias landscape of potential classifiers and trade-offs between bias and prediction performance.

A limitation of our current work is the limited availability of public datasets related to bias in the labor market. To evaluate our method more effectively, we used a human capital dataset where bias was manually introduced through sampling, as well as a second dataset commonly used in fairness research. Additional fairness research datasets could be used to further test the method.

The maximum accuracy obtained by the models in our experiment is slightly lower than what could be attained by a sophisticated conventional learning algorithm, but this slight difference could probably be overcome with some further engineering. In this context, it is worth reiterating that conventional machine learning algorithms result in only a single model and no opportunities for bias characterization.

While there are approaches that seek to alleviate biases rather than characterize them, they all have various shortcomings. For example, dominant classes could be undersampled, but that means that not all of the dataset will be used, likely impacting classifier performance and generalization; it also requires that an analysis of bias is done in the first place to understand which instances to undersample.

6   Conclusion and Future Work

We showed that CMA-ME can generate multiple models of varying bias, thereby allowing us to visualize the distribution of the accuracy of the models and its relation with the learned bias. In other words, we can understand the bias-performance trade-off. The result of the method is a set of trained models, where the user can choose any of them for deployment. The accuracy of the best-classifying model is comparable, though slightly inferior to, the output of state of the art conventional machine learning methods. The method is fast and would scale to networks of at least tens of thousands of parameters [30].

We had different bias-performance trade-off levels, going from 0.0431 to 0, while the difference in deviation goes from 0.5832 to 0. The correlation between the trade-off and the difference in the deviation between the most accurate model and the model with best accuracy within the fair zone (best fair) is 0.9150.

The method we have proposed here can be used as-is for training relatively small-size classifiers. These are useful for prediction from a small set of features. The kind of tabular data that is often used in credit scoring systems or hiring support systems are good examples of this. However, bias is common also in ML systems that work on much higher-dimensional, non-tabular data. A good example is face recognition systems, which are ubiquitous but can be highly biased, for example by more readily recognizing people of certain ethnicities [17]. These systems typically build on neural networks with millions of parameters, where evolutionary methods such as CMA-ME largely break down. However, there are QD methods capable of effectively training very large networks, such as the recently proposed Differentiable Quality Diversity (DQD) algorithm [9]. It would be interesting to apply the methodology proposed here to facial recognition systems.

Fairness is not the only concern when using ML models for important or sensitive decision; another important concern is explainability. Neural networks, while powerful, are unfortunately some of the least explainable ML models. Decision trees, on the other hand, are in principle human-interpretable and can be highly accurate on tabular data. The approach presented in this paper could easily be applied to finding interpretable decision trees by exchanging the CMA-ME algorithm for a version of MAP-Elites suited to tree-structured representations.

References

1. Albemarle Paper Co. v. Moody, 422 U.S. 405 (1975)
2. Becker, B., Kohavi, R.: Adult. UCI Machine Learning Repository (1996). https://doi.org/10.24432/C5XW20
3. Black, E., Raghavan, M., Barocas, S.: Model multiplicity: opportunities, concerns, and solutions. In: Proceedings of the 2022 ACM Conference on Fairness, Accountability, and Transparency, pp. 850–863 (2022)
4. Chen, Z., Zhang, J.M., Sarro, F., Harman, M.: A comprehensive empirical study of bias mitigation methods for machine learning classifiers. ACM Trans. Softw. Eng. Methodol. **32**(4), 1–30 (2023)

5. Code of Federal Regulations (CFR): Uniform guidelines on employee selection procedures (1978). https://www.ecfr.gov/current/title-29/subtitle-B/chapter-XIV/part-1607?toc=1 (1978), title 29, Subtitle B, Chapter XIV, Part 1607
6. Connecticut v. Teal, 457 U.S. 440 (1982)
7. Cully, A., Clune, J., Tarapore, D., Mouret, J.B.: Robots that can adapt like animals. Nature **521**(7553), 503–507 (2015)
8. Cully, A., Demiris, Y.: Quality and diversity optimization: a unifying modular framework. IEEE Trans. Evol. Comput. **22**(2), 245–259 (2017)
9. Fontaine, M., Nikolaidis, S.: Differentiable quality diversity. Adv. Neural. Inf. Process. Syst. **34**, 10040–10052 (2021)
10. Fontaine, M.C., Togelius, J., Nikolaidis, S., Hoover, A.K.: Covariance matrix adaptation for the rapid illumination of behavior space. In: Proceedings of the 2020 Genetic and Evolutionary Computation Conference, pp. 94–102 (2020)
11. Griggs v. Duke Power Co., 401 U.S. 424 (1971)
12. Hajian, S., Bonchi, F., Castillo, C.: Algorithmic bias: from discrimination discovery to fairness-aware data mining. In: Proceedings of the 22nd ACM SIGKDD International Conference on knowledge discovery and data mining, pp. 2125–2126 (2016)
13. Hansen, N., Ostermeier, A.: Completely derandomized self-adaptation in evolution strategies. Evol. Comput. **9**(2), 159–195 (2001)
14. Hellström, T., Dignum, V., Bensch, S.: Bias in machine learning-what is it good for? In: International Workshop on New Foundations for Human-Centered AI (NeHuAI) co-located with 24th European Conference on Artificial Intelligence (ECAI 2020), Virtual (Santiago de Compostela, Spain), September 4, 2020, pp. 3–10. RWTH Aachen University (2020)
15. Hoffman, S., Podgurski, A.: Artificial intelligence and discrimination in health care. Yale J. Health Pol'y L. & Ethics **19**, 1 (2019)
16. Huang, C., Zhang, Z., Mao, B., Yao, X.: An overview of artificial intelligence ethics. IEEE Trans. Artif. Intell. **4**(4), 799–819 (2022)
17. Jain, A., Memon, N., Togelius, J.: Zero-shot racially balanced dataset generation using an existing biased stylegan2. arXiv preprint arXiv:2305.07710 (2023)
18. Li, J., Li, G.: The triangular trade-off between robustness, accuracy and fairness in deep neural networks: a survey. ACM Comput. Surv. (2024)
19. Liem, C.C., et al.: Psychology meets machine learning: Interdisciplinary perspectives on algorithmic job candidate screening. In: Explainable and Interpretable Models in Computer Vision and Machine Learning, pp. 197–253 (2018)
20. Mehrabi, N., Morstatter, F., Saxena, N., Lerman, K., Galstyan, A.: A survey on bias and fairness in machine learning. ACM Comput. Surv. (CSUR) **54**(6), 1–35 (2021)
21. Meyer, A.P., Albarghouthi, A., D'Antoni, L.: The dataset multiplicity problem: How unreliable data impacts predictions. In: Proceedings of the 2023 ACM Conference on Fairness, Accountability, and Transparency, pp. 193–204 (2023)
22. Mouret, J.B., Clune, J.: Illuminating search spaces by mapping elites. arXiv preprint arXiv:1504.04909 (2015)
23. Möbius: Hr analytics: Employee promotion data (2019). https://www.kaggle.com/datasets/arashnic/hr-ana
24. Pessach, D., Shmueli, E.: A review on fairness in machine learning. ACM Comput. Surv. (CSUR) **55**(3), 1–44 (2022)
25. Pugh, J.K., Soros, L.B., Stanley, K.O.: Quality diversity: A new frontier for evolutionary computation. Front. Robot. AI, 40 (2016)

26. Schwartz, R., Schwartz, R., Vassilev, A., Greene, K., Perine, L., Burt, A., Hall, P.: Towards a standard for identifying and managing bias in artificial intelligence, vol. 3. US Department of Commerce, National Institute of Standards and Technology (2022)

27. Speicher, T., et al.: A unified approach to quantifying algorithmic unfairness: measuring individual &group unfairness via inequality indices. In: Proceedings of the 24th ACM SIGKDD International Conference on Knowledge Discovery & Data Mining, pp. 2239–2248 (2018)

28. Suresh, H., Guttag, J.: A framework for understanding sources of harm throughout the machine learning life cycle. In: Proceedings of the 1st ACM Conference on Equity and Access in Algorithms, Mechanisms, and Optimization, pp. 1–9 (2021)

29. Sweeney, L.: Discrimination in online ad delivery. Commun. ACM **56**(5), 44–54 (2013)

30. Tjanaka, B., Fontaine, M.C., Lee, D.H., Kalkar, A., Nikolaidis, S.: Training diverse high-dimensional controllers by scaling covariance matrix adaptation map-annealing. IEEE Robot. Autom. Lett. (2023)

31. Tjanaka, B., et al.: pyribs: A bare-bones python library for quality diversity optimization (2023)

32. Verma, S., Rubin, J.: Fairness definitions explained. In: Proceedings of the International Workshop on Software Fairness, pp. 1–7 (2018)

33. Wang, P., Weise, T., Chiong, R.: Novel evolutionary algorithms for supervised classification problems: an experimental study. Evol. Intel. **4**, 3–16 (2011)

34. Watson v. Fort Worth Bank i& Trust, 487 U.S. 977 (1988)

35. Watson-Daniels, J., Barocas, S., Hofman, J.M., Chouldechova, A.: Multi-target multiplicity: flexibility and fairness in target specification under resource constraints. In: Proceedings of the 2023 ACM Conference on Fairness, Accountability, and Transparency, pp. 297–311 (2023)

36. Zhang, Q., Liu, J., Yao, X.: Fairness-aware multiobjective evolutionary learning. IEEE Trans. Evolutionary Comput. (2024)

37. Zhang, Q., Liu, J., Zhang, Z., Wen, J., Mao, B., Yao, X.: Fairer machine learning through multi-objective evolutionary learning. In: Farkaš, I., Masulli, P., Otte, S., Wermter, S. (eds.) ICANN 2021. LNCS, vol. 12894, pp. 111–123. Springer, Cham (2021). https://doi.org/10.1007/978-3-030-86380-7_10

38. Zhang, Q., Liu, J., Zhang, Z., Wen, J., Mao, B., Yao, X.: Mitigating unfairness via evolutionary multiobjective ensemble learning. IEEE Trans. Evol. Comput. **27**(4), 848–862 (2022)

Genetic Programming with Co-operative Co-evolution for Feature Manipulation in Basal Cell Carcinoma Identification

Taran Cyriac John[1]([✉]), Qurrat Ul Ain[1], Harith Al-Sahaf[1,2], and Mengjie Zhang[1]

[1] Centre for Data Science and Artificial Intelligence, Victoria University of Wellington, P.O. Box 600, Wellington 6140, New Zealand
{taran.john,qurrat.ul.ain,mengjie.zhang}@ecs.vuw.ac.nz
[2] College of Information Technology Engineering, Al-Zahraa University for Women, Karbala, Iraq
harith.alsahaf@alzahraa.edu.iq

Abstract. As global mortality rates rise alongside an increasing incidence of skin cancer, particularly Basal Cell Carcinoma (BCC), the development of effective automated detection strategies gains urgency. Traditional diagnosis of BCC relies heavily on manual inspection of skin lesions, an approach limited by subjectivity, time constraints, and the invasive nature of biopsy procedures. To address these challenges, this study introduces the two-stage cooperative co-evolution (2SCC)-Criptor, a novel two-stage feature manipulation method that employs genetic programming and co-operative co-evolution for automated BCC identification. The first stage generates an ensemble of models that collaboratively extract discriminative features from decomposed colour channels of skin lesion images, while the second stage constructs additional features to enhance classification performance. The efficacy of the method is evaluated using standard machine learning classifiers, demonstrating statistically significant superiority over traditional and contemporary approaches in the field. Further analysis reveals that the model's success stems from the synergistic integration of colour channels rather than individual channel contributions, with remarkably uniform utilisation of LAB colour space components. The findings underscore 2SCC-Criptor's potential in enhancing BCC diagnostic processes while maintaining interpretability through evolved genetic programming individuals.

Keywords: Skin cancer · Genetic programming · Machine learning

1 Introduction

The global burden of skin cancer has reached alarming proportions, with both mortality rates and incidence surging across the world [1]. As one of the most prevalent forms of cancer [2], non-melanoma skin cancer (NMSC) accounts for

© The Author(s), under exclusive license to Springer Nature Switzerland AG 2025
P. García-Sánchez et al. (Eds.): EvoApplications 2025, LNCS 15612, pp. 254–269, 2025.
https://doi.org/10.1007/978-3-031-90062-4_16

the majority of cases, with Basal Cell Carcinoma (BCC) representing 80% to 85% of NMSC instances [3].

While the lower mortality rate of BCC might suggest reduced urgency for advanced detection methods, this perception is misleading. Left untreated, BCC can lead to severe complications, including extensive local tissue damage and potential disfigurement, particularly in sensitive areas like the face or neck [4]. The visual variability of BCC presentations and their potential to mimic other skin conditions highlights the importance of developing advanced, objective detection methods to support clinical diagnosis [5].

The integration of Computer Automated Diagnostic Systems (CADS) with Artificial Intelligence (AI) represents a promising direction in addressing these challenges [6]. However, two significant obstacles impede widespread implementation: the complexity of manual feature extraction algorithms and reluctance of medical practitioners to rely on AI-based recommendations due to lack of transparency. The extraction of interpretable features is paramount, as it enables clinicians to understand the underlying rationale of the decision-making process, thereby enhancing the potential for integration into clinical practice.

In the realm of computational problem-solving, Genetic Programming (GP) has emerged as a powerful paradigm. Inspired by biological evolution, GP cultivates populations of tree-structured programs that adapt and improve over generations to address specific challenges [7]. Since its inception, GP has demonstrated remarkable versatility, making significant inroads in diverse domains such as image processing [8] and feature detection/extraction [9].

The adaptability of GP is further enhanced when combined with the principle of co-operative co-evolution (CC). This synergistic approach dissects complex problems into more manageable "species" or sub-problems [10]. Each species undergoes independent evolution, with individual performance evaluated through interactions with high-performing members of other species, guided by a "context vector". This divide-and-conquer strategy allows GP to tackle intricate challenges with greater efficiency and precision.

GP has emerged as a prominent technique in Artificial Intelligence-based CADS for skin cancer detection. Ain et al. introduced a novel approach, "2SGP-W", for analysing ten types of skin lesions including malignant melanoma and BCC [11]. Their methodology incorporates a diverse array of features, including local, global, texture, frequency-based, and colour attributes, extracted via Local Binary Patterns (LBPs) and pyramid-structured wavelet decomposition. The GP process unfolds in two distinct phases: feature selection followed by feature construction. Comparative analyses revealed that 2SGP-W demonstrated superior performance relative to contemporary algorithms.

It must be noted that while the literature pertaining to GP has been primarily developed for the binary classification of malignant melanoma, there is significant potential to translate these methodologies to other domains of skin cancer detection, such as BCC. This translation forms a key aspect of the current study. However, it is important to recognise that the reliance on predefined crafted features may inadvertently overlook crucial discriminatory characteristics.

This limitation has provided motivation in the GP field to devise automated feature extraction algorithms. The capacity of GP to evolve feature extractors automatically presents several advantages over conventional methods such as LBP or deep learning architectures. Notably, GP facilitates concurrent feature extraction and selection, thereby enhancing model efficiency and mitigating errors commonly associated with systems employing separately optimised components [11].

In response to these challenges, John et al. proposed "CC-Criptor" [12], a novel feature extraction method for melanoma detection that demonstrated significant advancement in automated diagnostic capabilities. This methodology generates an ensemble of models working in conjunction to extract distinctive features from skin lesion images, with each model corresponding to a specific colour channel of the image. When evaluated using a publicly available dataset, CC-Criptor significantly outperformed most of the of established benchmarks, including traditional methods such as LBP, variants of Convolutional Neural Networks (CNNs), and other GP-based approaches.

Building upon these advancements, the inclusion of feature construction through the evolution of a GP program offers several potential advantages. Feature construction in GP involves evolving mathematical expressions that combine existing features into more sophisticated discriminators. While CC-Criptor effectively addressed the challenges associated with using manually defined features, it may be further improved by drawing inspiration from the feature construction component of Ain et al.'s algorithm [11]. This integration could yield three key benefits. Firstly, it can produce highly discriminatory features, potentially improving the overall classification capabilities of the extracted feature vector. Secondly, it provides valuable insight into the interactions between features, particularly the colour channels involved. Lastly, it may enhance the ability of the model to capture complex patterns and relationships within the data. This multi-faceted benefit not only may enhance the performance of the model but also may strengthen the understanding of the underlying patterns in dermatological image analysis, potentially leading to more robust and interpretable AI-based diagnostic tools for skin cancer detection.

1.1 Research Goals

The primary goal of this study is to develop an advanced, two-stage GP methodology incorporating CC for automated feature extraction and construction in the detection of BCC. To this end, the following objectives have been established:

1. Devising a new multi-model GP feature manipulation system utilising colour decomposition and CC principles, in which the extracted features will be utilised in a subsequent GP stage for feature construction.
2. Selecting appropriate fitness functions for each GP stage that integrate distance-based and wrapper-based measures to obtain the highest discrimination capabilities.
3. Conducting a comparative analysis of the performance of the proposed method against established and contemporary feature extraction techniques.

4. Elucidating the characteristics of the proposed model through an examination of colour channel interactions, feature importance, and interpretability of evolved GP individuals.

2 Background and Related Work

2.1 Background

2.1.1 GP-Criptor

Introduced in [13], GP-Criptor represents an evolutionary approach to automated feature extraction that parallels LBP methodology while eliminating the need for expert-designed pixel formulae. The algorithm operates through a three-phase process: pixel value extraction via a sliding window mechanism, application of arithmetic operations according to the evolved GP tree structure, and binary encoding of the resultant values. The final output manifests as a histogram-based feature vector, compatible with conventional machine learning classifiers such as k-nearest neighbour.

2.1.2 CC-Criptor

CC-Criptor [12] represents a significant evolution of GP-Criptor, specifically tailored for skin cancer detection, particularly melanoma. The algorithm leverages CC by decomposing colour images into their constituent channels, creating an ensemble of three specialised models. Each model evolves independently while maintaining coordination through a context vector, enabling the system to capture channel-specific characteristics of skin lesions. The final output manifests as a histogram-based feature vector, incorporating information from all three colour channels. This introduces colour-channel decomposition and cooperative evolution, allowing for more nuanced feature extraction specifically suited to dermatological image analysis.

2.2 Related Work

The landscape of CADS for skin cancer has been dramatically transformed by recent advances in AI. Deep Learning (DL) has emerged as a particularly influential technology in this domain, moving capabilities that once seemed futuristic into practical reality. A pioneering study by Codella et al. [14] demonstrated the potential of DL in skin cancer diagnosis by implementing the Caffe framework to extract discriminative features from dermoscopic images. The effectiveness of DL in this domain was further validated by Haenssle et al. [15], who conducted a comparative study between Google's Inception v4 CNN architecture [16] and 58 dermatology specialists. Their findings showed that even when dermatologists were provided with additional clinical information alongside dermoscopic images, the CNN model maintained superior specificity in classification tasks.

Ain et al. [17] developed an innovative approach that harmoniously combined domain-independent features, extracted through Local Binary Pattern descriptors, with domain-specific features derived from the 7-point dermatology checklist. Their research demonstrated that this integrated approach not only achieved

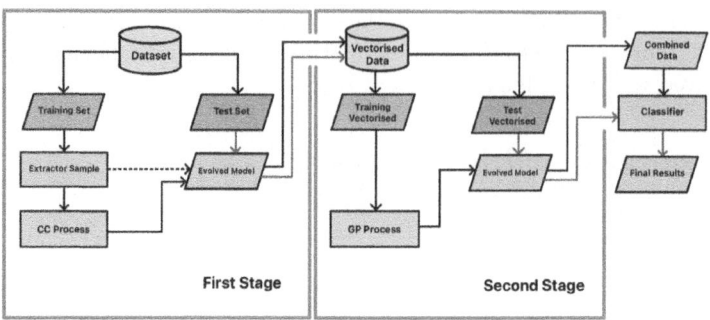

Fig. 1. The overall algorithm of the proposed method.

high classification accuracy but also generated interpretable models that highlighted the most discriminative features between benign and malignant lesions. The success of this methodology led to further developments, with the authors expanding their work to create sophisticated multi-tree GP systems [18].

Despite these advances, existing approaches face key limitations: deep learning methods operate as "black boxes," while feature-based approaches rely on predefined characteristics that may not capture all relevant BCC patterns. Furthermore, most work focuses on melanoma detection, with limited attention to the unique challenges of BCC identification.

3   Two Stage Co-operative Co-evolution Image Descriptor

3.1   Algorithm Overview

Demonstrated in Fig. 1 is an overview of the proposed algorithm: Two Stage Co-operative Co-evolution Image Descriptor (2SCC-Criptor). It should be noted that this algorithm consists of two distinct stages, with the first stage being feature extraction, followed by the second stage of feature construction, inspired by the works of John et al. [12] and Ain et al. [11], respectively. The first stage commences with the contents of the dataset being divided into training and test splits. During the training process, the *extractor sample* is randomly chosen, in which 20 samples from each class are selected. These extractor samples are fed to the CC evolutionary process, wherein a feature extraction model is obtained. This newly evolved model is then utilised to *vectorise* both the training and the test data, concluding the evolution of the first stage. The second stage commences by utilising the vectorised train data as input to the GP evolutionary process, where a GP program is employed to construct a highly discriminatory feature. This feature is concatenated to the end of the feature vector produced by the first stage. Following this GP evolution, the best-performing individual is evaluated to produce one constructed feature for both the training and test data, which is then concatenated to the existing feature vectors. The complete training feature vector is utilised to train a classifier, which is tested on the complete test vector to ascertain the final performance of the two-stage GP method.

3.2   Model Representation

The two distinct stages are composed of different representations, aligning with the works from which each was inspired. Thus, it would be prudent to detail each stage separately. Both stages, however, contain individuals that are structured according to Koza's tree-based GP representation [7], with terminal nodes derived from the terminal set and non-terminal nodes from the function set.

The evolved model comprises three distinct sub-trees, each representing the top-performing individual from a separate sub-population. As demonstrated by the findings of John et al. [12], the decomposition of colour channels into separate sub-populations proved to be efficacious in both performance efficacy and interpretability. Consequently, for this study, the LAB colour space was employed due to its perceptual uniformity and ability to separate luminance from chrominance information. Each image (data instance) was divided into three separate sub-populations corresponding to the L (Luminance), a^* (green-red), and b^* (blue-yellow) channels of the LAB colour space. This division adheres to cooperative co-evolution principles. Within each sub-population, individuals take the form of evolved mathematical formulae, which are employed to generate the feature vector. In juxtaposition to the first stage, the second stage of this algorithm evolves only one GP tree, in order to perform construction of a highly discriminatory feature. Thus, the final model consists of an ensemble of three distinct sub-trees which perform feature extraction, and another GP tree which solely performs feature construction.

3.3   Terminal Sets

3.3.1   First Stage

The terminal set for each GP tree in the first stage comprises the raw pixel values captured by each sliding window position. The size of this terminal set is determined by the dimensions of the sliding window (w). Additionally, the terminal set varies depending on the associated sub-population, as each tree encapsulates a distinct component of the feature space, specifically, one channel of the LAB colour space. For instance, if we consider a sliding window of size $w = 3$ (i.e. a 3×3 window), the terminal sets for the three colour channels would be represented as L_i, a_j^*, and b_k^*, where i, j, and k independently range from 1 to 9, corresponding to the indices of the flattened window matrix.

3.3.2   Second Stage

For the second stage, the terminal set is determined by the feature vector produced by the first stage, in which all features are treated as terminal nodes for this population. The terminal set for this stage is represented as F_m, where m ranges from 1 to the total number of features extracted in the first stage. For instance, if the output feature length from the first stage consists of 765 features, then the terminal set will comprise 765 nodes, each corresponding to a feature of this set. In this case, (m) would range from 1 to 765.

3.4   Function Sets

Both stages employ fundamental arithmetic operators $+$, $-$, $\times$, $\div$, where addition, subtraction, and multiplication maintain their standard arithmetic meanings. The division operator is specially modified to handle division by zero cases by returning 1 when such operations occur. The first stage combines these arithmetic operators with a specialised encode function, which is consistently positioned at the root of each tree and serves a crucial role in generating binary numbers at each location of the sliding window. The encode function connects to h child nodes, where h determines the length of the produced binary number, consequently establishing a range of 2^h possible representable values. The second stage expands upon the arithmetic operators by incorporating additional function types: trigonometric functions *sin, cos* and a conditional operator. The conditional operator, termed *"conditional_if"*, processes four input values and implements a simple decision mechanism: when the first input value exceeds the second, it returns the third value; otherwise, it returns the fourth value.

3.5   Feature Vector Extraction

In adherence to the method proposed by John et al. [12], the primary objective of feature extraction stage is to generate three GP-trees which can be utilised to convert a colour image into a one-dimensional vector. The system employs a sliding window that traverses the entire image horizontally and vertically, commencing at the top-left corner, to extract pixel values at each position.

- The pixel values of the current window are extracted and fed into the terminal set of the GP programme.
- The GP programme evaluates this terminal set, resulting in the *encode* node having h number of children (i.e. h integer values).
- The integer values of the children of the *encode* node are compared against a threshold (t). For each child, 1 is returned if the value exceeds the threshold, and 0 otherwise. This process generates a binary code of length h.
- This binary code represents a local pattern detected at the current window position. The code is converted to its decimal equivalent, and a histogram is constructed to track how frequently each unique pattern occurs in the image.
- The process is repeated for each colour channel (red, green, and blue), resulting in three separate histograms. Each histogram has 2^h bins, where each bin counts the occurrence of a specific pattern. These three histograms are concatenated to form a feature vector $\mathbf{x}$ of length 3×2^h.

3.6   Fitness Evaluation

Two distinct fitness evaluations are utilised in the first and second stages: a hybrid evaluation for the first stage and a wrapper evaluation for the second. Both stages, however, incorporate a balanced accuracy component, which is mathematically defined as:

$$\text{Balanced Accuracy} = \frac{1}{2}\left(\frac{TP}{TP+FN} + \frac{TN}{TN+FP}\right), \tag{1}$$

where TP, TN, FN, and FP denote true positive, true negative, false negative, and false positive values, respectively.

As described in [12], the first stage employs a hybrid fitness evaluation that simultaneously optimises two objectives. This fitness function aims to maximise the Balanced Accuracy on the training set while optimising class separation by maximising inter-class distances and minimising intra-class distances. The function can be mathematically defined as:

$$Fitness(\mathbf{x}) = (1 - \frac{1}{1 + e^{-5(D_b - D_w)}}) + (1 - \text{Balanced Accuracy}) \tag{2}$$

In this equation, D_w represents the mean intra-class distance (instances of the same class), while D_b denotes the mean inter-class distance (instances of different classes). The Balanced Accuracy is defined in Eq. (1). This formulation effectively balances the dual objectives of high classification performance and optimal feature space separation. The wrapper component assesses the ability of the evolved feature extractor to distinguish between classes. It employs the k-Nearest Neighbour (k-NN) algorithm to evaluate how effectively the extracted features discriminate between classes. To account for the limited number of instances in the dataset, the method uses stratified k-fold cross-validation ($k = 5$), which maintains the proportion of samples for each class in the training and validation folds. It is important to note that the fitness function uses the inverse of the mean accuracy from this evaluation, as the goal is to minimise the fitness value in this optimisation process.

Taking inspiration from [19], the second stage utilises a pure wrapper-based fitness evaluation, wherein the balanced accuracy is the sole component used to determine the fitness of the individual. While the original study employs decision trees, this study opts for Random Forest due to its enhanced robustness through ensemble learning and reduced susceptibility to overfitting. In a similar fashion to the first stage, stratified k-fold cross-validation is utilised.

4   Experiment Design

4.1   Dataset Overview

The University of Edinburgh in Scotland developed the Dermofit dataset [20], comprised of 1,300 skin lesion images captured using standard cameras without dermoscopy. The images vary in size and are accompanied by labels provided by expert dermatologists and dermatopathologists. Example images from this study are shown in Fig. 2. Figure 2(a) and 2(b) show examples of non-BCC cases, while Fig. 2(c) and 2(d) show examples of BCC cases. The distribution of skin lesion classes in the dataset is shown in Table 1. Additionally, the dataset includes binary masks for each image. For this work, which focuses on BCC detection, the dataset has been labelled in a binary manner. BCC instances are designated as the positive class, whilst all other instances are labelled as negative.

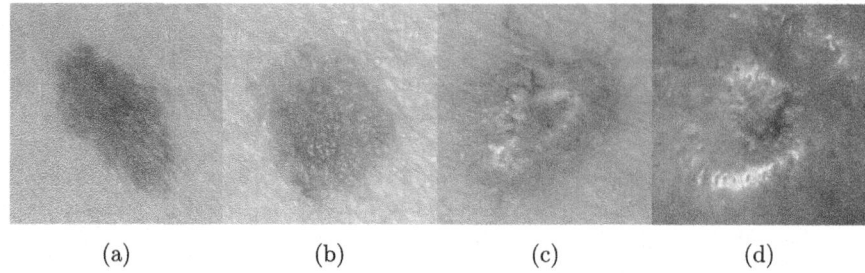

(a)   (b)   (c)   (d)

Fig. 2. Skin lesion image samples from Dermofit dataset.

Table 1. Distribution of skin lesion classes in the Dermofit dataset

Lesion Type	Instances	Lesion Type	Instances
Melanocytic Nevus (mole)	331	Intraepithelial Carcinoma	78
Seborrhoeic Keratosis	257	Haemangioma	97
Basal Cell Carcinoma	239	Dermatofibroma	65
Squamous Cell Carcinoma	88	Actinic Keratosis	45
Malignant Melanoma	76	Pyogenic Granuloma	24
		Total	1,300

4.2   Data Pre-processing

The dataset exhibited considerable variation in the size of the Region of Interest (ROI) relative to the overall image dimensions. To standardise the input and remove irrelevant background information, a cropping procedure was implemented. This involved using a bounding box to isolate the ROI in each image. Given that the primary objective is skin lesion classification, it was determined that only the information within the ROI was pertinent to the algorithm. To further refine the focus, the provided binary mask was applied to the cropped images. This ensured that the proposed algorithm would analyse solely the relevant skin lesion area, disregarding any remaining background elements.

4.3   Methods for Benchmark Comparison

The comparative framework draws inspiration from the benchmarking methodology established by John et al. [12], incorporating several key approaches. The implementation utilises LBP in two configurations: as a standalone feature extractor and in conjunction with GP for feature construction (denoted as $LBP_{constructed}$). This dual implementation facilitates evaluation of both extracted and constructed features against the proposed 2SCC-Criptor method.

Furthermore, to quantify the evolutionary advancement of the algorithm, the original CC-Criptor is included in the comparative analysis, enabling direct assessment of the performance enhancement achieved through the additional processing stage. Given the contemporary prevalence of deep learning in computer vision, benchmarking against the LeNet-5 CNN architecture as described in [21] was conducted, maintaining consistent methodological parameters including 5-fold cross-validation and preprocessing protocols.

4.4 Experiment Parameters

The parameter configurations for this study drew inspiration from [11,12], with some settings shared across both stages and others differing. Both stages employed the k tournament selection method ($k = 7$) for parent selection, coupled with a one-point crossover mechanism for offspring generation. Uniform mutation, involving random sub-tree generation and replacement within GP trees, was the chosen mutation technique. The evolutionary process was set to terminate after 50 generations in both stages. Genetic operator rates were consistent across stages, with crossover at 80%, mutation at 19%, and a 1% elitism rate. This elitism approach ensures the preservation of the fittest individuals and their genetic material throughout successive generations. Population initialisation utilised the ramped half-and-half method, incorporating both grow and full initialisation techniques. However, certain parameters diverged between stages. The first stage employed a population size (θ) of 50 for each species, totalling 150, while the second stage used a population of 1024. Although both stages maintained a maximum tree depth of 10, the initialisation tree sizes varied: 2 to 5 for the first stage, and 3 to 6 for the second.

4.5 Experiment Setup

Given the imbalanced nature of the dataset and the scarcity of positive samples, stratified 5-fold cross-validation was adopted. This approach seeks to minimise performance fluctuations arising from arbitrary data partitioning. To further counteract the potential impact of stochastic elements on algorithmic performance, 30 separate experiments were conducted, each utilising distinct random seeds for the stochastic methods, namely 2SCC-Criptor and LeNet-5.

To ascertain the efficacy of the complete feature vector extracted by 2SCC-Criptor, it was deemed prudent to utilise a range of machine learning classifiers for evaluation purposes. These included k-NN, Gaussian Naïve Bayes (GNB), Support Vector Machine (SVM), Decision Tree (DT), Random Forest (RF), Linear Discriminant Analysis (LDA), and Multi-layer Perceptron (MLP). It is worth noting that LeNet-5 diverged from this approach, instead employing fully connected neural network layers for classification, in accordance with the methodology outlined in [21].

5 Results and Discussions

Statistical comparisons between model variants were conducted using paired t-tests ($p < 0.05$). When comparing 2SCC-Criptor against other models, "↑" indicates 2SCC-Criptor performed significantly better, "↓" indicates significantly worse performance, and "=" denotes no statistically significant difference.

Table 2. Classification Performance of LBP$_{Constructed}$, CC-Criptor, and 2SCC-Criptor (%). Values represent the mean and standard deviation (mean ± std. dev.)

	Classifier	Precision	Recall	F1-Score	Balanced Accuracy
LBP$_{Constructed}$	k-NN	76.77 ± 1.24	59.42 ± 2.14	63.91 ± 1.91	61.68 ± 2.21
	GNB	78.65 ± 2.08	47.90 ± 5.78	49.82 ± 6.90	59.75 ± 2.95
	SVM	78.45 ± 1.36	50.42 ± 5.63	53.43 ± 7.03	61.00 ± 2.68
	DT	77.16 ± 1.37	61.22 ± 1.94	65.47 ± 1.72	62.57 ± 2.39
	RF	79.80 ± 0.76	66.80 ± 1.06	70.35 ± 0.89	66.76 ± 1.29
	LDA	79.98 ± 0.69	66.71 ± 1.09	70.33 ± 0.95	67.11 ± 1.26
	MLP	76.00 ± 6.04	60.20 ± 3.88	63.40 ± 4.87	63.20 ± 2.39
CC-Criptor	k-NN	82.85 ± 0.76	66.12 ± 2.25	71.65 ± 1.77	61.55 ± 2.08
	GNB	83.48 ± 0.78	73.18 ± 1.83	76.75 ± 1.27	63.18 ± 2.05
	SVM	83.48 ± 0.76	69.10 ± 2.80	73.68 ± 2.10	62.94 ± 2.00
	DT	83.48 ± 1.31	69.06 ± 2.44	73.92 ± 1.94	63.42 ± 3.65
	RF	85.83 ± 1.04	76.23 ± 2.10	79.40 ± 1.68	67.97 ± 2.83
	LDA	82.18 ± 1.13	62.99 ± 2.06	69.17 ± 1.69	59.55 ± 2.95
	MLP	84.98 ± 0.81	71.04 ± 1.51	75.55 ± 1.19	66.46 ± 2.20
2SCC-Criptor	k-NN	82.81 ± 0.83	65.69 ± 2.04	71.27 ± 1.61	61.37 ± 2.17
	GNB	81.85 ± 4.46	68.07 ± 7.35	70.87 ± 8.16	61.21 ± 2.33
	SVM	82.48 ± 3.22	67.35 ± 4.63	71.46 ± 4.54	61.15 ± 2.31
	DT	83.13 ± 1.53	69.31 ± 2.53	74.07 ± 1.98	62.51 ± 4.22
	RF	86.00 ± 0.89	76.84 ± 1.73	79.91 ± 1.37	**72.47 ± 2.46** ↑
	LDA	82.23 ± 1.29	62.99 ± 3.14	69.15 ± 2.56	59.67 ± 3.51
	MLP	84.48 ± 0.85	70.65 ± 1.57	75.16 ± 1.21	66.11 ± 2.24

5.1 Classification Performance of GP-Based Methods

The classification performance results presented in Table 2 compare three feature extraction methods: LBP$_{Constructed}$, CC-Criptor, and 2SCC-Criptor across seven standard machine learning classifiers. Random Forest (RF) consistently emerged as the top-performing classifier across all feature extraction methods, with 2SCC-Criptor achieving the highest balanced accuracy of 72.47% ± 2.46 on average, which was statistically significant compared to the other approaches. Both CC-Criptor and 2SCC-Criptor demonstrated notable improvements over LBP$_{Constructed}$, particularly in precision metrics, where they maintained scores above 82% across all classifiers. While CC-Criptor showed strong performance with GNB (76.75% ± 1.27 F1-Score), the 2SCC-Criptor RF implementation achieved the highest overall F1-Score of 79.91% ± 1.37, suggesting that the two-stage approach enhances feature discrimination capabilities while maintaining robust classification performance. This superior performance with Random Forest is expected, as the second stage fitness function specifically employs RF-based balanced accuracy to guide the evolutionary process of the GP individuals, effectively optimizing the features for RF classification.

5.2 Performance Comparison with Benchmark Methods

Table 3 presents a comprehensive comparison between traditional LBP approaches, deep learning (LeNet-5), and the GP-based methods. The 2SCC-

Table 3. Comparison against Benchmark Methods (%)

	Best Classifier	Precision	Recall	F1-Score	Balanced Accuracy
LBP	LDA	80.21 ± 0.81	66.79 ± 1.16	70.41 ± 1.00	67.49 ± 1.44
LBP$_{Constructed}$	LDA	79.98 ± 0.69	66.71 ± 1.09	70.33 ± 0.95	67.11 ± 1.26
LeNet-5	CNN	58.29 ± 2.74	59.51 ± 2.31	54.66 ± 3.92	59.51 ± 2.31
CC-Criptor	RF	85.83 ± 1.04	76.23 ± 2.10	79.04 ± 1.68	67.97 ± 2.83
2SCC-Criptor	RF	86.00 ± 0.89	76.84 ± 1.73	79.91 ± 1.37	**72.47 ± 2.46** ↑

Table 4. Performance of models on different colour decompositions (%).

	Classifier	Precision	Recall	F1-Score	Balanced Accuracy
Luminance Channel	k-NN	83.67 ± 3.42	83.67 ± 3.42	83.67 ± 3.36	54.60 ± 6.04
	GNB	85.81 ± 3.48	85.81 ± 3.48	85.71 ± 3.41	**54.59 ± 5.13** ↑
	SVM	87.99 ± 0.08	87.99 ± 0.08	87.79 ± 0.10	50.00 ± 0.06
	DT	78.56 ± 3.24	78.56 ± 3.24	78.58 ± 3.25	55.59 ± 6.44
	RF	84.49 ± 3.28	84.49 ± 3.28	84.48 ± 3.28	54.37 ± 5.68
	LDA	80.35 ± 5.54	80.35 ± 5.54	80.31 ± 5.49	56.54 ± 8.96
	MLP	87.13 ± 2.67	87.13 ± 2.67	87.05 ± 2.62	**52.42 ± 4.32** ↑
a* Channel	k-NN	83.36 ± 3.72	83.36 ± 3.72	83.35 ± 3.68	52.66 ± 4.71
	GNB	83.48 ± 4.20	83.48 ± 4.20	83.46 ± 4.11	51.94 ± 4.31
	SVM	87.99 ± 0.08	87.99 ± 0.08	87.79 ± 0.10	50.01 ± 0.19
	DT	78.89 ± 3.85	78.89 ± 3.85	78.92 ± 3.83	57.49 ± 8.17
	RF	86.76 ± 2.10	86.76 ± 2.10	86.77 ± 2.11	54.50 ± 5.10
	LDA	82.51 ± 4.31	82.51 ± 4.31	82.55 ± 4.28	57.03 ± 6.16
	MLP	87.21 ± 3.91	87.21 ± 3.91	87.13 ± 3.88	50.79 ± 1.90
b* Channel	k-NN	84.06 ± 3.46	84.06 ± 3.46	84.04 ± 3.39	**56.40 ± 6.27** ↑
	GNB	84.92 ± 3.44	84.92 ± 3.44	84.84 ± 3.39	53.66 ± 3.92
	SVM	87.99 ± 0.08	87.99 ± 0.08	87.79 ± 0.10	50.00 ± 0.06
	DT	80.42 ± 3.48	80.42 ± 3.48	80.41 ± 3.43	56.49 ± 6.45
	RF	87.81 ± 1.16	87.81 ± 1.16	87.79 ± 1.19	53.54 ± 3.55
	LDA	81.74 ± 9.10	81.74 ± 9.10	81.79 ± 9.06	54.68 ± 6.65
	MLP	87.74 ± 1.57	87.74 ± 1.57	87.66 ± 1.62	50.70 ± 2.08

Criptor method coupled with Random Forest demonstrates statistically significant superior performance, achieving the highest balanced accuracy of 72.47% ± 2.46 among all benchmark methods. Both CC-Criptor and 2SCC-Criptor substantially outperform the LeNet-5 CNN architecture, which achieved only 59.51% ± 2.31 balanced accuracy. While traditional LBP and LBP$_{Constructed}$ show comparable performance with balanced accuracies of 67.49% ± 1.44 and 67.11% ± 1.26 respectively, the GP-based methods exhibit notably higher precision and recall metrics. The superior performance of 2SCC-Criptor, particularly in F1-Score (79.91% ± 1.37), demonstrates the effectiveness of the two-stage GP approach in automatically evolving robust and discriminative features, while maintaining the interpretability advantages inherent to GP-based methods.

6   Further Analysis

6.1   Evolved Functions Analysis

To evaluate the contributory significance of each colour channel within the 2SCC-Criptor feature extraction algorithm, a systematic decomposition analysis was

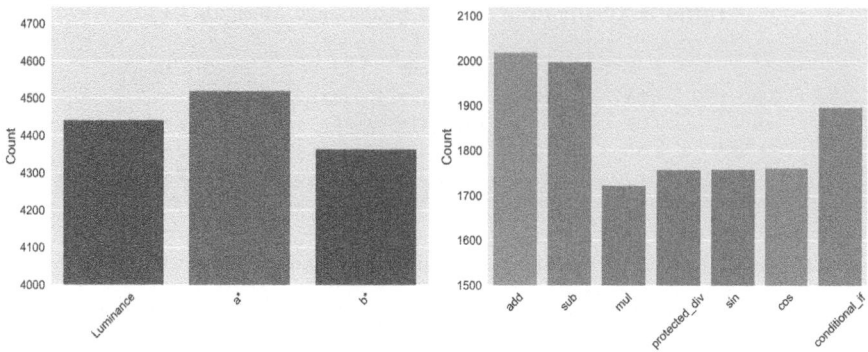

Fig. 3. Distribution of terminal nodes across LAB colour channels and function nodes in the evolved GP individuals.

conducted, demonstrated in Table 4. The methodology entailed the individual vectorisation of each colour channel (Luminance, a^*, and b^*), followed by the concatenation of these vectors with the constructed feature derived from the second stage of the algorithm. This methodological approach facilitated assessment of individual channel contributions to classification performance. The experimental results demonstrate that the utilisation of individual colour channels, even when combined with the constructed feature, yields performance metrics that are notably inferior to those obtained from the complete model. This observation suggests that the inherent discriminative power of the 2SCC-Criptor algorithm is substantially dependent on the synergistic integration of all colour channels, rather than the isolated contribution of any single channel.

Statistical analysis revealed that the Luminance channel achieved the highest balanced accuracy among all channels when using GNB (54.59%) and MLP (52.42%) classifiers, while the b* channel showed the highest balanced accuracy when using k-NN (56.40%). The predominance of the effectiveness of the Luminance channel suggests it may carry more discriminative information than the chromatic channels. However, the inferior performance of individual channels may be attributed to the co-operative co-evolutionary nature of 2SCC-Criptor, where individuals are evolved for collaborative rather than independent performance. Further investigation into evolving features for each channel independently before combination could provide deeper insights into their true discriminative capabilities.

6.2   Analysis of Constructed Features

Analysis of the node distribution across the evolved GP individuals reveals intriguing patterns in how 2SCC-Criptor utilises different colour channels and mathematical operations. As illustrated in Fig. 3, the distribution of terminal nodes across the three LAB colour channels demonstrates remarkable uniformity, with Luminance, a^*, and b^* channels being referenced 4440, 4519, and

4363 times respectively across all evolved individuals. This balanced utilisation (standard deviation of 78.5 nodes) indicates that the GP evolution process naturally converged to solutions leveraging the full colour space, despite the superior individual performance of the Luminance channel noted in the previous analysis.

The function node distribution similarly exhibits a balanced utilisation of different mathematical operations, with addition (2018) and subtraction (1997) operations being the most frequently employed, followed by conditional statements (1896). The trigonometric operations (sin: 1758 and cos: 1761) and protected division (1757) show remarkably similar frequencies, suggesting that the algorithm has found these operations equally valuable in constructing complex, non-linear relationships between colour components. This diverse yet balanced use of mathematical operations indicates that the evolved features capture sophisticated colour-space relationships, potentially explaining why the full model outperforms single-channel variants.

7   Conclusion and Future Work

Inspired by the effectiveness of GP in feature extraction, this study proposes a two-stage cooperative co-evolutionary approach for BCC detection. The comparative analysis reveals that the 2SCC-Criptor model, particularly when coupled with Random Forest, demonstrates statistically significant superior performance over traditional methods and deep learning approaches in binary classification tasks. The experimental results highlight that 2SCC-Criptor achieves a balanced accuracy of $72.47\% \pm 2.46$, substantially outperforming both conventional LBP approaches and the LeNet-5 CNN architecture. A distinctive feature of 2SCC-Criptor is its ability to automatically evolve GP individuals that synergistically integrate information across colour channels, as evidenced by the balanced distribution of terminal nodes across LAB channels. The systematic decomposition analysis reveals that the efficacy of the model stems from the collaborative interaction of colour channels rather than individual channel contributions, with the second-stage feature construction further enhancing discriminative capabilities through diverse mathematical operations. This comprehensive evaluation demonstrates 2SCC-Criptor's robust performance in automated BCC detection while maintaining interpretability through its evolved GP individuals.

Future research could explore extending the CC paradigm to the second stage, where multiple constructed features could be evolved simultaneously, potentially capturing different discriminative aspects of the extracted features while maintaining synergistic relationships. Additionally, investigating a multi-tree approach for the feature extraction stage, where trees evolve independently rather than through CC, could provide insights into whether the collaborative evolution is truly beneficial or if independent evolution could achieve similar results with reduced computational complexity.

References

1. Sung, H., Ferlay, J. Siegel, R., Laversanne, M., Soerjomataram, I., Jemal, A., Bray, F.: Global cancer statistics 2020: GLOBOCAN estimates of incidence and mortality worldwide for 36 cancers in 185 countries. CA Cancer Clin. **71**, 209–249 (2021). This report provides the latest global cancer statistics of incidence and mortality worldwide, 2022

2. Khazaei, Z., Ghorat, F., Jarrahi, A., Adineh, H., Sohrabivafa, M., Goodarzi, E.: Global incidence and mortality of skin cancer by histological subtype and its relationship with the human development index (HDI); an ecology study in 2018. World Cancer Res. J. **6**(2), e13 (2019)

3. Zambrano-Román, M., Padilla-Gutiérrez, J.R., Valle, Y., Muñoz-Valle, J.F., Valdés-Alvarado, E.: Non-melanoma skin cancer: a genetic update and future perspectives. Cancers **14**(10), 2371 (2022)

4. Rubin, A.I., Chen, E.H., Ratner, D.: Basal-cell carcinoma. N. Engl. J. Med. **353**(21), 2262–2269 (2005)

5. Altamura, D., et al.: Dermatoscopy of basal cell carcinoma: morphologic variability of global and local features and accuracy of diagnosis. J. Am. Acad. Dermatol. **62**(1), 67–75 (2010)

6. Esteva, A., et al.: Dermatologist-level classification of skin cancer with deep neural networks. Nature **542**(7639), 115–118 (2017)

7. Koza, J.R.: Genetic Programming: On the Programming of Computers by means of Natural Selection. MIT Press (1992)

8. Liang, J., Wen, J., Wang, Z., Wang, J.: Evolving semantic object segmentation methods automatically by genetic programming from images and image processing operators. Soft. Comput. **24**, 12887–12900 (2020)

9. Cano, A., Ventura, S., Cios, K.J.: Multi-objective genetic programming for feature extraction and data visualization. Soft. Comput. **21**, 2069–2089 (2017)

10. Potter, M.A., Jong, K.A.D.: A cooperative coevolutionary approach to function optimization. In: Proceedings of the International Conference on Parallel Problem Solving from Nature, pp. 249–257. Springer (1994)

11. Ain, Q.U., Al-Sahaf, H., Xue, B., Zhang, M.: Automatically diagnosing skin cancers from multimodality images using two-stage genetic programming. IEEE Trans. Cybern. **53**(5), 2727–2740 (2022)

12. John, T.C., Ain, Q.U., Al-Sahaf, H., Zhang, M.: Evolving feature extraction models for melanoma detection: a co-operative co-evolution approach. In: International Conference on the Applications of Evolutionary Computation (Part of EvoStar), pp. 413–429. Springer (2024)

13. Al-Sahaf, H., Zhang, M., Johnston, M., Verma, B.: Image descriptor: a genetic programming approach to multiclass texture classification. In: Proceedings of the 2015 IEEE Congress on Evolutionary Computation, pp. 2460–2467. IEEE (2015)

14. Codella, N., Cai, J., Abedini, M., Garnavi, R., Halpern, A., Smith, J.R.: Deep learning, sparse coding, and SVM for melanoma recognition in dermoscopy images. In: Proceedings of The International Workshop On Machine Learning in Medical Imaging, pp. 118–126. Springer (2015)

15. Haenssle, H.A., et al.: Man against machine: diagnostic performance of a deep learning convolutional neural network for dermoscopic melanoma recognition in comparison to 58 dermatologists. Ann. Oncol. **29**(8), 1836–1842 (2018)

16. Szegedy, C., Vanhoucke, V., Ioffe, S., Shlens, J., Wojna, Z.: Rethinking the inception architecture for computer vision. In: Proceedings of the IEEE Conference On Computer Vision And Pattern Recognition, pp. 2818–2826. IEEE (2016)

17. Ain, Q.U., Xue, B., Al-Sahaf, H., Zhang, M.: Genetic programming for skin cancer detection in dermoscopic images. In: Proceedings of The 2017 IEEE Congress on Evolutionary Computation, pp. 2420–2427. IEEE (2017)
18. Ain, Q.U., Al-Sahaf, H., Xue, B., Zhang, M.: A genetic programming approach to feature construction for ensemble learning in skin cancer detection. In: Proceedings of the 2020 Genetic and Evolutionary Computation Conference, pp. 1186–1194. Association for Computing Machinery (2020)
19. Loescher, L.J., Janda, M., Soyer, H.P., Shea, K., Curiel-Lewandrowski, C.: Advances in skin cancer early detection and diagnosis. In: Proceedings of Seminars in Oncology Nursing, vol. 29, pp. 170–181. Elsevier (2013)
20. Ballerini, L., Fisher, R.B., Aldridge, B., Rees, J.: A color and texture based hierarchical k-NN approach to the classification of non-melanoma skin lesions. Color Med. Image Anal. 63–86 (2013)
21. LeCun, Y., Bottou, L., Bengio, Y., Haffner, P.: Gradient-based learning applied to document recognition. Proc. IEEE **86**(11), 2278–2324 (1998)

Multi-objective Evolutionary Optimization of Virtualized Fast Feedforward Networks

Renan Beran Kilic(iD), Kasim Sinan Yildirim(iD), and Giovanni Iacca(✉)(iD)

Department of Information Engineering and Computer Science,
University of Trento, Trento, Italy
{renanberan.kilic,kasimsinan.yildirim,giovanni.iacca}@unitn.it

Abstract. Many embedded applications have strict energy, memory, and time constraints, making neural network (NN) inference particularly challenging. Recently, a novel NN architecture called Fast Feedforward Networks (FFFs) has been proposed to achieve inference with extremely lightweight computational demands and minimal latency. Yet, the memory footprint of such NNs remains a challenge. In this paper, we attempt to overcome this challenge by using a weight-sharing technique, called weight virtualization, proposing different virtualization methods that take advantage of the peculiarities of the FFFs' tree-based architecture. We further optimize the model's size (resulting from the virtualization configuration) and performance via multi-objective evolutionary optimization based on NSGA-II. Our experiments (https://github.com/DIOL-UniTN/MOE-VFFF) show that, in different benchmarks, leaf virtualization can reduce the memory footprint by up to 13x with negligible accuracy loss.

Keywords: Fast Feedforward Networks · Weight Sharing · Multi-Objective Optimization · In-memory Computing · Embedded intelligence

1 Introduction

The rise of Internet of Things (IoT) devices like smartphones, wearable devices, drones, and smart speakers, along with the huge amount of data they generate, has transformed how we interact with the world. By using their sensing, computing, networking, and communication capabilities, these devices can gather, process, and transmit different types of data such as images, audio, wireless signals, and texts from individuals and the environment. In recent years, advancements in AI, particularly in neural networks (NNs), have driven the integration of AI with IoT, creating the concept of AI of Things (AIoT) a reality [1]. However, IoT applications usually run on limited-hardware devices [2], while AI models are known to be, in most cases, resource-hungry. Therefore, there is a pressing need for AI models that are energy, memory, and computationally efficient [3].

© The Author(s), under exclusive license to Springer Nature Switzerland AG 2025
P. García-Sánchez et al. (Eds.): EvoApplications 2025, LNCS 15612, pp. 270–286, 2025.
https://doi.org/10.1007/978-3-031-90062-4_17

One promising approach to embedded intelligence is represented by a recently introduced NN architecture called Fast Feedforward Networks (FFF) [4]. FFFs divide the input space into distinct regions using a differentiable binary tree (built by optimizing these regions' boundaries) which uses shallow NNs on the leaves. FFFs provide significant advantages in terms of inference speed compared to vanilla Feedforward Networks (FF). However, due to their tree-based architecture, their memory requirement grows exponentially with the depth [4].

There are several widely used techniques to address the memory issue in NNs, such as quantization [5], knowledge distillation [6], pruning [7], and weight sharing [8,9]. Recently, an advanced weight-sharing technique called *weight virtualization* has been proposed to minimize the memory footprint and hardware requirements of embedded systems running multiple tasks with deep NNs [10]. One drawback of these methods is that the performance reduction introduced by compression is often not tolerable. Furthermore, compression methods involve several user-defined hyperparameters, such as the ranks and sparsities of weight matrices in pruning. Choosing the wrong values for these parameters can lead to a significant loss in performance. Hence, finding a well-performing compressed model is both complex and costly, as the process of tuning these hyperparameters is computationally intensive [11]. Multi-objective evolutionary algorithms can effectively search for optimal configurations in compression algorithms and help balance the trade-off between performance and compression levels. Additionally, they offer a range of Pareto-optimal models, which allows for the selection of the best option for specific devices.

In this paper, we specifically address the memory overhead issue in FFFs by 1) introducing weight virtualization in FFFs and 2) leveraging multi-objective optimization (based on NSGA-II [12]) to find the optimal hyperparameters related to the virtualization configuration. We experiment with three well-known datasets adopted in IoT research, MNIST [13], Human Activity Recognition (HAR) [14,15], and Google Speech Commands v2 (SC) [16]. We compare the results of the multi-objective evolutionary search with those achieved by random search and manual tuning; finding that NSGA-II can effectively find the optimal virtualization configurations for FFFs. Given the recent introduction of FFFs, to our knowledge, this is the first study to apply virtualization to this architecture and use evolutionary algorithms to optimize it.

The rest of the paper is structured as follows. In the next section, we briefly summarize the related work. Section 3 introduces the background concepts on FFFs and weight virtualization, along with the definition of Fisher information that is used in our approach. Then, Sects. 4 and 5 present the methods and experimental results, respectively. Finally, Sect. 6 concludes this work.

2   Related Work

This section reviews the related work on model compression and the application of evolutionary search to automate the optimization of compression techniques.

Model Compression. Quantization, pruning, and knowledge distillation are the main approaches to reduce the number of parameters in NN models and make them memory efficient. Quantization can reduce the size of the model and accelerate computation with hardware support by converting weights and activations from high precision to lower precision. This process can be applied either during training, known as Quantization-Aware-Training (QAT) [17–19], or after training, known as Post-Training Quantization (PTQ) [20]. PTQ is considered a very fast way to quantize NNs. Some QAT techniques aim to binarize or quantize Convolutional NNs (CNNs), or sometimes both [17,18]. Many researchers have proposed varying solutions for the accuracy loss of PTQ [21,22]. Yet, PTQ still tends to be less accurate than QAT [23].

Knowledge distillation is a compression strategy that instead includes a teacher-student architecture. The teacher network is a more complex pre-trained network, and the student network is a smaller, less complex one. The teacher network helps the student network by providing prior knowledge during training to improve its performance [23].

Finally, structured pruning is commonly used for accelerating NNs typically by removing entire channels, hidden layers, or neurons [24–26]. When a channel is pruned, the corresponding channels are also removed to maintain the model's consistency. The same process is applied to the pruned elements as well [27,28]. Many pruning techniques rely on the sum of absolute weights of the channels as a criterion for pruning [28,29]. These algorithms typically follow a train-compress-retrain approach to reduce the accuracy drop introduced by the compression while optimizing storage and computational demands by retaining only the most important parameters of the model [23,30].

Many of the above-mentioned compression strategies are designed for complex models, such as CNNs or Transformers, which are computationally intensive and less suited for resource-constrained applications compared to simpler models such as shallow FFs or FFFs. In contrast, weight sharing [8,9] reuses parameters across different parts of the model, rather than selecting which parameters to store, and as such it is somehow agnostic w.r.t. the underlying model architecture, hence being suitable also for FFFs. One special kind of weight sharing is *weight virtualization* [10], which was specifically introduced in the context of multi-task learning. While other compression strategies are widely used for reducing the size and complexity of NNs, weight virtualization can be highly effective in scenarios where multiple models are involved [23]. In this paper, we use this concept by treating the leaves of an FFF as sub-models.

Multi-objective Optimization. Many studies apply multi-objective optimization on several compression algorithms for IoT devices in order to reduce their memory and latency requirements [31] as well as their power consumption [32]. Most of the existing works, however, focus on convolutional models such as VGG, and MobileNet.

Among multi-objective approaches, Multi-Objective Evolutionary Algorithms (MOEAs) have been commonly employed for compression optimization and neural architecture search [33,34]. One important aspect that makes MOEAs particularly suitable for these tasks is that they provide a set of Pareto-optimal models which are characterized by different trade-offs between memory and performance [35,36].

For instance, in [11] the MOEA based on Decomposition (MOEA/D) [37] has been used for deep neural network compression. The advantage of MOEA/D is that it decomposes a multi-objective problem into several single-objective subproblems, making it an effective approach for multi-objective optimization. MOEA/D has also been used to train deep belief networks to use sparse output features and achieve small model reconstruction errors [38]. Recently, MOEA/D has been adopted for model pruning [39].

In [40], authors considered two objectives in the context of CNNs, i.e., the number of filters and accuracy, adopting NSGA-II [12] to find the optimal set of non-dominated solutions. The authors used a binary string encoding, with each bit representing a filter in CNN. In [41], instead, the authors use evolutionary search to design efficient 1-bit CNNs (BNNs).

Novelty of the Proposed Approach. With the exception of the aforementioned [41], which as said focuses on BBNs (which are lightweight, although their accuracy is low compared to FFFs [42]), most of the existing works dealing with multi-objective optimization applied to NNs focus on complex models such as CNNs rather than fast, lightweight models with conditional execution such as FFFs. On the other hand, these requirements are essential to reduce execution time and energy consumption during inference, which is crucial for IoT applications. Moreover, to the best of our knowledge, weight virtualization has not been explored in the context of FFFs. Beyond that, and despite the great potential of weight virtualization for model compression, no studies have explored using evolutionary search for this optimizing virtualization in FFFs, particularly considering the impact of hyperparameters on the process. Finally, only a few works include Fisher information to compress models, which as we will see below is another aspect of our approach.

3   Background

In this section, we present the key background concepts underlying our methods, specifically focusing on FFFs, weight virtualization, and Fisher information. As mentioned earlier, our framework indeed leverages FFFs as the core machine learning (ML) model, while we employ weight virtualization, a specialized form of weight sharing, as our compression technique. As we will see later, we utilize Fisher information in the virtualization pipeline to compute the informativeness of each parameter and leaf within FFF models.

3.1   Fast Feedforward Networks

Before discussing how virtualization can be applied to FFFs, it is important to understand the main characteristics of the FFF, as they have a direct impact on how we design the compression algorithm. In essence, an FFF is a balanced binary-tree structured NN, designed to benefit from the fact that different neurons are activated by different regions of the input space [4,42]. An FFF with a depth of d has two main types of computing blocks: N (inner) nodes and L leaves. Each node can be seen as a single neuron computing a weighted sum of its inputs and bias. The leaves are instead tiny FFs with a single hidden layer of size l (in the following, we refer to it as "leaf width") that are particularly trained to handle different parts of the input space.

We denote by $\mathcal{N} := \{N_{0,0}, N_{1,0}, N_{1,1}, \ldots, N_{d,2^{d-1}-2}, N_{d,2^{d-1}-1}\}$ the set of nodes, each one having a sigmoid activation on the output. We denote by $\mathcal{L} := \{N_{d,0}, N_{d,1}, \ldots, N_{d,2^d-2}, N_{d,2^d-1}\}$ the set of leaves. The notation $N_{i,j}$ indicates the j-th node (or leaf, when $i = d$) at depth i. Nodes and leaves form a balanced binary tree where $\{N_{i+1,2j}, N_{i+1,2j+1}\}$ are the children of $N_{i,j}$.

Exponentially Faster. Given an FFF with depth d and leaf width l, the model width w (i.e., the total number of hidden nodes across all leaves) is given by the number of leaves (2^d) multiplied by their width (l), i.e., $w = 2^d l$. During training, an FFF uses its $2^d l$ hidden neurons to learn; yet, at inference time it only uses one leaf with l neurons, with a complexity of $\mathcal{O}(2^d l) = \mathcal{O}(\log w)$. Hence, FFFs have a logarithmically reduced complexity, compared to an FF with the same model width w (i.e., no. of hidden nodes), which has a complexity of $\mathcal{O}(w)$.

Yet, Still Large. In FFFs, we need to store all weights even though we do not use all of them during inference. An FFF with depth d and leaf width l has a total number of parameters (γ) equal to:

$$\gamma = 2^{d-1}\gamma_N + 2^d l \gamma_L \tag{1}$$

where γ_N and γ_L are the total numbers of parameters in nodes and leaves, respectively. From here, it results that $\gamma \propto 2^d$, such that while the depth increases, the size of an FFF increases exponentially.

3.2 Weight Virtualization

As discussed earlier, there is a growing interest in compression techniques within the ML community. This is motivated on the one hand by application demands (e.g., in the context of IoT). On the other hand, recent evidence [43] showed that in deep NNs there is great redundancy in the number of parameters being used. In the context of FFFs, even more redundancy can be expected due to the repetition of similar operations on the same features across different leaves.

One way to reduce such redundancies is weight virtualization [10], a weight-sharing technique that works in a coarse-grained fashion. The idea behind weight virtualization is to map multiple weights (i.e., parameters of a model) to a smaller set of shared weight vectors. In the case of NNs, weight virtualization works by (1) using each block of memory to represent a block of weights shared across one or more NNs, and (2) mapping significant (if not all) weights from the NN onto a shared weight page, thereby reducing memory usage and improving efficiency. As we will see below, this approach can also be applied to FFFs, where, for instance, a weight block V may simultaneously represent k consecutive weights of a given leaf L_i as well as k consecutive weights of another leaf L_j.

3.3   Fisher Information

The Fisher information estimates how much information each parameter in a model f holds for a particular sample by analyzing the corresponding output probabilities [44]. In the context of FFFs, given a leaf L_i, we can compute Fisher information of its parameters Θ_{L_i} as follows:

$$I(\Theta_{L_i}) = \frac{1}{|\mathbf{X}_{L_i}|} \sum_{x \in \mathbf{X}_{L_i}, \theta \in \Theta_{L_i}} \left(\underbrace{\frac{\partial}{\partial \theta} \log f(x|\theta)}_{I(\theta)} \right)^2 \qquad (2)$$

where $\mathbf{X}_{L_i}$ is the set of samples that propagate through leaf L_i. As a result, for FFFs, Eq. (2) measures the importance of each leaf and the parameters therein.

4   Methods

Our framework consists of three main components; an FFF as the ML model, an optimal virtualization pipeline as a highly efficient compression algorithm, and

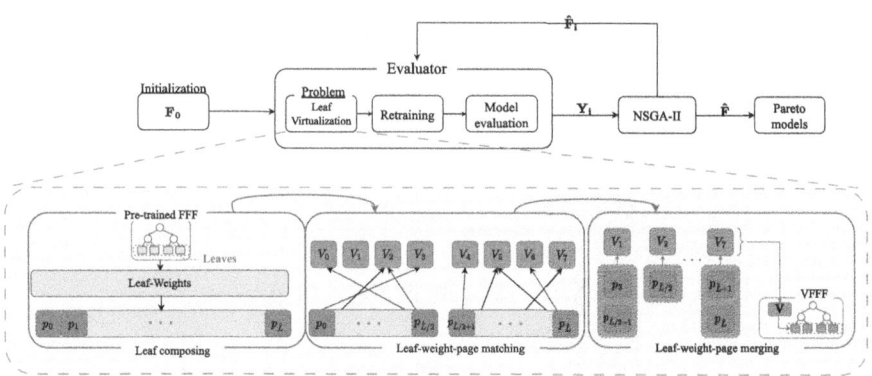

Fig. 1. Detailed blocks of the proposed leaf virtualization algorithm (bottom) where the split size is $n = 2$ and the number of leaf–pages $\tilde{L}$ is higher than the number of virtual pages $\tilde{V} = 8$, such that the model size is reduced by a factor of $\tilde{V}/\tilde{L}$, enabling the compressed model to fit within the memory limits of an edge device. The optimal virtualization pipeline (top) first initializes virtualization hyperparameters for a population of K individuals in $\mathbf{F_0} \in \mathbb{Z}^2$ and retrains the corresponding VFFFs. Then, it uses NSGA-II to iteratively generate new populations optimized for accuracy and model size. This process is repeated over multiple generations, producing Pareto-optimal FFF models that balance accuracy and model size.

multi-objective optimization via NSGA-II for optimally finding hyperparameters of the compression method with respect to two contrasting objectives, namely the model size and its accuracy. Figure 1 visually represents the overall framework.

Leaf Virtualization. We use weight virtualization to map similar parameters of FFFs into one parameter, to reduce the overall memory requirements of the model, hence attempting to solve the FFF memory footprint problem. We call the resulting FFFs *Virtualized Fast Feedforward networks* (VFFFs).

The proposed virtualization algorithm can be seen in the bottom part of Fig. 1. The goal of the algorithm is to have β fixed-sized groups of weights, that we call *leaf–pages* (P), each with page size α, where $\beta, \alpha \in \mathbb{Z}$. We aim to match those to ζ virtual pages (V) with the same size, where $\zeta < \beta$, and $\zeta \in \mathbb{Z}$. Note that in this virtualization pipeline, the parameters in the nodes are not compressed, since their memory footprint is negligible compared to that of the leaves, i.e., $\gamma_L >> \gamma_N$ (see Sect. 3.1).

Importance of the Leaves. For an FFF, since each leaf specializes on a part of the input space, we can calculate Fisher information of all leaves as:

$$\hat{I} = \frac{1}{L} \sum_{i=0}^{L-1} I(\Theta_{L_i}) \tag{3}$$

where L is the number of leaves and $I(\Theta_{L_i})$ is calculated as in Eq. (2).

Leaf–Page Matching. In order to apply leaf virtualization, the overall set of leaves' weights in the model must be split into groups and then matched to one of the $\beta \in \mathbb{Z}$ virtual pages. For this, we divide the weight vector into n splits, where $n = 2^k$ and $k \in \mathbb{Z}$ (this is equal to dividing the leaves into n pieces). Then, we perform a greedy search to find the optimal virtual page for each leaf–page using the following cost function:

$$C(p_i, p_j) = \underbrace{\sum_{(\theta_i \in p_i, \theta_j \in p_j)} (\theta_i - \theta_j)^2}_{C_m(p_i, p_j)} \underbrace{\sum_{(\theta_i \in p_i, \theta_j \in p_j)} (\hat{I}(\theta_i) + \hat{I}(\theta_j))}_{C_f(p_i, p_j)}$$

$$= \sum_{(\theta_i \in p_i, \theta_j \in p_j)} (\theta_i - \theta_j)^2 (\hat{I}(\theta_i) + \hat{I}(\theta_j)) \tag{4}$$

where p_i and p_j are two different leaf–pages, and θ_i and θ_j are single weights. The magnitude difference between pages helps group similar pages into virtual pages. Fisher information, on the other hand, adjusts the cost function, by ensuring that critical pages are matched with their closest counterparts.

Leaf–Page Merging. After leaf–page matching, we merge the matched pages into their virtual pages by taking the mean of each matched group. For optimization during training, we repeat this merging process at each iteration. In addition, we evaluate the leaf virtualization using three different search methods (referred to as Page, Leaf, and Leaf–Split, respectively) using as cost function the one shown in Eq. (4). The Page method sorts the leaf–pages in ascending order using C_f. The Leaf method sorts the leaves and each page in the leaves in the same way. Finally, the Leaf–Split method splits the leaves and then carries out the Leaf method for each chunk of pages.

Retraining and Model Evaluation. We retrain the model after compression to allow the virtualized model to adapt, recovering some of the accuracy lost during the compression process. We then evaluate the model based on (1) its validation error δ (with the validation accuracy being calculated as $1-\delta$) and (2) the number of model parameters γ, see Eq. (1). These two values are used as the objectives for the multi-objective optimization process carried out by NSGA-II.

4.1 Multi-objective Optimization via NSGA-II

Given our problem leaf virtualization $P(\alpha, \beta)$, NSGA-II takes the two hyperparameters α, β as input variables to optimize two objectives, namely accuracy and model size. We focused on optimizing both objectives jointly, rather than fixing the model size and optimizing only accuracy, because this approach has the advantage of providing multiple Pareto-optimal solutions for various scenarios involving devices with differing memory requirements. On the other hand, formulating the problem as single-objective constrained optimization would require multiple time-consuming experiments for each model size of interest.

We utilize the NSGA-II algorithm from the Pymoo library [45], employing its default operators for crossover and mutation, except for customizations made to the initialization process and the individual sampling strategy. During initialization, we randomly initialize the hyperparameter $\alpha_0 \in \{10, 11, \ldots, 10000\}$ since the algorithm gives a reasonable performance with a page size in this range [10]. Given a maximum number of leaf parameters $\hat{\gamma}_L$, the initial number of pages is randomly sampled such that $\beta_0 = \hat{\gamma}_L/\alpha_0$. During the optimization, each individual's page size α_i is sampled in the same way, while the number of leaf–pages β is sampled randomly such that $\beta_i \in \{1, 2, \ldots, \hat{\gamma}_L/\alpha_i\}$, where $i \in \{1, 2, \ldots, K\}$ where K is the population size.

5 Experimental Results

We evaluated the proposed method by testing it on three datasets, which have been selected for being representative of three different application domains. The three considered datasets are MNIST [13], Human Activity Recognition (HAR) [14,15], and Google Speech Commands v2 (SC) [16]:

- MNIST can be seen as an example of a simple, yet realistic image-based application. It includes 10 gray-scale hand-written digits with the size of 28×28 such that each sample has 784 features.
- HAR is an example of a wearable device application where the goal is to classify human activities based on sensor data. This dataset contains six activity classes (e.g., walking, standing, etc.). Data are collected by a gyroscope and

an accelerometer, which capture 3-axis angular velocity and 3-axis acceleration, respectively, at a sampling rate of 50 Hz. This results in 3 data points
for each axis per sensor. We use 1-second frames as input to our model for
both training and inference, where each frame contains 300 features, i.e., two
3-axis feature sets (one for each sensor) sampled over 50 timesteps.

- SC is an example of an audio-based application. Data consist of 1-second
speech samples, each sampled at 16 kHz. The dataset includes 35 distinct
word classes (e.g., yes, no, up, down, etc.). Ten of them are used as commands
by convention. The other words are considered to be auxiliary and labeled as
'unknown'. A subset of the 'unknown' class with a similar number of samples
w.r.t. the other classes is included to avoid data imbalance. The set also
contains noise samples of different lengths and a 1-second random frame is
taken from them during training. In the end, 12 classes were used during
training including 10 command classes, one unknown class, and one noise
class. We preprocessed each audio sample by extracting 13 Mel-Frequency
Cepstral Coefficients, obtained by applying a Short-Time Fourier Transform
with a window size of 25 ms and a hop size of 16 ms, resulting in 793 (61×13)
input features for each 1-second sample. For each set, we took 10% of the
training set as the validation set.

We used a pre-trained FFF model in all experiments. For training, we used
the Adam optimizer with a learning rate of 0.002 and a mini-batch size of 512.
Each model has been trained for 50 epochs. We repeated each dataset on each
experiment with 10 different seeds to ensure repeatability. Each experiment in
Subsect. 5.2 has been run for 10 generations (G) with a population size (K)
of 10. We took computing resource limitations into account when determining
these values. We set the maximum model size to 400kB, to ensure any model
can fit into a device with 512kB memory while still having some room for other
processes requiring memory. We have kept the FFF configuration fixed for all
models, with a depth of 4 and a leaf width of 16, since this configuration gives
acceptable performance for each dataset [42]. Please note that no additional
compression (e.g., based on pruning or quantization) is applied to the models,
and weights are maintained in full precision.

5.1   Preliminary Analysis

Before applying the proposed virtualization scheme and evolutionary search, we
performed two preliminary experiments to analyze the similarity between the
leaves and observe the feasibility of applying virtualization to them.

Leaves Grow Fast. Even though both the number of nodes and leaves increase
exponentially with the FFF depth, the fact that leaves have leaf width l results
in them being the size bottleneck, see Eq. (1). In Fig. 2, we can see how, for each
dataset, the model size increases exponentially as the depth increases, reaching
nearly 1MB at depth 4 for MNIST and SC. This exceeds edge devices with few
kB-sized memories.

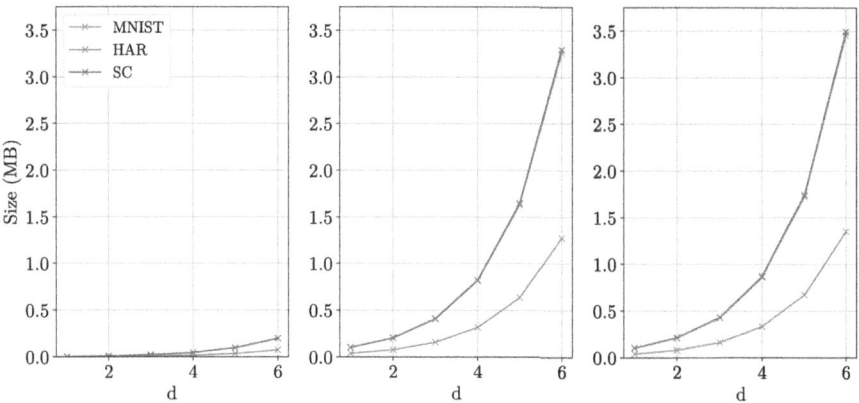

Fig. 2. Exponential increase of FFF's size when increasing the model depth. From left to right: memory required by nodes, leaves, and overall FFF model.

Leaves are Similar. We have investigated similarities between different leaves of an FFF with a leaf width $l = 16$ (for ease of representation) considering different models in terms of magnitude and Fisher information of the weights of each leaf (Fig. 3), and the sum of absolute weight differences as the model depth varies (Fig. 4). In Fig. 3, we can see that the distribution of the important (and unimportant) weights (i.e., the weights with high and low Fisher information) is very similar in each log-weight, especially between -1 and -6. In addition, in Fig. 4, we can see that there are several similar whole leaves and this behavior gets even more obvious while the depth increases. This means that we can compensate for the exponential increase of the weights by virtualizing them. Both findings indicate that the leaves are quite similar to one another, suggesting that sharing them would result in minimal loss.

Landscape Analysis. Finally, we have characterized the landscape of the validation error δ w.r.t. the parameters α and β, see Fig. 5. As expected, the performance on the different tasks varies with page size and number of pages. Given that the overall model size, which is proportionate to $\alpha \times \beta$, must be minimized along with the minimization of δ, the necessity of automated searches, such as evolutionary search, appears clear. Multi-objective optimization, in particular, allows one to find the optimal trade-offs between performance and model size, which in turn can be useful when dealing with different devices with different memory requirements and different performance requirements.

5.2 Optimal Compression

In the experiments, to compare the benefit of the different virtualization methods on especially FFFs, we also compressed FFs with a hidden width of 256 (chosen to be the same as the width of FFFs). We first compared the virtualized models in terms of performance, memory footprint, and complexity (MACs) against FF and FFF architectures.

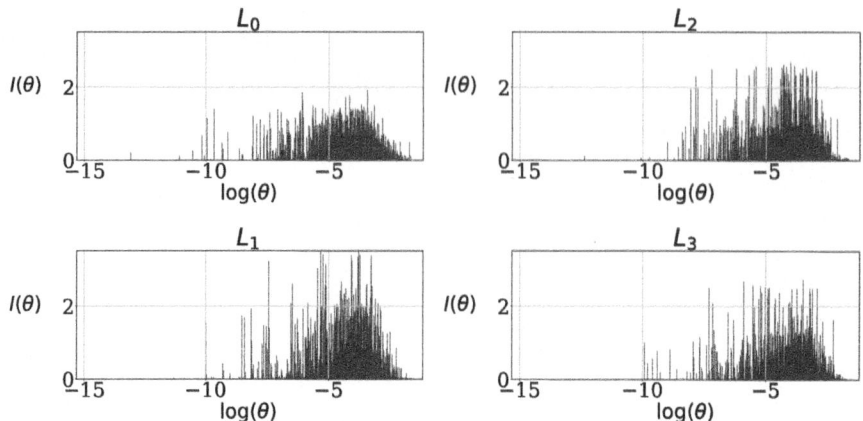

Fig. 3. Important and unimportant weights are similar. Fisher information $I(\theta)$ computed based on Eq. (2) vs. logarithm of each weight's magnitude $\log(\theta)$ across leaves of an FFF with $d = 2$ on the SC dataset.

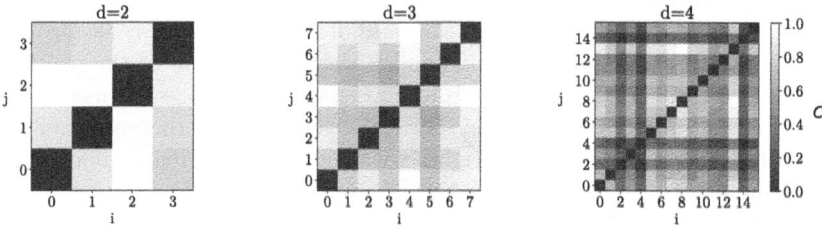

Fig. 4. Normalized matching cost $C = C(p_i, p_j)$ computed based on Eq. (4) of the leaves of a pre-trained FFF for different values of the leaf depth d on the SC dataset.

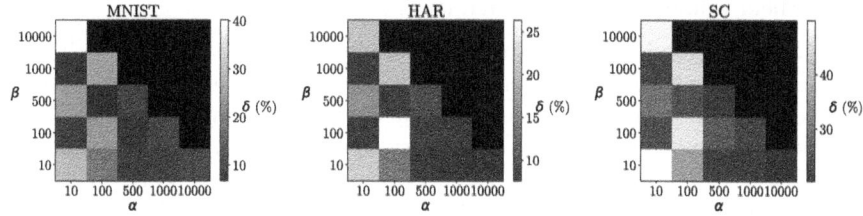

Fig. 5. Landscape of the validation error δ obtained via grid search. Black areas have not been considered due to their corresponding model size being too large.

In Fig. 6 (left), it can be seen that, with a very small accuracy drop, we achieve around half of the model size compared to FF and FFF (lower than 256 kB), which allows the model to fit into an extremely memory-constrained system, enabling in-memory computing. For extremely memory-constrained devices, we can achieve a model size smaller than 128 kB, yet, trading off some accuracy (∼2%). To provide a reference, the full-precision (uncompressed) FFF model

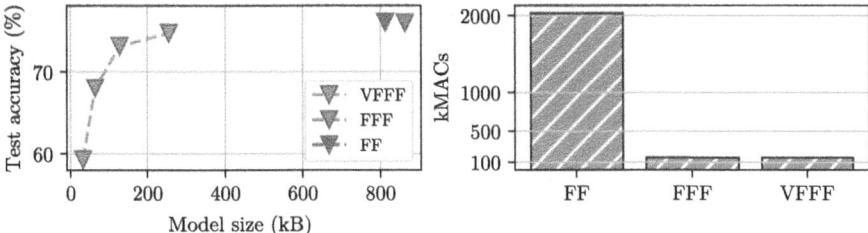

Fig. 6. Inference accuracy, model size, and model complexity of VFFFs, FFFs, and FFs with the same model width on the SC dataset for $n = 2$ and $\alpha = 100$.

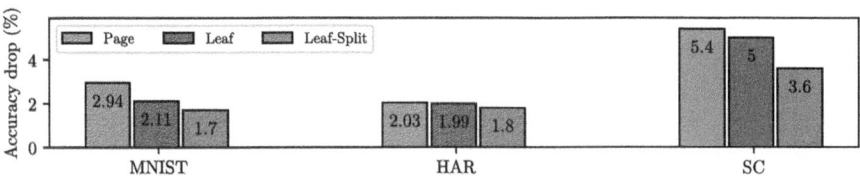

Fig. 7. Accuracy comparisons of the search methods for leaf–page merging on each dataset for $n = 2$ and $\alpha = 100$. For each dataset, the models are compressed to fit inside a 128 kB memory. The accuracy drop indicates the accuracy loss compared to the original accuracy before the compression.

sizes for MNIST, HAR, and SC are 857 kB, 327 kB, and 863 kB, respectively. In Fig. 6 (right), we can also observe how FFFs and VFFFs compare favorably w.r.t. FFs, achieving more than a 4x reduction in terms of complexity (MACs).

We have made further experiments to observe the performance drop using the different search methods for leaf–page merging described in Sect. 4, see Fig. 7. We can see that, for each benchmark, the Leaf–Split method has the smallest accuracy drop. We believe that one reason for this lies behind the tree-based architecture of FFF. Since the leaves that are close to each other are more similar (or recognize similar features), it is easier in this case that they are matched into the same virtual page.

We then decided to use the Leaf–Split method with $n = 2$ while performing virtualization based on the multi-objective evolutionary algorithm, since it has been observed that it is the best-performing method.

In Fig. 8, we see that, for each dataset, there are different best optimal solutions with specific page sizes α for different model sizes (which correlate with β) while trading off some validation accuracy so to meet specific memory requirements. For the MNIST dataset, models tend to perform better with smaller page sizes as the model size increases. Similarly, for the HAR dataset, the same trend holds, although the optimal page size is smaller. In the case of SC, smaller page sizes are generally preferred across most scenarios. For each dataset, we notice several trade-off points where a slight decrease in accuracy is accompanied by a significant reduction in model size.

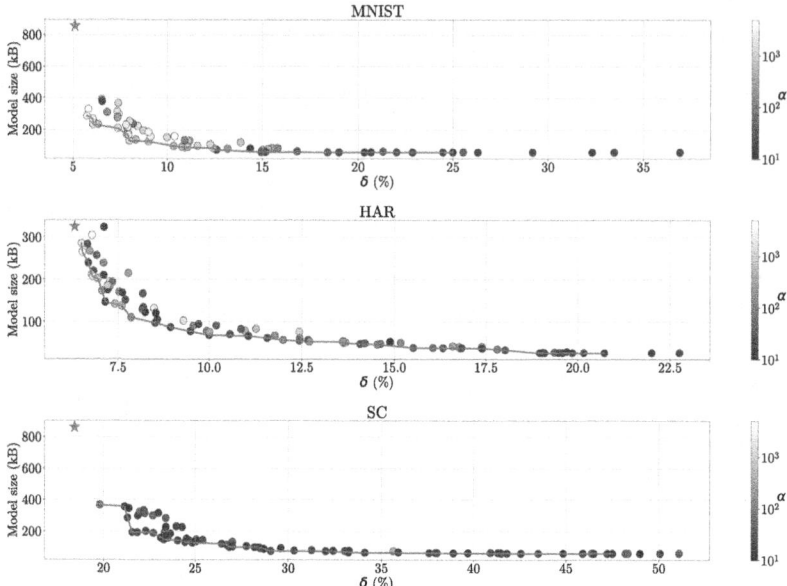

Fig. 8. Pareto fronts (solutions aggregated across 10 different seeds) for each different dataset. The trade-offs can be observed with each model's optimized page size α. For better visualization, the red line indicates the aggregated Pareto front interpolating the non-dominated solutions found across all runs. The red star represents the original FFF uncompressed model referred to in Fig. 6.

In the case of MNIST, we observe three distinct model sizes; 300 kB, 285 kB, and 225 kB, where all models have similar errors ($\sim$5%). A similar trend is observed in the other datasets as well: for HAR, the model sizes are 400 kB, 300 kB, and 225 kB, and for SC, they are 380 kB, 300 kB, and 200 kB, with all models showing nearly identical accuracy.

Additionally, for each dataset, there is a clear breaking point where the accuracy begins to decline sharply, indicating a steep improvement in accuracy as model size decreases. For MNIST, HAR, and SC, these points correspond to model sizes of 180 kB, 140 kB, and 200 kB, respectively. This could occur because the remaining pages are either highly important (with high Fisher information and small absolute weight differences), making them sensitive to even the slightest changes in magnitude, or they have significantly different absolute weight values (high magnitude differences, low Fisher information), which, despite being less important, lack comparable weights to match. It is also possible that both factors are present simultaneously, making the matching process even more lossy, causing the curve to become steeper.

We further compared VFFFs compressed via NSGA-II with those obtained by searching the page size either randomly or manually across three different model sizes, namely 256 kB, 128 kB, and 64 kB, on each dataset. The results of

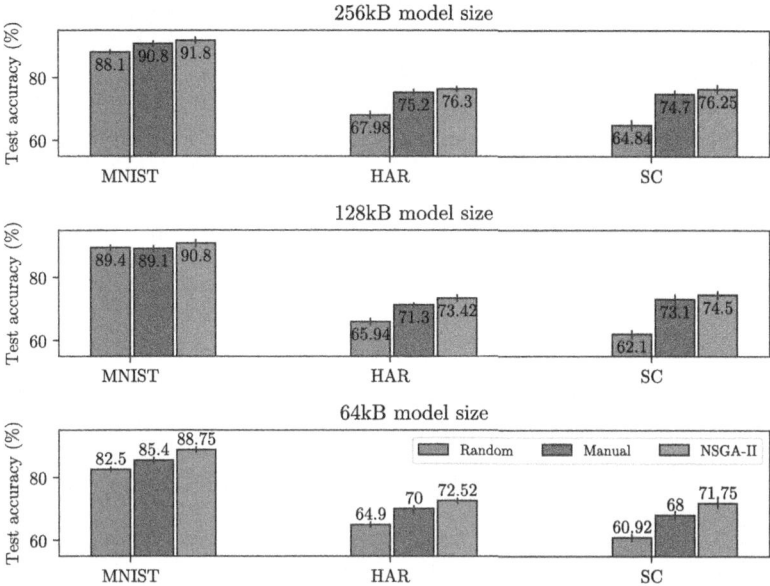

Fig. 9. Test accuracy on the different datasets comparing compression schemes via random search, manual tuning, and NSGA-II for different sizes of VFFFs.

this comparison are provided in Fig. 9. Each bar represents the mean of 10 different test seeds, with their standard deviation. For NSGA-II, we picked the model from the Pareto fronts shown in Fig. 8 with the closest model size to the desired size. For random search, we drew 100 discrete random variables—to match the complexity of evolutionary search $(G \times K)$—to sample page sizes $\alpha_i \in [10, 10000]$, and numbers of pages $\beta_i \in [10, 100000/\alpha_i]$, where $i \in \{1, \cdots, 100\}$. For manual search, we considered page sizes picked from $\{10, 100, 1000, 10000\}$ (as proposed in [10]) and took the page size that gives a virtualized model with the best performance. From the figure, for each model size, using NSGA-II is clearly superior to both random search and manual tuning, hence motivating the use of multi-objective optimization in this application. In addition, the performance difference becomes more significant when the model gets smaller. This indicates that for devices with more strict memory requirements, the usage of our method gets more crucial. Finally, while the advantage of using NSGA-II appears somehow more limited for a simpler task such as MNIST, it becomes more pronounced when the task gets more challenging, such as SC.

6 Conclusions

We have introduced VFFFs through leaf virtualization as a solution to the memory bottleneck of FFFs by compressing models up to 13x. Our approach involves experimenting with various methods to minimize the accuracy drop caused by compression, leveraging the unique characteristics of FFFs. To enhance our results, we have employed multi-objective optimization via NSGA-II to optimize the search for a more effectively compressed model that maintains the model's original accuracy, or even improves in some cases, which is significant while compressing such lightweight models. We also provided model options for different model sizes and accuracies thanks to evolutionary search.

Acknowledgments. Funded by the European Union (project no. 101071179). Views and opinions expressed are, however, those of the author(s) only and do not necessarily reflect those of the European Union or EISMEA. Neither the European Union nor the granting authority can be held responsible for them.

References

1. Siam, S.I., et al.: Artificial intelligence of things: a survey. ACM Trans. Sensor Netw. **21**(1), 1–75 (2024)
2. Boobalan, P., et al.: Fusion of federated learning and industrial Internet of Things: a survey. Comput. Netw. **212**, 109048 (2022)
3. Zhou, Z., Chen, X., Li, E., Zeng, L., Luo, K., Zhang, J.: Edge intelligence: paving the last mile of artificial intelligence with edge computing. Proc. IEEE **107**(8), 1738–1762 (2019)
4. Belcak, P., Wattenhofer, R.: Fast feedforward networks (2023). arXiv:2308.14711
5. Nagel, M., Fournarakis, M., Amjad, R.A., Bondarenko, Y., Van Baalen, M., Blankevoort, T.: A white paper on neural network quantization (2021). arXiv:2106.08295
6. Gou, J., Yu, B., Maybank, S.J., Tao, D.: Knowledge distillation: a survey. Int. J. Comput. Vis. **129**(6), 1789–1819 (2021)
7. Hoefler, T., Alistarh, D., Ben-Nun, T., Dryden, N., Peste, A.: Sparsity in deep learning: pruning and growth for efficient inference and training in neural networks. J. Mach. Learn. Res. **22**(241), 1–124 (2021)
8. Roth, W., Pernkopf, F.: Bayesian neural networks with weight sharing using Dirichlet processes. IEEE Trans. Pattern Anal. Mach. Intell. **42**(1), 246–252 (2020)
9. Ullrich, K., Meeds, E., Welling, M.: Soft weight-sharing for neural network compression. In: International Conference on Learning Representations (2017)
10. Lee, S., Nirjon, S.: Fast and scalable in-memory deep multitask learning via neural weight virtualization. In: International Conference on Mobile Systems, Applications, and Services, pp. 175–190. Association for Computing Machinery, New York (2020)
11. Huang, J., Sun, W., Huang, L.: Deep neural networks compression learning based on multiobjective evolutionary algorithms. Neurocomputing **378**, 260–269 (2020)
12. Deb, K., Pratap, A., Agarwal, S., Meyarivan, T.: A fast and elitist multiobjective genetic algorithm: NSGA-II. IEEE Trans. Evol. Comput. **6**(2), 182–197 (2002)
13. Deng, L.: The MNIST database of handwritten digit images for machine learning research. IEEE Signal Process. Mag. **29**(6), 141–142 (2012)
14. Ignatov, A.: Real-time human activity recognition from accelerometer data using Convolutional Neural Networks (2017)
15. Reyes-Ortiz, J.: Anguite: human activity recognition using smartphones. UCI Machine Learning Repository (2013)

16. Warden, P.: Speech commands: a dataset for limited-vocabulary speech recognition (2018). arXiv:1804.03209
17. Courbariaux, M., Bengio, Y., David, J.-P.: BinaryConnect: training deep neural networks with binary weights during propagations (2015). arXiv:1511.00363
18. Lin, X., Zhao, C., Pan, W.: Towards accurate binary convolutional neural network (2017). arXiv:1711.11294
19. Gysel, P., Pimentel, J., Motamedi, M., Ghiasi, S.: Ristretto: a framework for empirical study of resource-efficient inference in convolutional neural networks. IEEE Trans. Neural Netw. Learn. Syst. **29**(11), 5784–5789 (2018)
20. Ni, R., Chu, H.-M., Castañeda, O., Chiang, P., Studer, C., Goldstein, T.: WrapNet: neural net inference with ultra-low-resolution arithmetic (2020). arXiv:2007.13242
21. Cai, Y., Yao, Z., Dong, Z., Gholami, A., Mahoney, M.W., Keutzer, K.: ZeroQ: a novel zero shot quantization framework (2020). arXiv:2001.00281
22. Fang, J., Shafiee, A., Abdel-Aziz, H., Thorsley, D., Georgiadis, G., Hassoun, J.: Near-lossless post-training quantization of deep neural networks via a piecewise linear approximation (2020). arXiv:2002.00104
23. Li, Z., Li, H., Meng, L.: Model compression for deep neural networks: a survey. Computers **12**(3), 60 (2023)
24. Wang, X., Yu, F., Dou, Z.-Y., Darrell, T., Gonzalez, J.E.: SkipNet: learning dynamic routing in convolutional networks. In: European Conference on Computer Vision, pp. 420–436. Springer, Heidelberg (2018)
25. Zhou, H., Alvarez, J.M., Porikli, F.: Less is more: towards compact CNNs. In: Leibe, B., Matas, J., Sebe, N., Welling, M. (eds.) ECCV 2016. LNCS, vol. 9908, pp. 662–677. Springer, Cham (2016). https://doi.org/10.1007/978-3-319-46493-0_40
26. Luo, J.-H., Zhang, H., Zhou, H.-Y., Xie, C.-W., Wu, J., Lin, W.: ThiNet: pruning CNN filters for a thinner net. IEEE Trans. Pattern Anal. Mach. Intell. **41**(10), 2525–2538 (2019)
27. Yang, H., Ping, L., Ziwei, W., Yi, Y.: Pruning filter via geometric median for deep convolutional neural networks acceleration (2018). arxiv:1811.00250
28. Hao, L, Kadav, A., Durdanovic, I., Samet, H., Peter Graf, H.: Pruning filters for efficient ConvNets (2016). arxiv:1608.08710
29. Li, Q., Li, H., Meng, L.: Feature map analysis-based dynamic CNN pruning and the acceleration on FPGAs. Electronics **11**(18), 2887 (2022)
30. Han, S., Pool, J., Tran, J., Dally, W.J.: Learning both weights and connections for efficient neural networks (2015). arXiv:1506.02626
31. Zhuang, B., Pau, D.: A practical framework for designing and deploying tiny deep neural networks on microcontrollers. In: IEEE International Conference on Consumer Electronics, pp. 1–6 (2024)
32. Wang, Z., Luo, T., Li, M., Zhou, J.T., Goh, R., Zhen, L.: Evolutionary multi-objective model compression for deep neural networks. IEEE Comput. Intell. Mag. **16**(3), 10–21 (2021)
33. Zhang, C., Lim, P., Qin, A.K., Tan, K.C.: Multiobjective deep belief networks ensemble for remaining useful life estimation in prognostics. IEEE Trans. Neural Netw. Learn. Syst. **28**(10), 2306–2318 (2017)
34. Liu, J., Gong, M., Miao, Q., Wang, X., Li, H.: Structure learning for deep neural networks based on multiobjective optimization. IEEE Trans. Neural Netw. Learn. Syst. **29**(6), 2450–2463 (2018)
35. Zhou, Y., Yen, G.G., Yi, Z.: A knee-guided evolutionary algorithm for compressing deep neural networks. IEEE Trans. Cybern. **51**, 1626–1638 (2019)

36. Zhou, Y., Hu, B., Yuan, X., Huang, K., Yi, Z., Yen, G.G.: Multiobjective evolutionary generative adversarial network compression for image translation. IEEE Trans. Evol. Comput. **28**(3), 798–809 (2024)
37. Zhang, Q., Li, H.: MOEA/D: a multiobjective evolutionary algorithm based on decomposition. IEEE Trans. Evol. Comput. **11**(6), 712–731 (2007)
38. Gong, M., Liu, J., Li, H., Cai, Q., Su, L.: A multiobjective sparse feature learning model for deep neural networks. IEEE Trans. Neural Netw. Learn. Syst. **26**(12), 3263–3277 (2015)
39. Li, N., Ma, L., Yu, G., Xue, B., Zhang, M., Jin, Y.: Survey on evolutionary deep learning: principles, algorithms, applications, and open issues. ACM Comput. Surv. **56**(2), 1–34 (2023)
40. Zhou, Y., Yen, G.G., Yi, Z.: Evolutionary compression of deep neural networks for biomedical image segmentation. IEEE Trans. Neural Netw. Learn. Syst. **31**(8), 2916–2929 (2020)
41. Phan, H., Liu, Z., Huynh, D., Savvides, M., Cheng, K.-T., Shen, Z.: Binarizing MobileNet via evolution-based searching (2020). arxiv:2005.06305
42. Custode, L.L., Farina, P., Yildiz, E., Kilic, R.B., Yildirim, K.S., Iacca, G.: FastInf: ultra-fast embedded intelligence on the batteryless edge. In: Conference on Embedded Networked Sensor Systems, pp. 239–252. Association for Computing Machinery, New York (2024)
43. Denil, M., Shakibi, B., Dinh, L., Ranzato, M., de Freitas, N.: Predicting parameters in deep learning. In: Burges, C.J., Bottou, L., Welling, M., Ghahramani, Z., Weinberger, K.Q. (eds.) Advances in Neural Information Processing Systems, vol. 26. Curran Associates, Inc. (2013)
44. Tu, M., Berisha, V., Woolf, M., Seo, J.-S., Cao, Y.: Ranking the parameters of deep neural networks using the fisher information. In: International Conference on Acoustics, Speech and Signal Processing, pp. 2647–2651 (2016)
45. Blank, J., Deb, K.: pymoo: multi-objective optimization in Python. IEEE Access **8**, 89497–89509 (2020)

Variable-Size Genetic Network Programming for Portfolio Optimization with Trading Rules

Fabian Köhnke$^{(\boxtimes)}$ (iD) and Christian Borgelt

Department of Artificial Intelligence, University of Salzburg, Salzburg, Austria
fabian.koehnke@stud.plus.ac.at

Abstract. We present an extension of a graph-based evolutionary algorithm called Genetic Network Programming (GNP) by a novel mutation operator, which allows for a variable number of nodes and edges per individual. With this operator, the search space is significantly extended, but without the risk of incurring the bloat problem. The operator is fitness neutral and has no hyper-parameter. Due to higher flexibility, it is now possible for GNP to automatically adapt to the complexity of a given task and to find suitable features, especially for high dimensional data sets. We applied our mutation operator successfully in a GNP for a financial data set where it improved over standard GNP with an optimal network size while maintaining the interpretability of the solution candidates.

Keywords: Variable-Size Genetic Network Programming · Portfolio Optimization · Trading Rules · Time Series Analysis · Evolutionary Algorithms · Genetic Operators

1 Introduction

Genetic Network Programming (GNP) is an extension of Genetic Programming (GP), which has been developed and researched since around 2000 [1] and has been used to solve optimization problems with large search spaces. Due to the network structure of individuals and the possibility of reusing nodes, an implicit memory can develop. As a consequence, GNP targets dynamic environments and has been applied successfully in fields such as robotics and financial market analysis. In the course of time, several extensions have been proposed. Among these are time adapting GNP to distribute given capital onto a set of stocks [2], the addition of statistical models, with GNP generating trading rules and the statistical model distributing the capital [3], or an extension by reinforcement learning to obtain a trading model [4]. In other work, individuals were altered from directed to undirected graphs in which nodes represent particular stocks and edges capture the best relation between them [5]. Furthermore, recent studies combined GNP and neural networks to forecast stock returns [6].

All previous GNP models in the context of financial market analysis had a fixed number of nodes, because this prevents the bloat problem (i.e., a large number of superfluous, often useless nodes). However, a fixed size can also be a disadvantage. If the task is complex, it may be impossible to choose an appropriate size based on prior knowledge or to test sufficiently many different sizes. If the

© The Author(s), under exclusive license to Springer Nature Switzerland AG 2025
P. García-Sánchez et al. (Eds.): EvoApplications 2025, LNCS 15612, pp. 287–304, 2025.
https://doi.org/10.1007/978-3-031-90062-4_18

number of nodes is chosen too small, networks could lack expressivity/capacity. If the number of nodes is chosen too large, the search space may be too large, and individuals may overfit the data or may not find an appropriate solution.

Recently, a new type of GNP has been introduced that allows for a variable number of nodes by using a special crossover operator [7]. Although this method leads to variable-size GNP, the gene pool is still limited due to the initialization of the nodes for the following reasons:

1. The number of outgoing edges of a node always stays the same.
2. The number of different node functions (especially judgments) is limited.
3. The interval boundaries on the edges of numerical judgment nodes are limited.

In order to improve variable-size GNP w.r.t. these points, we developed a new GNP mutation operator that enhances flexibility by allowing for variable numbers of nodes and edges and simultaneously counters the bloat problem.

The remainder of this paper is structured as follows: Sect. 2 reviews the general idea of GNP and the encoding of individuals (candidate solutions), introduces multiple start nodes, and explains standard genetic operators. Section 3 explores the new mutation operator that enables variable-size networks. The deletion and addition of nodes, adjustments to the genetic operators, and the final algorithm are described in detail. Section 4 demonstrates the successful extension of GNP by simulation studies. Section 5 draws conclusions.

2   Genetic Network Programming

Genetic Network Programming works on network individuals like the one shown in Fig. 1, which are directed graphs with nodes and edges (phenotype). Such a network is executed on a multivariate time series of descriptive variables commencing at the start node (node 0, depicted as a rectangle) and following the connections. Apart from the start node there are two kinds of nodes: judgment nodes (J, depicted as hexagons) and processing nodes (P, depicted as circles). A judgment node has a conditional branch decision function and chooses

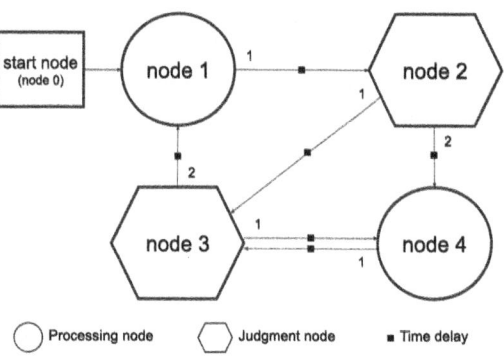

Fig. 1. GNP graph structure.
Source: own illustration according to [8]

between successor nodes. Hence it always has two or more outgoing connections. In this paper, a judgment node refers to a financial indicator that is to be evaluated for trading. For example, node 3 in Fig. 1 could evaluate the price/earnings ratio and determine the next node by a value between 0 and 15 leading to node 1, while price/earnings ratios greater than 15 lead to node 4. Processing nodes, on the other hand, produce outputs, which are buy or sell decisions in this paper.

Processing nodes have only one outgoing connection that simply determines the next node. On execution on a multivariate time series of financial indicators, judgment nodes maintain the current time step (and hence a sequence of judgment nodes can test multiple indicators before making a decision), while processing nodes advance it (that is, go to the next time step). The nodes visited in such an execution are called a *transition path*. The number and kinds of judgments on such a path are adapted and optimized by an evolutionary procedure [8].

2.1   GNP Encoding

The genotype of a GNP individual is shown in Fig. 2. It consists of $n+1$ node genes (encoding n nodes plus the start node 0), with type, function, and time delay, and $n+1$ connection genes, specifying successor nodes. The type t_i of the i-th node gene is either S: start node (only node 0), J: judgment node, or P: processing node. f_i is the node function identifier. For example, if $t_i = J$ and $f_i = 2$, then the node is a judgment node having function 2, where

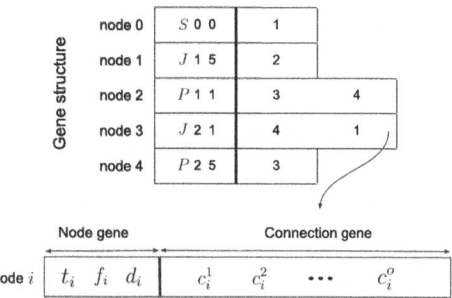

Fig. 2. GNP gene structure.
Source: own illustration according to [8]

function 2 means that the second feature of the data is processed (e.g. "compare the current value of variable X to threshold θ"). Two judgment nodes can have the same judgment function, but then they differ in their node index and can also differ in their connection genes [8].

Time delays d_i (which are *not* time steps) are included as hyper-parameters that model the (total) time spent on judgment, processing, and node transition and are relevant for real-world problems. For example, a moving robot needs some time to assess an obstacle (judgment node), to put the judgment into action (delay on transition), and to avoid the obstacle (processing node). An essential role of time delays is to detect deadlocks, which occur if a network execution never reaches a processing node because of a loop of judgment nodes, all of which examine the same row of the processed multivariate time series data over and over again. Network executions that exceed a user-specified time delay limit before reaching a processing node are stopped and typically removed from the population in the evolutionary process [8] (since the individual is dysfunctional). In an analogous manner, time delays can be used to constrain the number of indicators considered (judgment nodes visited) in a transition path [9].

Connection genes are lists $c_i^1, ..., c_i^{o_i}$, where o_i is the number of outgoing connections (or successor nodes). These genes indicate the destinations of transitions starting at node i. The result of a judgment function determines the superscript of the connection gene to be used. For example, if the result is $r = 3$, c_i^3 is retrieved. Note that the alleles of connection genes start at 1, because the transition path never returns to the start node (node 0). Furthermore, it is not possible to connect a node to itself, that is, $\forall k \in \{1, \ldots, o_i\} : c_i^k \neq i$.

The domain of a judgment function (feature) is split into as many equal-size intervals as the node has outgoing edges. The edge to follow is then chosen as the one associated with the interval into which the current value of the feature falls.

2.2 Multiple Start Nodes

In many cases, especially when processing financial market data, the data covers different assets or at least different stocks. For example, the data may contain the price records of multiple stocks for each day for a particular trading period. This gives rise to at least two different approaches:

1. Train one GNP for all stocks/assets (stocks/assets are "pooled").
2. Train one GNP for each stock/asset (stocks/assets are treated separately).

If stocks are treated separately, each stock needs its own GNP model, and thus a (potentially) vast number of models may be required. Moreover, similarities between the behaviors of different stocks cannot be exploited by the network. To render individuals able to handle several stocks, multiple start nodes were introduced [10]: each stock receives its own start node (different node function identifier f) and thus may follow a different transition path through the same network. As a consequence, one network can handle multiple stocks simultaneously, without treating them all in the same way.

The connections of the start nodes are initialized randomly so that they may point to different nodes in the network (or to the same node, thus allowing equivalent as well as different treatment). In a financial context, this can be motivated as follows: if one stock tracks the development of another, but with a certain time delay, the second stock could have a start node that enters the network transition path of the first stock a bit later. However, multiple start nodes also allow for more complex interdependent transition paths of different stocks. In addition, fitness functions that depend on aggregates over multiple stocks become possible. For example, later in this paper, the total fitness is the sum of all profits, while initially each stock has the same budget. However, because the same network handles all stocks, different budgets for each stock can be stored in its corresponding start node and evolve during the evolutionary process, which enables portfolio optimization.

2.3 Selection Method

Selection serves the purpose to choose the individuals that are used as parents in genetic operators (one for mutation, two for crossover). This selection should favor better individuals, giving them better chances to reproduce. We applied tournament selection, in which a certain number k of individuals is drawn randomly (with equal probability) from the current population, where $k \geq 2$ is the user-defined tournament size. The resulting k-tournament (duel for $k = 2$) is won by the individual with the highest fitness. The winning individual receives a descendant in the next generation. After the tournament, all participants are returned to the current population (including the winner). Since each tournament selects one individual, N tournaments have to be carried out to fill the next generation, where N is the size of the population. The selection pressure can be controlled by the tournament size parameter k [11].

2.4 Genetic Operators

Two genetic operators are applied in standard GNP. Following [12], we used standard mutation and so-called uniform crossover for our simulations.

Mutation randomly modifies a single individual.

1. Select a random individual (A) from the population (cf. Sect. 2.3).
2. Each connection c_i^r of each node g_i in A is selected with probability p_m.
3. If selected, it is replaced by a number $c_i^r \in \{1, \ldots, n\} - \{i\}$ (equal prob.).
4. The new individual may be subjected to crossover.

Crossover recombines genotypes of two parents to create two offspring individuals.

1. Select two individuals (A and B) from the population (cf. Sect. 2.3).
2. Each node g_i, $i \in \{1, \ldots, n\}$, is selected with probability p_c (but *never* g_0).
3. Exchange genes of the corresponding nodes (same node index), i.e., $g_i^A \leftrightarrow g_i^B$.
4. The generated individuals both become part of the next generation.

The crossover described above is valid in standard GNP with fixed-size networks in which there is a fixed mapping $i \mapsto \{S, J, P\}$ and therefore $\forall i : t_i^A = t_i^B$. Thus, the number of different node types (t_i) in each network is also fixed and is not changed by the crossover. Note that once we allow for variable-size individuals (cf. Sect. 3) these constraints no longer hold, which requires special treatment, like a repair mechanism that fixes edges that lost their destination.

3   Variable-Size Genetic Network Programming

A trade-off between simple and complex models for fitting the data is typical in data science. Complex models are more flexible and better able to fit the data, but also run the risk of overfitting, that is, fitting noise rather than only regularities. Occam's razor recommends to choose the simplest model that "explains" the data. Hence we want to find a model that fits the data well, but is as simple as possible [13]. With our novel mutation operator we try to attain this goal.

3.1   Basic Concept

The complexity of GNP can be measured by the size of the individuals (number of nodes). In order to find the optimal number (and thus complexity), Variable Size Genetic Network Programming (GNPvs) was proposed, which changes the size of the individuals and tries to find an optimal size during evolution. Proposed approaches differ in whether they allow only changing the number of nodes across generations, with all individuals of a generation having the same size [14], or whether individuals of the same generation can differ in size [15]. In the former, the decision how many (and which) nodes to add or to delete is based on a measure of how much a node contributes to performance. In the latter, different sizes are obtained by a modified crossover operator, which may exchange a certain number of nodes of one individuals with a different number from the other. However, the number of outgoing edges is still limited to the number possible at initialization. Neither of these approaches achieved variable sizes by a mutation operator, which is our approach.

Our novel mutation operator allows GNP individuals to grow and shrink individually—there is no fixed number of nodes, neither overall nor per generation. Compared to existing studies of GNP and GNPvs, this appears to be beneficial, because earlier approaches are constrained by

1. a limited number of outgoing edges (due to just exchanging existing nodes),
2. a limited number of nodes (g_i) and their functions (f_i).

Since our new operator is a mutation operator, it operates on only one solution candidate. In particular, it randomly chooses whether a node is either added or deleted in each generation, but not both. If nodes are added to or deleted from a GNP individual, several issues may arise. These are first listed here and then covered in detail in the following sections to finally end up with the new algorithm:

- Deletion of nodes can lead to huge fitness differences (epistasis, see below).
- If too many nodes are deleted, an individual may lack capacity to model the problem; in a worst case, all nodes are deleted, yielding an invalid individual.
- Edges pointing to a deleted node cause an invalid, non-executable graph.
- Crossover can cause problems if applied to individuals of different size.
- Too many inserted nodes can lead to very large networks and overfitting; furthermore, potentially limitless network growth can incur the bloat problem.

3.2   Deleting Nodes

Unrestricted deletion of nodes can lead to huge fitness differences, because it may alter the transition path significantly, thus affecting the trading policy and hence the fitness. This effect is known as epistasis: the expression of a gene is dependent on the expression of one or more other genes. To avoid epistasis, our mutation operator deletes a node only if it is *not used in the transition path* of the network (for the training data). Since unused nodes do not influence trading, our mutation operator is *fitness neutral* and hence does not "feel" any selection pressure. Deleting only unused nodes also helps with the problem of "dangling" edges, that is, edges that have no destination because they were deleted. These edges are randomly changed to an existing node in order to keep the number of outgoing edges the same. However, if only nodes are deleted that are never visited, these edges are not used in the transition path and thus can simply be altered to point to other nodes without affecting fitness.

3.3 Crossover Adjustment

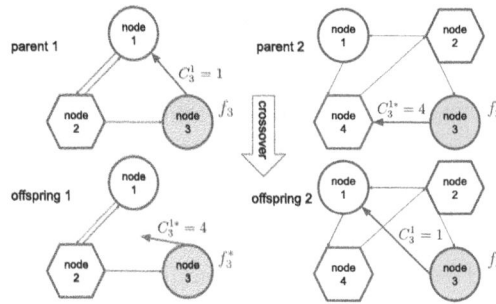

An individual's fitness can be changed only by the genetic operators discussed in Sect. 2.4. Standard mutation (which we also keep) changes successor nodes with probability p_m per edge, and this still works flawlessly if the number of nodes changes. However, a crossover of individuals of different size can cause two problems that are illustrated in Fig. 3, which shows the crossover of parents with 3 and 4 nodes, respectively.

Fig. 3. Faulty crossover example in which, after the exchange of node 3, an edge in offspring 1 no longer points to any node, leading to an invalid graph (due to an invalid, "dangling" edge). Note that node 4, only in parent 2, cannot be exchanged.

In the first place, not just any node of the parents may be selected for exchange. Even though exchanged nodes no longer need to be of equal type (since $\forall i : t_i^A = t_i^B$ no longer holds), if index $i = 4$ is chosen, this node only exists in the second parent, but not in the first, and hence cannot be exchanged with a node of the same index. To avoid this problem, we choose $i \in \{1, \ldots, \min\{n_A, n_B\}\}$, where n_A and n_B are the numbers of nodes of the parents A and B, for an exchange $g_i^A \leftrightarrow g_i^B$. Alternatively, one may choose an index $i \in \{1, \ldots, n_A\}$ and an index $j \in \{1, \ldots, n_B\}$ and exchange $g_i^A \leftrightarrow g_j^B$. However, this could lead to a loop (an edge connecting a node to itself), which would require an additional repair mechanism (an exploration of this alternative is left for future research).

Secondly, a crossover of individuals of different size can lead to "dangling" edges. This is shown in Fig. 3, where after the exchange of node 3 the outgoing edge of node 3 is "dangling" in offspring 1. This is fixed by simply choosing a new (random) destination from the nodes existing in offspring 1.

3.4 Adding Nodes

We now consider adding nodes to an individual. Especially by adding judgment nodes additional features may be used, which is particularly beneficial for high-dimensional data sets. In standard GNP, in order to enable exploiting information from all features, (at least) as many judgment nodes have to be

```
Algorithm: Mutation Operation Add/Delete Nodes

function mutation_add_delete_nodes (pop):
    for ind ∈ pop do                          (* traverse individuals of population *)
        mop ← random element of {add, delete} (* choose mutation operation *)
        for node ∈ ind.genes do               (* traverse nodes of individual *)
            if   mop = add                     (* if a node is to be added *)
            and ind.used ≥ len(ind.genes) then (* and there are no unused nodes *)
                dst ← new random node object   (* add a new node (with connections) *)
                ind.genes ← ind.genes ∪ {dst}  (* to the individual *)
            elif mop = delete                  (* if a node is to be deleted *)
            and len(ind.genes) − ind.used > 1  (* and more than one unused node *)
            and node ∉ used then               (* and the current node is not used *)
                ind.genes ← ind.genes − {node} (* delete the current node *)
                for nred ∈ ind.genes do        (* for all nodes in reduced individual *)
                    for con ∈ nred.connections do (* traverse the node's connections *)
                        if con points to node then (* if connection goes to deleted node *)
                            dst ← random element of ind.genes − {nred}
                            con ← index of dst (* choose a new destination randomly *)
                        else                   (* if connection goes to kept node *)
                            adapt connection destination index if necessary
```

Fig. 4. Mutation Operation Add/Delete Nodes

initialized as there are features. However, by allowing to add nodes in the evolution process, the initial number can be kept smaller, without precluding the use of any feature later.

The main issue when adding nodes is the bloat problem, which has (at least) two aspects: In the first place, unused nodes may occur, that is, nodes that are not visited on any transition path (on the training data) and are, therefore, fitness neutral. Such nodes are analogous to introns.[1] As a consequence, individuals bloated with unused nodes may persist during the evolution process.[2] The second aspect of bloating is redundancy: analogous transition paths may be needed in different situations. These may be coded by separate paths (i.e. with distinct nodes), even though it may be possible, in principle, to merge them, thus simplifying the individual. Another issue with adding nodes is that large individuals may overfit the training data, simply because they have the capacity to fit noise, rather than only regularities. In all of these respects, a (sufficiently small) fixed number of nodes in standard GNP is an advantage, because it counteracts bloat and overfitting. However, this comes at the price of a limited learning ability for more complex problems.

Note that newly added nodes are necessarily unused initially. Even though the outgoing connections of such new nodes are initialized to point to existing nodes in the individual, no existing nodes have outgoing connections pointing to *them*. Hence, adding nodes is not sufficient to make them part of any transition path (fitness neutral). However, a new node can evolve into a successor of another node with the standard genetic operators (cf. Sect. 2.4) and thus become part of a transition path. In addition, an added node can have exactly one more edge than the existing nodes, as the gene structure is increased by 1 ($n^* = n + 1$), which makes it possible to increase the number of outgoing edges of nodes.

To prevent the number of nodes from increasing excessively, nodes can be added by our novel mutation operator only if all existing nodes in the network are already used. Otherwise existing unused nodes should rather be put to use first. As a consequence, at most one node can be added to an individual per generation. On the other hand, multiple nodes can be deleted per generation, provided they are all unused. Thus, adding nodes is (potentially) less invasive than deleting nodes. This asymmetry is motivated by the intention that the algorithm should find the smallest suitable networks, grow them slowly, and only if the complexity of the optimization problem requires it. Nevertheless, both parts of the operator are constrained by the use of the current nodes in the transition path. To prevent a node from being inserted and then immediately (in the next generation) deleted again, only nodes in excess of one unused node are deleted. This ensures that one unused node in the network is protected from deletion by the mutation. The detailed algorithm is shown in Fig. 4. Note that it has the additional advantage of not having any hyper-parameter that needs tuning.

---

[1] In biology, introns are parts of the DNA sequence that do not carry any information in the sense that they do not code for any phenotypical characteristic [11].

[2] In biology, introns are kept in check, at least to some degree, by the metabolic costs of a larger genome. Especially bacteria often have a "streamlined" genome (few introns) in order to minimize the metabolic costs incurred in reproduction.

4   Simulation Study

We conducted a simulation study of several GNP with our novel mutation operator and compares the results to the standard GNP as well as a mean-variance analysis and a buy-and-hold strategy. Mean-variance analysis is a fundamental concept in modern portfolio theory that constructs portfolios by maximizing expected return for a given level of risk (measured by variance) and by selecting and weighing stocks differently [16]. In contrast, the buy-and-hold strategy involves buying a stock and just holding it, where all stocks in a portfolio are initially equal-weighted (unequal weights may only result from unequal performance).

We used a financial market data set that consists of the daily prices of 45 stocks between 02.01.2018 and 01.10.2022, which is uploaded as supplementary material. All stocks are part of the main index of the Eurozone, the so-called EURO STOXX 50. Not all stocks in the EURO STOXX 50 have a price history for the simulation period, which is why the data set only contains 45 of the 50 stocks. For the training of the GNP, the data were separated according to Sect. 2.2, where a starting node was initialized for each stock. Thus, each stock has its own transition path. The following data was retrieved from Yahoo Finance[3]: the nominal stock price, outstanding shares of the company, the traded daily stocks, the earnings (after-tax net income), the revenue, and the book value (difference between total assets and total liabilities) of the stock. The stock return and various indicators of financial market analysis were calculated from the stock price. All features are listed in the supplementary material in Table 3 where the features were calculated for $h \in \{5, 10, 15, 20, 50\}$, and where h represents the number of historical days used in the calculation from the reference day. In total, the data set consists of 28 features.

4.1   Fitness/Objective Function

We consider learning a trading strategy from given data with GNP, including that different stocks are weighted for portfolio optimization. A different weighting of stocks is possible by initializing several start nodes (see also Sect. 2.2). For this purpose, the fitness function is coded so that the budgets of each stock (stored in the start node) are allocated before the transition path of the multivariate time series is executed and optimized with each generation. During the execution of the transition path, the weights remain the same, and only trading decisions are executed (described below). The fitness function for training the GNP is defined as: $\mathrm{Fitness}(k) = \sum_{s \in S} \mathrm{Profit}(s, k)$, where s is the current stock and k the current generation. Profit is the difference between all sales and purchases during the trading period (transition path) with T time steps:

$$\mathrm{Profit}(s, k) = \sum_{t=1}^{T} \mathrm{Sell}(s, k, t) - \sum_{t=1}^{T} \mathrm{Buy}(s, k, t) \tag{1}$$

---

[3] https://finance.yahoo.com/.

Note that the final time step is treated as a sell to capture the total profit over the entire period. The profitability of each stock can be calculated based on the profit:

$$\text{Profitability}(s, k) = \frac{\text{Profit}(s, k)}{\text{Budget}(s, k)}, \tag{2}$$

where $\text{Budget}(s, k)$ is the initial budget of stock s in generation k. The initial budget in the first generation is calculated as: $\text{Budget}(s, 1) = \frac{1}{|S|} \cdot$ Total Budget, where "Total Budget" is the total capital of the portfolio. After that, the initial budget of each stock is weighted differently over the generations and each weight is determined by profitability:

$$\text{Budget}(s, k + 1) = \frac{\exp(\text{Profitability}(s, k))}{\sum_{s \in S} \exp(\text{Profitability}(s', k))} \cdot \text{Total Budget} \tag{3}$$

This coding of the stocks' fitness and weight is based on the approach presented in [2]. In our paper, each stock is weighted differently in the portfolio by Eq. 3 with the restriction that no stock must be weighted higher than 10%.

The actual target variable is the return computed from the stock prices. Based on the returns of the stocks, the profit (fitness) of each stock can be calculated. The processing node function is encoded as: $0 = buy$ and $1 = sell$ and a decision to hold a stock is not explicitly defined. However, if a stock was already bought and a buy signal occurs again, the stock is simply held (as no increase is possible). Stocks are bought and sold with the mentioned budget, and decisions always relate to the entire budget of a stock. During the execution of the transition path, the budget can only change because of the stock price performance, as the weights are optimized from generation to generation (after the execution of the transition path). In summary, the following rules are applied:

- processing function = 0 and stock not in portfolio → buy the stock
- processing function = 0 and stock already in portfolio → hold the stock
- processing function = 1 and stock not in portfolio → no action
- processing function = 1 and stock in portfolio → sell the stock

4.2 Results

We compared the results obtained by (classical) GNP and our new GNPvs on the stock data described above. For meaningful results, 24 consecutive test periods were cross-validated in each different setting. The validation procedure was the same as presented in paper [10]: the train and test periods are consecutive samples that were shifted by the size of the test period. The training data was used to select stocks to construct a portfolio and find trading rules for the stocks. All reported results are obtained from the test periods.

The data was divided into 80% training data and 20% test data. The 80% training data corresponds to 177 days, and the 20% test data corresponds to 44 trading days for each of the 24 validations. After this period, a new GNP/GNPvs was trained and tested, which resulted in a total test period of $24 \cdot 44 = 1056$

Table 1. Simulation Study of 7 different GNP/GNPvs with 24 validations each and the shown combination of judgment (J) and processing (P) nodes (column). The first five rows show the results with our novel mutation operator, the next five rows the results without. The bottom three rows list the results of standard approaches.

	J:1 P:2	J:5 P:5	J:10 P:10	J:15 P:15	J:25 P:25	J:50 P:50	J:100 P:100
Profit*	47.99	32.65	97.85	78.61	55.25	46.05	50.11
ProfitBH*	40.94	34.43	50.15	51.94	29.18	29.81	36.90
Mean Nodes*	22.08	27.06	32.48	31.88	24.50	58.28	70.79
Mean Connections*	193.00	350.12	578.33	567.73	285.28	2055.64	3946.27
Runtime (sec.)*	5799.26	5553.35	5963.53	6108.46	3537.80	12644.35	26209.81
Profit**	41.99	28.32	45.15	69.65	30.65	49.25	35.39
ProfitBH**	41.16	31.92	32.95	44.49	33.68	27.34	30.45
Mean Nodes**	3.00	10.00	20.00	30.00	50.00	100.00	200.00
Mean Connections**	4.00	20.43	101.32	242.19	623.68	2530.49	10144.33
Runtime (sec.)**	7924.60	6916.80	5078.94	5506.99	4989.51	15965.50	44543.03
ProfitBH all Stocks	34.83	34.83	34.83	34.83	34.83	34.83	34.83
Quadratic	35.96	35.96	35.96	35.96	35.96	35.96	35.96
Sharpe	21.65	21.65	21.65	21.65	21.65	21.65	21.65

* results with novel mutation (GNPvs) ** results without novel mutation (GNP)

trading days. Each network (portfolio) constructed by the GNP/GNPvs contains 25 stocks. A GNP/GNPvs was, therefore, initialized with 25 start nodes, while the corresponding stocks were selected at random (random distinct start node function identifier). This allowed the GNP/GNPvs to find an optimal subset of stocks from the 45 available stocks as the individuals with the more profitable stocks are favored in the evolutionary process. Each portfolio was initialized with a total budget of 100, and each stock was allocated an equally weighted budget, thus $\text{Budget}(s, 1) = \frac{100}{25} = 4$ per stock. To keep the results as realistic as possible, a fee of 0.05% was charged for each trade in a stock (buy as well as sell). The evolution of the GNP/GNPvs was simulated with the parameter setups shown in Table 4, using different numbers of initial judgment and processing nodes. This simulation study was repeated twice. Once with and once without the new mutation operator, to be able to discern its effect. Table 1 shows the results of the simulation study together with some characteristic quantities. The column labels indicate how many judgment and processing nodes have been initialized. Accordingly, one judgment node and two processing nodes were initialized in column "$J : 1 P : 2$" of Table 1.

It can be seen that the profit of (classical) GNP could be beaten in six of the seven simulations by our GNPvs with the new mutation operator (compare Profit* and Profit** in Table 1). Only with an initialization of 50 judgment nodes and 50 processing nodes was the profit slightly lower after completing all 24 cross-validations. Also, the GNP with *and* without the new mutation operator outperformed the buy-and-hold strategy in the majority of simulations (compare Profit*, Profit** with ProfitBH all Stocks). The better performance using GNP compared to the buy-and-hold strategy has been shown in previous papers, for example, in [2, 4, 17] and could be confirmed in our simulation study.

Furthermore, GNPvs also outperformed mean-variance analysis. Six of the seven simulations beat the performance of a portfolio optimized according to the quadratic utility model (compare Profit* and Quadratic). The quadratic utility model includes both the expected return and the risk (volatility) in the portfolio optimization and is based on Markowitz portfolio selection [16]. In our study, the risk tolerance was assumed to be maximally large, as the objective function of the GNP/GNPvs is profit-based only. In addition, an optimal portfolio based on the Sharpe ratio was constructed as a comparison [18]. In this case, our GNPvs performed better in all simulations. Detailed settings of both benchmark models can be found in Table 6 in the appendix.

ProfitBH shows the performances of the stocks selected as a subset by GNP/GNPvs without using trading decisions and stock weights (like the buy-and-hold strategy). Even if the operator does not directly affect the selection of stocks, the GNP was apparently more capable of identifying appropriate stocks through the variability of the judgment nodes J and processing nodes P and their node function identifier. It can be seen that GNPvs found better stocks in most of the cases using the new mutation operator, but furthermore, the additional profit is even greater than the out-performance of the picked stocks (difference between Profit* and ProfitBH*). In total, the GNPvs also performed better in six out of seven cases within trading decisions and stock-weighting and were more capable of optimizing the portfolios compared to the fixed-size GNP.

By using the new muta-tion operator, it is now pos-sible for the GNPvs individ-uals to grow and shrink in their number of nodes. This is shown in the rows "Mean Nodes", which compare the average number of nodes of all individuals at the end of each training (evolution) of the simulation study in Table 1. If only a few nodes were ini-tialized, it can be seen that the individuals have grown on average, whereas a high num-ber of initialized judgment and processing nodes led to a

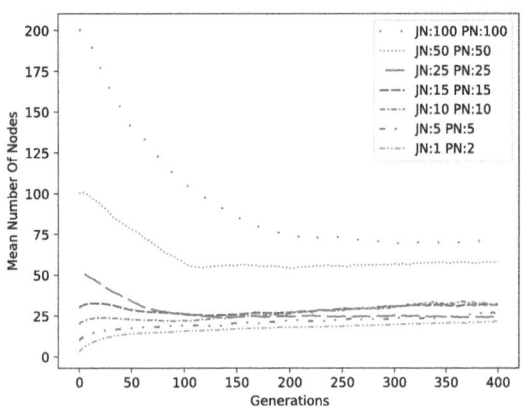

Fig. 5. Average number of nodes per individual per generation. Simulations converge to the average number of nodes in the population.

shrinking of the individuals. For a detailed illustration, Fig. 5 shows the average number of nodes of all individuals per generation. It can be seen that the mean number of nodes converges to a certain level in all simulations. If a number of less than or equal to 100 nodes per individual was initialized for the simulation, a similar level of convergence was always obtained. The mean number always reaches a level between 22.08 and 32.48 in these simulations. The two simula-tions with 100 and 200 nodes show a rapid decrease in the average number of nodes, but also a higher level of convergence. Apparently, large parts of the net-

Table 2. Mean number of used judgment nodes of the best individual and for all 24 training runs for each simulation.

	J:1 P:2	J:5 P:5	J:10 P:10	J:15 P:15	J:25 P:25	J:50 P:50	J:100 P:100
With Mutation	4.54	4.38	5.42	7.46	7.08	17.21	34.21
Without Mutation	1.00	5.00	9.17	12.75	20.62	42.21	82.17
Difference	3.54	−0.62	−3.75	−5.29	−13.54	−25.00	−47.96

work were used, which prevented the new operator from deleting nodes, and a convergence similar to the level between 22.08 and 32.48 was not reached.

Due to the variable number of nodes, another objective is also achieved: a variable number of connections. Especially when the GNPvs grow, it should be possible for the network to increase its number of edges to divide the features more finely (compare Sect. 3.4). This can be seen clearly in the rows "Mean Connections" of Table 1, which show the average number of connections of all networks. With an initialization of one judgment node and two processing nodes, for example, the judgment node in a fixed-sized GNP can have at most two outgoing edges. However, the network's growth leads to a mean number of connections of 193, allowing GNPvs to split value ranges much more accurately.

Because of excessive computing effort, trying various combinations of initial judgment and processing nodes is often not feasible. The simulation study emphasizes the advantage of initializing the GNPvs with a small number of nodes: *All simulations* with a total number of nodes up to 50 reached a similar level in the number of nodes in the last generation and achieved better results than GNP with fixed size individuals. In addition, these simulations were able to achieve a significant reduction in runtime compared to simulations with more than 50 initial nodes (see rows "Runtime (sec.)" in Table 1)

This is underscored the average number of utilized judgment nodes in the transition path of the best individual across all training phases, i.e., those available for validation, which can be seen in Table 2. As the networks grow, additional judgment nodes are integrated into the network, even with the minimal initialization of one judgment node (3.54 in row "Difference" in Table 2). For initialization with more than one node, fewer judgment nodes are used for the transition path in every case compared to the fixed-size GNP (compare Table 2), consequently resulting in the mentioned feature selection and in better results in all cases except for the run with 100 initial nodes. Consequently, it becomes evident that a suitable selection of features was found and that finding simple networks that "explain" the data was successful.

A core advantage of evolving trading rules with GNPvs is that the results are interpretable: In principle, a human expert can inspect a result network and check it for plausibility. As an example, an evolved, fully interpretable network is shown in Fig. 6 in the appendix. In contrast to this, neural network approaches, for example, which might produce good or even better results, are often "black boxes" that arrive at their results in a somewhat incomprehensible fashion.

4.3 Limitations and Future Research

24 cross-validations were performed for each of the simulations with different judgment and processing nodes. However, with extended computational capacity, the results could be further increased in significance by simulating the cross-validation multiple times for each setting. Furthermore, even though the feature selection worked well in the simulation study presented, it is not clear to what degree the algorithm reaches its limits with even higher-dimensional data. This could be tested in future research. Furthermore, it is not immediately apparent that the computation time of the algorithm could be improved overall. Clearly, the new mutation operator also adds computational costs. However, our recommendation to initialize small networks can prevent excessive computation time compared to initializations with a large number of nodes without observing the algorithm performing worse in its objective function. The computation time could be stabilized to a similar level regardless of the number of initialized nodes but is not better in every case (see rows "Runtime (sec.)") in Table 1.

The following topics could be of interest for future research:

- The newly developed mutation operator can add at most one node per generation. Therefore, growth is limited by the number of generations. The possibility of adding multiple nodes in each generation could lead to more suitable networks for complex problems.
- Also, the novel operator could add and delete start nodes to train GNPvs with different numbers of used stocks. Thus, portfolios with varying amounts of stocks could be created from the data. Alternatively, a mutation operator could change the node function identifier f of the start nodes, improving the discovery of optimal subsets of the stocks.
- Even if our GNPvs can now split the numerical value ranges into any granularity with the new mutation operator, the intervals remain of equal length. Future research could explore an operator that changes the boundaries of the intervals in such a way that different lengths are made possible.
- In order not to adapt too much to a certain training history, it would be interesting to investigate under "more difficult" conditions by shifting or changing the training intervals. This could enable the GNP to learn a wide variety of patterns and improve performance.

5 Conclusion

Our simulation studies demonstrate that our new mutation operator improves the flexibility and adaptability of a GNP. By growing and shrinking, GNP individuals can now precisely fit the complexity of given tasks. Despite the possibility of generating more complex network structures, the bloat problem could be counteracted by imposing a few, simple constraints on the operator. Our new mutation operator works without additional parameters, ensuring results remain fully interpretable. The results of the simulation studies have shown that the GNP with the new mutation operator outperforms the standard GNP on the financial data set and is better in portfolio optimization.

A   Appendix

A.1   Feature Calculation

Table 3. Outline of the Features

Feature	Description	Calculation
id	Stock id to separate the data (multiple start nodes)	–
MA_DIFFh	Percentage difference of the share price to the h days moving average	$\frac{c - ma_h}{ma_h}$
SDh	Standard deviation calculated from the last h days	$\sqrt{\frac{1}{N}\sum_i^N (x_i - \mu)^2}$
MOM_RELh	Relative momentum indicator calculated from the last h days	$\frac{c - c_h}{c_h}$
RSIh	Relative strength indicator calculated from the last h days	$100 - \frac{100}{1+RS}$ where, $RS = \frac{acp}{acn}$
VOL_RELh	RSI calculated with the tv from the last h days	RSIh(tv)
PE	historical Price-earnings ratio	$\frac{os \cdot c}{earnings}$
PS	Historical price-sales ratio	$\frac{os \cdot c}{sales}$
PB	Historical price-book ratio	$\frac{c}{\frac{ta - ia - tl}{os}}$

c = closing price
ma = moving average
acp = average c with positive changes
acn = average c with negative changes
os = number of outstanding shares
ta = total assets
ia = intangible assets
tl = total liabilities
tv = traded number of stocks (volume)

A.2   Simulation Parameter

Table 4. Parameter for Simulation

Fitness: Profit (cf. Sect. 4.1)
Generations: 400
Population size: 300
Judgment Nodes: see Table 1
Processing Nodes: see Table 1
Mutation Probability: 0.10
Crossover Probability: 0.10
Tournament Size: 2
Elitism: 2
Max. Time Delay on Nodes: 10

A.3   Simulation Resources

Table 5. Simulation Resources

Cloud Provider: Hetzner Online GmbH
Name: CCX33
vCPUs: 8 AMD (dedicated)
Model name: AMD EPYC-Milan Processor
Ram: 32 GB

A.4   Benchmark Settings

Table 6. Benchmark Settings

Library: PyPortfolioOpt
Used Class: EfficientFrontier
Quadratic: $max\_quadratic\_utility(risk\_aversion = 0.01)$
Sharp: $max\_sharpe()$

A.5   Example Network

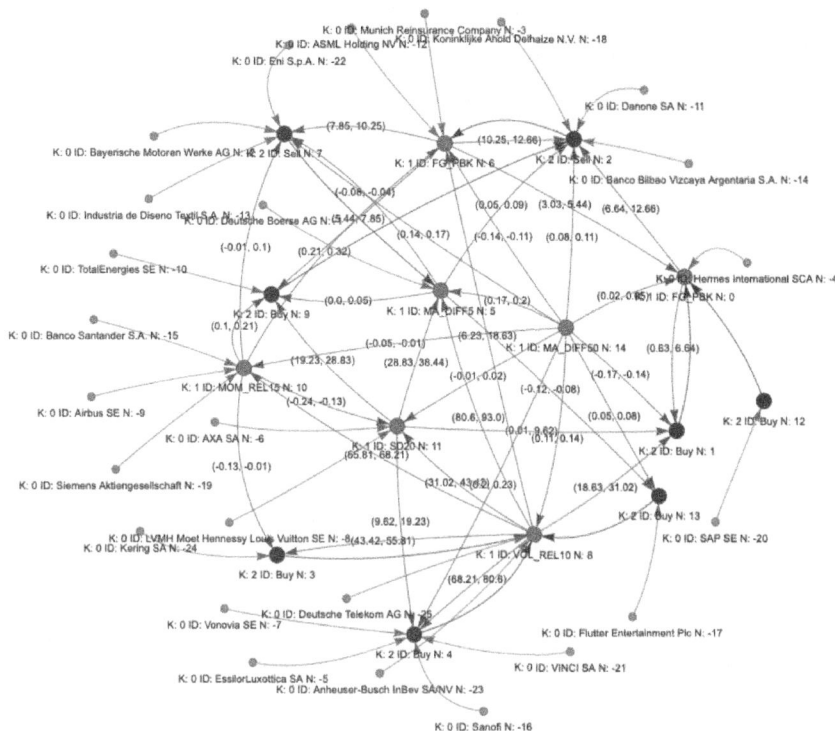

Fig. 6. Network Example of the Simulation with 1 initial judgment node and 2 initial processing nodes. It can be seen that the network is grown to a larger network with 8 processing nodes and 7 judgment nodes. The network can be interpreted in its decisions and is not a black box model. Yellow nodes are S : start nodes, blue nodes are P : processing nodes, and green nodes are J : judgment nodes. (Color figure online)

References

1. Katagiri, H., Hirasama, K., Hu, J.: Genetic network programming - application to intelligent agents. In: SMC 2000 Conference Proceedings. 2000 IEEE International Conference on Systems, Man and Cybernetics. 'cybernetics Evolving to Systems, Humans, Organizations, and Their Complex Interactions' (Cat. no.0), vol. 5, pp. 3829–3834 (2000). https://doi.org/10.1109/ICSMC.2000.886607
2. Chen, Y., Mabu, S., Hirasawa, K.: A model of portfolio optimization using time adapting genetic network programming. Comput. Oper. Res. **37**(10), 1697–1707 (2010). issn: 0305-0548. https://doi.org/10.1016/j.cor.2009.12.003, https://www.sciencedirect.com/science/article/pii/S0305054809003281
3. Chen, Y., Wang, X.: A hybrid stock trading system using genetic network programming and mean conditional value-at-risk. Eur. J. Oper. Res. **240**(3), 861–871 (2015). issn: 0377-2217. https://doi.org/10.1016/j.ejor.2014.07.034, https://www.sciencedirect.com/science/article/pii/S0377221714006006

4. Chen, Y., Mabu, S., Shimada, K., Hirasawa, K.: A genetic network programming with learning approach for enhanced stock trading model. Expert Sys. Appl. **36**(10), 537–546 (2009)

5. Chen, Y., Mabu, S., Hirasawa, K.: Genetic relation algorithm with guided mutation for the large-scale portfolio optimization. Expert Syst. Appl. **38**(4), 3353–3363 (2011). issn: 0957-4174. https://doi.org/10.1016/j.eswa.2010.08.120, https://www.sciencedirect.com/science/article/pii/S0957417410009280

6. Ramezanian, R., Peymanfar, A., Ebrahimi, S.B.: An integrated framework of genetic network programming and multi-layer perceptron neural network for prediction of daily stock return: an application in tehran stock exchange market. Appl. Soft Comput. **82**, 105:551 (2019). issn: 1568-4946. https://doi.org/10.1016/j.asoc.2019.105551, https://www.sciencedirect.com/science/article/pii/S156849461930331X

7. Bing, L.: Study on genetic network programming with variable size structure and genotype/phenotype mapping mechanism. Doctoral thesis (2013)

8. Mabu, S., Hirasawa, K., Hu, J.: A graph-based evolutionary algorithm: genetic network programming (GNP) and its extension using reinforcement learning. Evol. Comput. **15**(3), 369–398 (2007). https://doi.org/10.1162/evco.2007.15.3.369

9. Chen, Y., Hirasawa, K.: Generating trading rules on the stock markets with robust genetic network programming using variance of fitness values. Proc. SICE Annu. Conf. **2010**, 3095–3102 (2010)

10. Chen, Y., Mabu, S., Ohkawa, E., Hirasawa, K.: Constructing portfolio investment strategy based on time adapting genetic network programming. In: IEEE Congress on Evolutionary Computation, pp. 2379–2386 (2009)

11. Kruse, R., Borgelt, C., Braune, C., Klawonn, F., Moewes, C., Steinbrecher, M.: Computational Intelligence: Eine methodische Einführung in künstliche neuronale Netze, evolutionäre Algorithmen, Fuzzy-Systeme und Bayes-Netze (Computational Intelligence), 2., überarbeitete und erweiterte Auflage. Wiesbaden: Springer Vieweg (2015). isbn: 9783658109035

12. Katagiri, H., Hirasawa, K., Hu, J., Murata, J.: Comparing some graph crossover in genetic network programming. In: Proceedings of the 41st SICE Annual Conference. SICE 2002, vol. 2, pp. 1263–1268. IEEE (2002)

13. Berthold, M.R., Borgelt, C., Höppner, F., Klawonn, F.: Guide to Intelligent Data Analysis: How to Intelligently Make Sense of Real Data. Springer (2010)

14. Katagiri, H., Hirasawa, K., Hu, J., Murata, J.: Variable size genetic network programming. IEEJ Trans. Electron. Inf. Syst. **123**(1), 57–66 (2003)

15. Li, B., Li, X., Mabu, S., Hirasawa, K.: Variable size genetic network programming with binomial distribution. In: IEEE Congress of Evolutionary Computation (CEC), pp. 973–980. IEEE (2011)

16. Markowitz, H.: Portfolio selection. J. Finance **7**(1), 77–91 (1952). issn: 00221082, 15406261. http://www.jstor.org/stable/2975974 (visited on 10/21/2024)

17. Parque, V., Mabu, S., Hirasawa, K.: Enhancing global portfolio optimization using genetic network programming. Proc. SICE Annu. Conf. **2010**, 3078–3083 (2010)

18. Cornuejols, G., Peña, J., Tütüncü, R.: Optimization Methods in Finance. Cambridge University Press (2018)

Evolving Dynamic Fault Mitigation Strategies in a Robot Swarm for Collective Transport

Suet Lee[(✉)][iD] and Sabine Hauert[iD]

University of Bristo, Bristol Robotics Laboratory, Bristol, UK
{suet.lee,sabine.hauert}@bristol.ac.uk

Abstract. As robot swarms move to real-world deployment, safety will be a key factor in improving adoption and trust [45]. Robot swarms are composed of many robots: during real-world operation each individual may be susceptible to failure resulting in potentially degraded performance of the swarm overall. A necessary component for safety is then the ability to detect and mitigate faults in the swarm. In this paper, we present a novel approach to learning dynamic fault mitigation via neuroevolution, where mitigation actions are implemented by both faulty and non-faulty robots in a collective transport scenario. In particular, there is no explicit fault detection step and the evolved "mitigation module" maps between a set of locally observed metrics as input and mitigation actions as output. Our approach is able to learn effective mitigation for six types of fault independently. We show that by allowing robots of any state to freely apply actions, "loosely-coordinated" mitigation emerges improving on the baseline where no mitigation is applied.

Keywords: Neuroevolution · Fault Mitigation · Swarm Robotics

1 Introduction

Swarms have the potential for positive impact in real-world applications: they are distributed systems where individuals act according to local knowledge [37]. It is the local interactions between robots - and between robots and the environment - which give rise to emergent swarm behaviour. Examples of real-world swarm applications include search and rescue [5,10,23], construction [8,34,42], and space exploration [22,24,35,43]. However, many such applications are yet to be deployed in the real-world having been validated primarily in laboratory settings and in simulation [38]. In order to make the leap to reality, it will first be necessary to guarantee safe swarm operation through the development of methods for fault detection and mitigation.

Swarms have traditionally been considered to be robust through redundancy of a large number of individuals [37]. This assumption has now been demonstrated to be false depending on the scenario and type of fault [6,26]. Given the

Supplementary Information The online version contains supplementary material available at https://doi.org/10.1007/978-3-031-90062-4_19.

© The Author(s), under exclusive license to Springer Nature Switzerland AG 2025
P. García-Sánchez et al. (Eds.): EvoApplications 2025, LNCS 15612, pp. 305–322, 2025.
https://doi.org/10.1007/978-3-031-90062-4_19

occurrence of a fault, the most effective action to take is not obvious due to the complexity of swarm dynamics. In some situations it may be better to tolerate a fault and "do nothing" if a robot is still able to contribute positively to the swarm - e.g. a partial wheel fault causing a robot to move slower yet still being operational [26]. A fault and its criticality therefore depends on *context*. Context is key in a robot's decision-making process for applying any mitigation.

In this paper, we use neuroevolution, a gradient-free approach, to learn the most effective mitigation actions for individual robots to take based on their current context. Gradient-free approaches have been shown to be competitive with gradient-based reinforcement learning approaches [39]. The context of an individual robot can be encoded in local metrics observable through on-board sensors and components, such as the number of boxes delivered or objects perceived in proximity. Given a set of low-level mitigation actions, which can be implemented asynchronously and in a distributed manner, the goal is to automatically learn a mapping from the set of local metrics to the optimal mitigation action. We select an artificial neural network (ANN) as the substrate for learning as ANNs are, in theory, capable of representing any function [13,29]. The topology of the network is fixed whilst the weights are evolved to optimise for overall swarm performance, rather than that of an individual robot. Swarm performance is thus a global metric of fitness, evaluated at the end of each trial. The evolved ANN is labelled the *mitigation module* and serves as a controller for the mitigation. The module is copied over to all individuals in the swarm.

An important feature of this approach is that it does not depend on an explicit fault detection step, nor indeed a diagnosis step. Instead, robot states are implicitly learned by the ANN - they are encoded in the multi-dimensional space of local metrics. These states may be "faulty" or otherwise. An advantage of this approach is that the robot states are automatically determined in the learning process. An explicit definition of what constitutes a fault, or more specifically what states should trigger mitigation, is not required. This allows the mitigation module to freely learn the best action to take for a robot of any state, without predefined constraints. Such mitigation is dynamic as robots may switch between actions depending on context. The optimal action for an individual robot will be influenced by actions taken by other robots in the swarm.

This work aims to evolve a robot controller specifically for *reactive* fault mitigation in a swarm, using neuroevolution. Here, reactive mitigation refers to robots diverting from their default behaviour to implement mitigation. We focus on collective transport in an intralogistics scenario: robots are tasked with retrieving and delivering boxes to a designated deposit zone. Four robots are required to transport each box and explicit communication is necessary for team coordination. This presents a rich scenario for learning strategies where both the faulty and non-faulty robots may take action. Notably, the robots are not required to come to a consensus on the action to be taken - thus producing a "loosely-coordinated" strategy for the swarm overall.

2   Related Work

Fault mitigation for swarms is as yet an under-explored area of research. On the swarm level, work has been done in developing a safety assurance framework for the design of swarm algorithms, mitigating the potential risks and hazards of emergent behaviour of the collective [1]. Concerning fault mitigation implemented by robots in the swarm, current work largely takes one of two approaches: 1) embedded strategies where swarm behaviours are fault-tolerant by design, which do not require an explicit mitigation step, and 2) reactive strategies where robots divert from their default behaviour in order to execute mitigation. Embedded approaches include work on self-healing in shape formation [4,36], robust and adaptive control [3,11,17], and trial-and-error strategies in selecting the best behaviour from an archive of generated behaviours [7]. In the case of reactive strategies, work has been done in developing immune-inspired approaches for energy transfer to low-powered robots [40], reducing the influence of malicious agents [12], taking corrective actions based on a predefined rule-based dictionary [33], data-driven learning to select the best predefined actions [32], and automatic evaluation of the effectiveness of mitigation actions taken by faulty robots [26].

In the realm of evolutionary approaches to learning fault mitigation strategies, there are two embedded approaches: first, as previously mentioned, the generation of an archive of behaviours using MAP-Elites [30] from which the most effective may be selected through trial-and-error [7]. Another approach implements online lateral evolution where robots exchange genes to evolve resilient behaviours [31]. Beyond fault mitigation, there are many examples of evolutionary approaches to designing swarm controllers [14,25]. There has been a focus in particular on evolving artificial neural networks (ANN) as they are, in theory, capable of representing any function as well as having conceptual roots in, and taking inspiration from, biology [29]. In this paper we focus on fixed-topology neuroevolution. Work taking this approach includes evolving for collective transport behaviours [2,15], evolving for collective behaviour in a swarm of robots with limited sensing and communication abilities [41], and evolving sub-controllers in a hierarchical controller for task allocation [44]. A variety of evolutionary algorithms are implemented in this body of work: genetic algorithms, evolutionary strategies, and differential evolution.

There have been a wide range of ANN architectures considered in the literature: there is not a clear optimal choice and this may be task and scenario specific - similarly for the choice of evolutionary algorithm. There is evidence to suggest that a straightforward implementation of neuroevolution produces controllers competitive with more advanced approaches, both in simulation and real-world tests [16]. This motivates therefore the approach in this paper: we implement a genetic algorithm (GA) to evolve a fully connected ANN with a single hidden layer which acts as the fault mitigation module within the larger robot controller. We consider six fault types: modules are evolved for each type of fault independently, for varying numbers of fault. In particular, the output of the ANN is a vector of probabilities corresponding to low-level mitigation actions (e.g. bias to object, stop moving). In previous work, we found effective hardcoded actions for fault mitigation within this set, providing a good basis to learn dynamic mitigation strategies [26]. This differs from the majority of related work where action sets with a higher degree of granularity are considered: raw motor commands, as well as LED activations. In this paper, we are able to evolve effective dynamic strategies where both faulty and non-faulty robots are able to apply mitigation actions to improve swarm performance.

3   Method

3.1   Collective Transport Scenario

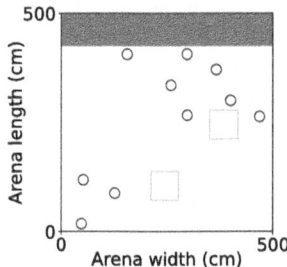

Fig. 1. Arena setup for collective transport: robots are represented by circles, boxes are represented by large orange squares. The deposit zone is the green-shaded strip at the top of the arena. (Color figure online)

We focus on a collective transport scenario: there is a high degree of interdependence between robots as they must explicitly communicate and coordinate in order to form a team, to pick up and transport a box. Robots operate in a 500×500cm bounded arena and deliver boxes to a 75cm-length horizontal strip extending along the upper boundary. They detect objects (robot, box or wall) via ArUco tags [18]. We run trials in a custom 2D simulator written in C++ and developed specifically for the intralogistics scenario[1]. The configuration and parameters for the scenario are selected to closely match the real-world arena and robot platform available at the Bristol Robotics Laboratory [19,20]. The DOTS robot (Distributed Organisation and Transport System) is a highly-developed platform with lifting capabilities in addition to cameras located on the perimeter and IR lasers used in obstacle avoidance.

In this task, robots employ a random walk behaviour only in the phase of searching for a box. Once detected, a robot positions itself under a box and waits for a complete team of four robots to be formed. Once complete, robots in the team broadcast their "vote" for the next collective action: either to *lift the box, move,* or *deposit the box.* A complete set of votes (one from each team member) is required to activate a team action. Upon depositing a box, the team

---

[1] https://bitbucket.org/hauertlab/2d_dots_simulator_cpp/src/master/.

is disbanded and robots search for a new box. We assume that one robot needs to be positioned at each corner of the box. Figure 1 visualises the arena setup. The configuration for the scenario and details for the random walk behaviour are provided in the supplementary material. Figure 2 shows the robot controller used in this task.

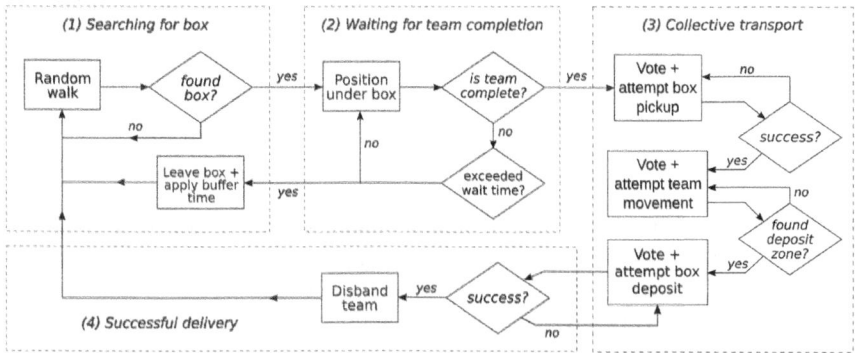

Fig. 2. Robot controller collective transport: the four main phases of robot behaviour are labelled (1)–(4) in the figure.

In collective movement, each robot chooses a heading at random and communicates the chosen heading to its teammates. The final heading chosen is the average of individually chosen headings. The team is assumed to move at the velocity of the slowest robot. Besides movement, there are two important parameters involved in the team formation mechanism. The first is a timeout variable, t_w, which acts as a threshold limit for the wait time for team formation. The wait time is tracked individually for each robot. If the threshold is exceeded and the team is not complete, the robot will leave its position under a box. A robot that leaves a team in this manner is not able to rejoin the team, or join a new team, for a given buffer time, t_b. This is the second parameter in team formation. In a parameter sweep, we select $t_w = 120$ seconds and $t_b = 4$ seconds to produce a high-performing controller for collective transport behaviour. Whilst the behaviour is not optimised against any benchmark and may be further fine-tuned, our focus is on selecting an algorithm where faults may have significant impact so that we can test our mitigation method.

3.2 Injected Faults

The faults selected take into consideration the sensors, actuators and communication capabilities of the robots. They are representative of the possible faults that could occur in the scenario:

F1	0% maximum speed	*F4* Perimeter camera fault
F2	10% maximum speed	*F5* Can't pick up boxes
F3	50% maximum speed	*F6* Can't receive messages

In the case of the perimeter camera fault, it is assumed that all cameras fail and the robot cannot detect objects in its surroundings. We also consider varying degrees of wheel failure causing robots to move at a range of reduced speeds. The inability to receive messages causes a robot to be unable to coordinate within a team for box transport. A single type of fault is injected at the beginning of each trial (varying in number) and persists for the duration of the trial.

3.3   Mitigation Actions

We consider the following set of mitigation actions which can be activated by individual robots in an asynchronous and distributed manner. In particular, there is the option to take "no action" which in effect, short circuits fault mitigation and causes a robot to perform the default behaviour of the controller:

A0	No action	*A8* Bias left
A1	Decrease speed 50%	*A9* Bias away from nearest robot
A2	Stop moving	*A10* Bias away from nearest box
A3	Don't join team	*A11* Bias away from nearest wall
A4	Leave team	*A12* Attract robots in range
A5	Bias towards nearest robot	*A13* Repel robots in range
A6	Bias towards nearest box	*A14* Stop receiving messages
A7	Bias towards nearest wall	*A15* Stop sending messages

The granularity of actions is lower than the typical actions considered in the literature for evolutionary approaches (usually motor commands and component activation, for example) which may provide a degree of readability for analysing the metric-to-action mapping. We make two assumptions: that 1) robots are able to switch between actions, and 2) both faulty and non-faulty robots may activate an action. Action selection is therefore dynamic and a new action can be selected per control cycle. The aim is to evolve dynamic mitigation strategies where robots can switch actions depending on the perceived context. Additionally, non-faulty robots may also apply actions to contribute positively to swarm performance. In this work, robots do not have a priori knowledge of faults, that is, we do not assume that they can self-detect nor self-diagnose faults.

3.4   Fault Mitigation Module

An ANN is implemented to learn a mapping between input metrics and output actions. Here, the ANN acts as a *fault mitigation module* which selects the appropriate action to take depending on context. The collective transport controller is augmented with the mitigation module which is activated at the beginning of the control loop. Figure 3 visualises the augmented controller. Actions selected by the module can modify behavioural parameters or trigger behaviours later on

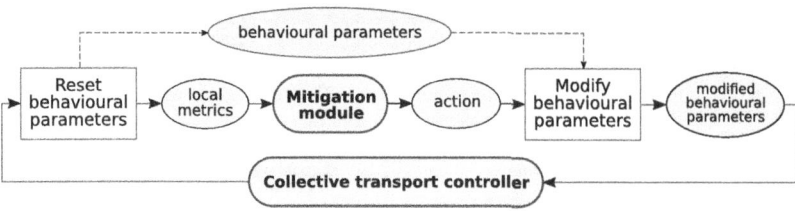

Fig. 3. The mitigation module is activated at the beginning of the control cycle: inputs are local metrics which are directly encoded, the output is a single action. The action is implemented by modifying behavioural parameters or by triggering a robot behaviour in the main controller. Behavioural parameters are reset before the module is activated.

in the control cycle. For example, selecting a *bias* movement action will set the next heading for movement, overriding the step of choosing a new random heading. The *attract* and *repel* actions trigger corresponding signals to be broadcast to robots in proximity. Robot controllers are hardcoded to respond as appropriate to such signals - a sort of predetermined "coordination protocol". The input metrics encode the context of a robot: its current state and its environment. We showed in previous work that a subset of these metrics is able to encode faulty (and thereby, non-faulty) states [27]. The metrics considered are:

M1	Robots in range	*M8*	Nearest wall distance
M2	Boxes in range	*M9*	Robot delivery count
M3	Walls in range	*M10*	Is robot in position under box
M4	Velocity	*M11*	Is robot in team
M5	Robot state	*M12*	Is robot lifting box
M6	Nearest robot distance	*M13*	Is robot moving box
M7	Nearest box distance	*M14*	Is robot depositing box

Here we define metrics which encode the stages of team formation and collective transport (M10-M14). The majority of metrics (M1-M8) and actions considered are task agnostic. Consequently, the ANN allows a robot to select an action based on the perceived context encoded in the local metrics. There is no explicit fault detection or diagnosis step as the neural network allows for a direct mapping between local metrics and local actions. The network therefore encompasses, implicitly, all steps of a *fault detection, diagnosis and recovery* (FDDR) pipeline.

The topology of the network is fixed and fully connected: there are 14 input nodes corresponding to the metrics, 1 hidden layer with 10 nodes, and 16 output nodes corresponding to the set of mitigation actions. A *tanh* activation function is applied to the nodes in the hidden and output layers. Finally, the nodes in the output layer are interpreted as a set of probabilities corresponding to the action set. Negative probabilities are discounted. The action corresponding to the highest probability is selected as the final output of the ANN. If there are no non-zero, non-negative probabilities, the default action is selected as "no action", A0. Figure 4 depicts the ANN implementation.

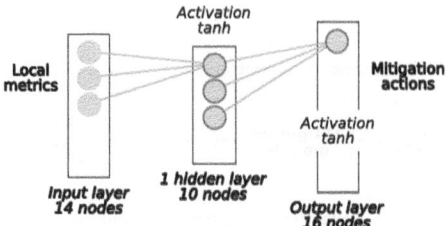

Fig. 4. The ANN consists of an input layer with 14 nodes, a single hidden layer with 10 nodes, and an output layer with 16 nodes. The nodes in the hidden and output layers are activated with a *tanh* function. The nodes are fully connected and the topology is fixed. Here, only a few nodes and their connections are depicted for illustrative purposes.

3.5 Evolutionary Algorithm

We apply a GA to evolve the weights of the network. As the ANN is fully connected, there are 300 weights in total. Each genome of the population is a direct-encoding for the weights of the network. Population size is selected to be $N = 192$ to optimise the computational resource available. The number of generations for evolution is $G = 50$. Fitness is evaluated as the mean performance over a batch of trials where a single fault type is injected in the collective transport scenario. Performance is evaluated as box delivery over time, computed as the integral under a step function $f(t)$ which describes the number of boxes delivered at time $t \in [0, T]$. The number of faults is varied in the batch of trials: there are 10 trials with no faults injected, and 5 trials for each of 2, 4, 6, 8, and 10 faults. The aim is to cover a range of possible numbers of fault in the swarm, whilst the fault type is fixed. In particular, 10 trials are chosen for the "no fault" trials in order to give higher weighting to performance in the default scenario. This is intended to ensure that the mitigation module does not disrupt swarm performance when no faults are present. For each genome in the initial population, the weights are initialised following a uniform distribution in the range of [-1, 1]. In the evolutionary process, two-point crossover and mutation operators are applied. The population is evolved for 50 generations. Table 1 lists the parameters of the GA and the steps in the implementation are as follows:

Table 1. Evolutionary parameters

Population size, N	192
Generations, G	50
Genome size	300
Elites, E	10
Crossover probability, p_x	0.2
Crossover-point selection probability, p_{xp}	0.05
Mutation probability, p_m	0.1

1. Each genome is evaluated for performance over the batch of trials. The fitness of the genome is computed as the mean performance score.

2. At the end of the evaluation step, the top $E = 10$ fittest individuals are selected as elites and copied directly to the next generation (no genetic operators applied).

3. From the remaining individuals, parents are independently selected through k-way tournament selection ($k = 3$) with replacement. Two-point crossover is applied with probability $p_x = 0.2$. Any two possible crossover points in the genome are independently selected with probability $p_{xp} = 0.05$. Otherwise in the case where no crossover is applied, the first selected parent is copied to generate the new individual.

4. Mutation is applied to the newly generated individual. Each element of the genome has $p_m = 0.1$ probability of mutation. The mutation value is sampled from a Gaussian distribution, parameterised $\mathcal{N}(0, 0.1)$.

5. Steps 3-4 are iterated to generate $N - E - 1$ individuals. The final individual to complete the population is randomly generated with weights sampled from a uniform distribution in the range of $[-1, 1]$.

4 Results

4.1 Evolutionary Fitness

We run six experiments, one for each fault type, the intention being to generate fault mitigation modules which target a specific type of fault as a first step. In preliminary experiments, evolving a controller over multiple fault types led to a generic, low-performing solution. By evolving distinct mitigation modules, our results show that we successfully avoid such local optima which may be encountered when evolving a generalised mitigation strategy.

Figure 5 shows the aggregated fitness scores of the population in each generation, across the six experiments. The fitness scores are normalised with respect to the maximum possible score (200,000). We use the terms fitness and performance interchangeably here, since fitness directly corresponds to the mean performance score over the batch of trials. Fitness is compared with a baseline performance: the baseline is the performance with no mitigation applied across the same batch of trials. Across all fault types, the fittest individuals in the final generation perform better than the baseline - the mean difference in performance over 600 trials (100 for each fault number $0, 2, 4, 6, 8, 10$) is as high as 0.15 (*10% speed*) and 0.14 (*camera fault*) compared to the lowest difference at 0.04 (*50% speed*).

We see similar fitness trajectories in the evolutionary process across all experiments. As the initial population is assigned random weights, the variation in fitness in the first generation of each experiment is large. The median score is 0 (or close to 0), suggesting that the majority of individuals in the first generations produce ANNs that disable the swarm from delivering boxes. Within 10 generations, the variation is already reduced and in all cases, the fittest individual performs better than the baseline. The fitness then tends to reach a plateau, suggesting that evolution has converged on a robust solution. The variation in

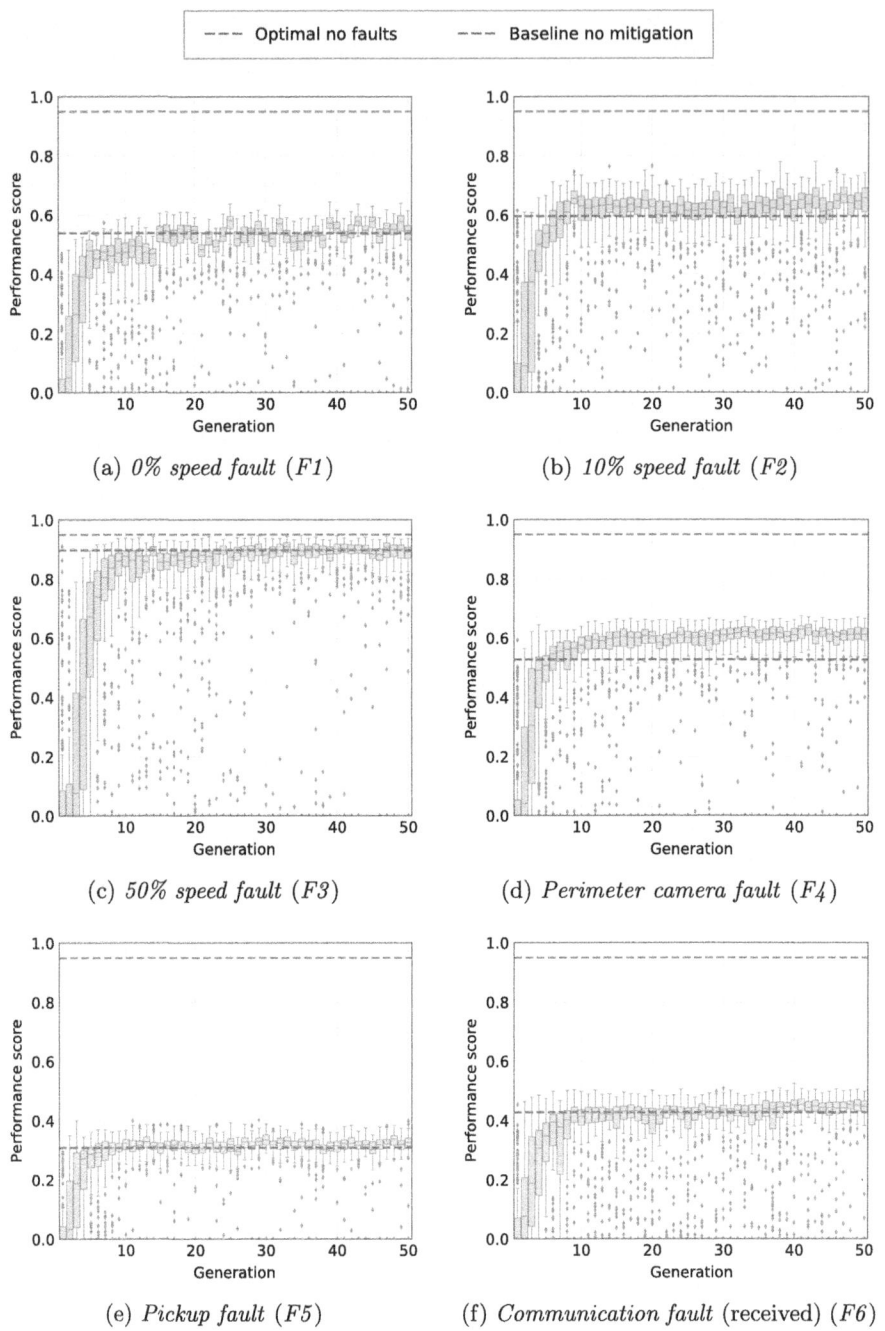

Fig. 5. Fitness of the population from generation 1 to 50, across all fault types. Here, fitness is the mean performance score over the batch of trials. For comparison, we also plot 1) the baseline performance with no mitigation, and 2) the "optimal" performance when no faults are present. For all fault types, the fittest individuals in the final generation (corresponding to the highest performance score) perform better than the baseline.

Table 2. Mitigation power of fittest individuals.

Fault type	Mitigation power (2 d.p.)	Mitigation power non-critical (2 d.p.) (0,2,4,6 faults)	Mitigation power 0 faults (2 d.p.)
F1: 0% speed	0.07	0.17	0.25
F2: 10% speed	0.11	0.19	0.28
F3: 50% speed	0.24	0.28	0.23
F4: camera fault	0.18	0.35	0.31
F5: pickup fault	0.02	0.04	0.45
F6: can't receive messages	0.05	0.08	0.26

performance of the final population across experiments is low compared to the first generation, which is further evidence of convergence. Notably, there are a number of outliers even in the final generation and this can be attributed to the single randomly generated individual in each iteration of the GA, which introduces novelty to the population.

The impact of a fault is dependent on the type of fault. The convergence point across experiments thus varies. As expected, fitness scores are highest for partial faults *10%* and *50% speed*, seen in Fig. 5b, 5c. In fact, for *50% speed*, the performance with mitigation is comparable to the default scenario when no faults are present. The "optimal" performance in the default scenario is plotted for comparison. In further analysis, we test the fittest individuals from each experiment over a fault range of $[0, 2, 4, 6, 8, 10]$, with 100 trials for each number of faults. Drawing from the statistical analysis implemented in previous work, we evaluate the effectiveness of the evolved mitigation modules - the *mitigation power* - compared to the baseline with no mitigation [26]. This is an application of the Mann-Whitney U test combined with effect size analysis - details are provided in the supplementary material. Table 2 shows the mitigation power of the fittest individuals for each fault type with 1 being the maximum power possible. In terms of *practical* significance, we may consider values greater than 0.1 to indicate a substantial improvement to swarm performance when the evolved mitigation module is implemented [28].

We also compute the mitigation power for a subset of trials with non-critical numbers of fault - i.e., when there are sufficient non-faulty robots (at least 4) to achieve non-zero performance in the task. The rationale here is that for a critical number of faults (for critical *types* of fault F4, F5, F6), performance for both mitigation and the baseline are indistinguishable (they score 0) and by excluding these scores, we can perform a more meaningful assessment of the effectiveness of the evolved strategy.

Fault types *50% speed* and *camera fault* have the highest mitigation power: 0.24 and 0.18. When considering non-critical numbers of fault, the camera fault has the highest power. Mitigation has the highest positive impact for these three fault types due to their critically negative impact on swarm performance when

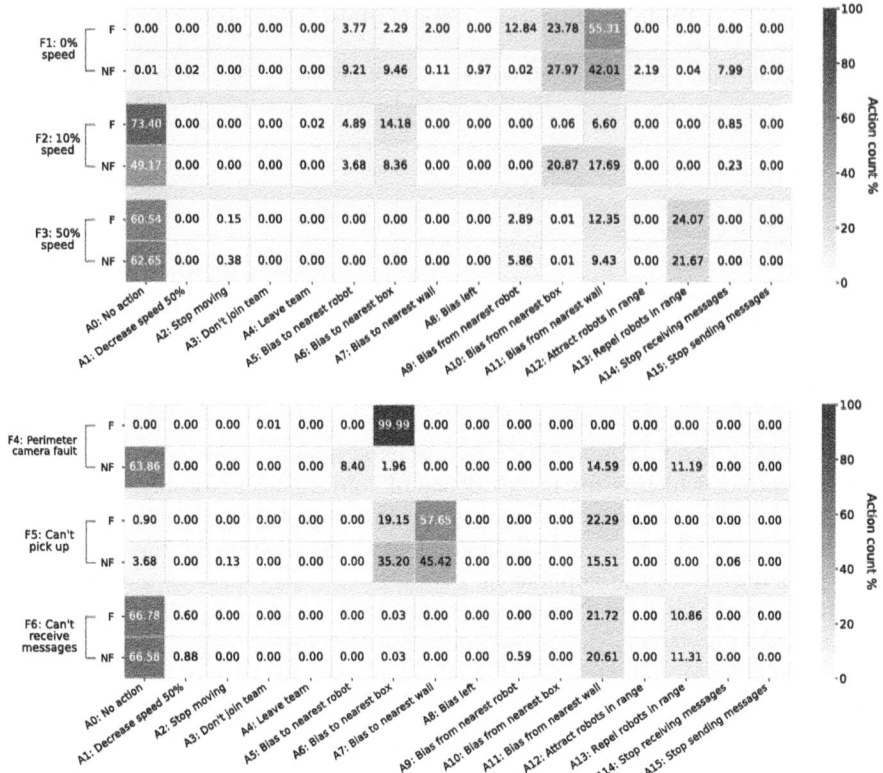

Fig. 6. The proportion of actions selected as output from the evolved mitigation modules, across experiments. The actions are counted separately for a faulty robot and a non-faulty robot. The counts are averaged over the batch of trials, from 0 to 10 faults.

left unchecked. Finally, we compute mitigation power when no faults are present: across all fault types, the power is higher than the evaluation on the whole range of numbers of fault, with the exception of the *50% speed* fault. This suggests that the evolved modules are optimising the baseline controller to some degree through the actions selected in the module.

4.2   Mitigation Analysis

We further examine the actions selected as the output from each ANN mitigation module, where the weights are taken from the fittest individual in generation 50. For each experiment, we compute the proportion of actions selected by 1) a faulty robot, and 2) a non-faulty robot. The proportion of actions is averaged over trials with different numbers of fault, $[0, 2, 4, 6, 8, 10]$. Figure 6 compares action outputs across fault types.

We make two key observations: firstly, an action "signature" is produced from the actions selected for a faulty and non-faulty robot across fault types. When comparing faulty versus non-faulty, some signatures are very similar, as in the case of *50% speed* and *can't receive messages*. This may be due to the difficulty of differentiating between fault and non-fault in these cases. For example, for the fault *can't receive messages*, the faulty robots can only differentiate themselves when interacting with a box - they are unable to join a team since they are unable to coordinate for team formation. Further, for the *50% speed* fault trials, non-faulty robots sometimes move at slower speeds making it hard to discriminate between fault and non-fault. However, we see a stronger difference in signatures for other fault types: for a *10% speed* fault, faulty robots choose "no action" 73% of the time compared to 50% for non-faulty robots. The *perimeter camera* fault shows the greatest difference - a faulty robot chooses to *bias to nearest box* close to 100% of the time whereas the non-faulty robot has a more varied action selection. The differences in action signature between faulty and non-faulty robots show that our approach evolves ANNs with some rudimentary specialization.

Secondly, bias actions are often selected - *bias from nearest wall* is selected across fault types, as well as other bias actions towards *robot, box, or wall*. The *bias to robot* action allows robots to congregate and when this happens in the proximity of a box, it can lead to faster team formation. Additionally, the *bias from wall* action produces a buffer between a robot and the walls of the arena, reducing time spent at the sides and corners. By applying these actions, non-faulty robots are able to contribute positively towards swarm performance by adapting their behaviour towards faster box delivery.

It is worth bearing in mind that some useful actions may not necessarily be selected frequently but may be effective when selected in the right context. For example, actions such as *don't join team* or *leave team* allow a robot to disengage from a team, which may be beneficial if the robot is faulty and these actions only need to be triggered within a single control cycle. The most distinct action signatures when comparing fault versus non-fault, are produced for the *perimeter camera* fault.

4.3   Camera Fault Mitigation in Action

Focusing on the *perimeter camera* fault, Fig. 7 shows the breakdown of action outputs in the case of a non-faulty robot. The action outputs vary with the number of faults. Whilst proportions vary, the *types* of selected actions remain the same. Differing proportions of selected actions may reflect the changing dynamics of the swarm over numbers of fault. This result suggests that our dynamic mitigation approach is able to adapt accordingly, especially as faults accumulate over the operational period of the swarm.

We demonstrate the mitigation module in action: Fig. 8 shows a series of snapshots for a single trial with four camera faults. Each snapshot demonstrates varying contexts in which actions are activated and robots are coloured accordingly. Faulty robots are represented by pentagons, non-faulty robots by circles. In the case where robots are not coloured, i.e. white, this indicates that "no action" has been selected. At 26 s, Fig. 8a, robots at (A) are biasing to the robot

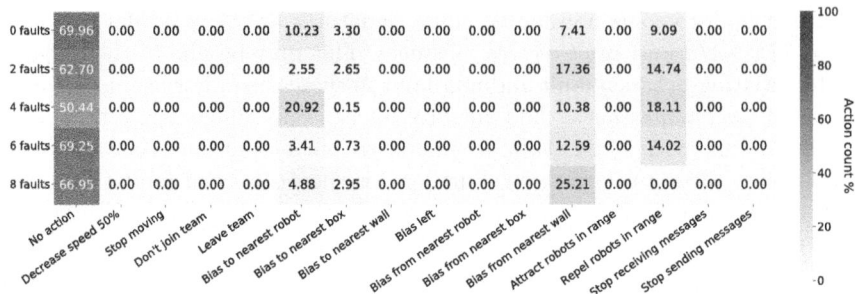

Fig. 7. Focusing on the *perimeter camera* fault: the proportion of mitigation actions selected for a non-faulty robot across numbers of fault (0 to 8).

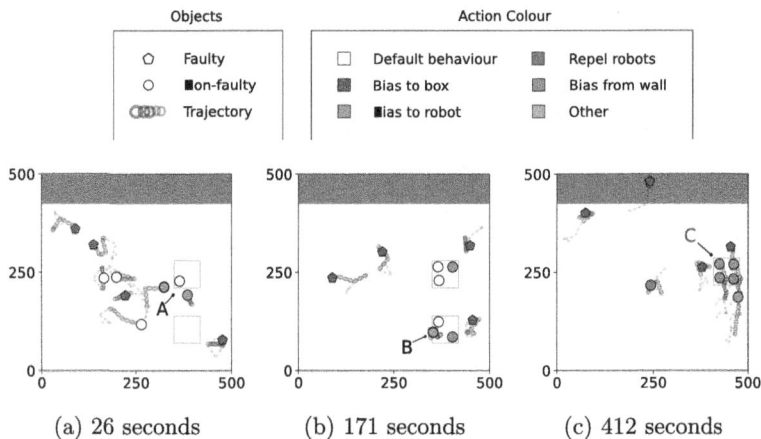

(a) 26 seconds   (b) 171 seconds   (c) 412 seconds

Fig. 8. Time series snapshots of dynamic mitigation for a trial with 4 faults (perimeter camera fault F4). Faulty robots are represented as pentagons, non-faulty robots as circles. The blue trail shows a robot's trajectory over the past 2 s. Action outputs are represented by six colours, labelled in the legend. Faulty robots select the *bias to box* action in all snapshots. Non-faulty robots select different actions depending on context: they select the *bias to robot* action which results in congregation near boxes; the *bias from wall* action is selected when robots can detect the walls of the arena.

positioned under the box. This clustering behaviour aids in team formation as robots gather near boxes. At 171 s, Fig. 8b, the robot at (B) has left the box after exceeding the wait time threshold. However, it is selecting the *bias to robot* action which effectively keeps it close to the box. Eventually, it repositions itself under the box. Finally at 412 s, Fig. 8c, the robots at (C) have successfully formed a team and are searching for the deposit zone. All four robots in the team are selecting the action *bias from wall* which keeps the team away from the wall where movement is restricted: this aids in the team's movement towards the deposit zone. Moreover, in the other snapshots, robots can be seen to select the *bias from wall* action when they are in proximity of a wall which suggests that, in general, it is useful to stay away from the walls of the arena where boxes are not located. In our examination of camera fault, we have thus demonstrated dynamic mitigation and the different contexts in which actions are triggered. A video demonstration can be found at https://youtu.be/YnG-R2_AseA.

5   Conclusion

This work has demonstrated a novel approach to evolving mitigation strategies for fault detection in swarms. Effective strategies are produced by evolving a mitigation module with local metrics as input and local actions as output. By mapping local metrics directly to the actions, an explicit fault detection step is not required as evolution learns the best action to take according to a robot's current state and context. The evolved mitigation modules result in better performance than the baseline controller without mitigation, across all fault types. This validates our approach which allows for dynamic action selection and for non-faulty robots to also take action. We see that improvement in performance, compared to no mitigation, is largely due to non-faulty robots acting to speed up team formation and to avoid the arena walls. Neuroevolution automatically learns the best action for each situation, for both faulty and non-faulty robots and depending on the number of faults. Whilst gradient-free learning has proven effective here, we aim to explore gradient-based approaches for comparison in future.

Our study has been grounded in a realistic scenario, based on the capabilities of the robot platform available [20]. Besides the faults considered in this work, noisy sensor measurements may impact the ability of robots to perceive context. To this end, we will explore online evolution in future work as it may allow for adaptation to noise and other perturbations which arise [9,21]. We will also consider actions where energy or modular components can be exchanged allowing robots to be repurposed to benefit other members of the swarm [32,40]. In addition to higher "readability" for metric-to-action mapping, selecting low-level behavioural actions may allow for a more robust transfer of the evolved mitigation module from simulation to the real-world [16].

Finally, mitigation modules are evolved for each fault type independently, in order to avoid local optima in a generalized mitigation strategy. In the real-world, there are likely to be multiple types of fault present in the swarm simultaneously. In the future, we will explore the combination of mitigation modules in a hierarchical controller, where an additional ANN can be trained to select between multiple module outputs.

Acknowledgements. This work was supported by the US Air Force Office of Scientific Research, European Office of Aerospace Research and Development, and the European Union under Grant Agreement 101070918 and UKRI grant number 10038942.

References

1. Abeywickrama, D.B., et al.: AERoS: assurance of emergent behaviour in autonomous robotic swarms. In: International Conference on Computer Safety, Reliability, and Security, pp. 341–354. Springer (2023)
2. Alkilabi, M., Narayan, A., Tuci, E.: Cooperative object transport with a swarm of e-puck robots: robustness and scalability of evolved collective strategies. Swarm Intell. **11**, 185–209 (2017)
3. Aniketh, R., Manohar, E., Yazwa, G.P.R., Nithya, M., Rashmi, M.: A decentralized fault-tolerant weights based algorithm for coordination of swarm robots for a disaster scenario. In: 2016 IEEE Annual India Conference (INDICON), pp. 1–5. IEEE (2016)
4. Arbuckle, D., Requicha, A.A.: Self-assembly and self-repair of arbitrary shapes by a swarm of reactive robots: algorithms and simulations. Auton. Robot. **28**, 197–211 (2010)
5. Arnold, R.D., Yamaguchi, H., Tanaka, T.: Search and rescue with autonomous flying robots through behavior-based cooperative intelligence. J. Int. Humanitarian Action **3**(1), 1–18 (2018)
6. Bjerknes, J., Winfield, A.: On fault tolerance and scalability of swarm robotic systems. In: Distributed Autonomous Robotic Systems: The 10th International Symposium, vol. 83, pp. 431–444. Springer (2013)
7. Bossens, D.M., Tarapore, D.: Rapidly adapting robot swarms with swarm map-based bayesian optimisation. In: 2021 IEEE International Conference on Robotics and Automation (ICRA), pp. 9848–9854. IEEE (2021)
8. Braithwaite, A., Alhinai, T., Haas-Heger, M., McFarlane, E., Kovač, M.: Tensile web construction and perching with nano aerial vehicles. Robot. Res. **1**, 71–88 (2018)
9. Bredeche, N., Haasdijk, E., Prieto, A.: Embodied evolution in collective robotics: a review. Front. Robot. AI **5**, 12 (2018)
10. Carrillo-Zapata, D., et al.: Mutual shaping in swarm robotics: user studies in fire and rescue, storage organization, and bridge inspection. Front. Robot. AI **7** (2020)
11. Chand, P.K., Kumar, M., Molla, A.R., Sivasubramaniam, S.: Fault-tolerant dispersion of mobile robots. In: Conference on Algorithms and Discrete Applied Mathematics, pp. 28–40. Springer (2023)
12. Chen, L., Ng, S.L.: Securing emergent behaviour in swarm robotics. J. Inf. Secur. Appl. **64**, 103047 (2022)
13. Doncieux, S., Bredeche, N., Mouret, J.B., Eiben, A.E.: Evolutionary robotics: what, why, and where to. Front. Robot. AI **2**, 4 (2015)
14. Francesca, G., Birattari, M.: Automatic design of robot swarms: achievements and challenges. Front. Robot. AI **3**, 29 (2016)
15. Groß, R., Dorigo, M.: Towards group transport by swarms of robots. Int. J. Bio-Inspired Comput. **1**(1–2), 1–13 (2009)
16. Hasselmann, K., Ligot, A., Ruddick, J., Birattari, M.: Empirical assessment and comparison of neuro-evolutionary methods for the automatic off-line design of robot swarms. Nat. Commun. **12**(1), 4345 (2021)
17. Hecker, J.P., Moses, M.E.: Beyond pheromones: evolving error-tolerant, flexible, and scalable ant-inspired robot swarms. Swarm Intell. **9**, 43–70 (2015)
18. Jones, S., Hauert, S.: Frappe: fast fiducial detection on low cost hardware. J. Real-Time Image Proc. **20**(6), 119 (2023)

19. Jones, S., Milner, E., Sooriyabandara, M., Hauert, S.: Distributed situational awareness in robot swarms. Adv. Intell. Syst. **2**(11), 2000110 (2020)
20. Jones, S., Milner, E., Sooriyabandara, M., Hauert, S.: DOTS: an open testbed for industrial swarm robotic solutions. arXiv preprint arXiv:2203.13809 (2022)
21. Jones, S., Winfield, A.F., Hauert, S., Studley, M.: Onboard evolution of understandable swarm behaviors. Adv. Intell. Syst. **1**(6), 1900031 (2019)
22. Kang, C.K., et al.: Marsbee-swarm of flapping wing flyers for enhanced mars exploration. Tech. rep. (2019)
23. Katada, Y., Hasegawa, S., Yamashita, K., Okazaki, N., Ohkura, K.: Swarm crawler robots using lévy flight for targets exploration in large environments. Robotics **11**(4), 76 (2022)
24. Kortman, M., Ruhl, S., Weise, J., Kreisel, J., Schervan, T., Schmidt, H., Dafnis, A.: Building block based iBoss approach: fully modular systems with standard interface to enhance future satellites. In: 66th International Astronautical Congress (Jerusalem), vol. 2, pp. 1–11 (2015)
25. Kuckling, J.: Recent trends in robot learning and evolution for swarm robotics. Front. Robot. AI **10**, 1134841 (2023)
26. Lee, S., Hauert, S.: A data-driven method to identify fault mitigation strategies in robot swarms. In: International Conference on Swarm Intelligence, pp. 16–28. Springer (2024)
27. Lee, S., Milner, E., Hauert, S.: A data-driven method for metric extraction to detect faults in robot swarms. IEEE Robot. Autom. Lett. **7**(4), 10746–10753 (2022). https://doi.org/10.1109/LRA.2022.3189789
28. McGraw, K.O., Wong, S.P.: A common language effect size statistic. Psychol. Bull. **111**, 361–365 (1992)
29. Mouret, J.B.: Evolving the behavior of machines: from micro to macroevolution. Iscience **23**(11) (2020)
30. Mouret, J.B., Clune, J.: Illuminating search spaces by mapping elites. arXiv preprint arXiv:1504.04909 (2015)
31. Neupane, A., Goodrich, M.A.: Learning resilient swarm behaviors via ongoing evolution. In: International Conference on Swarm Intelligence, pp. 155–170. Springer (2022)
32. Oladiran, O.O.: Fault recovery in swarm robotics systems using learning algorithms. Ph. D. thesis, University of York (2019)
33. Parker, L.E., Kannan, B.: Adaptive causal models for fault diagnosis and recovery in multi-robot teams. In: 2006 IEEE/RSJ International Conference on Intelligent Robots and Systems, pp. 2703–2710. IEEE (2006)
34. Petersen, K., Nagpal, R., Werfel, J.: TERMES: an autonomous robotic system for three-dimensional collective construction. In: Robotics: science and systems. vol. 7, pp. 257–264. Los Angeles, CA, USA (2011)
35. Post, M.A., Yan, X.T., Letier, P.: Modularity for the future in space robotics: a review. Acta Astronaut. **189**, 530–547 (2021)
36. Rubenstein, M., Shen, W.M.: A scalable and distributed model for self-organization and self-healing. In: Proceedings of the 7th International Joint Conference on Autonomous Agents and Multiagent Systems-Volume 3, pp. 1179–1182 (2008)
37. Şahin, E.: Swarm robotics: from sources of inspiration to domains of application. In: Şahin, E., Spears, W.M. (eds.) Swarm Robotics, pp. 10–20. Springer, Berlin Heidelberg, Berlin, Heidelberg (2005)
38. Schranz, M., Umlauft, M., Sende, M., Elmenreich, W.: Swarm robotic behaviors and current applications. Front. Robot. AI **7** (2020)

322   S. Lee and S. Hauert

39. Such, F.P., Madhavan, V., Conti, E., Lehman, J., Stanley, K.O., Clune, J.: Deep neuroevolution: genetic algorithms are a competitive alternative for training deep neural networks for reinforcement learning. arXiv preprint arXiv:1712.06567 (2017)
40. Timmis, J., Ismail, A.R., Bjerknes, J., Winfield, A.: An immune-inspired swarm aggregation algorithm for self-healing swarm robotic system. Biosystems **146** (2016)
41. Van Diggelen, F., Luo, J., Karagüzel, T.A., Cambier, N., Ferrante, E., Eiben, A.: Environment induced emergence of collective behavior in evolving swarms with limited sensing. In: Proceedings of the Genetic and Evolutionary Computation Conference, pp. 31–39 (2022)
42. Vasey, L., Felbrich, B., Prado, M., Tahanzadeh, B., Menges, A.: Physically distributed multi-robot coordination and collaboration in construction: a case study in long span coreless filament winding for fiber composites. Constr. Robot. **4**(1), 3–18 (2020)
43. Vassev, E., Sterritt, R., Rouff, C., Hinchey, M.: Swarm technology at NASA: building resilient systems. IT Professional **14**(2), 36–42 (2012)
44. Wei, Y., Hiraga, M., Ohkura, K., Car, Z.: Autonomous task allocation by artificial evolution for robotic swarms in complex tasks. Artif. Life Robot. **24**, 127–134 (2019)
45. Wilson, J., et al.: Trustworthy swarms. In: Proceedings of the First International Symposium on Trustworthy Autonomous Systems, pp. 1–11 (2023)

Inferring Reaction Elasticities from Metabolic Correlations in Cells Through Multi-objective Evolutionary Optimization

Arthur Lequertier[1] , Wolfram Liebermeister[1] , and Alberto Tonda[2,3]($\boxtimes$)

[1] Université Paris-Saclay, INRAE, MaIAGE, 78350 Jouy-en-Josas, France
{arthur.lequertier,wolfram.liebermeister}@inrae.fr
[2] UMR 518 MIA-PS, INRAE, Université Paris-Saclay, 91120 Palaiseau, France
alberto.tonda@inrae.fr
[3] UAR 3611 Institut des Systèmes Complexes de Paris Île-de-France (ISC-PIF), CNRS, Paris, France

Abstract. Parameter fitting in metabolic models can be challenging because experimental data are often noisy and sparse. In Bayesian estimation, prior knowledge about model parameters would be weighted against knowledge from data fitting. Since error bars and prior widths are often unknown, we explore a more flexible way of regulating this trade-off. We propose an evolutionary multi-objective approach to parameter estimation to find compromises between parameters matching the prior (prior loss) and yielding good data fits (likelihood loss). Our metabolic model describes an ensemble of steady states with correlated variation of all model variables. In the estimation, reaction elasticities are the parameters and the covariances of measurable state variables serve as measurement data. To evaluate our approach, we conduct two tests with artificial data and a known ground truth. We first consider a simple metabolic pathway with 3 reactions and 4 metabolites, where the correlated variation of variables can be understood intuitively. The second test involves a more complex real-world metabolic model of *Escherichia coli* bacteria with 62 metabolites, 57 reactions, and 234 elasticity coefficients to be fitted, where the results are almost impossible to guess even for domain experts. In both cases, the proposed method yields satisfactory results. This paves the way to studying biological objective functions unrelated to model fitting, including homeostasis or information transmission across metabolic networks.

Keywords: Bacteria · Covariance matrix · Metabolic model · Multi-objective optimization · Parameter estimation · Structural Kinetic Model

1 Introduction

Cell metabolism consists of a network of enzyme-catalyzed chemical reactions showing a complex dynamics. Metabolic models are based on reaction networks

© The Author(s), under exclusive license to Springer Nature Switzerland AG 2025
P. García-Sánchez et al. (Eds.): EvoApplications 2025, LNCS 15612, pp. 323–337, 2025.
https://doi.org/10.1007/978-3-031-90062-4_20

whose nodes and edges carry different types of variables – metabolite concentrations, enzyme levels, and metabolic fluxes – which depend on each other in complex ways. Physical laws, including mass balance relations and kinetic rate laws, govern the dynamics of this high-dimensional dynamical system. Experimental "omics" data, assigning values to model variables in different states of the cell, are usually scarce and noisy and often do not capture the absolute values of variables, rather measuring their relative variation between different states of the cell. This, together with limited data about model parameters, makes model parameterization a challenging task.

The Structural Kinetic Models (SKM) approach [17] is a way of formulating metabolic models that removes some of these difficulties. It describes metabolic systems in two steps, by first defining a steady reference state – a plausible set of all model variables, describing a viable state of the cell – and then modeling dynamic variations around this state with dynamics described as linear approximations and with reaction elasticities as parameters to be sampled or fitted. The reaction elasticities describe how individual reaction rates respond to changes in metabolite concentrations. A main advantage of the SKM formulation over traditional metabolic models is that it already starts from a steady state with plausible metabolic fluxes, which allows for constructing realistic models without the need for a brute-force parameter estimation.

Here we ask how the covariations of metabolic variables are shaped by network structure and details of enzyme kinetics and regulation. How much information is contained in observed correlations [16]? Focusing on covariations instead of a single steady state has different reasons: they are not only easier to measure in "omics experiments", but they also tell us more clearly how variables are dynamically related. Moreover, variance and covariances of variables may be important for cellular regulation, homeostasis, or adapted responses to changes in the cells' environment.

While covariation happens dynamically as cell variables fluctuate in time, we can also think of covariation across an ensemble of steady states, that is, states in which all variables remain constant in time, but differ between model instances, for example depending on cells' environments. To model such an ensemble of states, we may assume that some "external" variables are chosen from random distributions while all the remaining "internal" variables assume their steady-state values given those variables. The resulting variations and covariations of all variables are shaped by the structure of the metabolic network (which enforces, for example, covarying fluxes along linear metabolic pathways), but also by the reaction kinetics.

Here we study how SKM models can be fitted to covariance data. Given a network structure and a known reference state, the free parameters of an SKM are its reaction elasticities: in order to find their values, however, it is not enough to fit experimental data – in our case, the elements of its covariance matrix. It is also necessary to consider that elasticities should not differ too strongly from theoretical expectations, either because some of their values are approximately known from literature or because deviating too far from certain values might have costly side-effects for the cell. The two requirements naturally lead to a multi-objective problem, where each candidate solution will represent a trade-off between the two conflicting aims.

Below we propose a novel multi-objective evolutionary approach to parameter estimation for SKMs. In our model, we describe how covariances of model variables depend on reaction elasticities. Then we turn this question around: from a given covariance matrix of metabolite (or metabolite and enzyme) concentration data we infer the elasticities. This yields an estimate of the Jacobian matrix, an estimation problem that has been recently tackled in [10] using another methodology. In our method, the parameter estimation task is framed as an optimization problem with two conflicting aims: finding reaction elasticities that, on the one hand, fit a given covariance matrix of state variables and, on the other hand, are not too different from the theoretical expectations about approximately known or physiologically optimal parameters of the cell.

We test the approach on metabolic models following the SKM framework, first with a simple case study using a 3-reaction pathway (with 2 external and 2 internal metabolites); and then with a larger, more realistic network model describing the core metabolism of *Escherichia coli* bacteria. For simplicity, the elasticities with respect to external metabolites are assumed to be fixed and given. The results show that the proposed multi-objective optimization approach is able to find satisfying Pareto fronts for both cases. As expected, the candidate solutions match the ground truth when full data are available and deviate from the ground truth as more and more data are masked.

2   Background

This section briefly summarizes the methods used in this work: metabolic models, SKMs, and multi-objective evolutionary optimization.

2.1   Cell Metabolic Models

A metabolic model describes biochemical reactions that occur within a cell, enabling the cell to maintain its biological functions. Its nodes represent chemical species called metabolites and its edges represent the chemical reactions themselves. Here is an example of a 3-reaction linear pathway that will later serve as a toy model for the experimental evaluation:

$$A_{\text{ext}} \xrightarrow{\text{R}_1} B \xrightarrow{\text{R}_2} C \xrightarrow{\text{R}_3} D_{\text{ext}}$$

A_{ext} and D_{ext} are external metabolites (with concentrations treated as model parameters). The concentrations of internal metabolites B and C and the reaction rates v_1, v_2, and v_3 are state variables. Each reaction follows an unknown rate law of the form $v_i = e_i \, f_i(\mathbf{c})$. The metabolites consumed and produced in each reaction are described by a stoichiometric matrix $\mathbf{N}$ whose rows represent metabolites and whose columns correspond to reactions. Each matrix element represents the stoichiometric coefficient between a metabolite and a reaction, with negative elements for reaction substrates and positive elements for reaction products. Considering the mass balance for each metabolite, we can relate the temporal variation of internal metabolite concentrations (in the vector $\mathbf{c}$) to the reaction rate $\mathbf{v}$:

$$\frac{d\mathbf{c}}{dt} = \mathbf{N} \cdot \mathbf{v} \Rightarrow \frac{d}{dt}\begin{pmatrix} c_B \\ c_C \end{pmatrix} = \begin{pmatrix} 1 & -1 & 0 \\ 0 & 1 & -1 \end{pmatrix} \cdot \begin{pmatrix} v_1 \\ v_2 \\ v_3 \end{pmatrix}. \tag{1}$$

We now consider a steady state in which reaction fluxes and metabolite concentrations have reference values $\mathbf{v}^*$ and $\mathbf{c}^*$, satisfying the mass conservation equation $\mathbf{N} \cdot \mathbf{v}^* = \mathbf{0}$ and the rate laws $\mathbf{v}^* = \mathbf{v}(\mathbf{e}, \mathbf{c}^*)$ To describe how small changes in metabolite concentrations influence reaction rates near this steady state, we use elasticity coefficients. The reaction elasticity E_{ij} quantifies how sensitive a reaction rate v_i is to a small change in the concentration of metabolite c_j. Based on a given rate law $v_i = v_i(e_i, \mathbf{c})$, an elasticity is defined as:

$$E_{ij} = \frac{\partial v_i}{\partial c_j}\Big|_{\mathbf{c}^*, \mathbf{v}^*} \tag{2}$$

Applied to a reference state, it tells us how much the reaction rate v_i changes when c_j is slightly perturbed, assuming that all other system variables remain unchanged. Using these coefficients, we can approximate the direct effect of changes in metabolite concentrations $\delta \mathbf{c}$ on the reaction rates $\delta \mathbf{v}$:

$$\delta \mathbf{v} \approx \mathbf{E} \cdot \delta \mathbf{c}.$$

The elasticity matrix $\mathbf{E}$ contains the reaction elasticities for all the reactions and internal metabolites in the system.

2.2 Linearized Metabolic Model

In the Structural Kinetic Modeling (SKM) approach, we construct a linearized metabolic model in which unknown reaction elasticities are formally treated as model parameters [17]. The unscaled elasticity matrix can be written as

$$\mathbf{E} = \text{diag}(\mathbf{v}^*)\, \mathcal{E}\, \text{diag}(\mathbf{c}^*)^{-1}$$

where $\mathcal{E}$ is the scaled version of the elasticity matrix. The matrix $\mathcal{E}$ is sparse: it contains non-zero entries only for metabolites directly involved in reactions. The dimensionless coefficient $\mathcal{E}_{ij}$, for the ith reaction, measures the normalized degree of saturation of the catalyzing enzyme with respect to the jth metabolite. It has a known sign (1 for substrates and -1 for products), and its absolute value ranges between 0 (when the enzyme is fully saturated by the metabolite) and a maximum given by the absolute stoichiometric coefficient $|n_{ji}|$ (for a fully unsaturated enzyme). We can therefore write the matrix as

$$\mathcal{E} = -\mathbf{N}^\top \circ \mathbf{A} \tag{3}$$

with a matrix of saturation values a_{ij} in the range between 0 and 1. Here, $\circ$ denotes the component-wise multiplication (Hadamard product) of the two matrices. Below, for the optimization, we will represent this matrix by a vector $\mathbf{a}$ containing all the relevant (potentially non-zero) elements of the matrix. A given vector $\mathbf{a}$ of saturation values will, therefore, define a parameterized model.

To examine the dynamics of our metabolic model around a reference state, we study the Jacobian matrix, which characterizes the sensitivity of each metabolite to changes in its neighbor metabolites. The Jacobian determines how fluctuations in one metabolite can propagate across the network. For a given steady reference state, it is given by

$$\mathbf{J} = \mathbf{N} \, \mathbf{E} \tag{4}$$

where $\mathbf{N}$ is the (known) stoichiometric matrix and $\mathbf{E}$ is the unscaled elasticity matrix. By simulating the resulting metabolic dynamics, one can infer the overall behavior of the cell under different conditions and explore how the network responds to changes, despite uncertainties in the underlying kinetic details.

The SKM approach uses matrices $\mathbf{N}$ and $\mathbf{E}$, as well as fixed vectors $\mathbf{v}^*$ and $\mathbf{c}^*$, to represent the metabolism dynamic. While the stoichiometric matrix $\mathbf{N}$ represents the well-known topology of the metabolic network and plausible reference states can be guessed, the vector $\mathbf{a}$, defining the scaled elasticity matrix $\mathcal{E}$, is typically unknown. In SKM, one may circumvent this problem by sampling this vector at random [8,12,17]; here, instead, we will fit it to data.

2.3 Correlated Variation of Metabolic Variables

To model variability or uncertainties of metabolic variables, we describe them as random variables. Given random distributions of the external variables (external metabolite concentrations and enzyme levels), we obtain distributions of all the state variables (internal metabolite concentrations and fluxes). The model variables are described on a logarithmic scale and their variations are assumed to be small, allowing us to describe the system's dynamics in a linear approximation, using notions from Metabolic Control Analysis (MCA) [9,14]. Our aim is to compute a global linear response of all internal variables of the system (internal metabolite concentrations $\mathbf{c}$ and reaction fluxes $\mathbf{v}$) to external variables (enzyme levels and external metabolite concentrations) that serve as the sources of perturbations. This global linear response is captured by the unscaled response matrix [9]

$$R_p = \begin{pmatrix} R_p^c \\ R_p^v \end{pmatrix} \quad \text{where} \quad \begin{aligned} R_p^c &= -LJ_R^{-1}N_R \cdot E_p \\ R_p^v &= E \cdot R_p^c + E_p. \end{aligned}$$

In the formula., the reduced stoichiometric matrix N_R consists of a set of linearly independent rows of N, and the link matrix is defined to obtain N again via $N = LN_R$ [14]. The reduced Jacobian is given by $J_R = N_R \, E \, L$.

From now on we assume that all variables are described on a logarithmic scale. In analogy to the scaled elasticities, we define a scaled version $\mathcal{R}_p$ of the response matrix R_p, which is used in this case. Moreover, rather than considering specific perturbations, we consider random perturbation and treat all variables as random variables [7,11,18]. On a logarithmic scale, all variables are assumed to follow normal distributions around the given reference state. The (logarithmic) external variables (in a random vector P) are assumed to be independent, with a predefined, diagonal covariance matrix $\text{Cov}(P)$. Due to the linearized model, also the resulting (logarithmic) state variables will follow normal distributions. Altogether, we obtain a random vector Z comprising all the (logarithmic) external and internal variables, following a joint multivariate normal distribution with covariance matrix [11]

$$\text{Cov}(Z) = \begin{pmatrix} I_p \\ \mathcal{R}_p \end{pmatrix} \cdot \text{Cov}(P) \cdot \begin{pmatrix} I_p \\ \mathcal{R}_p \end{pmatrix}^T \tag{5}$$

where I_p is the identity matrix with a size equal to the number of external variables in p. Equation (5) is a compact representation of how uncertainty in the network's external environment leads to uncertainty in all variables. The covariance matrix $\text{Cov}(Z)$ depends on the covariance matrix $\text{Cov}(P)$ of external variables and on the scaled response matrix $\mathcal{R}_p$, which itself depends on the stoichiometric matrix N and the vector of saturation levels a.

2.4 Inferring Reaction Elasticities from Covariances in Model Variables

For a model with given reference state, stoichiometric matrix, and random distributions of the external variables, our aim is to estimate the unknown saturation levels in a based on data by assuming two different objectives. The first objective is to fit experimental data by minimizing the *negative* log likelihood between the model output and experimental data – in our specific case, a covariance matrix derived from experimental data in different steady states. Since in practice, data are limited, we assume that only some of the covariances are available for the estimation. The second objective reflects our prior knowledge about plausible saturation values in a, described by a prior distribution. For the prior, we assume a Gaussian distribution with given mean vector and standard deviations. The mean values represent our best guess for biologically reasonable levels of saturation. This second objective, effectively, penalizes deviations of the saturation values from their prior means.

Thus, our two-objective approach balances two key considerations: (1) the likelihood that the model accurately reflects experimental data, and (2) the prior knowledge or assumptions about plausible saturation values based on established biochemical principles. These objectives can be in conflict, as measurement data may suggest values that deviate significantly from prior expectations. This trade-off between empirical alignment and adherence to prior knowledge represents a central challenge in metabolic modeling.

2.5 Multi-objective Evolutionary Optimization

Machine learning and optimization algorithms have been applied to metabolic network models [2]. Among these methods, Evolutionary Algorithms (EAs) have emerged as a powerful class of optimization techniques [1]. These approaches also allow for multi-objective optimization and that the cell does not have a single stable state but rather a set of preferred modes, each optimizing, to a greater or lesser extent, key characteristics of homeostasis.

When dealing with multiple conflicting objectives, classic optimization techniques might employ a composite objective function defined as a weighted sum of the single objective functions. The idea behind multi-objective optimization [4], instead, is to not assign weights to the functions at all but to explore the space of trade-offs between the conflicting objectives. The result is not just one single best solution, but a set of candidate solutions that are not Pareto-dominated by others. Being population-based, EAs are well suited to multi-objective optimization and currently represent the state of the art in the domain [15]. In this work, we chose to employ the established Non-Sorting Genetic Algorithm II (NSGA-II) [5]: while not particularly recent, it is still extremely competitive in optimization problems with up to 3 objectives and has implementations in multiple programming languages, from C++ to Python.

3 An Approach for Model Fitting by Multi-objective Optimization

Given the two conflicting objectives of parameter estimation in metabolic models described above, we propose a novel approach based on multi-objective evolutionary optimization to find a set of good compromise solutions to be later analyzed for their biological relevance. A scheme of our estimation problem with the two objective functions is presented in Fig. 1.

3.1 Structure of a Candidate Solution

In our optimization problem, an individual represents a set of reaction elasticities, that capture the local sensitivities of reaction fluxes to changes in metabolite concentrations. The elasticity matrix is encoded by a saturation level vector $\mathbf{a}$, whose elements are real-valued numbers in $[0, 1]$. An individual $\mathbf{a}$ thus represents a configuration of the SKM model from which we compute the model's covariance

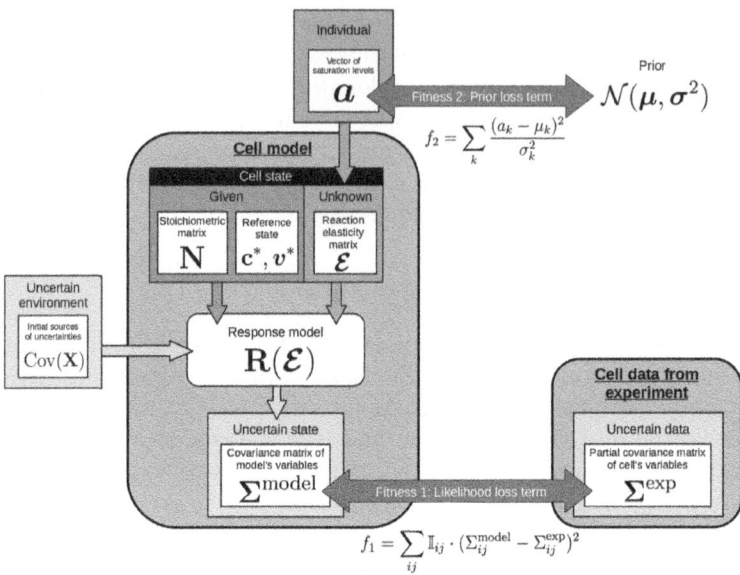

Fig. 1. Estimation problem for metabolic models. The aim is to estimate reaction elasticities, using covariances of state variables as data. A model instance is defined by a stoichiometric matrix $\mathbf{N}$, a known reference state ($\mathbf{v}^*$ and $\mathbf{c}^*$), and unknown reaction elasticities $\mathcal{E}_{ij}$, parameterized by saturation values in a vector $\mathbf{a}$. A model instance yields an input-output relationship described by a response coefficient matrix $\mathbf{R}$. Based on an assumed random distribution of external variables ("environment"), the model generates a covariance matrix of all the model variables. In our estimation procedure for saturation values, we consider two objectives. The first loss function f_1 compares the covariance results between our model and experimental data. The second loss function f_2 compares the elasticity matrix to a prior. Solutions are found by an evolutionary algorithm in which each individual represents a vector of saturation values, encoding an instance of our response model.

matrix, $\mathrm{Cov}(\mathbf{Z}^{\mathrm{model}})$. During our estimation procedure, this covariance matrix is then scored by one of our two objective functions.

Given the structure of a candidate solution – a simple numerical vector – the operators that will be used during the evolutionary optimization are a 1-point crossover and a polynomial mutation [6].

3.2 Objective Functions

Our first objective concerns the similarity between a covariance matrix $\mathbf{\Sigma}^{\mathrm{model}} = \mathrm{Cov}(\mathbf{Z}^{\mathrm{model}})$ obtained from a model and an "experimentally measured" covariance matrix $\mathbf{\Sigma}^{\mathrm{exp}} = \mathrm{Cov}(\mathbf{Z}^{\mathrm{exp}})$. The likelihood function (assuming a normal distribution for "measurement errors" of covariance values with a constant standard deviation σ), is given by

$$\mathcal{L}(\mathbf{a}^{\text{model}}|\mathbf{\Sigma}^{\text{exp}}) = \prod_{ij} \frac{1}{\sqrt{2\pi\sigma^2}} \exp\left(-\frac{(\Sigma_{ij}^{\text{model}}(\mathbf{a}^{\text{model}}) - \Sigma_{ij}^{\text{exp}})^2}{2\sigma^2}\right).$$

In our optimization, we consider the negative log-likelihood and ignore constant terms. Our resulting first objective function f_1, called "likelihood loss", can be written as follows:

$$f_1(\mathbf{a}^{\text{model}}) = \sum_{ij} \mathbb{I}_{ij} \cdot (\Sigma_{ij}^{\text{model}}(\mathbf{a}^{\text{model}}) - \Sigma_{ij}^{\text{exp}})^2 \qquad (6)$$

with $\mathbb{I}_{ij} \in \{0, 1\}$: 1 if a data point is accessible, 0 otherwise, reflecting the impossibility of observing specific matrix elements from biological data. A maximal likelihood corresponds to a minimal loss.

For the second objective, we score each possible vector $\mathbf{a}$ by a prior density $P_{\text{prior}}(\mathbf{a})$, an uncorrelated multivariate normal distribution with mean vector $\boldsymbol{\mu}$ and standard deviations in a vector $\boldsymbol{\sigma}$. With this prior, a saturation value vector $\mathbf{a}$ can be scored by how much it deviates from the prior mean. In analogy to our loss function f_1, we define our second objective function f_2, called "prior loss":

$$f_2(\mathbf{a}^{\text{model}}) = \sum_{k} \frac{(a_k^{\text{model}} - \mu_k)^2}{\sigma_k^2}. \qquad (7)$$

The function represents the negative log-prior, where constant terms are again ignored.

4   Experimental Evaluation

To validate our approach, we ran computer experiments on two different models. We first considered a simple 3-reaction pathway for which the results are easy to analyze. Then we considered the *E. coli* core model [13], a standard network model of *Escherichia coli* bacteria consisting of 62 metabolites and 57 reactions, and with a total of 234 reaction elasticities to be estimated.

In both cases, the artificial data used in the estimation were generated using a "true" instance of our model, taken to be our ground truth. The true saturation values were chosen randomly within biologically plausible ranges. In the models to be fitted, we kept all model parameters exactly the same except for the saturation values in $\mathbf{a}$ to be estimated. The prior mean values for all saturation values were set to $1/2$, describing a case in which enzymes are half-saturated with all the metabolites. Biochemically, this corresponds to metabolite concentrations matching their respective Michaelis-Mention constants K_{M}. All prior standard deviations were chosen to be equal, and also all data error bars (for covariance values) were chosen to be equal: the two numerical values do not play a role,

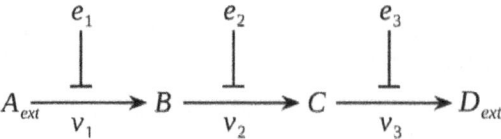

Fig. 2. Metabolic pathway of 3 reactions. Variability in external variables (enzyme levels and concentrations of external metabolites A_{ext} and D_{ext}) causes variability in internal variables (reaction fluxes and concentrations of internal metabolites B and C).

because their only effect is a linear scaling of the two objective functions, which will not change the shape of our Pareto front.

All the necessary code for reproducing the experiments is available in a public GitHub repository[1]; the scripts are implemented in Python 3, resorting to the pymoo library [3] for NSGA-II.

4.1   Simple 3-Reaction Pathway Model

As a first test for of our optimization algorithm, we studied a hypothetical 3-reaction linear pathway (Fig. 2). The internal variables are the fluxes v_1, v_2 and v_3 and the concentrations c_B and c_C, while the external variables (and therefore sources of uncertainty) are the external metabolite concentrations $c_{A_{ext}}$ and $c_{D_{ext}}$, as well as the enzyme activities e_1, e_2 and e_3. The elasticity matrix $\mathcal{E}$ (for internal metabolites B and C) contains only two columns, and only 4 of its elements are non-zero. Therefore, a candidate solution is represented by a saturation vector of length 4, $\mathbf{a}^{model} \in [0, 1]^4$, parameterizing the reaction elasticities with respect to internal metabolites. This minimal setup enables a rapid and easily interpretable analysis, which allowed us to identify and understand challenges during the implementation.

After a few trial runs, we ran NSGA-II with the following hyperparameters: population size $\mu_e = 1000$, offspring size $\lambda_e = 1000$, tournament selection with $\tau = 2$, probability of crossover $p_c = 0.9$, probability of polynomial mutation [6] $p_m = 0.9$, and a stop condition after $G_{max} = 100$ generations. The experiment took about 20 min to run on a server with 72 Intel Xeon w9-3475X CPUs and 128 GB of RAM, with one evaluation taking around 0.02 s on average.

The results are shown in Fig. 3 (left). The plots show how the algorithm progressively identifies high-quality, non-dominated points in each iteration, ultimately converging toward a satisfactory front. Since the objective functions are positive with optimal (minimal) values of 0, the fact that the front almost touches both axes indicates a good performance. The two red dots at the ends of the front represent the known extreme points; although they are shown here for reference, they were not used in computing the front.

---

[1] https://github.com/albertotonda/evolutionary-optimization-cell-models.

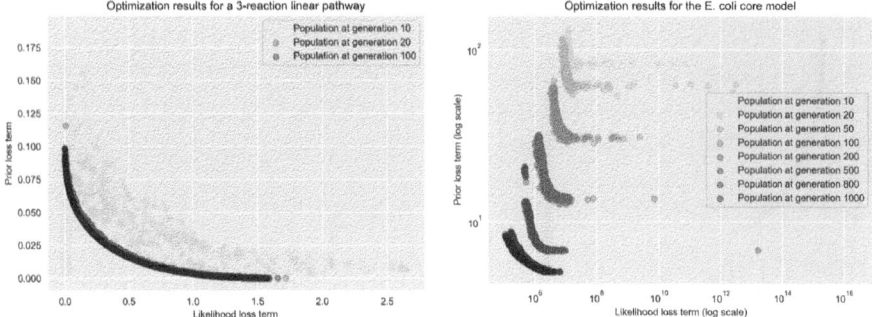

Fig. 3. Population of NSGA-II at selected generations for the 3-reactions pathway model (see Fig. 2) and the *E. coli* core model. Both objectives (prior and likelihood loss term) are minimized. Left: Experiment on the 3-reaction pathway. Right: Experiment on the *E. coli* core model. (Color figure online)

Our objective f_1 compares a covariance matrix produced by a candidate solution to our "true" the covariance matrix, replacing here covariances from biological data. In real biological experiments, only some of the cellular variables can be measured. We studied the effects of such incomplete data by considering a number of scenarios in which only some of the covariance data were used to compute the likelihood loss f_1. Figure 4 shows optimization results for different scenarios with such incomplete data. To indicate the varying quality of the estimation, the distance of each parameter set from the ground truth (the a vector of the "true model") is shown in color. The "true model" itself is represented by the red dot on the top left. The red point on the bottom right represents an individual whose values match the prior mean.

In Scenario 1 (top left), the whole covariance matrix is considered (a 10×10 matrix – but the symmetric shape of the matrix allows us to reduce it to 55 elements). Our Pareto front connects these two points. Distances between the individuals and the "true" saturation values are coherent compared with the values of the first objective function along the Pareto front. The individual that is the closest to the top-left red point has the lowest loss value, which suggests the uniqueness of the solution in this case. In Scenario 2 (top right), only the covariances of metabolites and enzymes (but not the fluxes) are considered, reducing to 13 the number of elements compared. Distances from the ground truth remain well sorted along the Pareto front. With more iterations, the front would probably extend to the two red points. In Scenario 3 (bottom left), only covariances between metabolites are considered, reducing to 7 the number of elements compared. The distances from the ground truth (in color) remain globally coherent, but do not fully match the first objective function. This means that an estimation based on metabolite covariances still works, but not fully reliably. In Scenario 4 (bottom right), only the covariance between internal metabolites was considered, reducing to 3 the number of elements compared. As expected,

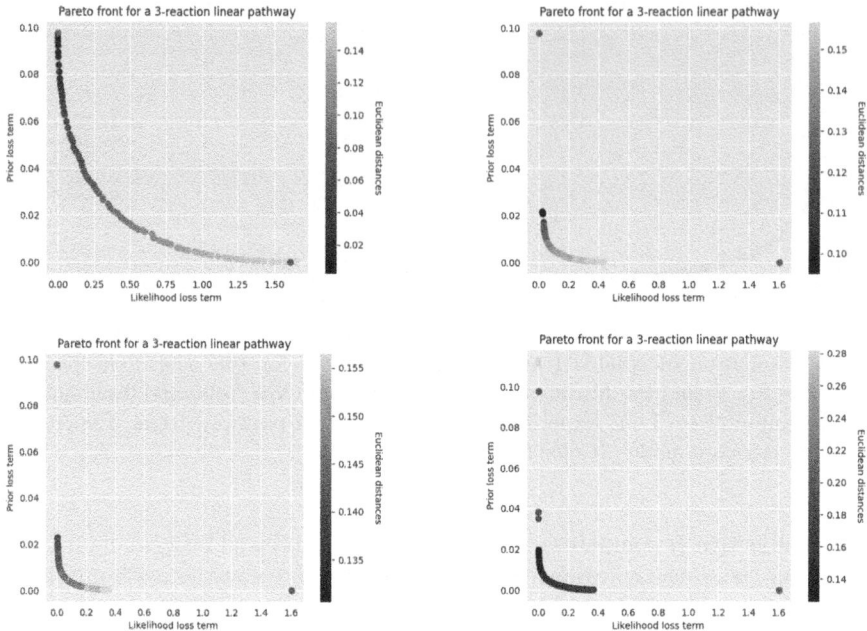

Fig. 4. Population of NSGA-II (last generation) after optimizing the 3-reaction pathway using different estimation scenarios. Top left (Scenario 1): Loss function f_1 based on covariances between all model variables (enzyme levels, metabolite concentrations, and fluxes). Top right (Scenario 2): f_1 based on covariances of 3 enzyme levels and 4 metabolite concentrations. Bottom left (Scenario 3): f_1 based on covariances of 4 metabolite concentrations. Bottom right (Scenario 4): f_1 based on covariances of 2 internal metabolite concentrations. (Color figure online)

with 3 data points and 4 parameters to be estimated, the estimation problem is ill-determined and the estimates along the entire front are partially shaped by the prior.

4.2 *Escherichia Coli* core model

Our second computer experiment targets the *E. coli* core model [13], which contains 62 metabolites, 57 reactions, and 234 considered reaction elasticities. A candidate solution is now represented by a vector **model** $\in [0, 1]^{234}$, again representing enzyme saturation values encoding reaction elasticities. As the problem is now more complex, a larger computational budget was allocated to NSGA-II, with hyperparameters: $\mu_e = 1000, \lambda_e = 1000, \tau = 2, p_c = 0.9, p_m = 0.9$, and $G_{max} = 1000$ generations. The experiment took about 96 h to run on a server with 72 Intel Xeon w9-3475X CPUs and 128 GB of RAM, with one evaluation taking on average 0.34 s.

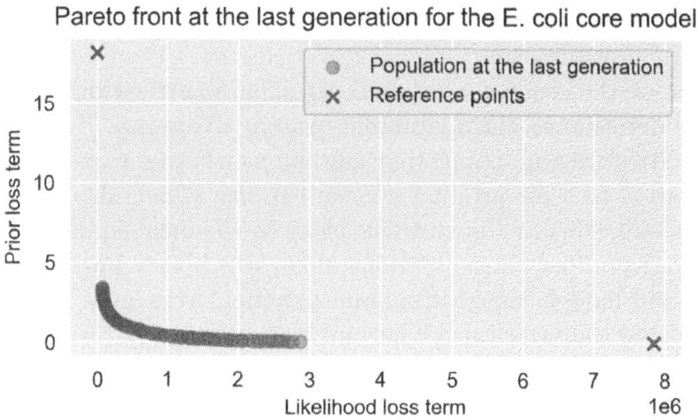

Fig. 5. Pareto front at generation 1000 for the *E. coli* core model.

The results from the evolutionary run on the *E. coli* core model are presented in Fig. 3 (right). Figure 5 shows the set of non-dominated points found at the last generation, along with the two reference points for the known optimal solutions. While the point at the bottom right represents the known prior mean, the point at the top left represents the "true" vector **a** behind the artificial data, which would not be accessible if the ground truth were not known. The results are satisfying: like with the smaller test model, the Pareto front comes very close to the optimal values of the single objectives.

5   Conclusions and Future Works

To explore how measured covariances in metabolite and enzyme data could be used for parameter estimation, we proposed a framework for multi-objective optimization in metabolic models. In our models, a know random distribution of external metabolite and enzyme concentrations leads to a simple formula for covariances between all model variables. We now used a given covariance matrix to infer some model parameters (the enzyme saturation values with respect to internal metabolites), assuming all other model parameters to be known.

In Bayesian estimation, a parameter set would be scored by its posterior density, a product of likelihood and prior. Since the relative importance of these terms may vary (depending on prior widths and data error bars), we treated them here as separate objectives. Each point on the Pareto front represents a model instance. Since the assumed "true" parameter set (our ground truth) differs from the prior mean, there is a trade-off between the objectives, resulting in an extended front that we recovered in our computer experiments. Points from the two ends of the front represent, respectively, solutions that are well supported by data (but with the risk of overfitting) versus more conservative solutions that stay close to our prior expectation about model parameters. By moving along the front, we can interpolate between these extremes and shift our focus of attention between prior and likelihood. This would not be possible in a standard Bayesian estimation.

Each point of the front corresponds to a parametrization of the Jacobian matrix which determines the metabolic system dynamics. Biochemically, the inferred elasticities are important: they can tell us whether reactions are forward-driven (insensitive to their product concentrations, which also makes them relatively enzyme-efficient) or, in contrast, close to chemical equilibrium (sensitive to substrate and product concentrations alike, which may imply a high enzyme demand). In our model, thermodynamic driving forces may also be included explicitly by using the Structural Thermokinetic Modeling variant of SKM [12].

In our optimization, we assumed that reference state and elasticities for external metabolites were known, and estimated only the elasticities for internal metabolites. This could be generalized to fit other model details, including reference fluxes or concentrations, thermodynamic forces, or the presence of regulatory arrows. Likewise, instead of covariances also other types of data could be considered, such as time series or sets of different steady states. With only minor modifications, the proposed framework can also be used to study biological objectives, for example, trade-offs between dynamic robustness and enzyme costs.

Acknowledgments. The authors would like to acknowledge funding from the ANR project Artificial Metabolic Networks (ANR-21-CE45-0021), a grant from the French National Agency for Research.

References

1. Ananda, R., Daud, K.M., Zainudin, S.: Non-dominated sorting differential search algorithm for optimizing regulatory-metabolic networks by using probabilistic approach. In: 2023 International Conference on Electrical Engineering and Informatics (ICEEI), pp. 1–6. IEEE (2023). https://doi.org/10.1109/iceei59426.2023.10346837
2. Bai, L., et al.: Advances and applications of machine learning and intelligent optimization algorithms in genome-scale metabolic network models. Syst. Microbiol. Biomanufacturing **3**(2), 193–206 (2022). https://doi.org/10.1007/s43393-022-00115-6
3. Blank, J., Deb, K.: pymoo: multi-objective optimization in python. IEEE Access **8**, 89497–89509 (2020)
4. Deb, K.: Multi-Objective Optimization Using Evolutionary Algorithms, vol. 16. John Wiley & Sons (2001)
5. Deb, K., Pratap, A.: A fast and elitist multiobjective genetic algorithm: NSGA-II. IEEE Trans. Evol. Comput. **6**, 182–197 (2002)
6. Deb, K., Sindhya, K., Okabe, T.: Self-adaptive simulated binary crossover for real-parameter optimization. In: GECCO '07, Proceedings of the 9th Annual Conference on Genetic and Evolutionary Computation, pp. 1187–1194. Association for Computing Machinery, New York, NY, USA (2007)
7. Elowitz, M.B., Levine, A.J., Siggia, E.D., Swain, P.S.: Stochastic gene expression in a single cell. Science **297**(5584), 1183–1186 (2002). https://doi.org/10.1126/science.1070919

8. Grimbs, S., Selbig, J., Bulik, S., Holzhütter, H.G., Steuer, R.: The stability and robustness of metabolic states: identifying stabilizing sites in metabolic networks. Mol. Syst. Biol. **3**, 146 (2007)
9. Hofmeyr, J.: Metabolic control analysis in a nutshell. In: ICSB 2001 Online Proceedings. http://www.icsb2001.org/toc.html (2001)
10. Li, J., Weckwerth, W., Waldherr, S.: Network structure and fluctuation data improve inference of metabolic interaction strengths with the inverse jacobian. npj Syst. Biol. Appl. 137 (2024)
11. Liebermeister, W., Klipp, E.: Biochemical networks with uncertain parameters. IEE Proc. Sys. Biol. **152**(3), 97–107 (2005)
12. Liebermeister, W.: Structural thermokinetic modelling. Metabolites **12**(5), 434 (2022). https://doi.org/10.3390/metabo12050434
13. Orth, J.D., Fleming, R.M.T., Palsson, B.O.: Reconstruction and use of microbial metabolic networks: the core escherichia coli metabolic model as an educational guide. EcoSal Plus **4**(1) (2010). https://doi.org/10.1128/ecosalplus.10.2.1
14. Reder, C.: Metabolic control theory: a structural approach. J. Theor. Biol. **135**(2), 175–201 (1988). https://doi.org/10.1016/s0022-5193(88)80073-0
15. Sharma, S., Kumar, V.: A comprehensive review on multi-objective optimization techniques: past, present and future. Arch. Comput. Methods Eng. **29**(7), 5605–5633 (2022). https://doi.org/10.1007/s11831-022-09778-9
16. Steuer, R., Kurths, J., Fiehn, O., Weckwerth, W.: Observing and interpreting correlations in metabolomic networks. Bioinformatics **19**(8), 1019–1026 (2003). https://doi.org/10.1093/bioinformatics/btg120
17. Steuer, R., Gross, T., Selbig, J., Blasius, B.: Structural kinetic modeling of metabolic networks. Proc. Natl. Acad. Sci. **103**(32), 11868–11873 (2006)
18. Uhlendorf, J., Bockmayr, A., Liebermeister, W.: Prediction of optimal enzymatic regulation architectures. Master's thesis, Department of Mathematics and Computer Science Bioinformatics Program (2009)

Trace-Elites: Better Quality-Diversity with Multi-point Descriptors

Harald M. Ludwig[1] , Ane Espeseth[2] , and Eric Medvet[3]([✉])

[1] Complexity Science Hub, Vienna, Austria
ludwig@csh.ac.at
[2] Center for Interdisciplinary Education, University of Oslo, Oslo, Norway
ake25@cantab.ac.uk
[3] Department of Engineering and Architecture, University of Trieste, Triest, Italy
emedvet@units.it

Abstract. MAP-Elites (ME) has been shown to be a successful way for solving quality-diversity (QD) optimization problems, *i.e.*, those where the goal is to obtain many diverse solutions of high quality, rather than just one single solution. ME achieves diversity by organizing the population in an archive, where solutions are indexed by a descriptor, *i.e.*, a function mapping each solution to a point in a p-dimensional space. There are however cases where mapping a solution to a single point is not enough for describing it: for instance, when the descriptor should capture the behavior of a robotic agent during a simulation and this behavior has many relevant facets. In this paper, we propose a simple modification of the standard ME which addresses this limitation by employing a descriptor which, in general, maps every solution to one *or more* points in $\mathbb{R}^p$. We call this novel extension Trace-Elites (TE), as the image of a solution in the descriptor space extends across several points, hence corresponding to a sort of trace left by the solution. We experimentally assess the effectiveness of TE on a set of QD problems consisting in the optimization of a controller of a simulated robot which is required to navigate an arena. We show that TE outperforms ME in effectiveness (in terms of both quality and diversity) and efficiency. We also show that, at least in the specific problem considered here, the visualization of the archives evolved by TE gives more insights about the problem than ME ones, potentially permitting a more informed choice of the solutions by the designer.

Keywords: MAP-Elites · Policy search · Neuroevolution

1 Introduction

Optimization problems often require not only identifying the best-performing solution but also discovering a diverse set of solutions that perform well under

---

H. M. Ludwig and A. Espeseth—Contributed equally to this work.

© The Author(s), under exclusive license to Springer Nature Switzerland AG 2025
P. García-Sánchez et al. (Eds.): EvoApplications 2025, LNCS 15612, pp. 338–353, 2025.
https://doi.org/10.1007/978-3-031-90062-4_21

various conditions. Quality-diversity (QD) algorithms, such as MAP-Elites (ME) [16], address this need by exploring the solution space and maintaining a collection of high-quality solutions spread across a behavior-performance map. ME seeks to illuminate the solution space by mapping solutions based on their behaviors and rewarding diversity alongside performance [16].

Traditional ME assigns each agent to a single location in the behavior-performance map, assuming that each solution excels at only one behavior. This simplification overlooks cases where an agent might benefit from showing multiple variations of a behavior, e.g., to encourage spacial exploration in navigation tasks, gait variation in locomotion tasks, or heterogeneous behaviors in task involving swarms rather than single agents. Pugh et al. [21] note that QD tends to improve when the chosen behavior characterization is *strongly aligned* with the implicit objective of the task. As their example they use a navigation task, where the behavior is defined by the agent trajectory in the arena: considering only the final point is in general less aligned with how well the agent navigates the arena than considering many points of the trajectory. However, with traditional ME, forcing this alignment soon runs into the curse of dimensionality. For instance, in the navigation case, Pugh et al. took three samples from each trajectory, hence obtaining a behavioral space in six dimensions: harder to fill, much less practical to visualize than one in two dimensions.

In the cases mentioned above, multiple complimentary aspects of a single solution behavior could potentially be extracted and compared in the same descriptor space. Such a method would retain low dimensions for the descriptor points, but improve representation by including more "samples" of how the solution fills the descriptor space. In the maze task, such "behavior sampling" could select the agent locations at certain time instants, without being limited to one final sample. For a robot swarm sharing the same controller [22], behavior could be sampled per individual or per cluster. Similarly, a Fourier transform could extract multiple frequency behaviors from a single gait solution [14], allowing exploration over individual components of the gait.

To address this limitation in behavior representation, we here introduce Trace-Elites (TE), an extension of ME that allows a solution to occupy multiple locations in the archive by representing complementary components of its behavior as a collection of points in the descriptor. This approach rewards solutions for excelling in diverse behaviors within a single evaluation, enabling a more nuanced representation of both quality and diversity.

We experimentally compare TE against ME on a set of navigation problems in which the solution is the policy of the agent and its behavior, a trajectory, can be conveniently mapped to many points in the descriptor space. We show that TE outperforms ME in both quality and diversity, as it finds a larger number of diverse successful solutions, the difference being much larger on harder problems. Moreover, TE finds successful solution earlier in the search process. We also analyze the impact of the key innovation of TE with respect to ME, *i.e.*, the number of points in the descriptor space assigned to each solution. We find that TE performance is robust with respect to this parameter and it slowly degrades to the performance of ME when approaching the extreme value of only one point per solution.

2 Related Work

When attempting to capture multiple factors of an agent behavior for QD, a few existing paradigms might come to mind [11]. *Multi-objective optimization* was first introduced to ME in 2022, and stores in each location a Pareto front hypervolume instead of a simple fitness value [19]. TE differs from multi-objective optimization in a subtle but important way: the fitness goal remains exactly the same, but agents are given the opportunity to reach that goal through diversifying their behavior—a preliminary version of this idea had been already explored in [13]. For example, an agent that solves a task well using both a running gait and a skipping gait would be rewarded as both the "best runner" and "best skipper". However, if another agent excels solely at running and outperforms the first in that behavior, it will take the top spot as the "best runner", while the original agent retains the title of "best skipper".

Although the low dimensionality of the archive in ME was one of its strongest selling points, allowing visualization and controlled exploration of the search space, *N-dimensional ME* was suggested already in the original paper [16]. Overall, the performance of ME reduces drastically with higher dimensionality [6]. CVT-ME improves on this by reducing the amount of niches in the archive hypercube, but suffers from low resolution in each dimension and does not improve on the performance of low-dimensional ME [24]. Some implementations sidestep high-dimensionality by maintaining several descriptor-spaces simultaneously [4,7,18]. Age-layered ME applies different "layers" of descriptors to different ages of the population [20]. Another approach lets the user alternate the use of different descriptor spaces while evolution is in progress [1,25], thereby emphasizing *interactive exploration* of the mapping from solution to descriptor. TE does not require higher dimensions in its descriptor space. In fact, by allowing one individual to occupy multiple niches, its archive fills faster than traditional low-dimensional ME.

Multi-task ME mimics the archive structure from ME, but uses task parameters (*e.g.*, environmental conditions, start-morphology) as its dimensions [2,17]. However, the authors note that this method requires at least 1072 tasks (user-defined or parameterized) to be feasible. There are many cases where just a few conditions exhaust the meaningful variation of the task. Behavior sampling of each condition could in theory incentivize multi-task specialization while maintaining behavioral diversity in the descriptor space.

For tasks in *noisy domains*, a ME solution can be sensitive to uncertainty in both the fitness function and the behavioural descriptors [10]. Repeated runs can reduce noise at the cost of additional resources. Solutions such as Deep-Grid ME [9] can reduce computation time at the expense of archive memory. Behavioral sampling within-run, however, comes at none of these costs, and could potentially de-noise the descriptor representation.

3   Trace-Elites (TE)

We here present our proposed algorithm, TE. Since TE is based on ME, we first describe how ME works, introducing its basic concepts, notations, and requirements. In this study, we employ a standard version of ME, taking inspiration from the work of Mouret and Clune [16], with some basic improvements, namely, the archive based on centroidal Voronoi tessellation [24] and the isoline variation operator [23].

MAP-Elites (ME). ME is an evolutionary algorithm (EA) which poses few requirements and has hence a broad applicability. It can be applied to problems defined by a pair $S, \prec$, where S is the search (or solution) space and $\prec$ is a partial order relation defined on S. The relation $\prec$ establishes the notion of the quality of solutions, and hence drives the optimization. We write that $s' \prec s''$ if the solution s' is better than the solution s''.

In order to work, ME requires the user to provide a probability distribution $p_{init} \in P_S$ over S, a crossover operator $c : S \times S \rightarrow P_S$, and a descriptor $d : S \rightarrow \mathbb{R}^p$. The probability distribution p_{init} gives, when sampled, a solution in $s \in S$: we denote this operation as $s \hookleftarrow p_{init}$. The crossover operator c is a function that, given two solutions s_1, s_2, returns a probability distribution $c(s_1, s_2) \in P_S$ over S: this distribution can be then sampled to produce another solution $s' \in S$: we denote this operation as $s' \hookleftarrow c(s_1, s_2)$. The descriptor is a function that, given a solution s, returns a point $\boldsymbol{d} = d(s)$ in the descriptor space $\mathbb{R}^p$. The descriptor d establishes the notion of diversity of solutions: *i.e.*, solutions are deemed different if they map to different points according to d.

Internally, ME employs an archive to store solutions during the iterative evolution process. The archive is organized in locations: at each location, at a given iteration of the evolution, there is at most one solution. ME put solutions at locations of the archive using the descriptor, possibly replacing already present solution if worse according to $\prec$.

Formally, we define as L the (discrete) set of locations. We define as $\ell : \mathbb{R}^p \rightarrow L$ the location mapping function: given a point $\boldsymbol{d}$ in the descriptor space $\mathbb{R}^p$, this function returns a location $l = \ell(\boldsymbol{d}) \in L$. Finally, we define the archive as a function $a : L \rightarrow S \cup \{\varnothing\}$ which, given a location $l \in L$, returns the solution $s \in S$ being stored at that location in the archive, if any, or $\varnothing$ otherwise; we denote as $a(L)|_{\neq \varnothing}$ the set being the image of L in S, *i.e.*, the image without empty locations, provided by a; $a(L)|_{\neq \varnothing}$ is hence the content of the archive. We remark that the role of ℓ is to "discretize" the descriptor, such that solutions can be mapped to discrete locations rather than to "continuous points" in $\mathbb{R}$, hence enforcing a stronger notion of diversity.

With the above definitions, the working of ME can be described concisely as follows (see Algorithm 1). Initially, ME generates n_{pop} solutions by sampling p_{init} n_{pop} times: each solution s is inserted in the archive at location $\ell(d(s))$ by setting $a(\ell(d(s)))$ to s if the location is empty or if the previous solution $s' = a(\ell(d(s)))$ at that location is worse than s, *i.e.*, if $s \prec s'$. Then the archive is updated for n_{iters} times by building an offspring from the current content and then inserting it in the archive. In detail, at each iteration ME builds the offspring $S_{\text{offspring}}$ by taking n_{pop} pairs of solutions in the archive (sampling it with uniform probability) and recombining them using c. Then, it inserts each generated solution s of the offspring $S_{\text{offspring}}$ in the archive at its location $\ell(d(s))$ if it is better then the currently contained one or if that location is empty. At the end, ME returns the content of the archive, $i.e.$, the set $a(L)|_{\neq\varnothing}$. The number of returned solutions is at most $|L|$ and at least one: moreover, they are all different based on the equivalence relation established by $\ell \circ d$, $i.e.$, for all pairs of different elements $s, s' \in a(L)|_{\neq\varnothing}$, it holds that $\ell(d(s)) \neq \ell(d(s'))$.

Algorithm 1. Pseudocode for ME. $\prec$ is the problem to be solved. p_{init}, m, and d are parameters depending on S. ℓ, n_{iters}, and n_{pop} are other parameters.

1: **function** EVOLVE($\prec; p_{\text{init}}, c, d, \ell, n_{\text{iters}}, n_{\text{pop}}$)
2:   **for** $l \in L$ **do**   ▷ initialize the archive
3:       $a(l) \leftarrow \varnothing$
4:   **end for**
5:   $S_{\text{init}} \leftarrow []$
6:   **for** n_{pop} times **do**
7:       $s \leftsquigarrow p_{\text{init}}$
8:       $S_{\text{init}} \leftarrow S_{\text{init}} \oplus [s]$
9:   **end for**
10:   UPDATE($A, S_{\text{init}}, \prec, d, \ell$)   ▷ update archive with initial population
11:   **for** n_{iters} times **do**   ▷ iterate
12:       $S_{\text{offspring}} \leftarrow []$ ▷ build offspring
13:       **for** n_{batch} times **do**
14:           $s_1 \leftsquigarrow U(a(L)|_{\neq\varnothing})$
15:           $s_2 \leftsquigarrow U(a(L)|_{\neq\varnothing})$
16:           $s' \leftsquigarrow c(s_1, s_2)$
17:           $S_{\text{offspring}} \leftarrow S_{\text{offspring}} \oplus [s']$
18:       **end for**
19:       UPDATE($A, S_{\text{offspring}}, \prec, d, \ell$)   ▷ update archive with offspring
20:   **end for**
21:   **return** $a(L)|_{\neq\varnothing}$
22: **end function**

1: **procedure** UPDATE($A, S', \prec, d, \ell$)   ▷ update the archive A with solutions in S'
2:   **for** $s \in S'$ **do**
3:       **if** $a(\ell(d(s))) = \varnothing \vee s \prec a(\ell(d(s)))$ **then**
4:           $a(\ell(d(s))) \leftarrow s$
5:       **end if**
6:   **end for**
7: **end procedure**

It can be seen from Algorithm 1 that ME requires, for solving the problem defined by $\prec$ operating on S, three parameters related to S (p_{init}, c, and d) and other three parameters not related to S (ℓ, n_{iters}, and n_{pop}).

For what concerns the location mapping function ℓ, the original ME employed a function discretizing the continuous descriptor space $\mathbb{R}^p$ by binning each coordinate of the descriptor independently: that is, with a p-dimensional descriptor, p intervals $[d_{1,\mathrm{inf}}, \ldots, d_{1,\mathrm{sup}}], \ldots, [d_{p,\mathrm{inf}}, \ldots, d_{p,\mathrm{sup}}]$, and p numbers of bins $h_1, \ldots, h_p$, L was defined as $\{1, \ldots, h_1\} \times \cdots \times \{1, \ldots, h_p\}$ and $\ell(\boldsymbol{d})$ returned the point where each j-th coordinate d_j of $\boldsymbol{d}$ was the index of the equal-width bin where d_j fell the corresponding j-th interval. Later, other variants for ℓ have been proposed, including the one used in this work, which is based on centroidal Voronoi tessellation [24]. With this location mapping function, given a p-dimensional descriptor and n_c centroids $\{c_i\}_{i=1}^{i=n_c}$, i.e., points in $\mathbb{R}^p$, L is $\{1, \ldots, n_c\}$ and $\ell(\boldsymbol{d})$ returns the index of the closest centroid which is the closest to $\boldsymbol{d}$, i.e., $\ell(\boldsymbol{d}) = \arg\min_{i \in \{1, \ldots, n_c\}} \|\boldsymbol{d} - c_i\|$.

Finally, concerning the crossover operator c, in this study we employ the isoline variation [23], which is applicable if S is $\mathbb{R}^m$, as in our case (see Sect. 4.1). Given two solutions $\boldsymbol{\theta}_1, \boldsymbol{\theta}_2 \in \mathbb{R}^m$, this crossover returns $\boldsymbol{\theta}_1 + \boldsymbol{\alpha} + \beta(\boldsymbol{\theta}_2 - \boldsymbol{\theta}_1)$, where $\boldsymbol{\alpha} \sim N(\boldsymbol{0}, \boldsymbol{I}\sigma_{\mathrm{point}})$, $\beta \sim N(0, \sigma_{\mathrm{line}})$, and $\boldsymbol{I}$ is the identity $m \times m$. We use $\sigma_{\mathrm{point}} = 0.005$ and $\sigma_{\mathrm{line}} = 0.05$.

Trace-Elites (TE). Our TE has the same broad applicability of ME. Indeed, its applicability is even broader, as it takes a descriptor d which is not constrained to return one single point in $\mathbb{R}^p$, but it takes one that can return *zero or more* points. Formally, $d : S \to \mathcal{P}(\mathbb{R}^p)$, with $\mathcal{P}(\mathbb{R}^p)$ being the power set of $\mathbb{R}^p$.

Internally, the only difference between ME and TE is in the way the archive is updated, i.e., in the UPDATE() procedure shown in Algorithm 2. Given a bag S' of solutions, for each s in S, TE obtains the set D of points in the descriptor space for s and, for each $\boldsymbol{d}$ in D, it inserts s in the archive at the corresponding location $\ell(\boldsymbol{d})$ if it is better then the currently contained one or if that location is empty.

Algorithm 2. Pseudocode for the UPDATE() procedure of TE.

1: **procedure** UPDATE($A, S', \prec, d, \ell$)   ▷ update the archive A with solutions in S'
2:    **for** $s \in S'$ **do**
3:        $D \leftarrow d(s)$
4:        **for** $d \in D$ **do**
5:            **if** $a(\ell(\boldsymbol{d})) = \varnothing \vee s \prec a(\ell(\boldsymbol{d}))$ **then**
6:                $a(\ell(\boldsymbol{d})) \leftarrow s$
7:            **end if**
8:        **end for**
9:    **end for**
10: **end procedure**

One direct consequence of the fact that the description $d(s)$ for one solution can consist of more than one points is that the uniqueness of solutions in the archive is not enforced anymore in TE (while it was in ME). We expect that this will impact on the trade-off between quality and diversity (and, more broadly, between exploitation and exploration) obtained by TE. In practice, we expect that TE will exhibit faster convergence to good quality than ME, because good solutions may be stored at several locations in the archive and can hence be selected with higher probability. At the same time, we expect to observe lower uniqueness: in the extreme case and depending on d, in TE one single solution might fill the entire archive.

Regardless on the impact on the overall search performance of the EA, we remark that we designed TE in order to be usable precisely in those scenarios where the user *wants* to have descriptors giving more points, rather than just one.

4   Experimental Evaluation

We compare TE against ME on four variants of a policy search problem with the aim of answering the following three research questions: (RQ1) Is TE more effective than ME? Does it find higher quality solutions? Does it find more diverse solutions? (RQ2) Is TE more efficient than ME? Does it reach faster the same quality of solutions? (RQ3) Are TE solutions different than ME ones? Moreover, we perform an experiment to assess what is the impact on TE effectiveness of the number of points the descriptor actually produces.

We implemented TE in the QDax framework [5] and conducted all experiments using the Kheperax simulator [12] for the navigation problem (see below). We set $n_{\text{iters}} = 1072$ and $n_{\text{pop}} = 100$ for both ME and TE: we remark that the two EAs do not significantly differ in computational complexity if the heaviest computation is the fitness evaluation, as in our case. In particular, they generate the same amount of new individuals at each iteration.

We run the experiments on a desktop PC equipped with an AMD Ryzen 7 3700X CPU and a NVIDIA GeForce RTX 3070 GPU. We observed no differences in execution times between the two EAs: both took $\approx 60\,\text{s}$ for each run.

4.1   Navigation Problems

We consider a policy search problem for an agent involved in a navigation task, *i.e.*, it has to reach a predefined target position in an arena with some obstacles.

The agent is a simulated differential drive circular-shaped robot equipped with three proximity sensors, two contact sensors, and two actuators (one for each wheel). The robot radius is $1.5\,\text{cm}$. The proximity sensors range is $20\,\text{cm}$ and they are directed at angles $-\frac{\pi}{4}$, 0, and $\frac{\pi}{4}$ with respect to the front of the robot. The contact sensors cover the front-facing quarters of the robot, from $\frac{-\pi}{2}$ to 0 (left) and 0 to $\frac{\pi}{2}$ (right). Proximity sensors perceive the distance to the closest obstacle (arena outer and inner walls), if within range, or -1 otherwise;

contact sensors perceive 1 when the robot is touching a wall and -1 otherwise. We normalize the sensor readings in order to make the agent observation be defined in $[-1, 1]^5$. The two actuators take values in $[-1, 1]$ each, hence giving an action to be defined in $[-1, 1]^2$.

We employ a multilayer perceptron (MLP) as the policy for the agent. We use an input layer with five neurons, two hidden layers with 1072 neurons each, and an output layer with two neurons. We use the ReLU activation function in the inner nodes and tanh in the output nodes, in order to ensure the output to be in the action space $[-1, 1]^2$. The policy is hence defined as the vector $\boldsymbol{\theta} \in \mathbb{R}^{4674}$ of the synaptic weights of the MLP, which is what we optimize.

For each candidate solution $\boldsymbol{\theta}$ we perform a simulation lasting at most 1072 time steps: we stop the simulation if the agent distance to the target becomes lower than 5 cm, $i.e.$, if the agent is $successful$. We associate with each candidate solution $\boldsymbol{\theta}$ a fitness score $f(\boldsymbol{\theta})$ being the negative duration of the simulation; if the agent does not reach the target (with the above criterion), we further add to the score $-100l$, with l being the distance at the last time step. This policy search problem hence corresponds to a search space $S = \mathbb{R}^{4674}$ and a partial order $\prec$ such that $\boldsymbol{\theta} \prec \boldsymbol{\theta}'$ if and only if $f(\boldsymbol{\theta}) > f(\boldsymbol{\theta}')$, $i.e.$, the order is actually total and the fitness has to be maximized.

We experiment with four equal size ($1\,\text{m} \times 1\,\text{m}$) arenas of increasing difficulty of exploration, because of the relative positions of agent starting point, target, and walls. We show them in Fig. 1.

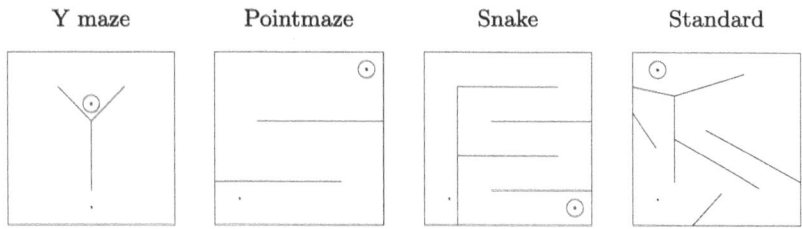

| Y maze | Pointmaze | Snake | Standard |

Fig. 1. The four arenas used in the experiments. The small marker denotes the starting point of the agent; the star enclosed in a black circle denotes the target point; the circle marks the "success" zone.

As descriptor, we use for both ME and TE a behavioral descriptor, $i.e.$, one that produces its output based on the trajectory of the agent in the arena. For ME we adopt the common choice of having d return the final agent position ($i.e.$, its two x and y coordinates) in the arena. For TE we make d return a fixed number n_d of points of the trajectory run by the agent and taken at regular time interval during the simulation, namely at time steps $\left\{ \left\lfloor k_{\text{final}} \frac{j}{n_d} \right\rfloor \right\}_{j=1}^{j=n_d}$, with k_{final} being the duration of the simulation (1072 for not successful agents and ≤ 1072 for successful agents). It can be noted that, with this descriptor and $n_d = 1$, our TE corresponds to ME. On the other hand, the greater the value of n_d, the more archive locations a single solution can fill. If not otherwise specified, we experimented with $n_d = 10$.

Finally, for both ME and TE we choose the centroidal Voronoi tessellation as location mapping function and we choose 1072 centroids as follows. First, we sample 1072 points uniformly distributed across the arena; then we apply k-means clustering with $k = 1072$ and use the cluster centers as centroids. We use the same centroids for all the experiments.

4.2   Results and Discussion

Quality. We applied both the EAs to each one of the arenas 1072 times, changing the seed for the random generator. Figures 2 and 3 show the fitness of the best solution in the archive (briefly, the best fitness), for each EA and arena. The former shows the progression of the best fitness during the evolution, plotting the median value across the 1072 repetitions and the interquartile range. The latter shows its distribution at the end of the evolution in the form of boxplots. Clearly, the best fitness accounts for the "quality part" of the ability of an EA to solve a QD problem.

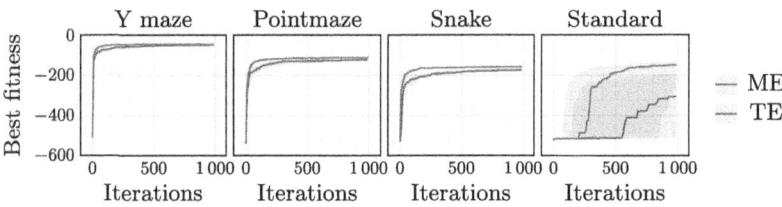

Fig. 2. Progression of the fitness of the best solution during the evolution: median and interquartile range across the 1072 repetitions for each pair EA/arena. TE converges faster than ME on all problems; moreover, the harder the problem, the larger the difference.

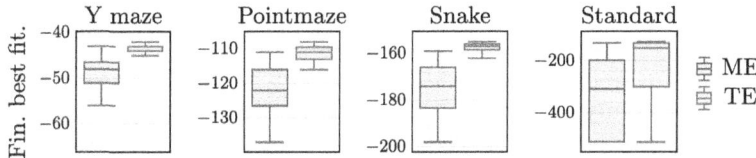

Fig. 3. Distribution of the final best fitness across the 1072 repetitions for each pair EA/arena. TE outperforms ME in all problems.

Three key observations can be made based on Fig. 2. First, three on four problems are solved by both EAs, *i.e.*, they both find at least one agent which reaches the target. This can be inferred from the fact that for Y maze, Pointmaze, and Snake, the fitness of the best solution converges to a value and does not further improve. Second, for one of the four problems, Standard, only TE appears to converge and the difference of best fitness values between TE and ME is much larger than in the other cases. Third, TE converges faster than ME on all problems: on the ones where both EAs find good solutions, TE finds them earlier in the optimization; in Standard, TE starts improving the initial unsuccessful solutions much earlier then ME. Summarizing, TE is at least as effective as ME on all problems and more effective than ME on the hardest problem, Standard. Indeed, by looking at the distribution of the final values for the best fitness in Fig. 3, we can see that TE is actually more effective and more consistent than ME on all problems. We performed a statistical significance test (Wilcoxon rank-sum test) and found that for all the problems the difference is significant with an $\alpha = 0.05$.

Quality and Diversity. When approaching a QD optimization problem, looking just at the quality of the highest quality solution (*i.e.*, the best fitness) achieved by an EA does not capture its ability to find many diverse solutions to the problem. To address this limitation, the QD-score metric has been introduced [21] which "sums" the quality measured for all the solutions in the archive. Here, we adjust the standard definition of the QD-score to take into account the fact that with TE the same solution may be stored at several locations in the archive: we hence consider only unique solutions in the summation. Moreover, since our fitness function takes negative values, we sum a predefined offset to make all fitness values positive while retaining the "the greater, the better" semantics.

Figures 4 and 5 show the progression and final values for our corrected QD-score with the same visual syntax of the corresponding figures for the best fitness.

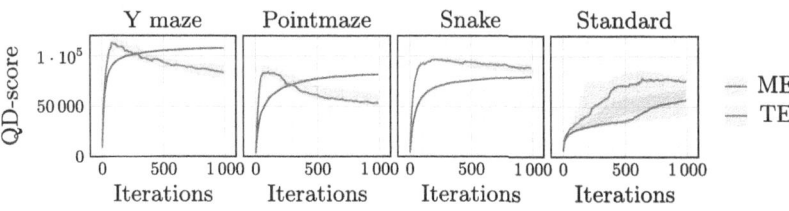

Fig. 4. Progression of the QD-score of the population during the evolution: median and interquartile range across the 1072 repetitions for each pair EA/arena. TE tends to decrease in QD-score after having reached an initial peak; for harder problem, however, it still outperforms ME.

Figures 4 and 5 reveal an interesting pattern: in simpler environments (Y maze and Pointmaze), TE shows a distinctive behavior where QD-score peaks early but then gradually declines. This decline occurs because better solutions found by TE tend to occupy multiple locations with slightly different trajectories,

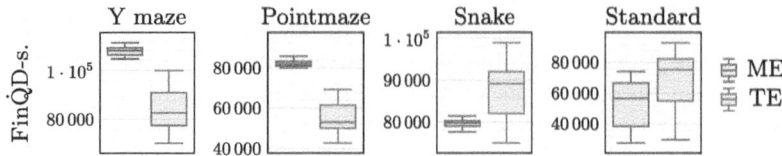

Fig. 5. Distribution of the final QD-score across the 1072 repetitions for each EA/arena pair. TE performs better than ME in harder problems and is outperformed in easier problems.

effectively reducing the diversity in the archive as measured by unique solutions. However, in more complex environments (Snake and Standard), this effect is less pronounced, and TE maintains higher QD-scores than ME. This suggests that in harder problems, the ability of TE to explore multiple trajectory variants becomes advantageous for maintaining both quality and diversity in the archive. More broadly, these results confirm that TE and ME find two different trade-off points between quality and diversity, with TE being better in quality and ME being better in diversity. However, while TE always outperform ME in best fitness, ME outperforms TE in QD-score only in two out of four problems, as shown in Fig. 5—the differences are always statistically significant.

Successful Diversity. For gaining deeper insights on the performance of TE and ME, we analyzed one evolutionary run for each arena in detail.

We first show in Fig. 6 a visualization of the archive at the end of the evolution for both EAs. We adopt the common visual syntax of the QD literature: we plot the two-dimensional descriptor space (here, the coordinates of the agent position in the arena) as square and we segment this square in regions which correspond to archive locations, dictated by ℓ. Finally, we fill each region with a color determined by the fitness of the solution stored in the archive at that location: the higher, the more yellow, the worse, the more blue.

The most apparent difference between TE and ME archives visible in Fig. 6 is in the amount of good locations, *i.e.*, yellow, or almost yellow, regions. However, we remark that with TE the same solution can be stored at different location, while with ME all solutions are unique by construction. Hence, by looking at the archive alone it is not possible to say at a glance "how much QD" the two algorithms found. We hence show in Table 1 the number of unique solutions and successful unique solutions found by the two EAs on the four problems (for the same evolutionary runs considered in Fig. 6). We use the criterion for success defined in Sect. 4.1, based on the distance of the agent to the target.

The figures in Table 1 makes apparent the fact that, for all the problems, TE is much better in finding many diverse *and* successful solutions than ME. The number n_{succ} of unique successful solutions is always at least one order of magnitude greater for TE. For the harder problem, Standard, TE finds 1072 successful solutions, whereas ME finds just two of them.

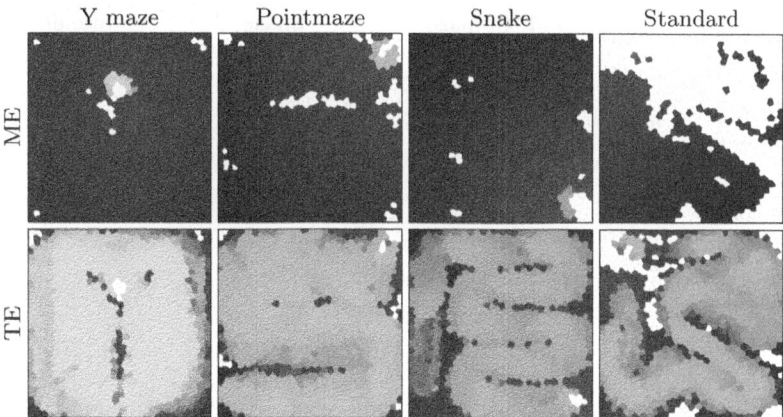

Fig. 6. Visualization of the archives for one randomly chosen evolutionary run for each EA/arena pair. Picking a successful solution with some desired descriptor value (*i.e.*, for this problem, one passing through some given point in the arena) is much easier in TE than in ME archives. (Color figure online)

Table 1. Number of unique (n_{unique}) and unique successful (n_{succ}) solutions in the final archive for the runs shown in Fig. 6.

EA	Y maze		Pointmaze		Snake		Standard	
	n_{unique}	n_{succ}	n_{unique}	n_{succ}	n_{unique}	n_{succ}	n_{unique}	n_{succ}
ME	1024	14	1024	13	1024	12	1024	2
TE	250	179	221	136	336	108	348	175

Besides the difference in raw numbers, we highlight a further consequence of the way TE fills the archive. Suppose the user running the optimization "in a QD way" wants to find a solution presenting some given traits, *i.e.*, being in some given point of the descriptor space, *and* having a good quality. This task can be performed much easier with TE archives than with ME archives. In the specific case of the navigation problems here considered, this means that (almost) whatever position in the arena the user considers, it is possible to find a successful agent which passes through that position in TE archives. This is simply not doable with ME archives. While we acknowledge that this practical advantage is strictly related with the specific descriptor space being employed, we remark that it can be obtained without paying any other cost, *i.e.*, without having neither a lower search effectiveness, nor a lower search efficiency.

Finally, to conclude the analysis, we show in Fig. 7 the trajectories of the best ten unique solutions for each run corresponding to the archives shown in Fig. 6. Note that, based on the numbers shown in Table 1, the ten best solutions found by TE are always successful, whereas for ME in the Standard problem, only two on ten best agents are actually successful.

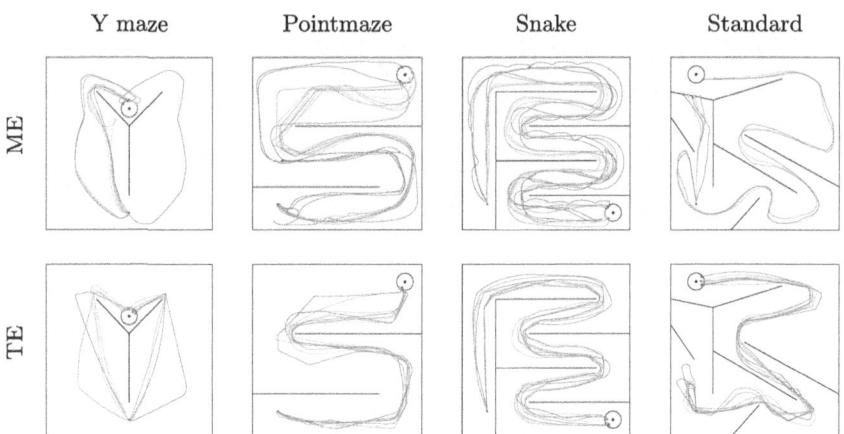

Fig. 7. The trajectories of the best ten solutions for one randomly chosen evolutionary run (the same of Fig. 6 and Table 1) for each EA/arena pair. TE trajectories are in general "more optimal", *i.e.*, shorter, than ME trajectories.

By looking carefully at Fig. 7 it can be seen that the trajectories of agents found by TE are in general better than the ones found by ME. That is, TE best agents, while running through diverse paths, all take shorter trajectories to the target, passing closer to walls which mark turning points. Indeed, this observation is consistent with the distributions shown in Fig. 3: while they were based just on the trajectory of the single best solution, they were showing that TE best solutions achieved in general better fitness values, *i.e.*, they reached earlier the target, hence running along a shorter trajectory.

Impact of the Number of Points in the Descriptor Space. The key difference between TE and ME is in the fact that the TE descriptor gives zero or more points in the descriptor space for each solution, rather that just one. In the problems considered in this study the actual number of points can be specified in advance by setting n_d: in the extreme case, TE with $n_d = 1$ corresponds to ME. We here look at the impact of n_d, *i.e.*, the size of the trace of each solution, on the performance of TE. For the sake of brevity, we consider only the Pointmaze arena. We show in Fig. 8 the progression of the best fitness for nine different values of n_d ranging from 2 to 10.

It can be seen that the search behavior of TE changes smoothly while decreasing the n_d. Lines corresponding to large values of n_d are quite close, while does corresponding to $n_d = 2$ or $n_d = 3$ are farther away, showing that TE behaves like ME in these conditions. Considering the simple modification of the algorithm that differentiates TE from ME, this finding is not particularly surprising. More broadly, the results of this experiment suggest that (a) TE is quite robust with respect to the size of the trace of solutions and, on the other hand, that (b) users can freely set this parameter to fit their needs without worrying too much about its impact on search effectiveness and efficiency.

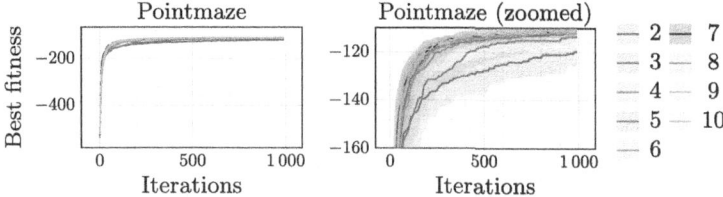

Fig. 8. Progression of the fitness of the best solution during the evolution for the Pointmaze arena with different values of n_d: median and interquartile range (only in the zoomed plot) across the 1072 repetitions for each n_d value. With decreasing n_d, TE search behavior smoothly degrades to ME one.

5 Concluding Remarks and Future Perspective

We presented a simple improvement of ME, a widely adopted EA for solving problems "in a QD way", $i.e.$, by finding many high quality and diverse solutions. Our extension, which we named Trace-Elites (TE), permits to adopt a descriptor which maps each solution not just to one point in the descriptor space, but to zero or more points. By design, TE is hence a generalization of ME which broadens its applicability. We performed an experimental evaluation of TE on a small set of policy search problems involving agents that have to navigate an arena. In this problems, diversity is useful both $per\ se$ and as a way for the EA to avoid falling in local optima. We found that TE is both more efficient and more effective than ME: it finds a larger number of diverse successful solutions in a shorter evolutionary time. Solutions found by TE are in general of higher quality than those found ME, with agents running along better trajectories to reach the target. Moreover, we showed that archives evolved with TE may be practically more useful if one has to pick "manually" one or more successful solutions which match some criterion based on their location in the descriptor space.

While we acknowledge that we here applied TE only on one kind of QD optimization problems, we do believe that it might be practically useful also on other kinds of problems. For example, one way to deal with policy search for swarm robotics [3,15] is to optimize one single controller that is deployed on every robot in the swarm: by using TE, one might employ a descriptor that takes each single robot behavior as input, rather than the swarm behavior as a whole. This way, the design of the descriptor might be easier and, at the same type, the archive would favor controllers which are able to exhibit different successful behaviors. A similar application might involve the case of modular robots, where the same controller is applied to each physically identical module but the behavior they actually exhibit depends on how they differently perceive the world [8] or on some plasticity available in the agent [22].

Acknowledgments. This study was carried out within the PNRR research activities of the consortium iNEST (Interconnected North-Est Innovation Ecosystem) funded by the European Union Next-GenerationEU (Piano Nazionale di Ripresa e Resilienza

(PNRR) - Missione 4 Componente 2, Investimento 1.5 – D.D. 1058 23/06/2022, ECS_00000043).

We thank the SPECIES Society and Anna Esparcia-Alcázar for organizing the SPECIES Summer School 2024, which brought us together and gave us the chance to start this collaboration.

References

1. Alvarez, A., Dahlskog, S., Font, J., Togelius, J.: Interactive constrained map-elites: analysis and evaluation of the expressiveness of the feature dimensions. arXiv.org (2020). https://doi.org/10.1109/TG.2020.3046133
2. Anne, T., Mouret, J.B.: Multi-task multi-behavior map-elites. In: Proceedings of the Companion Conference on Genetic and Evolutionary Computation, GECCO 2023, pp. 111–114. Companion, Association for Computing Machinery, New York, NY, USA (2023). https://doi.org/10.1145/3583133.3590730
3. Bossens, D.M., Tarapore, D.: Qed: using quality-environment-diversity to evolve resilient robot swarms. IEEE Trans. Evol. Comput. 25(2), 346–357 (2020)
4. Cazenille, L.: Ensemble feature extraction for multi-container quality-diversity algorithms. In: Proceedings of the Genetic and Evolutionary Computation Conference, pp. 75–83 (2021)
5. Chalumeau, F., et al.: Qdax: a library for quality-diversity and population-based algorithms with hardware acceleration. J. Mach. Learn. Res. 25(108), 1–16 (2024)
6. Chatzilygeroudis, K., Cully, A., Vassiliades, V., Mouret, J.B.: Quality-diversity optimization: a novel branch of stochastic optimization. In: Black Box Optimization, Machine Learning, and No-Free Lunch Theorems, pp. 109–135. Springer (2021)
7. Doncieux, S., Coninx, A.: Open-ended evolution with multi-containers qd. In: Proceedings of the Genetic and Evolutionary Computation Conference Companion, pp. 107–108 (2018)
8. Ferigo, A., Iacca, G., Medvet, E., Nadizar, G.: Totipotent neural controllers for modular soft robots: achieving specialization in body-brain co-evolution through hebbian learning. Neurocomputing, p. 128811 (2024)
9. Flageat, M., Cully, A.: Fast and stable map-elites in noisy domains using deep grids. IEEE Sympos. Artif. Life (2020). https://doi.org/10.1162/ISAL_A_00316
10. Flageat, M., Cully, A.: Uncertain quality-diversity: evaluation methodology and new methods for quality-diversity in uncertain domains. IEEE Trans. Evol. Comput. (2023). https://doi.org/10.48550/ARXIV.2302.00463
11. Gavrilescu, M., Leon, F., Ferariu, L.E., Buţincu, C.N.: A review of evolutionary optimization methods for information visualization and feature space exploration. In: 2024 International Conference on INnovations in Intelligent SysTems and Applications (INISTA), pp. 1–6 (2024). https://doi.org/10.1109/INISTA62901.2024.10683825
12. Grillotti, L., Cully, A.: Kheperax: a lightweight jax-based robot control environment for benchmarking quality-diversity algorithms. In: Proceedings of the Companion Conference on Genetic and Evolutionary Computation, pp. 2163–2165 (2023)
13. Jónsson, B.T., Erdem, Ç., Fasciani, S., Glette, K.: Towards sound innovation engines using pattern-producing networks and audio graphs. In: International Conference on Computational Intelligence in Music, Sound, Art and Design (Part of EvoStar), pp. 211–227. Springer (2024)

14. Methenitis, G., Hennes, D., Izzo, D., Visser, A.: Novelty search for soft robotic space exploration. In: Proceedings of the 2015 Annual Conference on Genetic and Evolutionary Computation, pp. 193–200 (2015)
15. Mkhatshwa, S., Nitschke, G.: Body and brain quality-diversity in robot swarms. ACM Trans. Evol. Learn. Optim. (2024)
16. Mouret, J.B., Clune, J.: Illuminating search spaces by mapping elites. arXiv preprint arXiv:1504.04909 (2015)
17. Mouret, J.B., Maguire, G.: Quality diversity for multi-task optimization. In: Proceedings of the 2020 Genetic and Evolutionary Computation Conference, GECCO 2020, pp. 121–129. Association for Computing Machinery, New York, NY, USA (2020). https://doi.org/10.1145/3377930.3390203
18. Nadizar, G., Medvet, E., Wilson, D.: Searching for a diversity of interpretable graph control policies. In: Proceedings of the Genetic and Evolutionary Computation Conference, pp. 933–941 (2024)
19. Pierrot, T., Richard, G., Beguir, K., Cully, A.: Multi-objective quality diversity optimization. In: Proceedings of the Genetic and Evolutionary Computation Conference, pp. 139–147 (2022)
20. Pozzuoli, A., Ross, B.: Increasing features in map-elites using an age-layered population structure. In: 2023 IEEE Congress on Evolutionary Computation (CEC), pp. 1–8 (2023). https://doi.org/10.1109/CEC53210.2023.10254093
21. Pugh, J.K., Soros, L.B., Szerlip, P.A., Stanley, K.O.: Confronting the challenge of quality diversity. In: Proceedings of the 2015 Annual Conference on Genetic and Evolutionary Computation, pp. 967–974 (2015)
22. Rusin, F., Medvet, E.: How perception, actuation, and communication impact the emergence of collective intelligence in simulated modular robots. Artificial Life, pp. 1–18 (2024)
23. Vassiliades, V., Mouret, J.B.: Discovering the elite hypervolume by leveraging inter-species correlation. In: Proceedings of the Genetic and Evolutionary Computation Conference, pp. 149–156 (2018)
24. Vassiliades, V., Chatzilygeroudis, K., Mouret, J.B.: Using centroidal voronoi tessellations to scale up the multidimensional archive of phenotypic elites algorithm. IEEE Trans. Evol. Comput. **22**(4), 623–630 (2017)
25. Wilson, J., Hauert, S.: Search space illumination of robot swarm parameters for trustworthy interaction. In: International Symposium on Distributed Autonomous Robotic Systems, pp. 173–186. Springer (2022)

Optimizing Camera Placement for Chicken Farm Monitoring

Kyriacos Mosphilis⬭ and Vassilis Vassiliades$^{(\boxtimes)}$ ⬭

CYENS Centre of Excellence, Dimarchou Lellou Demetriadi 1, 1016 Nicosia, Cyprus
{k.mosphilis,v.vassiliades}@cyens.org.cy

Abstract. Animal farming has transitioned from small-scale operations to large commercial ventures, raising concerns about animal welfare alongside productivity and profitability. Artificial intelligence technologies offer significant potential for enhancing welfare through improved monitoring. However, practical solutions for optimizing farm management decisions are still limited. This paper tackles the challenge of optimizing camera placement in chicken farms to achieve maximum coverage for effective monitoring. We employ the Covariance Matrix Adaptation Evolution Strategy (CMA-ES) and two Quality Diversity (QD) algorithms, MAP-Elites (ME) and CMA-ME, using two behaviour descriptors to identify optimal camera positions. Our findings show that algorithm-derived camera placements outperform human-designed configurations. Importantly, the QD algorithms offer a diverse set of high-quality solutions that can be selected without extra computation, in case the CMA-ES solution does not meet unforeseen constraints. Using our modelling, we have optimized the camera placement in a commercial farm in Cyprus offering maximum coverage surveillance.

Keywords: cma-es · quality diversity · map-elites · farm monitoring · optimization

1 Introduction

Animal welfare is a topic that has drawn much attention over the past years. Animal well-being and efficient farming has been associated with increased profits by reducing mortality, improving health, product quality, disease resistance; reduced medication; and increasing the ability to command higher prices from consumers [6]. Recent works have shown how Internet of Things (IoT) can revolutionize the agriculture industry with smart solutions for livestock monitoring, and precision farming with the assistance of Artificial Intelligence (AI) [8,20,25]. These works discuss how IoT-based data capture and AI models can improve livestock well-being and, consequently, producer profits. Other works have shown how AI and sensors can assist in creating smart poultry farms to empower welfare of egg-producing chicken flocks [1,12].

In this paper we present a farm monitoring optimization problem, which focuses on finding optimal camera positions and orientations. Similar works for

© The Author(s), under exclusive license to Springer Nature Switzerland AG 2025
P. García-Sánchez et al. (Eds.): EvoApplications 2025, LNCS 15612, pp. 354–369, 2025.
https://doi.org/10.1007/978-3-031-90062-4_22

placing cameras and other sensors have been carried out in the past. For example, a study took place for finding the optimal placement and orientation of visual sensors, like cameras, which proposed a new Field of View (FOV) model, which distributes the sensors similar to a human eye. Another related work has focused on finding optimal camera pose for video stitching using constrained greedy heuristic algorithms [24], and another work on finding optimal camera positions for overlapped coverage using 3D camera projections [16]. Other works have focused on finding optimal sensor placements in agriculture [15], agrohydrological systems [18], while others have used constrained optimization for sensor placement in nuclear reactors for the creation of nuclear digital twins [13]. Other, more directly related works, have shown optimal camera placement for bovine health monitoring [19], which focuses on a computer vision for livestock welfare monitoring, using Genetic Algorithms, and others have shown the usage of multi-objective genetic algorithms (Non-dominated Sorting Genetic Algorithm *NSGA-II* [7]) for camera localization [11].

Our contribution is the creation of a novel optimization problem, which focuses on placing cameras in a farm for chicken monitoring, using CMA-ES and Quality Diversity (QD) algorithms, which were not used before. This problem has an increasing amount of dimension as the camera requirements increase, marking this problem as a different optimization problem where the dimensionality can, it theory, be a variable. Through our experimentation, we have shown how QD algorithms can be as powerful as traditional single-objective evolutionary algorithms like CMA-ES, and provide more than one solutions. Whereas a traditional single-objective evolutionary algorithm would require changing of constraints; and/or change of the problem definition; and require a similar amount of evaluations to achieve a competitive solution, QD can achieve similar solutions with fewer evaluations; and as we will show in orders of magnitude. We will discuss how QD can be better that traditional evolutionary algorithms, and how CMA-ES can be more beneficial than QD in certain scenarios through defining specific constrains, either through a penalty function, or through a change in the problem configuration.

The structure of the paper is as follows: we will discuss the formulation of the problem in Sect. 3, provide information about the farm specifications and the experimental setup in Sect. 4, provide the necessary background information in Sect. 2, show our results in Sect. 5 and discuss our finding in Sect. 6, and finally provide a conclusion in Sect. 7.

2   Background

2.1 (μ_w, λ)-CMA-ES

The Covariance Matrix Adaptation Evolution Strategy (CMA-ES) samples λ solutions from a multivariate normal distribution $\mathcal{N}(m, \sigma^2 C)$, where m is the distribution weighted mean calculated from its best solutions μ_w, σ is the step-size, and C a covariance matrix of its population [10]. Given the updated population, the algorithm changes at each generation the mean and the covariance matrix, while using mechanisms such as step-size adaptation, and evolution paths to accelerate convergence to the optimum, but avoid converging prematurely.

2.2 Quality Diversity Optimization

Quality diversity (QD) optimization [4], is a family of algorithms that unlike traditional evolutionary algorithms that aim to find the best solution across the entire problem space, they aim to find a diverse set of high quality solutions.

MAP-Elites: Multi-dimensional Archive of Phenotypic Elites (MAP-Elites) [17], is an example of a QD algorithm that stores high quality, unique solutions (the elites). When the algorithm is deployed, the user defines n dimensions of interest, and a number m of discretization bins per dimension. This diverse solution space is called the *behaviour space*, which is a Cartesian space tessellated into an n-dimensional grid (or a Centroidal Voronoi Tesselation [23]) called the elite *archive* $\mathcal{E}$. Unlike a traditional evolutionary algorithm that only requires the *fitness* function $f(\mathbf{x})$ of a problem, MAP-Elites additionally requires a *behaviour function* $b(\mathbf{x})$ which will map each solution to its respective location in the created archive. Vanilla MAP-Elites uses an isotropic Gaussian mutation operator $\mathcal{N}(0, \mathbf{I})$, which creates perturbations to existing solutions (parents) in the archive, providing new solutions (offspring) to be evaluated $\mathbf{x}_i^{t+1} = \mathbf{x}_i^t + \sigma \mathcal{N}(0, \mathbf{I}) \mid \mathbf{x}_i \in \mathcal{E}$.

IsoLineDD: The IsoLineDD operator [22] combines two Gaussian perturbations, the original isotropic and a directional noise, from one parent to the other. The new offspring is defined as: $\mathbf{x}_i^{t+1} = \mathbf{x}_i^t + \sigma_1 \mathcal{N}(0, \mathbf{I}) + \sigma_2(\mathbf{x}_j^t - \mathbf{x}_i^t)\mathcal{N}(0, 1)$. With the directional noise, IsoLineDD, exploits linear correlations between the elites [5] and accelerates QD optimization.

CMA-ME: Covariance Matrix Adaptation MAP-Elites (CMA-ME) is an algorithm which integrates modified CMA-ES instances (called the emitters) into a MAP-Elites archive [9]. Each CMA-ES instance is modified by changing their ranking mechanism to randomly select parents from the elites, instead of the fittest solutions.

3 Problem Formulation

In the farm surveillance problem the goal is to maximize the area covered by $c \in \mathbb{N}_1$ cameras. For simplicity, the farm is modelled as a $2D$ rectangle with surface area $length \times width$, where $length, width \in \mathbb{R}_{\geq 0}$ are measured in *meters*. The farm is supported by *horizontal* and/or *vertical* beams, $hb, vb \in \mathbb{N}$, onto which cameras can be installed. The horizontal and the vertical beams are parallel and/or coincident to the X-*axis* and Y-*axis*, respectively. The location of each vertical beam is in the range $[0, width]$, and in the range $[0, length]$ for the horizontal beams. For each camera, the problem requires the desired depth capture d_i in *meters*, and the horizontal field of view (FOV) measured in degrees $\theta_i \in [0°, 360°]$. We represent farm coverage as a black and white image. White pixels indicate the uncovered areas by the cameras (blind spots), and black pixel indicate areas that are observed. Each pixel covers an area provided by the user in *squared meters*; by decreasing the area of each pixel, the resolution increases.

3.1 Genotype and Phenotype

The genotype of this problem comprises of 3 genes/dimensions per camera, or 1 chromosome per camera with 3 genes each, one for each characteristic that we would like to find: *(1)* x_i coordinate, *(2)* y_i coordinate, and *(3)* the orientation θ_i of the camera, where $|genotype| = 3c$, and i is the i_{th} camera. The orientation of the camera indicates where the middle part of the camera is positioned; for example, at $0°$ on a Cartesian coordinate system, a camera with a horizontal FOV angle of $100°$, at $50°$ it will coincide with the *X-axis*, having a span between $[310°, 50°]$. The value of every gene is in the range $[0, 1]$. The genotype is converted to the *phenotype* by scaling the value of each gene to the value it represents; i.e., the x_i coordinate (first gene of each chromosome) will be multiplied by the *width* of the farm, the y_i coordinate (second gene) will be multiplied by the *length* of the farm, and the orientation θ_i (third gene) by the max orientation $360°$.

3.2 Fitness Function

The fitness function measures the percentage of farm coverage by the cameras. The goal is to maximize the number of black pixels in the farm's image. We plot the farm's image by drawing the covered area by each cameras. We calculate that area, by modelling the camera's view as an isosceles triangle, with the camera at the apex, looking towards the base. To define the triangle, we need its three vertices. Given the camera's location at point $A_i(x_i, y_i)$ and the camera's depth d_i, we can find the other two vertices as follows. The height of the triangle is $h_i = d_i$, so the x-coordinates of the base points are $x^i_{1,2} = h_i + x_i$. To find the y-coordinates, we first calculate half of the base width using *half base_i* $= h_i \tan(\theta_i/2)$, where θ_i is the camera's viewing angle. The y-coordinates are then given by $y^i_1 = y_i + half\ base_i$, and $y^i_2 = y_i - half\ base_i$. We then apply the rotation matrix

$$\begin{bmatrix} x \cos\theta - y \sin\theta \\ x \sin\theta + y \cos\theta \end{bmatrix}$$

to the coordinates $B_i(x^i_1, y^i_1)$ and $C_i(x^i_2, y^i_2)$, adjusting their positions according to the orientation specified by the phenotype.

Next, we determine the rectangular bounding box by finding the minimum and maximum x and y coordinates of the triangle. We then check for pixels that lie within the triangle's covered area. To verify if a point D is inside the triangle, we calculate the area of the triangle $\triangle A_i B_i C_i$, and compare it with the sum of the areas of three smaller triangles formed with the evaluated point: $\triangle A_i B_i D$, $\triangle A_i C_i D$, and $\triangle B_i C_i D$. The area of each triangle is computed using the *Shoelace formula*:

$$T = \frac{|x_A(y_B - y_C) + x_B(y_C - y_A) + x_C(y_A - y_B)|}{2}.$$

If the area of $\triangle A_i B_i C_i$ matches the sum of the three smaller triangles, then the point D is within the triangle's area.

Finally, after drawing all the triangles, one for each camera, we calculate the percentage of black pixels in the image using the formula: $coverage = \frac{black\ pixels}{total\ pixels}$.

4 Experimental Setup

CMA-ES will be the traditional evolutionary algorithm of our choice (state-of-the-art Evolution Strategy), whilst MAP-Elites with the IsoLineDD operator and CMA-ME the QD algorithms. CMA-ES will be used to find optimal solutions of the entire farm on different runs, whereas the QD algorithms will be used to find optimal diverse solutions of the farm. We use existing implementations of these algorithms in JAX [2], a high performance array computing library, allowing GPU usage via just-in-time (JIT) compilations. More specifically, we use the evosax [14] library, which contains JAX-based Evolution Strategies, including CMA-ES, and the QDax [3] library, which contains JAX-based Quality Diversity algorithms, including IsoLineDD and CMA-ME. In addition, we use the behaviour space visualizing capabilities of the pyribs [21] library.

4.1 Farm and Camera Specifications

The farm dimensions were taken from a commercial farm in Cyprus, where the cameras have been deployed. The dimensions are $width = 20.5\,\text{m}$, and $length = 6.5\,\text{m}$, and we use a *pixel size* $= 0.1 \times 0.1\ m^2$. We use 6^1 *Raspberry Pi Camera Module 3* cameras, where the $FOV = 102°$, and we set $d = 5\,\text{m}$. The farm has in total 8 beams, where cameras can be installed; 5 vertical, and 3 horizontal. The horizontal beams are located at: $\{0\,\text{m}, 3.25\,\text{m}, 6.5\,\text{m}\}$, where $0m$ and $6.5m$, are the *outer frame beams* of the farm, and the vertical beams are located at: $\{0m, 5.125\,\text{m}, 10.25\,\text{m}, 15.375\,\text{m}, 20.5\,\text{m}\}$, where $0m$, and $20.5\,\text{m}$ are the *outer frame beams*. The genotype has 18 genes/dimensions in total; i.e., 3 for each camera.

4.2 Behaviour Descriptors

For the QD algorithms, we will use 2 different behaviour descriptors to find different solutions, which will assist with maintaining diversity: *(1)* **spread-xy** that describes the spread of the cameras, and *(2)* **used-beams** that describes the placement of the cameras on each beam.

---

[1] The total area that should be covered in the farm is $133.25\,\text{m}^2$. Each camera can cover approximately $\frac{102°}{360°}5^2\pi \approx 22.25\,\text{m}^2$. We can find the minimum required cameras, by ceiling the ratio of the camera's area to the farm's area: $\frac{133.25}{22.25} \approx \lceil 5.99 \rceil = 6$.

Spread-Xy Descriptor: The *spread-xy* descriptor is an *1D* behaviour descriptor, that measures how well spread the cameras are, given their (x, y) location in the farm. The descriptor's main goal, is to help us keep many solutions where the cameras are differently spread. The spread is in the range $[0, 1]$: 0 for the least spread (overlapping), 1 for the most spread (different locations). We calculate the descriptor, by finding the nearest neighbour of each camera using the ℓ^2 **norm** (Euclidean distance) to measure their distance, sum the nearest neighbour of each camera, and divide this over the square root of the sum of their squared maximum distance that each coordinate has (max distance of x and y), and multiplied by the number of total cameras, which will normalize the result:

$$b_{spread}(x) = \frac{\sum\limits_{c}^{\mathbf{C}} NN(c, \mathbf{C} \setminus c)}{|\mathbf{C}| \sqrt{max\ distance(x)^2 + max\ distance(y)^2}}$$

where $\mathbf{C}$ the set of the cameras. For simplicity, we use the genotype to calculate the spread, where the max distance is set to 1 for both coordinates; $x, y = 1$.

Used-Beams Descriptor: The *used-beams* descriptor indicates whether a beam is currently being used. Therefore, the dimensionality of this descriptor is equal to the number of beams. Each beam dimension has a grid size of 2: the first cell is filled if no cameras are deployed on the beam (unused), and the second cell is filled if at least one camera is deployed (used). The total cells that would be required to be filled are $2^{total\ beams}$. This approach allows for multiple solutions, where some beams may be left unused while others are selected for camera placement. Using this descriptor helps find a variety of high-performing solutions that involve different beams and beam combinations in a single run. In contrast, traditional optimization would require redefining the problem or adding specific constraints to achieve similar results. This flexibility is valuable in real-world scenarios, where unexpected situations might arise. With this descriptor, we can switch between solutions by querying which beams should be used without needing to re-optimize and/or modify the problem.

4.3 Methodology

For each algorithm we performed 30 experimental runs and their variations. For CMA-ES, we performed 3 different experiments targeting approximately 200 k evaluations to test the impact of its population size: *(1)* 16667 generations with 12 population (default setting) $\approx$ 200 k, *(2)* 2000 generations with 100 population, and *(3)* 1000 generations with 200 population. This will also show the impact of increasing the maximum total number of generations, and how it affects the convergence of the algorithm with a smaller population size.

To perform the *Quality Diversity* experiments, we opted for 1*M* evaluations for both IsoLineDD, and CMA-ME, over 1000 generations. Since both algorithms operate differently, we want to use the same amount of evaluations per generation for each one, hence we will be using 1 k evaluations for IsoLineDD, and 1 emitter with a batch size of $1k$ evaluations, resulting in a total of 1 k evaluations per generation for CMA-ME, which was the chosen combination of emitters and batch size per emitter, after hyperparameter tuning. Both algorithms will be tested using both behaviour descriptors.

For the *spread-xy* (1D) descriptor, we use 2 different grid sizes; 1 k and 10 k cells, to test the impact of archive capacity in algorithm performance. The *used-beams* behaviour descriptor, on the other hand, has a fixed capacity which depends on the pre-defined number of beams. In this scenario we have 8 beams; i.e., descriptor dimensions; with 2 cells each, which in total results in $2^8 = 256$ cells. Despite all the possible combinations, only 255 solutions can, theoretically, be found, as it is not possible to not place the cameras in a solution; at least one beam must be used. The source code of our experiments is available online[2].

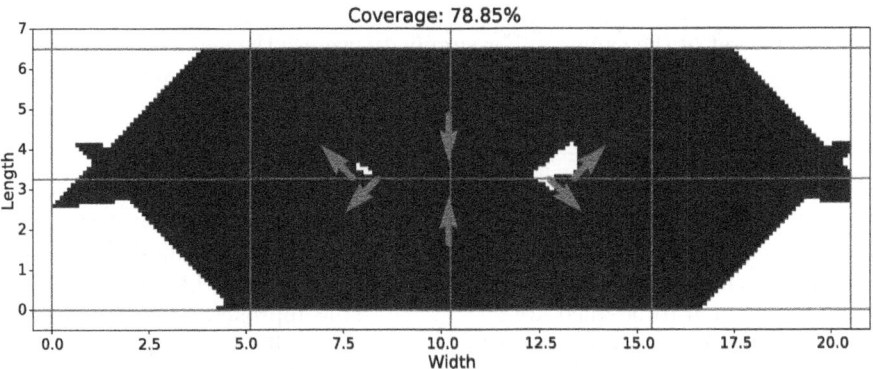

Fig. 1. Human created solution of the farm surveillance problem. The green lines indicate the available camera installation beams (8), the red points the cameras, and the red arrows indicate the focused location of the cameras. (Color figure online)

5   Results

5.1   Initial Human Design

Prior to conducting the experiments, we developed an initial design with the assumption that strong symmetry would contribute to an effective solution. However, our symmetric design; see Fig. 1; only achieved a fitness of 78.85% (coverage), which we will consider the base score that the two different approaches should have to compete with.

---

[2] https://github.com/CYENS/FarmCameraPlacementOptimizer.

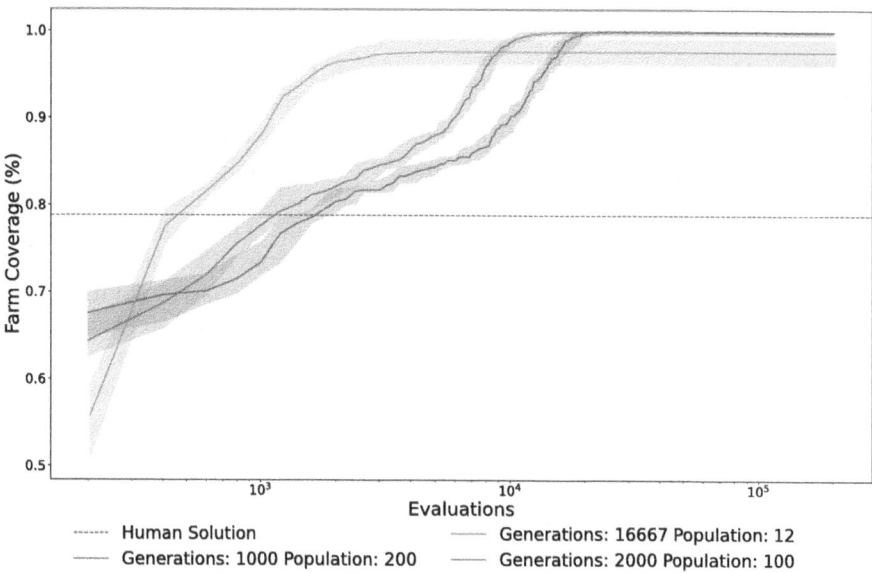

Fig. 2. CMA-ES results showing the coverage of the best solution provided at a given evaluation point. The shaded regions indicate the interquartile range (IQR) of the experiments, after 30 individual runs. The experiment with 1000 generations and 200 population seems to be the better choice of the experiments.

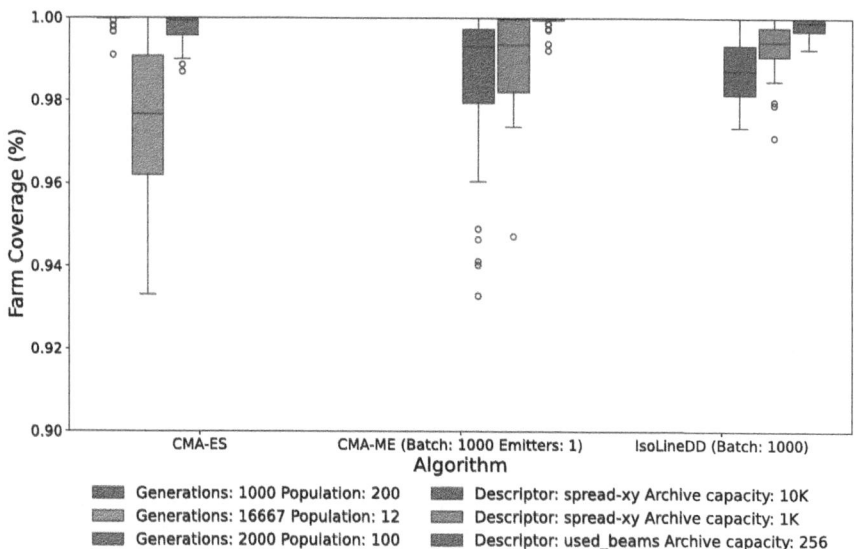

Fig. 3. The resulting box plots of all the experiments. Each box plot is created after 30 individual runs. The plot shows and compares versions of an algorithm (X-axis ticks), and also the algorithms with each other (X-axis). CMA-ES shows that higher population is better, and both QD algorithms show that the *used-beams* descriptor provides better, higher quality solutions than both versions of *spread-xy*.

5.2   CMA-ES Results

CMA-ES achieved complete coverage (100%) on this problem, after 30 runs. Following our 3 different experiments, we have noticed that only the higher population sized experiments were able to achieve 100%; see Fig. 2. Despite that the experiment with the 16667 generations and 12 (default) population size (orange line) was able to achieve better solutions using the same number of evaluations at the beginning, it could not achieve 100% coverage. This indicates that the algorithm can only find the best solution using multiple different generations, but the population limits the output, and thus the quality of the solution. The 2000 generations and 100 population size (green line), and the 1000 generations and 200 population size (blue line), were able to achieve 100% coverage using the same number of evaluations, and they both converge at approximately 30 k evaluations; both algorithms show that many generations are required for the algorithm to converge. However, the 1000 generations and 200 population size experiment's results have a smaller interquartile range, and a higher resulting median; see Fig. 3 under CMA-ES; making it the better choice. All approaches managed to outperform the initial design.

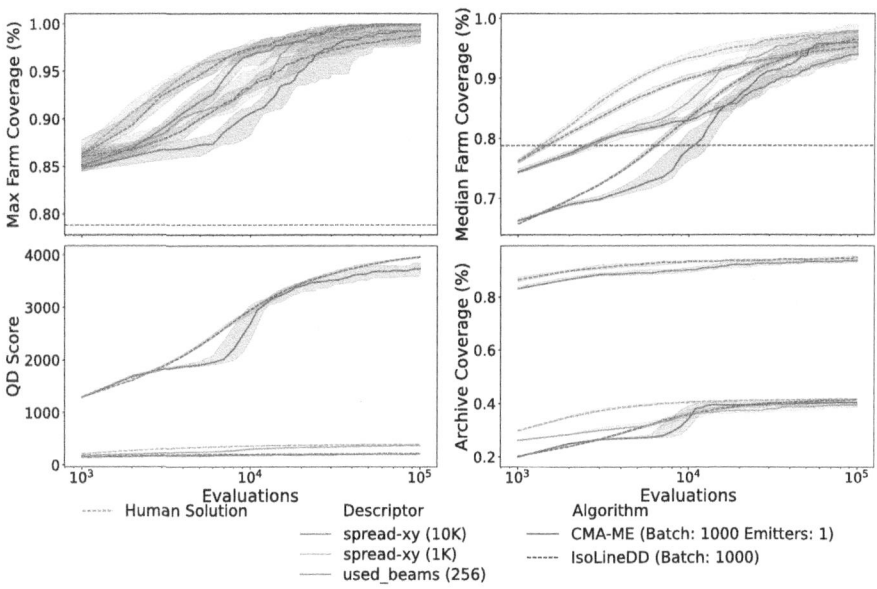

Fig. 4. Results of all the descriptor functions. The shaded regions indicate the IQR of the experiments after 30 individual runs. The top-left shows the Max Farm Coverage, which is the best elite from each archive, the top-right shows the Median Farm Coverage, which is the median elite from each archive, the bottom-left shows the QD-Score, which is the summation of the coverage of all the elites in a given archive, and the bottom-right shows the Archive coverage, which indicates how much of the archive is filled. All plots are drawn over a specific evaluation point.

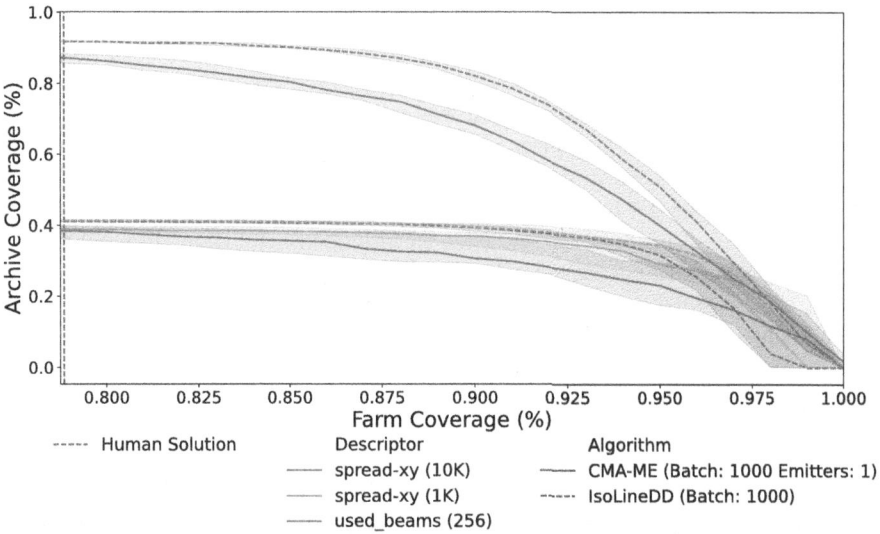

Fig. 5. Complementary Cumulative Distribution Function (CCDF) for all descriptor functions. Shows how many of the solutions in an archive are better than or equal to a given expected farm coverage.

5.3   QD Results

Just like the CMA-ES results, both QD algorithms and behaviour descriptors managed to outperform the human solution. We noticed that IsoLineDD managed to achieve a higher QD score and archive coverage, but CMA-ME outperformed IsoLineDD at the maximum fitness average of all archives; see Fig. 4. We have also noticed that CMA-ME managed to achieve more higher coverage solutions than IsoLineDD for all the descriptors; see Fig. 5 where the descriptors (colours) of the CMA-ME (lines) are always higher than those provided by IsoLineDD (dashed lines).

Despite that the two descriptors are trying to solve two different tasks, both have limitations. Whereas *used-beams* manages to achieve a much higher coverage than both versions of *spread-xy*, it is outperformed by them in the QD score, since they can find more solutions. However, this, is due to their nature, as *used-beams* can only have 256 (in theory 255) solutions, and *spread-xy* can have up to 1 k and 10 k solutions. We assume that this is the reason why both version of *spread-xy* have a higher median farm coverage, as they have a significantly bigger solution pool. On the other hand though, due to the limiting nature of *used-beams*, we notice in Fig. 5 that it has more solutions that achieve 100% coverage. Perhaps the reason that has caused this behaviour, is the ability to keep many solutions with the same spread, unlike *spread-xy*, which could have caused the elite hypervolume to be less scattered.

6   Deployment and Discussion

CMA-ES provided the solution shown in Fig. 6, which was consequently used for camera deployment. Another solution was provided by CMA-ES using a different configuration, where the outer beams were disabled for direct comparison of CMA-ES, with CMA-ES and constraints; see Fig. 8. This inspired the usage of QD algorithms and the behaviour descriptors, to provide us with different high quality solutions. IsoLineDD + *used-beams*, provided a solution (Fig. 9c) that is almost identical to the deployed solution (Fig. 6). This solution utilizes 5 beams in total (horizontal beams at: $\{0, 6.5\}$, vertical beams at: $\{0, 10.25, 20.5\}$). All of these solutions show a form of symmetry (bilateral), which was also observed in other runs of these algorithms; using these settings (descriptors and hyperparameters), and using different settings (Fig. 7).

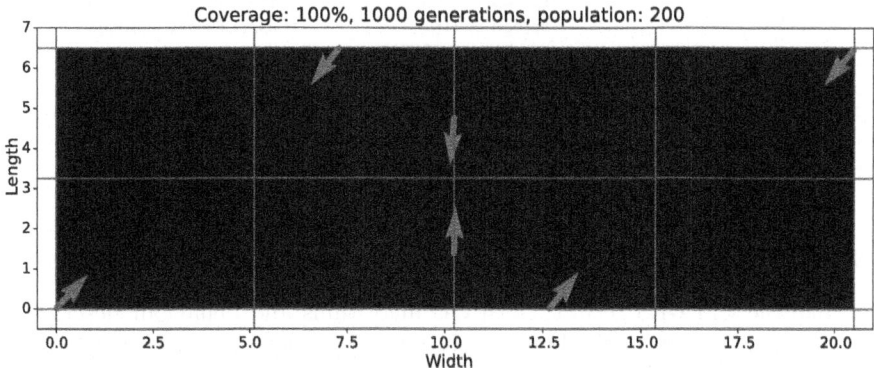

Fig. 6. Farm solution, provided by CMA-ES, that was set to deploy the cameras. Observe the symmetry.

In our exploration of potential solutions, we considered that a symmetric approach would be advantageous, which guided our decision-making process, which led us to create the solution at Fig. 1. Interestingly, we observed that some of the solutions presented, exhibited near-perfect symmetry. While this occurrence was coincidental, it is nonetheless an aspect of our findings.

Comparing the best results between two different approaches, CMA-ES with 200 population size for $1k$ generations and QD (CMA-ME and IsolineDD) with *used-beams*, we notice that CMA-ES and CMA-ME have minor differences, almost negligible ($p = 0.78$, Mann-Whitney U test), whereas CMA-ES and Iso-LineDD showed some difference ($p = 0.002$, Mann-Whitney U test); results in Fig. 3. Despite that we used a solution provided for the farm using CMA-ES, we have found almost the exact same solution using QD (IsoLineDD + *used-beams*. This case shows how QD can be at least as effective as traditional optimization, whilst providing many quality, and diverse solutions.

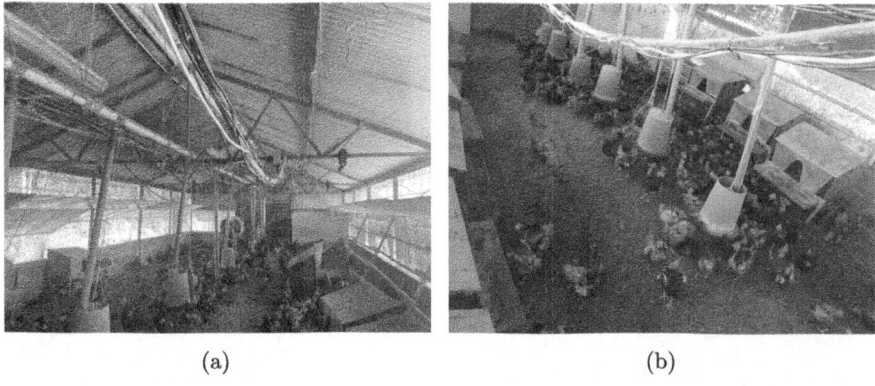

(a)   (b)

Fig. 7. Image of the deployed cameras (a) on the indicated metallic beams from Fig. 6, and an image of the farm view from a deployed camera (b).

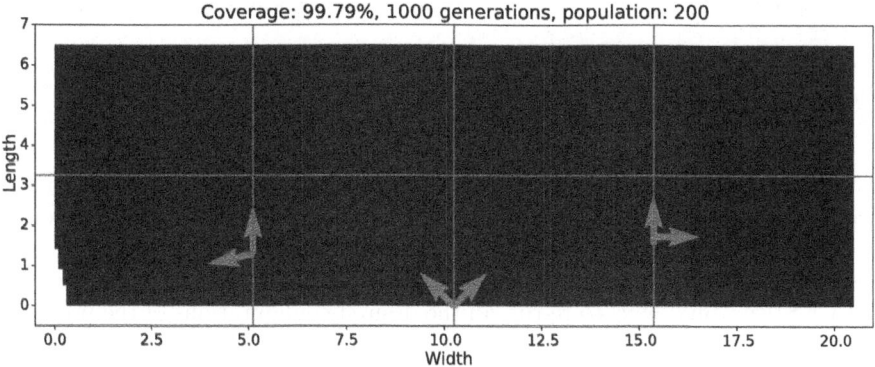

Fig. 8. Farm solution, provided by CMA-ES, that does not use the outer beams; note the missing outer green lines. Observe the symmetry. (Color figure online)

If we can picture scenarios where we have one solution, where the provided solution by CMA-ES was not applicable in the farm, we would have to run the experiment again, change the definition of the problem or add constraints using penalties.

One scenario could be the fact that the cameras are not close to each other, making their wiring or maintenance difficult, or the opposite, where the cameras are very close to each other making their placement difficult. Using QD and the *spread-xy* descriptor, we can always find the best solutions where their spread is limited. If we would like a different spread metric, e.g. spread on the X-axis, we can always change the descriptor. If we would like to do something similar with CMA-ES, we would have to create a penalty system, which would force the algorithm to find an optimal solution using a specific spread size.

Another scenario would be a limited availability on the installation beams. What if our solution uses all the beams, but we have found a problem during

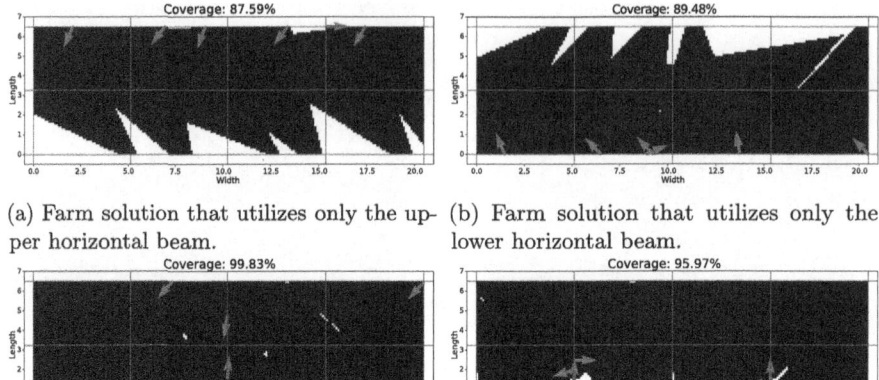

(a) Farm solution that utilizes only the upper horizontal beam.

(b) Farm solution that utilizes only the lower horizontal beam.

(c) Farm solution similar to the deployed solution (Figure 6).

(d) Farm solution that only utilizes two vertical inner beams, but not the middle.

Fig. 9. Results provided by different IsoLineDD + *used-beams* experiments over 1000 generations with 1 k evaluations per generation.

installation, forcing us to not use the outer beams of a farm? What if we are forced to not use specific beams due to other obstacles like wires, of other equipment on specific beams? With QD, we can have many such solutions, by only using *used-beams*, which can find the solution to many similar problems. For CMA-ES, we would have to re-run all the desired configurations of the problem, providing only the available installation beams.

Both of these scenarios can be performed by CMA-ES, but each run would require at least $30k$ evaluations (convergence of CMA-ES at Fig. 2). If we wanted to find the solutions provided by *used-beams*, we would require at approximately $7M$ evaluations (90% coverage of the archive; 256 cells $\times$ 30 k evaluations). Whilst CMA-ES results could be fitter, it would need at least 6M evaluations more. To emulate the results of *spread-xy* using 1k and 10 k archive capacity, we would require approximately 12M and 120M evaluations respectively (40% coverage of the archives $\times$ 30 k evaluations), which once again could be better, but require significantly more evaluations.

There are, however some considerations when using the current QD algorithms. If a farm is provided with a larger pool of installation beams, and one or more solutions using specific combinations of beams are required, CMA-ES would be a more suitable choice. CMA-ES would find the best solution(s) of (a) restricted version(s) of a farm with fewer evaluations, whereas QD + *used-beams* would require more evaluations to fill the archive, as the archive size would increase exponentially because its dimensionality increases; dimensionality = total beams = $2^{total\ beams}$, and to find equally good solutions for each beam combination. Another possible solution would be to disable the unwanted beams, and use *spread-xy*, to find multiple solutions on that specific configuration. Also, by disabling the unwanted beams and using *used-beams* to find solutions, can also be considered as a good alternative, as each removal of a beam reduces the exponentially increasing archive size.

Another reason to use CMA-ES, would be to find solutions for specific combinations of beams, and avoid all the "redundant" solutions that *used-beams* would provide. Finding solutions with a specific camera spread can also be a reason to not use QD, especially in a scenario were the farm is big, and the camera/sensor wiring is an issue. In this case a penalty function can be provided/created instead, which will result in one or more focused solutions, without the "redundant" solutions that *spread-xy* would find.

7   Conclusion and Future Work

We have presented a real life animal monitoring problem. We showed how the problem was solved and showed some of the possible solutions, using CMA-ES (a state-of-the-art Evolution Strategy) and the QD algorithms, MAP-Elites with IsoLineDD and CMA-ME. We presented and discussed how both of these algorithms can be used for providing problem solutions, discussing the ideal use case scenario for each one of them. We conclude that CMA-ES can provide a single high quality solution for any scenario, as long as the goal matches the problem description (problem configuration and/or fitness function penalties/restrictions), and QD can provide many high quality diverse solutions, as long as a reasonable dimensionality and archive capacity are provided.

Future works could see the creation of a 3D version of the problem, where the problem would feature the *height* of a farm. This way, we will also need to account for the height of the farm and require the vertical or diagonal FOV of a camera. Another change would be to convert the problem from a maximization problem (i.e., maximizing coverage) to a minimax or multi-objective problem, where the objective would also involve minimizing overlapping areas monitored by multiple cameras. Finally, we would also like to experiment with other farm shapes: squares; polygons; non-convex areas, and other camera types with different specifications.

Acknowledgments. The work was funded by the European Union Recovery and Resilience Facility of the NextGenerationEU instrument, through the Research and Innovation Foundation (CODEVELOP-ICT-HEALTH/0322/0061), as well as the European Union's Horizon 2020 Research and Innovation Programme under Grant Agreement No. 739578, and the Government of the Republic of Cyprus through the Deputy Ministry of Research, Innovation and Digital Policy. The authors would like to thank Marios Thoma and Pieris Panagi for their feedback during this project.

References

1. Astill, J., Dara, R.A., Fraser, E.D., Roberts, B., Sharif, S.: Smart poultry management: smart sensors, big data, and the internet of things. Comput. Electron. Agric. **170**, 105291 (2020)
2. Bradbury, J., et al.: JAX: composable transformations of Python+ NumPy programs, v0. 3.13 (2018)
3. Chalumeau, F., et al.: QDax: a library for quality-diversity and population-based algorithms with hardware acceleration. J. Mach. Learn. Res. **25**(108), 1–16 (2024)
4. Chatzilygeroudis, K., Cully, A., Vassiliades, V., Mouret, J.B.: Quality-diversity optimization: a novel branch of stochastic optimization. In: Pardalos, P.M., Rasskazova, V., Vrahatis, M.N. (eds.) Black Box Optimization, Machine Learning, and No-Free Lunch Theorems, vol. 170, pp. 109–135. Springer, Cham (2021). https://doi.org/10.1007/978-3-030-66515-9_4
5. Christou, K., Christodoulou, C., Vassiliades, V.: Quality diversity optimization using the IsoLineDD operator: forward and backward directions are equally important. In: Proceedings of the Companion Conference on Genetic and Evolutionary Computation, pp. 639–642. ACM, Lisbon Portugal, July 2023. https://doi.org/10.1145/3583133.3590737
6. Dawkins, M.S.: Animal welfare and efficient farming: is conflict inevitable? Animal Prod. Sci. **57**(2), 201–208 (2016)
7. Deb, K., Pratap, A., Agarwal, S., Meyarivan, TAMT.: a fast and elitist multiobjective genetic algorithm: NSGA-II. IEEE Trans. Evol. Comput. **6**(2), 182–197 (2002)
8. Farooq, M.S., Sohail, O.O., Abid, A., Rasheed, S.: A survey on the role of IoT in agriculture for the implementation of smart livestock environment. IEEE Access **10**, 9483–9505 (2022). https://doi.org/10.1109/ACCESS.2022.3142848
9. Fontaine, M.C., Togelius, J., Nikolaidis, S., Hoover, A.K.: Covariance matrix adaptation for the rapid illumination of behavior space. In: Proceedings of the 2020 Genetic and Evolutionary Computation Conference, pp. 94–102. ACM, Cancún Mexico (Jun 2020). https://doi.org/10.1145/3377930.3390232
10. Hansen, N., Ostermeier, A.: Completely derandomized self-adaptation in evolution strategies. Evol. Comput. **9**(2), 159–195 (2001)
11. Heyns, A.M.: Optimisation of surveillance camera site locations and viewing angles using a novel multi-attribute, multi-objective genetic algorithm: A day/night antipoaching application. Comput. Environ. Urban Syst. **88**, 101638 (2021)
12. Karatsiolis, S., Panagi, P., Vassiliades, V., Kamilaris, A., Nicolaou, N., Stavrakis, E.: Towards understanding animal welfare by observing collective flock behaviors via AI-powered analytics (2024)
13. Karnik, N., et al.: Constrained optimization of sensor placement for nuclear digital twins. IEEE Sens. J. **24**(9), 15501–15516 (2024). https://doi.org/10.1109/JSEN.2024.3368875
14. Lange, R.T.: Evosax: JAX-based evolution strategies, December 2022. https://doi.org/10.48550/arXiv.2212.04180
15. Lee, S.Y., Lee, I.B., Yeo, U.H., Kim, R.W., Kim, J.G.: Optimal sensor placement for monitoring and controlling greenhouse internal environments. Biosyst. Eng. **188**, 190–206 (2019)
16. Malhotra, A., Singh, D., Dadlani, T., Morales, L.Y.: Optimizing camera placements for overlapped coverage with 3D camera projections. In: 2022 International Conference on Robotics and Automation (ICRA), pp. 5002–5009. IEEE (2022)

17. Mouret, J.B., Clune, J.: Illuminating search spaces by mapping elites, April 2015. https://doi.org/10.48550/arXiv.1504.04909
18. Sahoo, S.R., Yin, X., Liu, J.: Optimal sensor placement for agro-hydrological systems. AIChE J. **65**(12), e16795 (2019). https://doi.org/10.1002/aic.16795
19. Sourav, A.A., Peschel, J.M.: Visual sensor placement optimization with 3D animation for cattle health monitoring in a confined operation. Animals **12**(9), 1181 (2022). https://doi.org/10.3390/ani12091181
20. Tedeschi, L.O., Greenwood, P.L., Halachmi, I.: Advancements in sensor technology and decision support intelligent tools to assist smart livestock farming. J. Anim. Sci. **99**(2), skab038 (2021)
21. Tjanaka, B., et al.: Pyribs: a bare-bones python library for quality diversity optimization. In: Proceedings of the Genetic and Evolutionary Computation Conference, pp. 220–229. ACM, Lisbon Portugal, July 2023. https://doi.org/10.1145/3583131.3590374
22. Vassiliades, V., Mouret, J.B.: Discovering the elite hypervolume by leveraging interspecies correlation. In: Proceedings of the Genetic and Evolutionary Computation Conference, pp. 149–156. ACM, Kyoto Japan, July 2018. https://doi.org/10.1145/3205455.3205602
23. Vassiliades, V., Chatzilygeroudis, K., Mouret, J.B.: Using centroidal Voronoi tessellations to scale up the multidimensional archive of phenotypic elites algorithm. IEEE Trans. Evol. Comput. **22**(4), 623–630 (2018). https://doi.org/10.1109/TEVC.2017.2735550
24. Watras, A.J., Kim, J.J., Liu, H., Hu, Y.H., Jiang, H.: Optimal camera pose and placement configuration for maximum field-of-view video stitching. Sensors **18**(7), 2284 (2018)
25. Wolfert, S., Ge, L., Verdouw, C., Bogaardt, M.J.: Big data in smart farming-a review. Agric. Syst. **153**, 69–80 (2017)

Adaptive Local Search for Real-World Multi-echelon Inventory Control

Agathe Métaireau Manche[1,2]($\boxtimes$) (iD), Clarisse Dhaenens[1] (iD),
Nadarajen Veerapen[1] (iD), and Manuel Davy[2]

[1] Univ. Lille, CNRS, Centrale Lille, UMR 9189 CRIStAL, 59000 Lille, France
ametaireau@vekia.fr
[2] Vekia, 16 rue Faidherbe, 59000 Lille, France

Abstract. This paper proposes an efficient solving method for a multi-echelon inventory control optimization problem applicable to real-world contexts. The method is tested on a real-world case study. We implement a metaheuristic based on a local search algorithm and develop custom neighborhood operators. To validate the proposed method and assess its performance in a realistic setting, we develop a simulation tool that reproduces the inventory control process over several time periods. This simulation tool is used to evaluate the performance of the multi-echelon approach and to compare it to existing single-echelon order engines used in production. The results of the experiments show the significant benefits of the proposed multi-echelon inventory control optimization method, including an improvement of between 13% and 40% in average service rate and a profit increase between 23% and 37% depending on the type of instance considered.

Keywords: Inventory Control · Multi-Echelon Replenishment · Local Search · Simulation

1 Introduction

Inventory control and planning is a central element of supply chain operations management, as these processes help companies ensure the availability of their products while adhering to budget constraints. In fact, one of the main functions of inventory management optimization is to plan the replenishment operations across the supply chain in order to satisfy an uncertain demand while minimizing the remaining inventory levels. To describe the different levels of a supply chain network, we refer to each one of them as an *echelon*.

Inventory control optimization is usually conducted by decomposing the supply chain into independent entities and computing the best orders for each of these entities. Research into this method, called the *single-echelon approach*, began with the seminal paper of Harris [7]. However, a supply chain is a network of interconnected entities that share common constraints. For example, if we consider a network of the One-Warehouse Multiple-Retailer (*1WnR*) type

© The Author(s), under exclusive license to Springer Nature Switzerland AG 2025
P. García-Sánchez et al. (Eds.): EvoApplications 2025, LNCS 15612, pp. 370–384, 2025.
https://doi.org/10.1007/978-3-031-90062-4_23

(Fig. 1), every retailer places orders to the same warehouse, which means they share its available inventory. If we use the single-echelon approach and compute the best order for each retailer separately, we may obtain a final solution with a total order quantity exceeding the warehouse's available inventory, resulting in unfeasible solutions that require a repair procedure. Such examples are plenty and justify the need to consider the supply chain network as a whole, rather than separating it into entities. This method is called the *multi-echelon approach*.

In this paper, we propose to develop a multi-echelon inventory control optimization method for industrial applications. This research is part of an industrial collaboration with Vekia, a company specializing in replenishment optimization and automation software. The software provides a probabilistic forecast to estimate future demand in different inventory control contexts, and computes optimized replenishment ordering plans. Given that the problem is too difficult to solve exactly [9], our research aims at developing a metaheuristic for multi-echelon inventory control in a One-Warehouse Multiple-Retailers supply-chain, in order to add a multi-echelon order optimization method in Vekia's software. This metaheuristic, based on a local search with custom neighborhood operators, is detailed in the paper and our research question is to see how it behaves and compares against existing single-echelon methods in Vekia.

Real-world complex systems, such as supply chains, operate in conditions that depend on multiple random factors, making their analysis in real-life situations difficult [16]. According to Alstrøm and Madsen [1], simulation tools are useful to analyze decisions in a framework that is similar to the reality. They are also known for their capacity to accurately capture and explain the actual behavior of the systems under study [3]. As a result, we developed a bespoke simulator to answer the aforementioned research question. To the best of our knowledge, the only published research explicitly comparing multi-echelon and single-echelon approaches are Hausman et al. [8] and Ekanayake et al. [4]. Consequently, research on comparing the performances of single-echelon and multi-echelon inventory control methods is still very sparse.

The remainder of this paper is organized as follows. We first detail the problem we want to solve, and describe its specific constraints in Sect. 2. We focus on the multi-echelon order engine and its resolution method in Sect. 3. In Sect. 4, we provide an overview of the inventory control process we want to reproduce, and a description of the simulation tool developed for this study. We also detail the different order engines compared in the simulations. Then, in Sect. 5, we describe the experimental protocol used for the simulations. We also provide and analyze the results of the experiments carried out. Finally, we provide a conclusion and propose future directions for our work in Sect. 6.

2   Problem

As we mentioned in the introduction, we want to investigate the multi-echelon approach in an industrial context. As a result, we choose to base our work on a real-world case study and use its data for our experiments.

The case-study concerns a company that provides spare parts and finished products for domestic heating systems. This company is a customer of Vekia. Its supply network consists of a warehouse, supplied by external suppliers, and a set of retailers. External suppliers provide goods, or *items*, to the warehouse, which then redistributes these goods to the retailers. This is a classic One-Warehouse Multiple Retailers (*1WnR*) network. In addition, the retailers also have the possibility to place orders directly to the suppliers, but the warehouse remains the privileged supplying source. This is referred to as *dual sourcing*. We provide a schematic representation of this network in Fig. 1.

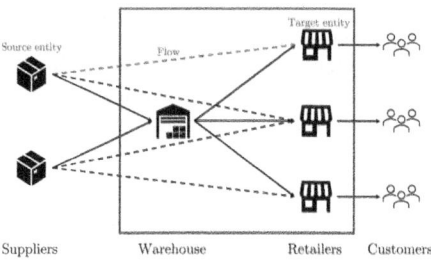

Fig. 1. Representation of the network under study. Solid lines represent the classic *1WnR* flows, and dotted lines the dual sourcing flows. The square defines the perimeter of the entities for which we manage the inventory.

In the following, we present the different features included in the proposed model. The case study provides a framework for representing an industrial inventory control context and its constraints, and serves as a basis to model the multi-echelon inventory control problem at hand. In order to make the model more generic, we also include additional constraints that are frequently encountered in practice.

Problem Specifications. We consider the suppliers to be external to the supply network under study, and we do not manage their inventory. We assume all the suppliers to be *perfect*, which means they have infinite inventory, and no uncertainty on supply and lead times.

The customer demand is not known in advance, and we have to rely on an estimated forecast provided by Vekia's software. Although a probabilistic forecast is available, we simplify the model by using the mean of this forecast as the estimated demand.

In our optimization model, we adopt a multi-echelon approach, calculating quantities transported throughout the whole network at once. This approach helps to avoid several arbitrary decision rules, for example on supplier selection or on inventory allocation in the warehouse.

The ordering process under study is also subject to multi-item costs and constraints. First, the suppliers impose a *Franco* penalty, which is an additional fixed cost applied to every order whose purchase total is under a given threshold.

Moreover, every order induces a *fixed ordering cost*, which is added to the total purchasing price.

The model developed is also aimed at anticipating the further behavior of the customer demand, as we are provided with forecast for several time periods. As a result, we propose a resolution method that provides an ordering plan for a multi-period planning horizon.

Modeling Choices. We have to account for in-progress orders, as the lead-times are not null in the system under study. To do so, we use the modeling technique, described by Boulaksil [2], which divides the total planning horizon into two parts. The first part, called the *frozen horizon*, represents the orders already placed several time periods ago, and that cannot be modified. The second part is the actual planning horizon, where we want to compute orders.

The problem we want to solve includes several costs and penalties that consolidate within a global objective function to be minimized. It contains the transportation cost, that includes the purchasing price of items to the external suppliers and the shipment cost of goods within the system. It also contains penalties for lost sales, holding stock, exceeding capacity, fixed ordering cost, and the Franco cost. The assumption of a lost sales customer behavior is motivated by the inventory control contexts we handle, *e.g.* retail or maintenance companies.

Mathematical Model. The full model is described in Métaireau et al. [9], where we show that exact methods are not suited to solve the problem in reasonable time. The decision variables are, for every flow-item couple (i.e. source entity, target entity and item) of the system, and for every period of the planning horizon, the quantity to order at this time period for this flow-item. The solutions of this problem can be described as the set of variables $(q_{t,f})_{t\in H, f\in F}$, with q representing the quantity to order, F the set of flow-items that forms the supply network, and H the planning horizon. The model aims at minimizing a cost function z, described in Eq. 1. S is the set of entity-item couples, E the set of entities, and L the set of flow *i.e.* source-entity and target-entity couples.

$$
\begin{aligned}
z = &\sum_{f\in F, t\in H} \mathcal{T}_f q_{t,f} + \sum_{s\in S, t\in H} \mathcal{H}_s i_{t,s} + \mathcal{LS}_s y_{t,s} \\
&+ \sum_{e\in E, t\in H} \mathcal{A}_e a_{t,e} + \sum_{l\in L, t\in H} \mathcal{K}_l \alpha_{t,l} + \mathcal{FR}_l \beta_{t,l}
\end{aligned}
\tag{1}
$$

The first sum represents the transport cost of items in the network. $\mathcal{T}$ is the cost including the purchase and carrying cost of the items. The second sum includes the holding cost, with i being the number of items staying in the inventory overnight, and $\mathcal{H}$ being the holding penalty. It also includes the lost sales cost, with y being the unmet demand level and $\mathcal{LS}$ the lost sale penalty. The third sum represents the cost of exceeding the entity's holding capacity, with a representing the excess inventory and $\mathcal{A}$ the exceeding capacity penalty. The last sum includes the multi-item costs. First, $\mathcal{K}$ is the fixed ordering cost, and α a boolean stating if a positive order is placed on the flow. Then, $\mathcal{FR}$ is the

Franco penalty, and β a boolean stating if the total purchase price of orders on the flow is under the Franco penalty.

3 Metaheuristic Resolution Method

The problem under study leads to a large and complex model, even for smaller instances. Consequently, exact methods are not able to provide feasible solutions in a reasonable time. This motivates the need to use metaheuristic methods, that are known to be one of the most practical approaches for solving many real-world problems [11]. We choose to take a local search approach.

3.1 Neighborhood Operators

There already exist plenty of neighborhood operators that are relevant to supply chain management and inventory control can be used in a local search. However, there is little work on operators tailored for the very specific nature of our multi-echelon multi-period problem. As a result, we choose to develop custom neighborhood operators for our problem. We develop a set of neighborhood operators of different kinds. Each operator is designed to influence a specific part of the cost function. In the following, we describe the neighborhood operator set we selected.

Random Addition. – This neighborhood operator is intended to reduce lost-sale costs by adding quantities to the solution. It randomly selects a flow-item from supplier to store, and a time period, then increments the corresponding variable by one, allowing quantity increases while keeping the solution feasible.

Guided Addition and Repair. – Building on the previous operator, it targets the variables leading to high lost sale costs. It first computes the lost sales associated to a given solution and arranges them in descending order. It then selects one variable randomly within the top-k lost-sales variables. We obtain an entity-item s and a time period t. It then selects $q_{t_1,f}$ such that the source entity of f is the warehouse, the target entity and the item correspond to the entity and item of s, ant $t_1 = t - LT_f$, with LT being the lead-time. We then select $q_{t_2,f'}$ with the item of f' corresponding to the item of s, the target entity being the warehouse, the source entity being the supplier with the lowest cost for this item, and $t_2 = t_1 - LT_{f'}$. If t_2 is in the planning horizon, we then add the lost sales quantity to $q_{t_1,f}$ and to $q_{t_2,f'}$. Otherwise, it selects $q_{t_3,f''}$ such that the target entity and item of f'' correspond to the entity and item of s, the source entity is the supplier with the lowest cost for this item. If t_3 is in the planning horizon, we add the lost sales quantity to $q_{t_3,f''}$. Otherwise, the solution is rejected.

Destruction. – This neighborhood operator aims to reduce the fixed costs and the unnecessary purchasing costs. It randomly selects one flow originating from the warehouse or from a supplier and targeting a store, and a time period. It deletes all orders placed on this flow at this time period.

Preponing. – This neighborhood operator is intended to reduce fixed costs and Franco costs by grouping orders. It selects one flow originating from a supplier and targeting the warehouse or a store, and a time period. It moves all orders to the previous period.

Period Preponing. – This neighborhood operator is an extension of the previous one, as it applies to a wider range of variables. It randomly selects a time period in the planing horizon, and moves all orders placed at this period to the previous one. The neighbor is kept only if the solution remains feasible.

Saturate Franco. – This neighborhood operator targets Franco costs specifically. It first lists all flows and time periods where the Franco threshold is not reached. It then randomly selects one flow from this group and randomly picks out a flow-item corresponding to this flow. The corresponding quantity is increased to reach the Franco threshold.

3.2 Adaptive Local Search

Preliminary experiments show that using the neighborhood operators individually in a classic local search do not give satisfactory results. However, each operator is designed to target different parts of the cost function, and using them jointly seems to be a promising avenue to improve the resolution method. Moreover, work on variable neighborhood search (VNS) [6] and adaptive operator selection [5,15] has shown that the usefulness of neighborhood operators is not always the same throughout the search process. For example, in our case, the *Saturate Franco* neighborhood operator may be very efficient on a solution with high Franco cost, but may also have an empty neighborhood if all Franco thresholds are met. These considerations lead to the implementation of a local search metaheuristic that uses several neighborhood operators. Each operator is assigned a score that reflects its current usefulness and that is used to select the next operator for the search. Once a neighborhood operator is selected, it is applied until an improving solution is found or some termination criterion is met. Its immediate performance is then used to make the neighborhood operator's score evolve. We call this algorithm *Adaptive Local Search*, the pseudo-code is provided in Algorithm 1.

For the sake of simplicity, and because our goal is to assess the suitability of such an approach for our real-world problem, we keep the operator selection process and the score computation fairly simple. In particular, we use *probability matching* [5], also known as *roulette wheel* selection, to select operators. In opposition to VNS algorithms, the neighborhoods are not explored sequentially. This allows for a more dynamic exploration of the search space. In the following, we describe the different algorithm design decisions.

Neighborhood Operator Score. – The scores associated to the neighborhood operators are intended to represent their usefulness. We compute it using the gap between the initial solution of the neighborhood search and the final solution

---

Algorithm 1. Adaptive Local Search

1: **Input:** Set of neighborhood operators O, Scores of neighborhood operators $G^* = \{G_o^*\}_{o \in O}$, Global termination criterion TC, Local termination criterion tc, Initial solution S
2: Compute selection probability distribution
3: **while** not TC **do**
4:     Select a neighborhood operator $u \in U$ following the selection probabilities
5:     **while** not tc **do**
6:         Generate a candidate neighbor S'
7:         **if** S' is better than S **then**
8:             $S \leftarrow S'$
9:             $tc \leftarrow True$
10:        **end if**
11:    **end while**
12:    Compute immediate efficiency of u
13:    Update G_u
14:    Update selection probabilities
15: **end while**

---

found by the neighborhood operator, and the number of neighbors explored to find this solution. This score can be described by $G = \frac{z^0 - z^*}{V}$, with z^0 being the fitness of the initial solution, z^* the fitness of the final solution, and V the number of neighbors evaluated. To initialize this score, we run a local search with a termination criterion of L evaluations for every operator at the beginning of the algorithm, with L being a hyper-parameter of our adaptive local search.

Neighborhood Operator Selection. – Based on the set of operator scores, we compute a probability corresponding to each score. This creates a probability distribution that is used to select the neighborhood operators at each step of the search.

Score Evolution. – At the end of every neighborhood exploration, the immediate performance score of the neighborhood operator is computed following the previous equation. This score represents the most up-to-date performance of the neighborhood operator, but it is also the result of a random neighborhood exploration. As a result, it may be the result of an arbitrarily advantageous, or disadvantageous, exploration. To make the score more representative, we have the score evolve in a way that is not too fluctuating but stays adaptive. In this respect, we set up an immediate score history, that we use to compute the neighborhood operator's score. This history is filled at every neighborhood exploration operator, up until a given limit K, which is a hyperparameter of the algorithm. This history is a sliding window, as is often the case in adaptive operator selection, which means that, at each new neighborhood exploration, the oldest immediate score of the history is erased, and the new immediate score is added to it. Based on this history, the new score of the neighborhood operator o

is computed as follows. Let G_o^* be the final score of the neighborhood operator, G_o its immediate performance score, and SH_o the scores stored in the sliding history.

$$G_o^* = \frac{1}{K+1}\left(G_o + \sum_{i=1}^{K} SH_o(i)\right) \qquad (2)$$

The global termination criterion TC, that ends the search, and the local termination criterion tc, that ends a neighborhood exploration if no improving neighbor is found before, are also hyper-parameters of the algorithm. The hyper-parameter set is the following : $\{L, K, TC, tc\}$, with L being the exploration limit for the initialization of the scores and K the size of the score sliding history. In the next sections, we detail the protocol used to assess the performance of the proposed resolution algorithm.

4   Performance Assessment Through Simulation

One of the main motivations of our work is to determine whether multi-echelon inventory is a good strategy for inventory control in a real-world business setting, and to compare it to single-echelon strategies. The goal is to evaluate the performance of our proposed algorithm in a realistic setting, and to compute business-oriented indicators. To achieve this, we want to reproduce the inventory control process in a framework that is the closest possible to reality. We therefore develop a simulation tool that aims at replicating the inventory control process on several time periods. This simulation tool is dedicated to the problem under study, and can be used to compute the performance for different order engines.

4.1   Inventory Control Process Simulation

Let us describe the inventory control process for a given network entity and item. Each period of time is divided into several steps. For a given period t, the entity starts with an initial inventory of the item $I0$. From this point, the order engine computes an order q_t based on the current state of the system and the demand forecast. The inventory then starts to evolve. First, the entity receives the orders placed beforehand. With LT being the total time to process an order, the inventory level after reception is the following: $I^R = I0 + q_{t-LT}$. The next step is to meet the customer demand. We operate under a *lost-sale* assumption, which means that every unsatisfied customer demand is lost, and not postponed to the next time period. With D being the demand on the period, the resulting inventory is then $I^+ = \left[I^R - D\right]$. This value corresponds to the final inventory of the period, which is the initial inventory for the next time period $t + 1$.

This process is reproduced with our simulation tool, developed especially for this research. This tool is aimed at recreating an inventory control environment and reproducing every step of a time period. Moreover, it recreates the inventory dynamics and its evolution throughout several time periods. The simulation tool begins a time period by initializing the inventory control environment with the necessary date, such as the initial inventory level and the demand forecast for example. Then, the optimization process is called at each time period in order to compute the orders. After this step, the orders are stored as in-progress shipment, and a demand scenario is generated. Based on the current inventory, the in-progress shipment and the demand, the inventory dynamic is computed and all selected performance criteria are assessed. Finally, the tool increments the time period and the data is carried forward to the next period.

4.2   Rolling Horizon

One important concept of our simulation is the rolling horizon. Since we propose a multi-period order engine, we provide orders for a finite planning horizon containing several periods. As the forecast estimating the incoming demand is rarely the actual customer demand, we cannot be sure that the ordering plan computed at the beginning of the planning horizon will stay relevant along the time periods. To tackle this problem, we choose to use the *rolling-horizon strategy*, introduced by Reiklaitis [12]. Let H be the planning horizon of the order optimization tool, and TH the total simulation horizon, such that $H \ll TH$. At each simulation period t, the order engine's planning horizon will be $[\![t, t + H]\!]$. Once the order engine computed the ordering plan, only the orders supposed to be placed on time period t are saved for the simulation. The simulation step then continues until the end of the period, and the next planning horizon is $[\![t + 1, t + H + 1]\!]$, and so on until the last simulation period $t = TH - H$.

4.3   Single-Echelon Inventory Control Order Engines

The simulation protocol is used to assess the performance of our proposed multi-echelon local search (ME-LS) order engine, but also to compare these performances to the ones of the already existing order engines in Vekia. These decision agents, $SPSE$ and MP^2, are two single-echelon order engines. Both of them are currently in use in production by Vekia and compute replenishment orders on a daily basis.

The $SPSE$ (*Single-Product Single-Echelon*) order engine is a single-echelon, single-item and single-period order engine. Its main feature is to compute an order for a given entity-item and for the period in which it is called. This order engine is used on every entity-item pair, for which we selected a supplying source beforehand (in the given supplying sources list, in our case the warehouse and all available external suppliers). The detailed process and optimization methods of this order engine is further explained by Sahu [13, Chapter 5]. In the simulator, we use the calculated orders as is.

We also work with the MP^2 (*Multi-Product Multi-Period*) order engine, further described in [10]. This order engine is still single-echelon, but multi-product and multi-period. Accordingly, this order engine is designed to compute the orders for an item group of a given entity and for a given number of time periods. For each item of the system, we pre-select a supplying source, in the same list as for *SPSE*, and group these items by entity and supplying source. Then, *SPSE* is used as a pre-solving tool to compute all order possibilities we want to take into account and their associated single-item cost (i.e. purchasing, holding and shortage cost). Then, the MP^2 order engine computes the best combination of these order possibilities on the given time horizon and returns the corresponding order planning. Like for the *ME-LS* order engine, we use the rolling horizon strategy for this order engine.

5   Experiments

In this section, we provide details on the experiments conducted with the simulation tool to validate the multi-echelon method, and to compare it to existing single-echelon order engines. The objective of this experimentation phase is twofold. First, we want to assess the performance of the proposed multi-echelon order engines from an industrial point of view, and record business-oriented criteria. Second, we want to quantify the benefits of using a multi-echelon method over a single-echelon one.

5.1   Instances

For our experiments, we use data from the case study that is the base of the problem under study and that comes from a company that provides spare parts and finished products for domestic heating systems. For this study, we need to create several instances to validate our proposed method in different situations.

To do so, we use the case study's data to extract reduced instances intended to be used in the simulations in order to evaluate how the proposed approach performs in different scenarios. The initial instance used to create the reduced one corresponds to the supply network of the case study. It includes one warehouse and 150 stores, which manage an inventory of 900 items. One should note that a large part of the item set is made up of very slow moving items, as 68% of items have an average forecast under 1. To create these reduced instances, we use the following procedure.

We first list all of the items of the case study, and sort them according to a preset sorting rule. This sorting rule is what we call the *generation mode* of the instance. Then, we select the n first items of this sorted list and the corresponding supply network to create the instance. n is the instance size.

For this experimentation, we use instances generated with two sorting rules. The first one is the *random order*, or `rand` generation mode, where items are sorted randomly. The second is the *average forecast order*, or `fora` generation mode, where items are sorted by decreasing average forecast. This is the average on the whole forecast history provided. Both generation modes are used with instances of the following size set: $\{15, 25, 40, 100\}$. The number of suppliers and entities in our network remains the same as in the original case study.

While using a reduced number of items will allow us to maintain a runtime within acceptable business limits, it also translates the fact that, in practice, we can separate the items into two groups. In the first group, we consider the items with high forecast, and solve the instance using our multi-echelon order engine. In the second group, the remaining items have a very low forecast. We can compute a satisfying ordering plan using a one-for-one, or $(S - 1, S)$, inventory control policy, which is known to be very efficient for items with an infrequent demand [14]. As a result, it is acceptable for our use case to test our proposed method on instances with a maximum size of 100 items, and this is especially relevant for the `fora` generation mode.

5.2   Experimental Protocol

In order to assess the performance of the order engines and to compare these performances, we launch simulation experiments with realistic instances of the problem. The main idea is to compare the three order engines on the same inventory control contexts so that the comparison is the fairest possible. For a given instance and order engine, we launch N simulations with a set of given seeds, and for a simulation horizon TH. TH is the number of time periods that are simulated. For our experiments, we set $TH = 45$, and $N = 10$. As we work with three order engines, this means that, for each instance, we launch a total of 30 simulations of 45 time periods. For the experiments, we set a global termination criterion of 2 h for the multi-echelon order engine, which corresponds to the real-life business constraint.

During the simulations, we monitor a set of performance criteria. These criteria allow us to assess the performance of the order engines with an industrial point of view. The criteria are chosen because they are the ones traditionally monitored by companies. In the following, we provide a list of the different criteria and their description.

Inventory levels: This represents the sum of all items remaining in the inventory after the receptions and the demand have been treated. It is split between warehouse inventory and store inventory.

Lost sales: This represents the sum of all demand that was not satisfied during the period.

Total costs: This represents the total monetary costs induced in the period. It includes the transport costs for the warehouse, the transport cost for the stores, the fixed costs and the Franco costs.

5.3   Results and Analysis

Based on the results obtained with the simulations experiments, we can assess and compare the performance of the different order engines under study. Out of concern for space, we only display the results for the instances of size 40 and 100 items in the paper, but all instance sizes have a similar behavior.

The first aspect we focus on is the inventory level assessment, that we represent on Fig. 2. For the single-echelon order engines, we can note that the warehouse inventory levels are rather stable along the simulations. The store inventory levels however, seem to significantly increase with time. On the other hand, the inventory levels obtained with the multi-echelon order engine are stable and lower for both the warehouse and the stores.

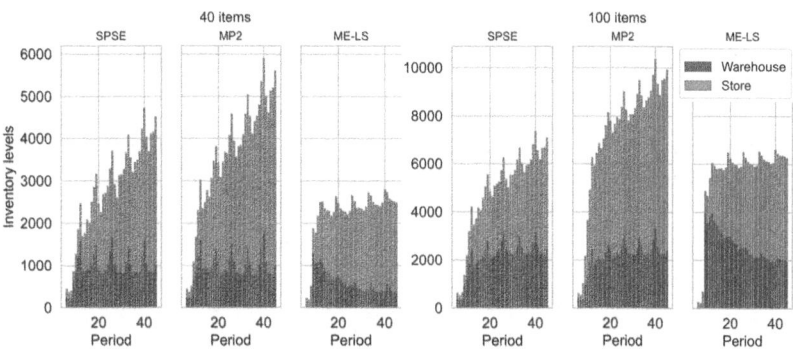

Fig. 2. Inventory levels at each simulated period averaged on all instances and seeds, for every problem size. Values are split for warehouse and store inventory levels.

Then, we confront the inventory level results with the lost sales results, that are displayed in Fig. 3. We notice that, although the lost sales levels of single-echelon order engines tend to decrease along the simulation, those obtained with the multi-echelon order engine are globally lower. If we combine results obtained for inventory levels and lost sale levels, we can conclude that the multi-echelon strategy allows to better answer the demand with lower inactive inventory. This means that the order strategy is more effective to manage the inventory versus demand satisfaction trade-off. If we compare the service rate obtained with *SPSE* and *MP²* with the one obtained with *ME-LS* on every instance, we notice an average service rate increase of 40% for the instances with high forecast, and 13% for the random instances with the multi-echelon method.

Next, we focus on the monetary costs induced by the ordering strategy proposed by the order engines. The results for this criterion are displayed in Fig. 4. For the single-echelon order engines, we can observe a stable tendency after the first simulated week. The total cost is lower and mainly due to the warehouse transport and purchase cost. With the multi-echelon order engine, the total cost is higher and mainly due to the transport and purchase cost of the stores. This is explained by the fact that this ordering strategy results in a higher amount of sales (which is obviously a good thing for the company), but implies a higher amount of items transported within the system, and consequently a higher monetary cost.

To balance the results obtained for the monetary costs, we also compute a profit performance criterion. For each run, we compute the total sales for all

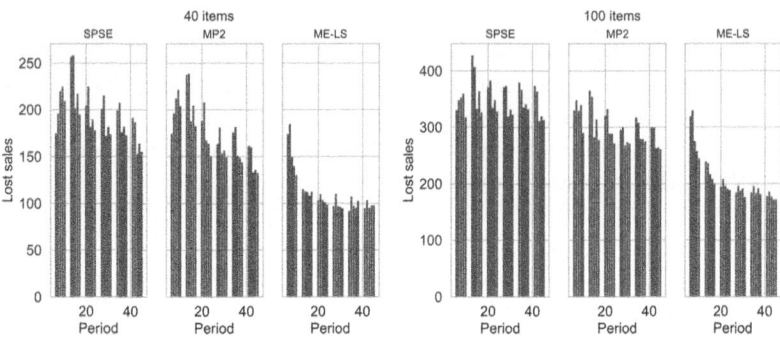

Fig. 3. Lost sales levels at each simulated period averaged on all instances and seeds, for every problem size.

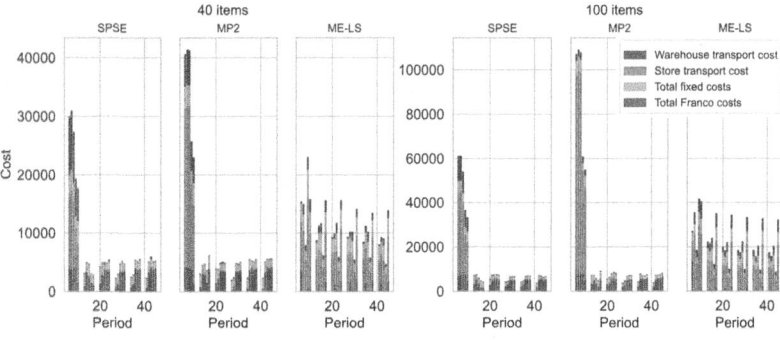

Fig. 4. Total monetary costs at each simulated period averaged on all instances and seeds, for every problem size.

simulation periods. We then subtract the total monetary costs from the sales value to obtain the profit. On average for every run and instance size, we obtain a profit increase with the multi-echelon engine of around 37% for the instances with high forecast and around 23% for the random instances.

To conclude on the simulation results, the use of the proposed multi-echelon order engine shows significant advantages for the inventory control problem under study. Although it implies a higher investment for inventory management, it allows for a better customer satisfaction with a lower inventory level. This better inventory versus lost-sales balance results in a profit increase with the use of the multi-echelon engine. This leads us to the conclusion that our proposed multi-echelon inventory control method achieves a better quality inventory control strategy.

6   Conclusion and Perspectives

In this paper, we propose to develop a multi-echelon inventory control optimization for industrial applications. The research is based on a real-world case study, that is used as a foundation for the definition of the problem under study, and that provides data for our experiments. To define the problem, the case study is enriched with constraints that are usually encountered in real-world inventory control contexts. This allow us to study an optimization problem that is close to cases encountered in practice, and makes the solution applicable to more general industrial situations.

In order to solve the problem under study, we develop a metaheuristic method aiming at giving acceptable solutions in a reasonable time. It is based on local search, and uses custom neighborhood operators. The main contribution in the development of this resolution method is the design of the neighborhood operators, that are created to target specific elements of the objective function of our model. The algorithm presented in this paper is a prototype and proof-of-concept and is aimed at being included in Vekia's software offering.

Finally, in order to assess the performances of our proposed validation method from an industrial point of view, we set up a simulation-based experimentation. We develop a simulation tool that allows us to reproduce the inventory control process of the problem under study on several time periods. The experiments conducted with the simulator have two main goals. First, we want to evaluate the advantages of our proposed method in a realistic setting. Second, we also want to compare the proposed multi-echelon method with existing single-echelon ones, in order to validate the benefits of a multi-echelon inventory control strategy. The results of this experimentation phase show significant benefits with the use of our proposed multi-echelon inventory control method, as we observe a service rate increase of up to 40%. This higher service level is accompanied by a profit increase of up to 37%.

In the future, there are a number of research avenues that seem interesting to explore. The main goal of our planned further research is to bring the model even closer to a wider variety of real-world contexts. This can be achieved with the inclusion of additional constraints, like packaging constraints for example. We also want to include supplier uncertainty in the model, so that it is more representative of the real-world. Moreover, we also want to investigate several research leads to make our proposed resolution method more efficient. The generation of the initial solution for the local search is one of our leads, and we also want to use strategies to reduce the search space, in order to make the convergence more efficient. We are also keenly aware that there has been much work in adaptive algorithm selection that our approach could benefit from, even though our current fairly simple *probability matching* approach already works well for our industrial context. In particular, we wish to explore the Multi-Armed Bandit paradigm [5] and focus on the exploration-exploitation properties of our operators [15].

References

1. Alstrøm, P., Madsen, P.: Simulation of inventory control systems. Int. J. Prod. Econ. **26**(1–3), 125–134 (1992). https://doi.org/10.1016/0925-5273(92)90054-B

2. Boulaksil, Y.: Safety stock placement in supply chains with demand forecast updates. Oper. Res. Perspect. **3**, 27–31 (2016). https://doi.org/10.1016/j.orp.2016.07.001

3. Chu, Y., You, F., Wassick, J.M., Agarwal, A.: Simulation-based optimization framework for multi-echelon inventory systems under uncertainty. Comput. Chem. Eng. **73**, 1–16 (2015). https://doi.org/10.1016/j.compchemeng.2014.10.008

4. Ekanayake, N., Joshi, N., Thekdi, S.A.: Comparison of single-echelon vs. multi-echelon inventory systems using multi-objective stochastic modelling. Int. J. Logist. Syst. Manage. **23**(2), 255–280 (2016). https://doi.org/10.1504/IJLSM.2016.073971

5. Fialho, Á.: Adaptive operator selection for optimization. Ph.D. thesis, Université Paris Sud-Paris XI (2010)

6. Hansen, P., Mladenović, N., Brimberg, J., Pérez, J.A.M.: Variable Neighborhood Search, pp. 57–97. Springer, Cham (2019). https://doi.org/10.1007/978-3-319-91086-4_3

7. Haris, F.W.: How many parts to make at once. Factory, the magazine of management (1913)

8. Hausman, W.H., Erkip, N.K.: Multi-echelon vs. single-echelon inventory control policies for low-demand items. Manage. Sci. **40**(5), 597–602 (1994). https://doi.org/10.1287/mnsc.40.5.597

9. Métaireau, A., Dhaenens, C., Veerapen, N., Davy, M.: Multi-echelon inventory planning for a one-warehouse multiple-retailer supply chain. In: Proceedings of the 7th International Conference on Logistics Operations Management, GOL'24, LNNS, vol. 1104, pp. 280–290. Springer, Cham (2024). https://doi.org/10.1007/978-3-031-68628-3_27

10. Métaireau, A., Sahu, R.K., Delecourt, S., Gerussi, A., Davy, M.: Multi-periodic joint replenishment planning method for various all-unit discounts. In: Proceedings of the 11th International Conference on Operations Research and Enterprise Systems - ICORES, pp. 86–93. SciTePress (2022). https://doi.org/10.5220/0010986800003117

11. Onar, S.Ç., Öztayşi, B., Kahraman, C., Yanık, S., Şenvar, Ö.: A literature survey on metaheuristics in production systems. Metaheuristics Prod. Syst. 1–24 (2016). https://doi.org/10.1007/978-3-319-23350-5_1

12. Reklaitis, G.V.: Review of scheduling of process operations. AIChE Symp. Ser. **78**(214), 119–133 (1982)

13. Sahu, R.K.: General framework and optimization methods for stochastic replenishment planning in industrial contexts. Ph.D. thesis, Université de Lille, August 2020

14. Schultz, C.R.: On the optimality of the (s–1, s) policy. Naval Res. Logist. (NRL) **37**(5), 715–723 (1990)

15. Veerapen, N., Maturana, J., Saubion, F.: An exploration-exploitation compromise-based adaptive operator selection for local search. In: Proceedings of the 14th annual conference on Genetic and evolutionary computation (GECCO 2012), pp. 1277–1284. ACM (2012). https://doi.org/10.1145/2330163.2330340

16. Wiśniewski, T.: Simulation study of inventory management in supply chains. Logist. Transp. **37**, 41–48 (2018)

Evolutionary Computation for Causality-Driven Feature Selection: A Preliminary Study

Emanuele Nardone[✉][iD], Tiziana D'Alessandro[iD], Claudio De Stefano[iD], and Francesco Fontanella[iD]

Department of Electrical and Information Engineering (DIEI), University of Cassino and Southern Lazio, Via G. Di Biasio 43, 03043 Cassino, FR, Italy
{emanuele.nardone,tiziana.dalessandro,destefano,fontanella}@unicas.it

Abstract. Selecting optimal features in high-dimensional spaces remains challenging due to their complexity and the focus on correlational rather than causal relationships. While evolutionary computation algorithms for feature selection show promising results, they often face challenges in identifying feature subsets that are both interpretable and causally relevant. In this paper, we present a preliminary study in which we investigate how causality affects the search capability of feature selection based on evolutionary computation. We tested Genetic Algorithms and Particle Swarm Optimization to this aim and compared their performance with wrapper-based approaches. Comprehensive experiments across multiple benchmark datasets reveal that our methods consistently identify features with stronger causal relationships and superior interpretability than traditional approaches. Our results demonstrate the significant potential of integrating causality to enhance evolutionary computation algorithms for feature selection.

1 Introduction

In recent years, the increasing availability of data has led to a strong and rapid growth in the number of available features in machine learning (ML) applications [4,6]. However, not all of these features may be relevant or informative for the target concept of a given problem. The presence of redundant or irrelevant features can cause the curse of dimensionality [26], where the sparsity of data in high-dimensional spaces negatively affects models. Those features typically slow down the learning process, make the learned model more complex, and eventually cause a deterioration of the classification performance. This problem can be mitigated by using feature selection techniques, which reduce the number of features to use by identifying and eliminating those that may have a detrimental effect [14]. The overall process of building ML-based applications is improved by feature selection, which speeds up the training process and leads to better models in terms of performance, as well as interpretability and explainability. Reducing the number of features makes models easier to understand and explain. Focusing

© The Author(s), under exclusive license to Springer Nature Switzerland AG 2025
P. García-Sánchez et al. (Eds.): EvoApplications 2025, LNCS 15612, pp. 385–401, 2025.
https://doi.org/10.1007/978-3-031-90062-4_24

on the most relevant features allows users to identify key variables. Furthermore, explainability techniques, e.g. LIME and SHAP, also benefit from such reduction by providing clearer insights into how predictions are made.

Feature selection can be viewed as the problem of finding the best feature subset (minimizing the cardinality and maximizing the relevance) in the search space made of all possible feature subsets. Therefore, feature selection implies the definition of a search procedure and an evaluation function. In principle, the optimal subset of features can be found by exhaustively evaluating all the possible solutions. Unfortunately, this strategy is not suitable in most situations because the number of possible solutions grows exponentially with the number of available features. Therefore, many search techniques have been applied to feature selection, such as complete search, greedy search, and heuristic search [18]. However, those algorithms do not consider complex interactions among the features, and their effectiveness is often limited by high computational costs and the tendency to get stuck in local optima.

The evaluation functions can be divided into two broad classes: filter and wrapper. The wrapper approach is a method of evaluating the quality of a given subset by measuring the performance of a classification algorithm. This approach is particularly costly in terms of computation, especially when a large number of evaluations are required and when the dataset is extensive. Furthermore, the outcomes of the wrapper approach are less generalisable since the results are dependent on the specific model used, which also limits their interpretability. On the other hand, filter approaches use statistical measures to evaluate the intrinsic properties of the data; this makes them independent of any classification algorithm and, in most cases, computationally less expensive and more general than wrapper algorithms. Regarding interpretability, filter approaches provide solutions that are not tied to a specific algorithm and are easier to interpret because they depend on the metrics used to evaluate the solutions.

Although filter approaches have shown to be effective, they still have some limitations. One key issue is that they often rely on correlation-based measures to evaluate the relevance of features, which may not capture the underlying causal relationships between features and the target variable. This can lead to selecting predictive features not causally related to the outcome, which could negatively impact the interpretability and generalisation of the resulting models. To address this limitation, researchers have started to explore the integration of causal inference techniques in feature selection algorithms [28]. Causal inference aims to identify the causal relationships between variables rather than mere correlations. By incorporating causal knowledge into the feature selection process, algorithms can potentially identify feature subsets with strong causal relationships with the target variable, improving model performance and generalisation. Focusing on causality rather than correlation proves particularly valuable in domains where understanding the underlying mechanisms is crucial, such as healthcare or finance, where decisions based on causal relationships tend to be more robust and reliable than those based on correlative patterns [20].

Evolutionary computation (EC) techniques have emerged as promising alternatives for feature selection due to their global search capabilities [9,10,27]. Genetic Algorithms (GAs) are among the most commonly used EC techniques since GA's binary chromosomes can represent feature subsets without any modification [11]. Also, Particle Swarm Optimization (PSO), inspired by bird and fish swarm behaviour, is widely used for feature selection [17]. PSO can use either binary vectors or real-valued numbers for feature selection tasks. In the real-valued approach, a feature is selected when its corresponding particle value exceeds a threshold θ. Although promising, few studies have proposed causal-based enhancements to EC-based algorithms for feature selection. Jiang et al. [15] used a causal graph with ant colony optimization, while our approach employs information-theoretic causality estimation that doesn't require explicit graph learning. Sun et al. [25] used NSGA-II with direct causal strength measurement, whereas we introduce a novel framework that combines normalized mutual information with pairwise feature interactions to estimate Average Causal Effects (ACE).

In this paper, we present a study in which we have investigated how causality affects the search capability of EC-based algorithms for feature selection. We tested GA and PSO. To this aim, we compared the performance of causality-driven PSO and GA with the results achieved by their wrapper-based version and those achieved by the recursive feature elimination (RFE) algorithm [12], a widely used wrapper feature selection algorithm. Furthermore, we also investigated the impact of causality on the stability of those algorithms, i.e. their capability to provide a limited subset of features when multiple runs are carried out. The experimental results confirm that causal analysis allows the evolutionary search process to identify feature subsets with stronger and more comprehensive causal relationships with the target variable, potentially improving model interpretability, generalisation, and performance.

The remainder of this paper is organized as follows: Sect. 2 details the proposed EC-based feature selection approaches based on causality. Section 3 provides an overview of the dataset and experimental setup. Additionally, Sect. 4 comprises an in-depth analysis of the results. Finally, Sect. 5 presents the conclusions and discusses possible directions for future work.

2   Methods

As mentioned in the previous section, feature selection methods often rely on correlation-based measures, which can lead to the selection of spurious features that are merely statistically associated with the target variable but do not represent genuine causal relationships. For instance, consider two features X_1 and X_2 where X_1 causes both X_2 and the target Y. While both features may strongly correlate with Y, only X_1 represents the true causal mechanism. This distinction is crucial for developing robust and interpretable ML models.

We propose integrating causal inference principles into evolutionary feature selection algorithms to identify features that correlate with and potentially cause changes in the target variable. EC-based algorithms are particularly well-suited for causal feature selection for several reasons: (i) They can efficiently explore the vast search space for potential causal relationships without making strong assumptions about the underlying causal structure. (ii) Their population-based nature allows simultaneous evaluation of multiple causal hypotheses. (iii) They can naturally handle the multi-objective nature of causal feature selection, balancing the trade-off between minimizing feature set size and maximizing causal influence. (iv) Their iterative improvement process aligns well with the concept of causal discovery, where understanding causal relationships is refined over time. We introduce a causality-based evaluation framework focused on the ACE to quantify causal relationships. The ACE represents the expected change in the outcome when intervening on a variable, as defined in Pearl's do-calculus:

$$\text{ACE}(X \to Y) = \mathbb{E}[Y|do(X = x)] - \mathbb{E}[Y] \tag{1}$$

where $do(X = x)$ is Pearl's do-operator, indicating intervention rather than mere observation. Since direct interventional data is often unavailable, we approximate causal strength using Normalized Mutual Information (NMI). The theoretical justification for this approximation stems from the Information Flow Framework (IFF) developed by Ay and Polani [2], which demonstrates that mutual information bounds causal influence under certain conditions. Specifically, when the system satisfies the causal Markov condition and faithfulness assumption, the mutual information between X and Y provides a lower bound on their causal strength:

$$\text{NMI}(X, Y) = \frac{2 \cdot I(X; Y)}{H(X) + H(Y)} \leq \text{ACE}(X \to Y)^2 \tag{2}$$

This inequality holds because mutual information captures both direct and indirect relationships while maintaining computational tractability with complexity $O(N \log N)$ for N samples. Under the Causal Markov condition (variables are conditionally independent of non-descendants given their parents) and faithfulness assumption (conditional independencies reflect the causal structure), NMI provides a reliable lower bound on causal strength. While these assumptions may not always hold perfectly in practice, empirical evidence shows NMI remains effective for causal feature selection [29]. To capture both direct and interaction-based causal effects, we define the collective causal effect (CCE) as a weighted sum of individual feature contributions and their pairwise interactions. This additive formulation follows the principles of information theory, where mutual information is combined independently. The parameter $\alpha \in [0, 1]$ controls the trade-off between individual and interaction effects, weighting the relative importance of pairwise causal relationships:

$$\text{CCE}(\mathbf{X_s}, Y) = \frac{1}{|\mathbf{X_s}|} \sum_{X_i \in \mathbf{X_s}} \text{NMI}(X_i, Y) + \alpha \sum_{i \neq j} \text{NMI}(X_i \otimes X_j, Y) \tag{3}$$

where α controls the importance of interaction effects and $\otimes$ represents feature interaction. The computational complexity of CCE is $O(|\mathbf{X_s}|^2 \cdot N \log N)$ due to the pairwise interactions. Finally, we introduce a causal relevance score that balances individual and collective causal effects:

Algorithm 1. Causality-Driven Genetic Algorithm (CDGA)

Require: X: feature matrix, y: target variable, N: population size, G: generations
Ensure: P: Pareto-optimal feature subsets maximizing causal relationships
 1: Initialize population P_0 of N binary strings representing feature masks
 2: **for** $g = 1$ to G **do**
 3:     Evaluate population P_{g-1} with bi-objective fitness function:
     – Minimize number of selected features for model simplicity
     – Maximize ACE using pairwise NMI
 4:     Apply adaptive crossover and mutation to create offspring Q_g
 5:     Form combined population $R_g = P_{g-1} \cup Q_g$ for elitism
 6:     Apply NSGA-II non-dominated sorting on R_g for Pareto optimization
 7:     Apply grid-based niching to maintain feature subset diversity
 8:     Select top N individuals from R_g to form next generation P_g
 9: **end for**
10: **return** final Pareto front P_G containing optimal feature subsets

---

$$\mathrm{CRS}(\mathbf{X_s}, Y) = \beta \cdot \mathrm{CCE}(\mathbf{X_s}, Y) - (1 - \beta) \cdot \frac{|\mathbf{X_s}|}{n} \tag{4}$$

where $\beta \in [0, 1]$ is a trade-off parameter and n is the total number of features. The overall complexity of computing CRS for each solution in the algorithms is $O(|\mathbf{X_s}|^2 \cdot N \log N + |\mathbf{X_s}|)$, where the second term accounts for the feature set size calculation. While theoretically $|\mathbf{X_s}|$ could equal N, leading to $O(N^3)$ complexity in the worst case, in practice $|\mathbf{X_s}| \ll N$ as the objective of feature selection is to identify a minimal subset of relevant features. This practical constraint keeps the computational cost manageable even for large datasets.

2.1   Causality-Driven Genetic Algorithm (CDGA)

CDGA extends the traditional GA by integrating causality principles into its core mechanisms. The algorithm represents solutions as binary strings $\mathbf{s} = s_1, ..., s_n$, where each bit indicates the inclusion (1) or exclusion (0) of a feature. While Eqs. 3 and 4 provide our theoretical framework for causal feature selection, we implement them through a computationally efficient bi-objective fitness function:

$$\mathbf{F}(\mathbf{s}) = [\min \sum_{i=1}^{n} s_i, -\mathrm{ACE}(\mathbf{X_s}, Y)] \tag{5}$$

The algorithm incorporates adaptive genetic operators where crossover and mutation rates decrease linearly ($0.7 \to 0.5$ and $0.2 \to 0.02$ respectively) across generations. A grid-based niching mechanism maintains diversity by partitioning the feature space into hypercubes of size $\sqrt{n}$, where n is the number of features, with a maximum of 2 solutions per cell. When a cell exceeds capacity, the solution with lower ACE is removed. For the case where $|\mathbf{X_s}| = 0$, we set ACE $= 0$. The method simultaneously minimizes the number of features and maximizes

---

Algorithm 2. Causality-Driven PSO (CPSO)

Require: X: feature matrix, y: target variable, N: particles, G: generations
Ensure: S: repository of best feature subsets
 1: Initialize positions $\mathbf{X}_0$, velocities $\mathbf{V}_0$, personal bests $\mathbf{P}$, and global best $\mathbf{g}$
 2: **for** $g = 1$ to G **do**
 3:     Calculate adaptive inertia $w(g)$
 4:     **for** each particle i **do**
 5:         Convert position to binary feature selection
 6:         Evaluate fitness with multi-objective function:
   – Minimize selected features
   – Maximize causal impact via ACE using NMI
 7:             Update personal best $\mathbf{p}_i$ if improved
 8:             Update global best $\mathbf{g}$ if improved
 9:             Update velocity using causal-aware equation
10:             Update position within $[-1, 1]$
11:     **end for**
12:     Apply grid discretization for diversity
13:     Update repository S with non-dominated solutions
14: **end for**
15: **return** S

---

the ACE. The ACE quantifies the collective causal effect of selected features on the target variable, implementing $\mathrm{CCE}(\mathbf{X_s}, Y)$ from Eq. 3 with regularization parameter $\lambda \in [0, 1]$:

$$\mathrm{ACE}(\mathbf{X_s}, Y) = \frac{1}{m} \sum_{i=1}^{m} \mathrm{NMI}(X_i, Y) + \frac{\lambda}{m(m-1)} \sum_{1 \le i < j \le m} \mathrm{NMI}(X_i \oplus X_j, Y) \quad (6)$$

where $m = |\mathbf{X_s}|$ and $\oplus$ denotes feature concatenation for interaction effects. This formulation provides an efficient balance between computational cost and causal discovery effectiveness in the evolutionary process. To enhance causal discovery, CDGA incorporates several key modifications. First, it employs adaptive genetic operators where crossover and mutation rates evolve over generations, allowing broad exploration of causal structures initially and focused refinement later. Second, a grid-based niching mechanism maintains population diversity by partitioning the objective space into hypercubes and limiting the number of solutions in each niche, ensuring the exploration of different causal pathways. The algorithm implements an efficient caching mechanism for ACE calculations, significantly reducing computational overhead when evaluating similar feature subsets. This is particularly important as causal estimation through NMI is computationally intensive. CDGA's output is a Pareto-optimal solution representing different trade-offs between feature set size and causal influence. Each solution represents a distinct hypothesis about the data's causal structure, allowing the selection of feature subsets based on their requirements for interpretability and causal fidelity.

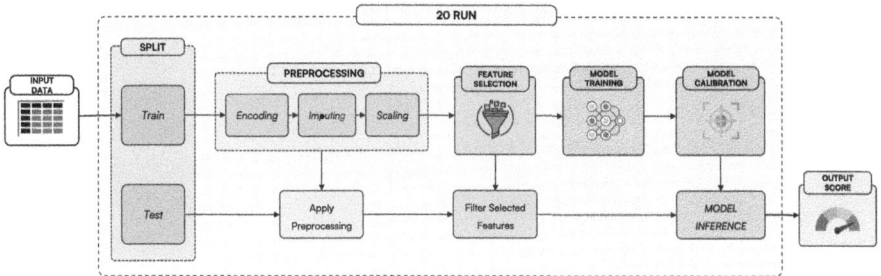

Fig. 1. Experimental workflow.

2.2 Causality-Driven Particle Swarm Optimization (CPSO)

CPSO adapts the classical PSO framework for causal feature selection by redefining particle dynamics to explore causal relationships. Each particle represents a potential feature subset through a continuous position vector $\mathbf{x} = x_1, ..., x_n$ in n-dimensional space, where each dimension corresponds to a feature. The position values are bounded in $[-1, 1]$ and are transformed to binary selections through a threshold function for feature selection: if $x_i > 0$, then $s_i = 1$; otherwise, $s_i = 0$.

The particle velocity update incorporates causal awareness through a modified update rule that balances individual and collective causal discoveries:

$$\mathbf{v}_i^{t+1} = w(t)\mathbf{v}_i^t + c_1 r_1(\mathbf{p}_i^{\text{best}} - \mathbf{x}_i^t) + c_2 r_2(\mathbf{g}^{\text{best}} - \mathbf{x}_i^t) \tag{7}$$

where $w(t)$ is an adaptive inertia weight that decreases linearly from w_{start} to w_{end} to transition from exploration of causal structures to exploitation. The cognitive (c_1) and social (c_2) components guide particles based on individual and swarm-wide causal discoveries. Personal best p_i and global best g updates follow dominance rules: a position is better if it achieves a higher ACE (with 10^{-10} threshold) or equal ACE with fewer features. The fitness evaluation uses the ACE metric from CDGA, which is adapted for PSO's continuous nature. Grid-based position discretization maintains diversity in causal hypothesis exploration, preventing premature convergence. An efficient caching mechanism for ACE calculations benefits PSO's tendency to explore similar search space regions. CPSO's velocity-based exploration provides a different perspective on causal feature selection than CDGA, allowing smoother transitions between causal hypotheses. The social learning component enables effective sharing of discovered causal relationships. The algorithm produces a repository of high-quality solutions balancing minimal feature sets and causal influence.

3   Experimental Setup

We used six datasets originating from various application domains. We employed datasets with diverse characteristics regarding sample sizes, features, and classes. In detail, we considered HAND [5], focused on handwriting features

Table 1. Datasets used in the experiments.

Dataset	#Samples	#Features	#Classes
HAND	174	90	2
Madelon	2000	500	2
Ovarian	216	2190	2
Ucihar	561	10299	6
Mfeat	2000	216	10
GCM	190	16063	14

for Alzheimer's disease diagnosis; Madelon [13], an artificial dataset created for the NIPS 2003 feature selection challenge; Ovarian [21], a proteomics dataset for ovarian cancer detection; UCIHAR [1], containing smartphone sensor data for human activity recognition; Mfeat [19], a subset of the Multiple Features dataset containing Fourier coefficients of character shapes; and GCM [23], a gene expression dataset for cancer type classification. For further details, please refer to Table 1.

We utilized two complementary metrics to evaluate the performance: accuracy, which reflects the overall correctness and the F1_score, the harmonic mean of precision and recall, which provides a balanced measure for imbalanced datasets. Given that we conducted 20 runs, we present the mean of both metrics and their standard deviations.

For our experiments, we employed four well-established ML models: Decision Trees (DT) [22], Multi-Layer Perceptron (MLP) [24], K-Nearest Neighbors (KNN) [7], and Random Forest (RF) [3]. These models were chosen for their diverse learning approaches and proven effectiveness across various domains.

3.1   Baseline Experiment

The initial experimental setup was a reference point for comparing its performance to the subsequent experiments. This comparison helps determine whether more advanced experiments' added complexity and effort lead to better performance. Figure 1 shows the complete experimental workflow. The baseline case doesn't have the feature selection block. The system takes a dataset and splits it into training (70%) and test (30%) sets. After, it performs a preprocessing step on the training set, which ensures data quality and aligns the dataset with the standards expected by the classification algorithms. The procedure involves three operations: encoding categorical features, handling missing values, and scaling all features with the RobustScaler standardization technique. The preprocessing steps remain the same for every experimental setting. We then train the classification models (DT, MLP, KNN, RF) on the training data. To ensure reliable probability estimates, we calibrate each model using isotonic regression [16]. Model calibration transforms the raw model outputs into well-calibrated probabilities, which is crucial when the model must provide accurate predictions and reliable confidence estimates for decision-making. The calibrated models make predictions on the test set. Table 2 shows the results for each dataset and model combination.

3.2 Causality-Driven Feature Selection

The scheme in Fig. 1 describes the second experiment. This includes adding a feature selection module (in orange) before the model training with respect to the baseline experiment, In particular, this experiment compares the traditional GA and PSO for feature selection with those proposed in our approach based on causality. Further technical details about the GA and PSO are explained in [8]. Table 3 and Table 4 reported the parameters of the algorithms and their fitness. It is important to note that the standard versions of the algorithms are multi-objective and wrapper-based, relying on the accuracy of the classifier. Furthermore, it should be highlighted that the classifier used to assess the accuracy of the EC-based algorithms is the same type as the one subsequently used for the final classification. The results obtained are reported and discussed in the next section.

4 Comparison Findings

This section discusses the outcomes of the experimental procedures detailed in the previous sections and assesses the effectiveness of the proposed approach in comparison to baseline or established methods in the field. Table 2 shows the results of the baseline experiment, described in Sect. 3.1, in which ML algorithms were applied without feature selection. Performance is reported for each dataset (first column) and each classification algorithm (first row), and the best outcomes are highlighted in bold. The table shows no overall best classifier; however, KNN and RF generally outperform the others, with KNN displaying major stability across runs.

Table 5 and 6 present the result of the experimental approaches described in Sect. 3.2, where GA and PSO are tested and compared with their causality-driven variants. Table 5 shows the results of applying GA and CDGA as feature selection methods (second column). Datasets are listed in the first column, while

Table 2. Baseline experiments results for different classifiers.

Dataset	DT		MLP		KNN		RF	
	Accuracy	F1	Accuracy	F1	Accuracy	F1	Accuracy	F1
HAND	57.08 (4.10)	56.03 (4.09)	51.13 (5.23)	50.38 (5.44)	37.74 (0.04)	37.74 (0.04)	**58.21** (2.95)	**57.47** (3.23)
Madelon	**73.90** (0.78)	**73.89** (0.78)	56.14 (1.65)	56.09 (1.68)	54.00 (0.03)	53.47 (0.03)	65.92 (1.10)	65.88 (1.11)
Ovarian	87.23 (1.87)	87.04 (1.89)	88.15 (3.00)	87.96 (3.02)	**90.77** (0.05)	**90.50** (0.05)	90.00 (1.90)	89.75 (1.95)
Ucihar	95.47 (0.38)	95.44 (0.37)	97.01 (0.28)	97.04 (0.28)	89.15 (0.06)	88.62 (0.06)	**97.98** (0.18)	**97.90** (0.18)
Mfeat	91.53 (0.52)	91.48 (0.52)	95.04 (0.63)	95.05 (0.63)	**96.83** (0.02)	**96.84** (0.02)	95.41 (0.28)	95.40 (0.28)
GCM	48.60 (3.78)	36.59 (5.08)	**66.67** (5.72)	**55.88** (8.06)	61.40 (0.07)	46.46 (0.07)	64.91 (3.12)	54.99 (4.08)

Table 3. Comparison of GA Parameters for Feature Selection Methods

Parameter	CDGA	Standard GA
Common Parameters		
Population Size	50	
Number of Generations	100	
Mutation Probability	0.1	
Crossover Probability	0.8	
Selection Method	NSGA-II	
Mutation Method	Flip Bit	
Method-Specific Parameters		
Crossover Method	Uniform	Two-Point
Niche Capacity	2	–
Fitness Function Objectives	(ACE, # Features)	(Accuracy, # Features)

the number of features chosen (third column) is also reported. The subsequent columns display the performance of ML algorithms, with the best values highlighted in bold. Across all cases, the causality-driven variant selects more features, resulting in improved outcomes with a lower or similar standard deviation. As for the baseline experiment, the RF algorithm typically outperforms others, though KNN demonstrates the lowest standard deviation values.

Tables 6 follows the same structure as Table 5, comparing results from applying PSO and CPSO as feature selection methods. Like CDGA, CPSO generally enhances performance, though to a lower percentage, and often with greater

Table 4. Comparison of PSO Parameters for Feature Selection Methods

Parameter	CPSO	Standard PSO
Common Parameters		
Number of Particles	50	
Number of Generations	100	
Initial Inertia Weight (w_{start})	0.9	
Final Inertia Weight (w_{end})	0.4	
Cognitive Coefficient (c_1)	2.0	
Social Coefficient (c_2)	2.0	
Position Bounds	$[-1, 1]$	
Velocity Bounds	$[-1, 1]$	
Method-Specific Parameters		
Grid Cell Capacity	2	–
Fitness Function Objectives	(ACE, # Features)	(Accuracy, # Features)

Table 5. GA vs CDGA experiments results for different classifiers.

Dataset	Method	#Feat	DT		MLP		KNN		RF	
			Accuracy	F1	Accuracy	F1	Accuracy	F1	Accuracy	F1
HAND	GA	29.25	61.98 (4.38)	61.79 (4.52)	48.77 (5.85)	47.93 (6.53)	54.72 (0.01)	53.91 (0.01)	62.83 (2.68)	62.24 (2.71)
	CDGA	63.40	63.22 (4.41)	63.03 (4.50)	49.75 (5.80)	48.89 (6.48)	55.81 (0.01)	54.99 (0.01)	**64.09** (2.70)	**63.48** (2.73)
Madelon	GA	199.65	73.77 (0.85)	73.76 (0.85)	55.90 (1.52)	55.83 (1.53)	54.00 (0.01)	53.47 (0.01)	65.93 (0.89)	65.89 (0.89)
	CDGA	320.85	**75.25** (0.84)	**75.24** (0.84)	57.02 (1.50)	56.95 (1.51)	55.08 (0.01)	54.54 (0.01)	67.25 (0.88)	67.21 (0.88)
Ovarian	GA	999.30	87.77 (1.69)	87.57 (1.75)	87.77 (2.47)	87.57 (2.44)	90.50 (0.01)	90.50 (0.01)	89.85 (1.61)	89.57 (1.67)
	CDGA	1344.40	89.53 (1.67)	89.32 (1.73)	89.53 (2.45)	89.32 (2.42)	**92.59** (0.01)	**92.31** (0.01)	91.65 (1.59)	91.36 (1.65)
Ucihar	GA	229.50	95.36 (0.43)	95.33 (0.43)	96.98 (0.30)	97.02 (0.32)	89.15 (0.01)	88.62 (0.01)	98.00 (0.13)	97.92 (0.13)
	CDGA	359.00	97.27 (0.42)	97.24 (0.42)	98.92 (0.29)	98.96 (0.31)	90.93 (0.01)	90.39 (0.01)	**99.96** (0.12)	**99.88** (0.12)
Mfeat	GA	84.45	91.50 (0.37)	91.45 (0.38)	95.21 (0.63)	95.21 (0.63)	96.83 (0.01)	96.84 (0.01)	95.50 (0.28)	95.50 (0.28)
	CDGA	133.20	93.33 (0.36)	93.28 (0.37)	97.11 (0.62)	97.11 (0.62)	**98.77** (0.01)	**98.78** (0.01)	97.41 (0.27)	97.41 (0.27)
GCM	GA	7811.85	48.86 (3.52)	36.77 (4.23)	66.67 (4.48)	55.79 (6.15)	61.40 (0.01)	46.46 (0.01)	66.93 (4.35)	57.79 (4.99)
	CDGA	9640.90	49.84 (3.49)	37.51 (4.20)	68.00 (4.45)	56.91 (6.12)	62.63 (0.01)	47.39 (0.01)	**68.27** (4.32)	**58.95** (4.96)

stability. However, an exception is observed with the Mfeat dataset, where PSO outperforms CPSO. In contrast to the GA results, the causality-driven variant of PSO selects a smaller number of features.

Comparing Table 2 (baseline approach) with Table 5 (GA approach) and Table 6 (PSO approach) provides several interesting insights. First, the baseline approach achieves performance levels comparable to those obtained with PSO and GA feature selection. However, adding causality to the evolutionary feature selection methods improves results. Notably, CDGA selects more features than GA, while CPSO selects fewer features than PSO. Overall, CDGA achieves the best performance among all methods.

Furthermore, Table 7 compares the proposed feature selection techniques, CDGA and CPSO, which achieved the best results in our experiments, with a standard and well-validated method, such as RFE. CDGA achieves the highest performance by selecting a number of features similar to RFE. While RFE shows the lowest performance, its results are comparable to CPSO's, which selects the fewest features. RFE's lower performance relative to CDGA and CPSO could be due to its sequential nature, which may lead it to overlook complex, non-linear

Table 6. PSO vs CPSO experiments results for different classifiers.

Dataset	Method	#Feat	DT		MLP		KNN		RF	
			Accuracy	F1	Accuracy	F1	Accuracy	F1	Accuracy	F1
HAND	PSO	43.65	56.98 (4.35)	56.11 (4.22)	50.57 (5.79)	49.74 (6.00)	37.74 (0.01)	37.74 (0.01)	58.49 (2.87)	57.84 (3.03)
	CPSO	35.10	58.02 (4.19)	56.97 (4.36)	49.53 (5.37)	48.88 (5.66)	37.74 (0.01)	37.74 (0.01)	**58.87** (3.16)	**58.08** (3.47)
Madelon	PSO	252.00	73.85 (0.85)	73.84 (0.85)	56.69 (1.40)	56.63 (1.40)	54.00 (0.01)	53.47 (0.01)	65.63 (1.05)	65.60 (1.04)
	CPSO	243.35	**73.97** (0.81)	**73.97** (0.81)	55.76 (1.44)	55.71 (1.45)	54.00 (0.01)	53.47 (0.01)	65.55 (1.00)	65.50 (0.99)
Ovarian	PSO	1076.40	88.00 (2.37)	87.82 (2.43)	88.92 (3.89)	88.73 (3.92)	90.77 (0.01)	90.50 (0.01)	90.08 (1.27)	89.81 (1.32)
	CPSO	991.40	88.38 (1.17)	88.22 (1.17)	89.62 (2.11)	89.43 (2.14)	90.77 (0.01)	**90.50** (0.01)	**90.00** (1.69)	89.72 (1.72)
Ucihar	PSO	282.90	95.34 (0.47)	95.31 (0.48)	97.08 (0.34)	97.12 (0.35)	89.15 (0.01)	88.62 (0.01)	97.96 (0.20)	97.88 (0.20)
	CPSO	264.80	95.33 (0.43)	95.30 (0.43)	96.99 (0.24)	97.02 (0.25)	89.15 (0.01)	88.62 (0.01)	**97.99** (0.20)	**97.91** (0.20)
Mfeat	PSO	110.25	91.20 (0.55)	91.15 (0.56)	95.00 (0.76)	95.00 (0.75)	**96.83** (0.01)	**96.84** (0.01)	95.47 (0.45)	95.46 (0.45)
	CPSO	133.60	91.48 (0.53)	91.43 (0.53)	95.15 (0.63)	95.15 (0.63)	96.83 (0.01)	96.84 (0.01)	**95.57** (0.33)	**95.56** (0.33)
GCM	PSO	8008.25	48.07 (2.51)	35.38 (3.44)	67.63 (3.24)	57.31 (4.33)	61.40 (0.01)	46.46 (0.01)	65.53 (3.57)	56.38 (5.05)
	CPSO	3782.35	48.60 (3.17)	36.01 (3.75)	**67.89** (4.03)	**57.57** (5.11)	61.40 (0.01)	46.46 (0.01)	64.91 (2.73)	55.10 (3.91)

Table 7. Comparison of CDGA, CPSO, and RFE experiments results for different classifiers. In bold are indicated the best values. * indicates the highest accuracy values that demonstrate statistical significance.

Dataset	Method	#Feat	DT		MLP		KNN		RF	
			Accuracy	F1	Accuracy	F1	Accuracy	F1	Accuracy	F1
HAND	CDGA	63.40	**63.22* (4.41)**	63.03 (4.50)	49.75 (5.80)	48.89 (6.48)	**55.81* (0.01)**	54.99 (0.01)	**64.09* (2.70)**	63.48 (2.73)
	CPSO	35.10	58.02 (4.19)	56.97 (4.36)	49.53 (5.37)	48.88 (5.66)	37.74 (0.01)	37.74 (0.01)	58.87 (3.16)	58.08 (3.47)
	RFE	73.00	59.25 (4.64)	58.28 (4.66)	**50.85* (4.64)**	49.52 (4.60)	37.74 (0.01)	37.74 (0.01)	58.40 (2.77)	57.71 (3.16)
Madelon	CDGA	320.85	**75.25* (0.84)**	75.24 (0.84)	**57.02* (1.50)**	56.95 (1.51)	**55.08* (0.01)**	54.54 (0.01)	**67.25* (0.88)**	67.21 (0.88)
	CPSO	243.35	73.97 (0.81)	73.97 (0.81)	55.76 (1.44)	55.71 (1.45)	54.00 (0.01)	53.47 (0.01)	65.55 (1.00)	65.50 (0.99)
	RFE	204.00	73.67 (0.63)	73.66 (0.63)	56.29 (1.11)	56.25 (1.11)	54.00 (0.00)	53.47 (0.01)	65.59 (0.78)	65.56 (0.77)
Ovarian	CDGA	1344.40	**89.53* (1.67)**	89.32 (1.73)	89.53 (2.45)	89.32 (2.42)	**92.59* (0.01)**	92.31 (0.01)	**91.65* (1.59)**	91.36 (1.65)
	CPSO	991.40	88.38 (1.17)	88.22 (1.17)	89.62 (2.11)	89.43 (2.14)	90.77 (0.01)	90.50 (0.01)	90.00 (1.69)	89.72 (1.72)
	RFE	1533.00	87.62 (1.54)	87.43 (1.56)	**89.85* (2.61)**	89.69 (2.61)	90.77 (0.01)	90.50 (0.01)	90.92 (0.99)	90.69 (1.03)
Ucihar	CDGA	359.00	**97.27* (0.42)**	97.24 (0.42)	**98.92* (0.29)**	98.96 (0.31)	**90.93* (0.01)**	90.39 (0.01)	**99.96* (0.12)**	99.88 (0.12)
	CPSO	264.80	95.33 (0.43)	95.30 (0.43)	96.99 (0.24)	97.02 (0.25)	89.15 (0.01)	88.62 (0.01)	97.99 (0.20)	97.91 (0.20)
	RFE	358.00	95.28 (0.38)	95.25 (0.38)	96.98 (0.32)	97.02 (0.33)	89.15 (0.01)	88.62 (0.01)	97.99 (0.16)	97.90 (0.16)
Mfeat	CDGA	133.20	**93.33* (0.36)**	93.28 (0.37)	**97.11* (0.62)**	97.11 (0.62)	**98.77* (0.01)**	98.78 (0.01)	**97.41* (0.27)**	97.41 (0.27)
	CPSO	133.60	91.48 (0.53)	91.43 (0.53)	95.15 (0.63)	95.15 (0.63)	96.83 (0.01)	96.84 (0.01)	95.57 (0.33)	95.56 (0.33)
	RFE	172.00	91.45 (0.59)	91.42 (0.59)	95.02 (0.74)	95.03 (0.74)	96.83 (0.01)	96.84 (0.01)	95.37 (0.38)	95.36 (0.38)
GCM	CDGA	9640.90	**49.84* (3.49)**	37.51 (4.20)	**68.00* (4.45)**	56.91 (6.12)	**62.63* (0.01)**	47.39 (0.01)	**68.27* (4.32)**	58.95 (4.96)
	CPSO	3782.35	48.60 (3.17)	36.01 (3.75)	67.89 (4.03)	57.57 (5.11)	61.40 (0.01)	46.46 (0.01)	64.91 (2.73)	55.10 (3.91)
	RFE	11244.00	49.30 (3.55)	36.74 (5.41)	64.91 (4.80)	53.40 (6.30)	61.40 (0.01)	46.46 (0.01)	66.75 (3.19)	57.35 (4.39)

dependencies among features that EC-based algorithms, like GA and PSO, are better equipped to capture. The fact that CDGA and CPSO optimize feature sets in a more global, population-based manner enables them to explore a broader solution space. We performed a Friedman test to evaluate the statistical significance of the differences among CDGA, CPSO, and RFE approaches across all datasets and classifiers. Since this non-parametric test revealed significant differences among the methods, we needed to conduct post-hoc pairwise comparisons to identify specific differences between approaches. Due to multiple comparisons (three pairwise comparisons: CDGA vs. CPSO, CDGA vs. RFE, and CPSO vs. RFE), we applied a Bonferroni correction by adjusting the significance level $\alpha = 0.05$ to $\alpha' = 0.05/3 \approx 0.0167$. In our plots (Fig. 2), the x-axis represents the possible number of occurrences (ranging from 0 to 20).

Occurrences. The primary objective of this experimental analysis is to evaluate the stability of feature selection across multiple runs of EC-based algorithms, which is crucial given their stochastic nature. While these algorithms may yield different solutions across different runs, we aim to demonstrate that certain features are consistently selected, thereby establishing algorithmic stability. This aspect is particularly beneficial for interpretability purposes. To quantify this, we employed Kernel Density Estimation (KDE)[1] to visualize the distribution of feature occurrences across the twenty runs. In our plots, the x-axis represents

---

[1] KDE creates smooth probability curves from discrete data points by placing Gaussian kernels at each point and summing them, following the formula $\hat{f}(x) = \frac{1}{nh}\sum_{i=1}^{n} K(\frac{x-x_i}{h})$, where h controls the smoothing bandwidth. This smoothing technique helps visualize and compare feature selection patterns.

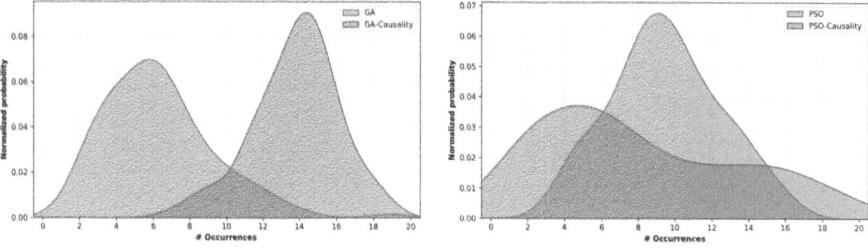

(a) Distribution of feature occurrences for HAND dataset.

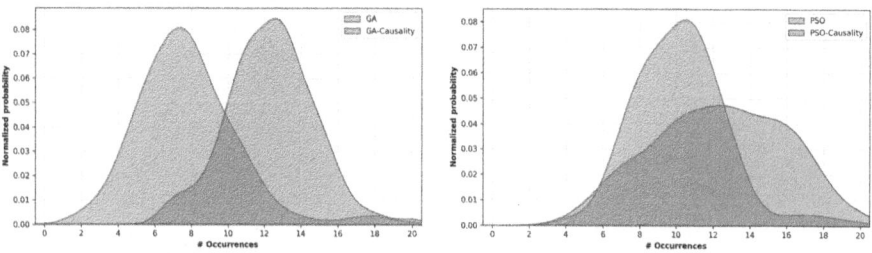

(b) Distribution of feature occurrences for Mfeat dataset.

Fig. 2. Distribution of feature occurrences across different datasets using DT.

the possible number of occurrences (ranging from 0 to 20), while the y-axis depicts the normalized probability of each occurrence count. Stable algorithms are characterized by distributions exhibiting higher peaks on the plot's right side, indicating features frequently selected across runs. Conversely, higher probabilities in the lower occurrence range suggest less stability, as features are selected inconsistently.

Fitness. In this second analysis, we investigated the behavior of CDGA and CPSO during the evolution. To this aim, we plotted the average fitness and the average number of selected features (computed over the whole population) as a function of the generation number. Figure 3 shows those plots for the HAND and Mfeat datasets. From the plot, we can observe that CDGA and CPSO demonstrated consistent evolutionary behaviour for the HAND dataset, successfully minimizing the number of selected features while maximizing the ACE metric. However, when applied to the Mfeat dataset, we observed divergent patterns between the two algorithms. While CDGA maintained its consistent feature minimization and ACE maximization performance, CPSO exhibited notably different behaviour. This divergence can be attributed to the inherent characteristics of PSO's particle dynamics, particularly when dealing with complex multiclass datasets like Mfeat. Unlike GA, which maintains population diver-

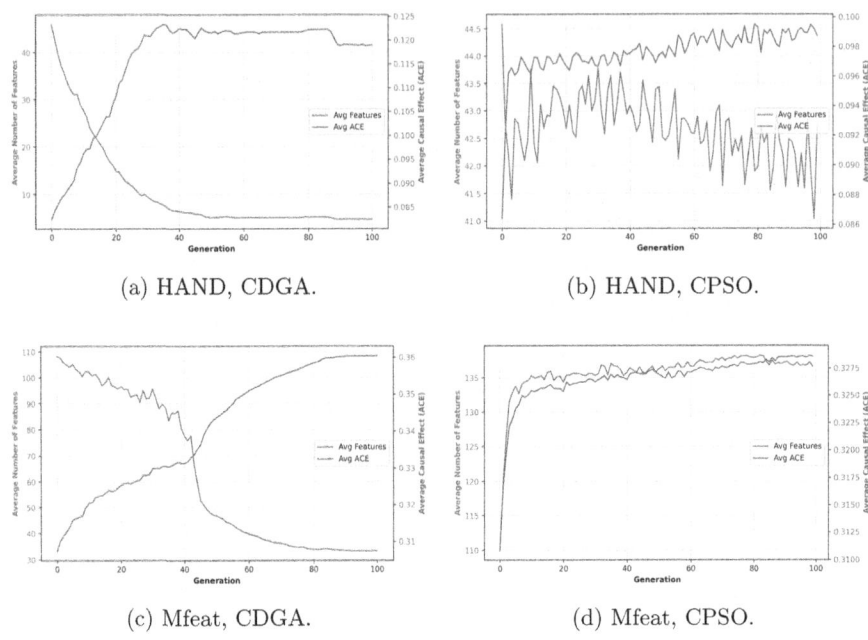

(a) HAND, CDGA.   (b) HAND, CPSO.

(c) Mfeat, CDGA.   (d) Mfeat, CPSO.

Fig. 3. Average fitness trend across different methods and datasets.

sity through crossover and mutation operators, PSO's velocity-based update mechanism may become susceptible to local optima trapping across different regions of the search space. This sensitivity is particularly pronounced in multiclass causal relationships, where the pairwise mutual information calculations underlying the ACE estimation may not adequately capture higher-order causal dependencies. The phenomenon becomes especially evident in high-dimensional feature spaces, where the interplay between the particle swarm's social and cognitive components drives the optimization process toward suboptimal trade-offs between feature count minimization and causality estimation accuracy.

5   Conclusions and Future Work

The results of this study highlight the advantages of incorporating causality principles into evolutionary feature selection methods like GA and PSO. By focusing on features with genuine underlying influences on outcomes rather than mere correlations, causality-driven approaches (CDGA and CPSO) improve model stability and generalizability across datasets. CDGA, in particular, tends to select a broader set of influential features than standard GA, likely due to its capacity to capture a wider range of causally relevant information while preserving only those features that impact outcomes. This balanced selection enhances both the robustness and interpretability of the model. PSO, by design, tends to explore smaller subsets of the feature space compared to GA, as it relies on swarm intelligence to navigate the search space efficiently. CPSO's causality filter further refines this process, selecting fewer but highly relevant features. Among the classification algorithms evaluated, RF consistently demonstrated the best performance, likely due to its ensemble nature, which combines multiple decision trees, reduces variance, and enhances robustness against feature selection variability. A key objective of this analysis was to assess feature selection stability across multiple runs, given the variable nature of EC-based algorithms. Using KDE, we found CDGA to be the most stable method, consistently selecting the same features and enhancing interpretability. Additionally, we analyzed the fitness dynamics of CDGA and CPSO. Both methods showed stability on some datasets, like HAND, minimizing feature count and maximizing ACE metric. However, for some datasets, CDGA remained consistent while CPSO varied due to its reliance on PSO's particle dynamics. Unlike GA, which uses crossover and mutation to maintain diversity, PSO's updates can lead to local optima, particularly in complex multiclass relationships, impacting the balance between feature minimization and causality accuracy. Future research could explore enhancing the causal component of evolutionary-based methods to improve performance on different high-dimensional and multiclass datasets. Further studies will investigate stability measures across other datasets and feature selection algorithms. Expanding the application of causal feature selection to different data types and diverse problem domains would help validate proposed approaches. Finally, incorporating advanced causality models and exploring interpretability frameworks aligned with domain-specific insights would deepen understanding of causally driven feature selection's role in AI and ML.

References

1. Anguita, D., Ghio, A., Oneto, L., Parra, X., Reyes-Ortiz, J.L.: Human activity recognition on smartphones using a multiclass hardware-friendly support vector machine. In: International Workshop on Ambient Assisted Living, pp. 216–223. Springer (2012)
2. Ay, N., Polani, D.: Information flows in causal networks. Adv. Complex Syst. 11(01), 17–41 (2008). https://doi.org/10.1142/S0219525908001465
3. Breiman, L.: Random forests. Mach. Learn. 45(1), 5–32 (2001)
4. Cilia, N., De Stefano, C., Fontanella, F., Scotto di Freca, A.: Variable-length representation for ec-based feature selection in high-dimensional data. Lecture Notes in Computer Science (including subseries Lecture Notes in Artificial Intelligence and Lecture Notes in Bioinformatics) 11454 LNCS, 325–340 (2019)
5. Cilia, N.D., De Stefano, C., Fontanella, F., Molinara, M., Scotto Di Freca, A.: Handwriting analysis to support alzheimer's disease diagnosis: A preliminary study. Lecture Notes in Computer Science (including subseries Lecture Notes in Artificial Intelligence and Lecture Notes in Bioinformatics) 11679 LNCS, 143–151 (2019)
6. Cordella, L.P., De Stefano, C., Fontanella, F., Marrocco, C., Scotto di Freca, A.: Combining single class features for improving performance of a two stage classifier. In: 2010 20th International Conference on Pattern Recognition, pp. 4352–4355 (2010)

7. Cover, T., Hart, P.: Nearest neighbor pattern classification. IEEE Trans. Inf. Theory **13**(1), 21–27 (1967)
8. D'Alessandro, T., De Stefano, C., Fontanella, F., Nardone, E.: Integrating data augmentation in evolutionary algorithms for feature selection: a preliminary study. In: Smith, S., Correia, J., Cintrano, C. (eds.) Applications of Evolutionary Computation, pp. 397–412. Springer Nature Switzerland, Cham (2024)
9. De Falco, I., Della Cioppa, A., Fontanella, F., Tarantino, E.: An innovative approach to genetic programming-based clustering. In: Abraham, A., Köppen, M. (eds.) 9th Online World Conference on Soft Computing in Industrial Applications (2004)
10. De Falco, I., Tarantino, E., Della Cioppa, A., Gagliardi, F.: A novel grammar-based genetic programming approach to clustering. In: SAC' 05: Proceedings of the 2005 ACM symposium on Applied computing, pp. 928–932 (2005)
11. De Stefano, C., Fontanella, F., Marrocco, C.: A ga-based feature selection algorithm for remote sensing images. Lecture Notes in Computer Science (including subseries Lecture Notes in Artificial Intelligence and Lecture Notes in Bioinformatics) **4974 LNCS**, 285–294 (2008)
12. Guyon, I., Weston, J., Barnhill, S., Vapnik, V.: Gene selection for cancer classification using support vector machines. J. Mach. Learn. Res. **46**, 389–422 (2002)
13. Guyon, I.: Design of experiments for the NIPS 2003 variable selection benchmark. In: NIPS 2003 Workshop on Feature Extraction and Feature Selection (2003)
14. Guyon, I., Elisseeff, A.: An introduction to variable and feature selection. J. Mach. Learn. Res. **3**, 1157–1182 (2003)
15. Jiang, X., Jia, X., Wang, L.: Feature selection based on ant colony optimization and causal graphical models. In: International Conference on Intelligent Computing, pp. 544–557. Springer (2020)
16. Kull, M., Silva Filho, T., Flach, P.: Beyond temperature scaling: obtaining well-calibrated probabilities with dirichlet calibration. Adv. Neural. Inf. Process. Syst. **30**, 12316–12326 (2017)
17. Li, A.D., Xue, B., Zhang, M.: Multi-objective particle swarm optimization for key quality feature selection in complex manufacturing processes. Inf. Sci. **641**, 119062 (2023)
18. Li, J., et al.: Feature selection: a data perspective. ACM Comput. Surv. (CSUR) **50**(6), 1–45 (2017)
19. van der Maaten, L., Postma, E.O., van den Herik, H.J.: Comparing discriminant analysis, neural networks and statistics for hand written digit classification. Pattern Recogn. **42**(6), 1388–1392 (2009)
20. Moraffah, R., Sheth, P., Vishnubhatla, S., Liu, H.: Causal feature selection for responsible machine learning (2024). https://arxiv.org/abs/2402.02696
21. Petricoin, E.F., et al.: Use of proteomic patterns in serum to identify ovarian cancer. The lancet **359**(9306), 572–577 (2002)
22. Quinlan, J.R.: Induction of decision trees. Mach. Learn. **1**(1), 81–106 (1986)
23. Ramaswamy, S., et al.: Multiclass cancer diagnosis using tumor gene expression signatures. Proc. Natl. Acad. Sci. **98**(26), 15149–15154 (2001)
24. Rumelhart, D.E., Hinton, G.E., Williams, R.J.: Learning representations by back-propagating errors. Nature **323**(6088), 533–536 (1986)
25. Sun, J., Fang, W., Liu, M., Xue, B.: A multi-objective evolutionary algorithm for causal feature selection in classification. Inf. Sci. **528**, 18–37 (2020)
26. Trunk, G.V.: A problem of dimensionality: a simple example. IEEE Trans. Pattern Anal. Mach. Intell. **1**(3), 306–307 (1979)

27. Xue, B., Zhang, M., Browne, W.N., Yao, X.: A survey on evolutionary computation approaches to feature selection. IEEE Trans. Evol. Comput. **20**(4), 606–626 (2016)
28. Yu, K., Guo, X., Liu, L., Li, J., Wang, H., Ling, Z., Wu, X.: Causality-based feature selection: methods and evaluations. ACM Comput. Surv. **53**(5) (2020)
29. Zhang, K., Hyvärinen, A.: On the identifiability of the post-nonlinear causal model. In: Proceedings of the Twenty-Fifth Conference on Uncertainty in Artificial Intelligence, pp. 647–655. AUAI Press (2009)

A Coach-Based Quality-Diversity Approach for Multi-agent Interpretable Reinforcement Learning

Erik Nielsen⑩, Andrea Ferigo⑩, and Giovanni Iacca$^{(\boxtimes)}$⑩

Department of Information Engineering and Computer Science, University of Trento,
Trento, Italy
{erik.nielsen,andrea.ferigo,giovanni.iacca}@unitn.it

Abstract. Thanks to the advances in deep Reinforcement Learning (RL) and its demonstrated capabilities to perform complex tasks, the field of Multi-Agent RL (MARL) has recently undergone major developments. However, current MARL approaches based on deep learning still suffer from a general lack of interpretability. Recently, hybrid models combining Decision Trees (DTs) with simple leaves running Q-Learning have been proposed as an alternative to achieve high performance while preserving interpretability. However, efficient search strategies are needed to optimize such models. In this paper, we address this challenge by proposing a novel Quality-Diversity evolutionary optimization approach, based on MAP-Elites. We test the method on a team-based game, on which we introduce a coach agent, also optimized via evolutionary search, to optimize the team creation during training. The proposed strategy is tested in conjunction with three different evolutionary selection methods and two different mappings between MAP-Elites archives and team members. Results demonstrate how the proposed approach can effectively find high-performing policies to accomplish the given task, while the coach pushes even further the team optimization, hence improving the algorithm's overall performance.

Keywords: Quality-Diversity · Genetic Programming · Multi-Agent Reinforcement Learning · Explainability · Interpretability

1 Introduction

Multi-Agent Reinforcement Learning (MARL) has received significant attention in recent decades [1], as it extends the applicability of traditional (single-agent) RL to problems where different agents must interact with each other and with a given environment. Typically, this results in a significantly more complex learning problem (compared to single-agent contexts), as in MARL the interactions between each single agent and the environment depend not only on the actions performed by that agent but also on what the other agents, which simultaneously interact with the same environment, do.

© The Author(s), under exclusive license to Springer Nature Switzerland AG 2025
P. García-Sánchez et al. (Eds.): EvoApplications 2025, LNCS 15612, pp. 402–418, 2025.
https://doi.org/10.1007/978-3-031-90062-4_25

The state of the art in MARL currently adopts deep RL models to carry out multi-agent tasks. This is mainly due to the fact that deep learning has shown high performance in various fields of applications, resulting in an effective tool also for MARL. However, most deep RL approaches use, by construction, black-box models, which makes it hard to interpret their functioning. On the other hand, interpretability is a crucial issue in many cases of safety-critical or high-stake decision-making, such as robotics, finance, and healthcare [2], where each decision taken by the algorithm should be human-verifiable and readable [3–5]. In the case of MARL, being able to understand why an agent takes an action is even more important, as it is necessary to understand not only how a single agent behaves, but also how agents adapt their decisions based on the other agents' behavior.

Previous work addressed this challenge by employing Decision Trees (DTs) combined with RL [6] in the context of MARL [7]. Yet, this approach was limited to the fact that only one solution (i.e., one policy obtained as a DT with RL) was generated. Another recent work, instead, showed that, in the case of single-agent RL tasks, applying a Quality-Diversity (QD) [8] algorithm, specifically MAP-Elites (ME) [9], can be an effective way to discover multiple DTs characterized by different depths and with different behaviors [10,11]. In this work, we combine these two approaches and apply ME to derive multiple interpretable models for a MARL task. To the best of our knowledge, this is the first attempt in the research community that goes in this direction. Furthermore, we propose the use of a "coach" agent to optimize team composition. This coach-based approach aims to enhance the trade-off between exploitation and exploration created by ME in order to fully exploit the characteristics of the evolutionary process. Our goal is to verify if applying ME to MARL can be a viable solution, how different selection and mapping approaches impact the performance of the algorithm, and ultimately analyze the interpretability of the obtained DTs.

The rest of the paper is structured as follows. The next Sect. 2 presents related works. Then, Sect. 3 describes the employed methods. Section 4 presents the results of the experimentation. Finally, Sect. 5 concludes this work.

2   Related Works

In this section, we report some of the most relevant studies related to MARL and QD algorithms and highlight the novelty of our approach with respect to the related works.

Multi-agent Reinforcement Learning First explorations in the MARL field [12] discussed the main advantages w.r.t. the traditional single-agent RL case. Later on, further research in MARL compared table-based policies and Recurrent Neural Networks (RNNs) in the iterated prisoner's dilemma [13], finding that RNNs tend to be less cooperative in this kind of task. Another work [14] showed how combining Minimax Q with Q-Learning returns more robust policies compared with state-of-the-art Q-Learning.

Finally, Tan [15] proposed a scalable and decentralized version, called Independent Q-learning (IQL), a version of Q-Learning specifically designed for MARL. However, IQL is not free of drawbacks. For example, due to the absence of the replay buffer, it is not possible to use NNs as approximator functions.

Several versions tried to address this problem, e.g., by using centralized critics [16], management agents [17], reward shaping [18], hysteretic Q-learning [19], or hysteretic deep recurrent networks [20]. Other works instead focused on the policy functions, switching the point of view to an actor-critic model. Chu and Ye, in [21], proposed a method based on sharing parameters between agents in MARL. In [22], the authors investigated the impact of the latency created in the communication between agents. Moreover, in [23] it was proposed to derive a novel MARL algorithm based on duality theory, while in [24], the authors used a value-decomposition network. Lately, the authors in [25] studied a multi-stage approach, where the agent structure is capable of performing multi- and single-agent tasks.

An alternative, non-NN-based approach consists of using coevolutionary settings and Genetic Programming (GP) to evolve the agents. In [26], these methods were shown to be more effective than the strategies handcrafted by the authors. Furthermore, GP-based approaches have the additional advantage of producing interpretable models, an aspect that has become a pressing need in the research community [5,27], where interpretability measures are often associated with the mathematical complexity [28] and structure [29] of the model. To reproduce interpretable-by-design models applicable to RL tasks, recent works [6,30] also proposed the use of evolutionary algorithms, combining Grammatical Evolution (GE) with RL in the context of single-agent RL tasks. A first attempt to extend this approach to MARL settings was made in [7], comparing solutions obtained by GE and GP.

Finally, it is worth mentioning that earlier works proposed to enhance performance in MARL by employing a coach supervising the training process [31,32].

Quality-Diversity Algorithms. QD optimization is a recently introduced para-digm in stochastic optimization whose focus is the search for diverse, optimal solutions [33]. Among QD algorithms, ME [9] represents a rather simple yet effective approach that has been successfully applied in various tasks, including RL [10,11]. One of the drawbacks of ME, however, is the need to define the solution descriptors. Recently, Kent et al. [34] attempted to address this issue by combining ME with Bayesian Optimization. Another recent work [35] introduced an alternative QD approach called Diverse Policy Optimization, achieving enhanced performance in different RL tasks.

Novelty of the Proposed Approach. In this paper, we propose an approach that applies ME in a MARL setting by evolving a set of solutions in the form of DTs. Moreover, we introduce a coach agent to perform optimal team formation. To the best of our knowledge, this is the first attempt to use ME to optimize DTs MARL settings and investigate the use of a coach in this context.

3   Methods

In this section, we describe how we evolve agents able to solve MARL tasks efficiently and cooperatively with an external coach agent whose purpose is to handle the team creation and evolutionary process. Figure 1 shows the whole process, illustrating how the selection methods and mapping strategies (see below for details) are set up in the *tell* and *ask* methods, which are important parts of our ME-based proposal. Our codebase is made publicly available on GitHub[1].

Following the approach previously proposed in [36], each agent is formed by a DT combined with simple leaves running Q-Learning. The structure of the DT is optimized with GA or ME, while IQL learns the optimal action for each leaf. Given that our aim is to apply this method to multi-agent settings, we need to define a mechanism to assemble the team involved in the task. In this regard, we compare three distinct processes for selecting agents, discussed in Sect. 3.3, each one featuring two separate mapping procedures, described in Sect. 3.4.

We use the GA-based approach from [7] as a baseline. In addition, we follow two different initialization methods, one that randomly initializes the population and one that injects a handcrafted solution into the initial population.

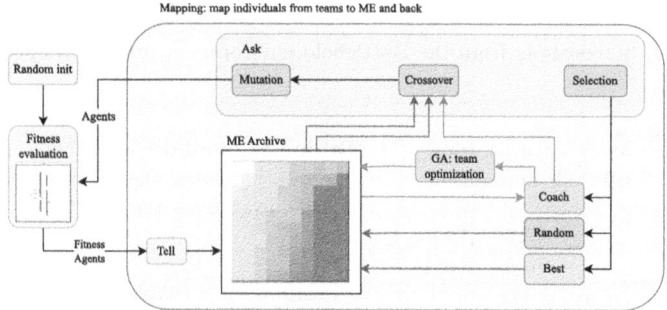

Fig. 1. Structure of the proposed algorithm workflow, showing how the selection and mapping mechanisms are employed within the ME optimization process.

3.1   Environment

Battlefield is a game environment specifically developed for testing MARL algorithms, available in the PettingZoo library [37]. Figure 2 gives an example of the game's stages. The game is based on two teams, blue and red, composed of 12 agents each. The goal of each team is to beat the other team by "killing" (i.e., removing from the environment) each member of it. In this work, we follow the same setting of [7], as we optimize the policy of the blue team against a red team performing random actions at each step. The environment is an 80x80

---

[1] https://github.com/DIOL-UniTN/Coach-QD-MARL.

grid, closed by outer walls that prevent the agents from exiting the game area; moreover, to make the task more challenging for the agents, there are 4 inner walls as obstacles. Table 1 reports the input features (preprocessed as in [7]) for this task. The agents have the option to perform 21 different actions, namely taking no action, moving to one of the 8 nearby cells, attacking one of those cells, or moving two cells on either left, right, up, or down (see Table 2 and Table 3).

Lastly, the reward is given by the performances during the task: each kill corresponds to 5 points; each attack carried out against an opponent corresponds to 0.9 points. The agent is penalized by −0.1 points for receiving hits, being killed, or attacking without hitting the enemy. In addition, at each step without other rewards, the agent loses −0.5 points.

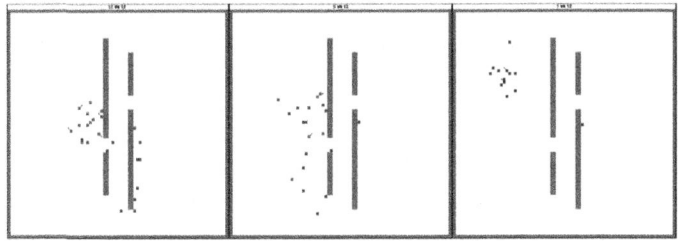

Fig. 2. Screenshots from the Battlefield environment during an episode.

Table 1. Extracted features, their abbreviation, and their domain.

Feature	Abbr.	Domain
Obstacle	o_{pos}	$\{0,1\}$
Allied global density	ag_{pos}	$[0,1]$
Enemies global density	eg_{pos}	$[0,1]$
Enemies local density	el_{pos}	$[0,1]$
Enemy presence	e_{pos}	$\{0,1\}$

Table 2. Available actions and their abbreviation. Considering the positions (Table 3), there are 21 actions.

Action	Abbreviation
Move	m_{pos}
Attack	a_{pos}

Table 3. Position abbreviations: n indicates the number of cells away from the agent (1 or 2).

Position	Abbreviation
Above	$X_{n,a}$
Above-left	$X_{n,al}$
Above-right	$X_{n,ar}$
Left	$X_{n,l}$
Right	$X_{n,r}$
Below	$X_{n,b}$
Below-left	$X_{n,bl}$
Below-right	$X_{n,br}$
Same quadrant	X_s

3.2   MAP-Elites

ME is a QD algorithm that explores a given descriptor domain to generate diverse, high-performing solutions. ME stores the best solution found for a given cell in a grid archive. The ME workflow starts with the initialization of the archive. Initially, $p_{initial}$ solutions are randomly generated. Then an iterative process starts: first, a batch of p solutions is retrieved (*ask* procedure) from the archive; the solutions are then updated using mutation and crossover in p'. After evaluation, the p' solutions are mapped back into the archive based on their descriptor and fitness (*tell* procedure).

Similar to [10, 11], the chosen features to describe each DT are the depth of the tree and the entropy of its leaves. The first feature measures the complexity of the tree's structure, as the deeper the tree, the more complex it is, while the entropy is correlated with the distribution of choices made by the leaves, i.e., with how many times each i-th action in the tree is executed.

3.3   Selection Strategies

ME retains, for each cell in the archive, the best solution found, based on its descriptor. The conventional approach is to select solutions to evolve stochastically. This solution may be ideal in certain scenarios, but for difficult tasks characterized by a significant degree of complexity, this approach may yield suboptimal solutions. Hence, we compare here three selection strategies: two baselines, random and best selection, and the proposed selection strategy based on a coach's decision.

Random Selection. The *Random* selection approach defines each team of agents by randomly picking m solutions, where m is the batch size, for k times. This approach serves as a baseline, which in principle provides the highest exploration of the given search space. However, it may not fully exploit the most suitable solutions because it does not consider the solution's fitness.

Best Selection. The second baseline is the *Best* selection method. In this case, each batch is formed by the m best solutions stored in the archive. This should favor the highest exploitation of the fittest solutions, but it may limit the ability to fully explore the search space.

Coach-Based Selection. Finally, we propose a selection strategy that involves a *Coach* agent, that actively oversees the ME archive and chooses the solutions to form the teams. The selection process is based on a GA (in our experiments, we resort to the implementation contained in the inspyred library [38]). The overall workflow of the coach can be described as follows:

1. The *Coach* collects the descriptors of the solutions with their corresponding fitness;
2. It optimizes k teams formed by m solutions via GA (with a population size of 100, uniform crossover with probability 1.0, Gaussian mutation with probability 0.1, and a maximum of 40000 generations), using the descriptors as "genes" and the n-percentile fitness of the solutions forming the teams as objective value;

3. For each optimized team, the *Coach* asks the corresponding solutions to the ME archive;
4. ME evolves the solutions and returns them to the *Coach*, to start the evaluation process.

The goal of using GA to select the optimal team is to efficiently achieve a trade-off between exploration and exploitation of solutions in order to increase the probability of having a higher number of optimal solutions stored in ME.

3.4   Mapping Approaches

To increase the likelihood of having a large number of high-performing solutions, it is crucial to devise methods to map each team with ME. Here, we use two different mapping approaches. The first one is based on the Fully Coevolutionary mapping [7], while the second one, which we call Singular, uses a single map that keeps track of each team member.

Fully Coevolutionary (FC). The Fully Coevolutionary technique [7] relies on the evolution of multiple, distinct archives, each containing its own population. This methodology should result in diverse solutions, as each member comes from a different evolutionary process. A graphical representation of this approach is reported in Fig. 3a, where it can be seen that FC requests a batch of k solutions, where k is the number of trained teams simultaneously, from m archives, where m is the number of team members.

Singular (S). The second approach is based on a single population that collects all the generated solutions. To ensure diversity, we employed ME, which, as described in Sect. 3.2, stores the solutions based on the predefined descriptors, hence exploiting the whole genetic material of the entire population. Figure 3b

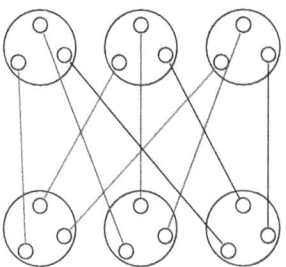

(a) Fully Coevolutionary mapping

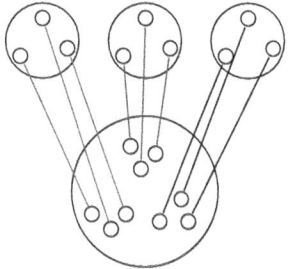

(b) Singular mapping

Fig. 3. Graphical representation of the proposed mapping schemes. The outer circles on the top represent teams that have to be evaluated, with the inner circles being the team members (i.e., the agents). The outer circles on the bottom are evolutionary processes, with the inner circles being the evolved solutions.

shows a visual representation of this methodology, where it can be seen that in this case, a batch of m solutions is requested for k times from a single archive.

3.5 Injection of a Handcrafted Solution

The random creation of the initial DTs employed in ME may lead to a cold start, as it may take several generations to identify good solutions. As a result, in [7], authors proposed to inject a handcrafted solution into the initial population at the beginning of the evolutionary process. To avoid cold starts, we test the two mapping approaches mentioned earlier by injecting the same solution into ME after translating their genotype into the corresponding DTs. We indicate the process *With Injection* as "W-I", and the process with *No Injection* as "N-I".

3.6 Evaluation Methods

We evaluate our setup in terms of ME performance, fitness function, and performance of the final teams achieved. We describe the corresponding evaluation metrics below.

MAP-Elites Evaluation. ME metrics are defined to highlight the differences in how each fills the archive and exploits the solutions therein. Several metrics can be used to measure the ME performances [9]. The metrics that are most suitable for the purposes of our experiments concern the archive coverage and the archive precision, which are defined as follows:

- *Archive coverage* It represents the percentage of the number of cells covered by the algorithm out of the maximum number of cells in the grid.
- *Archive precision* Given the highest fitness found across all cells in a given run of ME, it represents the average error between the best fitness found in each cell and the highest fitness overall. Since the fitness values, in this case, are continuous variables, we consider the root mean square error (RMSE):

$$RMSE = \sqrt{\frac{\sum(fitness_{cell,i} - \max fitness_{cell})^2}{n_{cell,i}}},$$

where n_{cell} indicates the number of cells and $fitness_{cell,i}$ indicates the best fitness in each i-th cell.

Fitness Evaluation. During the evaluation phase of the evolutionary process, a team consists of m agents that engage in a certain number of episodes, i.e., simulations of the task, denoted as N_{ep}. Once the simulation is over, the fitness score is based on the agents' rewards returned from the environment. To aggregate the rewards across episodes, we use $n\%$-percentile, which was shown to achieve a trade-off between considering the maximum and the average performance across episodes [7].

Final Team Evaluation. In the end, the algorithm stores the final optimal solutions to form the optimal team of agents. To evaluate the final teams, we test each formation derived from 10 executions of each approach in 100 new additional episodes. In this case, we evaluate these teams using three key metrics, namely the average agent reward, the average number of kills, and the average task completion rate (along with the overall fitness of the team).

4 Experimental Evaluation

With our experiments, we are interested in answering the following research questions:

RQ1 Is ME a viable approach to solving MARL tasks?
RQ2 What is the effect that different selection strategies have on the evolution?
RQ3 What is the difference between the Fully Coevolutionary and Singular mappings?
RQ4 Is it possible to interpret the optimized agents?

To answer these questions, we configure the methods described in Sect. 3 setting the hyperparameters as reported in Table 4 and Table 5, respectively for the common parameters and for those that differ depending on the approach. The parameters for the training process are reported in Table 6.

Table 4. Common hyperparameters for the FC and S mappings.

Parameter	Value
No. random seeds	10
$P_{crossover}$	0.4
Map bounds	$[0, 10] \times [0, 10]$
Descriptor dimensions	2
Conditions' features	$[0, 34]$
Conditions' threshold	$[0, 1]$
Actions	$[0, 21]$
Entropy bounds	$[0.8, 1]$
Max DT's depth	10
ϵ	1
ϵ decay	0.99
min ϵ	0.05
$n\%$-percentile	70%
No. teams per generation	60

Table 5. Specific hyperparameters for the FC and S mappings.

Parameter	FC	S
No. archives	12	1
Initial population size	60x12	120x1
Batch population size	60	12

Table 6. Training settings.

Parameter	Value
No. teams	60
Episodes	500
Generations	40

Note that, with this parametrization, FC has 60 initial solutions for each population, with 12 populations in total (one per agent), for a total of 720, while S uses a single map that is initialized with 120 solutions. These values have been selected to allow for a sufficient initial filling of the maps. During the evolutionary process, FC has a maximum capacity of 1200 solutions (i.e., a 10x10 map for each of the 12 agents in the teams), while S has a maximum capacity of a single map of 10x10 solutions. This means S utilizes only 1/6 of the initial population of FC, while its maximum population capacity during training is only 1/12 of FC's population.

4.1   RQ1: MAP-Elites Performance

Our first set of experiments aims to assess whether the use of ME can achieve high performance in MARL tasks. In Fig. 4, we report the fitness trends during the generations of 10 independent runs of each approach tested in the experiments. While the baseline algorithm (based on GA) seems incapable of finding high-performing solutions (with fitness values below 10), the approaches using ME reach fitness values above 20, similar to performances reported in [7]. We report the final fitness values achieved by each algorithm in Table 7. We confirmed that the differences across these values are statistically significant with the Wilcoxon signed-rank test ($\alpha = 0.05$), showing that the GA-based baseline is always worse than the ME-based methods.

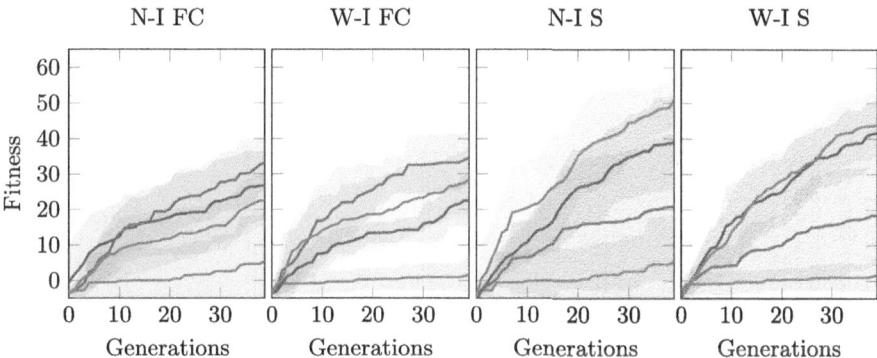

Fig. 4. Evolutionary trends (average ± std. dev. across 10 independent runs) of the best fitness value achieved at each generation by each approach tested in the experiments.

Table 7. Average ± std. dev. of best fitness values (to be maximized) achieved across 10 independent runs by each approach tested in the experiments.

Algorithm	GA	ME					
Mapping	FC	FC			S		
Selection	Random	Random	Best	Coach	Random	Best	Coach
N-I	5.4 ± 0.3	26.6 ± 9.8	33.2 ± 7.7	22.4 ± 7.3	38.9 ± 13.2	20.8 ± 20.40	51.0 ± 6.2
W-I	1.7 ± 3.4	22.4 ± 7.3	34.6 ± 7.5	28.5 ± 9.6	41.6 ± 9.1	18.3 ± 14.11	44.0 ± 9.2

4.2   RQ2: Selection Strategies

Our second analysis aims to evaluate ME using the three different selection mechanisms to determine whether our proposal, i.e., the *Coach* selection, can indeed improve the trade-off between exploration and exploitation. In Fig. 5, we show the best archives achieved across the 10 runs of each ME-based approach, where it is visible how the different approaches are characterized by different levels of exploration-exploitation trade-off. In particular, *Random* shows a better exploration of the archive; *Best* achieves more exploitation with the Singular mapping; whereas, *Coach* produces several high-performing solutions in a more dense area than *Random*.

In Fig. 6, we report the archive coverage and archive precision across the 10 runs, where it can be seen that *Random* and *Coach* selections provide better performances in terms of exploration of the search space. In particular, Fully Coevolutionary achieves the worst archive coverage and precision with the *Best* selection, as it is excessively exploitative. As for the Singular mapping, it achieves the best results with *Coach* selection, which allows for higher exploration. This is confirmed by the fact that *Random* (which is highly explorative) achieves higher coverage than *Best* in the case of Singular mapping. Overall, the best results in terms of archive coverage and precision are achieved by the Singular mapping with *Coach* and *Random* selection. Also in this case, we applied the Wilcoxon

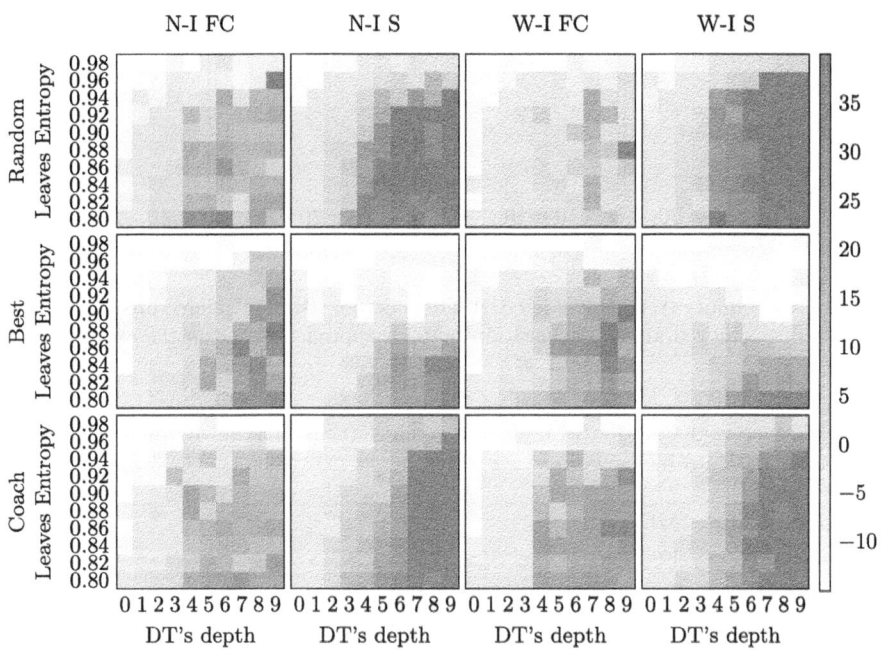

Fig. 5. Best archives achieved by ME with all the tested mapping and selection approaches (With Injection). Darker cells indicate better (i.e., higher) fitness.

signed-rank test ($\alpha = 0.05$) to statistically confirm these findings, revealing that in all mapping settings *Coach* selection outperforms *Best* and *Random* with statistical significance.

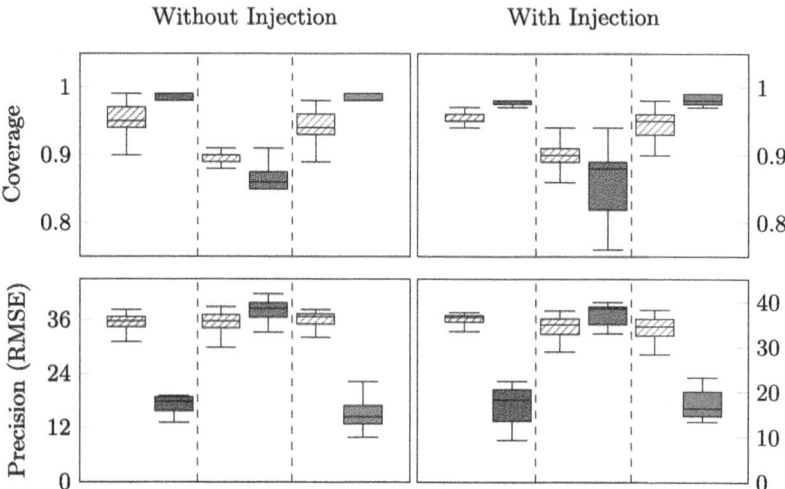

Fig. 6. Archive coverage and archive precision achieved by the ME-based approaches (Without Injection) across 10 independent runs.

Further evaluation considers the best teams (as described in Sect. 3.6) identified during the evolutionary process by all the settings of ME. Figure 7 shows the distribution of rewards achieved by each team identified by the different approaches. As can be seen, each combination of selection strategy and mapping approach yields different outcomes, indicating differences in terms of team composition and performance, with the *Coach* selection with Singular mapping consistently outperforming (in all the team-related metrics of interest) all the other approaches, showing a more effective optimization of the DT population.

4.3   RQ3: Mapping Approaches

Our third analysis focuses on comparing the two mapping approaches, Fully Coevolutionary and Singular. The fitness trends shown in Fig. 4 indicate that in both cases (With or Without Injection) the results obtained with the Fully Coevolutionary methods are lower than those found with Singular mapping. In particular, the best Singular case (*Coach* selection) outperforms the best Fully Coevolutionary approach (*Best* selection). However, the Singular mapping with *Best* selection performs worse than the worst Fully Coevolutionary case, confirming how all these methods combined affect the exploration-exploitation trade-off. Moreover, the injection of the solution does not benefit the evolutionary process, as it tends to achieve slightly worse results. This is primarily due to the increased exploitation of the algorithm.

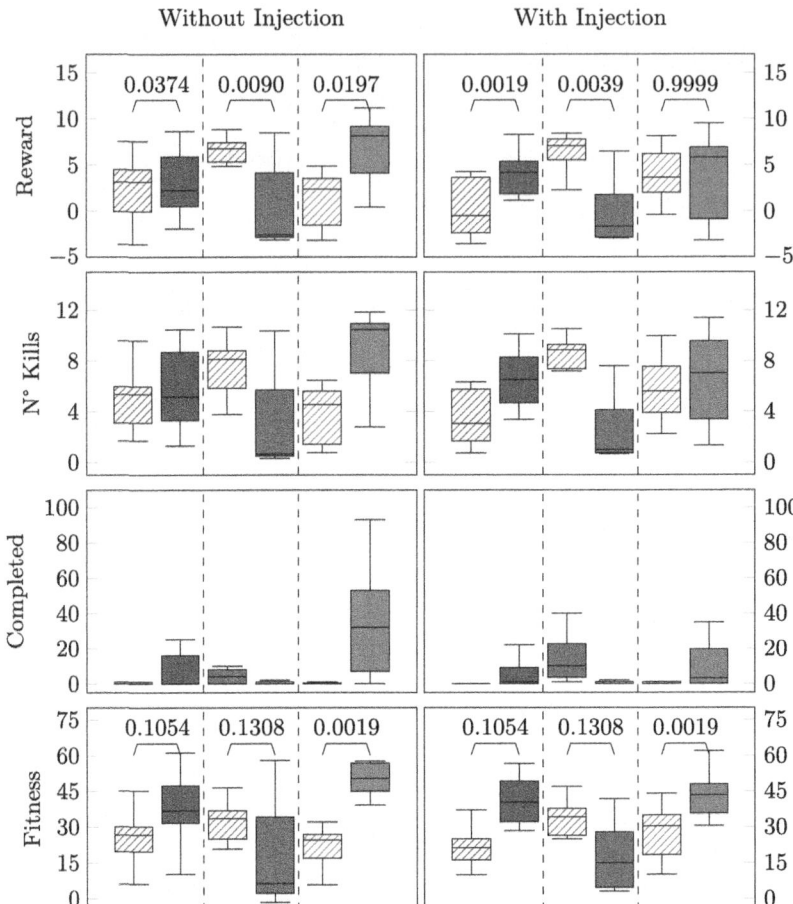

Fig. 7. Evaluation metrics of the final teams achieved by all the ME-based approaches across 10 independent runs.

4.4   RQ4: Interpretability Analysis

Finally, we conclude our investigation with an analysis of interpretability. As discussed earlier, all the evaluated optimization approaches are based on the evolution of the same type of agent (DTs with leaves running Q-Learning).

For the sake of simplicity, we focus here on an analysis of the structure of the best DT obtained across the 10 runs with Singular mapping and *Coach* selection, as this is the case that obtains the best performances. Figure 8 represents the structure of the DT, where the number in parenthesis in each node/leaf indicates the node id. Inner nodes make checks on the input features, while leaves correspond to actions. It is possible to describe the full sequence of actions taken by the agent and which environment feature led to those decisions. In this regard, we consider one of the quickest episodes observed in the experiment as an example for interpreting the agent's policy. The sequence explains how the agent first navigates around the obstacle, which is initially below-left (id 15) and then above (id 16), by moving first left, then below, and finally above (id 20, 22, 24). Then it follows the enemy's presence (id 14) by moving above (id 19). Occasionally, it adjusts the position to the right (id 19, 21). Then, it reaches the enemy from below (id 9), and it attacks 4 or 5 times (id 9). The episode concludes with the agent adjusting its position (id 5, 19, 24) until it detects the presence of an enemy above (id 6). Then, it moves above (id 22) to attack the last remaining enemy three times (id 9).

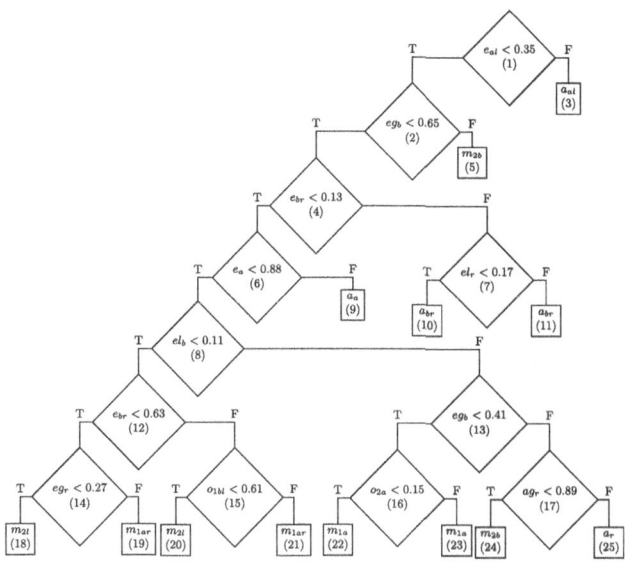

Fig. 8. Best DT found across the 10 runs of ME with Singular mapping and *Coach* selection.

5   Conclusions

In this paper, we developed an evolutionary algorithm based on ME that optimizes the creation of teams to deal with MARL tasks. Considering agents based on a DT structure with leaves running Q-Learning, the proposed QD approach optimizes the structure of the DT, while RL learns the optimal actions to perform. Hence, it maintains high interpretability while returning high-performing solutions.

In our experiments, we considered three selection mechanisms: two baselines, namely *Random* and *Best* selection, and then the proposed *Coach* selection, which uses a coach agent to supervise team formation. We also considered two ME mappings, namely Fully Coevolutionary and Singular.

The proposed *Coach* selection with Singular mapping revealed several advantages in terms of performance and computational complexity. In fact, it requires a reduced number of populations compared to Fully Coevolutionary mapping: the latter uses 12 populations, while Singular uses only one, besides the population for the *Coach* agent. In terms of performance, this configuration outperformed all the other approaches in all evaluation metrics, still finding interpretable models.

In future research, we intend to extend our model in several ways. Firstly, we intend to further explore the hyperparameter space, to find the best configuration and observe how the parametrization influences performance. We also intend to include in our studies alternative optimization algorithms, such as Bayesian optimization. Finally, it would be interesting to explore Quality-Diversity optimization of alternative model architectures, e.g., based on Hebbian Learning [39] or mixtures of simple experts [40, 41].

References

1. Oroojlooy, A., Hajinezhad, D.: A review of cooperative multi-agent deep reinforcement learning. Appl. Intell. **53**(11), 13677–13722 (2023)
2. Yu, C., Liu, J., Nemati, S., Yin, G.: Reinforcement learning in healthcare: a survey. ACM Comput. Surv. (CSUR) **55**(1), 1–36 (2021)
3. Degrave, J., et al.: Magnetic control of tokamak plasmas through deep reinforcement learning. Nature **602**(7897), 414–419 (2022)
4. Zheng, S., Trott, A., Srinivasa, S., Parkes, D.C., Socher, R.: The AI economist: taxation policy design via two-level deep multiagent reinforcement learning. Sci. Advances **8**(18), eabk2607 (2022)
5. Barredo Arrieta, A., et al.: Explainable artificial intelligence (XAI): concepts, taxonomies, opportunities and challenges toward responsible AI. Inf. Fusion **58**, 82–115 (2020)
6. Custode, L.L., Iacca, G.: Interpretable pipelines with evolutionary optimized modules for reinforcement learning tasks with visual inputs. In: Genetic and Evolutionary Computation Conference Companion, New York, NY, USA, Association for Computing Machinery, pp. 224–227 (2022)
7. Crespi, M., Ferigo, A., Custode, L.L., Iacca, G.: A population-based approach for multi-agent interpretable reinforcement learning. Appl. Soft Comput. **147**, 110758 (2023)
8. Pugh, J.K., Soros, L.B., Stanley, K.O.: Quality diversity: a new frontier for evolutionary computation. Front. Robot. AI **3**, 40 (2016)
9. Mouret, J.B., Clune, J.: Illuminating search spaces by mapping elites (2015) arXiv:1504.04909
10. Ferigo, A., Custode, L.L., Iacca, G.: Quality diversity evolutionary learning of decision trees. In: Symposium on Applied Computing, New York, NY, USA, Association for Computing Machinery, pp. 425–432 (2023)
11. Ferigo, A., Custode, L.L., Iacca, G.: Quality-diversity optimization of decision trees for interpretable reinforcement learning. Neural Comput. Appl. 1–12 (2023)
12. Stone, P., Veloso, M.: Multiagent Systems: A Survey from a Machine Learning Perspective. Technical report, Defense Technical Information Center, Fort Belvoir, VA (1997)

13. Sandholm, T.W., Crites, R.H.: On multiagent Q-learning in a semi-competitive domain. In: Adaption and Learning in Multi-Agent Systems, vol. 1042. Springer, Berlin, pp. 191–205 (1996)
14. Littman, M.L.: Markov games as a framework for multi-agent reinforcement learning. In: Machine Learning Proceedings 1994. Morgan Kaufmann, San Francisco, CA, USA, pp. 157–163 (1994)
15. Tan, M.: 1. In: Multi-Agent Reinforcement Learning: Independent vs. Cooperative Agents. Morgan Kaufmann, San Francisco, CA, USA, pp. 330–337 (1997)
16. Lauer, M., Riedmiller, M.A.: An algorithm for distributed reinforcement learning in cooperative multi-agent systems. In: International Conference on Machine Learning, San Francisco, CA, USA, Morgan Kaufmann, pp. 535–542 (2000)
17. Fuji, T., Ito, K., Matsumoto, K., Yano, K.: Deep multi-agent reinforcement learning using DNN-weight evolution to optimize supply chain performance. In: Hawaii International Conference on System Sciences, Honolulu, HI, USA, HICSS, pp. 1278–1287 (2018)
18. Tampuu, A., et al.: Multiagent cooperation and competition with deep reinforcement learning. PLoS ONE $12(4)$, e0172395 (2017)
19. Matignon, L., Laurent, G.J., Le Fort-Piat, N.: Hysteretic Q-learning: an algorithm for decentralized reinforcement learning in cooperative multi-agent teams. In: International Conference on Intelligent Robots and Systems, New York, NY, USA, IEEE/RSJ, pp. 64–69 (2007)
20. Omidshafiei, S., Pazis, J., Amato, C., How, J.P., Vian, J.: Deep decentralized multi-task multi-agent reinforcement learning under partial observability. In: International Conference on Machine Learning, Sydney, NSW, Australia, JMLR, pp. 2681–2690 (2017)
21. Chu, X., Ye, H.: Parameter sharing deep deterministic policy gradient for cooperative multi-agent reinforcement learning (2017). arXiv:1710.00336
22. Singh, A., Jain, T., Sukhbaatar, S.: Learning when to communicate at scale in multiagent cooperative and competitive tasks (2018). arXiv:1812.09755
23. Macua, S.V., Tukiainen, A., Hernández, D.G.O., Baldazo, D., de Cote, E.M., Zazo, S.: Diff-DAC: distributed actor-critic for average multitask deep reinforcement learning (2019). arXiv:1710.10363
24. Sunehag, P., et al.: Value-decomposition networks for cooperative multi-agent learning based on team reward. In: International Conference on Autonomous Agents and MultiAgent Systems, Stockholm, Sweden, pp. 2085–2087 (2018)
25. Yang, J., Nakhaei, A., Isele, D., Fujimura, K., Zha, H.: CM3: Cooperative multi-goal multi-stage multi-agent reinforcement learning (2020). arXiv:1809.05188
26. Haynes, T., Wainwright, R.L., Sen, S., Schoenefeld, D.A.: Strongly typed genetic programming in evolving cooperation strategies. In: International Conference on Genetic Algorithms, San Francisco, CA, USA, Morgan Kaufmann, pp. 271–278 (1995)
27. Bacardit, J., Brownlee, A.E., Cagnoni, S., Iacca, G., McCall, J., Walker, D.: The intersection of evolutionary computation and explainable AI. In: Genetic and Evolutionary Computation Conference Companion, New York, NY, USA, Association for Computing Machinery, pp. 1757–1762 (2022)
28. Barceló, P., Monet, M., Pérez, J., Subercaseaux, B.: Model interpretability through the lens of computational complexity. In: Advances in Neural Information Processing Systems, vol. 33 (2020)
29. Virgolin, M., De Lorenzo, A., Medvet, E., Randone, F.: Learning a formula of interpretability to learn interpretable formulas. In: Parallel Problem Solving from Nature, Springer, Cham, pp. 79–93 (2020)

30. Custode, L.L., Iacca, G.: A co-evolutionary approach to interpretable reinforcement learning in environments with continuous action spaces. In: IEEE Symposium Series on Computational Intelligence, pp. 1–8 (2021)
31. Riley, P.F., Veloso, M.M.: Coach planning with opponent models for distributed execution. Auton. Agent. Multi-Agent Syst. **13**, 293–325 (2006)
32. Liu, B., Liu, Q., Stone, P., Garg, A., Zhu, Y., Anandkumar, A.: Coach-player multi-agent reinforcement learning for dynamic team composition. In: International Conference on Machine Learning, PMLR, pp. 6860–6870 (2021)
33. Chatzilygeroudis, K., Cully, A., Vassiliades, V., Mouret, J.B.: Quality-diversity optimization: a novel branch of stochastic optimization (2020)
34. Kent, P., Gaier, A., Mouret, J.B., Branke, J.: Bayesian optimisation for quality diversity search with coupled descriptor functions. IEEE Trans. Evol. Comput. (2024)
35. Li, W., Wang, B., Yang, S., Zha, H.: Diverse policy optimization for structured action space (2023). arXiv:2302.11917
36. Custode, L.L., Iacca, G.: Evolutionary learning of interpretable decision trees. IEEE Access **11** (2023)
37. Terry, J., et al.: Pettingzoo: gym for multi-agent reinforcement learning. Adv. Neural. Inf. Process. Syst. **34**, 15032–15043 (2021)
38. Tonda, A.: Inspyred: bio-inspired algorithms in python. Genet. Program Evolvable Mach. **21**(1), 269–272 (2020)
39. Ferigo, A., Cunegatti, E., Iacca, G.: Neuron-centric hebbian learning. In: Genetic and Evolutionary Computation Conference, New York, NY, USA, Association for Computing Machinery, pp. 87–95 (2024)
40. Custode, L.L., Farina, P., Yildiz, E., Kilic, R.B., Yildirim, K.S., Iacca, G.: Fast-Inf: Ultra-Fast Embedded Intelligence on the Batteryless Edge. In: Conference on Embedded Networked Sensor Systems, New York, NY, USA, Association for Computing Machinery, pp. 239–252 (2024)
41. Vincze, M., Ferrarotti, L., Custode, L.L., Lepri, B., Iacca, G.: SMoSE: sparse mixture of shallow experts for interpretable reinforcement learning in continuous control tasks (2024). arXiv:2412.13053

FedGP: Genetic Programming for Evolutionary Aggregation in Federated Learning with Non-IID Data

Elia Pacioni[1,2]([✉]) [ID], Francisco Fernández De Vega[1] [ID], and Davide Calvaresi[2] [ID]

[1] Universidad de Extremadura, Av. Santa Teresa de Jornet, 38., 06800 Mérida, Spain
{eliapacioni,fcofdez}@unex.es
[2] University of Applied Sciences and Arts of Western Switzerland (HES-SO Valais/Wallis), Rue de l'Industrie 23, 1950 Sion, Switzerland
{elia.pacioni,davide.calvaresi}@hevs.ch
https://www.unex.es , https://www.hevs.ch

Abstract. Federated Learning (FL) represents a distributed, privacy-preserving machine learning (ML) paradigm that enables decentralized model training across multiple clients. While traditional aggregation techniques, such as Federated Averaging (FedAVG), have demonstrated effectiveness, they often struggle in Not Independent and Identically Distributed (non-IID) scenarios, where data distributions vary significantly among clients. To address these limitations, this study introduces FedGP, a novel aggregation strategy based on Genetic Programming (GP). FedGP dynamically evolves aggregation functions, enabling adaptive and personalized model updates that better capture the heterogeneity inherent in distributed data. The proposed method is evaluated on the PathMNIST dataset, employing a comprehensive experimental design comprising 24 configurations, including 8 setups with FedAVG and 16 with FedGP. The comparative analysis highlights FedGP's superior generalization capabilities and reduced biases, outperforming FedAVG in terms of accuracy. These results position FedGP as a robust and scalable solution for real-world FL applications, particularly in environments characterized by data heterogeneity.

Keywords: Federated Learning · Genetic Programming · Model Aggregation · FedGP · FedAVG

1 Introduction

FL was introduced by Google in 2016 as a collaborative ML paradigm that trains models across decentralized data sources while keeping data localized on client devices [1]. Designed to address growing privacy and confidentiality concerns, FL enables the development of large-scale models without requiring data transfer to a central server [2]. A notable application is Google's Gboard, which uses data from Android devices to train predictive text models locally, enhancing

© The Author(s), under exclusive license to Springer Nature Switzerland AG 2025
P. García-Sánchez et al. (Eds.): EvoApplications 2025, LNCS 15612, pp. 419–434, 2025.
https://doi.org/10.1007/978-3-031-90062-4_26

user experience while maintaining privacy [3]. Beyond mobile applications, FL has seen significant adoption in sectors such as healthcare [4] and finance [5,6], where sensitive data demands robust privacy measures. At the core of FL is the aggregation of models, a pivotal step in the distributed training process. During aggregation, client devices send locally trained models (or their weights) to a central server, which synthesizes them into a global model using a predefined aggregation strategy. This process maintains data locality, reducing communication overhead and ensuring client privacy. In 2017, McMahan et al. [1] proposed the de-facto standard approach, namely Federated Averaging (FedAVG), that combines client updates through weighted averaging based on local data sizes. However, FedAVG faces significant challenges in non-IID settings, where data distributions vary across clients [7,8]. In such cases, it struggles to effectively capture local data peculiarities, potentially resulting in a global model with reduced generalization capability [9,10]. Simplified variants of FedAVG, which eliminate client weighting, have been explored to address privacy concerns, but they often fail to address data heterogeneity effectively. Improving model aggregation in FL remains a critical challenge. Recent studies highlight the need for techniques that accommodate client-specific variability and adapt to the dynamic nature of distributed systems [9,10]. Enhancing aggregation mechanisms is essential for boosting global model accuracy and robustness, particularly in complex, heterogeneous environments.

This paper introduces FedGP (Federated Genetic Programming aggregation), a novel approach that leverages GP to address the limitations of traditional aggregation methods [11]. GP, a well-known ML evolutionary algorithm, is well-suited for adaptive and complex problems. During the last decades, it has shown flexibility to adapt to many different tasks, and has demonstrated its capabilities for addressing symbolic regression problems. This particular kind of problem is the inspiration for the problem we address: the dynamic evolution of aggregation strategies in the context of FL. As we show below, FedGP reduces biases and enhances generalization, surpassing the constraints of averaging-based methods.

The remainder of this paper is organized as follows: Sect. 2 outlines the theoretical background and technologies relevant to FL and GP. Section 3 details the methodology, including the FedGP framework, its evolution process, and selection criteria. Experimental results on the PathMNIST dataset [12], focusing on non-IID scenarios, are discussed in Sect. 4, comparing FedGP with FedAVG. Finally, Sect. 5 summarizes the findings, highlights contributions, and proposes avenues for future research.

2   State of Art

Since its introduction, FL has enabled decentralized ML across mobile and IoT (Internet of Things) systems, utilizing the vast amounts of locally generated data without requiring centralization. This paradigm has revolutionized distributed learning, with applications in mobile systems (e.g., GBoard for typing prediction [3], Siri for text prediction and speech recognition [13]), healthcare (e.g., clinical data aggregation [14]), finance, and beyond. FL operates through an iterative process where client devices train local models on private data. Instead of sharing raw data, only model updates (weights or gradients) are sent to a central server for aggregation, preserving privacy [2]. FL can follow either a centralized or decentralized architecture [15]. Centralized FL relies on a server to aggregate local patterns, while decentralized FL distributes this task among devices, enhancing robustness against server failures or attacks. FL can also utilize horizontal or vertical data partitioning [16,17]: in horizontal partitioning, clients hold data with similar attributes but different examples, while in vertical partitioning, clients have different attributes but the same examples, requiring complex integration techniques. Model updates in FL occur synchronously or asynchronously [17]. Synchronous FL waits for all clients to finish training before aggregation, which can cause delays in heterogeneous networks. Asynchronous FL updates models as client contributions arrive, improving efficiency but introducing challenges in stability. Security and privacy in FL are bolstered by differential privacy, secure multiparty computation (SMC), and homomorphic encryption, which protect user data during model updates [18,19]. Techniques like gradient compression and quantization further optimize communication in low-bandwidth environments. Aggregation is at the core of FL. Federated Averaging (FedAVG) is the most widely used method, combining local updates through a weighted average based on client data sizes [20]. However, FedAVG struggles in non-IID scenarios where data distributions vary across clients, leading to suboptimal global models. Alternative approaches, such as local customization and meta-learning, have been explored to address these limitations [21]. Another method, FedSGD, updates models using stochastic gradients instead of averaging but is less commonly adopted [22]. Model adaptability remains a critical challenge in FL, particularly in non-IID environments and resource-limited settings. Meta-learning has shown promise in enabling global models to quickly adapt to new contexts with minimal data [21]. Additionally, bio-inspired algorithms have been explored to optimize transmission efficiency and model performance [23].

2.1 Genetic Programming and Its Role in FL

Introduced by Koza et al. [24], GP excels in optimizing non-linear functions and dynamically generating solutions. Since its inception, various versions of GP have been introduced, each contributing unique methodologies and insights to the field. These include (i) Linear GP [25], which structures programs as linear sequences of instructions, thereby enhancing their interpretability and optimizing performance; (ii) PushGP [26], a variant that employs the Push programming language, specifically designed to handle complex data structures and streamline the manipulation of evolving programs; (iii) Cartesian GP [27], which represents programs as acyclic directed graphs, affording greater flexibility in structural representation and facilitating the modeling of complex functions and systems; (iv) Geometric Semantic GP [28], which incorporates geometric operators to improve solution-finding efficiency and yield more robust outcomes than conventional methodologies; (v) Grammatical Evolution [29], which leverages formal grammars to direct program evolution, thus enabling the generation of solutions in specific programming languages through the encoding of grammatical rules; and (vi) Genetic Improvement [30], which utilizes evolutionary techniques to enhance existing software, thereby augmenting performance, energy efficiency, or adaptability to new platforms while preserving the software's original functionality.

GP has demonstrated success across diverse domains, including creative applications (e.g., designing a commemorative coin in Portugal [32]), music transcription [33], learning the behaviors and generate controllers in the gaming field [34], in healthcare to improve medical image analysis [35], and to predict stock option prices in finance [36].

Notably, GP addresses a variety of problems, with the symbolic regression [31] being particularly salient, as it seeks to develop mathematical models that accurately represent a given set of data points defining the model. As described before, in FL, we need a mathematical function -the aggregation function- to work with multiple data, which somehow resembles the symbolic regression approach frequently addressed by means of GP. GP can be thus leveraged to evolve aggregation strategies tailored to heterogeneous data and varying client resources.

Its ability to adapt to complex scenarios makes it well-suited for enhancing FL aggregation methods. Despite advances in aggregation, traditional approaches like FedAVG show clear limitations [2], particularly in non-IID settings [5,6]. This paper proposes a novel use of GP to address these challenges, focusing on two key research questions: RQ1: How can GP improve model aggregation in FL, particularly in non-IID contexts? RQ2: What is the impact of parameter and hyperparameter configurations on the performance of FedGP compared to FedAVG (e.g., batch size, aggregation frequency)?

3   Methodology

To address challenges related to the aggregation limitations of traditional methods, such as FedAVG, in FL scenarios characterized by non-IID distributions, we developed FedGP, a GP method designed to improve model aggregation. This approach dynamically optimizes aggregation functions to accommodate data variability across clients. In addition, to evaluate the performance of FedGP compared to FedAVG, we explore the effect of different parameter configurations and hyperparameters, such as batch size and aggregation frequency, by analyzing their impact on the generalization ability of the overall model.

The methodology, illustrated in Fig. 1, is structured into six main blocks in agreement with the project's design phases. In particular, selection of the dataset study (Sect. 3.1); technology stack definition (Sect. 3.2); aggregation methods selection (Sect. 3.3); FL and GP's parameters setup (Sect. 3.4); environments configuration and tests execution (Sect. 3.5); and results aggregation and analysis (Sect. 3.6).

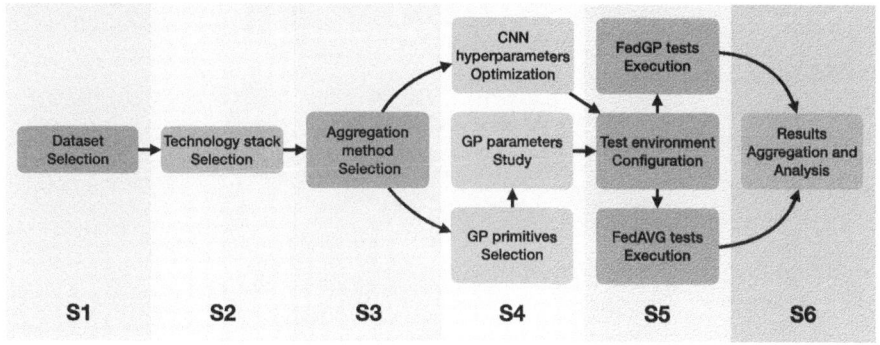

Fig. 1. Methodology organized in steps

3.1 S1: Dataset Selection

The first step concerns selecting a dataset suitable for FL experiments. Among FL's main application areas is the medical field, so the PathMNIST dataset from the MedMNIST collection [12], designed for medical image analysis, was chosen. Specifically, PathMNIST contains 107,180 colon histopathological images, each 28×28 pixels in size, it is important to note that in this study, we work directly with images and not with features pre-extracted in tabular data. The dataset is divided into nine classes representing different tissue types and histological structures. Derived from pathology slides, the dataset is intended for classification tasks in medical research, supporting automated diagnosis and analysis. The compactness and variety of classes make PathMNIST ideal for FL experiments. Figure 2 shows a sample montage of images from the dataset. The classes present are: (i) adipose, (ii) background, (iii) debris, (iv) lymphocytes', (v) mucus', (vi) smooth muscle, (vii) normal colon mucosa, (viii) cancer-associated stroma, (ix) colorectal adenocarcinoma epithelium.

3.2 S2: Technology Stack Selection

The entire project is developed using Python with PyTorch libraries to handle neural network (NN) training, operations on tensors, and DEAP [37] for GP. The model adopted for training the PathMNIST dataset is a convolutional neural network (CNN) suggested by the creators of the dataset. It has five convolutional layers followed by batch normalization, ReLU (rectified linear unit) activations, max-pooling, and a final section of fully connected layers. During the training process, accuracy and loss will be calculated for both training and validation data.

Individuals in FedGP are represented as expression trees, where each node corresponds to an operation or value. The tree structure is ideal for representing aggregation functions, as it allows mathematical and aggregation primitives to be dynamically combined to create complex functions. We describe below the functions and terminals used to build these expression trees.

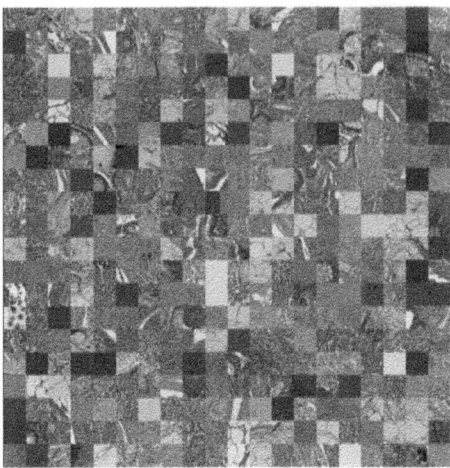

Fig. 2. PathMNIST image collage. Source: [12]

3.3   S3: Aggregation Method Selection

Figure 3 shows the FL workflow. We have the central server and the user clients. Initially, the global model is trained on the server and sent to the clients. Once in the clients, local training begins on the user data. Using the synchronous, centralized version of FL, the server waits to receive the clients' models before aggregating them into a new global model.

In these expressions, one will always work with model weights. However, there are aggregation methods that work with gradients.

The aggregation step represents a crucial moment in the FL process.

As analyzed in the literature review, there are several methods. However, since this is the first step toward introducing a new aggregation method, unweighted FedAVG was chosen as the reference method for comparison. FedAVG is a logical choice since it represents the standard for aggregation in FL. The choice of the unweighted version is intended to provide privacy to users, so the server does not need to know how much data each client has processed and what type it is. FedAVG averages the weights of the clients and thus creates a new model to propagate. Overall, this study proposes the introduction of FedGP. The latter, unlike FedAVG, does not simply average but generates an expression tree that will then be applied to the clients to aggregate the weights and generate the new model to be propagated.

In the FL workflow, these operations are repeated cyclically unless a stop condition is established, such as a threshold on the metrics.

3.4   S4: FL and GP Configurations

To ensure the proper functioning of the system, a prior study is carried out on the hyperparameters to be used for CNN. Specifically, we have:

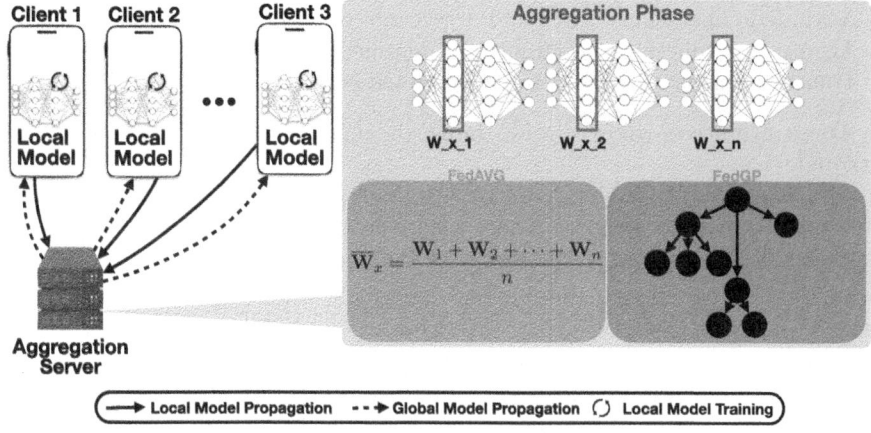

Fig. 3. FL workflow with explanation of FedAVG and FedGP aggregation methods.

- Learning rate: 0.0005. The optimizer's learning rate controls the weights' update rate during training.
- Momentum: 0.9. Momentum factor applied to improve convergence by reducing oscillations during optimization.
- An SGD optimizer is used to update CNN weights during training.
- Cost function: unweighted CrossEntropyLoss.

Next, the parameters for FL are set: (i) 5 clients; (ii) 2 epochs of global model training; (iii) 15 epochs of local training.

Before addressing the GP parameters, it is essential to define the primitives and the terminals that the GP can use during the evolutionary process. FedGP uses a specific set of primitives to work with PyTorch tensors. The functions torch.sum, torch.sub, torch.mul, and torch.abs are comprise in the PyTorch library, and torch_protected_div, torch_mean, torch_median, torch_protected_sqrt have been implemented for this study. For examples torch_median and torch_mean receive an arbitrary number of tensors, and place them in a stack. In turn, the averaging function is applied, so a new tensor with the same shape as the starting tensors and whose values are the mean and median of the tensors are obtained.

To handle data variability and prevent numerical errors, protected functions, such as protected division and protected square root, were introduced to avoid problematic situations such as division by zero and square roots of negative numbers.

$$F = \left\{ \begin{array}{l} \text{torch.sum, torch.sub, torch.mul,} \\ \text{torch\_protected\_div, torch\_mean,} \\ \text{torch\_median, torch.abs, torch\_protected\_sqrt} \end{array} \right\} \tag{1}$$

Equation 1 shows the available primitives, groupable into categories:

- Arithmetic operations: torch.add, torch.sub, torch.mul, torch_protected_div
- Aggregation functions: torch_mean, torch_median
- Transformations: torch.abs, torch_protected_sqrt

These primitives allow a wide range of aggregation configurations to be explored.

As terminals, placeholders are set up to represent the clients (Eq. 2) that have submitted their weights. Thus, when the tree is processed by the fitness function, the placeholder is replaced with the appropriate tensor.

$$T = \{\text{CLIENT}_1, \text{CLIENT}_2, \ldots, \text{CLIENT}_n \quad \} \tag{2}$$

A study is also carried out over GP parameters, choosing: (i) Generations: 10; (ii) Elitism: 1; (iii) Mutation rate: 40%; (iv) Crossover rate: 60%.

To balance the complexity of the generated aggregation functions and their generalization, a minimum depth of 1 and a maximum depth of 5 were set for the trees. These depth limits were also chosen to avoid trees that might be computationally burdensome. Trees are combined through a one-point crossover, in which a subtree of one individual is swapped with another, maintaining the syntactic validity of the trees. Mutation is done by generating a subtree and replacing it with a subtree of the individual. The elitism of size 1 is adopted, preserving the individual with the best fitness in each generation to ensure that the best solutions are preserved.

The fitness function performs model validation on a test dataset. Model validation consists of making predictions on the GP aggregated model, and the fitness value is the accuracy value obtained as a result of the inference. The test dataset used for inference comprises a subset of PathMNIST images not used in the training phase, with an IID distribution to evaluate the entire model equally.

3.5   S5: Tests Performed

The FL process was implemented in a virtualized environment, leveraging threads to simulate distributed execution rather than employing physically distributed devices. This approach was adopted to avoid introducing additional variables that were not instrumental to the study. All tests were conducted on a server running Ubuntu 20.04.6, equipped with two Intel® Xeon® Silver 4310 processors, 512 GB of RAM, and four NVIDIA A100 PCIe GPUs (40 GB memory each), using CUDA version 12.4 and Python version 3.11.10. The GP algorithm was executed on the CPU, while the training and validation of FL models, as well as the evaluation of GP individuals, were performed on the GPU. To emulate the federated setting, model training executions were conducted in isolation on the same machine, with no physical distribution of devices. This methodology allowed for an accurate investigation of FL dynamics in a controlled environment. Each configuration was executed 30 times to ensure the statistical robustness and reliability of the results.

All experiments are performed with horizontal data partitioning and centralized and synchronous FL architecture. Given the nature of the proposed methodology, the results can be easily generalized to asynchronous decentralized architecture.

Table 1 shows the variable parameters used in the experiments. Combining the available parameters between FedAVG and FedGP, taking into consideration that the number of individuals is only relevant for GP. The batch size represents the amount of data processed at each iteration in the NN. The Weights Sending Frequency determines every how many epochs model weights are sent from clients to servers, thus when the aggregate model is created. Then, the GP algorithm will run whenever the model needs to be aggregated. The *iid* parameter represents the data distribution; if the *iid* variable is false, data will have a non-IID distribution; otherwise, it will IID distribution. Figure 4 shows an example of IID and non-IID data distribution. In the non-IID case, it is not guaranteed that all clients have at least one example for each class; this condition promotes the divergence of local models from the global model trained on all classes. Independent of data distribution, a test dataset containing data from each class is used when the model is evaluated. For FedGP, the use of 20 or 50 individuals for the population is proposed. Finally, we find the chosen aggregation methods: FedGP and FedAVG.

Table 1. Possible values for each variable parameter in the experiment configurations.

Parameter	Values
Batch Size	64, 128
Weights Sending Frequency	3, 5
IID	TRUE, FALSE
Individuals (FedGP only)	20, 50
Aggregation Method	FedAVG, FedGP

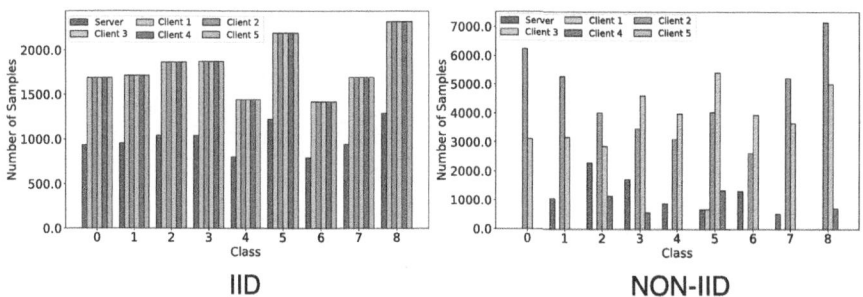

Fig. 4. Example of IID and non-IID data distribution for the PathMNIST dataset.

Table 2 summarizes the configurations used for the tests. Notably, the server is initialized with 10% of the dataset data to create the base model. The remaining 90% is distributed among the clients. This design choice is intentional and not a fixed requirement; it allows the model to be heavily influenced by client contributions, enabling a clearer evaluation of aggregation effects. The experiments are divided into two groups: the first uses a batch size of 64, while the second uses a batch size of 128.

Table 2. Experiments' Configurations (Aggregation frequency is expressed in epochs).

ID	Batch Size	Agg. Freq.	IID	Ind.	Method	ID	Batch Size	Aggr. Freq.	IID	Ind.	Method
1	64	3	TRUE	-	FedAVG	13	128	3	TRUE	-	FedAVG
2	64	3	TRUE	20	FedGP	14	128	3	TRUE	20	FedGP
3	64	3	TRUE	50	FedGP	15	128	3	TRUE	50	FedGP
4	64	3	FALSE	-	FedAVG	16	128	3	FALSE	-	FedAVG
5	64	3	FALSE	20	FedGP	17	128	3	FALSE	20	FedGP
6	64	3	FALSE	50	FedGP	18	128	3	FALSE	50	FedGP
7	64	5	TRUE	-	FedAVG	19	128	5	TRUE	-	FedAVG
8	64	5	TRUE	20	FedGP	20	128	5	TRUE	20	FedGP
9	64	5	TRUE	50	FedGP	21	128	5	TRUE	50	FedGP
10	64	5	FALSE	-	FedAVG	22	128	5	FALSE	-	FedAVG
11	64	5	FALSE	20	FedGP	23	128	5	FALSE	20	FedGP
12	64	5	FALSE	50	FedGP	24	128	5	FALSE	50	FedGP

3.6   S6: Results Aggregation and Analysis

For each run, accuracy values for each epoch of the server and all clients are saved, along with data for each aggregation step. With the information obtained, mean and standard deviation are then calculated for server, client, and aggregate. So, the average among the 5 clients of all runs is averaged for each configuration. Afterward, the data is prepared for analysis and is eventually processed to generate a graph. Initially, a graph is created for each configuration, and subsequently, additional graphs are produced to directly compare two configurations in order to determine which one performs best.

4   Results and Analysis

The results of the experiments are shown in Figs. 5 and 6 and represent the comparison between FedGP and FedAVG, in terms of accuracy, in the above configurations.

For FedGP to improve the compactness of the narrative, only the experiments involving 50 individuals are shown since they performed slightly better than the version with 20 individuals in each case.

4.1 IID Data Distribution

Figure 5 deals with the experiments with IID data distribution; experiments 1–3 and 7–9 have a batch size of 64, while experiments 13–15 and 19–21 have a batch size of 128. In this case, a larger batch size helps to decrease the results' variability by presenting a reduced standard deviation. Examining the comparison between FedGP (orange color) and FedAVG (blue color), we see that FedGP always obtains higher accuracy results than FedAVG. Results are always better regardless of the frequency of aggregation of the weights. In some cases, FedGP performs higher than the maximum value obtained by the clients; in each case, it obtains higher accuracy than the average of the clients.

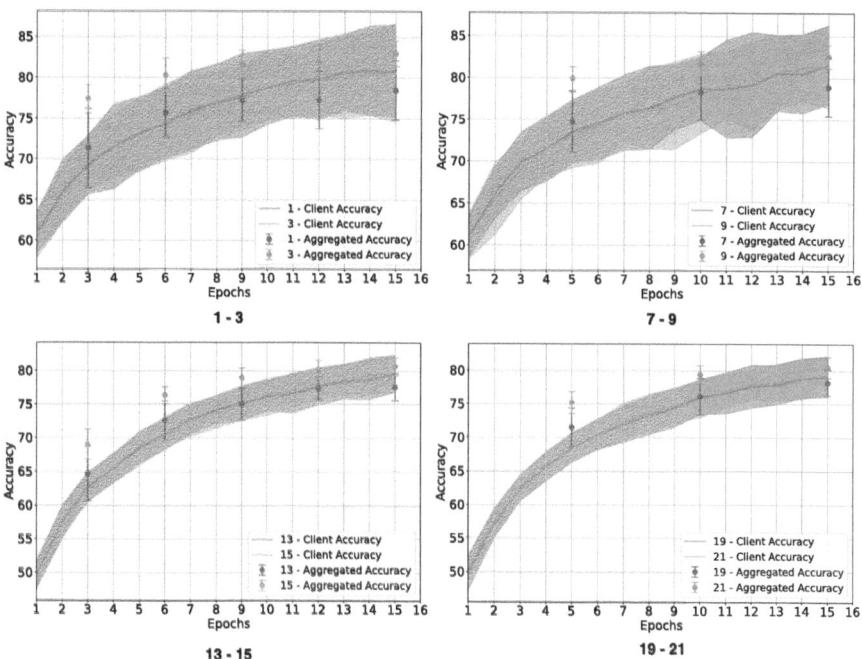

Fig. 5. Experiments with distribution of IID data. The blue represents the FedAVG, while the orange represents the FedGP. (Color figure online)

4.2 Non-IID Data Distribution

Focusing on the analysis of the experiments with non-IID data distribution, Fig. 6, we notice that the standard deviation takes significant values; this is because clients are not guaranteed to have examples of every class in the dataset, so at the time of evaluation, clients are obtained that perform very well and others of poor quality. This scenario is critical because it more closely represents the reality of FL application. Also, it is evident how FedGP always gets superior results than FedAVG. Moreover, with non-IID data, FedGP aggregates models with accuracy higher than the best of clients; this highlights the robustness and quality of the proposed aggregation method.

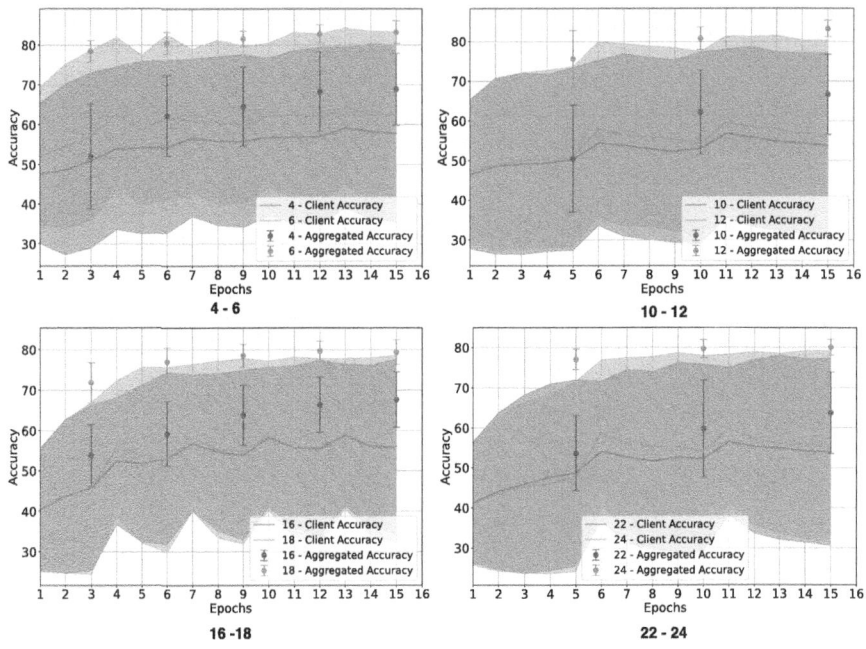

Fig. 6. Experiments with distribution of non-IID data. The blue represents the FedAVG, while the orange represents the FedGP. The numbers below each graph represent the configurations shown. (Color figure online)

Thus, regardless of batch size, aggregation frequency, and data distribution, FedGP represents a promising aggregation method in FL.

Figure 7 represents an expression tree produced by GP. It shows how it combined the primitives of mean and median and selected the clients by effectively discarding client3. Therefore, it is interesting to study how, in addition to aggregating the weights, GP can select the clients that bring quality to the final solution while directly ignoring the others.

However, occasionally, GP produces higher quality but also bigger individuals than the one presented in Fig. 7. For instance, Fig. 8 shows the largest individual produced by GP, respecting the algorithm's depth limits. In the future, we plan to analyze the way GP produces the aggregation of clients which may be of help to design new strategies.

Although only one network type was used for this experiment, FedGP is designed to be model-agnostic, thus applicable to any network architecture.

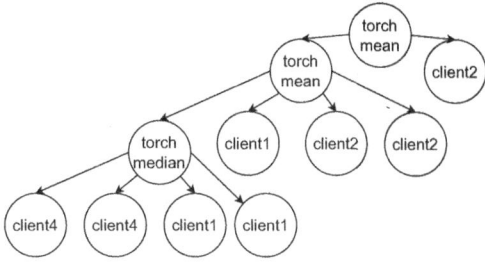

Fig. 7. Representation of an individual produced by the FedGP at the end of the evolutionary process.

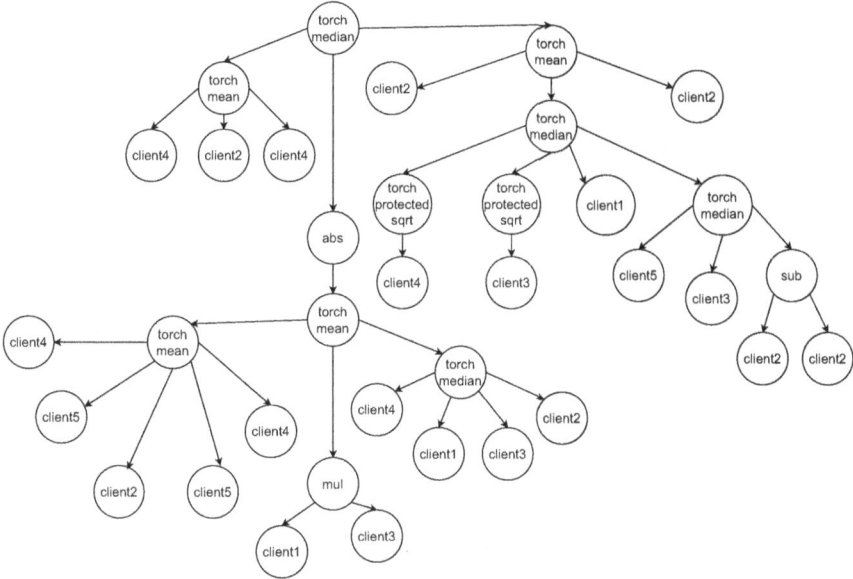

Fig. 8. Representation of the biggest individual produced by the FedGP.

5   Conclusions

This paper introduced FedGP, a novel GP-based aggregation method for FL. To evaluate its effectiveness, FedGP was compared against FedAVG, the most widely adopted aggregation method in FL. The results demonstrate that FedGP consistently outperforms FedAVG and, in some cases, even surpasses the best-performing individual client. These findings highlight the robustness and superior performance of GP-driven aggregation compared to traditional global models.

A key insight from the experiments is that FedGP is minimally impacted by changes in aggregation frequency. Aggregation intervals were varied between every three epochs and every five epochs, yet FedGP outperformed FedAVG in both configurations, showcasing its adaptability to diverse operational settings.

Further testing in both IID and non-IID data environments confirmed FedGP's efficacy in addressing challenges associated with heterogeneous data distributions. While the experiments focused on a CNN, the model-agnostic design of FedGP indicates its potential applicability to other types of NNs.

This study marks a significant step forward in advancing aggregation techniques for FL through GP. To enhance the proposed method's validation, future work includes (i) comparing FedGP with alternative aggregation methods beyond FedAVG, (ii) exploring a gradient-based version of FedGP in place of the current weight-based approach, (iii) investigations into FedGP's resilience in privacy-preserving scenarios by analyzing its behavior in the event of noise injection; (iv) The study of the impact of dynamic variation in client data on FedGP; (v) Finally, expand the range of GP primitives and incorporating random ephemeral constants could further refine its ability to handle tensor data transformations.

In summary, FedGP offers a compelling improvement over FedAVG, paving the way for the development of innovative aggregation operators through evolutionary algorithms. This work underscores promising opportunities for enhancing the future of Federated Learning.

Acknowledgements. This work was partially supported by the HES-SO RCSO ISNet HARRISON grant (WP2), the Spanish Ministry of Economy and Competitiveness (PID2020-115570GB-C21, PID2023-147409NB-C22), funded by MCIN/AEI/10.13039/501100011033, and the Junta de Extremadura (GR15068).

References

1. McMahan, H.B., Moore, E., Ramage, D., Hampson, S., Agüera y Arcas, B.: Communication-efficient learning of deep networks from decentralized data. In: Proceedings of the 20th International Conference on Artificial Intelligence and Statistics (AISTATS) (2017). http://arxiv.org/abs/1602.05629
2. Kairouz, P., et al.: Advances and Open Problems in Federated Learning. Now Found. Trends (2021)
3. Hard, A., et al.: Federated learning for mobile keyboard prediction. arXiv preprint (2019). https://arxiv.org/abs/1811.03604
4. Rieke, N., et al.: The future of digital health with federated learning. NPJ Digit. Med. 3(1), 119 (2020). https://doi.org/10.1038/s41746-020-00323-1
5. Liu, Y., Ai, Z., Z., Sun, Z., Zhang, S., Liu, Z., Yu, H.: FedCoin: a peer-to-peer payment system for federated learning. In: Lecture Notes in Computer Science (including subseries Lecture Notes in Artificial Intelligence and Lecture Notes in Bioinformatics), vol. 12500. Springer, Cham (2020). https://doi.org/10.1007/978-3-030-63076-8_9
6. Long, G., G., Tan, G., Jiang, J., Zhang, C.: Federated learning for open banking. In: Lecture Notes in Computer Science (including subseries Lecture Notes in Artificial Intelligence and Lecture Notes in Bioinformatics), vol. 12500. Springer, Cham (2020). https://doi.org/10.1007/978-3-030-63076-8_17

7. Nie, W., Yu, L., Jia, Z.: Research on aggregation strategy of federated learning parameters under non-independent and identically distributed conditions. In: Proceedings of the 2022 4th International Conference on Applied Machine Learning (ICAML), Changsha, China, pp. 41–48 (2022). https://doi.org/10.1109/ICAML57167.2022.00016

8. Reguieg, H., Hanjri, M.E., Kamili, M.E., Kobbane, A.: A comparative evaluation of FedAvg and Per-FedAvg algorithms for Dirichlet distributed heterogeneous data. In: Proceedings of the 2023 10th International Conference on Wireless Networks and Mobile Communications (WINCOM), Istanbul, Turkiye, pp. 1–6 (2023). https://doi.org/10.1109/WINCOM59760.2023.10322899

9. Li, X., Huang, K., Yang, W., Wang, S., Zhang, Z.: On the convergence of FedAvg on non-IID data. arXiv preprint (2020). https://arxiv.org/abs/1907.02189

10. Wang, J., Liu, Q., Liang, H., Joshi, G., Poor, H.V.: Tackling the objective inconsistency problem in heterogeneous federated optimization. In: Proceedings of the 34th International Conference on Neural Information Processing Systems (NIPS '20) (art. no. 638, pp. 1–13). Curran Associates Inc., Red Hook, NY, USA (2020)

11. Pacioni, E., Fernández De Vega, F., Calvaresi C.: Towards a meaningful communication and model aggregation in federated learning via genetic programming. In: ICAART 2024 (2020)

12. Yang, J., et al.: MedMNIST v2: a large-scale lightweight benchmark for 2D and 3D biomedical image classification. Sci. Data 10(1), 41 (2023). Nature Publishing Group UK London

13. Granqvist, F., Seigel, M., van Dalen, R., Cahill, Á., Shum, S., Paulik, M.: Improving on-device speaker verification using federated learning with privacy (2020). https://arxiv.org/abs/2008.02651

14. du Terrail, J.O., et al.: Collaborative federated learning behind hospitals' firewalls for predicting histological response to neoadjuvant chemotherapy in triple-negative breast cancer. medRxiv, https://doi.org/10.1101/2021.10.27.21264834 (2021)

15. Martínez Beltrán, E.T., et al.: Decentralized federated learning: fundamentals, state of the art, frameworks, trends, and challenges. IEEE Commun. Surv. Tutor. 25(4), 2983–3013, Fourthquarter 2023 (2023). https://doi.org/10.1109/COMST.2023.3315746

16. Yang, Q., Liu, Y., Chen, T., Tong, Y.: Federated machine learning: concept and applications. ACM Trans. Intell. Syst. Technol. 10(2), Article 12, 19 p. (2019). https://doi.org/10.1145/3298981

17. Xu, C., Qu, Y., Xiang, Y., Gao, L.: Asynchronous federated learning on heterogeneous devices: a survey. Comput. Sci. Rev. 50, 100595 (2023). ISSN 1574-0137, https://doi.org/10.1016/j.cosrev.2023.100595

18. Mugunthan, V., Polychroniadou, A., Byrd, D., Balch, T.H.: SMPAI: secure multi-party computation for federated learning. In 33rd Conference on Neural Information Processing Systems (NeurIPS 2019), Vancouver, Canada (2019). https://www.jpmorgan.com/content/dam/jpm/cib/complex/content/technology/ai-research-publications/pdf-9.pdf

19. Madi, A., Stan, O., Mayoue, A., Grivet-Sébert, A., Gouy-Pailler, C., Sirdey, R.: A secure federated learning framework using homomorphic encryption and verifiable computing. In: 2021 Reconciling Data Analytics, Automation, Privacy, and Security: A Big Data Challenge (RDAAPS), Hamilton, ON, Canada, pp. 1–8 (2021). https://doi.org/10.1109/RDAAPS48126.2021.9452005

20. Qi, P., Chiaro, D., Guzzo, A., Ianni, M., Fortino, G., Piccialli, F.: Model aggregation techniques in federated learning: a comprehensive survey. Futur. Gener. Comput. Syst. 150, 272–293 (2024). https://doi.org/10.1016/j.future.2023.09.008

21. Fallah, A., Mokhtari, A., Ozdaglar, A.: Personalized Federated Learning: A Meta-Learning Approach (2020). https://arxiv.org/abs/2002.07948
22. Yuan, H., Ma, T.: Federated accelerated stochastic gradient descent. In: Advances in Neural Information Processing Systems, vol. 33, pp. 5332–5344 (2020)
23. de Souza, M.M., Holm, A., Biczyk, M., de Castro, L.N.: A systematic literature review on the use of federated learning and bioinspired computing. Electronics 13(16), 3157 (2024). https://doi.org/10.3390/electronics13163157
24. Koza, J.R.: Genetic Programming: On the Programming of Computers by Means of Natural Selection. MIT Press, Cambridge. http://mitpress.mit.edu/books/genetic-programming
25. Brameier, M., Banzhaf, W.: Linear Genetic Programming. Springer, New York (2007). https://doi.org/10.1007/978-0-387-31030-5
26. Spector, L.: Autoconstructive evolution: push, PushGP, and Pushpop. In: Proceedings of the Genetic and Evolutionary Computation Conference (GECCO) (2001)
27. Miller, J.F., Thomson, P.: Cartesian genetic programming. In: Lecture Notes in Computer Science, vol. 1802, pp. 121–132 (2000). https://doi.org/10.1007/978-3-540-46239-2_9
28. Moraglio, A., Krawiec, K., Johnson, C.G.: Geometric semantic genetic programming. In: Coello, C., Cutello, V., Deb, K., Forrest, S., Nicosia, G., Pavone, M. (eds.) PPSN 2012. LNCS, vol. 7491, pp. 21–31. Springer, Heidelberg (2012). https://doi.org/10.1007/978-3-642-32937-1_3
29. O'Neill, M., Ryan, C.: Grammatical evolution. IEEE Trans. Evol. Comput. 5(4), 349–358 (2001). https://doi.org/10.1109/4235.942529
30. Petke, J., Haraldsson, S.O., Harman, M., Langdon, W.B., White, D.R., Woodward, J.R.: Genetic improvement of software: a comprehensive survey. IEEE Trans. Evol. Comput. 22(3), 415–432 (2018). https://doi.org/10.1109/TEVC.2017.2693219
31. Augusto, D.A., Barbosa, H.J.C.: Symbolic regression via genetic programming. In: Proceedings, vol. 1, Sixth Brazilian Symposium on Neural Networks, Rio de Janeiro, Brazil, pp. 173–178 (2000). https://doi.org/10.1109/SBRN.2000.889734
32. Machado, P., et al.: Designing coins with evolutionary computation. SIGEVOlution 17(2), Article 1, 9 p. (2024). https://doi.org/10.1145/3695933.3695934
33. Miragaia, R., Fernández, F., Reis, G., Inácio, T.: Evolving a multi-classifier system for multi-pitch estimation of piano music and beyond: an application of cartesian genetic programming. Appl. Sci. 11(7), 2902 (2021). https://doi.org/10.3390/app11072902
34. Wilson, D.G., Luga, H., Cussat-Blanc, S., Miller, J.F.: Evolving simple programs for playing Atari games. In: GECCO 2018 - Proceedings of the 2018 Genetic and Evolutionary Computation Conference, pp. 229–236 (2018). https://doi.org/10.1145/3205455.3205578
35. Langdon, W.B., Modat, M., Petke, J., Harman, M.: Improving 3D medical image registration CUDA software with genetic programming. In: GECCO 2014 - Proceedings of the 2014 Genetic and Evolutionary Computation Conference, pp. 951–958 (2014). https://doi.org/10.1145/2576768.2598244
36. Hsu, C.-M.: A hybrid procedure for stock price prediction by integrating self-organizing map and genetic programming. Expert Syst. Appl. 38(11), 14026–14036 (2011). https://doi.org/10.1016/j.eswa.2011.04.210
37. Fortin, F.-A., De Rainville, F.-M., Gardner, M.-A., Parizeau, M., Gagné, C.: DEAP: evolutionary algorithms made easy. J. Mach. Learn. Res. 13, 2171–2175 (2012)

A Genetic Algorithm Approach for Aggregation of Residential Electricity Prosumers' Flexibility

Vahid Rasouli$^{(\boxtimes)}$ ⓘ, Álvaro Gomes ⓘ, and Carlos Henggeler Antunes ⓘ

Department of Electrical and Computer Engineering, INESC Coimbra, University of Coimbra, Pólo II, R. Silvio Lima, 3030-290 Coimbra, Portugal
{vahid.rasouli,agomes,ch}@deec.uc.pt

Abstract. A genetic algorithm in combination with a mixed integer linear programming solver has been developed to deal with a bilevel (hierarchical) optimization model that represents the interaction of an aggregator and prosumers in flexibility aggregation. The proposed approach is used to assist the aggregator to define the optimal rewards to be given to residential prosumers to characterize their flexibility in the utilization of their energy resources. The rewards are determined in each generation of the genetic algorithm (upper-level problem) aiming to maximize the aggregator's profit; the rewards are then fed into the prosumers' problem (lower-level problem). The flexibility responsiveness of the prosumers is calculated using the mixed integer linear programming model at lower-level problem to reschedule the operation of their energy resources and minimize their cost, which in turn impacts the aggregator's profit. The aggregator can thus characterize the flexibility of prosumers and define the bids to be traded in the ancillary services market. The illustrative results show the performance of the proposed approach to determine the optimal rewards.

Keywords: Bilevel Optimization · Genetic Algorithm · Demand-side Flexibility Aggregation · Ancillary Services Markets · Demand Response

1 Introduction

1.1 Background and Motivation

The energy transition towards low-carbon generation promoted by national governments and international organizations has made power system management more complex due to the increasing integration of variable renewable energy sources (RES) into the generation matrix [1]. To assist in tackling the imbalances in the supply-demand chain, a priority of system operators is to make demand-side energy resources responsive to the fluctuations in generation [2].

Demand response (DR) enables the prosumers to adjust their consumption and utilization of other energy resources such as storage, local generation and bidirectional exchanges with the grid in response to stimuli such as rewards, lower prices or emergency signals [3]. These consumption changes, making the most of the prosumers'

© The Author(s), under exclusive license to Springer Nature Switzerland AG 2025
P. García-Sánchez et al. (Eds.): EvoApplications 2025, LNCS 15612, pp. 435–451, 2025.
https://doi.org/10.1007/978-3-031-90062-4_27

flexibility in their consumption patterns, can be utilized and traded as ancillary services in markets, thus aiding in the overall management of the power system. However, since the amount of flexibility of residential households is typically too small to be traded on their own and due to the technical requirements and market participation criteria [4], the prosumers' flexibility generally needs to be aggregated and dispatched by an intermediary entity between the prosumer resources and the markets/system operator, known as aggregator [5].

The aggregator uses DR programs to collect and utilize residential prosumers' flexibility to offer market bids. The price/incentive signals stimulate the prosumers' dynamic response regarding the management of their energy resources. However, the prosumers' preferences and resource availability, in which comfort is determinative, have significant impact on flexibility provision. Therefore, the characterization of the prosumers' flexibility by the aggregator is essential and helps the aggregator to perform DR programs with economic efficiency [6].

The characterization of prosumer flexibility has been performed through quantitative models and methodologies in the literature [7]. The prosumer's flexibility is accounted for by optimizing the operation scheduling of the demand-side resources in response to changes in time-of-use electricity prices [8]. Incentives over time-of-use tariffs can also stimulate changes in the consumption patterns of different types of prosumers [9]. In these settings, the power consumption of the prosumer is optimized twice, once before receiving the reward information and once after. By subtracting the reoptimized power consumption from the reference consumption, the amount of flexibility during a given response period can be calculated.

The quantification of flexibility at the aggregator level is more challenging and requires more complex models and methodologies to enable the aggregator to perform DR programs efficiently. This is due to the complexity raised from different prosumers' consumption profiles and comfort preferences that lead to different responsiveness to the same monetary stimuli. Paying higher fixed-value rewards does not necessarily lead to higher flexibility being provided due to the limitations in responsiveness of the prosumers' demand-side energy resources and comfort preferences [10].

In this paper, the aggregated flexibility is defined as the sum of all prosumers' increase/reduction in consumption stimulated by a rewarding mechanism defined by the aggregator. The operation of demand-side energy resources is modeled using physical-based models and the prosumer's problem is framed as a mixed integer linear programming (MILP). The aim of this study is to propose an aggregation algorithm that can define the reward signal to optimize the utilization of the prosumers' flexibility to maximize the aggregator's profit.

1.2 Literature Review

Rewarding mechanisms have been used by the aggregator to leverage the flexibility of the prosumers. Fairness plays a key role in designing an appropriate reward mechanism. An incremental rewarding mechanism was proposed in [11], where highly flexible prosumers receive higher rewards. Two different reward strategies were introduced and compared in [12]. In one strategy, the aggregator rewards only the selected prosumers whose flexibility maximizes the aggregator's profit, while the other strategy rewards all prosumers, regardless of their flexibility. The purpose of the second approach is to maintain prosumers' motivation for future participation in DR programs by providing a minimum reward to the non-selected prosumers.

In addition to fairness, the effectiveness of the rewarding mechanism is also relevant. Rewarding mechanisms can be defined in the context of a direct load control (DLC) DR program as simple and direct monetary payment/discount. In DLC programs, utility companies/aggregators control home equipment through IoT technology to meet the grid needs [13]. Security and privacy issues may reduce the motivation of prosumers to participate in DLC programs. Instead, they may be more willing to have their own home energy management system (HEMS) behind their smart meter to control their appliances [14]. The only information communicated with the aggregator will be the reward and flexibility amounts. The rewarding mechanism can have different values at different periods of the prosumer's load operation scheduling. An optimization tool is crucial for enabling the aggregator to define the reward signal, which can enhance the mobilization of prosumers' flexibility and improve the aggregator's profitability.

The aggregation problem of residential prosumers' flexibility has a hierarchical structure in which the aggregator leads the decision process by setting the rewards, while the prosumers respond by quantifying their available flexibility. Therefore, it can be formulated as a bilevel optimization model (BOM) [15]. However, solving BOM is not easy due to its intrinsic nonlinearity, in which an optimization problem appears in the constraints of another optimization problem. There are several methods in the literature that have been used to solve BOM. If the lower-level (LL) problem is convex, then it allows the BOM to be reformulated as a single level program by means of the Karush-Kuhn-Tucker (KKT) conditions of the optimality of the LL problem [16]. The interaction between the retailer and the prosumer has been modeled as a BOM in [17] to define the profit-maximizing dynamic prices. The BOM is reformulated as a single level MILP problem, and it is solved using KKT conditions of the LL model. Depending on what demand-side resources have been used in modeling the LL problem, the complexity of the BOM can be different thus requiring different algorithms. Photovoltaic (PV), heat production, and storage are the demand-side resources used in modeling the retailer-prosumer interaction in [18], which builds a BOM to define the optimal tariffs aiming to maximize the retailer's revenue. Although the physical-based model presented for the demand-side resources adds computational complexity to the problem, it still maintains a linear structure at LL. This allows the BOM to be solved by leveraging the strong duality of the LL problem. A discrete finite set of optimality conditions in [19] and mathematical program with equilibrium constraints (MPEC) in [20] are two other approaches to reformulate the BOM of prosumer's interaction with the supplier and with the distribution system operator (DSO), respectively. A hybrid approach composed of a particle swarm optimization (PSO) algorithm at the upper-level (UL) problem and an exact solver at the LL problem has been proposed in [21] to solve a bilevel mixed-integer nonlinear optimization model that represents the retailer-prosumer interaction and defines the profit-maximizing dynamic prices. This hybrid approach has been developed to cope with the computational complexity of the BOM.

In the literature, most of the studies have considered the retailer-prosumer interaction with limited types of demand-side resources that make it easier to solve the bilevel problem. The interaction of the aggregator with the prosumers is different from the retailer-prosumer interaction. That is because the aggregator uses the prosumer's resources to provide services and participate in markets on behalf of the prosumers while the retailer only sells electricity to the prosumers. An algorithm to define the profit-maximizing real-time prices for an aggregator of a residential energy community has been proposed in [22]. The demand-side resources are only distributed generation and battery storage system. Considering a diverse range of demand-side resources in the aggregator-prosumer hierarchical interaction and a computationally efficient algorithm that can solve the resulting BOM are the research gaps in literature.

1.3   Contributions

In this study, the aggregator-prosumers interaction is formulated as a BOM: at the UL problem the aim is to determine the optimal rewards that maximize the aggregator's profit, while the LL problem aims to compute the optimal flexibility to minimize the prosumers' energy cost accounting for the reward due to flexibility provision. At the LL problem various demand-side energy resources are considered, such as shiftable, interruptible, and thermostatically controlled loads, battery storage systems (BSS), electric vehicles (EV), and local generation, which have specific physical-based operation and control models. Different prosumer profiles are included to represent diverse comfort preferences, resource ownership and consumption patterns. The objective function of the aggregator's problem is nonlinear due to the multiplication of the decision variables related to reward and flexibility. In addition, due to the physical-based models used at the prosumer's problem, the LL problem is formulated as a MILP model, which makes the LL problem non-convex. Exact methods to solve BOM are very demanding from a computational point of view and, in general, require specific model characteristics for algorithm convergence. Metaheuristic approaches are an alternative to solve nonlinear BOM [23]. However, since they cannot guarantee the optimality of the LL problem, the feasibility of the overall solution cannot also be guaranteed which could be misleading. In our study, we use a genetic algorithm (GA) at the UL problem and an exact solver at the LL problem to ensure that the solutions are optimal to the LL problem and thus feasible to the BOM.

The main contributions of this paper are:

- proposing an algorithmic approach combing a GA at the UL problem to determine the rewards with an exact solver at the LL problem to guarantee the optimal solutions for the prosumers' cost minimization problems for each reward instantiation;
- considering a diverse range of demand-side energy resources and prosumer profiles;
- evaluating the performance of the hybrid approach using the GA for the UL search to determine the profit-maximizing rewards using LL optimality information (prosumers' reaction).

1.4   Organization of the Paper

The remainder of the paper is organized as follows. The mathematical formulation of the aggregator-prosumer BOM is described in Sect. 2. The GA algorithm adapted for this study is explained in Sect. 3. The data and parameters used for running the experiments along with illustrative results are given in Sect. 4. The conclusion and future research avenues are drawn in Sect. 5.

2   Bilevel Formulation of the Aggregator-Prosumer Interaction

The aggregator-prosumer interaction can be formulated as a BOM stated as (1)–(46). Due to the lack of space, the nomenclature is presented in a companion file available at https://data.mendeley.com/datasets/mr7c4p5zns/1. The objective of the aggregator (leader), OF_{Agg}, is to maximize the profit, i.e. revenue (reserve services sold to the market/system operator) minus the aggregation cost (rewards paid to the prosumers), by defining the optimal rewards, Rew_t, to be sent to the prosumers and optimal reserve bids, $P_{t'}^{Res}$, to be sent to the market/system operator at price of $C_{t'}^{Sell}$. The time index for the aggregation revenue is denoted by t', which corresponds to the market time-frame for trading reserve power. The time index for the aggregation cost is denoted by t, which pertains to the operation scheduling problem of prosumers, by which they receive rewards for providing flexibility, $Flex_{t,n}$. Equation (2) calculates the aggregated flexibility, $Flex_t^{Agg}$, of all prosumers participating in DR. Equations (3) to (5) link the aggregated flexibility provided by the prosumers to the minimum amount of aggregated flexibility, $Flex_t^{min}$, that can be made available at each time interval of the prosumer's problem. Equation (6) relates the minimum flexibility available at each interval of the prosumers' problem to the minimum reserve bid capacity that can be guaranteed at each interval of the market. The aggregator communicates the reward, the response period, $[T_{rs}, T_{re}]$, , and the response type (decrease, $RT = 1$, or increase, $RT = -1$, in demand) to the prosumers. The prosumers respond to the reward by reoptimizing their operation scheduling and defining the amount of flexibility they can make available. Each prosumer aims to minimize the cost also by increasing the reward through maximizing the flexibility during the response period defined by the aggregator. Therefore, the solution to the prosumer's problem is a new optimal scheduling of the demand-side resources during the whole planning period, satisfying the prosumer's comfort preferences while providing the required flexibility. Equation (7) represents LL (prosumer's) objective function as a constraint of the UL (aggregator's) optimization problem.

The objective function of prosumer's problem, OF_{EU_n}, is obtained by subtracting the total reward received, as the product of reward and flexibility, from the total energy cost, as the product of energy consumption and energy prices, C_t^{Buy}. The amount of flexibility at each interval of the response period is obtained by subtracting the reoptimized scheduled consumption, $P_{t,n}^{Con}$, from the reference consumption, $P_{t,n}^{Ref}$, which was optimized without inclusion of the reward impact thus only considering the energy prices impact (8). Equation (9) represents the power balance, stating the power requested from the grid plus local generation, such as photovoltaic (PV), $P_{t,n}^{PV}$, and wind energy, $P_{t,n}^{W}$, must be equal to the power requested by non-controllable loads, $L_{t,n}^{NC}$, and controllable loads at each interval of the prosumer's problem. Equation (10) ensures that the total power consumption requested from the grid for each prosumer remains within the limits defined by the contracted power with the supplier company, P_n^{Gmax}.

$$OF_{Agg} = \max_{P_{t'}^{Res} \text{ and } Rew_t}$$

$$\left[\sum_{t'=1}^{T'} C_{t'}^{Sell} \times P_{t'}^{Res} \times \Delta t' - \left(RT \times \sum_{n=1}^{N} \sum_{t=T_{rs}}^{T_{re}} Rew_t \times Flex_{t,n} \times \Delta t \right) \right] \quad (1)$$

Subject to:

$$Flex_t^{Agg} = \sum_{n=1}^{N} Flex_{t,n}, t \in [T_{rs}, T_{re}] \quad (2)$$

$$T' = \frac{(T_{re} - T_{rs} + 1)}{\Delta t'} \times \Delta t \quad (3)$$

$$Flex_t^{min} \le Flex_t^{Agg}, t \in \left[T_{rs} + \frac{(t'-1) \times \Delta t'}{\Delta t}, T_{rs} + \frac{t' \times \Delta t'}{\Delta t} - 1 \right],$$
$$t' - 1, \ldots, T' \quad (4)$$

$$Flex_{t+1}^{min} = Flex_t^{min}, t \in \left[T_{rs} + \frac{(t'-1) \times \Delta t'}{\Delta t}, T_{rs} + \frac{t' \times \Delta t'}{\Delta t} - 2 \right],$$
$$t' - 1, \ldots, T' \quad (5)$$

$$P_{t'}^{Res} = Flex_t^{min}, \quad t = T_{rs} + \frac{(t'-1) \times \Delta t'}{\Delta t}, \quad t' - 1, \ldots, T' \quad (6)$$

$$OF_{EU_n} = \min_{P_{t,n}^{Con}, Flex_{t,n}}$$

$$\left[\sum_{t=1}^{T} C_t^{Buy} \times P_{t,n}^{Con} \times \Delta t - \left(RT \times \sum_{t=T_{rs}}^{T_{re}} Rew_t \times Flex_{t,n} \times \Delta t \right) \right], n = 1, \ldots, N \quad (7)$$

Subject to:

$$Flex_{t,n} = P_{t,n}^{Ref} - P_{t,n}^{Con}, t = T_{rs}, \ldots, T_{re}, n = 1, \ldots, N \quad (8)$$

$$P_{t,n}^{Con} + S_n^{PV} P_{t,n}^{PV} + S_n^{W} P_{t,n}^{W} = L_{t,n}^{NC} + P_{t,n}^{AC} + P_{t,n}^{EWH} + \left(P_{t,n}^{H2B} - P_{t,n}^{B2H} \right) +$$

$$\left(P_{t,n}^{H2V} - P_{t,n}^{V2H} \right) + \sum_{j_n=1}^{J_n} P_{t,j_n,n}^{Sh}, t = 1, \ldots, T, n = 1, \ldots, N \quad (9)$$

$$0 \le P_{t,n}^{Con} \le P_n^{Gmax}, t = 1, \ldots, T, n = 1, \ldots, N \quad (10)$$

Equations (11) through (17) represent the operation of shiftable loads, allowing each load to function in various time periods based on the prosumer's comfort preferences. The power requested by each shiftable load at each cycle of its operation is defined by $P_{j_n,n,r_{j_n}}^{ShReq}$ for each prosumer. An auxiliary binary variable, $w_{t,j_n,n,r_{j_n}}^{Sh}$, is used to guarantee that once a shiftable load begins operating it continues until its cycle is complete.

$$P_{t,j_n,n}^{Sh} = \sum_{r_{j_n}=1}^{R_{j_n}} P_{j_n,n,r_{j_n}}^{ShReq} \times w_{t,j_n,n,r_{j_n}}^{Sh}, t = 1, \ldots, T, j_n = 1, \ldots, J_n, n = 1, \ldots, N \quad (11)$$

$$\sum_{r_{j_n}=1}^{R_{j_n}} w^{Sh}_{t,j_n,n,r_{j_n}} \leq 1, t = 1, \ldots, T, j_n = 1, \ldots, J_n, n = 1, \ldots, N \qquad (12)$$

$$w^{Sh}_{(t+1),j_n,n,(r_{j_n}+1)} \geq w^{Sh}_{t,j_n,n,r_{j_n}}, \ t = 1, \ldots, T-1, j_n = 1, \ldots, J_n$$

$$r_{j_n} = 1, \ldots, R_{j_n} - 1, n = 1, \ldots, N \qquad (13)$$

$$\sum_{t=1}^{T} w^{Sh}_{t,j_n,n,r_{j_n}} = 1, j_n = 1, \ldots, J_n, r_{j_n} = 1, \ldots, R_{j_n}, n = 1, \ldots, N \qquad (14)$$

$$\sum_{t=1}^{T-R_{j_n}+1} w^{Sh}_{t,j_n,n,r_{j_n}} = 1, j_n = 1, \ldots, J_n, r_{j_n} = 1, n = 1, \ldots, N \qquad (15)$$

$$w^{Sh}_{t,j_n,n,r_{j_n}} \in \{0,1\}, t = 1, \ldots, T, j_n = 1, \ldots, J_n, r_{j_n} = 1, \ldots, R_{j_n}, n = 1, \ldots, N \quad (16)$$

$$P^{Sh}_{t,j_n,n} \geq 0, t = 1, \ldots, T, j_n = 1, \ldots, J_n, n = 1, \ldots, N \qquad (17)$$

Equations (18) through (24) represent the operation of the BSS. At each interval, the state of charge (SoC) of the BSS, $E^B_{t,n}$, is a function of the previous SoC and the current amount of energy charged/discharged considering the charging/discharging efficiency, $\eta^{Bch}_n / \eta^{Bdch}_n$, (19). The SoC of the BSS must be always within a predefined range, $\left[E^{Bmin}_n, E^{Bmax}_n\right]$, (20). Using the binary variables, $S^{H2B}_{t,n}$ and $S^{B2H}_{t,n}$, it can be guaranteed that at each interval, the BSS is only in charging, discharging, or idle mode and its charging/discharging power stays within a range, $\left[0, P^{Bchmax}_n\right]/\left[0, P^{Bdchmax}_n\right]$, (21)–(24).

$$E^B_{t,n} = E^B_{0,n}, t = 1, n = 1, \ldots, N \qquad (18)$$

$$E^B_{t,n} = E^B_{(t-1),n} + \left(\eta^{Bch}_n \times P^{H2B}_{t,n} \times \Delta t\right) - \left(\left(1/\eta^{Bdch}_n\right) \times P^{B2H}_{t,n} \times \Delta t\right), t = 2, \ldots, T, n = 1, \ldots, N \qquad (19)$$

$$E^{Bmin}_n \leq E^B_{t,n} \leq E^{Bmax}_n, t = 1, \ldots, T, n = 1, \ldots, N \qquad (20)$$

$$0 \leq P^{H2B}_{t,n} \leq P^{Bchmax}_n \times S^{H2B}_{t,n}, t = 1, \ldots, T, n = 1, \ldots, N \qquad (21)$$

$$0 \leq P^{B2H}_{t,n} \leq P^{Bdchmax}_n \times S^{B2H}_{t,n}, t = 1, \ldots, T, n = 1, \ldots, N \qquad (22)$$

$$S^{H2B}_{t,n} + S^{B2H}_{t,n} \leq 1, t = 1, \ldots, T, n = 1, \ldots, N \qquad (23)$$

$$S^{H2B}_{t,n} \in \{0,1\}, S^{B2H}_{t,n} \in \{0,1\}, t = 1, \ldots, T, n = 1, \ldots, N \qquad (24)$$

Equations (25) through (33) represent the operation of the EV. At each interval, the SoC of the EV, $E_{t,n}^V$, is a function of the previous SoC, and the current amount of energy charged/discharged considering the charging/discharging efficiency, $\eta_n^{Vch}/\eta_n^{Vdch}$, (26). The SoC of the EV must be always within a predefined range, $\left[E_n^{Vmin}, E_n^{Vmax}\right]$, (27). The EV is being charged/discharged only during a charging cycle defined by $\left[T_n^{Vs}, T_n^{Ve}\right]$. A minimum desired level of SoC, E_n^{VminAv}, is defined by the prosumer by which the EV must reach that level by the end of the charging cycle (28). Using the binary variables, $S_{t,n}^{H2V}$ and $S_{t,n}^{V2H}$, it can be guaranteed that, at each interval, the EV is only in charging, discharging, or idle mode and its charging/discharging power stays within a range, $\left[0, P_n^{Vchmax}\right]/\left[0, P_n^{Vdchmax}\right]$, (29)–(33).

$$E_{t,n}^V = E_{0,n}^V, t = 1, n = 1, \ldots, N \tag{25}$$

$$E_{t,n}^V = E_{(t-1),n}^V + \left(\eta_n^{Vch} \times P_{t,n}^{H2V} \times \Delta t\right) -$$

$$\left(\left(1/\eta_n^{Vdch}\right) \times P_{t,n}^{V2H} \times \Delta t\right), t = 2, \ldots, T, n = 1, \ldots, N \tag{26}$$

$$E_n^{Vmin} \leq E_{t,n}^V \leq E_n^{Vmax}, t = 1, \ldots, T, n = 1, \ldots, N \tag{27}$$

$$E_{t,n}^V \geq E_n^{VminAv}, t = T_n^{Ve}, n = 1, \ldots, N \tag{28}$$

$$0 \leq P_{t,n}^{H2V} \leq P_n^{Vchmax} \times S_{t,n}^{H2V}, t = 1, \ldots, T, n = 1, \ldots, N \tag{29}$$

$$0 \leq P_{t,n}^{V2H} \leq P_n^{Vdchmax} \times S_{t,n}^{V2H}, t = 1, \ldots, T, n = 1, \ldots, N \tag{30}$$

$$S_{t,n}^{H2V} + S_{t,n}^{V2H} \leq 1, T_n^{Vs} \leq t \leq T_n^{Ve}, n = 1, \ldots, N \tag{31}$$

$$S_{t,n}^{H2V} = S_{t,n}^{V2H} = 0, t < T_n^{Vs} \vee t > T_n^{Ve}, n = 1, \ldots, N \tag{32}$$

$$S_{t,n}^{H2V} \in \{0,1\}, S_{t,n}^{V2H} \in \{0,1\}, t = 1, \ldots, T, n = 1, \ldots, N \tag{33}$$

Equations (34) through (40) represent the heating cycles of each prosumer's electric water heater (EWH). By multiplying the nominal power, P_n^{EWHnom}, and an auxiliary binary variable, $w_{t,n}^{EWH}$, the power consumption can be defined as summation of different heating intervals. Minimum and maximum time intervals for a heating cycle, $NT_{k_n,n}^{EWHmin}$ and NT_n^{EWHmax}, are defined by each prosumer according to its preferences and requirements for hot water consumption.

$$P_{t,n}^{EWH} = P_n^{EWHnom} \times w_{t,n}^{EWH}, t = 1, \ldots, T, n = 1, \ldots, N \tag{34}$$

$$\sum_{t=T_{k_n,n}^{EWHs}}^{T_{k_n,n}^{EWHe}} w_{t,n}^{EWH} \geq NT_{k_n,n}^{EWHmin}, k_n = 1, n = 1, \ldots, N \tag{35}$$

$$\sum_{t=T_{k_n,n}^{EWHs}}^{T_{k_n,n}^{EWHe}} w_{t,n}^{EWH} \leq NT_n^{EWHmax}, k_n = 1, n = 1, \ldots, N \tag{36}$$

$$\sum_{k_n'=1}^{k_n} \left(\sum_{t=T_{k_n'n}^{EWHs}}^{T_{k_n'n}^{EWHe}} w_{t,n}^{EWH} \right)$$
$$- \sum_{k_n'=1}^{k_n-1} \left(NT_{k_n',n}^{EWHmin} \right) \geq NT_{k_n,n}^{EWHmin}, k_n = 2, \ldots, K_n, \quad n = 1, \ldots, N \tag{37}$$

$$\sum_{k_n'=1}^{k_n} \left(\sum_{t=T_{k_n'n}^{EWHs}}^{T_{k_n'n}^{EWHe}} w_{t,n}^{EWH} \right)$$
$$- \sum_{k_n'=1}^{k_n-1} \left(NT_{k_n',n}^{EWHmin} \right) \leq NT_n^{EWHmax}, k_n = 2, \ldots, K_n, \quad n = 1, \ldots, N \tag{38}$$

$$w_{t,n}^{EWH} \in \{0,1\}, t = 1, \ldots, T, n = 1, \ldots, N \tag{39}$$

$$P_{t,n}^{EWH} \geq 0, t = 1, \ldots, T, n = 1, \ldots, N \tag{40}$$

Equations (41) through (46) represent the operation of the air conditioning (AC) system. The indoor temperature at each interval, $\theta_{t,n}^{in}$, is a function of the indoor and outdoor temperatures, $\theta_{(t-1),n}^{out}$, of the previous time interval and the current amount of heat/cold energy provided by the AC (42). The plus/minus sign of the third term is used in the heating/cooling mode. The AC can be off or operating at any power level within the nominal range of power, P_n^{ACnom}, (43)–(44). The AC is operating to satisfy the temperature comfort of the prosumer that is defined as $\left[\theta_{t,n}^{min}, \theta_{t,n}^{max} \right]$ (45).

$$\theta_{t,n}^{in} = \theta_{0,n}^{in}, t = 1, n = 1, \ldots, N \tag{41}$$

$$\theta_{t,n}^{in} = \alpha_n \theta_{(t-1),n}^{in} + \beta_n \theta_{(t-1),n}^{out} \pm \gamma_n P_{t,n}^{AC}, t = 2, \ldots, T, n = 1, \ldots, N \tag{42}$$

$$P_{t,n}^{AC} \leq P_n^{ACnom}, t = 1, \ldots, T, n = 1, \ldots, N \tag{43}$$

$$P_{t,n}^{AC} \geq 0, t = 1, \ldots, T, n = 1, \ldots, N \tag{44}$$

$$\theta_{t,n}^{min} \leq \theta_{t,n}^{in} \leq \theta_{t,n}^{max}, t = 1, \ldots, T, n = 1, \ldots, N \tag{45}$$

$$\theta_{t,n}^{in} \in \mathbb{R}, t = 1, \ldots, T, n = 1, \ldots, N \tag{46}$$

3   Workflow of the Proposed GA

A GA has been developed to deal with the nonlinearity of the UL problem to define the optimal reward to be communicated to the LL problem. An exact MILP solver model is then used to obtain the optimal solution to the LL problem. In each generation of the GA, a population of different reward signals is defined at the UL problem and is inserted into the LL problem as parameter to determine the optimal individual flexibilities of the prosumers. The GA customized for this problem has the following steps:

Step 1 – Initial Population. The initial population is created by generating P reward individuals $Rew^p = \left(Rew^p_{T_{rs}}, Rew^p_{T_{rs}+1}, \ldots, Rew^p_{T_{re}}\right)$, $p = 1, \cdots, P$ by randomly selecting values within a predefined range (47). To insert the reward individuals into the LL problem as a parameter, Eq. (48) is used to transform the reward individuals into the standard format of the optimization problem. For each individual Rew^p, the LL problems ($n = 1, \ldots, N$) are solved to find the optimal flexibility $Flex^p_n = \left(Flex^p_{T_{rs},n}, Flex^p_{T_{rs}+1,n}, \ldots, Flex^p_{T_{re},n}\right)$. Thereafter, the calculated individual flexibilities are inserted into the UL problem to determine the aggregator's profit.

$$Rew^{min} \leq Rew^p_t \leq Rew^{max}, t \in [T_{rs}, T_{re}] \tag{47}$$

$$Rew_t = \begin{cases} Rew^p, t \in [T_{rs}, T_{re}] \\ 0, t < T_{rs} \vee t > T_{re} \end{cases} \tag{48}$$

Step 2 – Fitness of the Initial Population. By inserting all individual flexibilities received from LL problem into (1)–(7), the fitness function (UL objective function) is calculated for each reward individual of the initial population using (49). The reward individuals are then sorted in descending order according to their fitness value.

$$OF^p_{Agg} = \sum_{t'=1}^{T'} C^{Sell}_{t'} \times P^{Res}_{t'} \times \Delta t'$$
$$- \left(RT \times \sum_{n=1}^{N} \sum_{t=T_{rs}}^{T_{re}} Rew^p \times Flex^p_n \times \Delta t\right), \quad p = 1, \cdots, P \tag{49}$$

Step 3 – New Generation (Generating P New Reward Individuals). A specific percentage of the best performing individuals from the previous population are identified as the elite set. The new generation is composed of the elite set and offsprings.

The elite set of the current population is passed to the next generation without any changes. The rest of the population is composed of offsprings obtained using one-point crossover and mutation operators as follows:

- Assign different selection weights to the individuals of the current population according to their fitness value.
- Select two parents from the current population using the roulette wheel selection method.
- Define the crossover point by randomly selecting a time interval within the response period

$$T_{rs} + 1 \leq cp \leq T_{re} - 1 \tag{50}$$

- Apply a one-point crossover to the reward individuals

$$Rew^{child_{p,1}} = \left(Rew^{p1}_{T_{rs}}, \cdots, Rew^{p1}_{cp}, Rew^{p2}_{cp+1}, \cdots, Rew^{p2}_{T_{re}}\right), p = 1, \cdots, P \quad (51)$$

$$Rew^{child_{p,2}} = \left(Rew^{p2}_{T_{rs}}, \cdots, Rew^{p2}_{cp}, Rew^{p1}_{cp+1}, \cdots, Rew^{p1}_{T_{re}}\right), p = 1, \cdots, P \quad (52)$$

- Apply mutation to one randomly selected gene of the selected individual with selection probability mp

$$Rew^{child_p}_t = \delta, Rew^{min} \leq \delta \leq Rew^{max}, t \in [T_{rs}, T_{re}] \quad (53)$$

Step 4 – Fitness of the New Generation. Each individual of the new generation is inserted into the LL problem to compute optimal flexibility. Then, the optimal flexibility of each individual is inserted into the UL problem to calculate their fitness value (54). The individuals of the new generation (including the elite set) are then sorted in descending order according to their fitness value. Steps 3 and 4 are repeated until all generations, G, are performed.

$$OF^{child_p}_{Agg} = \sum_{t'=1}^{T'} C^{Sell}_{t'} \times P^{Res}_{t'} \times \Delta t'$$
$$- \left(RT \times \sum_{n=1}^{N} \sum_{t=T_{rs}}^{T_{re}} Rew^{child_p} \times Flex^{child_p}_n \times \Delta t\right), \quad p = 1, \cdots, P \quad (54)$$

Step 5 – Stop. After the realization of all generations, G, the best individual (reward) of the last generation (Rew^P with the highest OF^P_{Agg}) is defined as the best solution of the BOM of the flexibility aggregation problem.

4 Illustrative Results

4.1 Data and Parameters

A group of 20 types of prosumers with different consumption profiles is considered as representative of a broad number of prosumers. The profiles are different considering three factors: possession of demand-side energy resources, technical characteristics, and comfort preferences. The parameters that were used to create the prosumer profiles are accessible via https://data.mendeley.com/datasets/wsy43pjnxn/1. The GA has been implemented in MATLAB2022a. The MILP model of the prosumer's problem given by (7)–(46) has been implemented in Open Programming Language (OPL) using IBM ILOG CPLEX Optimization Studio version 12.8.0. And solved by the CPLEX solver. The flows of input/output data and results are managed using the interaction mechanisms between MATLAB and IBM ILOG CPLEX Optimization Studio. An Intel® Core™i7-12700H, 14 Cores, 2700 MHz computer with 32 GB of RAM running Microsoft Windows 11 Pro was used to compute the illustrative results.

The planning period for all 20 prosumers is one day with 15-min time discretization. The non-controllable base load is equal for all prosumers. There are three shiftable loads:

washing machine (WM), dishwasher (DW), and clothes dryer (CD). The operation cycle of each shiftable load is the same for all prosumers. Some prosumers may have BSS, EV, local generation either by PV or wind energy generation that are modeled using Beta and Weibull probability density functions, respectively. Some prosumers may own an AC that operates in heating mode. The data related to the non-controllable base load and local generation models are accessible via https://data.mendeley.com/datasets/r79 cnyp8j6/1. The outdoor temperature data used in this study are real data recorded on a winter day. The data are accessible through https://data.mendeley.com/datasets/894vmn g8cj/1 with the time discretization of one minute. The data were transformed into 15-min time discretization to be compatible with the model in this study. A range of 0.01 €/kWh to 0.10 €/kWh was considered for the reward in each market period. Demand reduction (upward flexibility, $RT = 1$) was considered in this study for simplicity. However, the model can accommodate demand increment (downward flexibility, $RT = -1$) or a combination of both at different intervals by indexing the response type to the time (RT_t).

4.2 Results

Simulations were conducted using 100 and 300 generations, each with a population size of 30 reward signals, and two values for mutation probability, $mp = 0.04$ and $mp = 0.5$. The number of generations and mutation probability were defined based on experimental tests. Due to the high computational cost, every combination of generation and mutation probability was run only 10 times to provide a statistical analysis of the performance. The best objective function value (aggregator's profit), the average, and the standard deviation for each run are displayed in Tables 1, 2, 3 and 4. The average (Mean), standard deviation (STD), median, and interquartile range (IQR) of the 10 independent runs were calculated for each experiment to evaluate the sufficiency of the number of generations and the mutation probability. The best optimal objective value out of 10 runs is signaled in bold and underlined and the worst optimal value is in bold.

Comparing the results in Tables 1 and 2, higher mutation probability can lead to higher objective function values. The best value increases from 7.48 € to 11.24 € while the lowest value, 4.67 €, stays the same. The same pattern can be seen by comparing the results in Tables 3 and 4 for 300 generations: the highest objective function value increases from 8.71 € to 9.03 € while the lowest value, 4.82 €, remains unchanged. The average of the best objective values of 10 runs increases from 5.44 € to 6.83 € for 100 generations and from 6.47 € to 7.11 € for 300 generations.

Figures 1 and 2 show the aggregated flexibility, reserve power, and the reward signal associated with two different solutions. The former shows the best solution and the latter shows the 5th highest solution of the 10th run for 100 generations, with the mutation probability equal $mp = 0.5$. Comparing the two figures, the reward signal's structure directly impacts the mobilization of the individual flexibilities that shape the aggregated flexibility. The aggregator's profitability also depends on the reward signal's structure.

Table 1. Best objective function value, mean, and standard deviation along with mean, standard deviation, median, and interquartile range (IQR) of the best objective values of all runs ($G = 100, P = 30, mp = 0.04$)

Run	R1	R2	R3	R4	R5	R6	R7	R8	R9	R10
Best Objective (€)	5.92	4.82	5.24	**7.48**	**4.67**	4.82	6.00	4.82	5.83	4.82
Mean (€)	1.18	1.29	1.71	2.08	−0.61	1.31	2.88	0.93	0.56	0.80
STD (€)	3.02	2.82	2.88	2.07	2.67	3.12	2.06	3.22	2.06	3.20

Mean of best objectives	STD of best objectives	Median of best objectives	IQR (Q3–Q1)
5.44 €	0.84 €	5.03 €	1.07 €

Table 2. Best objective function value, mean, and standard deviation along with mean, standard deviation, median, and interquartile range (IQR) of the best objective values of all runs ($G = 100, P = 30, mp = 0.5$)

Run	R1	R2	R3	R4	R5	R6	R7	R8	R9	R10
Best Objective (€)	**11.24**	7.26	5.09	10.11	6.91	**4.67**	6.49	6.87	4.82	4.82
Mean (€)	0.25	0.99	0.69	1.5	0.8	0.44	0.54	0.79	1.58	1.92
STD (€)	4.43	3.1	1.66	3.55	3.37	2.16	3.07	2.2	2.34	2.59

Mean of best objectives	STD of best objectives	Median of best objectives	IQR (Q3–Q1)
6.83 €	2.15 €	6.68 €	2.28 €

Table 3. Best objective function value, mean, and standard deviation along with mean, standard deviation, median, and interquartile range (IQR) of the best objective values of all runs ($G = 300, P = 30, mp = 0.04$)

Run	R1	R2	R3	R4	R5	R6	R7	R8	R9	R10
Best Objective (€)	6.73	5.94	5.58	7.12	5.45	6.76	**8.71**	**4.82**	4.86	**8.71**
Mean (€)	0.63	0.39	1.32	3.57	0.52	0.73	1.00	0.13	1.04	0.89
STD (€)	3.20	2.32	2.37	2.16	3.11	3.22	3.97	3.65	2.73	2.69

Mean of best objectives	STD of best objectives	Median of best objectives	IQR (Q3–Q1)
6.47 €	1.34 €	6.33 €	1.55 €

Table 4. Best objective function value, mean, and standard deviation along with mean, standard deviation, median, and interquartile range (IQR) of the best objective values of all runs ($G = 300, P = 30, mp = 0.5$)

Run	R1	R2	R3	R4	R5	R6	R7	R8	R9	R10
Best Objective (€)	6.99	7.37	8.86	7.07	4.83	6.77	**9.03**	8.38	6.97	**4.82**
Mean (€)	0.25	0.82	1.45	0.3	0.74	0.04	-0.25	0.86	1.02	0.01
STD (€)	3.41	3.46	3.15	2.63	3.25	3.61	3.37	2.45	2.84	2.41

Mean of best objectives	STD of best objectives	Median of best objectives	IQR (Q3–Q1)
7.11 €	1.38 €	7.03 €	1.31 €

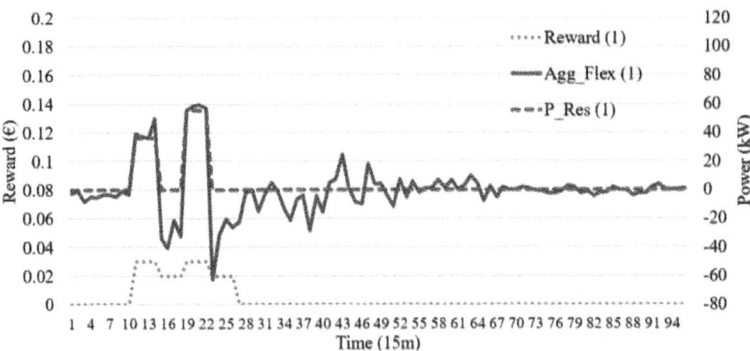

Fig. 1. Aggregated flexibility, reserve power, and reward signal of the best solution of the 10th run ($G = 100, P = 30, mp = 0.5$)

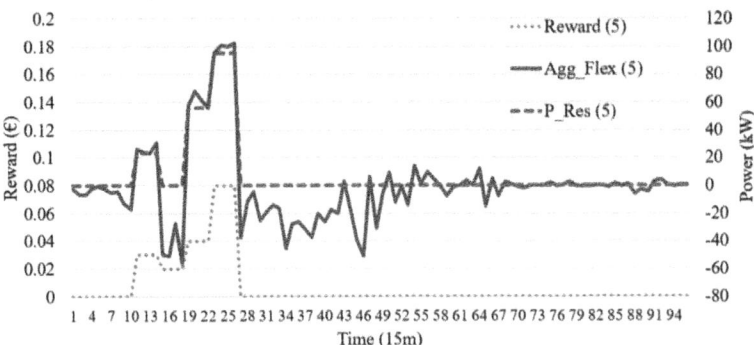

Fig. 2. Aggregated flexibility, reserve power, and reward signal of the 5th solution of the 10th run ($G = 100, P = 30, mp = 0.5$)

5   Conclusions

A GA has been developed to deal with a nonlinear BOM of flexibility aggregation. The GA was used at the UL problem in combination with an exact solver tackling the MILP model at the LL problem. The GA is responsible for finding the optimal reward that the aggregator (leader) sends to the prosumers (followers) to make their flexibility available during the response period. The GA was run for 100 and 300 generations, with a population of 30 individuals (rewards), for two different values of the mutation probability, 0.04 and 0.5. The GA was run 10 times for each combination of number of generations and mutation probability to evaluate the performance of the proposed algorithm to find the best aggregator's profit. It was shown that increasing the number of generations and mutation probability have an impact on the performance of the algorithm. The results revealed that the proposed approach was able to assist mobilizing the prosumers' flexibility and guarantee profitability for flexibility aggregation. The reward signal's structure has a significant impact on the aggregator's profitability. Extensions to this work will consist of comparing the proposed algorithm with other approaches to solve the nonlinear BOM considering the computational effort.

Acknowledgment. This work was supported by the Portuguese Foundation for Science and Technology (FCT) under doctoral grant SFRH/BD/151359/2021 (https://doi.org/10.54499/SFRH/BD/151359/2021), within the European Social Funds (FSE) under PORTUGAL2020, through the Regional Operational Program of the Center (Centro 2020) under MIT-Portugal Program (MPP2030), INESC Coimbra Pluriannual Funding Program (https://doi.org/10.54499/UIDB/00308/2020), the Portuguese Recovery and Resilience Plan (RRP – project 56, ATE – Alliance for Energy Transition), and the Energy for Sustainability Initiative of the University of Coimbra.

Disclosure of Interests.   The authors do not have any known competing financial interests or personal relationships to declare that are relevant to the content of this article.

References

1. International agreements. Commonwealth Law Bull. **3**(3), 452–456 (1977). https://doi.org/10.1080/03050718.1977.9985485
2. Qadir, S.A., Al-Motairi, H., Tahir, F., Al-Fagih, L.: Incentives and strategies for financing the renewable energy transition: a review. Energy Rep. **7**, 3590–3606 (2021). https://doi.org/10.1016/j.egyr.2021.06.041
3. Cruz, C., Palomar, E., Bravo, I., Aleixandre, M.: Behavioural patterns in aggregated demand response developments for communities targeting renewables. Sustain. Cities Soc. **72**, 103001 (2021). https://doi.org/10.1016/j.scs.2021.103001
4. Parvania, M., Fotuhi-Firuzabad, M., Shahidehpour, M.: Optimal demand response aggregation in wholesale electricity markets. IEEE Trans. Smart Grid **4**(4), 1957–1965 (2013). https://doi.org/10.1109/TSG.2013.2257894

5. Kovacevic, M., Vasak, M.: Aggregated representation of electric vehicles population on charging points for demand response scheduling. IEEE Trans. Intell. Transp. Syst. **24**(10), 10869–10880 (2023). https://doi.org/10.1109/TITS.2023.3286012

6. Tostado-Véliz, M., Rezaee Jordehi, A., Icaza, D., Mansouri, S.A., Jurado, F.: Optimal participation of prosumers in energy communities through a novel stochastic-robust day-ahead scheduling model. Int. J. Electr. Power Energy Syst. **147**, 108854 (2022). https://doi.org/10.1016/j.ijepes.2022.108854

7. Pedram, O., Asadi, E., Chenari, B., Moura, P., Gameiro da Silva, M.: A review of methodologies for managing energy flexibility resources in buildings. Energies **16**(17), 6111 (2023). https://doi.org/10.3390/en16176111

8. Heleno, M., Matos, M.A., Lopes, J.A.P.: Availability and flexibility of loads for the provision of reserve. IEEE Trans. Smart Grid **6**(2), 667–674 (2015). https://doi.org/10.1109/TSG.2014.2368360

9. Rasouli, V., Gomes, Á., Antunes, C.H.: Characterization of aggregated demand-side flexibility of small consumers. In: 3rd International Conference on Smart Energy Systems and Technologies, pp. 1–6. IEEE, Istanbul, Turkey (2020). https://doi.org/10.1109/SEST48500.2020.9203476

10. Rasouli, V., Gomes, Á., Antunes, C.H.: An optimization model to characterize the aggregated flexibility responsiveness of residential end-users. Int. J. Electr. Power Energy Syst. **144**, 108563 (2022). https://doi.org/10.1016/j.ijepes.2022.108563

11. Liu, D., Qin, Z., Hua, H., Ding, Y., Cao, J.: Incremental incentive mechanism design for diversified consumers in demand response. Appl. Energy **329**, 120240 (2023). https://doi.org/10.1016/j.apenergy.2022.120240

12. Rasouli, V., Gomes, Á., Antunes, C.H.: Impact of energy price scheme and rewarding strategies on mobilizing the flexibility of residential end-users and aggregator's profit. Int. J. Electr. Power Energy Syst. **158**, 109985 (2024). https://doi.org/10.1016/j.ijepes.2024.109985

13. Chai, Y., Xiang, Y., Liu, J., Gu, C., Zhang, W., Xu, W.: Incentive-based demand response model for maximizing benefits of electricity retailers. J. Mod. Power Syst. Clean Energy **7**(6), 1644–1650 (2019). https://doi.org/10.1007/s40565-019-0504-y

14. Shakeri, M., et al.: Implementation of a novel home energy management system (HEMS) architecture with solar photovoltaic system as supplementary source. Renew. Energy **125**, 108–120 (2018). https://doi.org/10.1016/j.renene.2018.01.114

15. Zhu, J., Alharthi, Y.Z., Wang, Y., Fatemi, S., Ahmarinejad, A.: A hierarchical structure for harnessing the flexibility of residential microgrids within active distribution networks: advancing toward smart cities. Sustain. Cities Soc. **106**, 105398 (2024). https://doi.org/10.1016/j.scs.2024.105398

16. Yang, J., Zhao, J., Wen, F., Dong, Z.Y.: A Framework of customizing electricity retail prices. IEEE Trans. Power Syst. **33**(3), 2415–2428 (2018). https://doi.org/10.1109/TPWRS.2017.2751043

17. Zugno, M., Morales, J.M., Pinson, P., Madsen, H.: A bilevel model for electricity retailers' participation in a demand response market environment. Energy Econ. **36**, 182–197 (2013). https://doi.org/10.1016/j.eneco.2012.12.010

18. Grimm, V., Orlinskaya, G., Schewe, L., Schmidt, M., Zöttl, G.: Optimal design of retailer-prosumer electricity tariffs using bilevel optimization. Omega **102**, 102327 (2021). https://doi.org/10.1016/j.omega.2020.102327

19. Besançon, M., Anjos, M.F., Brotcorne, L., Gomez-Herrera, J.A.: A bilevel approach for optimal price-setting of time-and-level-of-use tariffs. IEEE Trans. Smart Grid **11**(6), 5462–5465 (2020). https://doi.org/10.1109/TSG.2020.3000651

20. Askeland, M., Burandt, T., Gabriel, S.A.: A stochastic MPEC approach for grid tariff design with demand-side flexibility. Energy Syst. **14**(3), 707–729 (2023). https://doi.org/10.1007/s12667-020-00407-7

21. Soares, I., Alves, M.J., Antunes, C.H.: A deterministic bounding algorithm vs. a hybrid metaheuristic to deal with a bilevel mixed-integer nonlinear optimization model for electricity dynamic pricing. Comput. Oper. Res. **155**(C), 106195 (2023). https://doi.org/10.1016/j.cor.2023.106195

22. Sarfarazi, S., Mohammadi, S., Khastieva, D., Hesamzadeh, M.R., Bertsch, V., Bunn, D.: An optimal real-time pricing strategy for aggregating distributed generation and battery storage systems in energy communities: a stochastic bilevel optimization approach. Int. J. Electr. Power Energy Syst. **147**, 108770 (2023). https://doi.org/10.1016/j.ijepes.2022.108770

23. Carrasqueira, P., Alves, M.J., Antunes, C.H.: Bi-level particle swarm optimization and evolutionary algorithm approaches for residential demand response with different user profiles. Inf. Sci. **418–419**, 405–420 (2017). https://doi.org/10.1016/j.ins.2017.08.019

Algorithm Selection with Probing Trajectories: Benchmarking the Choice of Classifier Model

Quentin Renau$^{(\boxtimes)}$ and Emma Hart

Edinburgh Napier University, Edinburgh, Scotland, UK
{q.renau,e.hart}@napier.ac.uk

Abstract. Recent approaches to training algorithm selectors in the black-box optimisation domain have advocated for the use of training data that is 'algorithm-centric' in order to encapsulate information about how an algorithm performs on an instance, rather than relying on information derived from features of the instance itself. *Probing trajectories* that consist of a sequence of objective performance per function evaluation obtained from a short run of an algorithm have recently shown particular promise in training accurate selectors. However, training models on this type of data requires an appropriately chosen classifier given the sequential nature of the data. There are currently no clear guidelines for choosing the most appropriate classifier for algorithm selection using time-series data from the plethora of models available. To address this, we conduct a large benchmark study using 17 different classifiers and three types of trajectory on a classification task using the BBOB benchmark suite using both leave-one-instance out and leave-one-problem out cross-validation. In contrast to previous studies using tabular data, we find that the choice of classifier has a significant impact, showing that *feature-based* and *interval-based* models are the best choices.

Keywords: Algorithm Selection · Black-Box Optimisation · Algorithm Trajectory

1 Introduction

Per-instance algorithm selection (AS) that uses a machine-learning (ML) model to choose the most appropriate solver from a portfolio has shown much promise in both continuous and combinatorial optimisation domains [18]. Designing a selector has two important facets: determining what kind of input will be used to the selector, and the choice of the ML model itself.

With regard to the former, many approaches rely on an input vector that captures either features derived from the description of instance [1] or features derived from the landscape induced by the instance and an objective function, e.g. Exploratory Landscape Features (ELA) [25]. These methods can be considered as *instance-centric* as they are independent of the choice of solver. On the

© The Author(s), under exclusive license to Springer Nature Switzerland AG 2025
P. García-Sánchez et al. (Eds.): EvoApplications 2025, LNCS 15612, pp. 452–468, 2025.
https://doi.org/10.1007/978-3-031-90062-4_28

other hand, recent approaches to AS have taken an *algorithm-centric* approach, in which input to a model is derived from the performance of an algorithm on an instance: for example, Renau *et al.* [33,34] use time-series data as input to a classifier, in what is termed a trajectory-based approach. The trajectory contains a short sequence of the objective values obtained at each evaluation of a solution over a short time period.

The choice of ML model used in the selector is dependent on the type of input data. Feature-based approaches to AS have often used tree-based classifiers such as Random Forests [16], and there is a growing trend of exploiting new deep-learning architectures based on transformers, e.g. [9]. However, a different type of classifier is required to cope with time-series data when using trajectory-based approaches. In [34], a time-series forest classifier was used [7] as the model, but alternative approaches were not evaluated.

Kostovska *et al.* [20] assessed the influence of the ML model used in AS on single-objective black-box problems in the context of feature-based classifiers. They find that 'the choice of model has only a minor impact on the AS performance, as long as it is a method that demonstrates good performance on tabular data in general settings'. However, given that time-series data is fundamentally different to tabular data in that it is an ordered sequence of values and that specialised time-series classifiers fall into a wide range of categories, it is unclear whether the result from [20] holds when considering trajectory-based AS.

Contribution: In this paper, we address the question raised above, i.e. *"To what extent does the choice of ML model used in algorithm selection matter when using trajectory-based data?"*. We conduct an extensive evaluation of 17 time-series-based classifiers using the BBOB test suite as a benchmark, with a portfolio of three solvers. Experiments are conducted both in the Leave-one-instance-out (LOIO) and Leave-one-problem-out (LOPO) settings, where the latter is known to be harder [8]. We find that in contrast to tabular data, the choice of ML model in the time-series setting has a significant impact on the accuracy of the model. We improve the gain over ELA features obtained in [34] from 3% to 7% for similar budgets and we show that we can use even lower budgets of function evaluations and still obtain a 2% gain over ELA features.

Additionally, in the LOPO setting, we show that some functions are extremely difficult to classify correctly regardless of the model chosen, while a small set of four models obtain an accuracy of $\geq 90\%$ on 11 out of 24 functions.

The outline of this paper is as follows. Section 2 gives an overview of the background and related work. Section 3 describes the data used, the methods for obtaining probing-trajectories, describes the models benchmarked and experiments conducted in this paper. Section 4 describes the results obtained with the probing-trajectories on an algorithm selection task. Section 5 provides insights into the results and describes some limitations of our work. Finally, Sect. 6 highlights concluding remarks and future work.

2   Background and Related Work

The majority of previous work in algorithm selection is performed using information describing an *instance* as input to a selector. In the continuous optimisation domain, it is common to use Exploratory Landscape Analysis (ELA) [25] to extract features used as training data [14,29], while in combinatorial optimisation there has been a recent trend towards feature-free approaches that use information that directly describes an instance: in the bin-packing domain, the sizes of each item to be packed are directly used as input [1], while in the TSP domain, images that directly indicate city locations have been used [36]. However, these approaches all take an *instance-centric* view of a problem: that is, the input to a selector describes instance data only and is independent of the execution of any algorithm. Intuitively, incorporating some measure of algorithm performance into the input to a selector would seem beneficial.

Recognising this, recent work has begun to address this, using information derived from running a solver as input to a selector. For example, Jankovic *et al.* [15] propose using ELA features extracted from the search trajectory of an algorithm as training data. Their approach gave encouraging results but was outperformed by classical ELA features computed on the full search space. In [16], features obtained from algorithm trajectories are successfully used to determine whether to switch an algorithm during the course of solving an instance. In [5], standard statistics (mean, standard deviation, minimum, maximum) are extracted from function evaluations at each generation and used as input to a classifier that predicts which of the 24 BBOB function the trajectory belongs to. Renau *et al.* directly use algorithm trajectories to train a classifier to predict the best of three solvers using the BBOB test-suite, showing that these approaches outperform classifiers trained on ELA features, and later show that trajectories can also be used to train a classifier to detect whether an instance is easy or hard for a portfolio of solvers in [31]. Although not specifically concerned with algorithm selection, the work of [28] is also worthy of mention in taking an algorithm perspective by utilising information incorporated in CMA-ES *state variables* to train a surrogate model to predict performance.

Regardless of the type of input used in an algorithm selector, it is important to consider what type of machine learning (ML) model is best suited to the task. Kostovska *et al.* [20] study the impact of the choice of ML model when classifying tabular data, evaluating four different ML models (tree-based and deep-learning-based) on three AS approaches: regression, classification, and pair-wise classification. They find while per-instance algorithm selection has impressive potential, the ML technique is of minor importance.

Given that time-series data obtained from algorithm trajectories is significantly different to tabular data and requires a different family of ML model to be used as a classifier, we conduct a similar benchmarking exercise to that described in [20] to understand the influence of the choice of ML models when training an algorithm selector on trajectories.

3   Methods

We benchmark a suite of 17 different ML models that use time-series data as input to a classifier to better understand what type of classifier facilitates algorithm selection. The methods for obtaining the trajectory data and the models selected for evaluation are described below.

3.1   Trajectories

We use the same data as described in [34] to benchmark 17 classifiers on the noiseless Black-Box Optimisation Benchmark (BBOB) from the COCO platform [10] using data from three algorithms: CMA-ES [11], Particle Swarm Optimisation (PSO) [17], and Differential Evolution (DE) [37]. As in [34], all data is obtained from [38].

Input to each classifier is a *probing-trajectory*. This consists of a time-series consisting of the first n function evaluations from a run of an algorithm. Three types of trajectories are defined.

- *Best*: At each function evaluation, the best objective value seen so far is recorded per algorithm;
- *Current*: the objective value obtained after every evaluation is recorded per algorithm;
- *All*: the trajectories from each algorithm are concatenated to form a single trajectory (for either 'Best' or 'Current'), for example, in this paper ALL best refers to the concatenation of the *Best* trajectories from CMA-ES, DE, and PSO.

For a single instance, four trajectories can therefore be obtained per instance (one from each algorithm and one that concatenates the individual trajectories). Three of these trajectories contain data from one run of an algorithm; the final trajectory contains data from one run of each of three algorithms. Each trajectory is treated as a time-series and is used as input to an algorithm selector. The length of this time-series depends on the number of concatenated algorithms and the number of generations used by each algorithm. We match the experiments performed in [34] by testing trajectories in which the number of generations $g \in \{2, 7\}$. We use data from 5 instances of each of the 24 BBOB functions in dimension 10, and perform 5 runs per instance. Therefore, we obtain a dataset of $24 \times 5 \times 5 = 600$ trajectories.

3.2   Algorithm Selection

In order to replicate the experiments in [34], we consider an algorithm selection task as a classification task, i.e., given a time-series representing an instance, the output is the algorithm to use on that particular instance. The winning algorithm per instance is defined as the algorithm having the best median target value after 100,000 function evaluations and is used as the label for the classifier.

No single algorithm outperforms the others on all 24 functions: CMA-ES is the best performing algorithm for 11 functions, DE for 7, and PSO for 6.

Classification Models. Probing-trajectories consist of a time-series and thus require the use of a specialised time-series classifier. In [34], the authors used a Rotation Forests [35] from the *sktime* package [22][1]. In other work using trajectory-based input to a classifier [31] a Long Short Term Memory (LSTM) network [13] was used: these networks have been shown to deal well with sequential information. In this paper, our goal is to benchmark time-series classifiers to understand how the choice of classifier influences performance. As such, we evaluated 17 time-series classifiers, 16 from the *sktime* package and the LSTM from [31]. We excluded other classifiers that gave errors on our data or had an inference computation time that would exceed the running time of an optimisation algorithm. The list of classifiers is composed of 3 Deep Learning (DL) models, 3 distance-based (D) models, 2 feature-based (F) models, 3 interval-based (I) models, 3 kernel-based (K) models, 1 shapelet-based (S) model, 1 sklearn-based (sk) model, and 1 dummy model. Default parameters are initially used for all mentioned classifiers. The classifiers are defined as follows:

- **LSTM (DL)** as defined in [31]. Code for this model can be found at [32];
- **Time Convolutional Neural Network (CNN) (DL)** as defined in [41];
- **Multivariate Time Series Transformer for Classification (MVTST) (DL)]** as defined in [40];
- **K-nearest neighbors (D)** adapted version of the scikit-learn package [27] for time-series data;
- **Proximity Stump (D)** models a decision stump, i.e., a one-level decision tree, which uses distance to partition the data;
- **ShapeDTW (D)** as defined in [42];
- **Catch22 (F)** as defined in [24];
- **Summary (F)** extracts statistics on the data and builds a Random Forest [3] classifier;
- **Random Interval Spectral Ensemble (I)** as defined in [21];
- **Supervised Time Series Forest (I)** as defined in [4];
- **Time Series Forest Classifier (I)** as defined in [7];
- **Support Vector Classifier (K)** adapted version of the scikit-learn [27] SVC for time series data;
- **Arsenal (K)** as defined in [26];
- **Rocket (K)** as defined in [6];
- **Shapelet Transform Classifier (S)** as defined in [12];
- **Rotation Forests (sk)** as defined in [35];
- **Dummy** which simply outputs the most represented class in the training data.

Automated Configuration of Models. We perform automated configuration of the parameters of the best models. If not stated otherwise, we use *irace* [23][2]

---

[1] Version 0.33.0.

[2] Version 3.5.

to tune the models parameters. We use *irace*'s default parameters. The number of evaluations used is 5,000 for cheap models to evaluate and 1,000 for more expensive models. As CMA-ES has the smallest population size, 10, we choose to tune models using the best trajectory of CMA-ES for 2 generations. This choice is motivated by the computation time of models, i.e., shorter time-series minimises the training and inference times of models. We then transfer the results found for CMA-ES best trajectory to all other trajectories.

Validation Procedure. We perform two types of validations. As performed in [19], we perform a *leave-one-instance-out (LOIO) cross-validation* and we compute *the overall accuracy*. We train classifiers using runs from all 24 functions on all except one instance. Data from the left out instance is used as the validation set. Overall, $24 \times 4 \times 5 = 480$ inputs are used to train the model while the remaining $24 \times 1 \times 5 = 120$ inputs are used for validation.

As performed in [8], we also perform a *leave-one-problem-out (LOPO) cross-validation* and we compute *the overall accuracy*. We train classifiers using runs from 23 functions on all instances and data from the left out function is used as validation set. Overall, $23 \times 5 \times 5 = 575$ inputs are used to train the model while the remaining $1 \times 5 \times 5 = 25$ inputs are used for validation.

4   Results

In this section, we first present the results obtained with default models on a LOIO cross-validation (Sect. 4.1), followed by results obtained with the tuning of models on a LOIO cross-validation (Sect. 4.2). Finally, we present results obtained on the LOPO cross-validation (Sect. 4.3). We discuss the results in Sect. 5.

4.1   Default Models on LOIO Cross-Validation

In this section, we compare 17 default time-series classifiers on the algorithm selection task presented in Sect. 3.2. We use a LOIO cross-validation setting, i.e., models are trained on all functions for four instances and tested on the remaining instance. The 17 models are trained for best and current trajectories for all four trajectories described in Sect. 3.1, resulting in a total of 136 classifiers. As similar tendencies can be observed for all trajectories used as input, we will only present results for PSO trajectories for 2 generations and CMA-ES trajectories for 7 generations. We provide results for the other trajectories in the supplementary materials [30].

Figure 1 compares the classification accuracy of the classifiers trained using PSO trajectories for 2 generations (Fig. 1a) and CMA-ES trajectories for 7 generations (Fig. 1b). We can observe the same general pattern in Fig. 1. Classifiers that perform well when trained using PSO trajectories also perform well when trained with CMA-ES trajectories. This observation can be extended to all other studied trajectories.

Renau *et al.* [34] observed that training a classifier using the *best* trajectory is better than training using the *current* trajectory if running for a small number of generations, and vice versa for a large number of generations. However this result is not seen in our data—it only holds for 2 of the 17 models: *Rotation Forest* and *MVTSTransformer*. Moreover, we observe that the Rotation Forest classifier used in [34] is never the best performing model for any of the trajectories studied.

Furthermore, for both CMA-ES trajectories and PSO trajectories, we observe that some models perform on par or even worse than the Dummy classifier (recall this simply outputs the dominant class in the training data): this applies to 5 models trained using PSO trajectories and 3 for the CMA-ES trajectories. Two of these poor-performing models are kernel-based (*Arsenal* and *Rocket*), two are Deep Learning-based (*CNN* and *MVTSTransformer*) and one is distance-based (*ProximityStump*). On the contrary, all feature-based and interval-based models perform amongst the best performing classifiers. These two categories of classifier are the only ones in which all models from the category obtain performances above the Dummy classifier. Other models such as *ShapeletTransform* or *TimeSeriesSVC* display average performance. They outperform the Dummy classifier but are outperformed by at least 10% median accuracy by the best performing models.

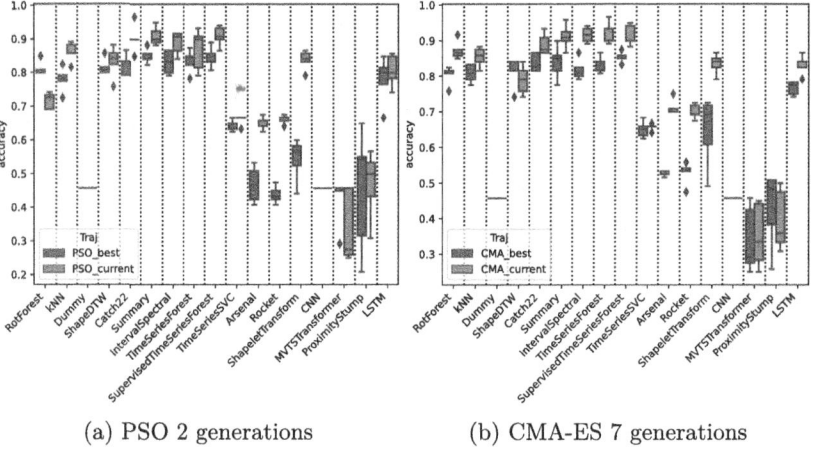

(a) PSO 2 generations   (b) CMA-ES 7 generations

Fig. 1. Accuracy of classification on the LOIO cross-validation for best-so-far and current probing-trajectories for 2 generations (PSO) and 7 generations (CMA-ES).

4.2   Tuning of Models on LOIO Cross-Validation

In order to improve the performance of models, we configure the parameters of the best performing models seen in Fig. 1. From these results, we identify

9 candidate models to tune. After removing models that are too expensive to tune and models where altering some parameters caused errors in the Python package, we are left with 6 models to compare. These models and the parameters that can be tuned are described below:

- **kNN**: number of neighbours (between 1 and 30), weight function (uniform or distance), and distance measure between time series (ten possibilities);
- **ShapeDTW**[3]: number of neighbors (between 1 and 30) and the descriptor function (six possibilities);
- **Time Series Forest**: number of estimators (between 10 and 500) and the minimum length of an interval (between 3 and 30);
- **Summary**: statistics functions used (8191 possible combinations) and the quantiles to compute (eight possibilities);
- **Rotation Forest**(see footnote 3): number of estimators (between 10 and 500), minimum and maximum size of an attribute group (between 3 and 30), and the proportion of cases to be removed per group;
- **Supervised Time Series Forest**: number of estimators (between 10 and 500). The tuning of this model was performed using a grid search dividing the search space in 50 values.

As mentioned in Sect. 3, tuning is performed using the *best* trajectory from runs of CMA-ES. The tuned parameters are then transferred to all trajectories.

We find that all *Supervised Time Series Forest* configurations have similar performances and thus we discard this classifier from the rest of the study on the configuration of parameters. Distributions of accuracies for all parameters tested overlap. We performed Kolmogorov-Smirnov tests on pairs of distributions and we cannot reject any null hypothesis, i.e., the two samples tested come from the same distribution (see supplementary material [30] for the plot). For all other models, we find that tuning improves performance. Details of the tuned configurations can be found in Table 1.

Figure 2 compares the results obtained from tuned/default configurations of the 5 models described above for models trained using (a) DE trajectories and (b) the concatenated ALL trajectories, i.e., we transfer the parameters found when tuning for CMA-ES best trajectory to DE and ALL trajectories. Results are provided using both best and current trajectories in each case. We observe that transferring the parameters found for models tuned with CMA-ES best trajectories improves the accuracy of classification in most settings, i.e., only kNN trained on the DE best trajectory (Fig. 2a), kNN on the ALL current, and Time Series Forest on the ALL best trajectory (Fig. 2b) do not improve in performance when compared to the default parameters.

The configured *Summary* classifier on current trajectories is the best performing model in Fig. 2. This result holds for models trained on 6 of the 8 types of trajectories used to train the models in this paper, i.e. from CMA-ES, DE, PSO, and ALL trajectories, calculated over 2 or 7 generations. Using trajectories from CMA-ES at both 2 and 7 generations, we observe that the tuned *Summary*

---

[3] 1,000 evaluations used by *irace*.

Table 1. Tuned parameters for the six models.

Models	Parameters
kNN	4 neighbors, uniform weights, two distance
ShapeDTW	4 neighbors, raw descriptor
Time Series Forest	460 estimators, 3 minimum interval
Summary	mean, min, max, kurtosis, variance, nb unique and count statistics, 0.25 quantile
Rotation Forest	367 estimators, min group of 10, max group of 19, remove proportion of 0.2364

classifier is outperformed by the tuned *Time Series Forest* classifier but obtains the rank of second best in terms of performance.

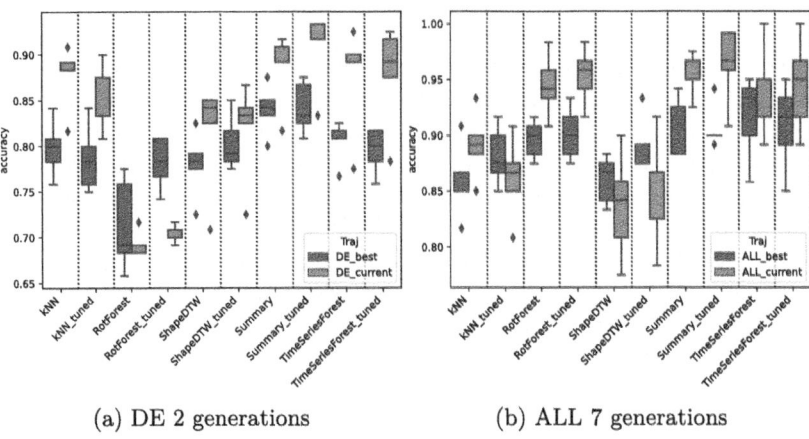

(a) DE 2 generations   (b) ALL 7 generations

Fig. 2. Accuracy of classification on the LOIO cross-validation for best-so-far and current probing-trajectories for 2 generations (DE) and 7 generations (ALL) for default and tuned models.

4.3   Default Models on LOPO Cross-Validation

In this section, we consider a LOPO cross-validation. Models are trained on all but one function and validated on instances from the remaining function. Given the construction of BBOB, this task is much harder than a LOIO cross-validation as functions are purposely designed to be diverse.

Figure 3 displays the results for all models on the LOPO cross-validation where the x−axis represents the function left out for validation, e.g., column F24 represents a training set composed of functions F1 to F23 and a validation set composed of F24. The trajectory used to train the models in each case is the CMA-ES best trajectory over 2 generations.

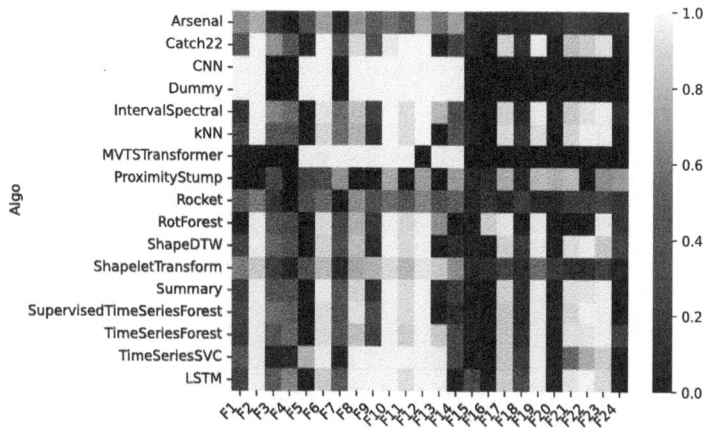

Fig. 3. Heatmap of classification accuracy on the LOPO cross-validation for CMA-ES best-so-far probing-trajectories for 2 generations for default. x−axis represents the functions left out for validation.

Insights into Model Performance. As expected, the performance obtained in the LOPO setting is poorer than using LOIO cross-validation, with a maximum average accuracy of 61.3% across all functions obtained by the LSTM model, which is the best model on average across all functions.

Figure 4 displays the number of functions where the accuracy of a model is below 10% or above 90%. Four models obtain an accuracy greater or equal to 90% on 11 functions: *CNN, Dummy, LSTM*, and *Summary*. The high performance of *Dummy* can be explained by the frequency of winning algorithms in the training data, i.e., CMA-ES wins on 11 functions and Dummy outputs the majority class in training, which is therefore CMA-ES. Looking at the behaviour of *CNN*, we observe that the model learns the same behaviour as *Dummy*, i.e., it always outputs CMA-ES as an answer. Thus, the only two models that learn to predict the correct algorithm rather than simply outputting the majority class are *LSTM* and *Summary*, obtaining at least 90% on 11 functions.

The lowest number of functions where models obtain an accuracy lower than 10% is 5. Three models obtain an accuracy lower than 10% on 5 functions: *LSTM, Time Series Forest* and *Random Interval Spectral Ensemble*. These three models also achieve at least 90% on 11, 8 and 9 functions respectively indicating that these three models can perform well and rarely perform poorly. Along with *LSTM*, the other best performing model, *Summary*, obtains a lower than

10% accuracy on 7 functions indicating that it does rarely perform poorly, even though it does not match the best three models in that matter. Overall, the best performing model on a LOPO cross-validation is *LSTM* followed by *Summary* for the number of well predicted functions.

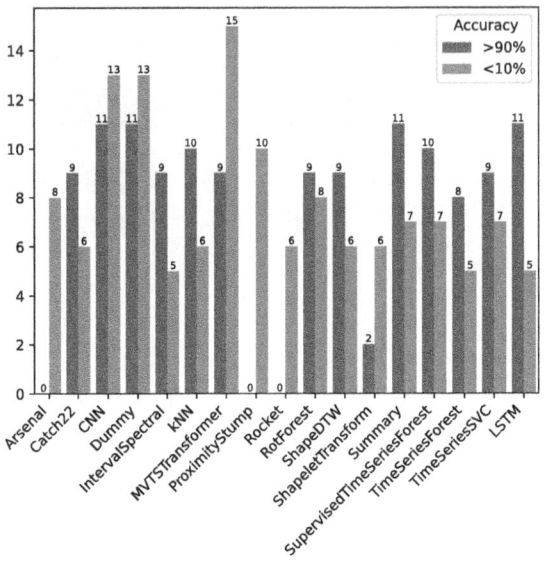

Fig. 4. Number of functions with accuracy above 90% or below 10% for each model.

Insights into the Difficulty of Predicting Functions. We observe that for some functions, it is relatively straightforward to train a high-performing selector, while for others, it is very difficult.

Figure 5 displays the number of models where the accuracy on a function is below 10% or above 90%. Well predicted functions are F10, F12, F2, and F6 with 14, 13, 12, and 12 models obtaining more than 90% accuracy on them respectively. No model performs below 56% on F10 and 32% on F6; only one model performs below 10% on F12 while two perform below this value on F2.

On the contrary, the most difficult functions to perform LOPO on are F15, F20, and F16 with 16 models, 14 models, and 12 models respectively performing below 10% accuracy. All models perform poorly on F15 with a maximum median accuracy of 24% (LSTM). On F20, the best median accuracy obtained is 80% from one model (*Proximity Stump*). Only one model performs above 90% accuracy on F16: *Rotation Forest*.

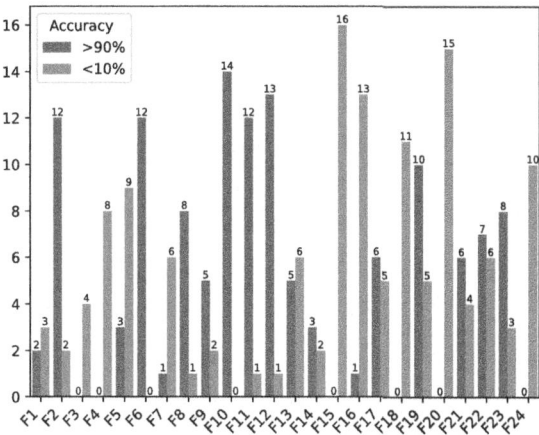

Fig. 5. Number of models with accuracy above 90% or below 10% for each function.

5   Discussion

In Sect. 4 we showed results obtained from training different models of algorithm selectors using two cross-validation strategies (LOIO, LOPO) using sixteen different types of training data as input: i.e. the best and current trajectories obtained from three algorithms, the ALL trajectories obtained by concatenating three best or three current trajectories, with trajectories collected over both 2 or 7 generations. From these experiments, we make a general observation that models performances in terms of accuracy are not linked to the type of cross-validation used and generally is invariant to the type of trajectory, i.e., the best performing model for one trajectory is likely to perform well on another trajectory. *Kernel-based* and *deep learning-based* classifiers (except LSTM) consistently perform poorly on the algorithm selection task. In fact, their performance is often comparable to those of the *Dummy* classifier which always predicts the majority class. Hence, we suggest that these types of models may not be suited for algorithm selection using time-series data. Concerning *deep learning* models, it is well-known in the machine-learning literature that they often require a lot of fine tuning to perform well, in terms of both the architecture and setting weights [2]. In our experiments, we only used off-the-shelf architectures for these models, hence fine tuning the architectures of these models may improve their performance.

We observe that *feature-based* and *interval-based* models consistently perform well. One feature-based model (*Summary*) and one interval-based model (*Time Series Forest*) are often ranked as the best performing model in an experiment. Hence, we recommend that one of these two models is used as a default when training algorithm selectors on raw time-series data.

Limitations: The study used a single test suite (BBOB). Although this is used extensively in the black-box optimisation community [19,31], it is known that it does not generalise well [39], especially in the LOIO cross-validation setting. We compensate for this by also performing a LOPO cross-validation, but extending the study to other test suites would be beneficial to understand better whether the results we obtained generalise across other benchmark suites, in both continuous and combinatorial optimisation. In addition, our tuning study only using CMA-ES trajectory data in tuning model parameters (Sect. 4.2). Although the tuned model clearly transfers well to models that use trajectories from the other algorithms as input, tuning a model on the specific trajectory type of interest might further improve results.

6 Conclusion

This article addressed the question of the extent to which the choice of machine-learning model used in the context of algorithm selection with time-series input influences the accuracy of the learned classifier. A large study benchmarked 17 classifiers from multiple families of time-series classifiers using the BBOB test suite, in both LOIO and LOPO settings.

Overall, we observe that classifiers that perform well in the LOIO setting also perform well in the LOPO setting. Additionally, the best performing classifiers for short trajectories of 2 generations are also the best performing classifiers for longer trajectories composed of 7 generations and this for all considered trajectory types. Two classes of models are consistently ranked as the best performing models for all settings considered: *Summary* and *Time Series Forest*. These models perform similarly despite the fact the classifiers operate in a very different manner. *Summary* is a feature-based model that extracts statistics on the time-series and builds a Random Forest classifier on these statistics while *Time Series Forest* is an interval-based model building an ensemble of decision trees on random intervals of the time-series. Hence we recommend the use of one of these two models to perform algorithm-selection using time-series input. We also performed an automated configuration of model parameters using a single trajectory type, and then transfer the tuned configurations to models trained on different trajectories. We show that tuned version of the models outperform the default configurations even when parameters learned on one type of trajectory are directly transferred to models that are trained on trajectories. Nevertheless, we expect that tuning the models using the trajectory type that will eventually be used to perform algorithm selection could bring further improvements.

In Renau *et al.* [34], the authors used a LOIO setting using a model trained using a *Rotation Forest* classifier. With this setting, the authors obtained a 3% gain in accuracy over a classifier trained using ELA features and using a similar budget (560 and 500 function evaluations). By tuning this classifier using automated algorithm configuration, we show that we can further increase the gain in accuracy to 5%. Interestingly, in a low budget setting using a model trained on the CMA-ES current trajectory with 70 function evaluations and using tuned *Time Series Forest* classifiers, we also obtain a 2% gain over ELA features but using more than 7 times fewer function evaluations. However, using the tuned *Summary* classifier, we further increase the gain to 6% using a similar function evaluation budget.

Obvious next steps include extending the benchmarking approach to other datasets both in continuous and combinatorial domains. Moreover, in this paper, we perform algorithm-selection from a classification perspective. As it is also common to train regressors as algorithm-selectors, benchmarking regressors for time-series inputs should also be undertaken in future.

Acknowledgments. Authors are supported by funding from EPSRC award number: EP/V026534/1.

Disclosure of Interests. The authors have no competing interests to declare that are relevant to the content of this article.

References

1. Alissa, M., Sim, K., Hart, E.: Automated algorithm selection: from feature-based to feature-free approaches. J. Heurist. **29**(1), 1–38 (2023). https://doi.org/10.1007/s10732-022-09505-4

2. Baratchi, M., et al.: Automated machine learning: past, present and future. Artif. Intell. Rev. **57**(5), 122 (2024). https://doi.org/10.1007/S10462-024-10726-1

3. Breiman, L.: Random forests. Mach. Learn. **45**(1), 5–32 (2001). https://doi.org/10.1023/A:1010933404324

4. Cabello, N., Naghizade, E., Qi, J., Kulik, L.: Fast and accurate time series classification through supervised interval search. In: 20th IEEE International Conference on Data Mining, ICDM 2020, Sorrento, Italy, 17–20 November 2020, pp. 948–953. IEEE (2020). https://doi.org/10.1109/ICDM50108.2020.00107

5. Cenikj, G., Petelin, G., Doerr, C., Korosec, P., Eftimov, T.: Dynamorep: trajectory-based population dynamics for classification of black-box optimization problems. In: Proceedings of the Genetic and Evolutionary Computation Conference, GECCO 2023, Lisbon, Portugal, 15–19 July 2023, pp. 813–821. ACM (2023). https://doi.org/10.1145/3583131.3590401

6. Dempster, A., Petitjean, F., Webb, G.I.: ROCKET: exceptionally fast and accurate time series classification using random convolutional kernels. Data Min. Knowl. Discov. **34**(5), 1454–1495 (2020). https://doi.org/10.1007/S10618-020-00701-Z

7. Deng, H., Runger, G.C., Tuv, E., Martyanov, V.: A time series forest for classification and feature extraction. Inf. Sci. **239**, 142–153 (2013). https://doi.org/10.1016/J.INS.2013.02.030

8. Derbel, B., Liefooghe, A., Vérel, S., Aguirre, H., Tanaka, K.: New features for continuous exploratory landscape analysis based on the SOO tree. In: Proceedings of Foundations of Genetic Algorithms (FOGA) '19, pp. 72–86. ACM (2019). https://doi.org/10.1145/3299904.3340308

9. Gorishniy, Y., Rubachev, I., Khrulkov, V., Babenko, A.: Revisiting deep learning models for tabular data. In: Advances in Neural Information Processing Systems, vol. 34, pp. 18932–18943 (2021)

10. Hansen, N., Auger, A., Ros, R., Mersmann, O., Tusar, T., Brockhoff, D.: COCO: a platform for comparing continuous optimizers in a black-box setting. Optim. Methods Softw. **36**(1), 114–144 (2021). https://doi.org/10.1080/10556788.2020.1808977
11. Hansen, N., Ostermeier, A.: Completely derandomized self-adaptation in evolution strategies. Evol. Comput. **9**(2), 159–195 (2001). https://doi.org/10.1162/106365601750190398
12. Hills, J., Lines, J., Baranauskas, E., Mapp, J., Bagnall, A.J.: Classification of time series by shapelet transformation. Data Min. Knowl. Discov. **28**(4), 851–881 (2014). https://doi.org/10.1007/S10618-013-0322-1
13. Hochreiter, S., Schmidhuber, J.: Long short-term memory. Neural Comput. **9**(8), 1735–1780 (1997). https://doi.org/10.1162/neco.1997.9.8.1735
14. Jankovic, A., Doerr, C.: Landscape-aware fixed-budget performance regression for modular CMA-ES variants. In: Proceedings of the Genetic and Evolutionary Computation Conference, GECCO '20 (2020). https://doi.org/10.1145/3377930.3390183, to appear
15. Jankovic, A., Eftimov, T., Doerr, C.: Towards feature-based performance regression using trajectory data. In: Applications of Evolutionary Computation - 24th International Conference, EvoApplications 2021, Held as Part of EvoStar 2021, Proceedings. LNCS, vol. 12694, pp. 601–617. Springer, Cham (2021). https://doi.org/10.1007/978-3-030-72699-7_38
16. Jankovic, A., Vermetten, D., Kostovska, A., de Nobel, J., Eftimov, T., Doerr, C.: Trajectory-based algorithm selection with warm-starting. In: IEEE Congress on Evolutionary Computation, CEC 2022, Padua, Italy, 18–23 July 2022, pp. 1–8. IEEE (2022). https://doi.org/10.1109/CEC55065.2022.9870222
17. Kennedy, J., Eberhart, R.: Particle swarm optimization. In: Proceedings of ICNN'95 - International Conference on Neural Networks, vol. 4, pp. 1942–1948 (1995). https://doi.org/10.1109/ICNN.1995.488968
18. Kerschke, P., Hoos, H., Neumann, F., Trautmann, H.: Automated algorithm selection: survey and perspectives. Evol. Comput. **27**(1), 3–45 (2019)
19. Kostovska, A., et al.: Per-run algorithm selection with warm-starting using trajectory-based features. In: Parallel Problem Solving from Nature - PPSN XVII - 17th International Conference, PPSN 2022, Dortmund, Germany, 10–14 September 2022, Proceedings, Part I. LNCS, vol. 13398, pp. 46–60. Springer, Cham (2022). https://doi.org/10.1007/978-3-031-14714-2_4
20. Kostovska, A., Jankovic, A., Vermetten, D., Džeroski, S., Eftimov, T., Doerr, C.: Comparing algorithm selection approaches on black-box optimization problems. In: Proceedings of the Companion Conference on Genetic and Evolutionary Computation, pp. 495–498 (2023)
21. Lines, J., Taylor, S., Bagnall, A.: Time series classification with hive-cote: the hierarchical vote collective of transformation-based ensembles. ACM Trans. Knowl. Discov. Data **12**(5) (2018). https://doi.org/10.1145/3182382
22. Löning, M., Bagnall, A.J., Ganesh, S., Kazakov, V., Lines, J., Király, F.J.: Sktime: a unified interface for machine learning with time series. CoRR abs/1909.07872 (2019). http://arxiv.org/abs/1909.07872
23. López-Ibáñez, M., Dubois-Lacoste, J., Pérez Cáceres, L., Birattari, M., Stützl, T.: The irace package: iterated racing for automatic algorithm configuration. Oper. Res. Perspect. **3**, 43–58 (2016)
24. Lubba, C.H., Sethi, S.S., Knaute, P., Schultz, S.R., Fulcher, B.D., Jones, N.S.: catch22: Canonical time-series characteristics - selected through highly comparative time-series analysis. Data Min. Knowl. Discov. **33**(6), 1821–1852 (2019). https://doi.org/10.1007/S10618-019-00647-X

25. Mersmann, O., Bischl, B., Trautmann, H., Preuss, M., Weihs, C., Rudolph, G.: Exploratory landscape analysis. In: Proceedings of the Genetic and Evolutionary Computation Conference, GECCO '11, pp. 829–836. ACM (2011). https://doi.org/10.1145/2001576.2001690

26. Middlehurst, M., Large, J., Flynn, M., Lines, J., Bostrom, A., Bagnall, A.J.: HIVE-COTE 2.0: a new meta ensemble for time series classification. Mach. Learn. **110**(11), 3211–3243 (2021). https://doi.org/10.1007/S10994-021-06057-9

27. Pedregosa, F., et al.: Scikit-learn: machine learning in Python. J. Mach. Learn. Res. **12**, 2825–2830 (2011)

28. Pitra, Z., Repický, J., Holena, M.: Landscape analysis of Gaussian process surrogates for the covariance matrix adaptation evolution strategy. In: Proceedings of the Genetic and Evolutionary Computation Conference, GECCO '19, pp. 691–699 (2019). https://doi.org/10.1145/3321707.3321861

29. Renau, Q., Dréo, J., Doerr, C., Doerr, B.: Towards explainable exploratory landscape analysis: extreme feature selection for classifying BBOB functions. In: Applications of Evolutionary Computation - 24th International Conference, EvoApplications 2021, Held as Part of EvoStar 2021, Proceedings. LNCS, vol. 12694, pp. 17–33. Springer, Cham (2021). https://doi.org/10.1007/978-3-030-72699-7_2

30. Renau, Q., Hart, E.: Algorithm Selection with Probing Trajectories: Benchmarking the Choice of Classifier Model - Data (2024). https://doi.org/10.5281/zenodo.14163833

31. Renau, Q., Hart, E.: Identifying easy instances to improve efficiency of ML pipelines for algorithm-selection. In: Parallel Problem Solving from Nature - PPSN XVIII - 18th International Conference, PPSN 2024, Hagenberg, Austria, 14–18 September 2024, Proceedings, Part II. LNCS, vol. 15149, pp. 70–86. Springer, Cham (2024). https://doi.org/10.1007/978-3-031-70068-2_5

32. Renau, Q., Hart, E.: Identifying easy instances to improve efficiency of ml pipelines for algorithm-selection - code and data (2024). https://doi.org/10.5281/zenodo.10590233

33. Renau, Q., Hart, E.: Improving algorithm-selectors and performance-predictors via learning discriminating training samples. In: Proceedings of the Genetic and Evolutionary Computation Conference, GECCO 2024, Melbourne, VIC, Australia, 14–18 July 2024. ACM (2024). https://doi.org/10.1145/3638529.3654025

34. Renau, Q., Hart, E.: On the utility of probing trajectories for algorithm-selection. In: Applications of Evolutionary Computation - 27th European Conference, EvoApplications 2024, Held as Part of EvoStar 2024, Aberystwyth, UK, 3–5 April 2024, Proceedings, Part I. LNCS, vol. 14634, pp. 98–114. Springer, Cham (2024). https://doi.org/10.1007/978-3-031-56852-7_7

35. Rodríguez, J., Kuncheva, L., Alonso, C.: Rotation forest: a new classifier ensemble method. IEEE Trans. Pattern Anal. Mach. Intell. **28**(10), 1619–1630 (2006). https://doi.org/10.1109/TPAMI.2006.211

36. Seiler, M., Pohl, J., Bossek, J., Kerschke, P., Trautmann, H.: Deep learning as a competitive feature-free approach for automated algorithm selection on the traveling salesperson problem. In: International Conference on Parallel Problem Solving from Nature, pp. 48–64. Springer, Cham (2020)

37. Storn, R., Price, K.: Differential evolution - a simple and efficient heuristic for global optimization over continuous spaces. J. Global Optim. **11**(4), 341–359 (1997). https://doi.org/10.1023/A:1008202821328

38. Vermetten, D., Hao, W., Sim, K., Hart, E.: To Switch or not to Switch: Predicting the Benefit of Switching between Algorithms based on Trajectory Features - Dataset (2022). https://doi.org/10.5281/zenodo.7249389

39. Vermetten, D., Ye, F., Bäck, T., Doerr, C.: Ma-bbob: many-affine combinations of bbob functions for evaluating autoML approaches in noiseless numerical black-box optimization contexts. In: Proceedings of the Second International Conference on Automated Machine Learning. Proceedings of Machine Learning Research, vol. 224, pp. 7/1–14. PMLR (2023). https://proceedings.mlr.press/v224/vermetten23a.html

40. Zerveas, G., Jayaraman, S., Patel, D., Bhamidipaty, A., Eickhoff, C.: A transformer-based framework for multivariate time series representation learning. In: Proceedings of the 27th ACM SIGKDD Conference on Knowledge Discovery and Data Mining. KDD '21, pp. 2114–2124. Association for Computing Machinery, New York, NY, USA (2021). https://doi.org/10.1145/3447548.3467401

41. Zhao, B., Lu, H., Chen, S., Liu, J., Wu, D.: Convolutional neural networks for time series classification. J. Syst. Eng. Electron. **28**(1), 162–169 (2017). https://doi.org/10.21629/JSEE.2017.01.18

42. Zhao, J., Itti, L.: Shapedtw: shape dynamic time warping. Pattern Recogn. **74**, 171–184 (2018). https://doi.org/10.1016/j.patcog.2017.09.020

Real Application Challenges in Evolutionary Optimization? People!

Tobias Rodemann[1]([⊠])[iD] and Christiane Attig[2][iD]

[1] Honda Research Institute Europe, Carl-Legien-Strasse 30,
63073 Offenbach/Main, Germany
`tobias.rodemann@honda-ri.de`
[2] Institut für Multimediale und Interaktive Systeme, Universität zu Lübeck,
Ratzeburger Allee 160, 23562 Lübeck, Germany
`christiane.attig@uni-luebeck.de`

Abstract. The application of evolutionary optimization methods for real world problems is often far less straight-forward than expected. One of the main challenges are the people involved in real business situations that decide on whether optimization projects are successful or not. In this work we present a few insights from 20+ years of applying Evolutionary Algorithms (mostly multi- and many-objective) with in-house customers and point to some key psychological insights that can explain several of the major issues that we encountered in our work but are also reported repeatedly in application sessions on major conferences. We argue that more convincing sales messages are needed; and explain, why trust, not better objective values, might be the ultimate goal of any optimization process. We can't provide numbers or new algorithms, but maybe a different perspective on some of the most persistent practical challenges.

Keywords: Application · Evolutionary Algorithms · User Interaction · Many-objective optimization

1 Introduction

Application sessions in conferences on evolutionary algorithms (EAs) like EvoStar, GECCO, or CEC often appear to be separated from other sessions, in the sense that few of the big topics in fundamental research are actually represented in application work and many of the practical challenges are not addressed in research. In this work we want to outline some of the problems we experienced many times in our own projects. The reason these problems are not properly discussed in EA research is that they are not computer science related but social and psychological in nature. Real-world problems are not just characterized by different objective functions but by the need to closely interact with different groups of people. In this paper we want to look at the three main groups in an optimization project:

© The Author(s), under exclusive license to Springer Nature Switzerland AG 2025
P. García-Sánchez et al. (Eds.): EvoApplications 2025, LNCS 15612, pp. 469–481, 2025.
https://doi.org/10.1007/978-3-031-90062-4_29

1. The optimization specialists and how their job is different in industrial projects compared to academic projects.
2. The application team who provide application knowledge, had the idea for the optimization project, and who need to specify the details of the project.
3. The decision maker who needs to be convinced to finance the optimization project and to realize the results of the optimization process.

We will argue that the current trends in research do not improve but rather weaken the communication between these groups and make an efficient cross-group cooperation more difficult. The main practical aspects we want to address in this work are:

1. In most real world problems, there is initially no clear definition of the project target. Most of the time and budget needs to go into extracting the actual task from the application specialists, leaving little time for testing and tuning many algorithms.
2. Commercial optimization specialists are orders of magnitude more expensive than students in academic projects. Many optimization projects struggle to finance even the costs of the optimization experts.
3. Some managers require a good explanation for choices made in optimization projects, such as why a certain optimization algorithm was chosen. The explosive growth of new algorithms especially for many-objective optimization has made any justification for a specific algorithm very challenging. Increasingly complex methods (for example the integration of Machine Learning methods in surrogates [9,13]) further increase the difficulties.
4. Decision makers typically not only consider numerical optimization results, but also require a certain level of trust in the methods and any choices made. Building this trust is often ignored and leads to mutual frustration at the end of the project when optimized solutions might not be implemented, due to low trust in the way they were generated.

Some of the issues above are actually well known from psychological research as fundamental human characteristics that need to be considered in any interaction. One of the key aspects is trust—without trust, no optimization project can be successful. Unfortunately, how to build and maintain trust has largely been ignored in computer science so far. Cognitive sciences have identified transparency and explainability as key elements for building trust. It is also known that there is threshold for the maximum cognitive load decision makers can handle and the negative impact on rational decisions and trust that results from exceeding this limit.

This article is based on a keynote speech presented at the IEEE WCCI conference 2024 in Yokohama, with several additions and modifications based on audience and reviewer comments.

2   Psychological Background

From the perspective of cognitive psychology, the optimization process can be understood as a complex human problem-solving process [14]: The status quo has been identified as an undesirable initial state by the application team, then the optimization team chooses from a number of operators (i.e., algorithms) that transform the initial state into the goal state. However, there exist barriers that impede the smooth transformation process from initial to goal state: (1) The precise goal state is unclear (i.e., ill-defined) or unknown to the optimization team, (2) the choice of operator is challenging due to the high and ever-growing number of applicable algorithms (i.e., the problem space is too vast to be fully comprehensible, leading to the necessity of using heuristics to avoid cognitive overload [21]), and (3) the algorithm selection is intransparent, likely causing low trust in the optimization process by the decision maker (see Fig. 1). A similar set of issues arises for the way the selected algorithm works and produces final candidates.

Solutions to ill-defined problems such as the described optimization problem typically are not correct or incorrect but fall on a spectrum from low to high acceptability [3,10]; moreover, solutions are context-dependent and need to be evaluated (e.g., through simulations). Consequently, the described optimization problem is characterized by at least three socio-cognitive challenges beyond algorithm quality: Insufficient goal state representation, risk of cognitive overload due to an overstraining number of possible solutions, and low trust in the optimization process. What is more, on a purely social level, communication challenges in the social interaction between the three involved groups can further complicate the optimization process during all stages.

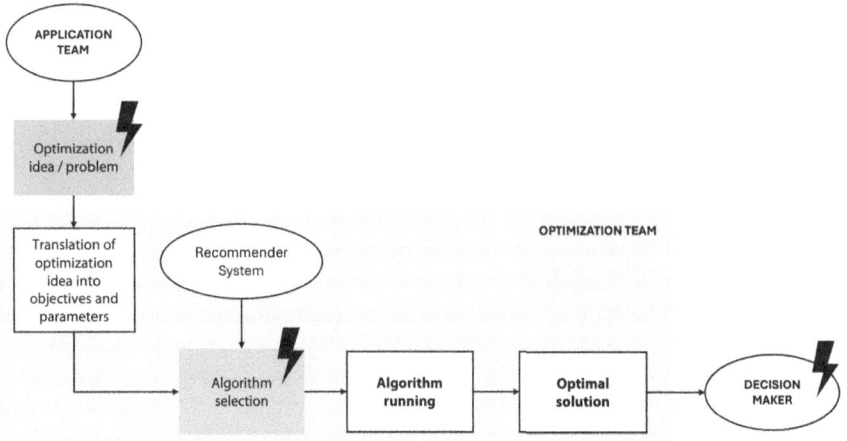

Fig. 1. Protoypical optimization process and socio-cognitive challenges in real-world applications.

In the following, we will briefly describe some of the socio-cognitive challenges with respect to the decision maker's information processing in more detail. To this end, we utilized classical control-theoretic models (e.g., [2]) to describe the psychological mechanisms in the decision maker that impede smooth optimization processes (see Fig. 2). Imagine the optimization team presenting their results (i.e., final objective values) to the decision maker. The decision maker first has to perceive and understand this information to be able to compare the technical output with the initial optimization goal (Feedback Loop 1). Only if the final objective values are sufficiently close to the optimization goal, the decision maker might decide to make investments. However, even if the reported objective values and the optimization goal are aligned, investment decision can be impeded or suspended. In fact, the decision maker not only evaluates the reported numerical values, but also the optimization process on a higher-order level (regarding e.g., number and comprehensibility of presented output values, length of the optimization process, comparison with prior optimization processes). Hence, the decision maker compares their expected optimization process performance with the actual optimization process (Feedback Loop 2). The result of this comparison is a certain subjective experience. For instance, if the presented technical output is too extensive or not adapted to the decision maker's state of expertise, cognitive overload and frustration might occur. In addition, the decision maker evaluates the traceability of the optimization process (i.e., algorithm selection and technical output; Feedback Loop 3). If they perceive the traceability as too low—maybe because the algorithm selection is based on past projects and not on the objective best fit to the optimization problem (see Sect. 4.3)—trust in the optimization process or team can be impaired.

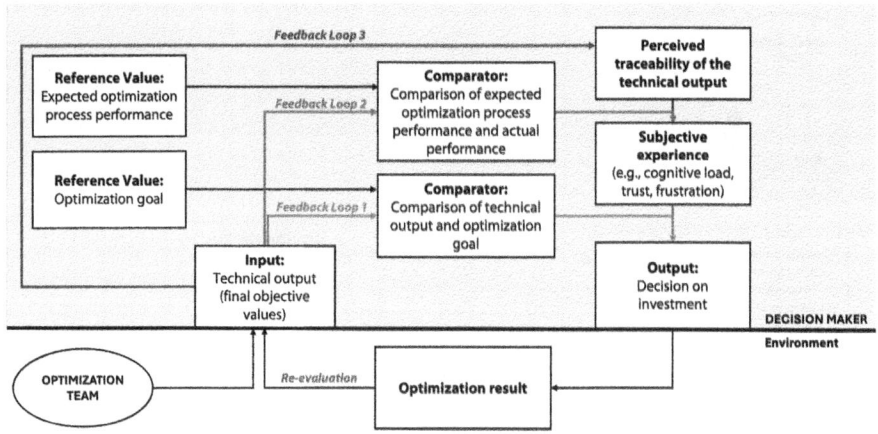

Fig. 2. Control-theoretic framework depicting psychological mechanisms in the decision-maker after presentation of the technical output by the optimization team.

Finally, the negative subjective experience of the decision maker may increase the likelihood of using heuristics that can delay or even suspend the investment decision [19]. Just to mention two well-known examples: The *feeling-as-information heuristic* [20] describes the behavioral tendency to use affective reactions (e.g., frustration, annoyance) as a basis of judgment. Furthermore, the *status quo bias* [5] is the tendency to stick with the status quo (i.e., the undesirable state from which the optimization idea arose in the first place) even though other options are feasible and better (as demonstrated by the optimization team). Particularly in instances with low trust in the optimization process, the status quo bias increases in probability because human decision makers are usually more driven to avoid losses than to make equal-sized gains [6]. When accepting the inherent socio-cognitive aspects of an optimization project, it becomes clear that a deep understanding of human psychology and cognitive limits is essential to improve the chances of a successful project.

3   Referenced Application Work

Our research group has worked intensively on different types of optimization problems in the past more than 20 years with company internal customers from different divisions. In this article we will focus on two key application areas, evolutionary design optimization (see for example [11]) and many-objective optimization of energy systems (equipment sizing optimization), see e.g., [16]. While from an optimization perspective the two applications are not too different they had a very different project setting.

In the design optimization one of the main challenges was a proper representation of shapes (e.g., turbine blades) that provides a high flexibility with few parameters. Another challenge was the evaluation of potential solutions via computational fluid dynamics (CFD) simulations. Especially in the beginning simulators had their problems and engineers did not fully trust the results. The road to success was a long-term commitment on both sides, slowly building trust in both simulation as well as optimization methods. We also put a focus on software transfer, maintenance and upskilling to ensure a long-term usability.

For the energy system optimization, the design parameters are obvious (dimensions of potential hardware purchases), but again the simulation was a cause of concern. In this application validating optimized solutions beyond a simplified simulation is very difficult and trust becomes an ever bigger issue. Even technically very reasonable results (simulations indicate strong and robust positive return on investment) did not lead to quick decisions, partially due to limited trust in the simulation and optimization process. We already elaborated on this in a prior article [18].

4   Overview of Challenges

This article has been motivated by a few occasions where cooperation partners asked questions or made comments that were difficult to answer or caused some consternation. As with many problems of this type one tends to initially see this as distraction from the real challenges but later realizes the fundamental issues behind those questions. The remainder of this article will elaborate more on these topics, trying to explain the social issues and the people involved. In this article we want to focus on the following real challenges that we encountered:

1. Cognitive overload in multi-objective optimization: Most (not all) people don't like Pareto fronts, they don't want to choose from 100 s of solutions. The extra degrees of freedom provided by a multi-objective optimization are quite often seen as a problem not a flexibility. One potential reason is the additional selection required for the decision maker (which of the many Pareto set solutions are selected). Quite obviously adding more objectives will further complicate matters.

2. Getting decisions: Requesting decisions from partners outside their domain of expertise is generally difficult, but even when staying inside the application domain, getting decisions might be challenging and lead to lower trust in the results. One example is the priorization of multiple objectives, for example as weights or reference vectors. In our experience many customers don't think in terms of explicit trade-offs and don't know how to express these in concrete numbers.

3. Concise optimization project summary: "Please summarize the results of your project in one line". In project reporting one typically needs to compress the results of the project into a few words (like "performance improved by X%"). For multi- and especially many-objective optimization the typical performance indicators like hypervolume or IGD are not well suited as they are hard to interpret.

4. Justification of algorithmic choices: "Why did you choose this method?". Decision makers often require a justification for the choice of algorithms. Especially in many-objective optimization there are so many algorithms (the popular PlatEMO [23] tool contains more than 270 EMOAs) that any choice is somewhat arbitrary, so it is difficult to come up with a good justification for any selection made.

5. Maintainability: In many cases, the project can only be called a success, if the customer is able to repeatedly employ the used optimization methods, perform minor adaptations and correctly interpret the results. Beyond software-related aspects (like DevOps) also know-how transfer has to be organized, considering the available skill-set at customer side.

The key underlying theme in most of these issues is **trust**. In the end, the decision maker, who is typically not an expert for optimization, has to evaluate not only the technical outcome of the project (the final objective values) but also the trust in the overall process. It is known [1] that trust is closely connected to transparency and explainability, meaning the ability to understand how optimal solutions were found (the process level) and why these solutions are good, respectively. These factors are increasingly considered in the field of ML and will also require more attention in EAs.

One issue often ignored is the problem of cognitive load or cognitive capacity like Miller's 7 ± 2 rule [12] (a limited short-term memory capacity of around 7 items, for example objective values). Any decision setting that exceeds the cognitive limit of the decision maker (like selecting a subset of solutions out of a Pareto set with 100 s of solutions and many objectives each), could trigger low-level psychological processes like fundamental decision heuristics (risk aversion, status-quo bias, etc.) as described above and diminish the trust in the process. On the other hand, trust can be transferred to and from other human beings, like domain experts. Convincing an independent domain expert might be easier than convincing the final decision maker.

4.1 Project Initialization

In the academic world the optimization problem is assumed to be precisely specified, with given design variables including valid ranges, a mathematical objective function, definition of constraints, etc.

In practice, the problem is often only explained in application terminology. This means that the underlying specs of the optimization tasks need to be extracted in discussions with the application team, often leaving many aspects undefined. For example, after an initial optimization run, additional constraints or objectives might surface that were initially not considered. A substantial communication effort is required for the optimization specialist to collect all information before the "real" optimization can start.

A further problem is that in many projects the actual high-level target of the project is not well defined or not shared among team members. For example, the application team might be interested in seeing new (novel [8]) solutions as inspirations, while the optimization expert is focusing on improving objective values. It is therefore essential to clarify early on which part of the optimization results is used by whom.

Because the specification of the project objectives requires a decent amount of time, less time is available for tuning hyperparameters and benchmarking different algorithm candidates. As a result it seems that many practitioners only use a few standard algorithms out of the box.

4.2 Budget

Most optimization algorithms are developed and benchmarked by (groups of) students in university settings. In this specific situation, personnel costs are not a concern. In a business environment however, optimization experts need to be hired which can charge substantial fees, easily exceeding the monthly costs of university students in a single day.

Being able to quickly translate the business challenge into a well-defined optimization problem is essential otherwise the customer will run out of budget before the actual optimization has even started. This means that large-scale algorithmic exploration, especially if a manual supervision is required, is out of the question. Since most business applications have non-trivial fitness functions (expensive function evaluations), only few runs with few algorithms and standard hyperparameter settings can often be done.

As argued in [15], algorithm portfolio management is also costly. Even though increasingly more algorithms are available for download, ensuring that the software is bug-free and validating the results on relevant benchmarks costs a lot of time. With hundreds of new algorithms being proposed annually (see for example the list of implemented methods in [23]), keeping up-to-date with recent research developments is next to impossible.

In addition to the factor time also the computing budgets are limited. Academic research can often access high-performance computing hardware for free, while in industry expensive cloud services might have to be procured. This means that expert and budget availability in academia can be very different from industry, leading to different preferences especially for the exploration/exploitation trade-off.

Developers of new optimization algorithms should also consider the effort to understand, implement, and tune the algorithm. Marginal performance gains at the cost of massively increased algorithm complexity might need to be reconsidered.

4.3   Choices

Due to the large number of algorithms proposed by the academic community and the increasing number of hyper-parameters for more and more complex algorithms, many choices have to be made. These choices can be hard to motivate. Let's take as an example the choice of a (many-objective) optimization algorithm. Different motivations come to mind, from best to worst:

1. **Full comparison:** "All existing algorithms have been compared on the problem function and the best one chosen". As outlined above this is next to impossible.
2. **Only fitting algorithm:** In some cases there is only one algorithm which can handle the specific requirements of your project, but in general there should be many candidate algorithms.
3. **Newest algorithm:** "The latest algorithm was used". Again very difficult, as algorithms are proposed faster than they can be tested. Also newer algorithms are not guaranteed to be better (see [17]).
4. **Everybody uses this algorithm:** Not very likely, and there is no process to check. Most academic research is based on self-made algorithms, while applied work frustratingly often uses dubious metaphor-based algorithms like the Grey Wolf (see [22] for a critical analysis) (note that this algorithm has been cited more than 15,000 times as of September 2024).
5. **Benchmark winner:** The algorithm performed well on recent benchmark competitions. One of the best options, but the test functions are probably unrelated to the application problem and only a small subset of (order of 10 vs order of 100 or 1000) algorithms has been compared.
6. **Reputation:** "The algorithm was developed by a prestigious research group in a famous university". In practice, one of the best arguments (trust in an algorithm is taking from trust in experts and institutions).
7. **In-house development:** "We developed the algorithm ourselves and we believe it to be really well designed". Again this is replacing performance evaluation by trust in people.

8. **Tested for application:** The same algorithm has been used before for the same application (but often without a comparison to many algorithms). We found this often in application papers that used the same optimization algorithm that others used on the same domain before.
9. **Past projects:** The algorithm worked well in past projects and so we use it again. The obvious question is why the algorithm was chosen in the prior work.
10. **Marketing impact:** "It is quite famous! And featured in a Youtube video!". We hope that the minimal scientific merit of this statement does not require further explanation.
11. **Fatalism:** "The choice of algorithms does not matter anyways!" This is probably not true and would be very hard to digest leading to the question which problem the mass of publications on optimization algorithms is addressing.
12. **Easy to get:** "We could download it for free from the internet". This is unfortunately not so rare as a motivation, especially for users unaware of the number of alternative algorithms.

Without a good motivation for key algorithmic choices, decision makers might be hesitant to trust the results of the optimizer. If you think this is exaggerated, consider that multi-million Euro investments or human lives might depend on the output of the optimizer. Not being able to concisely argue for key algorithmic decisions can be a major obstacle. Unfortunately, even for articles in high-quality journals, the choice of algorithms for benchmarking is rarely motivated concisely.

Having well-established best practice solutions (standard algorithms to try first for an unknown application problem, accepted by the scientific community) would provide a stronger justification.

Obviously, if it is possible to cheaply and reliably assess the quality of an optimized solution and when reference solutions (like current best) are available, trust can be built directly from the evaluation and comparison to other methods. But in many cases, evaluating a solution might require building expensive prototypes (for example jet turbines) or is only feasible by building the final system (as for renewable energy systems).

4.4 Many-Objective Optimization and Performance Indicators

One of the hottest topics in evolutionary algorithms in the past decades is many-objective optimization (MAO). The potential to consider a larger set of objectives and analyze the trade-off between those objectives is very appealing. In our experience, however, customers' enthusiasm was limited. MAO, especially for more than two objectives, is hard to understand and selecting one of the Pareto set solutions more difficult than expected.

A special problem is performance indicators. In engineering projects numerical progress indicators are highly valued. For single objective optimization one can simply use the objective value (or some indicator derived from that). This is not possible for MAO, so special performance indicators like Hypervolume (HV) or IGD are used. For a computer scientist, these indicators are relatively easy to understand (but contain a number of non-obvious issues that might affect algorithmic performance, see e.g. [4]). For the application team and decision makers this is far more difficult and often outright confusing.

Let's imagine we want to optimize an engine for costs (in Euro) and CO_2 emissions (in g/km). Both objectives are easy to understand individually, but using HV in MAO would generate a unit-free number, say initially 0.02 and improving to 0.3 over the course of the optimization. Understanding this value requires multiple steps of thought. Firstly, one needs to understand that HV does not rate individual solutions but the quality of a population. Secondly, all values are relative to a (somewhat) arbitrary reference point. This leads to numbers that have no interpretability any more, there is no physical unit attached, and HV should increase even when we want to decrease the individual objectives. The numerical improvement in HV is also hard to interpret.

A potential remedy is to use desirabilities [24] instead of objective values. One could then report for example an increase in average desirability for all objectives over the entire population.

4.5   Maintenance

When finally a good solution has been found and accepted by the decision maker, a new question might arise: How to make the optimization process available to the customer on a continuous basis? For example customers might want to repeat the optimization once new data arrives or constraints and objectives change.

Again, one needs to consider the involved people. Which skills are available in the team, what complexity of software, algorithms and visualization can they handle? Using very sophisticated methods to derive optimal solutions only to discover that the customer is unable to maintain the software (assuming that at least some minor changes might be required), is ultimately a failure. Here again, discussing the project to the very end at the beginning is a key to success.

The Machine Learning (ML) community has come up with a variety of tools like MLOps [7], that formalize some of the mentioned challenges and provide supporting tools. With increasing contributions from ML like surrogate models this might become more relevant.

5   Recommendations

As stated initially, real world optimization projects are an inherently social activity. It is essential to clarify right from the beginning, who the involved people are. Especially, the decision maker can be opaque until quite late in the project. It is then necessary to clarify the skill level, goals, and interests of all involved people, identify required trade-off decisions (e.g., cost minimization vs. performance maximization) and propose an initial approach that aligns with all those constraints. There is very little research on the required cognitive skills for understanding different optimization approaches, for example single vs multi- or even many-objective optimization. It can be helpful to build a mock-up of the final optimization output (like a Pareto curve or a parallel coordinate plot) and check with the involved people, if this could be a useful project outcome. Simplifying the interpretation of results by for example using desirabilities should be discussed.

Another important topic is to clarify how optimization results can be validated. Is the objective function the real benchmark or just an approximative simulation? Can the results be measured independently (for example in a wind tunnel)? Maybe there is an expert who can judge the solutions derived from the optimizer based on years of practical, real-world experience to generate additional trust. A big drawback in evolutionary optimization is that in practice there are few guarantees that help to assess the current state of the optimization. Finding the optimal solution can't be guaranteed and even it is found, it can't be proven. Methods like MILP (Mixed Integer Linear Programming) provide theoretical guarantees and upper bounds, that help deciding on next steps. In evolutionary optimization, future progress is hard to estimate and it is unclear how much better solutions might get if additional resources are invested. For a customer, getting additional guardrails is very helpful.

For the academic community our suggestion is to invest more into understanding the psychological and cognitive aspects of optimization projects, working more closely with experts from these fields to improve transparency, explainability, and trust-related aspects. It would also be advisable from our experience to work more on the practical usability of optimization methods rather than excelling in benchmarks. Methods that are intuitive, with few elements and hyper-parameters, and robust (good performance over many different problems rather than top performance in some) are of high practical relevance.

Finally, a community effort to define a set of good first algorithms to try on any unknown problem would help to increase trust.

Acknowledgments. The authors would like to thank the anonymous reviewers for invaluable feedback. Summarizing a variety of observations and ideas into a short conference paper turned out to be much more difficult than (we) anticipated. Based on the reviews we tried to clarify our line of argumentation, but want to apologize for any remaining hardships when reading this article.

Disclosure of Interests. The authors have no competing interests to declare that are relevant to the content of this article.

References

1. Büscher, C., Sumpf, P.: "Trust" and "confidence" as socio-technical problems in the transformation of energy systems. Energy Sustain. Soc. **5**, 1–13 (2015)
2. Carver, C.S., Scheier, M.F.: On the structure of behavioral self-regulation. In: Handbook of Self-regulation, pp. 41–84. Elsevier, Amsterdam (2000)
3. Goel, V.: Comparison of well-structured & ill-structured task environments and problem spaces. In: Proceedings of the Fourteenth Annual Conference of the Cognitive Science Society (1992)

4. Guerreiro, A.P., Fonseca, C.M., Paquete, L.: The hypervolume indicator: computational problems and algorithms. ACM Comput. Surv. (CSUR) **54**(6), 1–42 (2021)
5. Johnson, J.G., Busemeyer, J.R.: Decision making under risk and uncertainty. Wiley Interdisc. Rev. Cogn. Sci. **1**(5), 736–749 (2010)
6. Kahneman, D., Tversky, A.: Prospect theory: an analysis of decision under risk. In: Handbook of the Fundamentals of Financial Decision Making, chap. 6, pp. 99–127. World Scientific (2013). https://doi.org/10.1142/9789814417358_0006
7. Kreuzberger, D., Kühl, N., Hirschl, S.: Machine learning operations (MLOps): overview, definition, and architecture. IEEE Access **11**, 31866–31879 (2023)
8. Lehman, J., Stanley, K.O.: Abandoning objectives: evolution through the search for novelty alone. Evol. Comput. **19**(2), 189–223 (2011)
9. Liu, Q., Lanfermann, F., Rodemann, T., Olhofer, M., Jin, Y.: Surrogate-assisted many-objective optimization of building energy management. IEEE Comput. Intell. Mag. **18**(4), 14–28 (2023)
10. Lynch, C., Ashley, K.D., Pinkwart, N., Aleven, V.: Concepts, structures, and goals: redefining ill-definedness. Int. J. Artif. Intell. Educ. **19**, 253–266 (2009)
11. Menzel, S., Olhofer, M., Sendhoff, B.: Application of free form deformation techniques in evolutionary design optimisation. In: Proceedings of 6th World Congress on Structural and Multidisciplinary Optimization, Rio de Janeiro, Brazil (2005)
12. Miller, G.A.: The magical number seven, plus or minus two: some limits on our capacity for processing information. Psychol. Rev. **63**(2), 81 (1956)
13. Pan, L., He, C., Tian, Y., Wang, H., Zhang, X., Jin, Y.: A classification-based surrogate-assisted evolutionary algorithm for expensive many-objective optimization. IEEE Trans. Evol. Comput. **23**(1), 74–88 (2019). https://doi.org/10.1109/TEVC.2018.2802784
14. Quesada, J., Kintsch, W., Gomez, E.: Complex problem-solving: a field in search of a definition? Theor. Issues Ergon. Sci. **6**(1), 5–33 (2005). https://doi.org/10.1080/14639220512331311553
15. Rodemann, T.: Industrial portfolio management for many-objective optimization algorithms. In: 2018 IEEE Congress on Evolutionary Computation (CEC) (2018). https://doi.org/10.1109/CEC.2018.8477693
16. Rodemann, T.: A many-objective configuration optimization for building energy management. In: IEEE (ed.) WCCI 2018 Conference. IEEE (2018)
17. Rodemann, T.: A comparison of different many-objective optimization algorithms for energy system optimization. In: Kaufmann, P., Castillo, P.A. (eds.) EvoApplications 2019. LNCS, vol. 11454, pp. 3–18. Springer, Cham (2019). https://doi.org/10.1007/978-3-030-16692-2_1
18. Rodemann, T., Attig, C.: How can digital twins help to accelerate the transition to a carbon-neutral energy system? In: Lecture Notes in Computer Science, LNCS, LNAI, LNBI. Springer, Cham (2024)
19. Samuelson, W., Zeckhauser, R.: Status quo bias in decision making. J. Risk Uncertain. **1**, 7–59 (1988)
20. Slovic, P., Finucane, M.L., Peters, E., MacGregor, D.G.: Risk as analysis and risk as feelings: some thoughts about affect, reason, risk and rationality. In: The Feeling of Risk, pp. 21–36. Routledge, London (2013)
21. Slovic, P., Peters, E., Finucane, M.L., MacGregor, D.G.: Affect, risk, and decision making. Health Psychol. **24**(4, Suppl), S35–S40 (2005). https://doi.org/10.1037/0278-6133.24.4.S35
22. Sörensen, K.: Metaheuristics–the metaphor exposed. Int. Trans. Oper. Res. **22**(1), 3–18 (2015)

23. Tian, Y., Cheng, R., Zhang, X., Jin, Y.: PlatEMO: a MATLAB platform for evolutionary multi-objective optimization [educational forum]. IEEE Comput. Intell. Mag. **12**(4), 73–87 (2017). https://doi.org/10.1109/MCI.2017.2742868

24. Wagner, T., Trautmann, H.: Integration of preferences in hypervolume-based multiobjective evolutionary algorithms by means of desirability functions. IEEE Trans. Evol. Comput. **14**(5), 688–701 (2010)

The Importance of Being Earnest: Multiple Heterogeneous Container Loading with a Simple Genetic Algorithm

Francesco Rusin[1]([⊠])(ID), Jan Fiala[2](ID), Julian Sanker[3], Suyesh Bhattarai[4], and Anikó Ekárt[5](ID)

[1] Department of Engineering and Architecture, University of Trieste, Trieste, Italy
francesco.rusin2@phd.units.it
[2] Institute of Automation and Computer Science, Brno University of Technology, Brno, Czechia
jan.fiala6@vut.cz
[3] Yale College, Yale University, New Haven, USA
julian@sankergroup.org
[4] Mechatherm International Limited, Cradley Heath, UK
suyesh.bhattarai@mechatherm.co.uk
[5] Aston Centre for Artificial Intelligence Research and Application, Aston University, Birmingham, UK
a.ekart@aston.ac.uk

Abstract. In this study we address the complex practical problem of multiple heterogeneous container loading with a simple genetic algorithm. We demonstrate that with a well-chosen representation including a heuristic and a suitable fitness function the other aspects of the genetic algorithm do not need extensive work for good results. Following systematic study of our method on synthetically generated data, we visually showcase the solution for a company-based problem instance.

Keywords: multiple heterogeneous container loading · genetic algorithm · heuristic

1 Introduction

Containerised transport of goods is an area of continuing interest for all modes of transport, road, maritime, rail, and air. According to the Pitney Bowes Global Parcel Shipping Index [15] in 2022, 161 billion parcels were shipped worldwide. They predict that the global volume will reach 225 billion parcels by 2028. Efficiency is key for both container space utilisation and loading of containers. Optimising container space utilisation is essential not only from a business and cost perspective, but also environmentally, as the increasing number of transport vehicles will ultimately lead to increased CO_2 emissions and thus have a negative environmental effect.

© The Author(s), under exclusive license to Springer Nature Switzerland AG 2025
P. García-Sánchez et al. (Eds.): EvoApplications 2025, LNCS 15612, pp. 482–495, 2025.
https://doi.org/10.1007/978-3-031-90062-4_30

The container loading problem is a complex real world problem characterised by many constraints [2] that can be generic, such as stability and distribution of weight or specific to transport companies, such as container types, sizes and loading options (via a side door for closed containers or the top for open top containers) or the cargo itself, such as non-rotatable, non-stackable, or fragile. The optimisation problem could be stated as maximising the volume or value of goods that can be transported with a single container or as minimising the number of containers required to transport all the goods (assuming availability of an unlimited number of containers) [10]. In the literature, most attention has been dedicated to single container loading [8]. Multiple container loading is considered less frequently and in many cases only for one container size [3,6].

We took inspiration from the practical case of Mechatherm International Limited that needs to transport parts that then need to be assembled at destination. A variety of containers are available, of mainly two sizes and also as closed or open top (in case there are items with one dimension exceeding the dimensions of the container).

2   State of the Art

The container loading problem (CLP) belongs to the same family as the stock cutting problem [17]. On the abstract level they are both concerned with fitting cuboids optimally into larger cuboids (fitting boxes into containers or cutting out required sizes from stock sizes). Their objectives are typically to minimise empty space or wasted stock material, respectively. They both received considerable attention in the operations research literature over the past 60 years, including a dedicated special issue [12]. For exact solutions of several types the reader is referred to the work of Kurpel et al. [10].

The CLP is not only NP-hard, but presents additional challenges due to the practical constraints. In order for the solutions to be useful for companies, their specific constraints need to be met [8]. Commercial software solutions such as easyCargo or pier2pier [5,13] do not allow personalised constraints.

Existing methods often do not concern themselves with the packing order, so in practice the found solutions may not be applicable directly in practice. As the items would need to be loaded via the door of the container and in the appropriate orientation, if there is not sufficient space to move an item past some already loaded items, those would need to be unloaded before this item could be loaded. Simulation software to visualise the loading order and the dynamics exist and these can help companies plan [11].

Various heuristic algorithms have been devised for the single CLP [14], and some for multiple CLP, with one type of container.

Genetic algorithms have been successfully employed to solve various cases of the single CLP, taking into consideration orientation, weight, stability and balance constraints, using tower sets [9], or layering [4,16]. Hybridisation of the genetic algorithm with complex data structures and complex coding-decoding mechanisms are often applied [4,7]. The multiple heterogeneous CLP to our knowledge has not been addressed by genetic algorithms.

3   Problem Definition

Assuming an unlimited supply of containers of specified types and given a set of cargo items, we define the multiple CLP as the problem of finding the ordered list of containers, together with the sequence of loading, orientation and positioning of all cargo items on these containers, so that the space utilisation on the containers is maximised and the weight and stability constraints are satisfied.

For the multiple heterogeneous CLP, we consider two types of containers that are cuboids with fixed dimensions, the so-called *20ft* and *40ft* containers named according to their lengths in the industry. In practice, they both have a base version (loaded through a door) and an open top one, with slightly different dimensions. For simplicity, for each type we consider the most restricting measures between the base and the open top versions. For packing purposes, only the internal dimensions are relevant and these are:

- the *20ft container*: $L = 5.9\,\text{m}$, $H = 2.39\,\text{m}$, $W = 2.35\,\text{m}$, and a maximum load capacity of $W = 28\,250\,\text{kg}$
- the *40ft container*, with $L = 12.04\,\text{m}$, $H = 2.39\,\text{m}$, $W = 2.35\,\text{m}$, and $W = 26\,760\,\text{kg}$

We consider cargo items with a cuboid shape and homogeneous density, which we will refer to as boxes. These will be defined by their height (h), width (w), length (l) and weight (w).

4   Candidate Solution Representation

A candidate solution will need to specify for each box its precise container placement, together with its orientation. The position of a box in the chosen container can be represented as a three-dimensional coordinates of its center of mass with respect to the system whose origin is on the side opposite to the entrance of the container, in the bottom-left corner. As common in both the literature and practice, we restrict placement of boxes orthogonally with respect to the walls of the container, which results in 6 possible orientations.

The ultimate goal is to produce a sequence of boxes for loading, however to have as simple a method as possible, we will focus on the feasible placement of boxes onto the containers (so all constraints are satisfied), acknowledging that the order produced by our algorithm may need to be post-processed. In the absence of any loading order constraints or cost information, we assume that all loading orders are feasible from the perspective of the boxes and the quality of solutions is not dependent on the order itself.

We do not consider the fitting of boxes through the door, but assume that an open top container will be employed in such a case. The container cost is not included in the optimization.

A candidate solution will need to include two parts:

- the loading of boxes onto containers: each box i is represented as $(c_i, x_i, y_i, z_i, o_i)$, $c_i \in \mathbb{N}$ is the number of the container the box is placed in (0 for the first one, 1 for the second one, and so on), $x_i, y_i, z_i \in \mathbb{R}^+$ are the coordinates of the center of mass, and $o_i \in \{0, 1, 2, 3, 4, 5\}$ represents the orientation and
- the order of containers: a bitstring b, where $b[i]$ is *true* if the ith container is a 40 ft container, *false* otherwise; b can be chosen of any length.

While this definition of a solution only describes the positions and orientations of the boxes inside the container, to actually find a feasible solution we split the representation into three parts:

1. the order of the boxes,
2. the type of ordered containers, and
3. the actual positioning of the boxes given the order.

We represent the order of the boxes as a permutation $p \in S_n$, where n is the number of boxes. The types of containers to be considered are represented with a bitstring b of length k, where the i-th container is 40 ft if and only if $b[i \bmod k]$ is *true*, and it is 20 ft otherwise. Finally, for a given ordering of the boxes and a given container order, the actual solution is produced via a heuristic, which we describe in Sect. 4.1. Thus, the genotype is the concatenation of permutation p and bitstring b, and the corresponding phenotype is obtained by applying the heuristic.

4.1 Heuristic

Algorithm 1 provides the pseudocode description of the heuristic. As trying each possible combination is too computationally expensive, we simplify the computations by employing the concept of "spaces": an empty container is a space and each box must be placed inside a space; however, when a box is placed in a space, said space is removed from the computations and the remaining volume is divided into three new spaces. We start from a single empty container and each box is subsequently placed into the smallest space it can fit into; if a box cannot fit into any space or all the spaces it could, would make their container exceed its maximum weight, we open a new container, whose type is dictated by the bitstring, and place the box there. Figure 1 shows a visual representation of the first steps of the heuristic. Each time we place a box inside a space, we make it adhere to the bottom left corner far away from the door. The solution would be identical if any other corner was chosen consistently.

4.2 Visualisation

Visual representation is crucial to provide an intuitive understanding of box placement, space utilization and adherence to dimensional boundaries. Our algorithm generates positioning of boxes with a starting position at the bottom-left corner of the container. This approach simplifies the association of containers

Algorithm 1. Heuristic to place the boxes (given as the ordered array *Boxes*), where the ordered array *Containers* represents the container order

```
 1: procedure HEURISTIC(Containers, Boxes)
 2:     Spaces ← [0 : [Containers[0]]] ▷ hashmap from the containers into their spaces
 3:     for box in Boxes do
 4:         min_space ← null
 5:         c ← −1
 6:         while min_space == null and c<size(Spaces)-1 do
 7:             c ← c + 1
 8:             min_space ← smallest space in Spaces[c] that can contain box
 9:         end while
10:         if min_space == null then
11:             c ← size(Spaces)
12:             Spaces[c] ← [Containers[c]]
13:             min_space ← Spaces[c][0]
14:         end if
15:         place box in min_space in the bottom left corner
16:         Spaces[c].remove(min_space)
17:         generate S1, S2, and S3 by cutting min_space along the axes
18:         for S in [S1, S2, S3] do
19:             if S is not too small then
20:                 Spaces[c].add(S)
21:             end if
22:         end for
23:     end for
24: end procedure
```

with different sizes and box orientations. For each solution, the 3D layout displays containers aligned side-by-side. Different colours distinguish container walls from boxes, and varying transparency levels indicate overlapping or nested boxes (this helps visually distinguish infeasible solutions, as boxes cannot overlap).

5   Evolutionary Algorithm

As the heuristic that we employ is deterministic, the combination of a box ordering and a container bitstring uniquely identifies a solution.

For solving the multiple heterogeneous CLP, we use a typical genetic algorithm:

1. randomly generate n_{pop} possible solutions
2. generate n_{pop} offspring from selected parents through crossover and mutation
3. from the parent and offspring populations select n_{pop} individuals to survive to the next generation
4. repeat 2 and 3 until n_{evals} solutions have been generated
5. return the best performing solution in the current population.

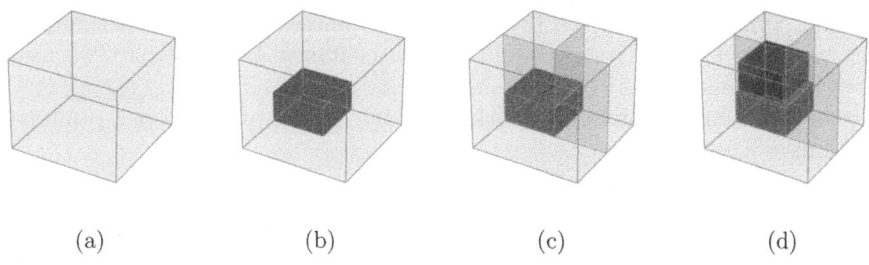

(a)   (b)   (c)   (d)

Fig. 1. A sketch of the heuristic working: starting from an empty container (a), a box is placed in the bottom left far away corner (b); each time a box is placed, the remaining space is sliced in three directions orthogonally to the sides of the box (c), then the following box is placed in the smallest space that it can fit (d), and the process is repeated until all boxes are placed. If a box cannot fit into any space, a new container is opened and the box is placed there.

For our study, we define a suitable crossover operator, two mutation operators, two fitness functions and two survivor selection methods. We summarise the parameters in Table 1.

Table 1. The genetic algorithm parameters.

Parameter	Value
Elitism	yes
Parent selection	tournament
Survivor selection	$\mu + \lambda$, generational (μ, λ)
Crossover	order and two-point
Mutation	(basic swap, roulette) and bit flip
Population size	100
Number of evaluations	10000
Tournament size	5
Mutation rate	20%, 50%, 80%

Initialisation. The randomly generated solutions will all be feasible due to the choice of representation. Any permutation p in the first part will correspond to a feasible order of boxes, including all boxes. The bitstring b of fixed length k can represent an ordered list of boxes of any length i through the application of the *mod* operator.

Crossover and Mutation. We define our crossover operator to be composed of two parts, corresponding to the two parts of the candidate solutions representing the box permutation and the bitstring of containers: for the box permutation part we use an order crossover and for the container bitstring part we use a two-point crossover.

Our mutation operators also work separately on the two parts of a candidate solution:

– for the permutation part, they swap two randomly chosen elements. To increase exploration, the mutations have a probability $p_p = 0.1$ to perform another random swap after the first one, and if this is successful, they keep performing additional swaps with the same probability until one swap fails. The difference between the two mutations lies in the way they select the elements to swap: "base mutation" uses uniform random selection, while "roulette mutation" takes into consideration the current stage of the evolutionary algorithm and applies tuning; Algorithm 2 shows the pseudocode for roulette mutation.

– in the bitstring part, bit flip mutation is applied, with a probability $p_b = 0.1$ of switching each bit's value; in the case of bitstrings of length 10, on average a single bit flips.

Algorithm 2. Roulette mutation at the s-th generation of an evolutionary algorithm that runs for a total of n_{gens} generations on n_{boxes} boxes identified by their number from 0 to $n_{\text{boxes}} - 1$; Table 1 shows the values of the evolutionary parameters in our experiments, while Table 2 shows the parameters related to the problems we tackled

---

```
1: function ROULETTE(n_boxes, s, n_gens)
2:    P ← {}          ▷ hashmap from each number to its probability (not normalized)
3:    for i in range(n_boxes) do
4:        P[i] ← min(i, ⌊n_boxes (s/n_gens)⌋)   ▷ the larger s is, the less boxes are uniform,
      with the last ones getting bigger P[i]
5:    end for
6:    rn ← U(1, ∑_{i=0}^{n_boxes} P[i])
7:    sum ← 0                        ▷ variable to identify the interval rn falls into
8:    box ← 0                        ▷ variable to identify the correct box number
9:    while sum < rn do
10:       box + +
11:       sum+ = P[box]
12:   end while
13:   return box
14: end function
```

Both crossover and mutation will always generate feasible candidate solutions.

Fitness Function. The goal is to load all the boxes with the best utilisation of space on the containers, which is equivalent to using as few containers as possible (in our case a $40ft$ container is double the size of a $20ft$ container). Therefore measuring the amount of "air transportation" as the unused space on the containers is a suitable measure. In the base case, which we call "empty space sum", given the volumes $v_0, \ldots, v_k$ we compute the "empty space sum" as

$$ESS = \sum_{j=0}^{k} v_j.$$

This measure accurately describes how much space is wasted; however, given that all boxes are included in every solution, any solution using the same containers will share the same fitness. We define an alternative fitness function that encourages tighter loading of earlier containers by gradually decreasing the contribution of empty space on later containers:

$$ESD = \sum_{j=0}^{k} \frac{v_j}{j+1}$$

which we will refer to as "empty space decsum". This can result in the last containers becoming empty, therefore leading to less containers needed overall.

Survivor Selection. We employ two survivor selection methods:

- $\mu + \lambda$: the parent and offspring populations (of equal size) are merged and the top half are kept; as the ESS fitness evaluation leads to many candidate solutions with equal fitness, to increase exploration we prioritize offspring in the case of ties;
- μ, λ or generational: only the offspring and the best performing individual (the elite) of the parent population survive.

6   Experiments

Before applying the genetic algorithms on the real data provided by the company, we wanted to analyse their performance on synthetic data of increasing complexity. We generated synthetic data with container dimension usage ranging from 70% to 100%, with a mix of $40ft$ and $20ft$ containers. Section 6.1 describes our synthetic data generation method and Fig. 2 provides some examples. We compared all combinations of fitness measures ESS and ESD, base and roulette mutation operators, $\mu + \lambda$ and μ, λ survivor selection. Applying the central limit theorem, we performed each experiment 30 times.

6.1   Synthetic Data Generation

Generating synthetic data simulating boxes with diverse dimensions and weights together with containers filled to varying degrees was essential for the tuning of

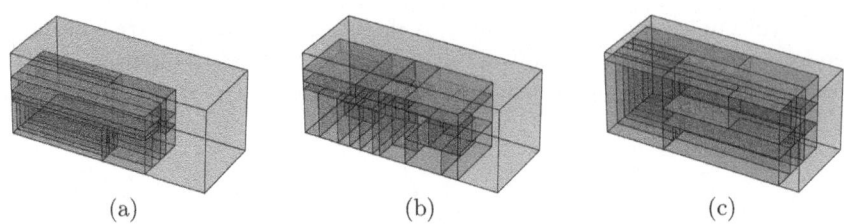

(a)   (b)   (c)

Fig. 2. Visualisation of cargo loading at different capacity levels: (a) 70%, (b) 80%, and (c) 90%.

the algorithm and for ensuring its robustness and suitability for application on real world data, where the optimum is not known. These boxes are created by dividing the container space using guillotine cuts, adhering to specified parameters such as split ratios and minimum dimensions. The generated data can be visualised, so that the containers and their designated load is viewed from any desired viewpoint.

The algorithm shown as Algorithm 3 follows a simple rule, whereby starting from the entire container space to be filled as the initial box, the largest box is iteratively split into two smaller boxes until the target number of boxes is reached (lines 1–9). Splits are performed along one of the three axes (length, width, height), ensuring that no dimension of any resulting box falls below a pre-specified minimum size. At each split, the original box is replaced with these new boxes in the collection. The process continues until the target number of boxes is achieved (lines 10–18). In Table 2 we summarise the parameters and the resulting box ranges.

Table 2. Synthetic data generation parameters and examples of resulting box size ranges. The data will be made available upon request.

Feature	Proportion of container space used			
	70%	80%	90%	100%
Number of containers	4	4	4	4
Container types	40 ft/20 ft	40 ft/20 ft	40 ft/20 ft	40 ft/20 ft
Total number of boxes	80	80	80	80
Weight per container (kg)	26760	26760	26760	26760
Initial box length (mm)	8428/4130	9632/4720	10836/5310	12040/5900
Initial box width (mm)	1645	1880	2115	2350
Initial box height (mm)	1673	1912	2151	2390
Box length range (mm)	307–5190	261–9632	495–10836	557–12040
Box width range (mm)	177–1645	287–1880	98–1386	161–1425
Box height range (mm)	125–1673	128–1287	255–2151	226–2390

Algorithm 3. Algorithm to generate boxes by guillotine splitting

```
 1: procedure GENERATEBOXES(Container, Target)
 2:     Boxes ← {Container}
 3:     while number of Boxes < Target do
 4:         box ← select largest box in Boxes
 5:         SPLITBOX(box, Boxes)
 6:     end while
 7:     return Boxes
 8: end procedure
 9: procedure SPLITBOX(box, Boxes)
10:     axis ← random axis (Length, Width, or Height)
11:     if box can be split along axis then
12:         location ← random % in given range
13:         [box1, box2] ← split box along axis at location
14:         Boxes.remove(box)
15:         Boxes.add(box1)
16:         Boxes.add(box2)
17:     end if
18: end procedure
```

7   Results

We found that on average ESD leads to less containers being used, proving its superiority, and therefore we only show the detailed results for the ESD fitness function in Fig. 3. Table 3 shows the average number of extra containers used in the end-of-run best solutions compared to the number of containers in the generated data for both ESS and ESD, counted in $20ft$ containers. Negative values indicate better solutions by the genetic algorithm. It is noticeable that the genetic algorithms perform better in the cases with lower capacity utilisation. The choice of mutation or survivor selection does not make a difference to the overall result. We believe that largely due to the effectiveness of the heuristic, a simple genetic algorithm is capable of solving the complex multiple CLP with multiple container sizes. Additional constraints or more sophisticated fitness functions can be incorporated easily, for example to account for cost models, where the cost of a $40ft$ container is different from the double of a $20ft$ container or for boxes that cannot be rotated.

We additionally compared our results with random search, using the Mann-Whitney U rank test with the null hypothesis of equality of the median, with $\alpha = 0.05$. The test confirmed that, with the exception of 20% mutation for the 70% volume case, the difference is significant.

Example solutions to the synthetically generated problems found by our genetic algorithms are visually displayed in Fig. 4. The improvement in the case of lower space utilisation in the synthetic data is visible for 70%. Following the results on synthetic data, the results on the dataset provided by the company are showcased in Fig. 5. Once again, ESD is superior. In this case the ground truth solution was not known in advance, therefore it is essential to display the

Table 3. Average number of extra $20ft$ containers in the end-of-run best solutions with respect to the original, with the ESS fitness function at the top and ESD at the bottom. "B" refers to "base" and "R" to "roulette".

		70%			80%			90%			100%		
		20%	50%	80%	20%	50%	80%	20%	50%	80%	20%	50%	80%
B	$\mu+\lambda$	−0.80	−0.77	−0.80	−0.07	−0.07	0.00	1.53	1.57	1.43	3.43	3.33	3.40
B	μ,λ	−0.77	−0.77	−0.80	−0.10	−0.03	−0.10	1.47	1.40	1.43	3.30	3.27	3.33
R	$\mu+\lambda$	−0.77	−0.73	−0.83	−0.10	−0.10	−0.10	1.47	1.50	1.43	3.17	3.17	3.20
R	μ,λ	−0.73	−0.77	−0.77	0.13	0.13	0.00	1.57	1.53	1.67	3.33	3.50	3.37
B	$\mu+\lambda$	−0.90	−0.90	−0.77	−0.30	−0.23	0.03	1.27	1.37	1.37	3.23	3.23	3.07
B	μ,λ	−0.80	−0.90	−0.80	−0.20	−0.27	−0.10	1.30	1.43	1.37	3.50	3.23	3.33
R	$\mu+\lambda$	−0.97	−1.00	−0.77	−0.30	−0.33	0.00	1.23	1.23	1.27	2.90	2.70	3.10
R	μ,λ	−0.93	−1.00	−0.83	−0.37	−0.33	−0.10	1.27	1.17	1.23	2.87	3.00	2.97

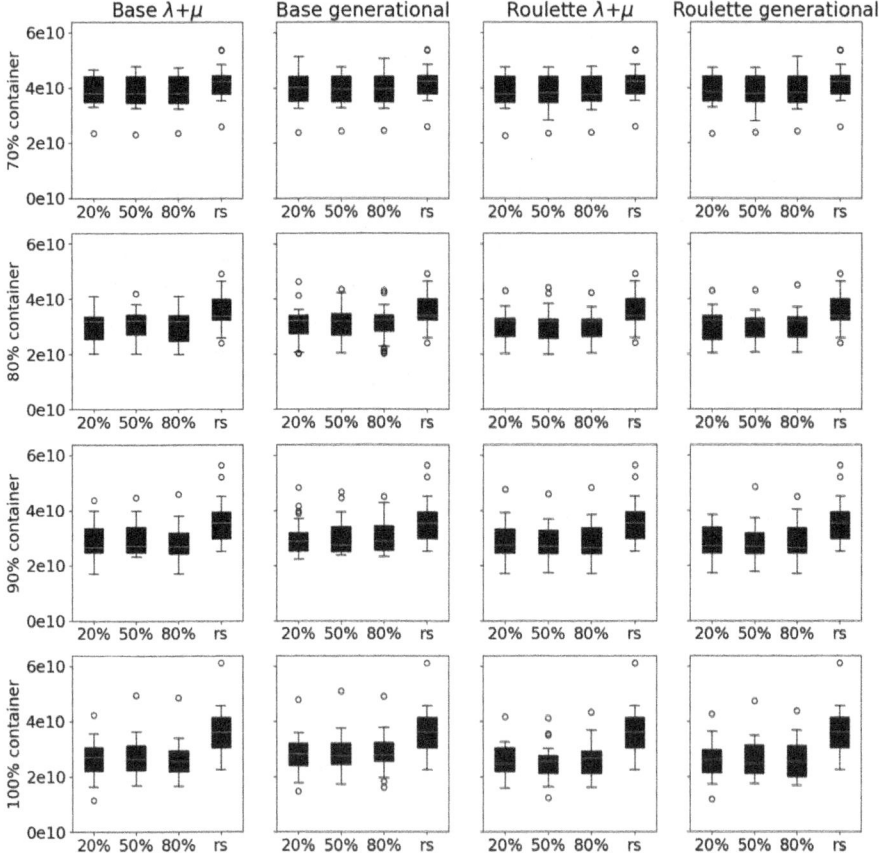

Fig. 3. Numerical results with the ESD fitness function.

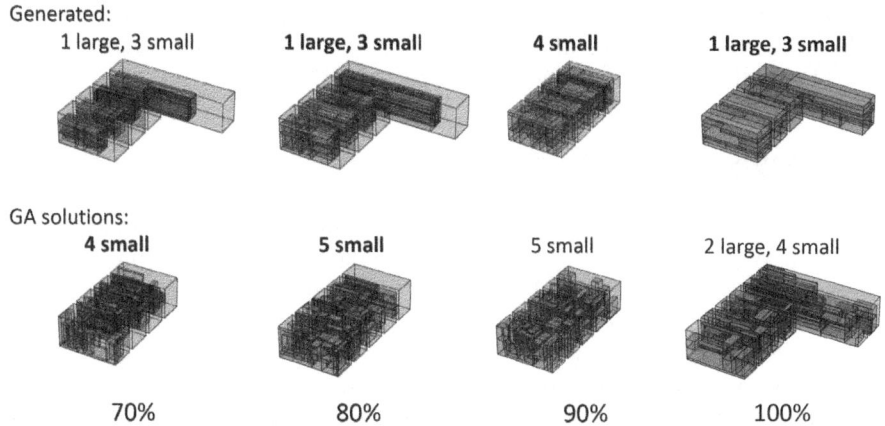

Fig. 4. Example generated containers and solutions found by the genetic algorithm for all four cases of 70%, 80%, 90%, and 100% capacity. The GA solution is better for 70%, equivalent for 80% and worse for 90% and 100%, considering one large container equivalent to two small containers.

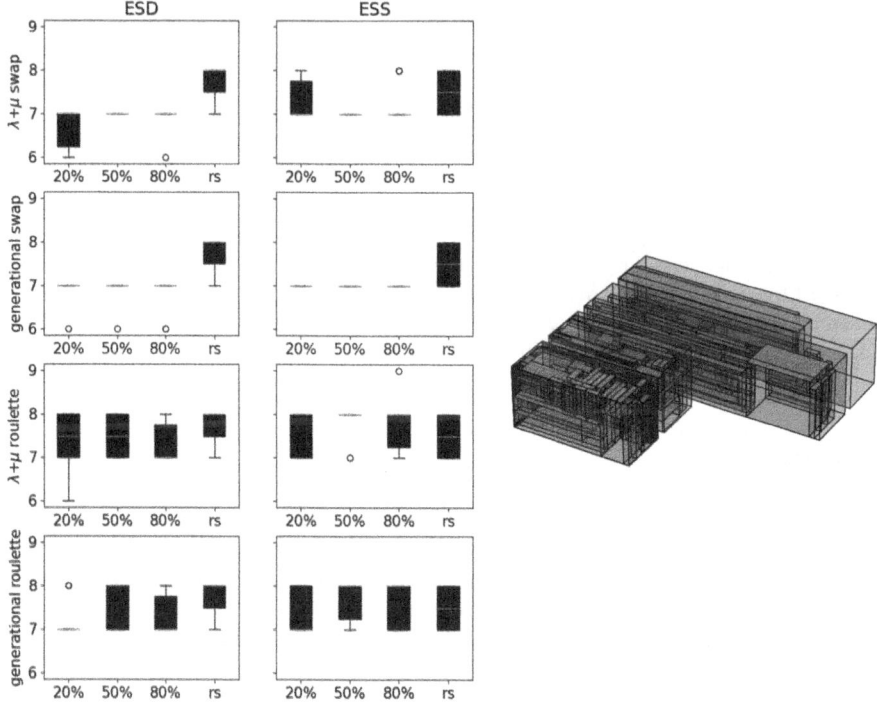

Fig. 5. Genetic algorithm results for the real data, reported as equivalent number of small containers (left). Example best solution found by the genetic algorithm for the real data, with two large and two small containers. (right).

solutions visually. By visual inspection we can establish that it is unlikely that a solution with better space utilisation would exist, in particular given the long boxes that can only fit into large containers.

8 Conclusions

In this study we addressed the multiple heterogeneous CLP with a simple genetic algorithm. We demonstrated how, with a suitable candidate solution representation, including a heuristic and a suitable fitness function, genetic algorithms offer a robust solution to this complex problem. In particular, if the problem involves a lot of 'transportation of air' (N.B. the 70% container dimension usage means 34.3% volume utilisation and the 80% container dimension usage means 51.2% volume utilisation), the simple genetic algorithm can find equally good or better solutions, even with modest resources (10000 evaluations). Following establishing the usefulness of the method on synthetic data, we applied it on a real dataset with good outcome that could be visually tested.

We are planning to continue our collaboration with the company and expand our method to handle additional constraints, such as requirements for subsets of boxes to be in the same container, as well as more diverse containers. We will introduce different objectives for the optimisation in line with company priorities and address performance improvement through increasing the resources.

Acknowledgements. The authors are grateful to Mechatherm International Limited for providing the practical problem and real data. This joint work and publication would not have been possible without the support of the Society for the Promotion of Evolutionary Computation in Europe and its Surroundings (SPECIES). This paper is the result of the participation of F. Rusin, J. Fiala, and J. Sanker as students at the SPECIES summer school in 2024 and their working together as a team on the challenge presented by A. Ekárt as mentor, with her supervision, and practical industry context and advice provided by S. Bhattarai. Jan Fiala was supported by project GACR No.24-12474S 'Benchmarking derivative-free global optimisation methods.'

References

1. Araya, I., Guerrero, K., Nuñez, E.: VCS: a new heuristic function for selecting boxes in the single container loading problem. Comput. Oper. Res. **82**, 27–35 (2017)
2. Bischoff, E.E., Ratcliff, M.: Issues in the development of approaches to container loading. Omega **23**(4), 377–390 (1995)
3. Bischoff, E., Ratcliff, M.: Loading multiple pallets. J. Oper. Res. Soc. **46**(11), 1322–1336 (1995)
4. Bortfeldt, A., Gehring, H.: A hybrid genetic algorithm for the container loading problem. Eur. J. Oper. Res. **131**(1), 143–161 (2001)
5. EasyCargo. https://www.easycargo3d.com/. Accessed Nov 2024
6. Eley, M.: Solving container loading problems by block arrangement. Eur. J. Oper. Res. **141**(2), 393–409 (2002)

7. Falkenauer, E.: A hybrid grouping genetic algorithm for bin packing. J. Heurist. **2**, 5–30 (1996)
8. Gajda, M., Trivella, A., Mansini, R., Pisinger, D.: An optimization approach for a complex real-life container loading problem. Omega **107**, 102559 (2022)
9. Gehring, H., et al.: A genetic algorithm for solving the container loading problem. Int. Trans. Oper. Res. **4**(5–6), 401–418 (1997)
10. Kurpel, D.V., Scarpin, C.T., Junior, J., Schenekemberg, C.M., Coelho, L.C.: The exact solutions of several types of container loading problems. Eur. J. Oper. Res. **284**(1), 87–107 (2020)
11. Martínez-Franco, J.C., Álvarez-Martínez, D.: Physx as a middleware for dynamic simulations in the container loading problem. In: 2018 Winter Simulation Conference (WSC), pp. 2933–2940. IEEE (2018)
12. Morabito, R., Arenales, M.N., Yanasse, H.H.: Special issue on cutting, packing and related problems. Int. Trans. Oper. Res. **16**(6), 659–659 (2009)
13. PIER2PIER. https://www.pier2pier.com/. Accessed Nov 2024
14. Pisinger, D.: Heuristics for the container loading problem. Eur. J. Oper. Res. **141**(2), 382–392 (2002)
15. PitneyBowes: Parcel shipping index (2023). https://www.pitneybowes.com/content/dam/pitneybowes/us/en/shipping-index/23-mktc-03596-2023_global_parcel_shipping_index_ebook-web.pdf. Accessed Nov 2024
16. Ramos, A.G., Oliveira, J.F., Gonçalves, J.F., Lopes, M.P.: A container loading algorithm with static mechanical equilibrium stability constraints. Transp. Res. Part B: Methodol. **91**, 565–581 (2016)
17. Wäscher, G., Haußner, H., Schumann, H.: An improved typology of cutting and packing problems. Eur. J. Oper. Res. **183**(3), 1109–1130 (2007)

Emergent Kin Selection of Altruistic Feeding via Non-episodic Neuroevolution

Max Taylor-Davies[1]([✉]), Gautier Hamon[2], Timothé Boulet[2], and Clément Moulin-Frier[2]

[1] School of Informatics, University of Edinburgh, Edinburgh, Scotland
m.taylor-davies@sms.ed.ac.uk
[2] Flowers team, Inria Center of the University of Bordeaux, Talence, France

Abstract. Kin selection theory has proven to be a popular and widely accepted account of how altruistic behaviour can evolve under natural selection. Hamilton's rule, first published in 1964, has since been experimentally validated across a range of different species and social behaviours. In contrast to this large body of work in natural populations, however, there has been relatively little study of kin selection *in silico*. In the current work, we offer what is to our knowledge the first demonstration of kin selection emerging naturally within a population of agents undergoing continuous neuroevolution. Specifically, we find that zero-sum transfer of resources from parents to their infant offspring evolves through kin selection in environments where it is hard for offspring to survive alone. In an additional experiment, we show that kin selection in our simulations relies on a combination of kin recognition and population viscosity. We believe that our work may contribute to the understanding of kin selection in minimal evolutionary systems, without explicit notions of genes and fitness maximisation.

Keywords: Kin selection · Neuroevolution · Multi-agent systems · Artificial ecosystems

1 Introduction

At first glance, it seems difficult to square the phenomenon of purely altruistic behaviour (acts which confer a benefit to the recipient at a cost to the actor) with the basic principle of natural selection: how can a gene be selected for when it decreases, rather than increases, the fitness of its host? One plausible account can be made through the theory of *inclusive fitness*. Key to this theory is the recognition that individual organisms within a shared social environment are not isolated from one another in terms of fitness, as the behaviour of any one individual can directly impact the survival of others. Individuals can therefore influence the propagation of their genes to future generations not only via their own reproductive success, but also via the reproductive success of others that share those genes. Whether a given gene is selected for is thus determined by

© The Author(s), under exclusive license to Springer Nature Switzerland AG 2025
P. García-Sánchez et al. (Eds.): EvoApplications 2025, LNCS 15612, pp. 496–509, 2025.
https://doi.org/10.1007/978-3-031-90062-4_31

its effect(s) on the fitness of any bearers of copies of that gene. Under this view, we can think of an altruistic act as an exchange of fitness from one agent to another. If the exchange is positive-sum and both sides are bearers of the gene in question, then from the gene's perspective the behaviour confers a fitness benefit–even while it decreases the fitness of the acting individual.

Hamilton's rule, published in 1964, offers a simple and elegant formalisation of this idea [8,9]. Imagine a particular gene produces some social behaviour which confers fitness benefit b to the recipient at a cost c to the actor. Given a measure r of the genetic relatedness between the two, the rule states that this gene should increase in frequency if $rb > c$. A key consequence of Hamilton's rule is that an altruistic behaviour that targets the actor's kin will undergo more positive selection than an equivalent behaviour which is indiscriminate or favours non-kin. This is referred to as 'kin selection' [16], and can be seen as a more narrow form of inclusive fitness. Targeting of behaviour towards kin can be facilitated either by mechanisms of kin recognition, or more simply by population viscosity (the tendency of individuals to remain close to their place of birth). Since the publication of Hamilton's rule, empirical studies across a range of different species [1] have confirmed its predictions for social behaviours such as guarding among female Allodapine bees and defence of dominant male Wild turkeys by non-dominant males [13,17]. Kin selection theory has also been used to explain insect eusociality [9,11] and allomothering in certain species of monkey [3], and has even been invoked in accounts of language evolution [4,5].

In this paper, we demonstrate an emergent kin selection mechanism within a population of continuously evolving artificial agents situated in an ecologically plausible environment. We place thousands of agents within a large gridworld filled with resources that regenerate through time (Fig. 1). Each agent is controlled by its own artificial neural network, mapping its local observation of the environment to its possible actions: moving in space, consuming a resource, feeding the agent next to it, or reproducing. We depart from the standard evolutionary computation paradigm: our simulations include no explicit fitness measure, instead relying on a 'minimal criterion' for individual agents' survival and reproduction, governed by a simple physiological model. In addition, we do not reset the agent population or environment state at any point during simulation (a framework referred to as non-episodic neuroevolution [10]). Importantly, the only kin-specific mechanism we introduce is the ability of agents to distinguish between adults, infants and their own offspring. The answer to our main research question is therefore far from trivial: in these simple, ecologically plausible conditions, will agents evolve the altruistic behaviour of feeding their own offspring? It is important to note that doing so will not favour their own survival and reproduction in any way. It will only favour the propagation of their own (mutated) genome, but without any explicit incentive to do so. To study this research question, we manipulate certain features of the environment, indirectly varying the fitness tradeoff $rb - c$. Our results show that altruistic feeding behaviour emerges as this differential increases. We also show that an increase in the prevalence of the altruistic feeding behaviour is associated with an increase in its selectivity towards agents' own offspring. To our knowledge, this is the first demonstration that kin selection naturally evolves in realistic ecological conditions where artificial agents continuously evolve in a non-episodic environment.

1.1 Related Work

While this work is to our knowledge the first to explore the kin selection of targeted altruism using large-scale continuous neuroevolution, there are some relevant lines of prior research which are worth highlighting. In particular, researchers in the fields of artificial life and evolutionary robotics have previously studied the emergence of cooperative and altruistic behaviour under different conditions. In [14], Montanier and Bredeche used a tragedy-of-the-commons setting to study the evolution of an indirect form of altruism where agents avoid pursuing a maximally greedy strategy of resource consumption; and in follow-up work [15] investigated the relationship between spatial dispersion and altruism. Floreano et al. [6] found that simulated robots in a hazardous foraging environment were more likely to evolve cooperative communication when selection took place at the level of colonies rather than individuals—which can be seen as an externally-imposed version of inclusive fitness. In [7], Frénoy et al. used the Aevol platform [12] to study the *robustness* of evolved cooperative behaviour in changing environments, finding that the evolutionary history of a given population has a significant effect on the future emergence of cooperative phenotypes. Closer to our work, Waibel et al. [18] used a system of simulated robots to explicitly test the predictions made by Hamilton's rule concerning the minimum genetic relatedness (r) required for kin selection, finding (as predicted) that consistent altruistic behaviour emerged only when $r > bc$.

2 Environment and Agents

2.1 Environment Dynamics

The environment we use is based on the non-episodic neuroevolution framework described in [10]. Simulations are run in a 100×100 gridworld which contains both food resources (plants) and agents, with each tile containing at most one agent, up to a maximum population size of 2500. Resources are generated and regenerated via simple spontaneous growth: at the start of simulation, each tile has probability $p = 0.1$ of containing a plant; at every subsequent timestep, each non-plant-containing tile has constant probability $p = 0.003$ of spontaneously spawning one. The environment is initialised with a starting population of 2000 agents, all with random neural network weights. Agents' survival and reproduction is governed entirely by simple physiological rules, to enable minimal-criterion neuroevolution [2]. Agents are born with an initial energy of 65. They lose 0.1 energy passively at each timestep, with a further loss of 1.0 for every non-idle action they perform. An agent dies when its energy falls below 0. Agents can increase their energy (up to a maximum of 200) by consuming plants, with

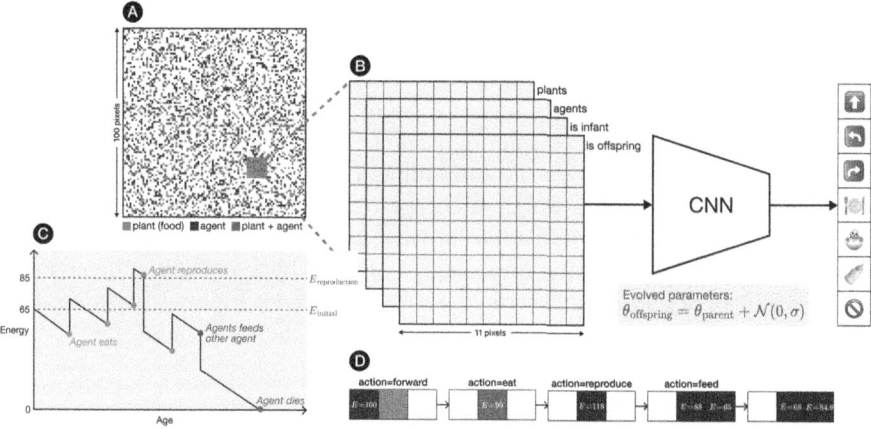

Fig. 1. (A) A snapshot of the simulation environment. (B) an illustration of the agent architecture, observation and action spaces. Each agent receives an $11 \times 11 \times 4$ observation of its local environment–this is fed through a convolutional neural network (CNN) to produce a vector of action probabilities. (C) an example illustration of the energy level over time for a single agent. (D) an illustration of a short action sequence in which an agent moves onto a food resource, eats the food resource, reproduces and then feeds their offspring.

each plant worth 20 units of energy. If an agent's energy is at or above a threshold of 85, and the population is less than the maximum of 2500, it can use the reproduce action to produce a single offspring (at an energy cost of 30). Offspring spawn into the nearest empty square and receive a copy of their parent's neural network weights randomly mutated by the addition of zero-mean Gaussian noise, i.e. $\theta_{\text{offspring}} := \theta_{\text{parent}} + \mathcal{N}(0, \sigma)$ (for all simulation runs reported in this paper we use $\sigma = 0.02$). Agents younger than 100 timesteps are considered to be 'infants'–they are unable to reproduce, and may also experience slightly different environment dynamics or action mechanics (see Sect. 3 for more detail). All of the parameters given here were chosen to produce environments where the difficulty of survival is neither trivial (leading to little selective pressure on agent behaviour) or too high (leading to population collapse or very low population sizes), and to give us sufficient 'room' to manipulate the cost-benefit tradeoffs involved in altruistic behaviour.

2.2   Agents

At each timestep, every agent in the population samples an action from a policy parameterised by an individual neural network (NN) controller, implemented as simple 3-layer convolutional neural network (CNN). The weights of each agent's controller remain constant throughout their lifetime–i.e. there is no learning, and behavioural adaptation can happen only through neuroevolution. Agents observe the environment within an 11×11 window centred at their current location, within which they see the locations of both plants and other agents. They can also see, for each other agent within this window, two binary values indicating whether that agent is A) an infant, and B) their own offspring. The action space is given by {forward, turn left, turn right, eat, reproduce, feed, idle}. The 'eat' action has the effect of consuming a single plant resource, and is only effective if the agent is occupying a tile containing a food resource. The 'feed' action transfers one food resource's worth of energy to a single other agent, and is only effective if the two agents are in adjacent tiles, and the feeder is directly oriented towards the feedee (facilitating some degree of selectivity in feeding behaviour). The 'reproduce' action is as described above (Sect. 2.1). Figure 1 gives an illustration of the environment, agent architecture, and energy dynamics.

3   Experiments

In our experiments, we seek to determine whether altruistic feeding behaviour (i.e. the zero-sum transfer of resources from one agent to another) can emerge in our neuroevolution framework as a result of kin selection. We focus on a narrow version of kin selection that considers only the relationship between parents and their direct offspring. Since in this case the genetic relatedness r is close to 1 (specifically, $1 - \sigma$), Hamilton's rule [8] tells us that in our environment, resource transfer from parent to child should be selected for when the fitness benefit b to the child of gaining the resource exceeds the fitness cost c to the parent of losing the resource. Our hypothesis is thus that in simulation environments where $b - c$ is greater, we should observe both more feeding behaviour *and* feeding behaviour that is more selectively directed towards agents' own offspring. If on the other hand there is no kin selection process,

While controlling this tradeoff directly is challenging, we can take an indirect approach by varying the relative effect of a marginal food resource on the continued survival of parent and child. We do this by manipulating certain environment conditions that make it harder for *infant* agents to survive to adulthood (and thus to reproductive age), while keeping the survival difficulty for adult agents unchanged. This should have the effect of increasing b while keeping c constant, and thus increasing the differential $b - c$. The parameters we manipulate for this are:

- **infant food energy proportion**: the proportion of available energy that infant agents gain from each food resource (where adults get 100%), defaulting to 1.0. For example, if this value is 0.5, then an infant's energy level will increase by 10 upon consuming a plant, whereas an adult's would increase by 20.
- **infant eat success probability**: the probability with which an infant's 'eat' action succeeds, assuming they are located at a food resource (where adults' 'eat' actions are deterministically successful), defaulting to 1.0. For example, if this value is 0.5, an infant would have to use the 'eat' action twice as often as an adult to get the same expected energy gain.

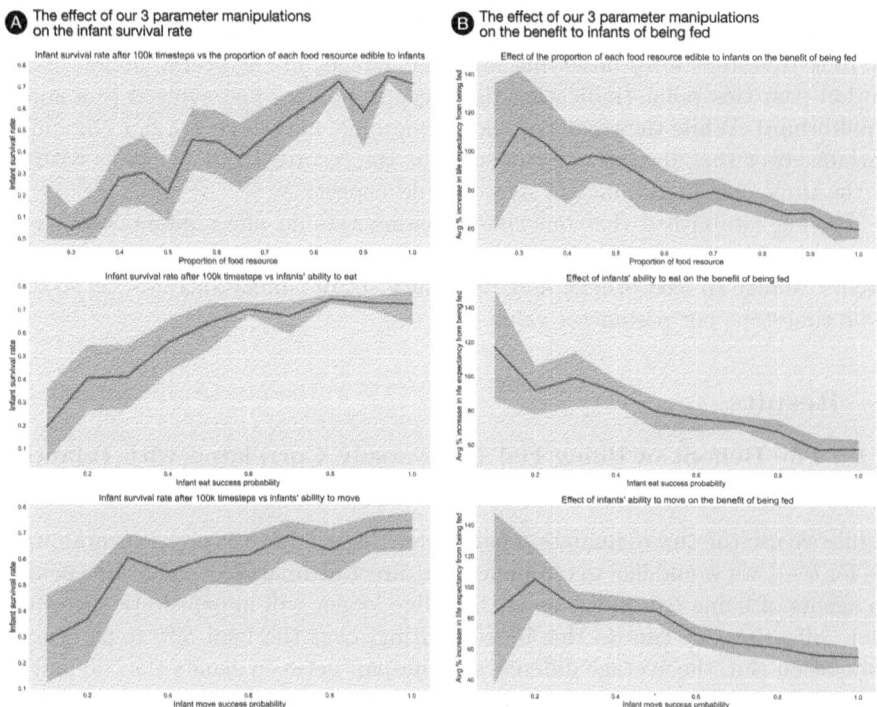

Fig. 2. The effect of varying the three chosen simulation parameters on (A) the infant survival rate (ISR) with feeding disabled, and (B) the estimated benefit of being fed as an infant ($\hat{b}$), given by the average percentage increase in lifespan for infants that are fed relative to those that aren't. Error bars in all plots represent bootstrapped 95% confidence intervals over 20 seeds per parameter value. ISR is measured over the final 10k steps of 100k-timestep simulation runs; $\hat{b}$ is measured over the final 50k steps of 500k-timestep simulation runs.

- **infant move success probability**: equivalent to **infant eat success probability** but for the 'move' action, defaulting to 1.0. For example, if this value is 0.5, an infant would move around the environment twice as slowly as an adult following the same action policy.

To establish the effect of each of these parameters on the environment survivability, we first run a set of simulations with **feeding behaviour disabled** (i.e. agents' action spaces do not contain the 'feed' action), and track the 'infant survival rate' (ISR) over time. For a window $[t_1, t_2]$, we compute this as

$$\text{ISR} = 1 - \frac{\text{num infant deaths in } [t_1, t_2]}{\text{total num deaths in } [t_1, t_2]} \tag{1}$$

Figure 2 (A) shows the ISR recorded over the final 10k steps of 100k-step simulations as a function of each of the three parameters listed above, with 20 seeds run per parameter value (and the other parameters held constant at their default value of 1.0). For these preliminary experiments, we chose to end simulations at 100k timesteps after observing minimal changes in the overall action distribution past this point (indicating that agents' policies had evolved to a stable equilibrium). While the trends are not monotonic, and there is a fair amount of variance over the 20 seeds, we can see that the recorded ISR increases with all of the three chosen parameters–as we would expect.

Having established that our chosen parameters do affect infants' ability to survive to adulthood (and thus in theory the excess fitness $b - c$), we re-enable agents' ability to feed others, and run a new set of simulations over 20 seeds x 500k timesteps per parameter value.

4   Results

4.1   The Benefit of Being Fed Is Inversely Correlated with Infants' Ability to Survive

While we use the three simulation parameters listed above as experimental proxies for $b - c$, we would like to obtain a more direct estimate $\hat{b}$ of the fitness benefit to agents of being fed. Since agents that live longer will in expectation produce more offspring, we can do this by computing, over the final 50k steps of each simulation run, the average difference in lifespan between agents that do and do not receive feeding as infants. Figure 2 (B) shows how $\hat{b}$ changes as a function of each of the three experimental parameters. Across all three parameters, we see that the benefit of being fed during infancy on agents' life expectancy (and thus ability to reproduce) is higher in environments that are more hostile to infants' survival.

4.2   Agents Engage in More Feeding and More Selective Feeding When the Benefit of Being Fed Is Higher

Having established an approximate measure of the benefit to reproductive fitness of being fed, we now test the relationship between this benefit and the actual feeding behaviour observed. If feeding behaviour is being evolved via kin selection, then we should expect two effects as we increase the importance of being fed to young agents' reproductive fitness:

– The amount of feeding behaviour observed in the population should increase,
– The selectivity of feeding behaviour towards agents' own offspring should increase.

We measure the amount of feeding as simply the number of times the 'feed' action was successfully used over the final 50k timesteps (10%) of each simulation run, divided by the average population size over the same period. To measure selectivity, we first take the proportion of these feeding events where the resource was transferred from a parent to their own offspring. But note that this proportion is not necessarily a reliable measure of selectivity—it could be

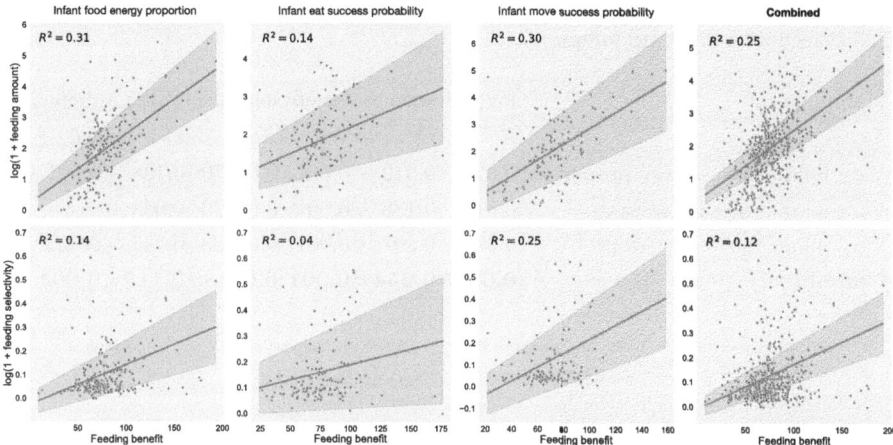

Fig. 3. The relationship between the estimated benefit to infants of being fed and both the amount and selectivity of feeding observed, shown separately for each of the three experimental parameters we varied (and combined in the rightmost column). Each scatter plot point represents a single 500k-timestep simulation run (with values averaged over the final 50k timesteps); regression lines (with 95% confidence intervals) are shown in green. Note that the y-axis shows log(measure) for both amount and selectivity. (Color figure online)

higher, for example, if offspring stay closer to their parents on account of being less able to move. To account for this, we track all instances where one agent (agent A) is directly facing another agent (agent B), and compute the proportion of times where B is A's offspring. We can treat this proportion as a baseline for the proportion of feeding towards offspring that we would expect if there were no 'real' selectivity; i.e. if agents were equally likely to use the feed action when facing any other agent. Our final measure of selectivity is then obtained by subtracting this baseline from the original proportion, to give a more robust measure:

$$\text{selectivity} = \frac{\#\text{ times offspring fed}}{\#\text{ all feeding events}} - \frac{\#\text{ times agent facing offspring}}{\#\text{ times agent facing any agent}} \quad (2)$$

To test for the two trends listed above, we carried out a linear regression analysis (by the OLS method). We found significant positive correlations between feeding benefit and both log(feeding amount) and log(feeding selectivity), indicating that the two trends are indeed observed, and have an exponential shape. The regression results are given in Table 1, and visualised in Fig. 3.

Table 1. Results of OLS linear regression analysis for feeding amount and selectivity against feeding benefit (to infants)

Manipulation	log(amount) ~ benefit			log(selectivity) ~ benefit		
	β	r^2	p-val	β	r^2	p-val
Vary infant food energy proportion	0.0225	0.312	<0.001	0.00170	0.138	<0.001
Vary infant eat success prob	0.0137	0.138	<0.001	0.00120	0.043	0.014
Vary infant move success prob	0.0297	0.297	<0.001	0.00320	0.247	<0.001
Combined	**0.0215**	**0.254**	**<0.001**	**0.00180**	**0.117**	**<0.001**

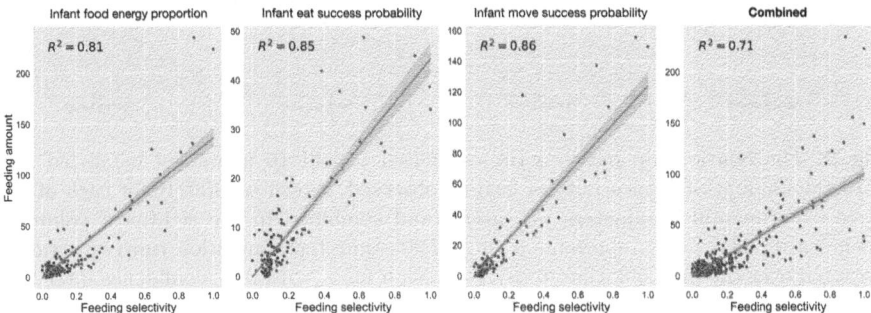

Fig. 4. The relationship between the amount of feeding behaviour observed and its selectivity towards agents' own offspring, shown separately for each of the three experimental parameters we varied (and combined in the rightmost column). Each scatter plot point represents a single 500k-timestep simulation run (with values averaged over the final 50k timesteps); regression lines (with 95% confidence intervals) are shown in green. (Color figure online)

4.3 The Amount of Feeding Observed Increases with the Selectivity of Feeding Towards Offspring

If a behaviour evolves through kin selection, then we should expect to see a positive correlation between the overall prevalence of the behaviour and the extent to which it is used selectively towards agents' own kin. As an additional check, we therefore test the relationship between the amount and selectivity of feeding observed in our simulation runs. As is shown in Table 2 and Fig. 4, we find a very strong positive correlation between the two measures, across all three experimental parameters.

4.4 Kin Selection Operates Through a Combination of Kin Recognition and Population Viscosity

As mentioned briefly in Sect. 1, kin selection is understood to operate via two possible mechanisms: kin recognition, and population viscosity [8, 9]. In the case of kin recognition, individuals have some faculty for recognising their own kin,

Table 2. Results of OLS linear regression analysis for feeding amount against feeding selectivity

Manipulation	amount $\sim$ selectivity		
	β	r^2	p-val
Vary infant food energy proportion	136.5	0.806	<0.001
Vary infant eat success prob	44.31	0.848	<0.001
Vary infant move success prob	123.8	0.862	<0.001
combined	**100.4**	**0.708**	**<0.001**

which allows them to target certain behaviours specifically towards them. In the absence of such a faculty, kin selection may still operate if the average movement of individuals from their birthplace is sufficiently slow. In such populations, the local social environment of an individual likely consists in large part of their relatives, simply by default–and so mere targeting of behaviour towards neighbours is sufficient for kin selection to occur.

As a final experiment, we seek to understand which of these two mechanisms is responsible for the kin selection we observe in our virtual agent populations. To do this, we perform two independent ablations to our existing setup, and re-run the full set of simulations with each. To test the importance of kin recognition, we simply disable the observation feature that tells an agent whether each other agent they can see is one of their own offspring (see Fig. 1; we retain the 'is infant' feature). Population viscosity is a little trickier to intervene on directly. Our approach was to modify the reproduction dynamics. Previously, each new offspring was born into the tile their parent was facing at the time of reproduction (meaning parent and child were initially directly adjacent to one another). Instead, we now randomly select the birth location of all new offspring agents, meaning that a newborn is no more likely to be near to its parent as to any other agent in the population.

Figure 5 shows the same regression analysis as performed previously for the two independent ablation runs. We can see that for both ablations, the relationships between feeding benefit (b) and the amount and selectivity of feeding are weakened. Interestingly, the effect is considerably stronger for randomised birth locations than for disabled kin recognition–suggesting that while both mechanisms play a role in the kin selection we observe, population viscosity is more important in facilitating the evolution of kin-selective feeding behaviour.

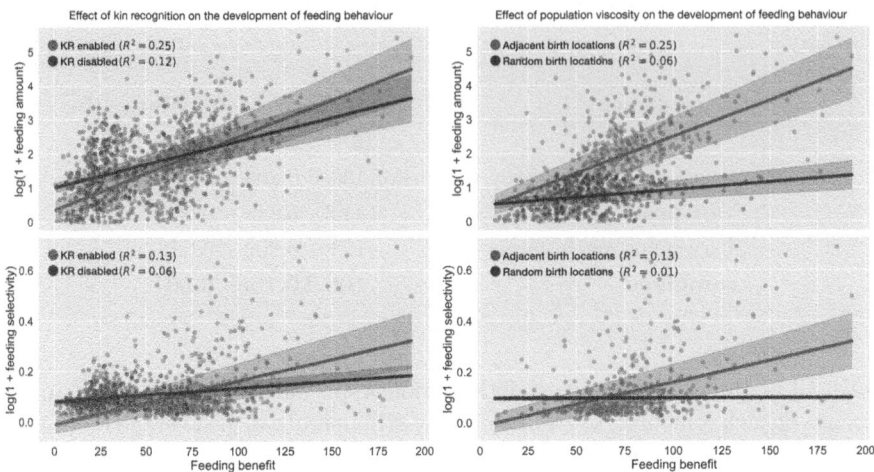

Fig. 5. The effect of disabling kin recognition (left) and reducing population viscosity (right) on the relationships between feeding benefit and feeding amount and selectivity. Each dot represents a single 500k-timestep simulation run. Simulations were run over the same values of the three experimental parameters as before, and as before we ran 20 seeds per parameter value. 95% confidence intervals are shown for all regression lines.

5   Discussion

In this paper, we have presented what is to our knowledge the first evidence of kin selection emerging from interactions between simulated agents undergoing neuroevolution in a simple but ecologically plausible environment. Focusing specifically on the phenomenon of agents feeding (donating energy to) their own infant offspring, we showed that this feeding behaviour saw greater adoption as a function of the fitness differential $rb - c$, as predicted by Hamilton's rule. We also found a very strong relationship between the prevalence of feeding behaviour observed and its selectivity towards agents' own offspring, supporting the idea that the use of the 'feed' action emerged principally through a mechanism that reinforced its use towards kin but not non-kin. Finally, we investigated the mechanisms by which this kin selection process occurred, and found that while both kin recognition and population viscosity played a role, the latter had a more significant impact.

As well as serving as a methodological demonstration for how kin selection and inclusive fitness can be studied using neuroevolution of simulated agents, we believe that our results may have some broader theoretical implications. The commonly accepted account of kin selection is made in terms of fitness maximisation from the perspective of individual genes shared by genetically related organisms. However, our simulations involve neither a proper notion of 'gene' (or any discrete genetic subunit), nor an explicit notion of fitness maximisation. Agents are governed by, and pass to their offspring, a set of continuous neural network weights with no particular structure; agents survive and reproduce as long as they can maintain their internal energy above a certain level. The fact that we were still able to observe a kin selection phenomenon emerging suggests the possibility of a more general framing of the theory for such systems. One possible interpretation of our results is that by increasing the difficulty for infants to survive without being fed (i.e. increasing $rb - c$), we increased the extent to which the feeding of offspring was necessary for an agent to propagate a lineage over several generations. That is, as $rb - c$ increased, the probability increased for the genome of any agent *not* engaging in offspring-feeding to disappear from the population, increasing the overall prevalence of the behaviour in the phenotype of the population.

While we do believe our results to be of some interest, we close by noting a number of limitations that could be addressed in future work that builds upon our framework. First, our agents' behaviour is governed by a very simple reactive architecture, in which actions are selected based purely on the current state observation, and there is no within-lifetime adaptation. This limits the complexity of the behavioural policies that agents can implement, and it would be interesting to explore a variant using (for example) recurrent architectures. For instance, an agent with memory could in principle 'remember' which other agents are its own offspring without needing access to an explicit marker; or could adapt their behaviour to focus more on caregiving in later life. Secondly, our kin recognition mechanism is also very simplistic, only allowing agents to distinguish their own direct offspring. For a more complete test of the quantitative predictions made by Hamilton's rule, we would need to provide a continuous measure of the genetic relatedness between agents, in terms of their distance in neural parameter space. Finally, while not specifically a limitation of the current work, we believe that another fruitful research direction could be in exploring the evolution of more complex social behaviours within the same non-episodic neuroevolution framework. For example, could environment dynamics that change over time (instead of being stationary) foster the emergence of teaching or demonstration by parents for the benefit of their newborn offspring?

Acknowledgments. This work was supported by the United Kingdom Research and Innovation (grant EP/S023208/1), EPSRC Centre for Doctoral Training in Robotics and Autonomous Systems (RAS). This work was also partially funded by the French National Research Agency (https://anr.fr/, project ECOCURL, Grant ANR-20-CE23-0006). This work also benefited from access to the HPC resources of IDRIS under the allocation 2023-[A0151011996] made by GENCI, using the Jean Zay supercomputer.

A   Appendix

All code is available at this Github repo.

References

1. Bourke, A.: Hamilton's rule and the causes of social evolution. Philos. Trans. Royal Soc. B: Biol. Sci. **369**(1642), 20130362 (2014)
2. Brant, J.C., Stanley, K.O.: Minimal criterion coevolution: a new approach to open-ended search. In: Proceedings of the Genetic and Evolutionary Computation Conference, pp. 67–74. GECCO '17, Association for Computing Machinery, New York, NY, USA (2017). https://doi.org/10.1145/3071178.3071186
3. Fairbanks, L.A.: Reciprocal benefits of allomothering for female vervet monkeys. Anim. Behav. **40**(3), 553–562 (1990)
4. Fitch, W.T.: Kin selection and "mother tongues": a neglected component in language evolution. In: Evolution of Communication Systems: A Comparative Approach. The MIT Press (2004). https://doi.org/10.7551/mitpress/2879.003.0022
5. Fitch, W.T.: Evolving Meaning: The Roles of Kin Selection, Allomothering and Paternal Care in Language Evolution, pp. 29–51. Springer London, London (2007). https://doi.org/10.1007/978-1-84628-779-4_2
6. Floreano, D., Mitri, S., Magnenat, S., Keller, L.: Evolutionary conditions for the emergence of communication in robots. Curr. Biol. **17**, 514–519 (2007)
7. Frénoy, A., Taddei, F., Misevic, D.: Robustness and evolvability of cooperation. In: ALIFE 2012: The Thirteenth International Conference on the Synthesis and Simulation of Living Systems, pp. 53–58. MIT Press (2012). https://doi.org/10.1162/978-0-262-31050-5-ch008
8. Hamilton, W.D.: The genetical evolution of social behaviour. I. J. Theor. Biol. **7**(1), 1–16 (1964)
9. Hamilton, W.: The genetical evolution of social behaviour. II. J. Theor. Biol. **7**(1), 17–52 (1964)
10. Hamon, G., Nisioti, E., Moulin-Frier, C.: Eco-evolutionary dynamics of non-episodic neuroevolution in large multi-agent environments. In: Proceedings of the Companion Conference on Genetic and Evolutionary Computation, pp. 143–146. GECCO '23 Companion, Association for Computing Machinery, New York, NY, USA (2023). https://doi.org/10.1145/3583133.3590703
11. Hughes, W., Oldroyd, B.P., Beekman, M., Ratnieks, F.: Ancestral monogamy shows kin selection is key to the evolution of eusociality. Science **320**(5880), 1213–1216 (2008)
12. Knibbe, C., Fayard, J.M., Beslon, G.: The topology of the protein network influences the dynamics of gene order: from systems biology to a systemic understanding of evolution. Artif. Life **14**(1), 149–156 (2008). https://doi.org/10.1162/artl.2008.14.1.149
13. Krakauer, A.H.: Kin selection and cooperative courtship in wild turkeys. Nature **434**, 69–72 (2005). https://doi.org/10.1038/nature03325
14. Montanier, J.M., Bredeche, N.: Surviving the tragedy of commons: emergence of altruism in a population of evolving autonomous agents. In: ECAL 2011: The 11th European Conference on Artificial Life. MIT Press (2011). https://doi.org/10.7551/978-0-262-29714-1-ch085
15. Montanier, J.M., Bredeche, N.: Evolution of altruism and spatial dispersion: an artificial evolutionary ecology approach. In: ECAL 2013: The Twelfth European Conference on Artificial Life, pp. 260–267. MIT Press (2013). https://doi.org/10.1162/978-0-262-31709-2-ch040
16. Smith, J.M.: Group selection and kin selection. Nature **201**, 1145–1147 (1964). https://doi.org/10.1038/2011145a0

17. Stark, R.E.: Cooperative nesting in the multivoltine large carpenter bee xylocopa sulcatipes maa (apoidea: Anthophoridae): do helpers gain or lose to solitary females? Ethology **91**(4), 301–310 (1992)

18. Waibel, M., Floreano, D., Keller, L.: A quantitative test of Hamilton's rule for the evolution of altruism. PLOS Biol. **9**(5) (2011). https://doi.org/10.1371/journal.pbio.1000615

Stalling in Space: Attractor Analysis for Any Algorithm

Sarah L. Thomson[1]([✉]) [iD], Quentin Renau[1] [iD], Diederick Vermetten[2] [iD],
Emma Hart[1] [iD], Niki van Stein[2] [iD], and Anna V. Kononova[2] [iD]

[1] Edinburgh Napier University, Edinburgh, UK
s.thomson4@napier.ac.uk
[2] LIACS, Leiden University, Leiden, The Netherlands

Abstract. Network-based representations of fitness landscapes have grown in popularity in the past decade; this is probably because of growing interest in *explainability* for optimisation algorithms. Local optima networks (LONs) have been especially dominant in the literature and capture an approximation of local optima and their connectivity in the landscape. However, thus far, LONs have been constructed according to a strict definition of what a local optimum is: the result of local search. Many evolutionary approaches do not include this, however. Popular algorithms such as CMA-ES have therefore never been subject to LON analysis. Search trajectory networks (STNs) offer a possible alternative: nodes can be any search space location. However, STNs are not typically modelled in such a way that models temporal *stalls*: that is, a region in the search space where an algorithm fails to find a better solution over a defined period of time. In this work, we approach this by systematically analysing a special case of STN which we name *attractor networks*. These offer a coarse-grained view of algorithm behaviour with a singular focus on stall locations. We construct attractor networks for CMA-ES, differential evolution, and random search for 24 noiseless black-box optimisation benchmark problems. The properties of attractor networks are systematically explored. They are also visualised and compared to traditional LONs and STN models. We find that attractor networks facilitate insights into algorithm behaviour which other models cannot, and we advocate for the consideration of attractor analysis even for algorithms which do not include local search.

Keywords: fitness landscape · local optima networks · search trajectory networks

1 Introduction

Understanding the landscape or structure of the search space associated with a problem instance is important for both predicting the performance of algorithms and improving algorithm design so that they can more efficiently traverse the landscape. In the past decade, several techniques for visualising landscapes have

© The Author(s), under exclusive license to Springer Nature Switzerland AG 2025
P. García-Sánchez et al. (Eds.): EvoApplications 2025, LNCS 15612, pp. 510–526, 2025.
https://doi.org/10.1007/978-3-031-90062-4_32

arisen that provide new insights into this task. Local optima networks (LONs) [1] capture the number, distribution and connectivity pattern of local optima in the form of a network where nodes represent local optima and edges capture the transitions between them. LONs have most often been used to understand landscapes in combinatorial settings [2–4]—mainly due to the fact that constructing a LON requires an iterated local search (ILS) algorithm to be run to identify local optima. More recently, LONs have been constructed for continuous spaces, mostly using monotonic basin-hopping [5,6], which is essentially an ILS framework for continuous optimisation. The LON model has been particularly useful identifying basins of attraction in a landscape [4,7]. Search trajectory networks (STNs) [8] were introduced as a more generalisable alternative to LONs. STN are directly constructed from data gathered while running an algorithm. Unlike LONs, there is no condition that nodes must be a local optima. STNs enable a user to visualise (for example) whether multiple algorithms or runs of an algorithm traverse the same locations, indicate termination points of an algorithm (both optimal and sub-optimal) and show the frequency with which algorithms pass through and escape from the nodes via an edge weight. However, STNs do not typically provide information as to **how long** an algorithm remained stalled at a node before managing to escape—the information they encode is valuable, but does not generally have this temporal aspect.

To address this, we put forward a new variant of an STN which is applicable to any type of algorithm (including those that do not easily permit local search) and which demonstrate the *stagnation behaviour* of an algorithm at various points in the search. We dub these regions *stall locations*: a period of the search process where the best value of the objective function does not change over a period of β evaluations. A stall location can hence be seen as an attractor of the search and can be detected using any type of algorithm (i.e. it does not rely on local search). This approach, which we call *attractor networks*, brings new insights into the operation of algorithms which are not evident in either LON networks or STN networks. We study the properties and illustrate the benefits of attractor networks by constructing them using three different algorithms on the 24 functions of the BBOB test suite [9]—comparing metrics and visualisations obtained from the networks to those obtained from both LON and STN. The contributions can be summarised as follows:

° The formal introduction of a method for visualising and analysing attractors in the genotype space (termed an *attractor network, or AN*) in a landscape that is applicable to algorithms from both the continuous and combinatorial optimisation domains and regardless of whether local search is applicable.
° A demonstration that ANs can provide new insights into intermediate attractors in the search process that are not detected by standard LON or STN analysis, obtained by evidence gathered by conducting a systematic analysis over suite of functions and algorithms.

2   Background

2.1   Network Models

Local optima network (LON). A local optima network is a directed and weighted graph comprised of: a) nodes $lo_i \in LO$ which belong to a set of local optima (this can be the complete set or a sample) and b) edges $e_{ij} \in E$ which connect pairs of local optima (nodes) lo_i and lo_j with a weight which denotes the frequency of transition w_{ij} iff $w_{ij} > 0$. Extensive descriptions can be found in [10].

Search trajectory network (STN). A search trajectory network is a directed and weighted graph comprised of: a) nodes $sl_i \in SL$ which belong to a set of search locations from algorithm trajectories (these do not need to be local optima) and b) edges $e_{ij} \in E$ which connect pairs of locations (nodes) sl_i and sl_j with a weight which denotes the frequency of transition w_{ij} iff $w_{ij} > 0$. Extensive descriptions can be found in [11].

2.2   LON and STN Manifestations

In combinatorial spaces, LONs are constructed such that a local optimum is the result of the application of a local search method [10]. In this way, the LON nodes satisfy the condition that they are the best solution within their [sampled or complete] neighbourhood with respect to a basic mutation operator. This is usually achieved either by using iterated local search as the construction algorithm [2–4] or occasionally by augmenting population-based approaches with local search, rendering them memetic [12,13].

Approaches in continuous spaces are typically similar; the Limited-memory Boyden-Goldfarb-Shanno (L-BFGS-B) [14] optimiser—a quasi-Newton method—has been used as local search to obtain LON nodes [5,6,15]. The Nelder-Mead downhill simplex algorithm has been used as an alternative method to identify local optima for LONs [16]; very recently, a greedy local sampling akin to $(1 + \lambda)$ Evolutionary Strategy has been proposed for discovering local optima for these purposes [17]. One contribution augmented differential evolution with local search so that LONs could be constructed [18]. The common thread with LON works in continuous spaces is that the network is constructed using algorithms which may not always closely resemble the type of algorithms actually used to search these kind of spaces in practice. For example, networks which reflect the dynamics encountered by monotonic basin-hopping may only give *limited insight* into how CMA-ES behaves on a problem.

STNs are more generalisable than LONs, but have their own considerations. One is the decision of when to log a new STN node. As it relates to population-based algorithms moving in continuous spaces, the convention is logging the best individual in the population every G generations; G has been set as (for example) 1 [11] or the dimension of the problem [8]. This could potentially have the effect that a location is logged even if the search is only there for as little as one generation.

Very recently, an article on Cartesian genetic programming (CGP) [19] took the alternative approach of logging STN nodes if the search was at a location in the behaviour space for at least one iteration. However, in practice, the search was usually stuck at locations in the behaviour space for many iterations. It is therefore the case that what resulted was essentially an attractor network (as it is defined in this paper), although there are some *key differences*: a) this was not in the genotype space, as we consider here; b) a single node in the behaviour network represented typically thousands of genotypically different solutions from the genotype space which map to the same program behaviour; and c) the condition for a behaviour location to be considered a stall point was a single iteration. Despite the differences with the networks we analyse here, in that work attractors were visualised and this helped give valuable comparative insights about the two studied CGP algorithms. This finding serves as a good motivator for our work in focussing on attractors.

Another choice related to STN modelling in continuous spaces is what the threshold in decision space should be defined to consider two solutions the same. For this, a *partition factor* has been used [8,11]. The same concept has also been used in LON analysis for continuous functions; the original paper used a threshold of 10^{-5} [5] and this is also the value used in the Python package PFLACCO [20]. With STNs, the partition factor has conventionally been decided according to a formula involving the search space bounds, x_{max} and x_{min} and dimension of the problem. They consider the largest integer n for which the following holds: $(x_{\max} - x_{\min}) \times D \geq 10^n$; with n, the expression $2 - n$ is then carried out to obtain the partition factor, which is equivalent to the exponent of 10. According to this formula, our partition factors would lead to coordinate precision $\epsilon = 10^{-1}$ for 2D and $\epsilon = 10^0$ for 10D. This is rather coarse when we consider that our region of interest is [-5, 5]; we therefore take 10^{-2} as our most coarse setting, but also consider other lower values which are specified in Sect. 4.

It is standard with visualisation of STNs to make the size of the nodes proportional to how many search runs reached that location [21] and the related notion of *attraction areas* has been considered recently [22]; however, these ideas are *not* equivalent to the notion of attractor which we use in this paper. The visual node sizes and identified attraction areas will show locations where search goes frequently, **but not how long it stays there**. For example: a node which is large in a standard STN visualisation could represent a location where search flows through often, but where it does not stay for longer than a single generation of the algorithm. There could also be a case where a node is small—due to having only been found in a single run—but search stalled there for a significant number of evaluations before finally moving on. This would not be captured with standard STN conventions, but is important to capture. In general, neither LONs nor STNs typically consider the temporal aspect of algorithm behaviour directly. That thought, alongside an appetite for applying LON analysis to real-world problems (where the relevant search algorithms may not mirror LON construction methods, but attractors may nevertheless be of great importance), motivates the idea of attractor networks introduced here.

2.3   The Notion of Attractors for CMA-ES and DE

As introduced in Sect. 4, experiments carried out in this study are based on variants of two popular heuristic algorithms for continuous setting: CMA-ES and DE. Therefore, we briefly examine the literature for the analysis of attractors and these two algorithm classes.

To the best of our efforts, no papers were identified that consider search behaviour in the sense of stalling moments or locations during runs of continuous optimizers. While not really discussing the concept of stalling/attractors per se, the heuristic (continuous) optimisation community has actively worked on mechanisms preventing such stalling. These include: restarts of stagnated [23] or redundant [24] runs with possible increase of population size (IPOP [23] and BIPOP [25] mechanisms in CMA-ES and adaptive population control [26,27] in DE), step-size thresholding [28] and control of covariance matrix eigenvalues in CMA-ES, various population diversity improving components for both algorithms, improved covariance adaptation mechanisms for high-conditioned landscapes, improved handling of boundary constraints to balance exploration of the whole domain [29].

3   Methodology

3.1   Attractor Networks

In the context of optimisation algorithms, the notion of attractors can be defined in a traditional sense inspired by [30,31] that is independent of the algorithm, as described in [32]. For a metric space X, the basin of attraction in discrete-time dynamics is defined for the system $(X, \varphi : X \to X)$, where the k-th iteration of the system corresponds to the k-fold composition $\varphi \circ \varphi \ldots \circ \varphi(\cdot)$. The application of φ to x is assumed to shift x towards a vector of smaller magnitude in the direction opposite to $\nabla f(x)$. If this iterative process converges to a single point x such that $\varphi(x) = x$, this point is referred to as an *attractor*.

Attractor. In the context of attractor networks, we define an *attractor* to be a location in the search space where at least one of the r algorithm runs used to construct the network stalled for at least β fitness evaluations. We are working in continuous spaces in this study; it follows that the notion of attractor also depends on the coordinate precision ϵ. This value mandates the cutoff threshold in decision space for two solutions to be considered the same attractor node in the network: if the two solutions differ in at least one variable by an amount equal to or over ϵ then they will not be represented by the same node.

Attractor Network Construction. Attractor networks can be constructed from a log taken during algorithm runs. The log must include all moments of improvement of the best-so-far solution; each logging event will consist of three things: the number of elapsed fitness evaluations, the fitness of the new solution, and the genotype of that solution. From this type of log, an attractor network can be built in mostly the same manner as an STN, except nodes (and edges) are

exclusively logged if the number of elapsed evaluations between two consecutive improvements of the best-so-far solution is greater than β. Mirroring LONs and STNs, an attractor network is comprised of the trajectories for multiple independent runs of the algorithm. The network is also defined according to a level of coordinate precision ϵ: the radius in decision space [computed with Manhattan distance] within which two solutions are considered the same node.

A Note on Attractor Networks. We note that standard LONs are a special case of STNs, essentially encoding search trajectories of monotonic basin-hopping (in continuous space) or iterated local search (combinatorial). The attractor networks described here may be seen as both LONs and STNs—although, instead of strict hill climbing local optima we consider generalised attractors. The perspective is non-specific enough that any algorithm can be studied using it, unlike LONs. STNs can also be used to understand any algorithm, but they have not been modelled in such a way that attractor structure in the genotype space is evident. Attractor networks could be viewed as a reduced STN, and essentially offer a coarser-grained view with a singular focus on search traps.

3.2 Approach to Validation

Naturally, the attractor networks should be compared with standard local optima networks and standard search trajectory networks. This brings to mind the question of what 'standard' means for these models; we now clearly define what we consider to be the standard models (for the purposes of the study). We define the standard LON as the method implemented in PFLACCO and described fully in previous literature [5]. The construction is based on repeated runs of monotonic basin-hopping with L-BFGS as the local search component. We consider the standard STN to be such that a node (location) is logged as the best individual in the population every k iterations, which is the setting used in a previous STN study on population-based algorithms for continuous domains [8]. For random search, this is every k evaluations; for the other algorithms, it is every $k \times popsize$ evaluations.

In addition to comparison with other network models, random search is included in our portfolio for validation. Random search should stall frequently and show approximately the same behaviour across all functions. That is, random search attractor networks should not look significantly different depending on the function. In addition, the attractor networks of random search should not resemble those of our other algorithms.

4   Experimental Setup

For our experiments, we use COCO's BBOB problem suite [33], consisting of 24 single-objective, noiseless, continuous minimisation problems, which we access via IOHEXPERIMENTER [34]. We make use of a single instance (IID 1) for each problem, in both problem dimensionality 2 and 10. We consider several settings for β: $\in [10, 20, 40, 80, 160, 320, 640]$ and ϵ for the attractor networks:

$p \in [0.01, 0.001, 0.0001, 0.00001]$. We construct attractor networks from 10 runs and 30 runs: the smaller networks are for visualisation, while the larger ones are used for computing statistics. For both the standard LON and standard STN, we consider the same ϵ for decision variables as the setting found in the standard LON implementation in the PFLACCO[1] package: 10^{-5}. LON construction is implemented through PFLACCO functions and the STN construction was written from scratch in Python (the code and data will be published upon acceptance of this work). LON extraction has a parameter which serves as the termination condition for an individual run: we use the default setting for this, which is 1000 iterations without improvement. To match the attractor networks, we also construct LONs and STNs from 10 and 30 runs, respectively for visualisation and statistics purposes. In the cases where networks relating to 10-dimensional functions are plotted, the position of nodes on the x-axis is obtained through SCIKIT-LEARN [35] multi-dimensional scaling on the decision vectors. Where search on 2-dimensional functions is visualised, the position of nodes is relative to their actual location in decision space.

Our algorithm portfolio consists of three algorithms: vanilla CMA-ES from the ModCMA package [36], DE rand/1/bin with uniform initialisation, $F = 0.5$, $Cr = 0.5$ and saturation corrections for infeasible solutions from ModDE [37] and RandomSearch taken from Nevergrad [38][2]. For both modular algorithm packages, we stick with the *default parameter settings*, as we aim to illustrate the attractor network methodology rather than explore the best-performing versions of these algorithms. In particular, this decision means that the population sizes are dimensionality-dependent, with population sizes 6 and 10 for the 2 and 10-dimensional problems, respectively (which is relatively low for DE, but ensures having an equal amount of generations for both algorithms). From now on, we will refer to CMA-ES as CMA, DE as DE and random search as RS. In addition, the monotonic basin-hopping algorithm used in LON construction is referred to as MBH.

5   Results

5.1   Attractor Networks: Characteristics

We begin by visualising and analysing attractor networks. Figures 1 and 2 show the change in CMA AN structure with increasing β (left-to-right) for *f21* and *f1*, respectively. Figure 1 reflects 2D function optimisation. The axes of the plots, and the placement of AN nodes, reflect the actual 2D coordinates; colour represents fitness. Figure 2 reflects search on 10D functions. In both Figures, we can see that increasing β leads to sparser attractor networks. It is interesting that for both, there are still attractors even at a high β—in the case of Fig. 1c, which represents optimisation with a population size of 6, we can see that there are still multiple CMA attractors even with β=320 fitness evaluations (which equates to

---

[1] version 1.2.2.
[2] Versions used: modcma 1.0.2 (C++ backend) modde 0.0.1, nevergrad 1.0.0.

53 generations). Similarly, Fig. 2b shows that 10-dimensional *f1*—the uni-modal sphere function—is associated with a sub-optimal attractor when $\beta=80$, which equates to 8 generations of CMA. We generated Figures for all combinations of [function, algorithm, dimension, β, and ϵ]: these can be found in the supplemental material [39].

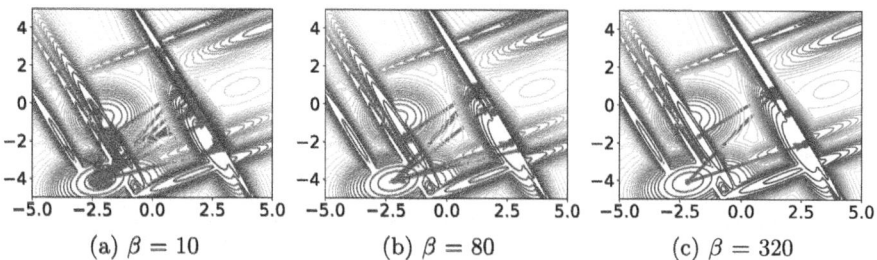

Fig. 1. Attractor networks for CMA on 2D Gallagher function *f21* across different values of β. Blue dots represents final points of the search while red dots represents attractors.

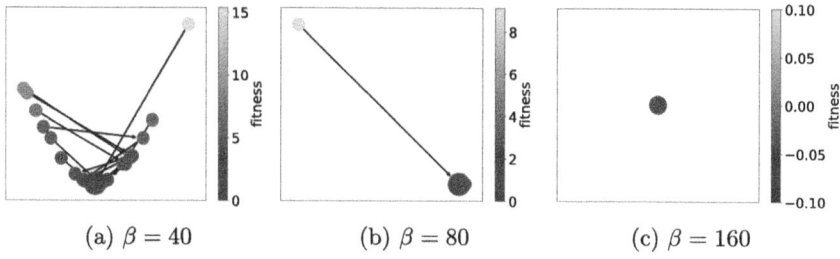

Fig. 2. Attractor networks [$\epsilon=10^{-5}$] for CMA on the 10-dimensional Sphere function *f1* across different values of β. Node size is proportional to the number of runs which reached that location

Table 1 presents summary statistics for the distribution over the 24 functions of evaluation differentials recorded on AN edges. This reflects networks constructed with 30 runs of the algorithms and $\epsilon=0.00001$; by *evaluation differential* we mean the elapsed fitness evaluations at the destination node of the edge minus the elapsed evaluations at the source node of the edge. For each network, the median differential *md* within it is recorded. The values in the table are the median and IQR value of *md* across the 24 functions is reported. Each row captures data for a given algorithm and setting for β. The trend is that a higher β leads to ANs with larger *md*. In the 2D case, DE has smaller *md* values with lower IQR value when compared to CMA. In 10 dimensions, the two have

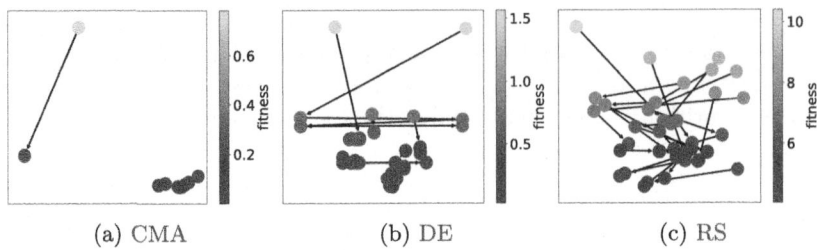

(a) CMA   (b) DE   (c) RS

Fig. 3. Attractor networks for 10D Schaffer function $f17$, $\beta = 320$, $\epsilon=10^{-5}$ for different algorithms.

Table 1. Median and IQR value [over 24 functions] for the median evaluations differential [over all edges in a given AN with $\epsilon=10^{-5}$ built from 30 runs]. Cells are shaded according to how often [out of 24 functions] the global optimum is present in the network: more vibrant green is more often.

model	2D		10D	
	median	IQR value	median	IQR value
DE AN [$\beta=40$]	52.25	6.13	55.5	67.88
DE AN [$\beta=80$]	101	15.5	119.25	93.13
CMA AN [$\beta=40$]	61.5	13.25	61.5	7.5
CMA AN [$\beta=80$]	168	153.75	108	18.75
RS AN [$\beta=40$]	509	128.63	484	126
RS AN [$\beta=80$]	732.25	132.88	659.5	191.5

similar md values but there is a larger dispersion for DE. Random search RS has, by far, the highest md values of the three algorithms.

Figure 3 shows, for a fixed β setting of 320, ANs for the three considered algorithms optimising 10-dimensional Schaffer function $f17$. All three algorithms have ANs which look noticeably different. Random search RS has the most dense (and most unstructured) network, and this is the trend across other functions as well (please see the supplemental material, where plots for all [function, algorithm, β, ϵ] combinations are available [39]). The DE network is smaller and a bit more structured: by following the arrows and the horizontal spacing, we can see that as search progresses there is movement towards a particular promising region of decision space. The CMA network is smaller still; we can see that at this setting of $\beta=320$ (that is, a stall of 32 generations), there is only a couple of sub-optimal attractors for this algorithm.

Figure 4 shows the change in network size (number of nodes) in ANs as β increases and for various ϵ, for CMA (left) and random search (right) on 10-dimensional functions. For each ϵ and β, there is a bar representing the median (over the 24 functions) and the variance is shown. Notice that within a coordinate precision level, nodes decrease approximately linearly with increasing β. In the case of CMA, a coarser ϵ leads to more nodes, and the variance decreases substantially with β. For random search, low β leads to many fewer nodes than in the CMA counterparts, but there are still rather a lot of nodes at high β (the decline in network size is much less dramatic in RS than we see with CMA).

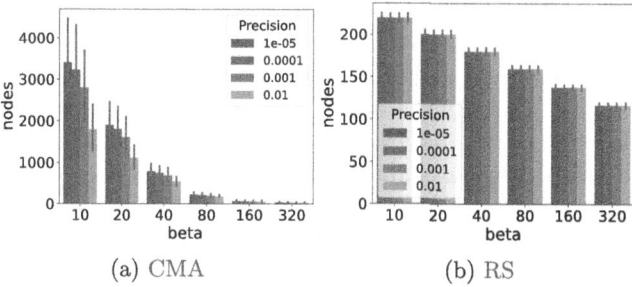

(a) CMA   (b) RS

Fig. 4. 10D ANs [constructed with 30 runs] number of nodes with increasing β and various ϵ.

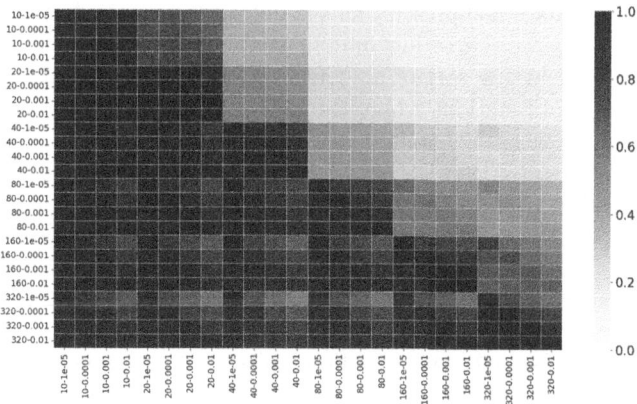

Fig. 5. The median proportion [over 24 functions] of node matches [i.e. fraction of mutual solutions] between CMA (2D) attractor networks with different network construction settings β and ϵ. In the plot labels, the integer part is the setting for β and the float is for ϵ.

Figure 5 focuses on the extent to which attractor networks constructed using different configuration settings (β and ϵ) for the same functions have an overlap in the node locations for CMA on 2D functions. We are focussing more heavily on CMA than DE; this is because CMA tended to produce more interesting attractor networks. However, the same plot for DE can be found in the supplemental material. In Fig. 5, pairs of AN configuration parameters are on each axis and the heat captures the median [across 24 functions] extent of mutual locations. For a given [vertical, horizontal] pairing, the heat value captures the fraction of the locations for the network type *on the vertical axis* which are also present in the network type *on the horizontal axis*. Squares with heat representing a value of 1.0 indicate that across the functions, that pair of differently-configured ANs have a complete node overlap. The purpose of this plot to further understand the properties of ANs better. Looking at the plot, we notice from the dark blue regions that ANs with the same β but different ϵ are often very similar. We can also observe that networks with a different (but similar) β seem to frequently have some relation—although this is to a lesser extent: in the region of 40–60% of nodes matching. For the higher levels of β (above 80), networks are also related in this way even to network counterparts which have a substantially different β—for example, pairs associated with the settings 80 and 320. In the top-right of the plot, we can notice that only a low proportion of nodes present in low-β settings are also present with high β, which fits with intuition.

5.2 Network-Based Landscape Models: a Comparison

Table 2 presents the median and IQR value [over 24 functions] for network nodes and edges with respect to LONs, STNs, and ANs with different β configurations—all of them constructed with a $\epsilon=10^{-5}$. Notice that networks associated with RS are not present: this is because we noticed during STN construction that they were growing unmanageably large and deduced that they were unlikely to yield any sort of meaningful insight. Cell shading captures the proportion of the 24 functions where the associated network contains the global optimum; vibrant green means more often.

Notice from the table that DE networks have fewer nodes and edges than their CMA counterparts. Comparing the LON metrics from the first row with other network types, we can see that on 2D functions, the LONs are of similar size to DE STNs; the CMA STNs are substantially larger than both of them. In fact, the 2D CMA STNs are larger even than the 10D networks. From looking at the raw data, we see that this is because the 2D CMA STNs are often comprised of many nodes with optimal or near-optimal fitness, but which are slightly different in decision space: at least 10^{-5} in one variable or more. There was a population size of six for this algorithm-dimension pair and the frequency of STN logging was every k generations, so there could be a maximum of 37500 (1250 per run for 30 runs) STN nodes if no node was ever seen twice. On 10-dimensional functions, we observe that low-β CMA ANs are larger than LONs, but that high-β ANs are smaller than LONs. Notice also that for most DE ANs, the IQR value for the nodes and edges is actually the same. Through investigation, we found that this happens when the number of nodes is exactly 30 more than the number of edges in the network; the reason for this phenomenon occurring is attributable to the 30 separate runs used to construct the networks, and takes place when each of the runs terminates in a different location.

Figure 6 is an illustrative comparison between different network model types: LONs, CMA STNs, and CMA ANs [with two β settings] are shown. Note that

Table 2. Number of network nodes and edges for networks [$\epsilon=10^{-5}$] constructed from 30 runs: median and IQR value [over 24 functions]. Cells are shaded according to how often [out of 24 functions] the global optimum is present in the network: more vibrant green is more often.

model	2D		10D	
	nodes	edges	nodes	edges
LON	67.5 (143.25)	73 (164.5)	233 (521)	210 (520)
DE STN	62 (94.5)	52.5 (94.75)	86.5 (52)	77 (56.5)
CMA STN	1367.5 (2166.75)	1585 (2268.25)	559 (1320)	579 (1361.5)
DE AN [β=40]	46.5 (8.25)	16.5 (8.25)	416.5 (141.25)	389 (140.25)
DE AN [β=80]	31 (3)	1.5 (3)	138 (223.25)	108 (223.25)
CMA AN [β=40]	104 (41.5)	96 (39.5)	860 (558.25)	830 (570)
CMA AN [β=80]	44.5 (25.5)	29 (20.25)	204.5 (216.25)	196 (195.25)

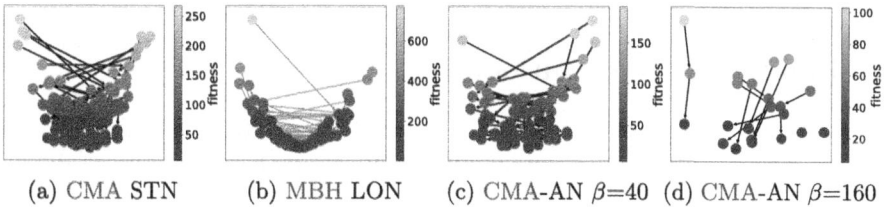

(a) CMA STN   (b) MBH LON   (c) CMA-AN β=40   (d) CMA-AN β=160

Fig. 6. Networks [$\epsilon=10^{-5}$] for 10D Rosenbrock function $f3$; the STN and AN show CMA behaviour, while the LON is constructed according to MBH (that is the convention).

visualisations for the other problem instances are available in the supplemental material [39]. In the case of Fig. 6, all plots reflect algorithms running on 10-dimensional Rastrigin function $f3$.

Surveying the figure, we notice that the STN, the LON, and the low β AN seem to reveal the overall structure of how algorithms move on the problem (forming a single-funnel shape). In the case of the LON, each node had been obtained through local search. In the case of the STN (Fig. 6a), nodes are not necessarily attractors. The low β AN resembles the STN. For the high β AN, there is a focus on exclusively attractors a.k.a. "local optima" for CMA. While each node in the LON (6b) is a local optimum for MBH, every node in Fig. 6d is an attractor for CMA. We can see that the latter is the sparsest of the four, and allows us to see a visualisation of search trajectories with a **temporal** component than none of the other models can show: we know that CMA stalled at each of these locations for at least 160 evaluations [16 generations]. Although an STN or LON can convey how often search passed through locations, they do not convey **how long** it spent there. The STN visualised here is rather crowded and it would not be possible to know which nodes are temporal attractors. While the LON does show local optima, there are two limitations: a) the temporal aspect is not captured and b) the local optima are with respect to MBH, rather than the algorithm under study (here: CMA).

The two plots in Fig. 7 show the proportion of CMA AN nodes which are also present in their corresponding STN (7a) and LON (7b) respectively. The networks were constructed with 30 runs and $\epsilon=0.01$ is used. Various β settings for the ANs are considered. We can observe that as β increases, the proportion of AN nodes increases as well. This is because there are less AN nodes and the ones which *are* still present are strong attractors; it makes sense that these nodes would also be present in the STN and LON. As β increases, though, the variance increases—particularly in the case of the STN matching in Fig. 7a. This implies that the degree of matching as it relates to high β ANs may depend on the nature of the particular function. The LON overlaps are less substantial than the STN overlaps; this makes sense, because the LON is constructed using MBH, while the STNs and ANs are constructed using CMA (in this case). It seems that MBH and for CMA traverse many *different* parts of the search space and for CMA attractors do *not* seem to be equivalent to LON local optima in most cases.

6   Limitations and Outlook

While attractor networks (ANs) offer an interesting way for understanding algorithmic stalling behaviour across different optimization landscapes, several limitations should be taken into consideration. Firstly, the construction and interpretation of ANs depend significantly on both the algorithm and the optimisation problem. The AN assumes a single population or individual traversing the search space. Methods that introduce diversity maintenance, such as niching and quality diversity algorithms, *may* result in misleading or overly complex ANs. For these approaches with multiple subpopulations, attractors might not represent stalling but rather indicate ongoing exploration within different niches. Therefore, alternative visualizations or constructions would be required to accommodate algorithms that are inherently multimodal or explicitly diversity-promoting. Another consideration is the sample dependency in AN construction. For problems with highly multimodal landscapes or higher dimensionality, the attractor

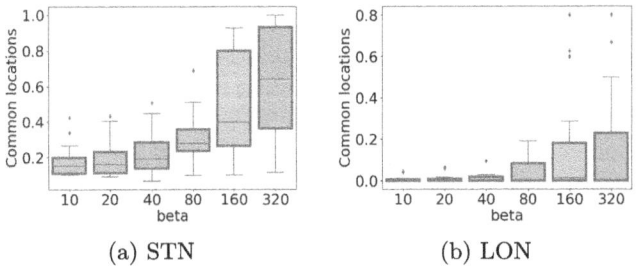

(a) STN   (b) LON

Fig. 7. CMA 10D proportion of matching locations between AN and other network models for all 24 functions [networks built with 30 runs and $\epsilon=10^{-2}$] across increasing AN β; represented as a proportion of AN size.

network structure *may* vary substantially across different runs. This necessitates additional runs to capture a more comprehensive network structure and to ensure robustness in AN-derived insights. The use of a fixed evaluation threshold for stalling also presents limitations in long-running or fine-tuned searches, where stagnation detection would benefit from adaptive thresholds. A future direction could be to explore an increasing window for evaluation that adapts to the progression of the search. Despite these limitations, the specificity of ANs to individual algorithms also gives an advantage. It allows for algorithm-level comparisons that go beyond performance metrics alone. This could enable a more nuanced understanding of algorithm dynamics, particularly in contexts where standard metrics are insufficient to capture structural differences. Looking forward, expanding the application of ANs across diverse algorithm classes and optimization scenarios could provide a better method to analyse algorithm behaviour and problem landscapes. The AN framework can be adapted for compatibility with multimodal algorithms or for discrete problem spaces. Additionally, future studies could explore hybrid network models that integrate insights from both traditional local optima networks and search trajectory networks.

7   Conclusions

In this work, we formalised and put forward the notion of attractor networks (ANs) as a novel framework for analysing the stalling behaviour of optimization algorithms. By focusing on attractor points—locations in the search space where algorithms experience prolonged stagnation—ANs provide a lens for examining algorithm dynamics beyond the reach of traditional local optima networks (LONs) and search trajectory networks (STNs). Unlike LONs, which are limited to hill-climbing algorithms, and STNs, which do not typically emphasise intermediate stalling behaviour, ANs facilitate a structured view of algorithm trajectories for any optimisation approach—including those that don't use local search. Through systematic analysis across 24 BBOB functions, we demonstrated that ANs reveal meaningful contrasts in how CMA-ES, differential evolution, and random search engage with search spaces. We show that ANs can give insights into intermediate attractors where alternative network-based models of algorithm behaviour would not. The AN model's flexibility in capturing unique behavioural characteristics presents a valuable direction for comparative analysis, enabling insights into algorithm-specific stalling and convergence patterns. Future studies may adapt the AN approach to accommodate more complex multimodal search strategies, broadening the applicability of ANs within optimization research. Code and data for this paper are available in a Zenodo repository [39].

Acknowledgments. Emma Hart and Quentin Renau are supported by funding from EPSRC award number: EP/V026534/1.

References

1. Ochoa, G., Tomassini, M., Vérel, S., Darabos, C.: A study of NK landscapes' basins and local optima networks. In: Proceedings of the 10th annual conference on Genetic and evolutionary computation, pp. 555–562 (2008)
2. Treimun-Costa, G., Montero, E., Ochoa, G., Rojas-Morales, N.: Modelling parameter configuration spaces with local optima networks. In: Proceedings of the 2020 Genetic and Evolutionary Computation Conference, pp. 751–759 (2020)
3. Ochoa, G., Veerapen, N., Daolio, F., Tomassini, M.: Understanding phase transitions with local optima networks: number partitioning as a case study. In: Hu, B., López-Ibáñez, M. (eds.) EvoCOP 2017. LNCS, vol. 10197, pp. 233–248. Springer, Cham (2017). https://doi.org/10.1007/978-3-319-55453-2_16
4. Ochoa, G., Chicano, F.: Local optima network analysis for max-sat. In: Proceedings of the Genetic and Evolutionary Computation Conference Companion, pp. 1430–1437 (2019)
5. Adair, J., Ochoa, G., Malan, K.M.: Local optima networks for continuous fitness landscapes. In: Proceedings of the Genetic and Evolutionary Computation Conference Companion, pp. 1407–1414 (2019)
6. Mitchell, P., Ochoa, G., Chassagne, R.: Local optima networks of the black box optimisation benchmark functions. In: Proceedings of the Companion Conference on Genetic and Evolutionary Computation, pp. 2072–2080 (2023)
7. Sánchez-Díaz, X.F., Masson, C., Mengshoel, O.J.: Regularized feature selection landscapes: an empirical study of multimodality. In: International Conference on Parallel Problem Solving from Nature, pp. 409–426. Springer (2024)
8. Ochoa, G., Malan, K.M., Blum, C.: Search trajectory networks of population-based algorithms in continuous spaces. In: International Conference on the Applications of Evolutionary Computation (Part of EvoStar), pp. 70–85. Springer (2020)
9. Hansen, N., Auger, A., Ros, R., Mersmann, O., Tusar, T., Brockhoff, D.: COCO: a platform for comparing continuous optimizers in a black-box setting. Optimization Methods Software **36**(1), 114–144 (2021)
10. Verel, S., Daolio, F., Ochoa, G., Tomassini, M.: Local optima networks with escape edges. In: Artificial Evolution: 10th International Conference, Evolution Artificielle, EA 2011, Angers, France, October 24-26, 2011, Revised Selected Papers 10, pp. 49–60. Springer (2012)
11. Ochoa, G., Malan, K.M., Blum, C.: Search trajectory networks: a tool for analysing and visualising the behaviour of metaheuristics. Appl. Soft Comput. **109**, 107492 (2021)
12. Veerapen, N., Ochoa, G., Tinós, R., Whitley, D.: Tunnelling crossover networks for the asymmetric TSP. In: Handl, J., Hart, E., Lewis, P.R., López-Ibáñez, M., Ochoa, G., Paechter, B. (eds.) PPSN 2016. LNCS, vol. 9921, pp. 994–1003. Springer, Cham (2016). https://doi.org/10.1007/978-3-319-45823-6_93
13. Thomson, S.L., Ochoa, G.: The local optima level in chemotherapy schedule optimisation. In: European Conference on Evolutionary Computation in Combinatorial Optimization (Part of EvoStar), pp. 197–213. Springer (2020)
14. Wright, S.J.: Numerical optimization (2006)
15. Contreras-Cruz, M.A., Ochoa, G., Ramirez-Paredes, J.P.: Synthetic vs. Real-World continuous landscapes: a local optima networks view. In: International Conference on Bioinspired Methods and Their Applications, pp. 3–16. Springer (2020)
16. Karatas, M.D., Akman, O.E., Fieldsend, J.E.: Towards population-based fitness landscape analysis using local optima networks. In: Proceedings of the Genetic and Evolutionary Computation Conference Companion, pp. 1674–1682 (2021)

17. Fieldsend, J.: Scalable local optima networks for continuous search spaces. University of Exeter (2024)
18. Homolya, V., Vinkó, T.: Leveraging local optima network properties for memetic differential evolution. In: Thi, H., Le, H.M., Dinh, T.P. (eds.) Optimization of Complex Systems: Theory, Models, Algorithms and Applications, pp. 109–118. Springer, Cham (2019)
19. De La Torre, C., Lavinas, Y., Cortacero, K., Luga, H., Wilson, D.G., Cussat-Blanc, S.: Multimodal adaptive graph evolution for program synthesis. In: International Conference on Parallel Problem Solving from Nature, pp. 306–321. Springer (2024)
20. Prager, R.P., Trautmann, H.: Pflacco: feature-based landscape analysis of continuous and constrained optimization problems in Python. Evol. Comput., 1–6 (2024)
21. Chacon-Sartori, C., Blum, C., Ochoa, G.: Search trajectory networks meet the web: a web application for the visual comparison of optimization algorithms. In: Proceedings of the 2023 12th International Conference on Software and Computer Applications, pp. 89–96 (2023)
22. Chacón Sartori, C., Blum, C., Ochoa, G.: Large language models for the automated analysis of optimization algorithms. In: Proceedings of the Genetic and Evolutionary Computation Conference, pp. 160–168 (2024)
23. Auger, A., Hansen, N.: A restart CMA evolution strategy with increasing population size. In: 2005 IEEE Congress on Evolutionary Computation, vol. 2, pp. 1769–1776 (2005)
24. de Nobel, J., Vermetten, D., Kononova, A.V., Shir, O.M., Bäck, T.: Avoiding redundant restarts in multimodal global optimization. In: Parallel Problem Solving from Nature - PPSN XVIII: 18th International Conference. PPSN 2024, Hagenberg, Austria, September 14–18, 2024, Proceedings, Part II, pp. 268–283. Springer-Verlag, Berlin, Heidelberg (2024)
25. Hansen, N.: Benchmarking a bi-population CMA-ES on the BBOB-2009 function testbed. In: Proceedings of the 11th Annual Conference Companion on Genetic and Evolutionary Computation Conference: Late Breaking Papers, pp. 2389–2396. GECCO '09, Association for Computing Machinery, New York, NY, USA (2009). https://doi.org/10.1145/1570256.1570333
26. Tanabe, R., Fukunaga, A.: Success-history based parameter adaptation for differential evolution. In: Proceedings of the IEEE Congress on Evolutionary Computation (CEC), pp. 71–78. IEEE, Cancún, Mexico (2013)
27. Tanabe, R., Fukunaga, A.: Improving the performance of success-history based parameter adaptation for differential evolution by linear population size reduction. In: Proceedings of the IEEE Congress on Evolutionary Computation (CEC), pp. 1658–1665. IEEE, Beijing, China (2014)
28. Hansen, N., Ostermeier, A.: Completely derandomized self-adaptation in evolution strategies. Evol. Comput. 9(2), 159–195 (2001)
29. Kononova, A.V., Vermetten, D., Caraffini, F., Mitran, M.A., Zaharie, D.: The importance of being constrained: dealing with constraints in evolution strategies using a new repair method. Evol. Comput. 32(1), 3–25 (2024)
30. Milnor, J.: On the concept of attractor. Commun. Math. Phys. 99, 177–195 (1985)
31. Collet, P., Eckmann, J.P.: Iterated maps on the interval as dynamical systems. Springer Science & Business Media (2009)
32. Antonov, K., Botari, T., Tukker, T., Bäck, T., van Stein, N., Kononova, A.V.: New solutions to Cooke triplet problem via analysis of attraction basins. In: Kress, B.C., Czarske, J.W. (eds.) Digital Optical Technologies 2023, vol. 12624, p. 126240T. International Society for Optics and Photonics, SPIE (2023). https://doi.org/10.1117/12.2675836

33. Hansen, N., Finck, S., Ros, R., Auger, A.: Real-Parameter Black-Box Optimization Benchmarking 2009: Noiseless Functions Definitions. Research Report RR-6829, INRIA (2009), https://hal.inria.fr/inria-00362633

34. de Nobel, J., Ye, F., Vermetten, D., Wang, H., Doerr, C., Bäck, T.: IOHexperimenter: benchmarking platform for iterative optimization heuristics. Evol. Comput. **32**(3), 205–210 (2024). https://doi.org/10.1162/evco_a_00342

35. Pedregosa, F., et al.: Scikit-learn: Machine learning in Python. J. Mach. Learn. Res. **12**, 2825–2830 (2011)

36. de Nobel, J., Vermetten, D., Wang, H., Doerr, C., Bäck, T.: Tuning as a means of assessing the benefits of new ideas in interplay with existing algorithmic modules. In: Krawiec, K. (ed.) GECCO '21: Genetic and Evolutionary Computation Conference, Companion Volume, Lille, France, July 10-14, 2021, pp. 1375–1384. ACM (2021). https://doi.org/10.1145/3449726.3463167

37. Vermetten, D., Caraffini, F., Kononova, A.V., Bäck, T.: Modular differential evolution. In: Silva, S., Paquete, L. (eds.) Proceedings of the Genetic and Evolutionary Computation Conference, GECCO 2023, Lisbon, Portugal, July 15-19, 2023, pp. 864–872. ACM (2023). https://doi.org/10.1145/3583131.3590417

38. Bennet, P., Doerr, C., Moreau, A., Rapin, J., Teytaud, F., Teytaud, O.: Nevergrad: black-box optimization platform. ACM SIGEVOlution **14**(1), 8–15 (2021)

39. Thomson, S.L., Renau, Q., Vermetten, D., Hart, E., van Stein, N., Kononova, A.V.: Code, data, and plots for the paper: Stalling in Space: Attractor Analysis for any Algorithm (2024). https://zenodo.org/records/14170241

Using Local Correlation Between Objectives to Detect Problem Modality

Tea Tušar[1,2]([✉]) [iD] and Jordan N. Cork[1,2]([✉]) [iD]

[1] Department of Intelligent Systems, Jožef Stefan Institute, Ljubljana, Slovenia
{tea.tusar,jordan.cork}@ijs.si
[2] Jožef Stefan Postgraduate School, Ljubljana, Slovenia

Abstract. Understanding the various characteristics of multiobjective optimization problems (MOPs) is crucial for designing and configuring optimization algorithms to efficiently solve them. This paper introduces a method that uses the estimation of local correlation between objectives to transform MOP landscapes into single-objective problem (SOP) landscapes. With this transformation, we make it possible to apply SOP landscape features to MOPs, thereby extracting valuable information about problem properties, such as modality. Our approach integrates both sample-based and search-based features, which are assessed for their ability to distinguish between unimodal, moderately multimodal, and highly multimodal MOPs. The proposed method is validated through a two-phase experimental setup. In the first phase, we select features that can reliably identify problem modality under ideal conditions with abundant data. The second phase evaluates their performance in more realistic scenarios with smaller samples and higher problem dimensions. The results show that features computed on the local correlation landscape achieve comparable or better performance than existing MOP features. These findings demonstrate the capability of SOP features to generalize to MOPs, showcasing their potential for characterizing MOP landscapes and inspiring future research on extending this approach to uncover additional problem properties.

Keywords: Multiobjective optimization · Correlation between objectives · Landscape features · Problem modality

1 Introduction

The efficiency of an optimization algorithm highly depends on the properties of the problem it is employed to solve. Being able to describe an optimization problem in terms of its characteristics, such as modality and separability, is therefore valuable as it enables one to choose and/or configure an algorithm to efficiently solve it. This work is concerned with continuous *black-box* multiobjective optimization problems, for which the objective function definitions are unknown to the optimizer. Because of this, most black-box problem properties are hard to detect. One way to assess them is by sampling the problem and using these solutions to

© The Author(s), under exclusive license to Springer Nature Switzerland AG 2025
P. García-Sánchez et al. (Eds.): EvoApplications 2025, LNCS 15612, pp. 527–542, 2025.
https://doi.org/10.1007/978-3-031-90062-4_33

compute *problem landscape features*, low-level numerical attributes that can be used to successfully predict certain high-level problem properties [22].

However, most research on landscape features is focused on Single-objective Optimization Problems (SOPs; see [11] for a collection of many works proposing such features), while Multiobjective Optimization Problems (MOPs) have received much less attention. A notable exception is a fairly recent paper proposing features for continuous (unconstrained) MOP landscapes [17]. Because of this gap, it would be particularly beneficial to be able to apply the many SOP features to MOP landscapes, thus acquiring additional features to further characterize MOPs.

There are several ways in which the landscape of a continuous MOP can be reduced to a single function and, therefore, viewed similarly to that of a continuous SOP—imagined as a terrain with peaks, basins, valleys and plateaus. Examples include the global dominance rank ratio [7], local dominance [6], optimal trade-offs [30], gradient length [4,10], local correlation [3] and Pareto [15] landscapes. We base our work on the recent local correlation landscapes due to their close tie to problem modality and the interpretability of the correlation coefficient values.

The main idea of this paper is thus to use a sample of solutions to construct an approximate local correlation landscape of an MOP and then apply 'single-objective' Exploratory Landscape Analysis (ELA) [22] to compute its features. As the local correlation landscape is highly related to problem modality, we devise two experiments, testing whether the resulting features are able to express this important problem property and compare their performance to that of existing MOP features from [17].

In the following, Sect. 2 presents some basic concepts, the local correlation landscapes and problem landscape features, while Sect. 3 explains how they are used to extract problem modality features. Next, Sect. 4 details the experimental evaluation of our approach and Sect. 5 concludes the paper with a summary and ideas for future work.

2   Background

2.1   Multiobjective Optimization Problems

We are interested in continuous multiobjective *minimization* problems that can be formally defined as:

$$\min_{x \in \mathbb{R}^d} F(x) = (f_1(x), f_2(x), \dots, f_m(x)),$$

where $\mathbb{R}^d$ is the *search space*, d is the problem dimension, and f_i, $i \in \{1, \dots, m\}$, are the m objective functions.

Solution $x \in \mathbb{R}^d$ *dominates* solution $y \in \mathbb{R}^d$, iff $f_i(x) \leq f_i(y)$ for all $i \in \{1, \dots, m\}$ and at least one of these inequalities is strict. Solutions which are not dominated by any other solution in the search space are *Pareto optimal*. All Pareto optimal solutions constitute the *Pareto set*. Its image in the *objective space* $\mathbb{R}^m$ is called the *Pareto front*.

A solution $x \in \mathbb{R}^d$ is *locally optimal* if it is not dominated by any other solution from its neighborhood N, $x \in N \subset \mathbb{R}^d$. If any locally optimal solution is also Pareto optimal, the problem in *unimodal*. Otherwise, it is *multimodal*.

In the remainder of the paper, we will be dealing with bi-objective problems, that is, $m = 2$ for all problems.

2.2 Local Correlation Landscapes

The concept that differentiates MOPs from SOPs is not the mere presence of multiple objectives, but the fact that they are typically *in conflict*, resulting in MOPs having multiple Pareto optimal trade-off solutions. In contrast, if the objectives would be in perfect *harmony*, i.e., completely equal, the MOP would be equivalent to the corresponding SOP. Therefore, when dealing with MOPs, we usually assume that they have conflicting objectives.

However, the conflict between two objectives exists primarily on the locally optimal sets and their 'vicinity', not the entire search space. In fact, it is a local problem property, not a global one, which is often ignored or disregarded.

To explain its local nature, we first need to formalize the relationship between two objectives. We can do this by considering their *correlation*. The correlation between two objectives can be estimated by the Pearson correlation coefficient [29], which measures the linear correlation between the objectives of a sample of solutions and takes a value between -1 (perfect linear anti-correlation that corresponds to conflicted objectives) and 1 (perfect linear correlation that corresponds to harmonious objectives). A zero value implies there is no linear dependency between the objectives, i.e., the objectives are neither conflicted nor in harmony.

Consider the simple example of the two-dimensional double sphere problem presented in Fig. 1. This is a bi-objective problem defined on the search space $[-5, 5]^2$, where each objective is a sphere function with the optimum located at a different point in the search space. The Pareto set of this function is the line segment connecting the two single-objective optima (shown in black[1] in Fig. 1a). The Pearson correlation coefficient visualized in Fig. 1b is computed for each grid point from a set of 100 solutions in its close proximity (see [3] for more details).

We can see that the Pearson correlation coefficient between the two objectives depends on the position in the search space. The correlation values along the Pareto set equal -1. This is to be expected as on the Pareto set, one cannot improve in one objective without deteriorating in the other. With increasing distance from the Pareto set in a direction perpendicular to it, the correlation coefficient increases, eventually becoming positive. On the parts of the line with the two single-objective optima that go beyond the Pareto set, the correlation coefficient takes on the value of 1, which is again understandable since at that location, a move in the direction toward the Pareto set results in simultaneous improvement in both objectives.

---

[1] Note that the black region in the plot is thicker than a line because of the discretization of the search space into a 501×501 grid for visualization purposes.

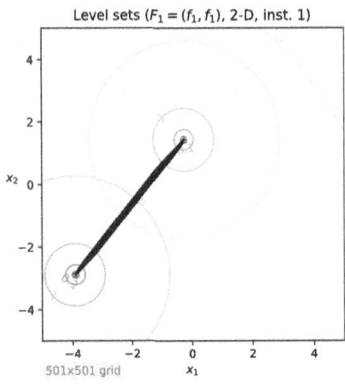
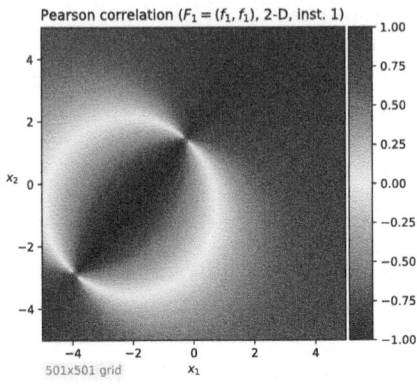

(a) Level sets for the two objectives in purple and green and the Pareto set approximation in black.

(b) Values of the Pearson correlation coefficient ranging from -1 in dark red to 1 in dark blue.

Fig. 1. Two grid-based visualizations of the search space of the first instance of the 2-D double sphere problem F_1 from the `bbob-biobj` suite [4] of the COCO platform [8].

To summarize, even for a simple unimodal problem such as the one from Fig. 1, the correlation between objectives is not constant, but depends on the position in the search space. On the Pareto set and in certain regions that are close to it, the correlation is negative, in others, it is zero or positive.

The relationship between objectives becomes even more complex when they are less regular or multimodal—see the examples from Fig. 2 in Sect. 4. There, we can see that some unimodal problems have anti-correlated objectives not only close to the Pareto set, but also far away from it. Additionally, visualizations of local correlations on multimodal problems demonstrate that many distinct anti-correlated regions can be located throughout the search space, surrounded by regions with correlated objectives.

The correlation between objectives is closely connected to problem modality and to the bi-objective gradient [10] as it equals -1 on any locally optimal set of solutions, not just the Pareto set. This is why the plots of multimodal problems in Fig. 2 contain many distinct regions with a negative correlation—one per locally optimal set of solutions.

Although these examples demonstrate that the concept of a 'global correlation between objectives' is effectively meaningless, the relationships between the objectives, as well as their mutual correlations, are almost always discussed solely on the global scale. Even a recently published book chapter [5] that provides an overview of the use of correlations among objectives in multiobjective optimization, explores several ways of estimating correlation in addition to the Pearson correlation coefficient and reviews the use of correlations for reducing redundant objectives, does not address their local nature. Similarly holds for the MOP feature `f_cor` from [17], which equals the (Spearman) correlation among objective values and is measured on the entire sample, i.e., is treated globally.

The local correlation landscape, therefore, provides an insightful view of the problem that is worth further exploration. However, this is a limited view as it does not contain enough information by itself to infer whether a locally optimal solution is also Pareto optimal. This is why any features computed solely from the local correlation landscape can meaningfully characterize only certain problem features, like (but not necessarily limited to) modality.

2.3 Problem Landscape Features

Most problem landscape features require only a set of solutions, called a *sample*, to be computed. Typically, the sample is generated with a procedure that tries to evenly cover the search space, such as Latin Hypercube Sampling (LHS) [21], and is evaluated beforehand. We will call such features *sample-based features*. However, there are also other problem features that require additional solution evaluations to be computed. For example, they can be based on a random walk [20], a hill climber run [1] or basin hopping iterations [2], to name a few. We will refer to these as *search-based features*. In real-world optimization scenarios, especially those with time-consuming evaluations, the latter might not always be retrievable. In this work, we use both sample- and search-based features, but only those search-based ones for which we can limit the number of additional evaluations.

The set of considered SOP features thus includes a total of 117 features that can be categorized into the following groups: dispersion features [19], classical ELA features (convexity, y-distribution, levelset, and meta model features) [22], fitness distance correlation features [9,24], cell mapping features (angle, convexity and gradient homogeneity features) [12], information content features [25], gradient features [20], nearest better clustering features [13], length-scale features [23], linear model features [14], and principal component features [14]. Of these, only the ELA convexity features, the gradient features and the length-scale features (with a total of 18) are search-based, the rest (99) are sample-based. All SOP features were computed with the `pflacco` Python library [27,28].

The set of MOP features used in the comparison comprises the 49 features from [17], which include global landscape features (among them, the global correlation between objectives), multimodality features, evolvability features and ruggedness features. All features are sample-based and were computed with the freely available `features.R` R script [16,17].

3 Detecting Problem Modality

The basic idea of this paper is to test whether features computed on the local correlation landscape can be used to detect problem modality. This is essentially done in three steps:

1. Approximate the local correlation landscape of the problem.
2. Compute SOP features of this landscape.
3. Measure the feature success in detecting problem modality.

Step 1: Approximating the Local Correlation Landscape. The execution of this step depends on the type of features—whether they are sample- or search-based, since search-based features guide the choice of solutions in the sample.

For sample-based features, the sample of solutions is retrieved independently from the features. This is done using LHS. Since we are interested in the *local* correlation of objectives, we need to define the neighborhood of solutions. For each solution in the sample, the neighborhood is comprised of n closest solutions to it, in terms of the Euclidean distance. This always includes the solution itself. The local correlation between objectives at each solution is then estimated by computing the Pearson correlation coefficient using the objective values for all solutions in its neighborhood.

Because search-based features use some inherent procedure to select the solutions to be evaluated, we cannot build the entire local correlation landscape upfront. Therefore, we first take a small ratio of the entire sample size s to produce an initial sample of solutions using LHS. Then, we construct the initial local correlation landscape using the same neighborhood definition as for sample-based features. Next, we let the search-based feature guide the choice of the subsequent solutions. For each, we find its current n closest neighbors and use them to approximate its local correlation value with the Pearson correlation coefficient. Note that the estimation of the local correlation for search-based features is less accurate at the beginning (when only a few solutions are available) than at the end.

Step 2: Computing Landscape Features. This step is straightforward—it requires computing the SOP feature values using the local correlation landscape instead of an objective landscape.

Step 3: Measuring Feature Success. Finally, feature success is measured by determining whether the feature can successfully differentiate between three groups of problems: unimodal, moderately multimodal and highly multimodal ones. We use clustering for this, because we are interested in the prediction capabilities of the feature. First, feature values are clustered into three clusters by k-means clustering with a fixed $k = 3$ [18, 26]. Then, we count the errors—number of problems that have not been clustered correctly[2]. The lower the error, the better the feature in detecting problem modality.

4   Experiments and Results

In this section we first explain the various problems used in the experiments. Then we present the two experiments and their results.

---

[2] This is not trivial to do because there is no fixed order in how k-means labels clusters and feature values can be increasing or decreasing with increasing problem modality. Therefore, we check all possible 2^3 orderings of the three clusters and use the one with the smallest error count.

Many of the features listed in Sect. 2.3 are not useful for detecting problem modality, which could diminish the predictive capability of the entire set of features. To avoid this, we split the study into two parts. In the first experiment, we identify individual features that are able to differentiate well between unimodal, moderately multimodal and highly multimodal problems when given a lot of data at their disposal. Then, in the second experiment, we use only these features to more comprehensively test their capabilities in a real-world-like scenario with less available data. Before detailing the two experiments, we present the problems used in both of them.

4.1 Problems

To test our idea, we need a selection of problems with diverse modality. While we first planned to use only problems from the bbob-biobj suite [4] of the COCO platform [8], they do not cover the modality range well enough, as they are either unimodal or highly multimodal. To fill this gap, we construct the moderately multimodal problems ourselves.

The set of moderately multimodal problems are Python implementations of Wessing's Multiple Peaks Model problems [31], here labeled multi-peak problems. For each objective, a multiple peak function is generated by taking the minimum value of a set of individual peak functions. Each individual peak function consists of a center point and a positive definite Hessian matrix. The separate problems within the set were generated by randomly configuring these center and matrix settings. The number of peaks per objective, however, were set manually, to provide different degrees of modality within the moderate range. The degree of modality is determined by the combinations of peaks between objectives, which each provide a basin of attraction.

Table 1 presents the 15 problems selected for this study. P_1–P_5 are unimodal bbob-biobj problems, P_6–P_{10} are moderately mutimodal multi-peak problems and P_{11}–P_{15} are highly multimodal bbob-biobj problems. While all problems can be instantiated in any dimension, $d \in \{2, 3, 5, 10\}$ is used in this work. The local correlation landscapes for all 15 2-D problems are shown in Fig. 2.

We can see that the multi-peak problems indeed represent the middle ground between the unimodal and highly multimodal bbob-biobj problems. We can also hypothesize that among all problems, P_{14} might be the hardest to categorize correctly, as its local optima are located in a relatively small region of the entire search space, which can be easily overlooked, especially with sparse sampling.

4.2 First Experiment

Experimental Setup. To find features with a potential for detecting problem modality, we simplify the task as much as possible. We use only 2-D problems and provide a large budget of $s = 10\,000$ solutions to compute the features. For sample-based features, all solutions are placed on the 100×100 grid, while for search-based features, the initial grid contains 32×32 solutions (which roughly equals 10% of the budget s) and the rest is made available to the method to

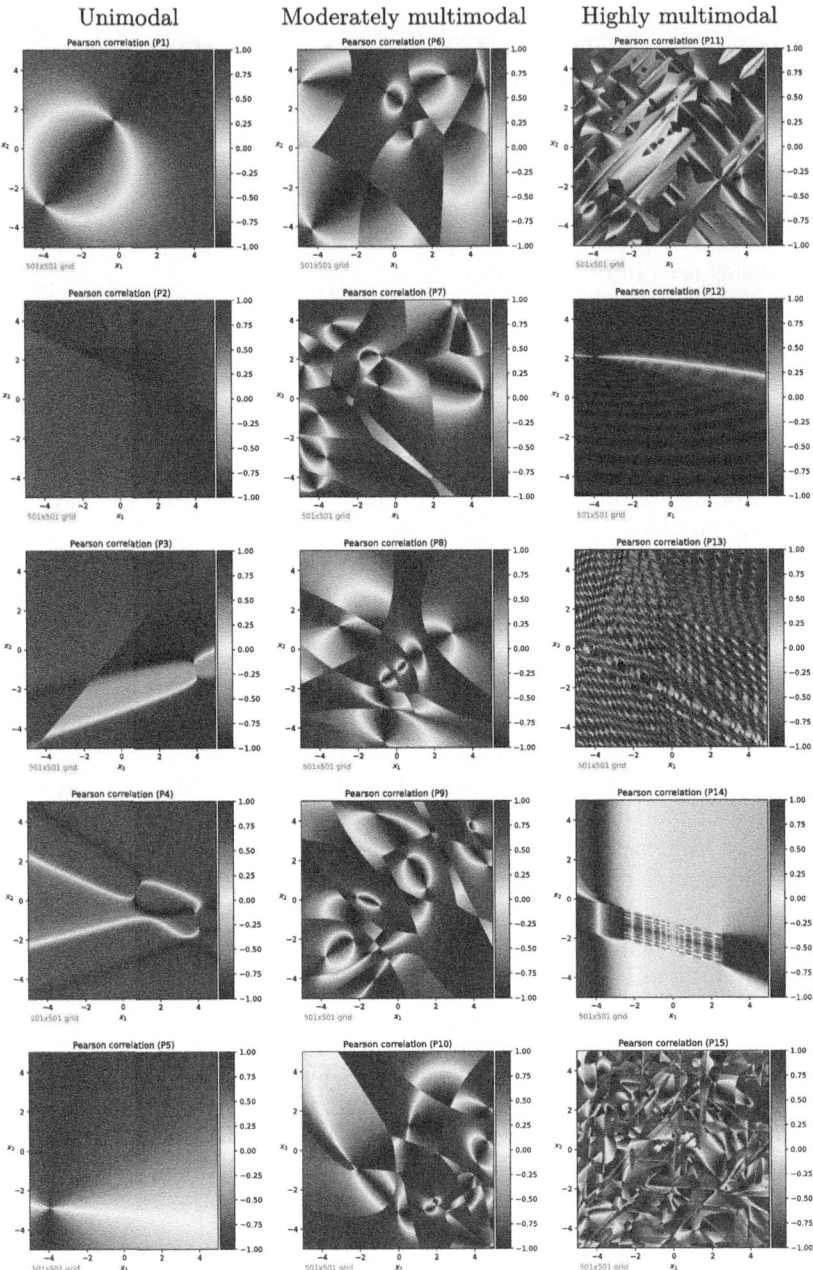

Fig. 2. Local correlation landscapes calculated with the Pearson correlation coefficient for 2-D unimodal problems P_1–P_5 (left column), moderately multimodal problems P_6–P_{10} (middle column) and highly multimodal problems P_{11}–P_{15} (right column). Red areas denote regions with negatively correlated objectives, while blue areas designate regions with positively correlated objectives.

Table 1. The 15 problems employed in this study. We always use only the first instance of a bbob-biobj problem. For multi-peak problems, the two numbers in brackets determine the number of peaks in the first and second objective.

Unimodal problems

P_1	bbob-biobj problem $F_1 = (f_1, f_1)$
P_2	bbob-biobj problem $F_{14} = (f_2, f_{13})$
P_3	bbob-biobj problem $F_{36} = (f_{13}, f_{14})$
P_4	bbob-biobj problem $F_{41} = (f_{14}, f_{14})$
P_5	bbob-biobj problem $F_{58} = (f_1, f_5)$

Moderately multimodal problems

P_6	multi-peak problem with $(2, 5)$ peaks
P_7	multi-peak problem with $(2, 10)$ peaks
P_8	multi-peak problem with $(4, 4)$ peaks
P_9	multi-peak problem with $(4, 8)$ peaks
P_{10}	multi-peak problem with $(6, 6)$ peaks

Highly multimodal problems

P_{11}	bbob-biobj problem $F_{10} = (f_1, f_{21})$
P_{12}	bbob-biobj problem $F_{17} = (f_2, f_{17})$
P_{13}	bbob-biobj problem $F_{24} = (f_6, f_{15})$
P_{14}	bbob-biobj problem $F_{33} = (f_8, f_{20})$
P_{15}	bbob-biobj problem $F_{55} = (f_{21}, f_{21})$

With f_i we denote a single-objective bbob function as follows:

f_1 sphere	f_8 original Rosenbrock	f_{17} Schaffers F7
f_2 ellipsoidal	f_{13} sharp ridge	f_{20} Schwefel
f_5 linear slope	f_{14} different powers	f_{21} Gallaghers Gaussian
f_6 attractive sector	f_{15} Rastrigin	101-medium peaks

sample the space according to its principle. In both cases, the neighborhood size n equals 9, which corresponds to the Moore neighborhood for the internal grid solutions.

Results and Discussion. In this experiment, we apply k-means clustering separately for each feature. The number of clustering errors committed on the 15 problems ranges from one to ten and is collected for all features in the histogram in Fig. 3a. The colors distinguish among features of the three different types (search- and sample-based SOP features, and MOP features). We can see that the distribution over error counts is roughly similar for all feature types with most features performing very badly (making six or more mistakes on 15 problems). Among the best features (making less than five mistakes) we have 22% of all sample-based

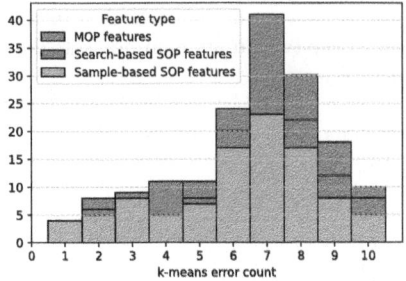
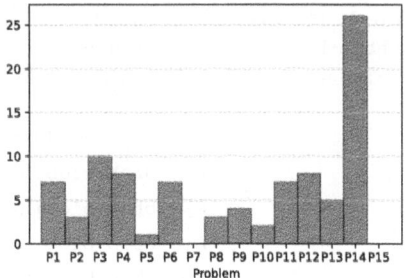

(a) Histogram of the clustering error count grouped by feature type.

(b) Number of clustering errors on each problem using only the best features.

Fig. 3. Results of the first experiment applying k-means clustering separately for each feature.

Table 2. The errors of the best SOP features computed on the local correlation landscape with 10 000 sampled solutions. There are 24 features with a clustering error lower than five. The only two search-based features are denoted by an asterisk (*), while the rest of them are sample-based.

Feature	Error	Feature	Error
cm_angle.y_ratio_best2worst_mean	1	cm_angle.dist_ctr2worst_mean	3
cm_grad.mean	1	cm_angle.dist_ctr2worst_sd	3
nbc.nb_fitness.cor	1	cm_conv.convex.hard	3
nbc.nn_nb.mean_ratio	1	disp.diff_median_02	3
cm_angle.angle_mean	2	disp.ratio_median_02	3
cm_conv.concave.hard	2	*ela_conv.lin_dev_abs	3
*gradient.g_avg	2	ic.eps_s	3
ic.costs_runtime	2	disp.diff_median_05	4
ic.eps_ratio	2	disp.ratio_median_05	4
limo.length_mean	2	ela_level.mmce_qda_50	4
cm_angle.dist_ctr2best_mean	3	ic.eps_max	4
cm_angle.dist_ctr2best_sd	3	ic.h_max	4

SOP features, 11% of all search-based SOP features and 14% of all MOP features. We set the threshold for 'good' features to four or fewer to discard features which clearly cannot distinguish among problems of different modality even when provided with plenty of data, but still keep enough to experiment with.

The complete list of the best 24 SOP features is given in Table 2. We can see that they come from various groups, with cell mapping, nearest better clustering, information content and dispersion features being the most well represented. These results are very positive as they show that we have a large number of SOP features that can be applied on local correlation landscapes to detect problem modality.

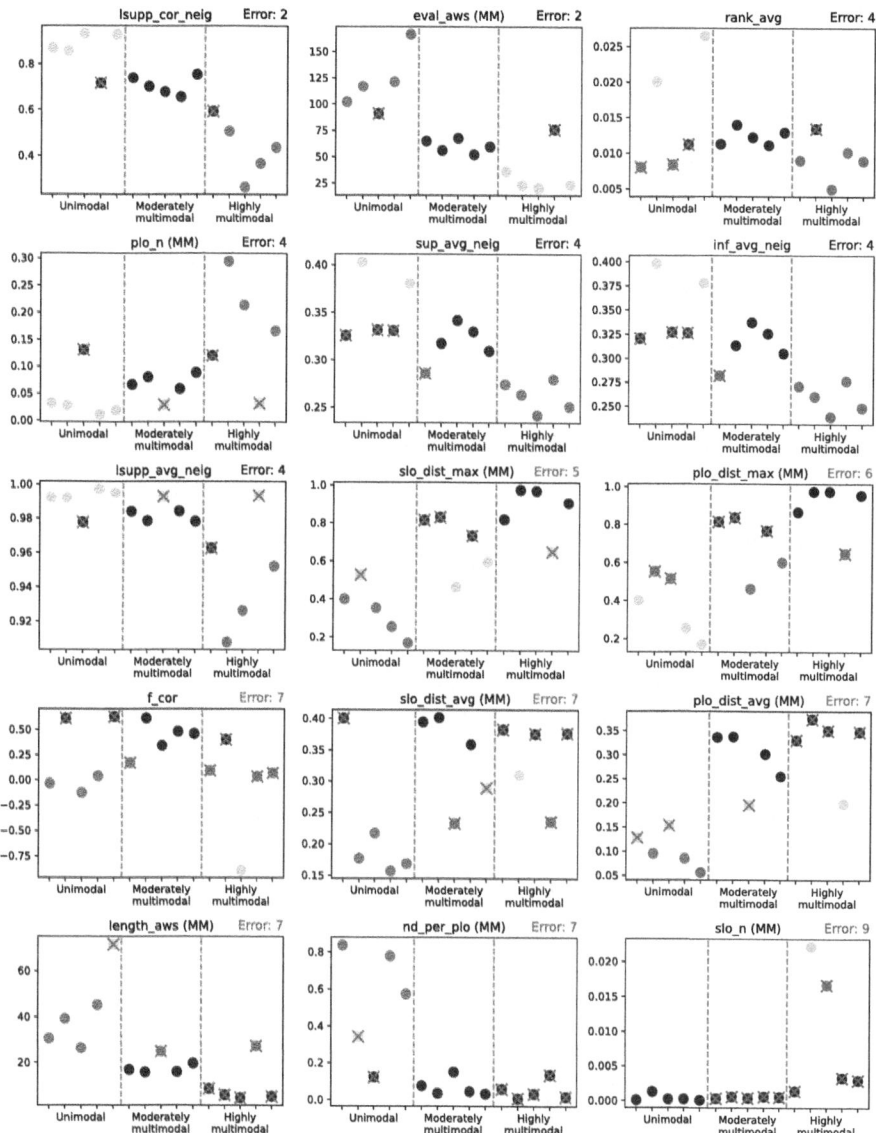

Fig. 4. Feature values (on the y-axis) for all 2-D problems (on the x-axis) computed on samples with 10 000 solutions for a selection of MOP features that is comprised by all seven features with an error smaller than five, all multimodal features (denoted by '(MM)' after their name) and the correlation feature f_cor. The features are sorted in ascending order of their error. The color of the dots represents the cluster determined by k-means and a red cross denotes every incorrectly categorized problem.

Next, the results for some chosen MOP features are presented in greater detail in Fig. 4. Every plot contains dots showing the feature value (y-axis) on each of the 15 problems (x-axis). Their colors denote the cluster assigned to that problem by k-means, while the red crosses represent wrongly categorized problems. This visualization comprises all seven MOP features with an error smaller than five, all nine multimodal (MM) features and the feature f_cor measuring correlation between objectives. The features are sorted in ascending error count. We see that only two of the nine multimodal features make less than five errors on 15 problems when detecting their modality and five other features outperform the rest of the multimodal ones. Also, we empirically show that the global correlation feature has very little meaning (see the first plot of the fourth row in Fig. 4). According to f_cor, most of the 15 problems have mildly correlated objectives (values between 0 and 0.5), with P_{13} being the only problem with highly anti-correlated objectives, which is incorrect.

Finally, Fig. 3b shows which of the problems were most often wrongly clustered by the 31 best features. We see a clear outlier—problem P_{14} that stands out from other highly multimodal problems in our set because its local optima are concentrated in a relatively small part of the search space.

4.3 Second Experiment

Experimental Setup. Only the 31 best features identified in the first experiment are included in the second part of this study. Here, we investigate how larger problem dimensions $d \in \{2, 3, 5, 10\}$ and smaller sample sizes $s \in \{200d, 1000d\}$ affect the capability of features to detect problem modality. In addition, to find a good neighborhood size n, we experiment with two settings, $n \in \{5, 10\}$. Similarly as before, for search-based SOP features, only 10% of the sample size s is created by LHS, while the rest is used to explore the search space according to the feature method. We repeat all the experiments five times, using different samples.

Results and Discussion. First, we discuss the results of using k-means clustering on separate features. Figure 5a shows how its error count depends on the type of feature, the problem dimension and sample size. The neighborhood size is not shown separately as it does not visibly affect the results. We can see that the sample size has a large effect on the feature capability to discern problem modality with the larger sample size ($1000d$) generally supporting better results than the smaller one ($200d$). The effect of problem dimension d is also visible—the error count typically (but not always) increases with higher dimensions. Both results are in line with expectations. Finally, the comparison among the three feature types shows that SOP features computed on the local correlation landscapes perform comparable to MOP ones. A visibly better performance is achieved by the two search-based SOP features only on 2- and 3-D problems with a large sample size.

This means that the excellent results achieved in the first experiment, where SOP features on local landscapes were outperforming MOP features, were not

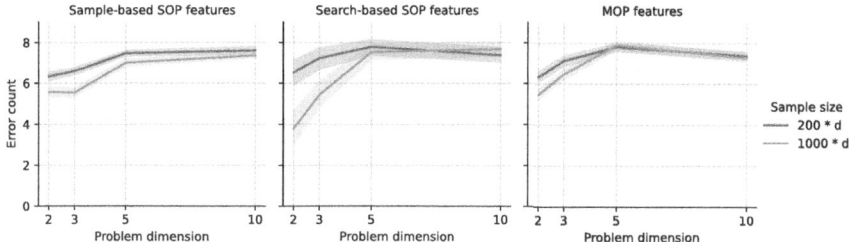

(a) k-means clustering results achieved using a single feature are here aggregated over all features of the same type, multiple samples and neighborhood sizes.

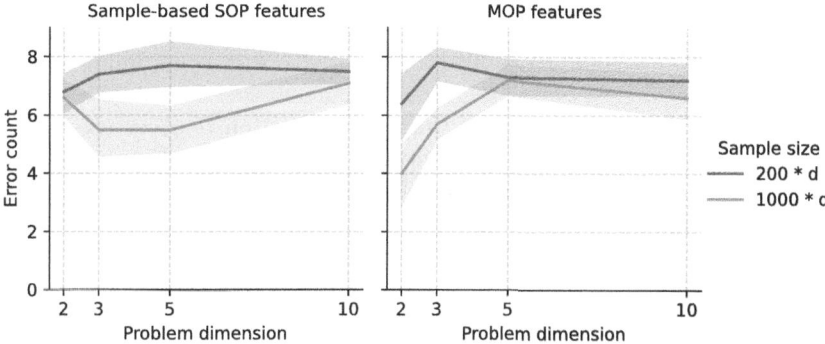

(b) k-means clustering results achieved using all features of the same type are here aggregated over multiple samples and neighborhood sizes. Search-based SOP features are missing because their aggregation is not meaningful.

Fig. 5. Results of k-means clustering on (a) separate features and (b) all features of the same type. The plots show how clustering error count (y-axis) depends on the problem dimension d (x-axis) and sample size s (color). The line represents the mean, while the shaded region corresponds to the 95 % confidence interval.

replicated in the more difficult scenario with higher problem dimensions and less available data. Still, the approach achieved results that are generally not worse than those by MOP features, meaning that it has established its merit.

However, the predictive power of features can be combined. Therefore, we present in Fig. 5b k-means clustering results using all features of the same type. The search-based SOP features are excluded from this analysis, because the two features of this type come from two different methods, meaning that their resulting samples are different and cannot be meaningfully combined. Similarly as before, we see a fairly reliable effect of the sample size and problem dimension (with the notable exception of sample-based SOP features with $1000d$ samples on dimension 2 that perform worse than expected). Surprisingly, combining the features does generally not help to (considerably) improve their separate results.

Finally, an analysis of the errors per problem (results not pictured) does not result in any stark outliers as the one from Fig. 3b. Rather, all problems are similarly difficult (or easy) to categorize, with slightly higher errors achieved on unimodal problems P_2 and P_4. A rather surprising result for which we cannot yet provide an explanation.

5   Conclusions

This paper demonstrates that estimating local correlation between objectives can effectively transform a multiobjective problem landscape into a single-objective one. This transformation enables the application of SOP features, including both sample-based and search-based features, to MOPs, facilitating the extraction of valuable information about problem characteristics—in this case, problem modality. Furthermore, this research paves the way for exploring alternative transformations that could be applied to similarly capture other important problem properties.

An important limitation of this work is its focus on bi-objective problems, as correlation can only be computed between two objectives. For problems with three or more objectives, only pairwise correlation values can be obtained, making it impossible to calculate a direct multi-way correlation. In future work, we would like to explore potential approaches to overcome this limitation.

Acknowledgments. We acknowledge financial support from the Slovenian Research and Innovation Agency (research core funding No. P2-0209 and projects No. GC-0001 "Artificial Intelligence for Science" and N2-0254 "Constrained Multiobjective Optimization Based on Problem Landscape Analysis").

Disclosure of Interests. The authors have no competing interests to declare that are relevant to the content of this article.

References

1. Abell, T., Malitsky, Y., Tierney, K.: Features for exploiting black-box optimization problem structure. In: Nicosia, G., Pardalos, P. (eds.) LION 2013. LNCS, vol. 7997, pp. 30–36. Springer, Heidelberg (2013). https://doi.org/10.1007/978-3-642-44973-4_4

2. Adair, J., Ochoa, G., Malan, K.M.: Local optima networks for continuous fitness landscapes. In: Companion Proceedings of the Genetic and Evolutionary Computation Conference (GECCO 2019). pp. 1407–1414. Association for Computing Machinery, New York, NY, USA (2019). https://doi.org/10.1145/3319619.3326852

3. Allmendinger, R., Fonseca, C.M., Sayin, S., Wiecek, M.M., Stiglmayr, M.: Multiobjective optimization on a budget (Dagstuhl Seminar 23361). Dagstuhl Reports **13**(9), 1–68 (2023). https://doi.org/10.4230/DAGREP.13.9.1

4. Brockhoff, D., Auger, A., Hansen, N., Tušar, T.: Using well-understood single-objective functions in multiobjective black-box optimization test suites. Evol. Comput. **30**(2), 165–193 (2022). https://doi.org/10.1162/EVCO_A_00298

5. Chugh, T., Gaspar-Cunha, A., Deutz, A.H., Duro, J.A., Oara, D.C., Rahat, A.: Identifying correlations in understanding and solving many-objective optimisation problems, pp. 241–267. Springer International Publishing, Cham (2023). https://doi.org/10.1007/978-3-031-25263-1_9

6. Fieldsend, J.E., Chugh, T., Allmendinger, R., Miettinen, K.: A feature rich distance-based many-objective visualisable test problem generator. In: Proceedigs of the Genetic and Evolutionary Computation Conference (GECCO 2019). pp. 541–549. Association for Computing Machinery, New York, NY, USA (2019). https://doi.org/10.1145/3321707.3321727

7. Fonseca, C.M.: Multiobjective genetic algorithms with application to control engineering problems. Ph.D. thesis, University of Sheffield (1995)

8. Hansen, N., Auger, A., Ros, R., Mersmann, O., Tušar, T., Brockhoff, D.: COCO: a platform for comparing continuous optimizers in a black-box setting. Optim. Methods Softw. **36**(1), 114–144 (2021). https://doi.org/10.1080/10556788.2020.1808977

9. Jones, T., Forrest, S.: Fitness distance correlation as a measure of problem difficulty for genetic algorithms. In: Proceedings of the International Conference on Genetic Algorithms (ICGA 1995), pp. 184–192. Morgan Kaufmann Publishers Inc., San Francisco, CA, USA (1995). https://doi.org/10.5555/645514.657929

10. Kerschke, P., Grimme, C.: An expedition to multimodal multi-objective optimization landscapes. In: Trautmann, H., et al. (eds.) EMO 2017. LNCS, vol. 10173, pp. 329–343. Springer, Cham (2017). https://doi.org/10.1007/978-3-319-54157-0_23

11. Kerschke, P., Preuss, M.: Exploratory landscape analysis. In: Companion Proceedings of the Genetic and Evolutionary Computation Conference (GECCO 209), pp. 1137–1155. Association for Computing Machinery, New York, NY, USA (2019). https://doi.org/10.1145/3319619.3323389

12. Kerschke, P., et al.: Cell mapping techniques for exploratory landscape analysis. In: Tantar, A.A., et al. (eds.) EVOLVE – A Bridge between Probability, Set Oriented Numerics, and Evolutionary Computation V. Advances in Intelligent Systems and Computing, vol. 288, pp. 115–131. Springer International Publishing, Cham (2014). https://doi.org/10.1007/978-3-319-07494-8_9

13. Kerschke, P., Preuss, M., Wessing, S., Trautmann, H.: Detecting funnel structures by means of exploratory landscape analysis. In: Proceedings of the Conference on Genetic and Evolutionary Computation (GECCO 2015), pp. 265–272. Association for Computing Machinery, New York, NY, USA (2015). https://doi.org/10.1145/2739480.2754642

14. Kerschke, P., Trautmann, H.: Comprehensive feature-based landscape analysis of continuous and constrained optimization problems using the R-package flacco, pp. 93–123. Springer International Publishing, Cham (2019). https://doi.org/10.1007/978-3-030-25147-5_7

15. Liang, Z., Cui, Z., Li, M.: Pareto landscape: visualising the landscape of multi-objective optimisation problems. In: Affenzeller, M., et al. (eds.) Parallel Problem Solving from Nature (PPSN XVIII), pp. 299–315. Springer Nature Switzerland, Cham (2024). https://doi.org/10.1007/978-3-031-70085-9_19

16. Liefooghe, A.: Landscape features for MO-ICOPs. https://gitlab.com/aliefooghe/landscape-features-mo-icops (2024). https://gitlab.com/aliefooghe/landscape-features-mo-icops, gitLab repository

17. Liefooghe, A., Verel, S., Lacroix, B., Zăvoianu, A.C., McCall, J.: Landscape features and automated algorithm selection for multi-objective interpolated continuous optimisation problems. In: Proceedings of the Genetic and Evolutionary Computation Conference (GECCO 2021), pp. 421–429. Association for Computing Machinery, New York, NY, USA (2021). https://doi.org/10.1145/3449639.3459353

18. Lloyd, S.P.: Least squares quantization in PCM. IEEE Trans. Inf. Theory **28**(2), 129–137 (1982). https://doi.org/10.1109/TIT.1982.1056489

19. Lunacek, M., Whitley, D.: The dispersion metric and the CMA evolution strategy. In: Proceedings of the Conference on Genetic and Evolutionary Computation (GECCO 2006), pp. 477–484. Association for Computing Machinery, New York, NY, USA (2006). https://doi.org/10.1145/1143997.1144085

20. Malan, K.M., Oberholzer, J.F., Engelbrecht, A.P.: Characterising constrained continuous optimisation problems. In: Proceedings of the Congress on Evolutionary Computation (CEC 2015), pp. 1351–1358. IEEE (2015). https://doi.org/10.1109/CEC.2015.7257045

21. McKay, M.D., Beckman, R.J., Conover, W.J.: A comparison of three methods for selecting values of input variables in the analysis of output from a computer code. Technometrics **42**(1), 55–61 (1979)

22. Mersmann, O., Bischl, B., Trautmann, H., Preuss, M., Weihs, C., Rudolph, G.: Exploratory landscape analysis. In: Proceedings of the Conference on Genetic and Evolutionary Computation (GECCO 2011), pp. 829–836. Association for Computing Machinery, New York, NY, USA (2011). https://doi.org/10.1145/2001576.2001690

23. Morgan, R., Gallagher, M.: Analysing and characterising optimization problems using length scale. Soft. Comput. **21**(7), 1735–1752 (2015). https://doi.org/10.1007/s00500-015-1878-z

24. Müller, C.L., Sbalzarini, I.F.: Global characterization of the CEC 2005 fitness landscapes using fitness-distance analysis. In: Di Chio, C., et al. (eds.) Applications of Evolutionary Computation. Lecture Notes in Computer Science, vol. 6624, pp. 294–303. Springer, Berlin, Heidelberg (2011). https://doi.org/10.1007/978-3-642-20525-5_30

25. Muñoz, M.A., Kirley, M., Halgamuge, S.K.: Exploratory landscape analysis of continuous space optimization problems using information content. IEEE Trans. Evol. Comput. **19**(1), 74–87 (2015). https://doi.org/10.1109/TEVC.2014.2302006

26. Pedregosa, F., et al.: Scikit-learn: machine learning in Python. J. Mach. Learn. Res. **12**, 2825–2830 (2011). http://jmlr.org/papers/v12/pedregosa11a.html

27. Prager, R.P.: Pflacco: feature-based landscape analysis of continuous and constrained optimization problems (2024). https://github.com/Reiyan/pflacco. Accessed 15 Nov 2024

28. Prager, R.P., Trautmann, H.: Pflacco: feature-based landscape analysis of continuous and constrained optimization problems in Python. Evol. Comput., 1–25 (2023). https://doi.org/10.1162/evco_a_00341

29. Rodgers, J.L., Nicewander, W.A.: Thirteen ways to look at the correlation coefficient. Am. Stat. **42**(1), 59–66 (1988). https://doi.org/10.1080/00031305.1988.10475524

30. Schäpermeier, L., Grimme, C., Kerschke, P.: One PLOT to show them all: visualization of efficient sets in multi-objective landscapes. In: Bäck, T., et al. (eds.) PPSN 2020. LNCS, vol. 12270, pp. 154–167. Springer, Cham (2020). https://doi.org/10.1007/978-3-030-58115-2_11

31. Wessing, S.: Two-stage methods for multimodal optimization. Ph.D. thesis, Technische Universität Dortmund, Fakultät für Informatik, Dortmund, Germany (2015). http://dx.doi.org/10.17877/DE290R-7804

Greater AI Design Control Aids Evolution of Computational Materials

Piper Welch[1(✉)], Monica Li[3], Shawn Beaulieu[1], Annie Xia[3], Dong Wang[3],
Medha Goyal[3], Atoosa Parsa[2], Corey S. O'Hern[3], Rebecca Kramer-Bottiglio[3],
and Josh Bongard[1]

[1] Department of Computer Science, University of Vermont, Burlington,
VT 05405, USA
{piper.welch,shawn.beaulieu,josh.bongard}@uvm.edu
[2] Department of Biology, Tufts University, Medford, MA 02155, USA
atoosa.parsa@tufts.edu
[3] Department of Mechanical Engineering and Materials Science, Yale University,
New Haven, CT 06520, USA
{monica.s.li,annie.xia,dong.wang,medha.goyal,corey.ohern,
rebecca.kramer}@yale.edu

Abstract. Unconventional computing may overcome some of the limitations of traditional silicon-based systems using alternative materials and computational mechanisms. However, due to their complex underlying dynamics and high-dimensional parameter space, the design of these materials such that they perform computation is non-intuitive, making AI-driven design attractive. It has been shown that evolutionary algorithms can tune the structural properties of grains within a granular material such that it computes logical functions. In recent years, programmable granular metamaterials have been developed so that multiple physical properties of individual grains can be altered independently. This raises the question of whether allowing evolutionary algorithms to tune more grain features within a granular material frustrates or facilitates its ability to embed computation. In this work, we show that the latter is the case, when grain sizes and stiffnesses are co-evolved to embed Boolean logic gates, compared to evolving just sizes or stiffnesses alone. We report physical verification of evolved designs, taking a further step toward the provision of alternatives to electronic computing.

Keywords: Granular Metamaterials · Mechanical Computing ·
Sim2Real

1 Introduction

For over fifty years, Moore's law has governed the rapid development of semiconductors as the density of transistors on silicon chips exponentially increased [11]. But as transistors shrink toward atomic scales, both theoretical and practical barriers are hindering this rate of growth [19]. The impending end of Moore's

© The Author(s), under exclusive license to Springer Nature Switzerland AG 2025
P. García-Sánchez et al. (Eds.): EvoApplications 2025, LNCS 15612, pp. 543–557, 2025.
https://doi.org/10.1007/978-3-031-90062-4_34

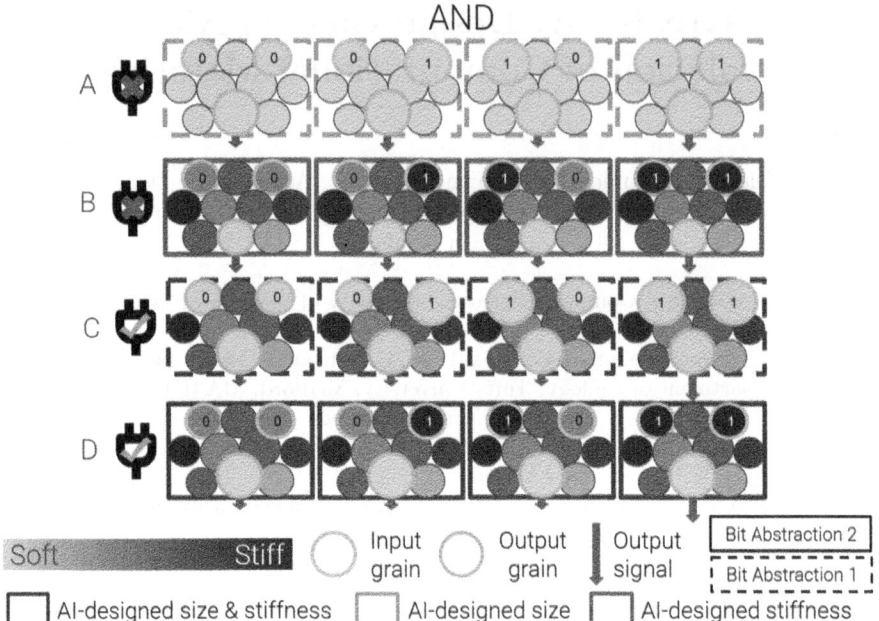

Fig. 1. Increasing AI design control over computational materials. (**A**): An evolutionary algorithm attempted to tune the sizes of 8 out of 10 grains in an *in silico* granular material to increase its "ANDness", but failed. Here, manually making the two input grains small or large provided the two input bits; the resulting force at the output grain (red arrow) was interpreted as the output. (**B**): In another material the evolutionary algorithm could tune only grain stiffnesses, and the two input grains were softened or stiffened to provide the input value of the logic gate. Again, the evolutionary algorithm failed. (**C**): In a third material the evolutionary algorithm could tune sizes and stiffnesses, and input values were provided via grain size change. Here AND behavior was achieved. (**D**): AND behavior was also achieved in a fourth material in which sizes and stiffnesses were evolved, but input was provided via grain stiffness change.

law signals an urgent need for alternative computational substrates capable of sustaining progress in computing beyond the limits of conventional silicon architectures. To this end, some investigators have expanded upon traditional semiconductor technologies, such as the work by Jayachandran *et al.*, who demonstrated the successful fabrication of 3D chips using non-silicon materials (MoS_2 and WSe_2) [7]. Meanwhile, others have explored entirely new paradigms, leading to the emergence of mechanical, chemical, neuromorphic, and quantum computing [4,10,18,23].

One promising novel computing strategy is mechanical computation using granular metamaterials, as there is currently no known upper limit on their computational density. This is due to the fact that these materials report the results of the computation as oscillatory signals at different frequencies simultaneously. The substrate of such a system, granular metamaterials, are materials with discrete units, or grains, whose strategic arrangement and local interactions can give rise to emergent properties not present in their constituent components. By tuning the spatial configuration and contact network of these grains, it is possible to engineer novel mechanical behaviors that arise solely from the collective dynamics of the system [3, 5, 6, 8, 15, 22].

Computational granular metamaterials (CGMMs) represent a class of granular metamaterials that have been designed, *in silico*, to execute logical operations [1, 12–14]. However, such materials are notoriously non-intuitive to understand and to design, so in all work to date, machine learning has been employed to tune the features of the grains within the materials such that they instantiate the desired computation. In all previous CGMM work, binary digits have been encoded as the presence ('1') or absence ('0') of mechanical vibration applied to manually specified grains representing the input ports. Then, the presence or absence of resulting vibration at a pre-specified output grain is interpreted as the value of the binary output. The benefit of encoding information as vibration is that one material can simultaneously act like more than one Boolean logic gate, with each gate operating at a different frequency. As the frequency spectrum has, in theory, infinite capacity, there is currently no known upper limit on CGMMs' computational density. The current experimental upper bound is 16 Boolean gates within a single material [1].

1.1   Physical Realization of Computational Materials

A profound limitation of vibrational CGMMs is the difficulty of their physical realization. This stems from, among other things, the sensitivity of the vibrational bit abstraction. Vibrational systems are sensitive to noise, and damping. Even minor perturbations such as deviation in resonance frequency or a slight phase misalignment can disrupt computation, significantly diminishing the robustness and reliability of the system's function.

Given this brittleness, we herein explore metamaterials that encode computation using mechanical phenomena other than vibration, and which do not change during the computing process. We refer to this class of CGMMs as *static* CGMMs. This facilitates the physical realization of AI designed, *in silico* computational materials, which we report below. Specifically, we test bit abstractions that involve changing a grain's bulk properties, such as size and stiffness, to encode binary bits. Computation is instantiated by the particular static force chains that arise within a given material. Static force chains operate by transmitting compressive or tensile forces through stable contact points between grains, forming reliable pathways for information flow [20]. The result of the computation is interpreted as the appearance of high ('1') or low ('0') forces arriving at a pre-specified output grain. By AI designing grain features within the material, different computations can be realized.

Although the abstraction of input and output bits differ, abstraction equality—and thus the cascading of multiple logic gates to compute more complex functions—could be achieved via hardware translation of input forces into input grain property change [9]. The advantage of our encoding is that static CGMMs are a non-volative computing system [23]. That is to say, they can sustain their computational states over time without requiring continuous energy input, unlike vibrational systems that rely on ongoing actuation. Unfortunately, in static CGMMs, the ability to operate multiple gates within a substrate at different frequencies is lost. However, we could alternatively utilize several bulk properties, such as force and shear modulus, to operate multiple gates simultaneously. This investigation of polycomputational static CGMMs is beyond the bounds of the present work.

1.2   Inverse Design of Computational Materials

Regardless of what physical phenomena encode information in a CGMM, as mentioned above, the inverse problem of designing a material that embodies a desired computation is non-intuitive and thus is difficult to tackle with hand-designed strategies. For this reason, machine learning has been employed to design CGMMs by allowing the optimization process to tune one grain feature: stiffness. However, new adaptive metamaterials are on the horizon in which multiple features of individual grains may be modified independently [9]. It is unknown whether broadening the design reach of machine learning to these new tunable grain features would frustrate design because of the increased dimensionality of the parameter space, or facilitate design as the larger space would contain more gradients to follow.

Specifically, in the case of static CGMMs, it remains unknown if independently varying multiple grain features could yield better computational results than tuning one feature per grain. To investigate this, here, we use evolutionary algorithms to evolve the properties of granular materials to behave as a variety of logic gates (AND, OR, XOR, and NAND). We find granular materials with AI designed grain sizes and stiffnesses are able to produce more distinguishable binary outputs for all tested logic functions, compared to materials with only AI designed grain sizes or stiffnesses. We then report the successful transfer of a designed logic gates from *in silico* to physical material. We conclude by providing an overview of the potential applications of this new computational platform and discussing future research directions.

2   Simulated Computational Materials

In this section, we first describe how simulated computational materials were designed (Sect. 2.1) and then investigate the quality of those designs (Sect. 2.2).

2.1   Methodologies

Several Boolean gates were AI designed into simulated granular materials with varying degrees of success.[1]

---

[1] This work's code and supplementary materials can be found at https://github.com/piperwelch/static_logic_gates/..

CGMM Setup. In this work, our materials are 2-D and comprised of 10 grains placed on a 3-4-3 hexagonal lattice. The lattice is inside a bounding box of fixed size in the x and y directions.

Experimental Design. To test the generality of our computational system, we optimize granular materials to perform as one of AND, OR, XOR, or NAND. For each logic gate, we conduct four experiments, testing two input bit abstractions, each under two different AI design conditions. Inputs are passed to a material in one of two ways: in *bit abstraction 1*, binary values were encoded by minimizing ('0') or maximizing ('1') the radial size of two input grains, while in *bit abstraction 2*, they were encoded through the minimization ('0') or maximization ('1') of grain stiffness. For each of the two input bit abstractions, we conducted two AI design conditions: first, optimization of the grain feature used for bit abstraction (only size or stiffness), and second, optimization of both grain features (size and stiffness). In all four cases, the output is measured as the amount of force exerted by the central grain in the bottom lattice row against the bounding wall.

Simulation. To simulate our granular system, we use the Discrete Element Method (DEM) [21]. All grains are frictionless and have the same mass ($m = 1$). Depending on the experimental setup, grains can have different radii or/and elastic moduli. Gravity is not included in our simulations. Particle-particle interactions are modeled as repulsive with linear spring potential. The Fast Inertial Relaxation Engine (FIRE) [2] is used for energy minimization.

Optimization. To design materials that behave as logic gates, we couple our physics simulator with an evolutionary algorithm. Specifically, we use Age-Fitness Pareto Optimization (AFPO) [17]. We selected AFPO as it requires no hyperparameters to promote diversity within the population. It is also well suited for problems with multiple local optima and prevents premature convergence.

Initialization. Each individual in the population represents the configuration of one granular material and is encoded as two float vectors with length 10, one assigning each grain's stiffness and another assigning their diameters. The initial population is instantiated with 200 randomly generated materials with grain properties drawn from a uniform distribution. Grains have variable stiffness ratios ($\in [0.5, 10]$) and diameters ($\in [1.0, 1.04]$). The stiffness ratio is chosen according to previous related work [1], while the diameter ratio was chosen as simulations can become unstable if all grains have a diameter above 1.04. The stiffness and sizes are unitless quantities and derive meaning from the ratios between grains. The locations of the grains chosen as the input and output of the logic gate are fixed to the locations shown in Fig. 1.

Evaluation. To assess a material's fitness, we measure the force, F_n, exerted by the output grain on the bottom wall of the simulation environment under each input case n ∈ {'00', '01', '10', '11'}. We use fitness functions that measure a given material's logical behavior ("GATEness") as a floating point number. This allows gradients to be followed during the evolutionary search. Equations 1-4 present the fitness functions for designing AND, OR, XOR, and NAND gates respectively.

$$\text{ANDness} = \ln \left(\frac{F_{11}}{(F_{01} + F_{10} + F_{00})/3} \right) \tag{1}$$

$$\text{ORness} = \ln \left(\frac{(F_{01} + F_{10} + F_{11})/3}{F_{00}} \right) \tag{2}$$

$$\text{XORness} = \ln \left(\frac{(F_{01} + F_{10})/2}{(F_{00} + F_{11})/2} \right) \tag{3}$$

$$\text{NANDness} = \ln \left(\frac{(F_{01} + F_{10} + F_{00})/3}{F_{11}} \right) \tag{4}$$

Each of these fitness functions is designed to maximize the force under the input conditions where a '1' output is expected, while minimizing the force under the input conditions where a '0' output is expected. In the above equations, ln denotes the natural logarithm.

Selection. After all 100 materials have been simulated and their behavior assessed, the materials with the lowest fitness values are removed. Survivor selection occurs by iteratively selecting two random individuals from our population, and discarding one if it is Pareto-dominated. Specifically, material i is removed if it is compared to another material j that demonstrates better performance ($f_j > f_i$) and belongs to a lineage l_j that is as young or younger than the lineage of material i ($l_j \leq l_i$). The age of each lineage is defined by the number of generations since it first emerged in the population. In the initial generation of randomly generated materials, each material begins a unique lineage with an age of zero. This process repeats until the population size reaches 100.

Reproduction. The materials allowed to reproduce are selected via tournament selection. In our implementation, two materials are randomly selected and the one with higher fitness is allowed to reproduce. This process repeats until the population contains 100 new materials.

Variation. Material mutation occurs by mutating the size, and/or stiffness of grains within a material. Specifically, we employ uniform mutation with a probability of 0.2 for each grain. The mutation size for grain diameter is ∈ ±0.005, while the mutation size for grain stiffness is ∈ ±0.5. We do not employ crossover.

Diversity. At each generation, one randomly generated material is injected into the population. This has the effect of introducing diversity into the evolving population of materials.

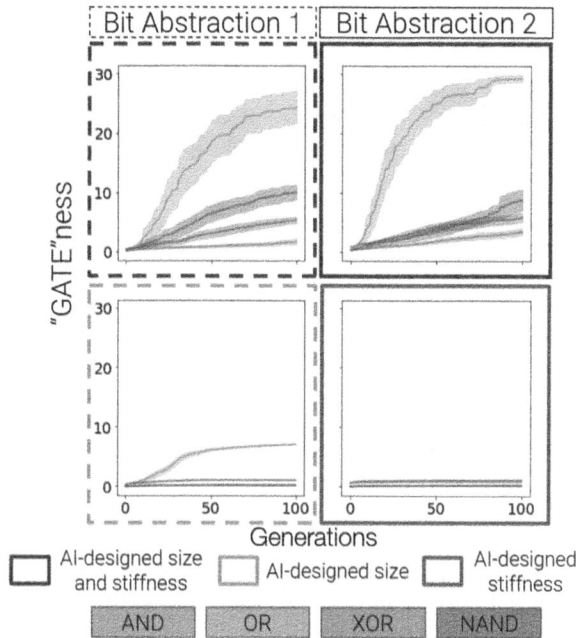

Fig. 2. Fitness over evolutionary time. This figure shows the fitness over evolutionary time for the 4 logic gates we investigated (AND, OR, XOR, NAND) across our 2 input bit abstractions, and 3 AI design conditions. Each line is the average of thirty experimental trials with a 95% confidence interval plotted in a lighter shade.

The process of simulating and evaluating new and freshly generated materials, removing low-fitness materials, duplicating and mutating surviving materials, and introducing a newly generated material is repeated for 100 generations of material evolution. We repeat this evolutionary process thirty times for each experimental condition. Each replicate uses a different random seed, and therefore each initial population is a novel random instantiation. We use the Mann-Whitney U test for all statistical comparisons and Bonferroni correction for multiple pair-wise comparisons.

2.2   Results and Discussions

The maximum fitness of the population over evolutionary time for each input bit abstraction, AI design condition, and logic gate is reported in Fig. 2. Here, each graph reports the average population maximum across 30 trials with the shaded region representing a 95% confidence interval. We find that (+AI designed size, +AI designed stiffness) granular materials have significantly higher fitness than those with only AI design for size or stiffness across both input bit abstractions and across all logic gates ($p < 0.005$ for all comparisons). When comparing the (+AI designed size, +AI designed stiffness) condition across bit abstraction

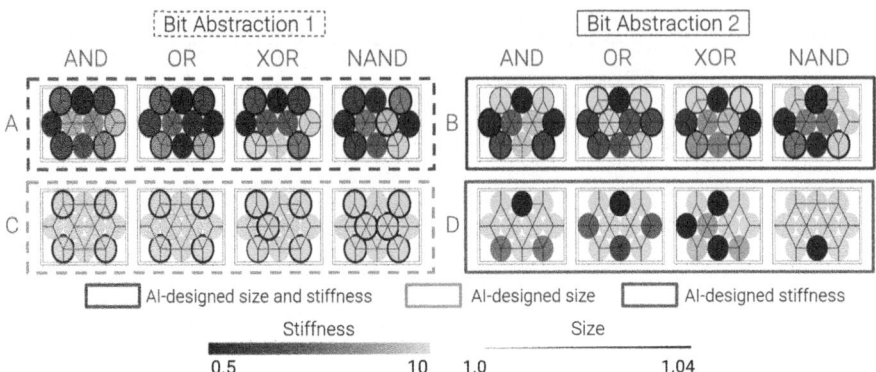

Fig. 3. Highest Fitness Solutions by AI-design Condition and Bit Abstraction Across Logic Gate. This figure shows the highest solution in the '11' input case for each logic gate across each AI-design condition and input bit abstraction.

1 and bit abstraction 2, we find that, after Bonferroni correction, there is no significant difference between any gates, save for XOR ($p < 0.05$), which is significantly higher for bit abstraction 2. In the (+AI designed size, +AI designed stiffness) conditions, for both bit abstraction 1 and bit abstraction 2, we find that the fitness from materials optimized for OR is the highest, followed by NAND, followed by AND, and ending with XOR. This pattern is not seen in the (+AI designed size, -AI designed stiffness) condition, which displays the highest fitness for materials optimized for OR, followed by AND, followed by XOR, and NAND. Similarly, the (-AI designed size, +AI designed stiffness) condition resulted in the materials optimized for AND having the highest fitness, followed by OR, NAND, and XOR. These results indicate that, while it is always beneficial to have AI design in both size and stiffness, some AI interventions are better suited for specific gates.

Top Performing Logic Gates. Figure 3 presents the configurations of the highest-performing materials at generation 100 for each logic gate, organized by AI design condition and bit abstraction. In this visualization, grain stiffness is denoted by color, while grain size is represented by both the plotted size and line width of each grain's border. Several intriguing structural and material property patterns emerge. We proceed with enumerating a few.

First, the AND and OR logic gates within the (+AI designed size, -AI designed stiffness) condition for bit abstraction 1 (Fig. 3C) share notable similarities in structure, though the OR gate exhibits a slight size increase of certain grains that are not found in the AND gate. The differences between the AND and OR gates for all other experimental conditions are more pronounced. We do not have any intuition regarding why this similarity has only appeared in the (+AI designed size, -AI designed stiffness) condition.

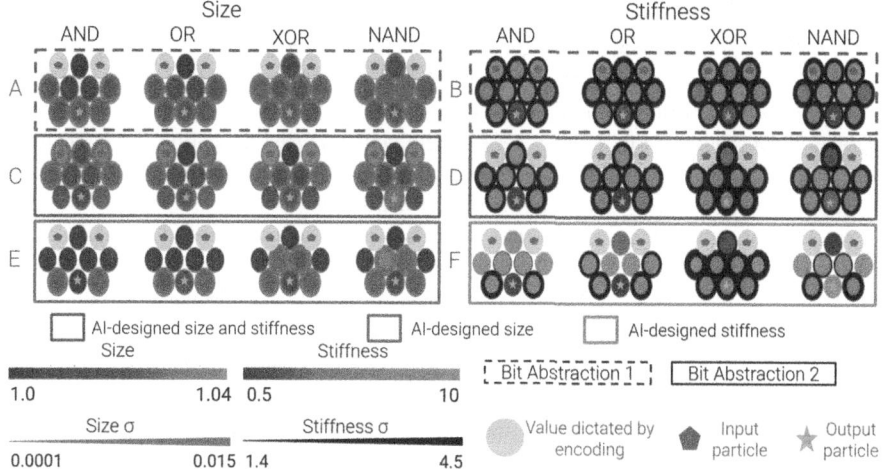

Fig. 4. Heat maps of mean size and stiffness across logic gates, input bit abstractions, and AI-design condition. This figure shows the mean size and stiffness across each input bit abstraction. **(A)**: Heat maps for mean size in materials with AI designed size and stiffness. **(B)**: Heat maps for mean stiffness in a materials with AI designed size and stiffness. **(C)**: Heat maps for mean size in materials with AI designed size and stiffness. **(D)**: Heat maps for mean stiffness in materials with AI designed size and stiffness. **(E)**: Heat maps for mean size in materials with only AI designed size. **(F)**: Heat maps for mean stiffness in materials with only AI designed stiffness.

Furthermore, when comparing the solutions in the (+AI designed size, -AI designed stiffness) condition (Fig. 3C) and the (-AI designed size, +AI designed stiffness) condition (Fig. 3D), grain property symmetry is observable along the central longitudinal axis in all materials. This pattern holds for all logic gates, except for the XOR gate, which remains distinctly asymmetric. This divergence in XOR may indicate a necessary structural asymmetry for achieving this logic function.

Another pattern we observe is that when stiffness is controlled by AI (Fig. 3A-B,D), the grain located between the input nodes consistently displays a low stiffness. The XOR in the (-AI designed size, +AI designed stiffness) (Fig. 3D) condition is the sole exception. This anomaly might stem from an alternative stress distribution pattern required to achieve the XOR functionality.

General Anatomy of a Logic Gate. Figure 4A-F presents heatmaps depicting the mean stiffness and size of individual grains extracted from the top-performing material across each of our thirty evolutionary runs. Each grain's color represents its relative mean stiffness or size, while the edge width indicates the standard deviation. Similar to Fig. 3, several interesting patterns have emerged across different input bit abstractions, AI design conditions, and logic gates. We proceed with enumerating a few.

The variation in mean stiffness, size, and standard deviation across different conditions highlights the complex ways in which the level of AI design control and choice of input bit abstractions shape the functionality of the granular materials. For instance, in the (+AI designed size, +AI designed stiffness) condition for bit abstraction 1 (Fig. 4A,B), the stiffness distributions for each logic gate appear to share a similar pattern, while the mean sizes display similarities between the AND and OR gates, and likewise between the XOR and NAND gates. This statement is correct for both the mean values and standard deviations. Similarly, under the (+AI designed size, -AI designed stiffness) condition (Fig. 4E), the AND and OR gates exhibit comparable configurations, while the XOR and NAND exhibit comparable configurations.

In the (-AI designed size, +AI designed stiffness) condition (Fig. 4F), the mean stiffness values for several grains are maximized to the upper limit (10) for the AND, OR, and NAND gates. Interestingly, many of these high-stiffness grains exhibit very low standard deviation. We can also see in this panel that the grains horizontally adjacent to the output grain maintain a relatively high standard deviation across all gates. In the case of XOR, the overall standard deviation across grains is higher than in other gates, indicating divergent solutions to this particular logic gate.

The distinct patterns under each level of AI design control condition emphasize that different combinations AI design of size and stiffness, along with input bit abstractions, lead to divergent internal structures. This exemplified the nuanced interplay between material properties and their computational outcomes.

3   Physical Computational Materials

In this section, we proceed with the physical validation of a granular material evolved *in silico* to act as an AND gate.

3.1   Experimental Setup

To design and construct a physical implementation of a computational granular material, we created a simple hardware with six grain types. The grain types are differentiated by two stiffnesses and three sizes, shown in Fig. 5A. These grains are made by casting silicone elastomer in molds laser-cut from a 0.25 in thick acrylic sheet. The grain diameters are 1.00 in (blue), 1.02 in (green), and 1.04 in (red), which span the 4% size change analogous to the simulations. The grain's material dictates its stiffness, which either has an elastic modulus at 100% strain of 10 psi (light, Smooth-On Ecoflex 30) or 86 psi (dark, Smooth-On Dragon Skin 30). We constructed a rudimentary force-sensing box experimental setup for these grains, shown in Supplementary Fig. 1. The grain packing was rigidly restricted from out-of-plane motion on both faces. The exterior in-plane bounding box was laser-cut from a 0.25 in thick clear acrylic sheet, similar to the negative molds for the grains. We designed the box's inner dimension to account for the laser beam width and to rigidly constrain the packed configuration of the

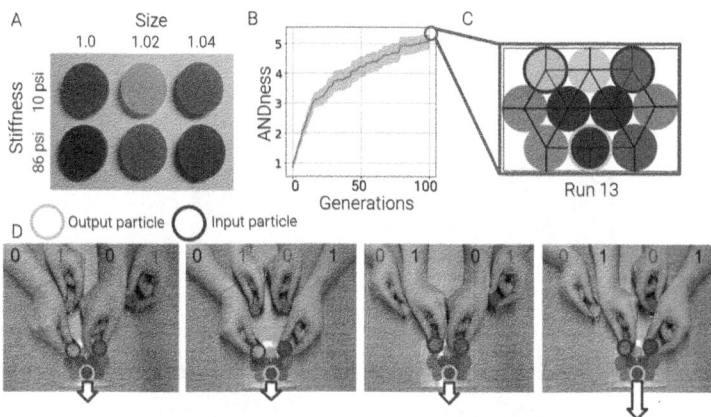

Fig. 5. Physical validation of an AND gate. (A): This panel shows the 6 grain types that we used to build the hardware implementation of our logic gates. Grains are available in 3 sizes and 2 stiffnesses. **(B)**: The evolution of ANDness over generations across 30 replicates with 95% confidence interval. In these evolutionary runs, the packings are AI-designed for size and stiffness and the logical input control policy is guided by stiffness. **(C)**: This panel shows the packing with the highest fitness value found by evolution. **(D)**: This panel shows how the 4 input cases ['11', '10', '01', '00'] are passed to the material. Here, both individuals have a '0' in their right hand (soft grain) and a '1' in their left hand (stiff grain). Because this gate was evolved to behave as logical AND, we expect a high force on the output grain in the '11' input case, and a low force on the output grain in the '10', '01', and '00' input cases.

smallest grain size. The stiffness of cast acrylic is 10,000 psi, so we assume the 0.5 in wide walls are rigid. To measure force output, the wall in contact with the output grain was rigidly linked to the force sensor (Vernier Go Direct®Sensor Cart) and unattached from the rest of the wall. This section of the rigid wall was a T-shape to keep the piece from slipping out completely and was free to move 0.02 in perpendicular to the wall face. Force measurements were sampled at 50 Hz and averaged over a 2 s collection interval.

To transfer the optimized designs from simulation to the reduced-degree-of-freedom physical material, we re-execute evolution using the aforementioned experimental setup. However, we constrain the grain properties to discrete values. Specifically, we use the non-dimensional stiffness values $\in \{0.11627, 1\}$ and sizes $\in \{1.0, 1.02, 1.04\}$, which are based on the physical grain ratios. We conduct 30 independent replicates for each logic gate, under both input bit abstractions. Here, we only test the (+AI designed size, +AI designed stiffness) condition as the prior simulation results revealed its superiority. Figure 5B shows the fitness over time for 30 replicates for bit abstraction 2, where logic is passed to the material based on the stiffness of input grains. Figure 5C shows the highest fitness evolved design across all replicates at generation 100. Figure 5D shows how the different input cases are passed into the material. In this visual metaphor, there are two individuals, each of whom has a soft ('0') and stiff ('1') grain in

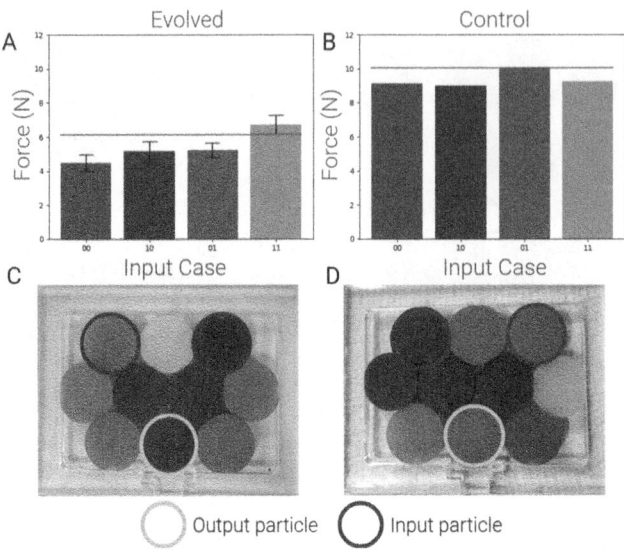

Fig. 6. Physical validation results. (A): Mean force recorded for each input case in the AND gate hardware implementation. The red line represents the magnitude of the threshold above which a force value is outputting a '1'. **(B)**: Mean force recorded for each input case for a control configuration. **(C)**: The configuration of the evolved AND gate when supplied with the '11' input case. **(D)**: The configuration of the control material when supplied with the '11' input case.

their hands. The varying hands that are extended represent four different input cases passed into the material.

To account for variability in grain positioning and contact networks, we systematically 1) unpacked, 2) repacked, and 3) remeasured the force, 10 times for each grain configuration. Consequently, we can assess the robustness and stability of the material's functional output under minor perturbations in grain arrangement. As the control experiment, we measured five randomized packing configurations. Each grain was independently and randomly chosen from the six size and stiffness combinations, shown in Fig. 5A. Figure 6D shows one of the randomized configurations tested, and images of the others are available in the supplementary material. It is worth noting that each random material was measured once. All data collected is also available in the supplementary material.

3.2   Design Transfer

We selected the highest-performing configuration for AND, with the input logic coded in the grain stiffness, for hardware deployment. To evaluate the performance of the physical implementation, we packed the optimized configuration into the bounding box and took force measurements on the output grain for the input cases $\in$ {'00', '01', '10', '11'}.

Figure 6A shows the mean force measured across each input case for the evolved configuration from Fig. 5B. Figure 6C shows the randomized reconfiguration of the evolved material. Figure 6B shows the mean force measured across each input case for the randomized configurations.

During evolution, we used a continuous metric-termed "GATEness" for GATE $\in$ {AND, OR, NAND, XOR}, to guide the optimization for materials that exhibit the desired logical behavior. However, post-evolution, this quantitative measure loses its meaning, as a gate either functions correctly or it does not. Therefore, we implement discrete thresholding to quantify the gate's logical behavior. Specifically, we define an observed output force above a given threshold as a '1', and an observed output force below a given threshold as a '0'. This approach is similar to that used in digital electronics [16]. We define a threshold based on the highest mean recorded for an input configuration that should produce a '0' output. For example, in the physical instantiation of an AND gate, we observe a mean maximum force of 5.965 N where a '0' output is expected. Consequently, this value serves as our threshold, meaning any output force greater than 5.965 N is interpreted as a '1'. For each randomized control, we set the threshold to the value of the force when the input is '00'. This value then becomes the control threshold for categorizing outputs.

Table 1 presents the observed frequencies of logical AND behavior across various materials, with the aforementioned universal threshold set for each material type. To evaluate whether the evolved configurations show a statistically higher proportion of AND behavior than random configurations, we use Fisher's exact test. We receive a p-value < 0.005, providing evidence that the evolved materials produce AND logic more frequently than their randomized counterparts.

Table 1. Evaluation of physical realization of the evolved and random configurations.

Metric	Evolved Configuration	Random Configuration
% trials acting as AND	90%	0%

4   Conclusions and Future Work

The shift away from traditional digital electronic computing will become increasingly salient in the coming years. To continue the pace of technological innovation, we must explore alternatives using novel computational substrates and paradigms. In this work, we introduced a new approach for using computational granular metamaterials as logic gates by leveraging static force chains that emerge within their internal contact network. By encoding binary inputs as physical properties like grain size and stiffness, our system translates these variations into force outputs, allowing the implementation of fundamental logic gates such as AND, OR, XOR, and NAND. We discovered that granular materials with AI design in both size and stiffness exhibit significantly more computational behavior compared to materials with only AI designed size or stiffness.

Our findings highlight the importance of material diversity in achieving emergent behaviors that are critical for logic operations. To ground our results in physical reality, we transfered an evolved AND gate configuration from simulation, demonstrating the first functional physical realization of a CGMM. Our sim2real transfer confirms the feasibility of using granular materials as next-generation computing technologies.

We plan to continue this research by improving upon our *in silico* design and optimization pipeline. Specifically, we hope to optimize our materials to display more complex computational behaviors. For example, one granular assembly could be designed to function as multiple logic gates simultaneously by using different bulk properties as inputs and outputs. Multiple gates could also operate simultaneously by using two sets of input grains and 2 output grains. Alternatively, one pair of input grains could have size and stiffness consecutively act as bit abstractions while outputs could be measured at different output grains, or different force angles on the same output grain (i.e. one in the x-direction and the other in the y-direction). We also plan to explore shape as an additional design parameter, allowing the evolution to optimize not only the size and stiffness of grains but also their geometry. It remains to be seen whether shape-varying CGMMs can further enhance the functionality of a material acting as a logic gate, but this direction holds promising potential for expanding the CGMM computational capacity.

Acknowledgements. We would like to thank Patrick Charron for lending the Vernier Cart used in the physical instantiation experiments. We would also like to thank Kameron Bielawski and Amanda Bertschinger for their assistance in creating Fig. 5. This material is based upon work supported by the National Science Foundation Graduate Research Fellowship Program under Grant No. 2235204. Any opinions, findings, and conclusions or recommendations expressed in this material are those of the author(s) and do not necessarily reflect the views of the National Science Foundation. We would like to acknowledge financial support from the National Science Foundation under the DMREF program (award number: 2118810).

Disclosure of Interests. The authors have no competing interests to declare that are relevant to the content of this article.

References

1. Beaulieu, S., Welch, P., Parsa, A., O'Hern, C., Kramer-Bottligio, R., Bongard, J.: Refractive Computation: parallelizing logic gates across driving frequencies in a mechanical polycomputer. In: ALIFE 2024: Proceedings of the 2024 Artificial Life Conference. MIT Press (2024)
2. Bitzek, E., Koskinen, P., Gähler, F., Moseler, M., Gumbsch, P.: Structural relaxation made simple. Phys. Rev. Lett. **97**(17), 170201 (2006)
3. Ciliz, D., O'Hern, C.S.: Granular metamaterials for soft robotic applications. J. Next Front. Life Sci. AI p. 48 (2021)
4. De Leon, N.P., et al.: Materials challenges and opportunities for quantum computing hardware. Science **372**(6539), eabb2823 (2021)

5. Fu, K., Zhao, Z., Jin, L.: Programmable granular metamaterials for reusable energy absorption. Adv. Func. Mater. **29**(32), 1901258 (2019)
6. Gantzounis, G., Serra-Garcia, M., Homma, K., Mendoza, J.M., Daraio, C.: Granular metamaterials for vibration mitigation. J. Appl. Phys. **114**(9) (2013)
7. Jayachandran, D., et al.: Three-dimensional integration of two-dimensional field-effect transistors. Nature **625**(7994), 276–281 (2024)
8. Kim, E., Yang, J.: Wave propagation in granular metamaterials. Funct. Composites Struct. **1**(1), 012002 (2019)
9. Li, M.S., Do, B.H., Le, C.L., O'Hern, C., Kramer-Bottiglio, R.: Variable stiffness and variable size particles for reconfigurable granular metamaterials. In: 2025 8th IEEE-RAS International Conference on Soft Robotics (RoboSoft) (2025)
10. McArdle, S., Endo, S., Aspuru-Guzik, A., Benjamin, S.C., Yuan, X.: Quantum computational chemistry. Rev. Mod. Phys. **92**(1), 015003 (2020)
11. Moore, G.E., et al.: Progress in Digital Integrated Electronics, vol. 21, pp. 11–13. Washington, DC (1975)
12. Parsa, A., Wang, D., O'Hern, C.S., Shattuck, M.D., Kramer-Bottiglio, R., Bongard, J.: Evolving programmable computational metamaterials. In: Proceedings of the Genetic and Evolutionary Computation Conference, pp. 122–129 (2022)
13. Parsa, A., Wang, D., O'Hern, C.S., Shattuck, M.D., Kramer-Bottiglio, R., Bongard, J.: Evolution of acoustic logic gates in granular metamaterials. In: International Conference on the Applications of Evolutionary Computation (Part of EvoStar), pp. 93–109. Springer (2022)
14. Parsa, A., Witthaus, S., Pashine, N., O'Hern, C., Kramer-Bottiglio, R., Bongard, J.: Universal mechanical polycomputation in granular matter. In: Proceedings of the Genetic and Evolutionary Computation Conference, pp. 193–201 (2023)
15. Pashine, N., et al.: Tessellated granular metamaterials with tunable elastic moduli. Extreme Mech. Lett. **63**, 102055 (2023)
16. Pedroni, V.A.: Digital electronics and design with VHDL. Morgan Kaufmann (2008)
17. Schmidt, M.D., Lipson, H.: Age-fitness pareto optimization. In: Proceedings of the 12th Annual Conference on Genetic and Evolutionary Computation, pp. 543–544 (2010)
18. Shrestha, A., Fang, H., Mei, Z., Rider, D.P., Wu, Q., Qiu, Q.: A survey on neuromorphic computing: models and hardware. IEEE Circuits Syst. Mag. **22**(2), 6–35 (2022)
19. Theis, T.N., Wong, H.: The End of Moore's Law: a new beginning for information technology. Comput. Sci. Eng. **19**(2), 41–50 (2017)
20. Tordesillas, A., Walker, D.M., Lin, Q.: Force cycles and force chains. Phys. Rev. E-Stat. Nonlinear Soft Matter Phys. **81**(1), 011302 (2010)
21. Wu, Q., Cui, C., Bertrand, T., Shattuck, M.D., O'Hern, C.S.: Active acoustic switches using two-dimensional granular crystals. Phys. Rev. E **99**(6), 062901 (2019)
22. Xia, A., Wang, D., Zhang, J., Shattuck, M.D., O'Hern, C.: Anisotropic shear response of 3D tessellated granular metamaterials. Bull. Am. Phys. Soc. (2024)
23. Yasuda, H., et al.: Mechanical computing. Nature **598**(7879), 39–48 (2021)

Scalable Evolution of Logically Independent Polycomputational Materials

Piper Welch[1]([✉]), Atoosa Parsa[2], Shawn Beaulieu[1], Corey S. O'Hern[3],
Rebecca Kramer-Bottiglio[3], and Josh Bongard[1]

[1] Department of Computer Science, University of Vermont, Burlington,
VT 05405, USA
{piper.welch,shawn.beaulieu,josh.bongard}@uvm.edu
[2] Department of Biology, Tufts University, Medford, MA 02155, USA
atoosa.parsa@tufts.edu
[3] Department of Mechanical Engineering and Materials Science, Yale University,
New Haven, CT 06520, USA
{corey.ohern,rebecca.kramer}@yale.edu

Abstract. Substrates that compute using vibration rather than electricity offer the potential of creating and deploying computers into electronics-denying environments. Another advantage of these materials is that some of them can compute multiple functions in the same place and at the same time, providing computational results at different frequencies. These so-called polycomputational materials may eventually compete with more traditional computers in terms of computational density because there is no currently known upper bound on how many functions can be simultaneously computed by a vibrational substrate. However, three challenges remain for polycomputational materials: how to ensure that the different functions are computed independently; developing evolutionary algorithms that allow for embedding increasingly more functions into these materials *in silico*; and validating the evolved *in silico* materials as physical materials. Here we report progress on all three of these issues.

Keywords: Granular Metamaterials · Mechanical Computing · Polycomputing

1 Introduction

The demand for faster and more efficient computation continues. Since the 1950 s, silicon-based computers have dominated. However, these conventional systems fail in harsh environments, rely on electricity, and degrade into electronic waste [8,19]. Unconventional computers are those which explore novel substrates and bit abstractions [5,7,9,15,16,20]. They offer alternatives that are more resilient in extreme conditions, operate without traditional electricity, or produce less electronic waste [4,7,11,21]. Despite these potential advantages, unconventional computers currently fall far short of the computational ability and density of digital computers. However exponential growth patterns such as

© The Author(s), under exclusive license to Springer Nature Switzerland AG 2025
P. García-Sánchez et al. (Eds.): EvoApplications 2025, LNCS 15612, pp. 558–572, 2025.
https://doi.org/10.1007/978-3-031-90062-4_35

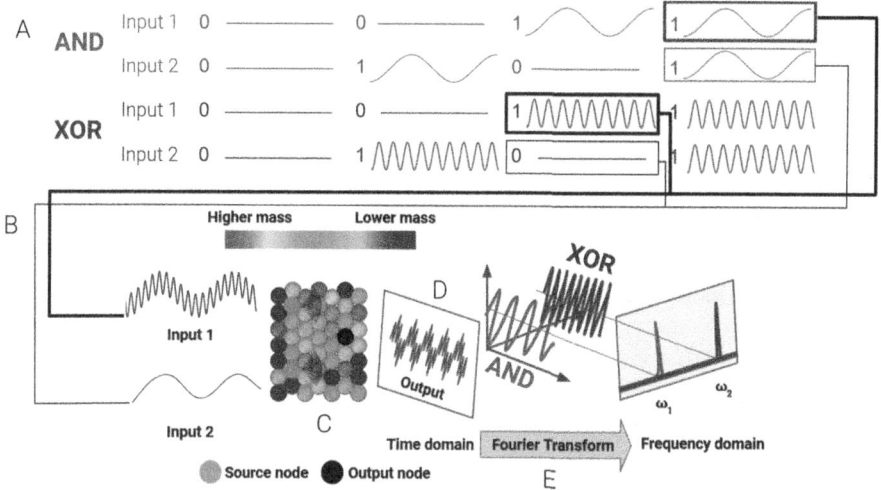

Fig. 1. CGMM inputs and outputs. This figure shows an overview of the system inputs and outputs for a CGMM where gate g_1 = AND and g_2 = XOR. **A)** This panel shows the possible system inputs for the AND gate operating at ω_1 Hz and the XOR gate operating at ω_2 Hz. In this example, the '11' input is passed into the AND gate, and the '10' input is passed to the XOR gate. **B)** The input signals are a combined superposition of the individual input cases. **C)** The CGMM is supplied with the selected input cases. In this cartoon, the input grains are marked in gray, while the output grain is marked in black. **D)** The output signal in the time domain. **E)** We perform a fast Fourier transformation on the time-domain oscillation of the output grain. We then analyze the power at the driving frequencies where power above a given threshold corresponds to an output value of '1' and power below a given threshold corresponds to an output value of '0'.

Moore's Law are not sustainable indefinitely. Transistors are nearing their physical limits, which impedes further miniaturization [17]. The particular class of unconventional computing material investigated herein, so-called polycomputational materials, can compute different logic gates at different frequencies [2,3]. Since there are infinite frequencies, in theory, such materials could have infinite computational density. In practice, no upper bound on computational density has been determined for these materials.

Polycomputational materials are designed using machine learning to engineer the properties, such as stiffness or mass, of grains in granular metamaterials [10]. Such materials are referred to as *computational granular metamaterials* (CGMMs) [1,12–14]. An overview of the inputs and outputs of a CGMM is displayed in Fig. 1. In addition to offering potentially unbounded computational density, CGMMs offer several advantages over traditional computational architectures. Notably, they could provide more robust and energy-efficient computation, as they would not rely on electronic components or specific substrates. These systems also have the potential to revolutionize future robotic systems, enabling sensing, control, and actuation to be driven by vibrational computation.

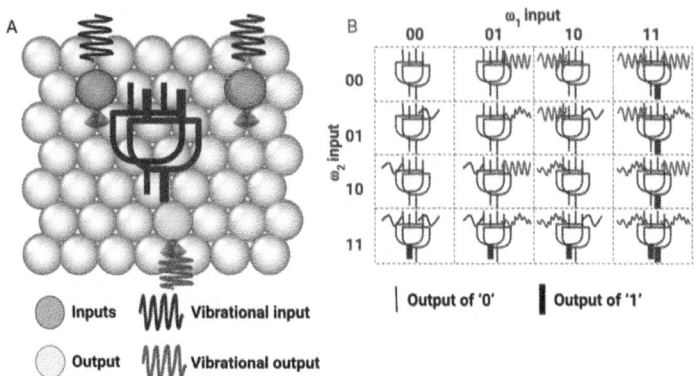

Fig. 2. Logical independence of a CGMM. A) Overview of a granular material with superimposed AND gates g_1 and g_2 operating at ω_1 and ω_2 hertz respectively. Bits are treated as the presence (1) or absence (0) of vibration at a given frequency. In panel a), both input grains are vibrated at ω_1 Hz, representing 1,1 supplied to gate g_1. Neither is vibrated at ω_2 Hz, representing the simultaneous application of 0,0 to gate g_2. The output grain is observed to vibrate in response, correctly, at ω_1 Hz (1,1→1), but not vibrate, correctly, at ω_2 Hz (0,0→0). **B)** For a given material, one can ask how many of the 16 possible cases (each of the four input cases, at both frequencies) results in the correct behavior of the output grain. Here, both the logic gates within the material are AND. An ideal material is logically independent: the gate at one frequency acts correctly regardless of which of the four input cases is being supplied to the other gate at the other frequency.

As polycomputational granular materials are an emerging technology, we have yet to explore many of their properties. One such unexplored yet important behavior is the relative *independence* of gates embedded within a given substrate. Here, independence refers to the ability of a logic gate to function autonomously, regardless of the input received by other gates within the same substrate. Prior studies have concentrated on the behavior of these materials across the input scenarios where source grains receive identical logical inputs *simultaneously* (e.g., the gate operating at frequency ω_1 receives the input case '10', while the gate operating at frequency ω_2 also receives the input case '10') [12–14]. Therefore, it remains unknown if gates g_1 and g_2 in a given material exhibit any degree of independence. That is to say, we do not have insight into how much the logical input into g_1 impacts the behavior of g_2 and vice versa.

One might think a simple solution to achieving logically independent CGMMs is to evolve for proper behavior at each of the 16 input case pairs (shown in Fig. 2B). However, explicitly evolving a granular material for independence has poor scalability: as logic gates n embedded into a material increases, the potential superpositions of frequencies expand exponentially with $O(2^{2n})$. Therefore, in this work, we focus on the following questions: 1) Can we evolve logically independent CGMMS, and 2) how do we do so in a scalable manner? To answer these questions, we evolve CGMMs using a variety of methods. We find that it is possible to evolve independent CGMMs and we can do so using scalable methods. This work presents the first CGMMs known to have this property and therefore highlights the potential of CGMMs to serve as dependable and robust computational systems.

In the following section, we describe our *in silico* model of granular materials, and their physical environments, followed by an outline of our evolutionary methods and conducted experiments. The Results section states our findings. The Discussion Section enumerates the behaviors of the best-evolved materials and their relative independence. The Hardware Implementation section presents a minimal physical implementation of a granular logic gate. We conclude with future directions for this work, including the transfer of our evolved designs *in situ*.

2   Methods

We use multi-objective optimization to evolve the masses of grains in a granular material. Each material has two logic gates, gates g_1 and g_2 operating at frequencies ω_1 and ω_2. For each material, either $g_1 = $ AND and $g_2 = $ AND, or $g_1 = $ XOR and $g_2 = $ AND. To explore scalable methods of evolving independent CGMMs, we test four methods of evaluating material fitness during evolution. Specifically, in place of evaluating materials for correct behavior in each of the 16 combinations of input case pairs, we evaluate gate behavior on various subsets of input case pairs for $input_1$ at $\omega_1 \in$ ['00', '01', '10', '11'] and $input_2$ at $\omega_2 \in$ ['00', '01', '10', '11']. These differing input case pair subsets used during evolution are referred to as *evaluation methods*. 30 independent evolutionary runs were collected for each experiment. We use the Wilcoxon rank-sum test and Bonferroni corrections for all statistical analyses with multiple pairwise comparisons.[1]

2.1   CGMM Definitions

In this work, a material consists of 49 3D grains arranged in a seven-by-seven triangular lattice. They are placed within a simulation box featuring fixed boundaries along the x and y directions. Each grain shares the same diameter D = 10cm. Logical inputs are sinusoidal vibrations supplied to two fixed input grains, and outputs are vibrations measured at a fixed output grain. The locations of output and input grains are arbitrarily fixed as the grains marked in Fig. 2. The logical input of True, or '1', is encoded as a sinusoidal input of amplitude $A = 1$ at driving frequency ω_n. Conversely, the logical input of False, or '0', is encoded as a sinusoidal input of amplitude $A = 0$ at driving frequency ω_n. To quantify output signals, we perform the fast Fourier transformation to convert output signals from the amplitude-time domain to the power-frequency domain. For a granular material to function as a logic gate, the relative magnitude of the

---

[1] The source code for the experiments in this paper is available here: https://github.com/piperwelch/logical_independence/.

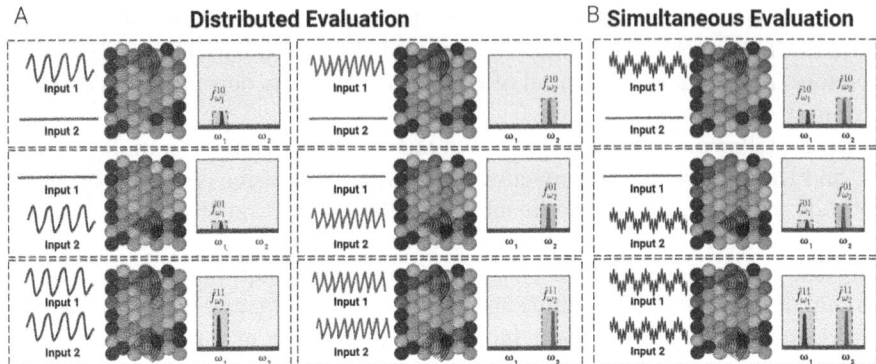

Fig. 3. Comparison of the distributed and simultaneous evaluation methods. A) This panel shows the inputs and outputs for the distributed evaluation method, in which the material is first supplied with the blue frequency, then the red frequency. **B)** This panel shows the inputs and outputs for the simultaneous evaluation method, in which the material is supplied with the red and blue frequencies simultaneously.

power at the driving frequency in each case must be consistent with the desired behavior of the gate. For example, when computing AND, we expect a relatively high power at the driving frequency when both input grains are supplied with vibration. In this work, gate g_1 is optimized to function at 15 Hz, while gate g_2 is optimized to function at 20 Hz.

2.2   Definitions

Notation. Herein, we will refer to a given input case pair supplied to a material as such: $F_x^{\omega_n}(ij_n)F_x^{\omega_m}(ij_m)$, where ω_n and ω_m are the driving frequencies, ij_n and ij_m are the logical inputs $\in$ ['11', '01', '10', '00'] supplied in the x-direction at ω_n and ω_m, respectively. For example, we represent the case when '11' is supplied at ω_1 and '00' is supplied at ω_2 as $F_x^{\omega_1}('11')F_x^{\omega_2}('00')$.

Independence. We define the independence of a material as the ability of its embedded gates to function correctly regardless of the input cases applied to other gates within the same material. The metric for quantifying independence is defined as follows:

$$I = \sum_{i=1}^{15} B_i \tag{1}$$

Here, B_i is 1 if behavior i is what we would expect in an ideal logical gate (e.g. if we observe a '1' where we expect it), and 0 otherwise. To determine whether an observed output signal aligns with the expected output, we binarize the material's output signals using a power threshold. This threshold is calculated as the midpoint between the lowest power magnitude corresponding to an output of 1 and the highest power magnitude corresponding to an output of 0. The independence I is a value $\in [0, 15]$, where 15 represents a material that behaves appropriately in each of the 15 input case pairs. The input case $F_x^{\omega_1}(\text{`00'})F_x^{\omega_2}(\text{`00'})$ is not considered as it is trivial.

2.3   Simulator

In this work, we use the open-source software LAMMPS (Large-scale Atomic/ Molecular Massively Parallel Simulator) [18] to model our granular materials. Both normal and tangential grain interactions adhere to a Hookean contact model. A Hookean model of contact also governs grain interactions with the system boundaries. We use a time step of 0.001 and run our simulations for 2000 steps. All reported units are expressed in SI units.

2.4   Optimization

We use a $(\lambda + \mu)$ multi-objective evolutionary algorithm, where $\lambda = 100$ and $\mu = 200$. The genome of each material is a 49-length array of floats, where each float represents the mass of one grain within the material. Each value in a material's genome is randomly initialized on a uniform distribution $\in [0.5\,\text{kg}, 1.3\,\text{kg}]$.

2.5   Fitness

Each material has a fitness associated with gate g_1, and a fitness associated with gate g_2. In practical applications, logic gates are either functional or non-functional; however, during evolution, we use a continuous fitness function to enable evolution to follow a fitness gradient. The metric we use to quantify logical AND behavior, or "ANDness" is:

$$\text{ANDness}_{g_n} = \frac{\hat{f}_{\omega_n}^{\text{`11'}}}{(\hat{f}_{\omega_n}^{\text{`10'}} + \hat{f}_{\omega_n}^{\text{`01'}})/2} \tag{2}$$

while the metric we use to quantify logical XOR behavior, or "XORness" is:

$$\text{XORness}_{g_n} = \frac{(\hat{f}_{\omega_n}^{\text{`01'}} + \hat{f}_{\omega_n}^{\text{`10'}})/2}{\hat{f}_{\omega_n}^{\text{`11'}}} \tag{3}$$

Here, $\hat{f}_{fn}^{ij}$ is the power at the driving frequency of the Fourier transform when logical input ij is supplied to the material. These metrics are maximized when the power for outputs that should be '1' is high relative to outputs that should be '0'. The '00' input case is not included in these equations since it is trivial. These metrics are subsequently referred to as GATEness.

Because power exists on a spectrum, while traditional logic gates operate in a binary manner, we establish a threshold of power post-evolution to binarize the behavior of our materials. Signals above this threshold are interpreted as

'1', while those below are interpreted as '0'. To encourage comparable power magnitudes for output signals from gates g_1 and g_2, we introduce a fitness term p. Here, p is designed to introduce pressure for similar output power when materials produce a '1' output. It is defined as:

$$p = \frac{1}{(1 + |\hat{f}_{f1} - \hat{f}_{f2}|)} \tag{4}$$

The fitness of a material for ω_n is then calculated as:

$$\text{fitness}_{g_n} = \text{GATEness}_{g_n} * p \tag{5}$$

2.6   Evaluation Methods

The varying subsets of input case pairs used during evolution to assess the fitness of g_1 and g_2 are subsequently enumerated.

Distributed Evaluation. The first evaluation method we test evaluates gate fitness when input cases are supplied to a material in a distributed manner. In this evaluation method, materials are evolved for correct logical behavior for gates g_1 and g_2 when the inputs for g_1 and g_2 are supplied during different simulations. This means a material is simulated six times to construct fitness_{g_1} and fitness_{g_2}. The input case pairs for distributed evaluation are represented visually in Fig. 3A. In closed form, fitness_{g_1} is constructed by examining the material's behavior when supplied with the $F_x^{\omega_1}$('11')$F_x^{\omega_2}$('00'), $F_x^{\omega_1}$('10')$F_x^{\omega_2}$('00') then $F_x^{\omega_1}$('10')$F_x^{\omega_2}$('00') input case pairs. Subsequently, fitness_{g_2} is constructed by examining the material's behavior when supplied with the $F_x^{\omega_1}$('00')$F_x^{\omega_2}$('11'), $F_x^{\omega_1}$('00')$F_x^{\omega_2}$('10') then $F_x^{\omega_1}$('00')$F_x^{\omega_2}$('01') input case pairs. The number of simulations this method requires scales as $O(n)$ with the n the number of gates embedded within a material.

Simultaneous Evaluation. The second evaluation method we test evaluates gate fitness *simultaneously*. That is, under this evaluation method, materials are evolved for correct logical behavior for gates g_1 and g_2 when the inputs for g_1 and g_2 are supplied during the same simulation. This means a material is simulated three times to construct fitness_{g_1} and fitness_{g_2}. The input case pairs for simultaneous evaluation are represented visually in Fig. 3B. In closed form, fitness_{g_1} and fitness_{g_2} are constructed by examining the material's behavior when it is supplied with the $F_x^{\omega_1}$('11')$F_x^{\omega_2}$('11'), $F_x^{\omega_1}$('10')$F_x^{\omega_2}$('10') and $F_x^{\omega_1}$('01')$F_x^{\omega_2}$('01') input case pairs. The number of simulations this method requires scales as $O(1)$ with the n the number of gates embedded within a material.

Input Case Varying Evaluation. The third evaluation method we test evolves for correct behavior as evaluated using three randomly selected input cases. That is, in this method of evaluation, granular materials are evolved for behavior at g_1 and g_2 when input cases for g_1 and g_2 are randomly selected $\in$ ['11', '01', '10', '00']. The input case pair $F_x^{\omega_1}$('00')$F_x^{\omega_2}$('00') and repeat selection of the same input case pair are excluded from consideration due to their redundancy. Figure 2B visually displays the 15 possible input case pair choices. A new set of three input case pairs is chosen for each generation.

Due to the shifting fitness function, this method employs several algorithmic adaptations from the aforementioned multi-objective optimization. Two notable changes have been implemented. First, the input case pairs needed to calculate GATEness may not be present in the three randomly selected input case pairs. Consequently, this method's relevant power values at a given driving frequency are amalgamated into a fitness metric for ω_1 and ω_2 as the product of three fitness terms for ω_1 and ω_2, respectively. Specifically, if a given input case should result in a '0' output then the fitness term is $\frac{1}{\hat{f}_{\omega_n}}$, while if a given input case is that should results in a '1', then the fitness term is $\frac{\hat{f}_{\omega_n}}{1}$. Similar to the GATEness metrics, this method rewards high fitness for materials that minimize power for any output that should be a '0' output and maximize power for any output that should be '1'. Another algorithmic adaptation required is that when a new set of three input case pairs gets selected at each generation, all individuals in the population must be re-evaluated by the latest fitness function. This is because the chosen input case pairs directly influence the magnitude of fitness. Without re-evaluation, there would be stagnation and evolutionary progress would not proceed. The number of simulations this method requires scales as $O(1)$ with the n the number of gates embedded within a material.

Tri-Objective. The final evaluation method we test follows the same process as the simultaneous method while introducing a third objective for independence on a subset of three input case pairs. By adding an objective for independence in a subset of input case pairs, we explicitly evolve for independence without having to perform $O(2^n)$ operations for each material. To construct the independence objective, we simulate material behavior in 3 randomly selected input case pairs. To ensure we can create a binarizing threshold by averaging the power for the lowest value of '1' and the highest value of '0', we enforce that at least one of the selected input case pairs results in a logical output of '1', and one results in a logical output of '0'. At every generation, we select a new subset of input cases on which to evaluate independence. The independence is then a value $\in$ [0,3]. The number of simulations this method requires scales as $O(1)$ with the n the number of gates embedded within a material.

2.7   Survivor Selection

After all the materials are evaluated, those allowed to maintain in the population are either selected based on the Pareto-front of fitness$_{g_1}$ and fitness$_{g_2}$ or, in experiments using the tri-objective evaluation method, off of the Pareto-front of fitness$_{g_1}$ and fitness$_{g_2}$ and independence. Survivor selection occurs by iteratively selecting two random individuals from our population, and discarding one if it is Pareto-dominated. This process repeats until the population size reaches λ. Each of the remaining materials is allowed to produce a mutated copy of itself until the population size reaches $(\lambda + \mu)$.

2.8   Mutation

Our mutation operator acts on each child with a 10% chance of mutating the mass of a given grain in the child's genome. The mutation size is $\in \pm 0.0524$ kg. Grains are limited to a minimum mass of 0.262 kg and a maximum mass of 5 kg. We repeat this process of creating, evaluating, and mutating materials until 100 generations have occurred.

3   Results

After evolution, we analyze each individual in the population at generation 100. We retrieve the material with the highest independence I from the population. This is repeated for all 30 evolutionary replicates. Table 1 shows the mean independence of these 30 top-performing materials in each evaluation method.

Table 1. Comparison of our four evaluation methods and random. The mean independence of materials evolved using our distributed, simultaneous, input case varying, and tri-objective evaluation methods. Here, big-O refers to how the number of physics simulations scales with n the number of logic gates embedded within a material.

Evaluation Method	Gate 1	Gate 2	Mean Independence	Big-O
Distributed	AND	AND	12.30 ± 1.77	$O(n)$
Simultaneous	AND	AND	12.60 ± 2.04	$O(1)$
Input Case Varying	AND	AND	10.80 ± 2.45	$O(1)$
Tri-Objective	AND	AND	12.86 ± 1.83	$O(1)$
Distributed	XOR	AND	14.70 ± 1.00	$O(n)$
Simultaneous	XOR	AND	14.03 ± 1.62	$O(1)$
Input Case Varying	XOR	AND	14.93 ± 0.35	$O(1)$
Tri-Objective	XOR	AND	14.46 ± 1.28	$O(1)$

To assess whether evolution produced materials with greater independence than randomly generated ones, we created a population of 1000 random materials and evaluated the fitness of each material. For all evaluation methods, evolved materials had significantly higher independence than the randomly generated ones ($p < 0.05$ for all comparisons).

AND & AND. When the logic gates g_1 and g_2 embedded within a material are logical AND, the tri-objective evaluation method produced the materials with the highest mean independence. In this method, 11 of the 30 replicates produced materials with perfect independence $I = 15$. The method that produced the second highest mean independence in the simultaneous method, followed by the distributed method. Respectively, using these methods, 10 and 6 replicates produced materials with perfect independence $I = 15$. The worst-performing evaluation method was the input case varying method. It resulted in materials with the lowest mean independence and only 5 replicates produced materials with perfect independence. Using the Mann-Whitney U test, we conclude that the independence of materials from the input case varying evaluation method is significantly less than material evolved using the other three methods ($p < 0.05$ for all comparisons). There is no significant difference between the distributed, simultaneous, and tri-objective methods ($p > 0.05$ for all comparisons). Figure 4 shows the behavior of sample $I = 15$ materials from each of the evaluation methods.

XOR & AND. When logic gate g_1 is XOR and logic gate g_2 is AND, the input case varying evaluation method produced the materials with the highest mean independence. Using this evaluation method, 29 of the 30 replicates produced materials with perfect $I = 15$ independence. The method that produced the second highest population mean fitness is the distributed method, which produced $I = 15$ materials in 27 of the 30 replicates. The tri-objective evaluation method produced the third highest population independence and found $I = 15$ materials in 25 of the 30 replicates. The worst-performing evaluation method was the simultaneous one, which produced the lowest mean population independence and found $I = 15$ materials in 22 of the 30 replicates. There is no significant difference between the independence of materials evolved from any of the methods.

4   Discussion

In this work, we have shown that it is possible to evolve entirely independent computational granular materials without explicitly selecting for independence. Our results unveil computationally efficient methods of embedding logic gates into granular metamaterials. We proceed with analyzing our findings.

4.1   Relative Independence of AND & AND Vs XOR & AND

Comparing within each valuation method, we find that materials where $g_1 =$ AND, $g_2 =$ AND have significantly lower independence than materials where $g_1 =$ XOR, $g_2 =$ AND (p < 0.05 for all comparisons). That is, materials with heterogeneous logic gates display higher independence in all evaluated methods tested. It is not clear why this is the case. One hypothesis is that in materials

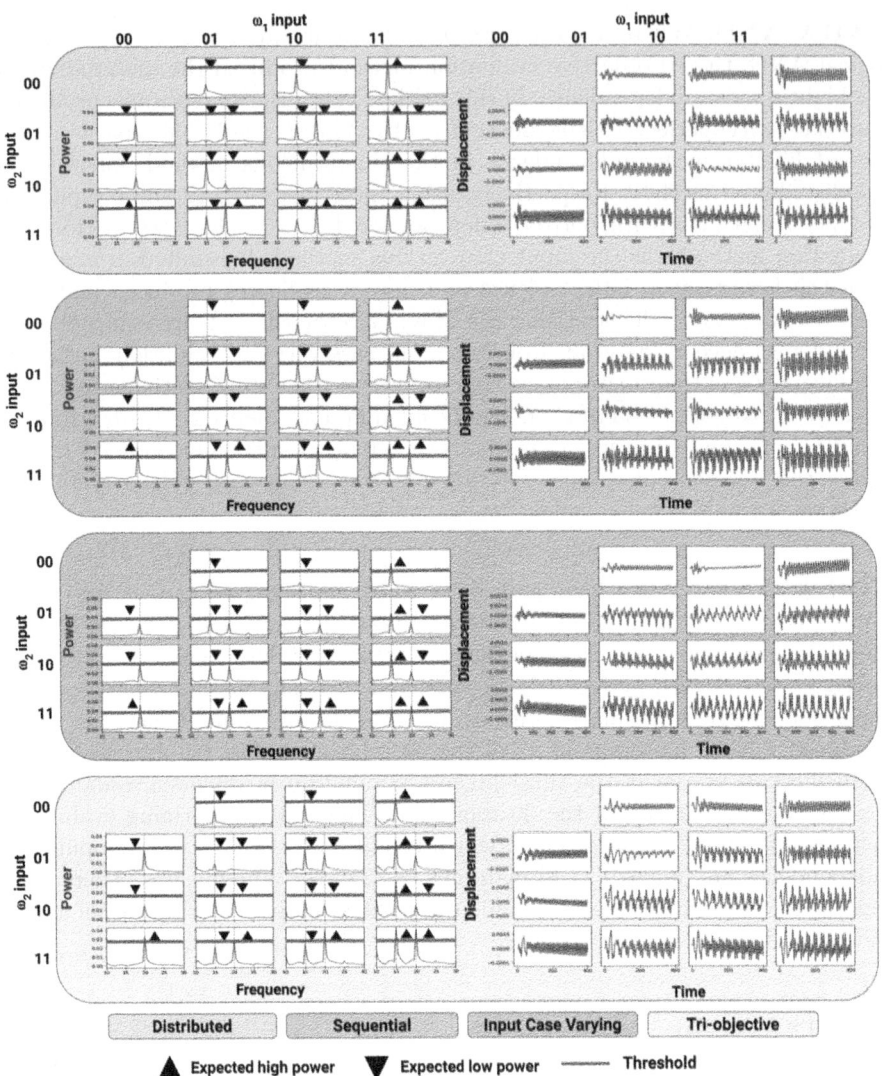

Fig. 4. Comparison of sample material behavior where $g_1 = \text{AND}$, $g_2 = \text{AND}$. This figure compares sample behavior for materials evolved across our four evaluation methods. The panels on the left side show behavior in the frequency domain, while the panels on the right side show behavior in the time domain. Each material here has a perfect independence score of $I = 15$.

with homogeneous logic gates, each gate might rely on the same physical phenomena to complete its logical function. For example, each logic gate could rely on constructive interference between the same pair of grains at a certain location within the material. Therefore, when the gates are supplied with different

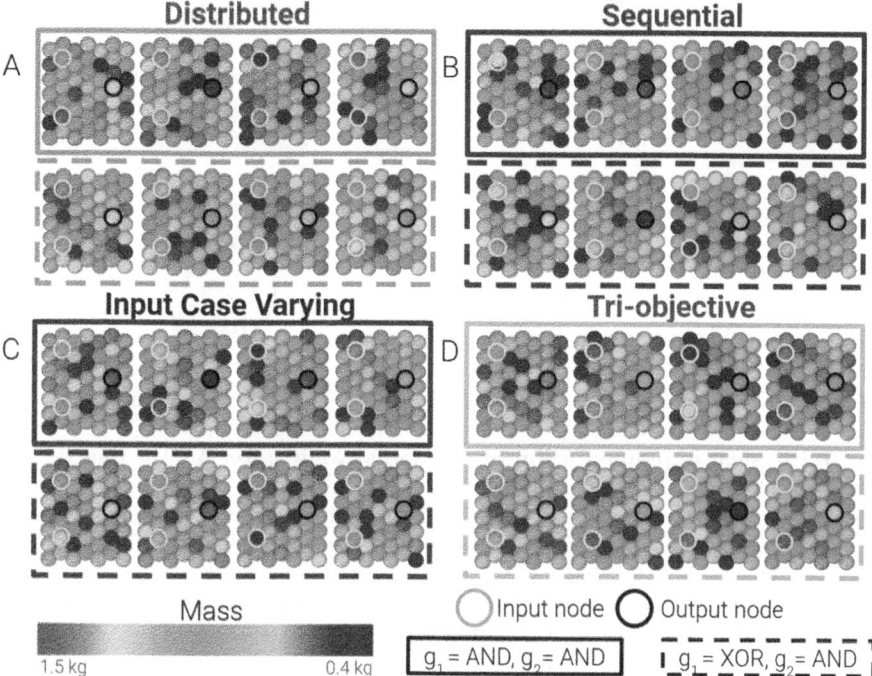

Fig. 5. Comparison of material configurations. This figure displays configurations from the four $I = 15$ materials from each of our evaluation methods both where g_1 = AND, g_2 = AND (materials with solid outline), and where g_1 = XOR, g_2 = AND (materials with dashed outline).

input cases, the internal pathways on which they each rely to complete their function are polluted by interference from the other gate. Further investigation is required to understand the relationship between logic gate choice and logical independence. Further investigation is also required to understand if this behavior is present for other logic gate choices, such as where g_1 = XOR and g_2 = XOR or g_1 = AND, g_2 = NAND. If this finding is ubiquitous across other logic gate pairings, it would have implications in the application of CGMMs. Specifically, it would guide CGMM design and implementation such that we would avoid embedding homogeneous logic gates into a CGMM, in preference for CGMMs with mixed logic gates.

4.2 Evolved Material Configurations

It is interesting to visually compare the configurations of materials both between and within each evaluation method. Figure 5 displays the configurations of four materials with $I = 15$ independence materials that were evolved using each of the tested evaluation methods, both where g_1 = AND, g_2 = AND, and where

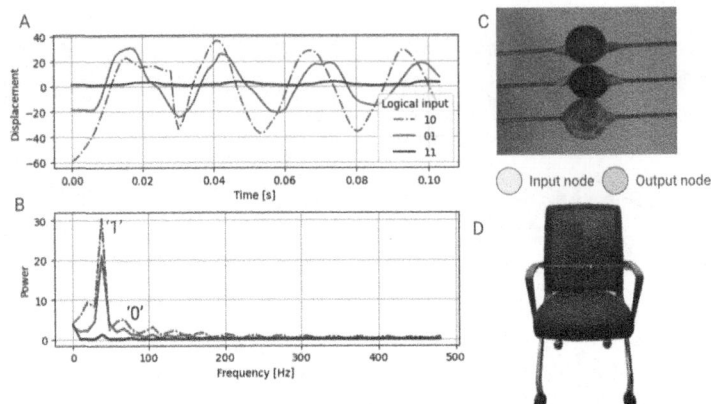

Fig. 6. Physical implementation of a granular logic gate. A) The time domain displacement of the input grain under each logical input. **B)** The frequency domain signal under each logical input. Here, we expect a high power at the driving frequency for each of the green lines, and a low power at the driving frequency of the blue line. The observed behavior here is consistent with the expected behavior of an XOR logic gate. **C)** This image shows a close-up of our 3-bead system. **D)** This photograph displays our entire chair-XOR system.

$g_1 = $ XOR, $g_2 = $ AND. We encourage the reader to closely examine the configurations of these materials. Interestingly, each of the materials depicted appears to have distinct grain-mass organizations. There are very few discernible patterns among the materials. This includes no discernible pattern in the properties of input grains, output grains, or border grains. This applies to both materials where here $g_1 = $ AND, $g_2 = $ AND, and where $g_1 = $ XOR, $g_2 = $ AND. The absence of clear patterns in material properties underscores the intricate and unintuitive relationship between form and function in CGMMs. This diversity of configuration reveals a breadth of functional material arrangements that can arise within the design space.

5   Hardware Implementation

To further the physical realizability of CGMMs, we proceed with implementing a simple physical computational granular XOR. We use everyday materials, including an office chair, guitar strings, adhesive putty, and beads. Our implementation consists of three grains strung onto parallel guitar strings placed across the armrests of an office chair. Figure 6C shows a close-up view of the system, while Fig. 6D shows a photo of the entire chair system. The two outer beads act as input grains and the center bead acts as an output grain. The grains' y-locations were adjusted such that the beads were in light contact when at rest. Their locations in the x-plane were aligned and then maintained by adhesive putty. In this system, an input of '1' is passed into the top grain as a hand-pluck with amplitude $\approx 2.5\,cm$ applied $90°C$ to the string, while an input of '1' is passed into the bottom grain as a hand-pluck with amplitude $\approx 2.5\,cm$ applied $270°C$ to the guitar string. This system is designed such that there is destructive interference from the input grains when the '11' input case is passed into the system, leading to little movement of the output grain.

To assess the function of this system, we recorded its behavior under the '01', '10', and '11' input cases. The '00' input case is not considered as it is trivial. The behavior of the output grain was captured on video with a frame rate of 960fps using Super Slo-mo mode on a Galaxy S21 Android phone. Each video was then processed to extract the displacement of the output grain under each logical input using trackR [6]. These videos are included in the supplementary materials. Figure 6A and 6B respectively show the behavior of the output grain over time, and across frequency. Both of these figures support that our system exhibits appropriate XOR behavior.

6 Conclusions & Future Work

This study presents scalable methods for evolving independent polycomputational granular materials without explicitly optimizing for independence. We report the first discovery of entirely logically independent CGMMs. Notably, we observed that evolution discovers more independent CGMMs when a material has mixed logic gates compared to those with homogeneous gates. However, the extent to which this trend generalizes to other gate pairings remains unclear. Resolving this uncertainty will be a focus of future investigations. We also implement a vibrational XOR gate in hardware. While not AI-designed, this gate serves as a physical proof-of-concept for computational functionality in vibrational granular assemblies. Expanding this work to include more complex, two-dimensional grain structures will be the next step to transferring evolved designs from simulation to reality.

Acknowledgements. We want to thank Frederic Sansoz for his guidance with LAMMPS and Emily Ertle for her assistance with the hardware implementation. This material is based upon work supported by the National Science Foundation Graduate Research Fellowship Program under Grant No. 2235204. Any opinions, findings, and conclusions or recommendations expressed in this material are those of the author(s) and do not necessarily reflect the views of the National Science Foundation. We would like to acknowledge financial support from the National Science Foundation under the DMREF program (award number: 2118810).

Disclosure of Interests. The authors have no competing interests to declare that are relevant to the content of this article.

References

1. Beaulieu, S., Welch, P., Parsa, A., O'Hern, C., Kramer-Bottiglio, R., Bongard, J.: Refractive computation: parallelizing logic gates across driving frequencies in a mechanical polycomputer. In: ALIFE 2024: Proceedings of the 2024 Artificial Life Conference. MIT Press (2024)
2. Bongard, J.: From rigid to soft to biological robots: how new materials are driving advances in the study of the embodied cognition. Artif. Life Robot., 1–5 (2023)
3. Bongard, J., Levin, M.: There's plenty of room right here: biological systems as evolved, overloaded, multi-scale machines. Biomimetics **8**(1), 110 (2023)
4. Córcoles, A.D., et al.: Challenges and opportunities of near-term quantum computing systems. Proc. IEEE **108**(8), 1338–1352 (2019)
5. van De Burgt, Y., Melianas, A., Keene, S.T., Malliaras, G., Salleo, A.: Organic electronics for neuromorphic computing. Nature Electron. **1**(7), 386–397 (2018)
6. Garnier, S.: trackR - Multi-object tracking with R (2022). https://swarm-lab. github.io/trackR/, r package version 0.5.3
7. Gorecki, J., et al.: Chemical computing with reaction-diffusion processes. Philos. Trans. Royal Soc. A: Math. Phys. Eng. Sci. **373**(2046), 20140219 (2015)
8. Hameed, S.A.: Controlling computers and electronics waste: toward solving environmental problems. In: 2012 International Conference on Computer and Communication Engineering (ICCCE), pp. 972–977. IEEE (2012)
9. Horowitz, M., Grumbling, E.: Quantum computing: progress and prospects (2019)
10. Kim, E., Yang, J.: Wave propagation in granular metamaterials. Funct. Compos. Struct. **1**(1), 012002 (2019)
11. Knill, E.: Scalable quantum computing in the presence of large detected-error rates. Phys. Rev. A **71**(4), 042322 (2005)
12. Parsa, A., Wang, D., O'Hern, C.S., Shattuck, M.D., Kramer-Bottiglio, R., Bongard, J.: Evolving programmable computational metamaterials. In: Proceedings of the Genetic and Evolutionary Computation Conference, pp. 122–129 (2022)
13. Parsa, A., Wang, D., O'Hern, C.S., Shattuck, M.D., Kramer-Bottiglio, R., Bongard, J.: Evolution of acoustic logic gates in granular metamaterials. In: International Conference on the Applications of Evolutionary Computation (Part of EvoStar), pp. 93–109. Springer (April 2022)
14. Parsa, A., Witthaus, S., Pashine, N., O'Hern, C., Kramer-Bottiglio, R., Bongard, J.: Universal mechanical polycomputation in granular matter. In: Proceedings of the Genetic and Evolutionary Computation Conference, pp. 193–201 (2023)
15. Paun, G., Rozenberg, G., Salomaa, A.: DNA computing: new computing paradigms. Springer Science & Business Media (2005)
16. Tanaka, G., et al.: Recent advances in physical reservoir computing: a review. Neural Netw. **115**, 100–123 (2019)
17. Theis, T.N., Wong, H.: The end of Moore's law: a new beginning for information technology. Comput. Sci. Eng. **19**(2), 41–50 (2017)
18. Thompson, A.P., et al.: LAMMPS - a flexible simulation tool for particle-based materials modeling at the atomic, meso, and continuum scales. Comp. Phys. Comm. **271**, 108171 (2022). https://doi.org/10.1016/j.cpc.2021.108171
19. Vega, A., Bose, P., Buyuktosunoglu, A.: Rugged embedded systems: computing in harsh environments. Morgan Kaufmann (2016)
20. Xu, X.Y., Jin, X.M.: Integrated photonic computing beyond the von Neumann architecture. ACS Photonics **10**(4), 1027–1036 (2023)
21. Yasuda, H., et al.: Mechanical computing. Nature **598**(7879), 39–48 (2021)

Author Index

© The Editor(s) (if applicable) and The Author(s), under exclusive license
to Springer Nature Switzerland AG 2025
P. García-Sánchez et al. (Eds.): EvoApplications 2025, LNCS 15612, pp. 573–576, 2025.
https://doi.org/10.1007/978-3-031-90062-4

The manufacturer's authorised representative in the EU is Springer
Nature Customer Service Centre GmbH, Europaplatz 3, 69115 Heidelberg,
Germany. If you have any concerns regarding our products, please
contact ProductSafety@springernature.com

Printed and bound by CPI Group (UK) Ltd, Croydon, CR0 4YY

28/04/2026

02098515-0012